THE KENTUCKY BREEDING BIRD ATLAS

THE KENTUCKY BREEDING BIRD ATLAS

Brainard L. Palmer-Ball Jr.

The University Press of Kentucky

Sponsored by

Kentucky State Nature Preserves Commission

and

Kentucky Department of Fish and Wildlife Resources
Nongame Wildlife Program

in cooperation with

Kentucky Ornithological Society

This project was funded in part using proceeds from taxpayer donations to the Nature and Wildlife Fund.

Cover illustration of a male Kentucky Warbler at the nest by Ron Austing.

Scholarly publisher for the Commonwealth,
serving Bellarmine College, Berea College, Centre College of Kentucky, Eastern Kentucky University,
The Filson Club, Georgetown College, Kentucky Historical Society, Kentucky State University,
Morehead State University, Murray State University, Northern Kentucky University, Transylvania
University, University of Kentucky, University of Louisville, and Western Kentucky University.

Editorial and Sales Offices: The University Press of Kentucky
663 South Limestone Street, Lexington, Kentucky 40508-4008

Library of Congress Cataloging-in-Publication Data

Palmer-Ball, Brainard L.
The Kentucky breeding bird atlas / Brainard L. Palmer-Ball, Jr.
p. cm.
Includes bibliographical references (p.) and index.
ISBN 0-8131-1965-0 (alk. paper)
1. Birds–Kentucky. 2. Birds–Kentucky–Geographical
distribution. I. Title.
QL684.K4P35 1996
598.29769–dc20 96-807

This book is printed on acid-free recycled paper meeting the requirements
of the American National Standard for Permanence of Paper for Printed Library Materials.

Manufactured in the United States of America

To all the birders, past and present, whose contributions made this book possible

CONTENTS

FIGURES

TABLES

FOREWORD

The *Kentucky Breeding Bird Atlas* is one of many similar works that have been published or are under way in many states and in other countries. Naming the creatures of the earth is older than civilization, and for centuries individuals have felt that discovering all the species of living things was important. Today, although new species are still being found, the emphasis has shifted to determining the numbers and distribution of known species.

Biological censuses such as this one help substantiate the changes in species frequency and distribution. Sound biological data can be used for intelligent decision-making by all sectors of society. Clearly there have been significant changes in avian populations concurrent with the significant land-use changes of the last two centuries, and these changes are still occurring.

The first of such comprehensive biological surveys conducted by teams of competent observers began in Great Britain. The completion of a survey of the flora of Britain and subsequent publication of the *Atlas of the British Flora* (Perring and Walters 1962) was followed by the *Atlas of Breeding Birds in Britain and Ireland* (Sharrock 1976). This work provided a method and an inspiration for other countries, including Third-World countries that realize the importance of their natural heritage and the necessity of having realistic data with which to assess environmental destruction and wildlife loss.

In most states birds are the logical first class of animals to be chosen as the basis of a biological census. They are large enough and colorful enough to be readily seen, and their calls and songs can be learned by skilled observers. For birds, a good deal of census and survey work has been done largely by amateurs. Many skilled volunteer hours have contributed to projects that would be almost impossible to complete if paid staff were necessary.

In the fall of 1984, Sherri Evans, biologist with the Nongame Wildlife Program of the Kentucky Department of Fish and Wildlife Resources (KDFWR), began discussions with representatives of the Kentucky Ornithological Society (KOS) and the Kentucky State Nature Preserves Commission (KSNPC) to determine their interest in a breeding bird atlas of Kentucky. With assurance of support, Evans began work soon thereafter. A change in employment forced her to resign her position as state coordinator two years into the field effort. In the latter years of the atlas, Brainard Palmer-Ball Jr. of KSNPC assumed responsibility for directing the fieldwork and overseeing the project through publication. Brainard's work on this project has been both an avocation and a work assignment as terrestrial zoologist for KSNPC.

The result of ten years of field work, data entry and analysis, and writing is now in your hands. Despite the limitations of the study, discussed in the introduction, the *Kentucky Breeding Bird Atlas* is the most thorough survey of Kentucky's breeding birds ever undertaken. Coming thirty years after Robert M. Mengel's landmark *The Birds of Kentucky*, the comparisons it makes between the present (1985-91) and the time of Mengel's most intensive field study (1948-52) are invaluable.

The survey resulting in this book produced numerous significant new records and range extensions. It also provides excellent status reports of relatively recent developments such as invasions of the Blue Grosbeak and House Finch. The results also clarify the status of many relatively obscure species, in some cases indicating surprisingly widespread occurrence of heretofore poorly known species. In contrast, the striking declines in breeding populations of other species that are indicated by the negative information generated will be critical to protection efforts undertaken by the sponsoring state agencies.

All of those involved with this project hope that you will find it fascinating and useful.

Robert M. McCance Jr.
Director, KSNPC

ACKNOWLEDGMENTS

The Kentucky Breeding Bird Atlas Project (KBBA) was sponsored by the Kentucky State Nature Preserves Commission (KSNPC) and the Nongame Wildlife Program of the Kentucky Department of Fish and Wildlife Resources (KDFWR), in cooperation with the Kentucky Ornithological Society (KOS). This effort could not have been undertaken without the support of Richard Hannan (former director) and Robert McCance Jr. (current director), of KSNPC, Lauren Schaaf (former director) and Roy Grimes (current director), of the Wildlife Division of KDFWR, and Don McCormick (former commissioner) and C. Tom Bennett (current commissioner) of KDFWR. In addition, without the thousands of hours contributed by volunteers, especially those from KOS, this project would not have been possible. Special appreciation is extended to those individuals who served as regional coordinators for the KBBA. Thanks is also extended to those individuals who were willing to donate their photographic skills to illustrate this publication: Ron Austing, Tom Bloom, Gene Boaz, Jeffrey A. Brown, Kathy Caminiti, Julian Campbell, Herbert Clay Jr. M.D., Marc Evans, Sherri A. Evans, Hal H. Harrison, Joseph O. Knight, Lewis Kornman, Landon McKinney, Maslowski Wildlife Productions, Gary Meszaros, James Parnell, Matthew Patterson, Mabel Slack, Philippe Roca, Bill Schoettler, Alvin E. Staffan, Bud Tindall, and Deborah White. In addition, several state and federal agencies provided photographic material for this book, including the Louisville District of the U.S. Army Corps of Engineers (Robert van Hoff), the Missouri Department of Conservation (Rochelle Renken), and the Ohio Department of Natural Resources (Thomas Schwartz). Assistance from Bruce Peterjohn, coordinator of the National Biological Service (formerly U.S. Fish and Wildlife Service) Breeding Bird Survey (BBS) in Laurel, Maryland, allowed for incorporation of seven years' worth of BBS data into the KBBA database. This project is also deeply indebted to Anne Stamm, state coordinator for the Cornell Nest-Card Program since its inception in 1965, for her assistance in incorporating a wealth of information from the thirty years' worth of nest cards for which she serves as curator. Thanks is also extended to Steve Phillips, of the U.S. Department of Agriculture Forest Service, London Ranger District, for providing information on the Red-cockaded Woodpecker in Kentucky. Likewise, Jeffery Sole, Ronald Pritchert, and George Wright of KDFWR provided access to KDFWR survey information for the Wood Duck, the American Woodcock, the Ruffed Grouse, and the Wild Turkey. Steve Kull of the Kentucky Department of Forestry provided historical and recent data on forest cover in the state.

Maintaining this project required a great deal of data entry and quality control, correspondence, and other similar work. Theresa Anderson, Lynda Andrews, Brenda Hamm, Danny Watson, and Traci Wethington (KDFWR) assisted in various aspects of data processing and correspondence. Sincere thanks is extended to David Yancy (KDFWR) for his diligence and patience in dealing with numerous administrative duties. Tom Bloom and Tim Clarke (KSNPC) made a tremendous contribution in computer programming for the project. Julie Smither and Melissa White (KSNPC) provided administrative assistance in publishing annual KBBA newsletters. I am indebted to Tom Bloom, Tim Clarke, Marc Evans, Martina Hines, Gary Libby, and Deborah White of KSNPC; Julian Campbell of The Nature Conservancy; George A. Hall; Lee McNeely; Bruce Peterjohn; Clell Peterson; Jeffery Sole of KDFWR; and Michael Stinson, for their critical reviews of portions of the manuscript. Other individuals who contributed in some manner include Mary Gustafson, Donald Harker, and George Rogers.

Financial support for the KBBA came from general funds of both KSNPC and KDFWR. Other funding came from the private sector, most notably through taxpayer donations to the Nature and Wildlife Fund. Members of KOS contributed funds through a general appeal as well. Special gratitude is extended to Kenneth Bates (director) and Steve Pigg (technician) of the State of Kentucky's Natural Resources and Environmental Protection Cabinet's Office of Information Services GIS Branch, for use of a Geographic Information System to map the KBBA data.

Finally, I wish to thank two individuals, Lee McNeely and Clell Peterson, whose energy and interest in the KBBA contributed immeasurably to coverage in their respective regions of the state. Both served above and beyond the call of duty as regional coordinators, and their dedication is greatly appreciated.

State Coordinators

Sherri A. Evans
John MacGregor
Brainard Palmer-Ball Jr.
David Yancy

Seasonal Atlas Blockbusters

Alan Barron (1988–89)
Richard Healy Jr. (1990–91)
Andrew Mullen (1989)
C. Michael Stinson (1988–91)

Regional Coordinators

Blaine Ferrell
Lee McNeely
Robert Morris
Brainard Palmer-Ball Jr.
Clell Peterson
Albert Powell

Volunteers

Mike Adam
Terry Anderson
Brad Andres
Sarah Andres
Lee Andrews
Lynda Andrews
Mary Bill Bauer
Rosemary Bauman
Brenda Bellamy
Joyce Bender
George Beringer
Morris Black
Gene Boaz
Earl Boggs
Ann Bradley
Eugene Bradley
David Brauer
Mike Brown
Robert Brown
W. Horace Brown
Vivian Brun
Mike Burns
Julia Burriss
Fred Busroe
Joe Caminiti
Kathy Caminiti
Denver Campbell
Frances Carter
Richard Cassell
Hap Chambers
Ron Cicerello
Kathryn Clay
Marcus Cope
Manton Cornett
Joseph Croft (BBS)
Wayne Davis
David Dister
Ronald Duncan
Brenda Eaden
Tony Eaden
Diane Ebel
Neil Eklund
Garry Elam
Mary Elam
George Elliott
Jackie Elmore
Joe Tom Erwin
Sherri A. Evans
Harold Eversmeyer
John Ferner
Blaine Ferrell
Jeff Finn
Stephen Figg
Mike Flynn
Tobin Foster
Jonathan Glixon
Darlena Graham
Vicki Griffin
Mark Gumbert
Wendell Haag
James Hancock
Robert Head (BBS)
Richard Hines
Gary Howard
Michael Hurst
Richard Hutson
Ramon Iles
Bill Jacoby
Carolyn Johnson
Kathy Johnson
Howard Jones
Jeff Jones

Marian Jones
Kell Julliard
J. William Kemper
Gerri Kennedy
Don King
Virginia Kingsolver
Wendell Kingsolver
Charles Kohler
Lewis Kornman
Mrs. R.K. Lane
Sally Leedom
Orville Litteral
Frederick Loetscher
Timothy Love
Henry Lowsma
John MacGregor
Scott Marsh
Wayne Mason
Ruth Mathes
William Mathes
Carolyn Mayo
Thomas Mayo
William McComb
Lee McNeely
Lynda McNeely
Robert McNeese
Melissa Mefford
Michael Miller
Burt Monroe Jr. (BBS)
Robert Morris
Carl Mowery Jr.
Tim Niehoff
Joan Noel
Doxie Noonan
Michael O'Hara
Patricia O'Hara
Don Parker
Jim Pasikowski
William Peeples (BBS)
Clell Peterson
David Peterson
Lawrence Philpot
Martha Pike
Joyce Porter
Albert Powell
Kerry Prather
James Prewitt
William Qualls
Glenn Raleigh
Lene Rauth
Art Ricketts
Tina Ricketts
Gary Ritchison
Michael Roberts
Emelene Rowland
Robert Rowland
Joe Russell
Mark Sanders
Carmen Schulte
Patricia Scott
Herbert Shadowen (BBS)
Larry Smith
Tony Smith
Carol Sole
Jeffery Sole
Donnie Spencer
Anne Stamm (BBS)
Russell Starr
Stephen Stedman (BBS)
Doug Stephens
Tommy Stephens
Thomas Stevenson
Allen Stickley Jr.
Tanya Stinson
Marie Sutton
Kathleen Tidwell
Tim Towles
Johnny Upton
Linda Ward
Sally Wasielewski
Leslie Welburn
Steve Welburn
Julianne Whitaker
Deborah White
William Wiglesworth Jr.
James Williams
Timothy Williams
Alita Wilson
James Woodring
Ben Yandell

INTRODUCTION

In the late 1970s and early 1980s, several northeastern states and Canadian provinces initiated breeding bird atlases. In the fall of 1984, the idea of undertaking a similar effort in Kentucky was initiated by the Nongame Wildlife Program of the Kentucky Department of Fish and Wildlife Resources (KDFWR). Along with representatives of the Kentucky Ornithological Society (KOS) and the Kentucky State Nature Preserves Commission (KSNPC), KDFWR made plans for a cooperative Kentucky Breeding Bird Atlas (KBBA). The goal of this project was to conduct a systematic survey of the state's breeding birds over a five-year period, beginning in 1985.

Methods

Preliminary Setup

During the winter of 1984–85 those of us involved in the project began to prepare materials, patterning most after those used by some other states already at work (Ohio, Pennsylvania, Vermont). We also recruited volunteers. As had been done in other states, U.S. Geological Survey (USGS) 7.5-minute topographic quadrangle maps were each divided into six equal units, each unit covering approximately ten square miles (Figure 1). The initial goal of the KBBA was to survey one unit, or "priority block," within each quadrangle covering the state. Priority blocks were designated to be inventoried in 727 topographic quadrangles wholly

A	B
C	D
E	F

Figure 1. Representation of USGS Topographic Quadrangle Map Showing Priority Block Designations

or partially in Kentucky (Figure 2), but 52 quadrangles lying along the state's borders had portions within Kentucky too small or isolated to be considered for atlas coverage. Initially, all 727 priority blocks were chosen randomly, but after the third field season, all unassigned priority blocks were switched to the central west "C" block. Designation of this block eliminated frequent clustering of priority blocks and allowed observers more easily to determine routes into and out of a priority block using a single quadrangle map. This midcourse change yielded the distribution of priority blocks seen in Figure 2: the more heavily populated areas (where more block assignments were made during the project's first three years) show a random priority block distribution, while other parts of the state show a uniform distribution. For quadrangles bordering other states, a priority block was often designated that included as much territory as possible within Kentucky's borders. For the 52 quadrangles that were not covered as part of the atlas, a few incidental observations were included, especially those of Notable Species.

Blocks other than the designated priority block on a given quadrangle that were surveyed were considered special blocks. During the atlas fieldwork, data were submitted for six special blocks. In addition, specific surveys were undertaken on a few special areas, that is, wildlife management areas, state parks, and so on. Data from special blocks and special areas were incorporated as incidental observations. In some other states, special block and/or special area contribution to the data set is significant, but we thought that an attempt to survey such areas in Kentucky would dilute the primary effort to cover priority blocks, and thus the survey of special areas was not emphasized during this project.

Figure 2. Location of KBBA Priority Blocks

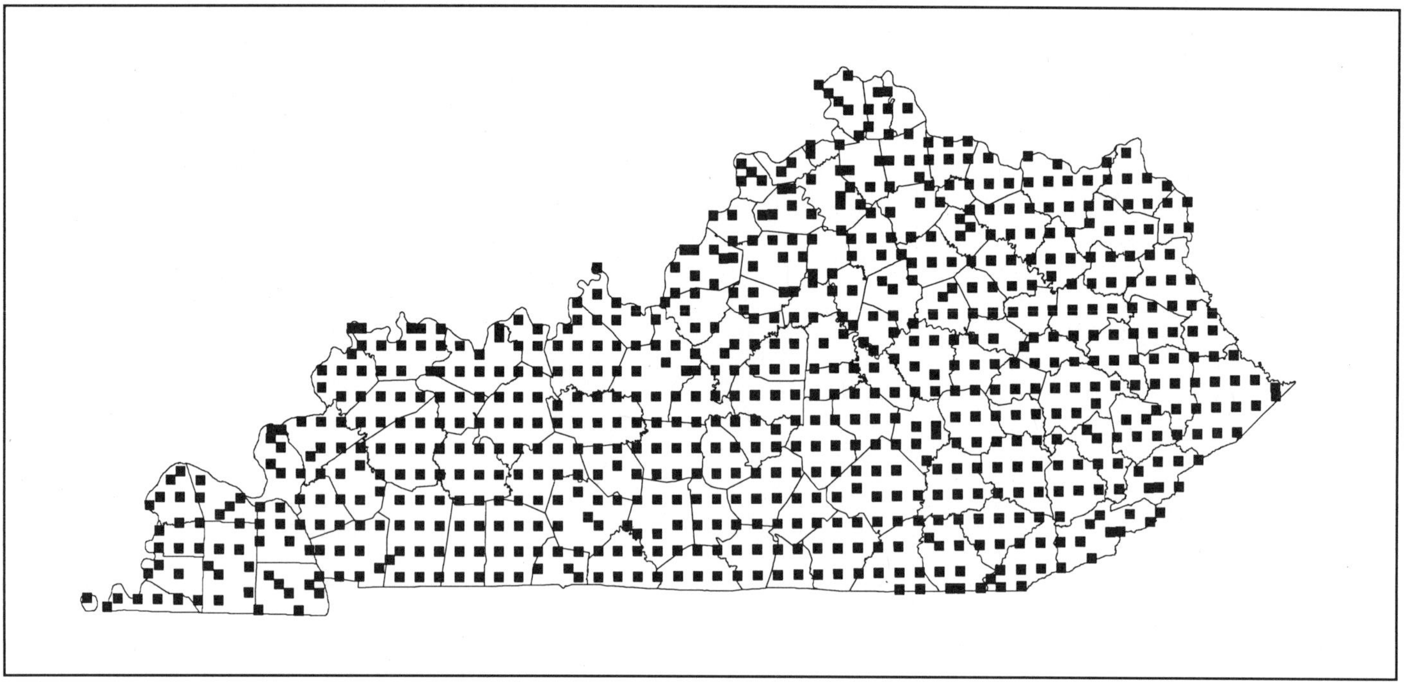

Materials

KBBA materials were similar to those used in other states. They included an Instruction Manual, a Field Card, a Summary Sheet, a Notable Species Form, an Incidental Observation Form, an Identification Card, and a Vehicle Identification Card. These materials were duplicated by KSNPC and KDFWR and distributed to volunteers. Priority blocks were marked on a set of USGS 7.5-minute quadrangle maps and were likewise sent to volunteers.

Data Recording

Observers recorded data using a standardized set of breeding codes based on observed behavior. The breeding codes and their definitions are presented in Table 1. Breeding codes and definitions used in Kentucky were similar to those used in other atlas projects, although a code relating the observation of a certain number of territorial birds to nesting evidence was not included. Some states have used such a code to indicate probable or confirmed breeding. Observers entered data onto Field Cards and transcribed them onto Summary Sheets at the end of each field season.

Abundance

As supplementary qualitative information, observers recorded the approximate abundance of all species encountered in each priority block using the following five-level system: only one individual or one pair observed in block; very small numbers restricted to a very small area in block; small numbers seen in scattered localities in block; fairly widespread in block; and one of the commonest species in block. Abundance values were entered into a separate column on the Field Card and later transferred to the Summary Sheet. Abundance data also were entered for incidental observations.

Table 1. Breeding Codes and Definitions Used

Observed (OB):

O	Species *observed* in the block but not believed to be breeding in the immediate vicinity of the block. These included birds seen before or after the normal nesting season dates for the species given in the *Instruction Manual,* probable migrants, vagrants, and other birds suspected to be nonbreeding.

Possible Breeding (PO):

X	Species heard or seen *in breeding habitat during the breeding season.* These included birds observed *during* the nesting season as given in the *Instruction Manual* but for which no further evidence of nesting was observed. Waterbirds observed in the immediate vicinity of suitable nesting habitat were included here. Most flyovers were also included in this category (e.g., foraging swallows or soaring vultures or hawks).

Probable Breeding (PR):

A	*Agitated* behavior or anxiety calls from adults. Parent birds respond to threats with distress calls or by attacking intruders. This does not include response to "squeaking," "pishing," or tape playing.
P	*Pair* observed in suitable habitat in breeding season.
T	Permanent *territory* presumed to be held through observed behavior (e.g., chasing other birds, or singing from the same location on two occasions at least a week apart).
C	*Courtship* or *copulation* observed. This includes displays and courtship feeding.
N	Visiting probable *nest* site. Primarily applied to cavity nesters. Also applies when a bird visits a site repeatedly but no further evidence of nesting is seen.
B	*Nest building* by wrens or excavation of a nest hole by woodpeckers. Not confirmed until young are detected in these circumstances because some wrens build dummy nests and woodpeckers excavate roost cavities.

Confirmed Breeding (CO):

DD	*Distraction display* or injury feigning (agitated behavior and/or anxiety calls are only a code A under Probable Breeding).
NB	*Nest building* by all except wrens and woodpeckers. Carrying of sticks is part of the courtship ritual of some species (code C).
UN	*Used nest* located. These must be carefully identified if they are to be counted as evidence. Some nests (e.g., that of the Northern Oriole) are persistent and characteristic, but many are difficult to identify correctly.
PE	*Physiological evidence* of breeding (i.e., highly vascularized incubation patch or egg in oviduct). This evidence is generally based on a bird in the hand, although brood patches may be visible in the field.

Table 1 continued

FL	*Recently fledged young* (altricial species) or *downy young* (precocial species) restricted to the natal area by dependence on adults or limited mobility. Young cowbirds confirm both cowbirds and host species. Do not forget to look for dead young on roads or elsewhere in the block.
FS	Adult bird seen carrying *fecal sac.* Nestlings of some species (mainly songbirds) void their excrement into a small sac, which the parents dispose of after leaving the nest site.
FY	Adult bird seen carrying *food for young.* Some birds (terns and raptors) continue to feed their young long after they have fledged, and even after they have moved considerable distances. Also, some birds (terns, egrets, herons, etc.) may carry food long distances to young in a neighboring block. Care should also be taken to avoid confusion with courtship feeding (code C).
ON	*Occupied nest.* Presumed by activity of parents (e.g., adults entering or leaving nest or cavity while exchanging incubation responsibility, carrying food to nest or cavity; or adult staying on nest for extended period, etc.). Primarily intended for high nests or nest holes where contents were not visible.
NE	*Nest with eggs.* Adult incubating eggs, or identifiable eggs or eggshells found beneath nest. Note: Cowbird eggs confirmed both cowbird and host.
NY	*Nest with young.* Once again, a young cowbird in a nest confirms both the cowbird and the host species.

Incidental Observations

During the fieldwork observers encountered a number of birds outside priority blocks. They were instructed to record such observations as "incidental records," noting the specific location or block of the sightings. Incidental observations were typically recorded on Incidental Observation Forms and turned in with priority block data at the end of each field season. At the end of the final field season, these observations were added to the atlas data set along with information from other sources.

Notable Species

Species listed for monitoring by KSNPC (Warren et al. 1986; KSNPC 1992) or otherwise considered rare or of unknown distribution in Kentucky were designated as Notable Species. Documentary details were required for all observations of these species to substantiate their validity. Details were submitted on a Notable Species Location Form and reviewed by regional and state coordinators before entry into the KBBA data set. Notable Species are listed in Table 2.

Forest Cover

The percentage of forest cover was estimated for each priority block by interpreting forest cover as depicted on the USGS quadrangle maps. Each priority block was assigned one of five forest cover values, based on the amount of forest cover indicated on the map: (1) no or very little forest cover; (2) some forest cover but predominantly open; (3) good mixture of open areas and forest cover; (4) predominantly forested but some open areas; and (5) virtually all forested. Whenever possible, priority blocks were field-checked to corroborate the values obtained from maps.

Table 2. Notable Species Arranged by Taxonomic Order (AOU 1983)

Species	KSNPC Status*	Federal Status**
Pied-billed Grebe	Endangered	—
Double-crested Cormorant	Endangered	—
American Bittern	Endangered	—
Least Bittern	Threatened	—
Great Blue Heron	Special Concern	—
Great Egret	Endangered	—
Little Blue Heron	Endangered	—
Cattle Egret	Special Concern	—
Black-crowned Night-Heron	Endangered	—
Yellow-crowned Night-Heron	Threatened	—
Blue-winged Teal	Endangered	—
Hooded Merganser	Endangered	—
Osprey	Threatened	—
Mississippi Kite	Special Concern	—
Bald Eagle	Endangered	Endangered
Northern Harrier	Threatened	—
Sharp-shinned Hawk	Special Concern	—
Cooper's Hawk	—	—
King Rail	Endangered	—
Common Moorhen	Threatened	—
American Coot	Endangered	—
Spotted Sandpiper	Endangered	—
Upland Sandpiper	Endangered	—
Least Tern (Interior race)	Endangered	Endangered
Black Tern	Extirpated	C2 Candidate
Black-billed Cuckoo	—	—
Barn Owl	Special Concern	—
Red-cockaded Woodpecker	Endangered	Endangered
Least Flycatcher	Threatened	—
Bank Swallow	Special Concern	—
Cliff Swallow	—	—
Fish Crow	Special Concern	—
Common Raven	Endangered	—
Brown Creeper	Endangered	—
Bewick's Wren	Special Concern	C2 Candidate
Sedge Wren	Special Concern	—
Veery	—	—
Bell's Vireo	Special Concern	—
Solitary Vireo	—	—
Golden-winged Warbler	Threatened	—
Chestnut-sided Warbler	—	—
Black-throated Blue Warbler	—	—
Blackburnian Warbler	Threatened	—
Canada Warbler	Special Concern	—
Rose-breasted Grosbeak	Special Concern	—
Bachman's Sparrow	Endangered	C2 Candidate
Vesper Sparrow	Endangered	—

Table 2 continued

Species	KSNPC Status*	Federal Status**
Lark Sparrow	Threatened	—
Savannah Sparrow	Special Concern	—
Henslow's Sparrow	Special Concern	C2 Candidate
Dark-eyed Junco	Special Concern	—
Bobolink	Special Concern	—

* KSNPC statuses are interpreted from KSNPC (1992). *Endangered*: A species in danger of extirpation and/or extinction throughout all or a significant part of its nesting range in Kentucky. *Threatened*: A species likely to become endangered within the foreseeable future throughout all or a significant part of its nesting range in Kentucky. *Special Concern*: A species that should be monitored because (a) it exists in a limited geographic area, (b) it may become threatened or endangered because of modification or destruction of habitat, (c) certain characteristics or requirements make it especially vulnerable to specific pressures, (d) experienced researchers have identified other factors that may jeopardize it, or (e) it is thought to be rare or declining but insufficient information exists for assignment to the threatened or endangered status categories.

** Federal statuses come from U.S. Fish and Wildlife Service (1991, 1993). Definitions come from U.S. Congress (1983); U.S. Fish and Wildlife Service (1991). *Endangered*: "Any species which is in danger of extinction throughout all or a significant portion of its range." *C2 Candidate*: Status review species for which information in the possession of the U.S. Fish and Wildlife Service "indicates that proposing to list as endangered or threatened is possibly appropriate, but for which conclusive data on biological vulnerability and threat are not currently available to support proposed rules."

Data Processing and Quality Control

At the end of each field season, atlas data were submitted on Summary Sheets for quality control and data entry. Records were hand-checked for obvious errors, then entered into a computer database. At the completion of each year's data entry, error-checking routines were run on all data, and printouts were generated. Printouts were sent to regional coordinators and volunteers for an additional level of quality control. Data from subsequent years were added to the database at the end of each field season and again checked for errors. A final check of the data set was accomplished before production of the atlas maps began.

Results and Discussion

Fieldwork

Early in the KBBA project, volunteer participation was high, but after the third field season it became clear that seasonal employees would be needed to complete the atlas project. In 1988 and 1989 KDFWR funded two seasonal positions for KBBA data collection. After the fifth field season, in 1989, many priority blocks remained unsurveyed, and it was determined that two additional field seasons would be required to complete the fieldwork. KDFWR again funded two seasonal positions in 1990 and 1991, and KSNPC made one staff person available during June and July of 1990 and 1991 to assist in completion of priority block surveys.

Supplemental Data Sources

After the final field season, in 1991, the KBBA data set was supplemented with records from three sources: records obtained from the National Biological Service Breeding Bird Surveys (BBS) for 1985–91; records obtained from KDFWR game bird surveys for 1985–91; and miscellaneous records acquired through personal communications and review of issues of *The Kentucky Warbler,* the journal of the Kentucky Ornithological Society, for 1985–91. The BBS data provided the greatest amount of information. Individual route stops were plotted on USGS quadrangle maps in consultation with BBS cooperators. Original BBS field sheets were compared with KBBA Summary Sheets for additional priority block and incidental observations. The Wildlife Division of KDFWR provided game bird survey data. Wood Duck brood survey counts, American Woodcock singing ground surveys, Ruffed Grouse drumming surveys, and some Wild Turkey gobbling surveys for 1985–91 yielded additional records for these species. As a final supplement to the data set, personal communications and a review of *The Kentucky Warbler* for the seven-year KBBA period provided various records of breeding birds. Although this latter effort yielded relatively few records, it helped fill in some gaps in the KBBA data.

Coverage

By the end of the seventh field season, at least some survey work had been undertaken in all 727 priority blocks. The average number of species reported per block statewide was 58.5. Priority block species totals ranged from 19 in a block dominated by urban streets and buildings to 90 in a diverse, rural part of Boone County (see Table 3 and Figure 3). The average number of species recorded per block varied noticeably between physiographic province sections (Table 4). Diversity was highest in the western part of the state and decreased eastward, being lowest in the Cumberland Plateau and the Cumberland Mountains. This variation was largely the result of the amount of forest cover.

Incidental reports increased the priority block data set by about 5 percent. When incidental observations were included, the average number of species reported per quadrangle statewide was 61.4, and four birds were added to the list of species that were found in priority blocks.

During the seven-year atlas project, 154 species were reported as at least possibly breeding. Of these, 146 species were confirmed as breeding. This set includes two species (Short-eared Owl and Brown Creeper) never before found breeding in the state, and one (Northern Harrier) that had not been reported as nesting since the early 1800s. The following species were recorded as possible or probable breeders only: Least Bittern, King Rail, Least Flycatcher, Fish Crow, Golden-winged Warbler, Canada Warbler, Bachman's Sparrow, and Vesper Sparrow.

Abundance

Abundance data for each species are plotted on maps in the individual species accounts. These data also were used for analysis of trends in average abundance for different physiographic province sections and degrees of forest cover. These trends are briefly discussed in the species accounts.

Table 3. Number of Species Recorded in Priority Blocks

Number of Species	Number of Blocks
19	1
20–29	2
30–39	20
40–49	105
50–59	253
60–69	256
70–79	70
80–89	19
90	1

Figure 3. Priority Block Coverage by Number of Species

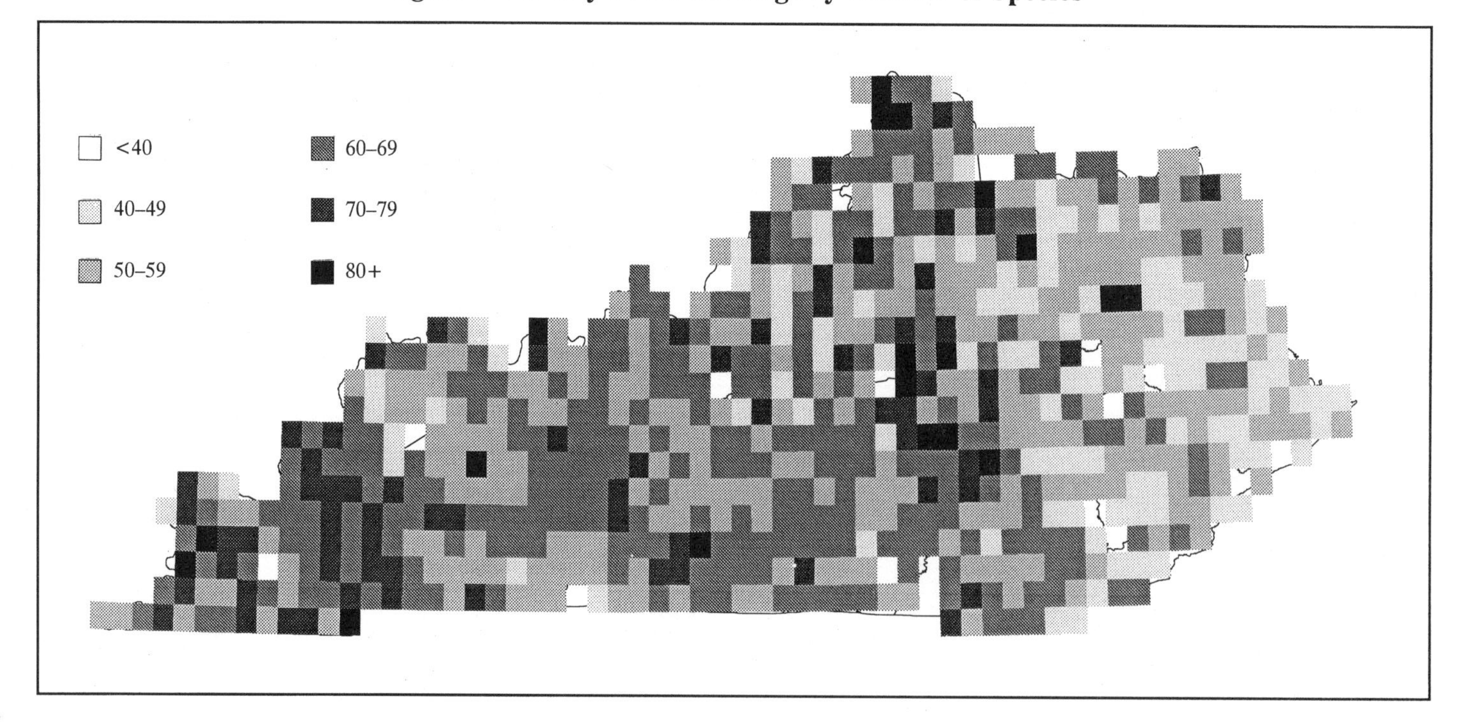

Table 4. Average Number of Species per Block by Physiographic Province Section

Physiographic Province Section	No. of Blocks in Section	Total No. of Species in Section*	Average No. of Species per Block
Mississippi Alluvial Plain	14	109	66.6
East Gulf Coastal Plain	36	120	64.0
Highland Rim	139	128	61.8
Shawnee Hills	142	133	59.8
Blue Grass	204	128	58.5
Cumberland Plateau	173	120	53.6
Cumberland Mountains	19	105	51.8
State Totals	727	154	58.5

*Includes incidental records

Abundance data also were used to compare overall relative abundance between species. The state's thirty most common breeding birds based on total abundance are presented in Table 5. Also included in Table 5 are the rankings for these species using the number of priority blocks, average abundance, and the National Biological Service BBS data for the period 1982–91. None of these ranking schemes are based on a comprehensive quantitative methodology, and the three ranking schemes using atlas data are biased by several factors, noted in the section "Limitations/Qualifications of Data." The BBS ranking is based on a more consistent methodology than was the atlas survey, but it is also biased by some of the same factors, as well as limitations in overall state coverage. Most BBS routes surveyed are in central and western Kentucky, where habitats are much more open; relatively few routes are run in the heavily forested eastern part of the state.

Forest Cover

Forest cover values were incorporated into the atlas data set and used in analysis of patterns in individual species' distribution and abundance, especially relative to physiographic province sections. These values are only approximations of the degree of forestation, and they are based on a crude interpretation of the USGS quadrangle maps. Comments on the influence of the amount of forest cover on individual species' distribution and abundance are presented in the species accounts.

Forest cover values for priority blocks are plotted in Figure 4 and arranged by physiographic province section in Table 6. Forest cover is highest in the Cumberland Plateau and Mountains and lowest in the East Gulf Coastal Plain. The Blue Grass section's average value is relatively high, because of the inclusion of the Knobs subsection with the predominantly cleared Inner and Outer Blue Grass subsections. Likewise, the Highland Rim's eastern transition to the Cumberland Plateau is more heavily forested than most of its central and western portions. Figure 4 and Table 6 are especially useful when compared with the maps and occurrence tables presented in the species accounts. Further discussion of characteristics of the state's major physiographic province sections can be found in Appendix A.

Table 5. Ranks of Most Common Kentucky Breeding Species

Species	Rank by*			
	Total Abund.	Number of Blocks	Average Abund.	BBS Data 1982–91
Indigo Bunting	1	1	1	6
Northern Cardinal	2	2	2	9
American Robin	3	3	3	5
Mourning Dove	4	11	4	8
Common Yellowthroat	5	10	9	14
Carolina Chickadee	6	6	10	35
Carolina Wren	7	4	13	27
Tufted Titmouse	8	7	15	21
Barn Swallow	9	15	11	11
House Sparrow	10	19	8	4
Blue Jay	11	5	19	17
Eastern Bluebird	12	14	16	20
American Crow	13	8	21	10
Red-winged Blackbird	14	25	7	3
European Starling	15	28	6	2
American Goldfinch	16	9	24	25
Rufous-sided Towhee	17	12	23	24
Eastern Meadowlark	18	32	4	7
Field Sparrow	19	17	20	16
Common Grackle	20	31	12	1
Eastern Wood-Pewee	21	13	25	31
Red-eyed Vireo	22	22	18	29
Chimney Swift	23	16	27	12
Song Sparrow	24	39	14	15
Chipping Sparrow	25	21	26	23
Eastern Kingbird	26	29	22	34
Northern Mockingbird	27	40	17	19
Blue-gray Gnatcatcher	28	20	29	41
Brown-headed Cowbird	29	27	28	18
Downy Woodpecker	30	18	33	45

* *Total Abund.* is the sum of all abundances recorded for the species in priority blocks; *Number of Blocks* is the number of priority blocks in which the species was recorded; *Average Abund.* is the average of all abundances for priority blocks in which the species was reported; *BBS Data, 1982–91* is the average number of individuals recorded per BBS route for the period 1982–91. For example, the House Sparrow ranked tenth according to total abundance, nineteenth by the number of priority block records, eighth by average abundance, and fourth according to BBS data for the period 1982–91.

Figure 4. Estimated Percentage of Forest Cover

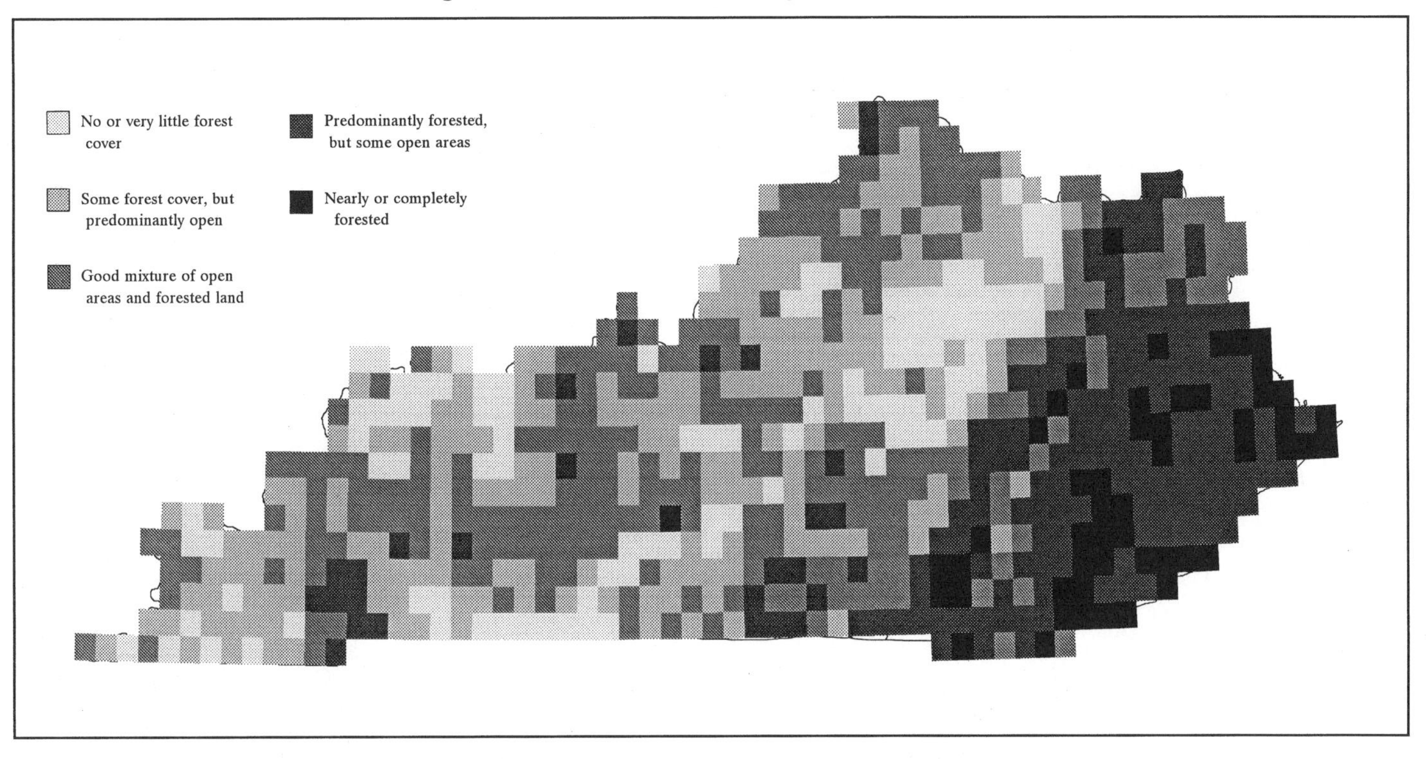

Based on interpretation of forest cover as depicted on USGS quadrangle maps. Values determined for each priority block were used to represent forest cover for the entire quadrangle.

Table 6. Forest Cover Values by Physiographic Province Section

Physiographic Province Section	Total Blocks	Percentage of Blocks in Section with Value =					Avg. Value
		1	2	3	4	5	
Mississippi Alluvial Plain	14	.14	.36	**.50**	—	—	2.36
East Gulf Coastal Plain	36	.22	**.58**	.14	.06	—	2.03
Highland Rim	139	.14	.30	**.51**	.04	—	2.46
Shawnee Hills	142	.17	**.35**	**.35**	.12	.01	2.42
Blue Grass	204	.23	**.34**	**.33**	.09	.01	2.30
Cumberland Plateau	173	—	.01	.16	**.59**	.24	4.05
Cumberland Mountains	19	—	—	.05	**.47**	**.47**	4.42
Total Blocks Statewide	727	.14	.26	.32	.21	.07	2.82

See page 4 for definitions of forest cover values; predominant values in each physiographic province section are shaded and appear in bold type. Physiographic province sections follow Fenneman 1938; see Appendix A and Figure 6.

Limitations/Qualifications of Data

Data Processing. Quality-control checks of the data set likely removed virtually all processing errors. Individual transcription errors by atlasers were believed to be infrequent; however, a few were detected by chance, and some degree of error must be assumed. Specific comments concerning the reliability of data are included in species accounts.

Coverage. Much fieldwork for the KBBA was conducted by volunteers, but seasonal employees of KDFWR and KSNPC completed a majority of priority blocks during the project's last four years. Most other atlas projects have used a goal-oriented system that allowed atlasers to determine when they had achieved sufficient coverage of their blocks. The KBBA did not use such a system, in large part because time constraints seldom allowed for priority blocks to be satisfactorily surveyed, especially by seasonal employees who typically devoted only half a day to each priority block. Surveying two priority blocks per day meant that one of the blocks was covered in the afternoon, when bird activity had declined dramatically. In contrast, blocks covered by volunteers were usually surveyed very well, and these blocks account for many of the highest species totals obtained during the atlas survey.

Largely as a result of time constraints, blocks covered by seasonal employees averaged only 57 species, while those covered by volunteers averaged more than 64. In contrast to Kentucky's atlas survey, in Ohio priority blocks were covered for at least an entire day, and much effort was expended to detect nocturnal species. The average number of species per block statewide in Ohio was 78 (Peterjohn and Rice 1991).

Time constraints also required that most survey work be conducted from roadsides, which biased results in favor of birds of more open habitats and reduced representation of woodland species. For example, woodpeckers were often difficult to record during atlas fieldwork. While most species of woodpecker are quite conspicuous in springtime, when territorial behavior is at its peak, calling drops off dramatically once nesting is under way, and the group as a whole is rather reclusive throughout the summer. Outside of a period of several hours in the morning, woodpeckers generally were difficult to detect, and it is likely that most were underreported both in occurrence and abundance. Other groups of birds that were underrepresented in the atlas results included waterbirds (e.g., waterfowl, rails), raptors (e.g., most hawks), early-season nesters (e.g., American Woodcock, Louisiana Waterthrush), and birds with soft or obscure songs (e.g., Cerulean Warbler, Henslow's Sparrow).

Although seasonal employees covered most priority blocks, volunteers surveyed a large percentage of blocks in certain parts of the state, most notably the Blue Grass and East Gulf Coastal Plain. This distinction biased the results obtained for certain species, most notably nocturnal ones. Time constraints did not allow seasonal employees to accomplish much nocturnal work, so the occurrence of most owls and nightjars usually appeared lower in areas surveyed primarily by seasonal employees. In contrast, volunteers typically visited blocks on several occasions and often included at least one night visit in their fieldwork.

Finally, such cursory coverage also meant that seasonal employees seldom had time to obtain substantive evidence of nesting, thus generating relatively few probable and confirmed breeding records. In this respect, data for Kentucky fall short of those for adjacent states, although this problem is also at least in part caused by the fact that the KBBA did not use a code for confirmed or probable breeding based on the presence of a certain number of territorial males. In Ohio, such a code was used to confirm breeding if seven or more territorial males were recorded in a block (Peterjohn and Rice 1991), and in Indiana (E. Hopkins, pers. comm.) and Tennessee (C. Nicholson, pers. comm.) a similar code was used to indicate probable breeding. In this regard Kentucky data are comparable to those of New York, where such a code was also not used (Andrle and Carroll 1988).

Breeding Codes. Although most breeding codes were used correctly, a few appeared to be misinterpreted by some observers. Comments on the misuse of specific codes are included in the species accounts when appropriate, but a brief discussion of a few of the most frequently misinterpreted codes is included here.

The FL code was used to confirm breeding based on the observation of recently fledged young; however, it was used too liberally in some cases, most notably when applied to family groups of full-grown young accompanying, but no longer dependent upon, their parents. Such sightings would have been more appropriately recorded as probable. Likewise, for some wide-ranging species, including most raptors, the FY code confirming breeding based on the observation of adults carrying food for young may have been used too liberally. For most species, however, the FY code was used appropriately. In fact, observations of adults carrying food were often accompanied by agitated action or distraction displays that served to reinforce the confirmation of the presence of young birds nearby.

Another code sometimes interpreted too broadly was the X code, indicating possible breeding based on the observation of a bird in appropriate breeding habitat during the breeding season. Although this code was used properly in most instances, occasionally it was applied to sightings that would have been recorded more appropriately as Observed, implying the probability that the bird was a vagrant, a late or early migrant, or a nonbreeder. For observations of rare or out-of-range, flyover raptors, the O code was usually applied. This was also the case for most observations of waterbirds and late summer sightings of passerines that were suspected of originating from other breeding areas. A nearly full-grown Black-throated Green Warbler observed during atlas work near Versailles, in Woodford County, on June 29, 1991, clearly represents a transient bird, but many less obvious cases confronted observers during the atlas fieldwork.

Abundance. The definitions given for the abundance codes were relatively straightforward, and most atlasers appeared to interpret them consistently. On the other hand, atlasers' impressions of abundance were subject to several biases, including completeness of block coverage, conspicuousness of individual species (size, song, habitat, etc.), and variability in levels of expertise. Most birds were recorded from roadsides during a day or less of time, and fairly widespread species were sometimes observed only once or twice. In general, woodland species were underrepresented in comparison with birds of open areas, where access was better, song carried farther, and visual sightings were more frequent. Good coverage of forested habitats often necessitated gaining access to private property, a step many times not taken during atlasing because of time constraints or reluctance to contact unknown landowners. Other groups of birds that likely were not recorded in actual abundance include nocturnal species, early-season nesters, those with crepuscular activity or song, those with soft or obscure songs, and those of special habitats, such as marshes.

Despite these limitations, the abundance data proved useful in comparing species' relative abundances between various regions of the state and degrees of forest cover. In addition, the abundance maps accompanying the species accounts provide a general picture of each species's statewide abundance and often accurately depict differences in relative abundance between species, especially within certain groups.

Forest Cover. Forest cover values were estimated by interpretation of USGS topographic maps. Thus they are only approximations of the degree of forestation. Field-checking usually corroborated the interpreted values, but it is possible that the data on some maps, especially older ones, were not completely accurate. Overall, the data as presented in Figure 4 and Table 6 correlate well with county-level data published by the U.S. Department of Agriculture (Alerich 1990).

Species Accounts

Most species accounts include the following information: a brief synopsis of the species's breeding occurrence in Kentucky, including references, particularly by Mengel (1965), to former status; a description of habitats used; a discussion of the species's probable distribution and abundance in Kentucky before settlement; a brief summary of the species's nesting biology, including peak dates of clutch completion and average clutch sizes; a summary of nest sites and a brief description of the nest, including average height and materials typically used in construction; a summary of atlas results, including discussion of variation in occurrence and abundance relative to physiographic province sections and percentage of forest cover; and any pertinent notes concerning trend data from the National Biological Service Breeding Bird Survey (BBS).

Some of these sections are occasionally omitted because of inadequate information. BBS data are not included for species with limited sample sizes. For very rare species, some or most of the above-mentioned sections are omitted in favor of a brief synopsis of breeding records, including historical information.

Robert M. Mengel's comprehensive monograph *The Birds of Kentucky* (1965) remains the benchmark for ornithological study in the state, and it was the primary resource for information on historical ranges and details of nesting season and behavior. Early references to the birdlife of Kentucky are relatively scarce compared with documentation in some surrounding states, and Mengel cited all of the substantial ones. In large part, specific references cited by Mengel and used by him to summarize distribution, occurrence, average dates of clutch completion, clutch sizes, and nest heights are not repeated herein. If more specific information was taken from these sources, their citation is included. Literature accounts published since *The Birds of Kentucky*—which was primarily completed by the late 1950s—are cited much more frequently, especially when they contain additions to nesting data presented by Mengel. Some data generated by the KBBA have already appeared in published form, especially in the Seasonal Reports feature of *The Kentucky Warbler.* As a general rule, for most observations generated by the KBBA that have already been published, citations are not included, especially if these observations appeared only in the Seasonal Reports; however, items that appeared in longer field notes and articles are cited, and all data obtained through independent field efforts during the atlas period (1985–91) are cited, even if they appeared only in the Seasonal Reports.

For most species observed in more than 10 percent of the priority blocks, a Forest Cover table is included that summarizes occurrence and average abundance relative to the percentage of forest cover in the priority block. This table is not included for most nocturnal species because of biased coverage. Only priority block data are included in the Forest Cover table. While this analysis is obviously based on generalized evaluations, it helps to illustrate the effect of one of the most influential factors on occurrence and abundance trends observed for many species.

Typically a section in each account is devoted to discussion of a species's status in Kentucky before European settlement. In some cases it is impossible to speculate about the status of individual species two centuries ago, but many times a reasonable guess is possible. Human alteration of Kentucky's landscape has been profound, and the avifauna of the state today is much different than that which European settlers first encountered in the latter part of the 18th century. Before that time, climatological changes had driven a slow but steady evolution of natural communities. In recent geologic times Native Americans had altered the landscape to some degree, and certain activities had probably affected many species. For example, the construction of villages and the clearing of land for crops likely resulted in local changes in avifauna. The rate of these changes, however, was very slow compared with the rate of changes that have occurred since European settlement. For this reason, it is worthwhile to reflect on the presettlement status of Kentucky's avifauna. An understanding of human influence on the landscape is a useful tool in recognizing some of the trends we see in bird populations today. It is also intriguing to recognize that some of our most common and widespread species may have been among the rarest only two centuries ago. The historical section focuses on a comparison of presettlement conditions with the current landscape. For this reason, it typically does not discuss short-term human-induced changes that have occurred during the 20th century, such as the severe deforestation that occurred during the early 1900s. Many forested areas were completely denuded during a period of intensive use of wood for construction and power. Kentucky's woodlands continue to recover from this era today, resulting in local avifaunal changes that are not usually addressed herein.

Discussion of BBS data typically includes results of trend analysis performed by the National Biological Service (B. Peterjohn, pers. comm.). For statistically significant trends, competence or p values included in the text are based on the National Biological Service analysis. Trends are not considered statistically significant for $p>0.1$.

Scientific and common names used herein conform to the latest revisions of the American Ornithologists' Union (AOU) *Check-list of North American Birds* (AOU 1983) and more recent supplements (AOU 1985, 1987, 1989, 1991, 1993). The species are presented in the taxonomic order specified by the same sources.

Maps

The tables and figures in the Introduction serve as references to assist the reader in interpreting data contained on the species maps. Figure 4 (p. 11) is especially useful for comparing species occurrence and abundance to percentage of forest cover. Figure 5 (p. 334) serves as a reference for county boundaries.

In each species account, the **Breeding** map plots the breeding codes recorded by atlasers during fieldwork. Square symbols indicate records obtained within priority blocks, and these are plotted at the true location of observation. Incidental records (including observations from special blocks) are plotted as triangles at the priority block location; thus they do not represent true locations of observation.

The **Abundance** maps show abundance values as recorded by atlasers during fieldwork. In addition to priority block data, these maps also include values for incidental reports. The symbols are plotted so that they represent abundance over the entire quadrangle for which either priority block or incidental abundance values were submitted. In cases where both a priority block and an incidental abundance value were submitted for a given quadrangle, the priority block value is plotted. Abundance values reported by atlasers often underrepresented true abundance in local areas, and they were subject to several biases, discussed above. For these reasons, the reader is cautioned not to attempt to apply the data site specifically or to compare local abundances critically. These maps are intended only to provide a generalized picture of each species's abundance across the state.

Tables

An **Analysis of Block Data by Physiographic Province Section** table provides a breakdown of atlas data by Kentucky's seven major physiographic province sections as defined by Fenneman (1938). The boundaries of these regions as interpreted for this work are illustrated in Figure 6 (p. 335). The **Summary of Breeding Data** table provides a breakdown of the breeding code data. Both tables include data from priority blocks only; incidental records are not included. Percentages in columns in both tables may not total 100.0 because of rounding. Average abundance is not included in the physiographic province section table for species for which a forest cover table is not provided in the species account.

Abbreviations

The following abbreviations are used in the text that follows:

AOU American Ornithologists' Union
BBS Breeding Bird Survey
KDFWR Kentucky Department of Fish and Wildlife Resources
KOS Kentucky Ornithological Society
KSNPC Kentucky State Nature Preserves Commission
NWR National Wildlife Refuge
USFWS United States Fish and Wildlife Service
WMA Wildlife Management Area

BREEDING SPECIES OBSERVED

Pied-billed Grebe

Podilymbus podiceps

The peculiar courtship calls of the Pied-billed Grebe are a rare sound in Kentucky today. This grebe has apparently declined as a nesting bird since the late 1950s, when Mengel (1965) described it as a fairly common summer resident, breeding locally west of the Cumberland Plateau. Mengel attributed most observers' impressions that the species was rare not necessarily to a scarcity of birds but rather to the rarity of nesting habitat. He further noted that Pied-billed Grebes occurred regularly in all suitable situations investigated. While the species is probably overlooked to some extent today, it certainly does not occur regularly in suitable nesting habitat.

Before the atlas fieldwork, Pied-billed Grebes had been reported nesting in six counties: Daviess (one record), Fulton (one), Hopkins (several at two localities), Jefferson (several at one locality), Trigg (two at one locality), and Warren (several at one or two localities). Despite the paucity of historical breeding records, it is likely that the species was somewhat more numerous in the past, especially before settlement. During the past two centuries the amount of suitable nesting habitat has declined significantly, as wetlands have been cleared and drained for development and agricultural purposes. It is not known to what extent the accumulation of residues from pesticides such as DDT may have contributed to a decrease in the species in the mid–20th century, but habitat destruction has more likely had a much greater effect. In contrast, it is possible that a recent resurgence of the beaver will eventually result in an increase in the number of nesting Pied-billed Grebes, because of the increase in available habitat created or enhanced by these mammals.

Pied-billed Grebes occur in marshy, shallow water habitats with an abundance of submerged vegetation. Although permanent bodies of water are often used, these small grebes also frequent temporary ponds and sloughs formed by heavy spring rains. The species is generally absent on larger open-water lakes, but when present it occurs along marshy or weedy shallow-water margins. The few recent records of summering birds have come from the temporary karst lakes of the Highland Rim, as well as seasonally flooded impoundments in the Shawnee Hills. In southern Warren County, Pied-billed Grebes linger and probably attempt to nest whenever water in the karst lakes lasts into May (Wilson 1962a). While it is likely that water does not remain long enough for nesting to be completed in most years, young appear to be raised successfully at least on an occasional basis.

The origin of the few Pied-billed Grebes that nest in Kentucky is uncertain, but most birds likely come from wintering areas farther south. Courtship activities may commence by the end of March, when transient grebes are still very much in evidence. Nest building has been observed as early as the first part of April, although breeding activity greatly depends upon water levels. Egg laying has been noted before the end of April, but clutch completion probably peaks during May (Mengel 1965). Young birds have been recorded most frequently in late May and June, and the observation of two small young in Trigg County on August 30, 1941, represents the latest observation of nesting activity (Mengel 1965). It is unclear whether late records indicate the raising of a second brood or attempts to renest following earlier failures. Mengel gave the average size of 14 clutches known or thought to be complete as 6.3 (range of 5–9). A nest observed during atlas fieldwork at Chaney Lake in Warren County on May 15, 1991, contained five eggs.

Pied-billed Grebes build a bulky nest of dead plant material and algae that is loosely attached to both submerged and emergent vegetation. Although the nest is partially submerged by its weight, it will float at the water's surface. Large fluctuations in water level, however, sometimes cause inundation or exposure of the nest and may result in abandonment. When leaving the nest to forage, the adult bird typically covers the eggs with nest material.

Philippe Roca

Atlas fieldwork yielded one priority block record of the Pied-billed Grebe, and two incidental observations were reported. Confirmed breeding was reported from only one location, the transient lakes near Woodburn in southern Warren County. In early July 1989 a single juvenile bird was observed at McElroy Lake (Palmer-Ball and Boggs 1991). Although an active nest was not located at the lake, adults were present throughout the season, and it is likely that the young bird was raised there. In mid-May 1991 two nests containing eggs were found at nearby Chaney Lake, and at least one young bird was observed being fed by adults there through mid-June (Palmer-Ball 1991b).

The only other evidence of nesting obtained during the atlas fieldwork came from the Sauerheber Unit of Sloughs WMA in western Henderson County, where territorial birds were heard and seen on at least three different impoundments in various years during the atlas period. Four additional reports were submitted without evidence of nesting; they have not been included in the atlas results because of the possibility that these birds were migrants or nonbreeders. Only one of these reports originated from suitable breeding habitat, in a cattail marsh along Mayfield Creek north of Cunningham, in Carlisle County.

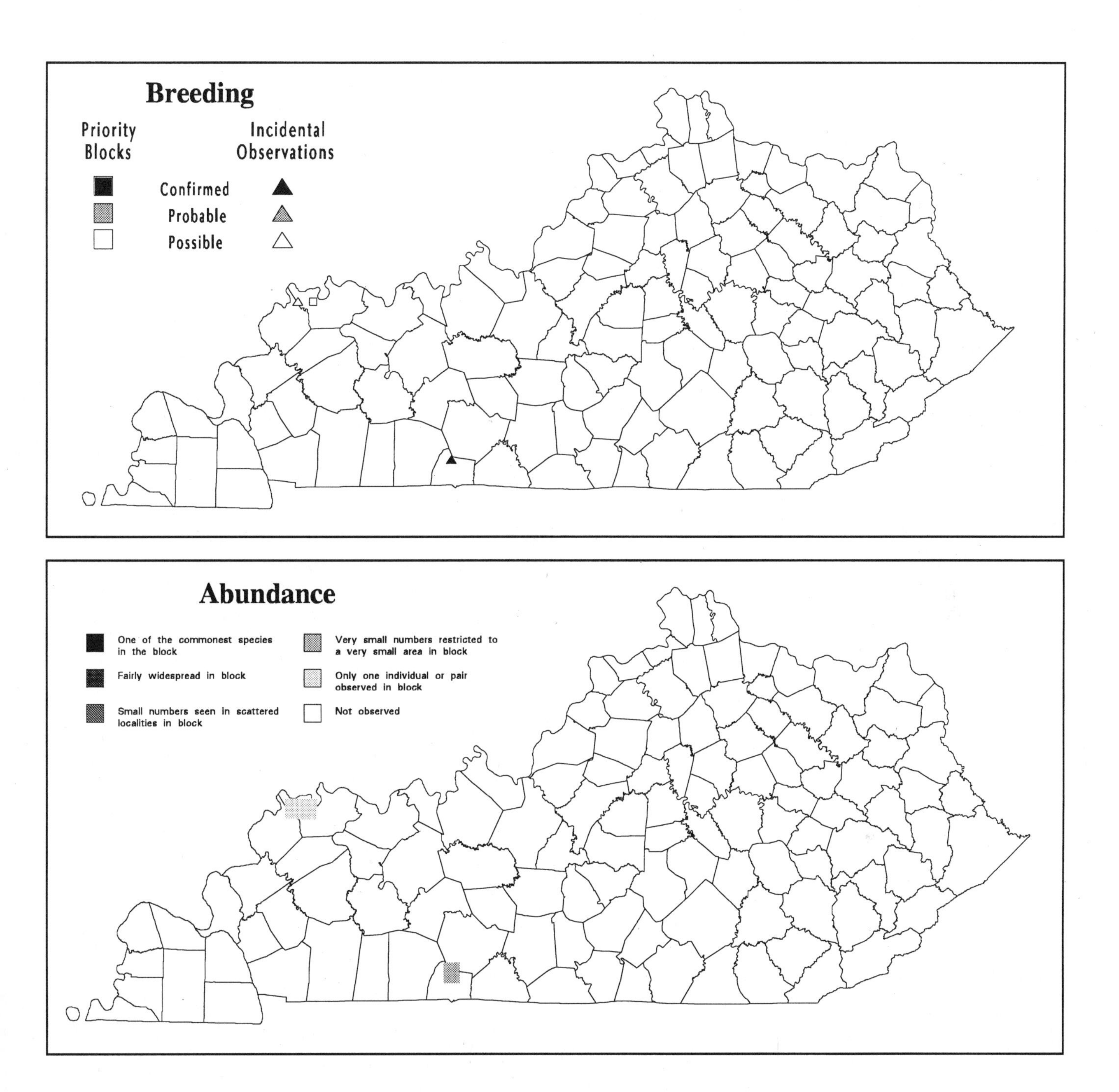

Analysis of Block Data by Physiographic Province Section

Physiographic Province Section	Total Blocks Surveyed	Blocks with Data	% with Data	Section's % for State
Mississippi Alluvial Plain	14	-	-	-
East Gulf Coastal Plain	36	-	-	-
Highland Rim	139	-	-	-
Shawnee Hills	142	1	0.7	100.0
Blue Grass	204	-	-	-
Cumberland Plateau	173	-	-	-
Cumberland Mountains	19	-	-	-

Summary of Breeding Status

Number of Blocks in Which Species Was Recorded		
Total	**1**	**0.1%**
Confirmed	-	-
Probable	-	-
Possible	1	100.0%

Pied-billed Grebe

Least Bittern

Ixobrychus exilis

Least Bitterns nest across much of the eastern United States and southern Canada, including a large part of Kentucky (AOU 1983). As is the case throughout much of the species's range, however, these small wading birds are very locally distributed, being reported regularly at only a few localities and nowhere in substantial numbers. Mengel (1965) said that the species, in summer, was distributed widely but locally in favorable situations west of the Cumberland Plateau.

The Least Bittern's breeding status in Kentucky has always been unclear. Only a few historical records exist, and the majority of these originated from one small marsh in Jefferson County that has been destroyed. Documented nesting records have been published for the counties of Carroll (Webster 1951; Lovell 1951, 1952), Daviess (Gray 1968b; Stamm 1969), Hopkins (Hancock 1954), Jefferson (Monroe 1935; Mengel 1965), and Nelson (Stamm and Croft 1968). Additional summer sightings have been published for Ballard (Mengel 1965; Palmer-Ball and Barron 1982), Calloway and Crittenden (Wilson 1923, 1942), Clinton (Ganier 1935), Edmonson (Wilson 1962a), and Fayette and Marshall (Mengel 1965) Counties. The extent to which the species has been overlooked is unknown, but its reclusive behavior and the inaccessibility of its preferred habitat must account in large part for the paucity of records.

Least Bitterns are encountered primarily in marshy habitats dominated by herbaceous aquatic vegetation such as cattails, burreeds, bulrushes, and sedges. In Kentucky they have been found nesting in both natural and artificial situations, including marshes, natural lakes and ponds, reservoirs, waterfowl management impoundments, and fish hatchery brood ponds. Unlike the American Bittern, which is regularly found in drier habitats, the Least Bittern is typically found in association with an aquatic environment.

Least Bitterns winter in the southern United States, Mexico, and Central America (AOU 1983). Summer residents arrive on their Kentucky breeding grounds during the first two weeks of May, and nesting activities may be under way before the end of the month. According to Mengel (1965), clutches are completed from about May 20 to July 2, with a peak in clutch completion about June 1–10. Mengel gives the average size of eight clutches as 4.4 (range of 3–6). Two nests have been reported in detail since Mengel: one in a marshy pond in Nelson County containing six eggs on June 27, 1967 (Stamm and Croft 1968), and one at Carpenter's Lake, in Daviess County, containing five eggs on July 13, 1968 (Stamm 1969).

The nest is typically concealed among rank herbaceous growth and consists of a bulky platform constructed of whatever dead plant material is readily available. It is usually placed just above the water, but one nest in Jefferson County was situated about four feet above the water in the crown of a fallen willow tree (Mengel 1965). Pairs may nest in rather dense colonies; 19 nests were found in one small marsh in Jefferson County in 1935 (Mengel 1965).

The atlas survey yielded five records of Least Bitterns, three of which were incidental observations. Although none of the records were for confirmed breeding, observations of more than one bird on several occasions during the nesting season constituted probable breeding at three localities: a cattail marsh along Mayfield Creek, in Carlisle County, where four birds were seen on June 17, 1990 (Stamm and Monroe 1990a); the Sauerheber Unit of Sloughs WMA, in Henderson County, where birds were observed on several occasions in late spring and summer; and the Minor Clark Fish Hatchery, in Rowan County, where up to three birds were seen on several occasions in June and July 1985 (Stamm 1985) and twice in May 1988 (Stamm 1988c).

Gary Meszaros

In addition, possible breeding was reported at two localities. Up to seven birds were reported on a cattail marsh and pond on a reclaimed surface mine near Cool Springs, in Ohio County, in July and early August 1989 (Stamm 1989a). These observations were only reported as possible breeding because they occurred so late in the summer. The species was also observed in suitable nesting habitat in the summer of 1985 at Pond Bayou in western Henderson County.

From the results of the atlas fieldwork, it appears that the Least Bittern remains a widely distributed, although sporadic, summer resident in Kentucky. Perhaps in the future the species will benefit from wetland protection projects being undertaken through public and private efforts.

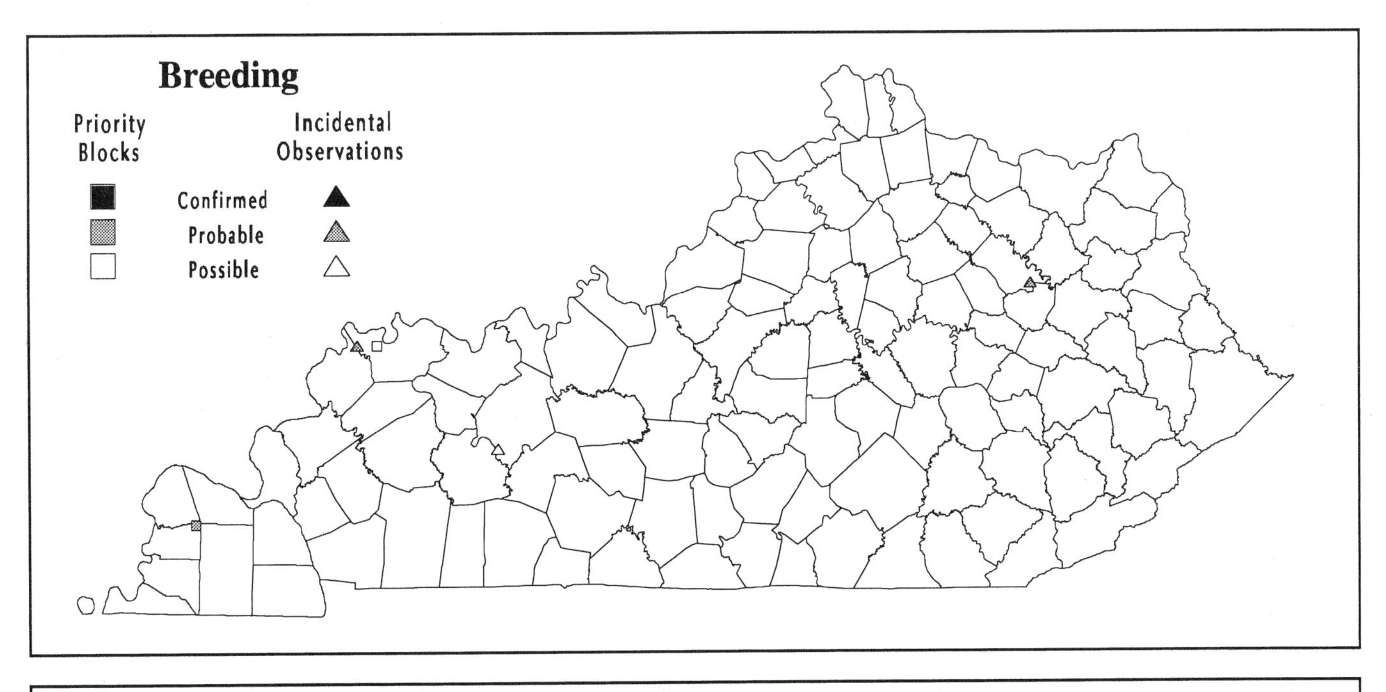

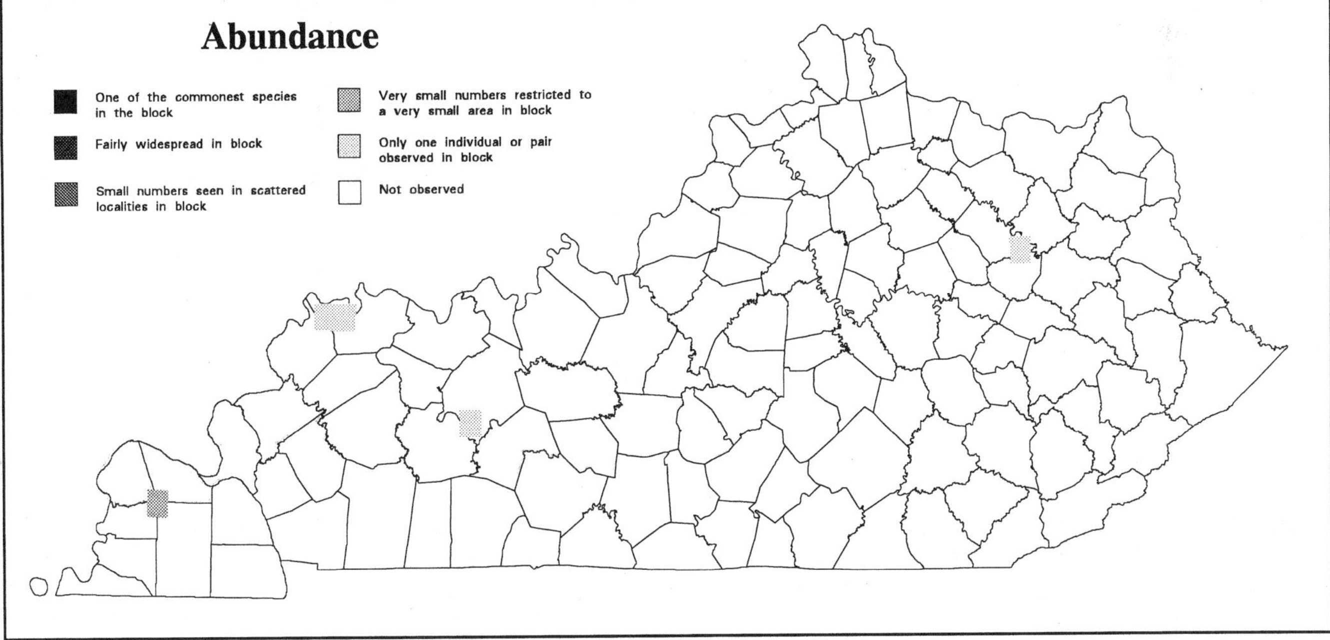

Analysis of Block Data by Physiographic Province Section

Physiographic Province Section	Total Blocks Surveyed	Blocks with Data	% with Data	Section's % for State
Mississippi Alluvial Plain	14	-	-	-
East Gulf Coastal Plain	36	1	2.8	50.0
Highland Rim	139	-	-	-
Shawnee Hills	142	1	0.7	50.0
Blue Grass	204	-	-	-
Cumberland Plateau	173	-	-	-
Cumberland Mountains	19	-	-	-

Summary of Breeding Status

Number of Blocks in Which Species Was Recorded		
Total	**2**	**0.3%**
Confirmed	-	-
Probable	1	50.0%
Possible	1	50.0%

Least Bittern

Great Blue Heron

Ardea herodias

The Great Blue Heron is the most widely distributed large wading bird in Kentucky. Although the species decreased significantly during the 1950s and 1960s, it has recovered remarkably in the last two decades and at present appears to be more plentiful than at any previous time.

Little is known about the Great Blue Heron's historical status in Kentucky. It may have been affected less dramatically in the late 1800s and early 1900s by hunting for the millinery trade than the intensely sought egrets. As of the early 1950s, Great Blues had been reported nesting in eight distinct colonies, primarily along the Mississippi and lower Ohio Rivers from Henderson County southwestward. These heronries each contained from 12 to 300 nests (Mengel 1965).

It seems that some time in the 1950s or early 1960s, the numbers of nesting Great Blues declined dramatically in Kentucky. The ornithological literature lacks published reports of nesting birds during the 1960s and early 1970s, and according to some accounts (e.g., Wiley 1964), some colonies disappeared. The species's disappearance coincided with a period when residues from chlorinated hydrocarbon pesticides (primarily DDT) were first being recognized as causing dramatic declines in some other waterbirds and raptors (Anderson et al. 1969; Cooke 1973). After the banning of DDT and other harmful pesticides in the United States in 1972, the species began to recover, and reports of nesting reappeared in the ornithological literature in the mid-1970s (Dodson 1976, 1977). Since that time Great Blue Herons have continued to become more common, and additional nesting colonies have been discovered. As of 1984 the species was reported nesting in at least five colonies of 77 to 262 active nests (S. Evans, unpub. rpt.). By 1990 more than 1,300 breeding pairs were known from 16 heronries in 12 counties (Palmer-Ball 1991a).

Great Blue Herons occur in a great variety of habitats, from small creeks and farm ponds to the shores of the largest lakes and rivers. Although these large waders formerly nested only in the vicinity of the larger rivers and floodplain swamps of western Kentucky, they have begun to nest on reservoirs during the last two decades.

Many Great Blues remain in the state throughout the winter, and the breeding season typically begins early. It is not uncommon for nest building to be under way by early March in the heronries of far western Kentucky. Farther north and east, nesting may be delayed for several weeks. Most clutches are probably completed by the end of March, although later clutches may not be completed until late April. Incubation requires nearly a month (Ehrlich et al. 1988), and most young hatch during April. By the middle of June the young have fledged from most nests, and by mid-July most heronries are nearly empty.

Great Blue Herons are colonial nesters. Kentucky heronries have included as few as 3 nests (Stamm 1990) and as many as 432 (Stamm 1986a). Colonies are typically situated in tracts of large trees along riparian corridors or within swamps, but sometimes they occur in bottomland or upland forest relatively far from water. Nests have been observed in a variety of deciduous trees; living trees are usually used, but some colonies have been located in stands of dead or dying timber. Kentucky data on clutch size are lacking, but clutches of 3–5 eggs have been noted at Reelfoot Lake, just south of the state line (Ganier 1933; Gersbacher 1939).

Nests are bulky platforms, constructed of sticks and branches and lined with finer twigs and green leaves (Harrison 1975). They are typically placed within a fork of an outer branch, rather high up in the tree and just inside the outer crown of foliage. More than a dozen nests may be constructed in a single tree.

Bud Tindall

The atlas survey yielded records of Great Blue Herons in nearly 20% of priority blocks statewide; however, most of these reports were known, or believed, to pertain to nonbreeding birds, especially in central and eastern Kentucky. For this reason, the only records included in the atlas data set were those of active nesting colonies and those involving adult birds observed repeatedly during the late spring and early summer.

Five priority block records and 18 incidental observations of confirmed breeding were reported during the atlas survey, including one from the extreme eastern Highland Rim. All 22 records represented the observation of active nesting colonies. Before 1991 Great Blues were only known to nest as far east as Ohio County (Palmer-Ball and Barron 1990), but in 1992 small nesting colonies were discovered near Lake Cumberland (Hodges 1992) and Laurel River Lake (J. MacGregor, pers. comm.). According to local reports, the species used the Lake Cumberland site in 1991, the final year of the atlas survey. If the current trend continues, most heronries should be expected to grow, and additional colonies should become established, especially near larger rivers and reservoirs.

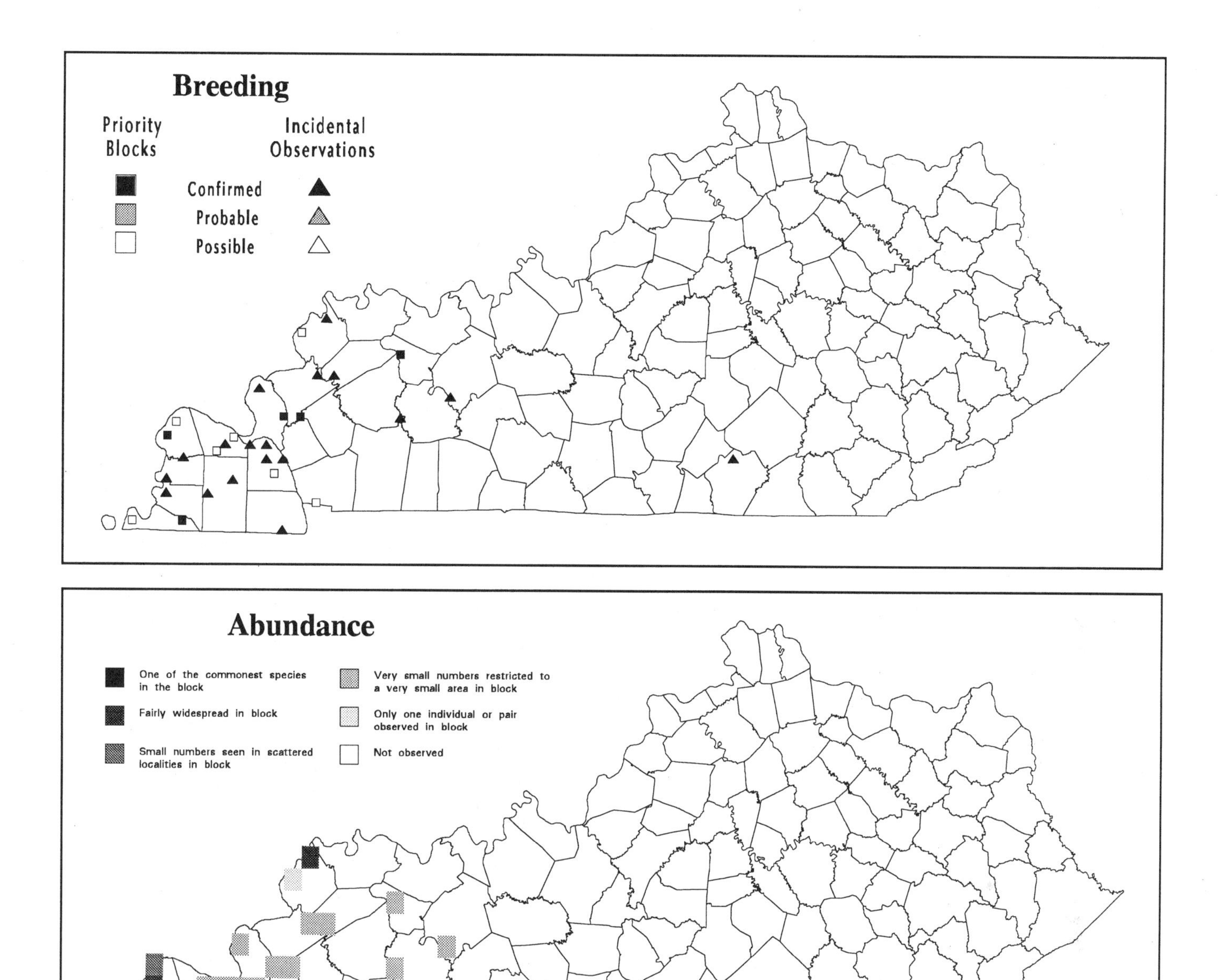

Analysis of Block Data by Physiographic Province Section

Physiographic Province Section	Total Blocks Surveyed	Blocks with Data	% with Data	Section's % for State
Mississippi Alluvial Plain	14	4	28.6	33.3
East Gulf Coastal Plain	36	3	8.3	25.0
Highland Rim	139	3	2.2	25.0
Shawnee Hills	142	2	1.4	16.7
Blue Grass	204	-	-	-
Cumberland Plateau	173	-	-	-
Cumberland Mountains	19	-	-	-

Summary of Breeding Status

Number of Blocks in Which Species Was Recorded		
Total	**12**	**1.7%**
Confirmed	5	41.7%
Probable	-	-
Possible	7	58.3%

Great Blue Heron

Great Egret

Casmerodius albus

Although nowhere more abundant than in the coastal regions of the southeastern United States, the Great Egret is a locally common but sporadically distributed breeding bird throughout the Great Lakes and the Mississippi Valley (AOU 1983). In Kentucky the species has been documented nesting only in the western third of the state, where it is typically associated with the large river floodplains.

It appears that numbers of Great Egrets have fluctuated widely since early ornithological work was undertaken in the Midwest. Regional accounts suggest that the species was common and widely distributed in the Mississippi Valley during the 1800s (Nelson 1877; Langdon 1879; Butler 1897; Widmann 1907), but reports of its occurrence in Kentucky at that time are limited. During the second half of the 19th century, some species of wading birds, including the Great Egret, were pushed to the brink of extinction by overhunting for the millinery trade. Fortunately, these birds were given legal protection in the early 1900s, and their numbers began to rebound soon afterward (Terres 1980). During the 1920s reports of Great Egrets began increasing in central and western Kentucky, and by the mid-1930s the species was documented as breeding for the first time (Mengel 1965).

As of the early 1950s the species had been documented nesting with Great Blue Herons at five mixed-species colonies in western Kentucky. These heronries were located along the Ohio River in western Henderson County; at Axe Lake, in Ballard County; near Fish Lake, in Carlisle County; at Murphy's Pond, in Hickman County; and on the northern end of Reelfoot Lake in what is now the Reelfoot NWR, in Fulton County. Numbers of nesting pairs varied from only a few at most sites to more than 150 in Fulton County (Mengel 1965).

Published accounts of nesting Great Egrets after the mid-1950s are lacking, and it seems that the species disappeared from Kentucky as a breeding bird. Numbers of birds decreased dramatically in nonbreeding areas during this time as well. Stamm et al. (1967) observed a decrease in postbreeding migrants at the Falls of the Ohio through the early 1960s. They noted that fewer birds appeared each year, until in 1966 no more than six were observed on any one occasion. In contrast, as many as 60 to 80 birds had been seen there in 1959 (Stamm et al. 1960). This decline occurred during a period when the harmful effects of residues from chlorinated hydrocarbon pesticides (primarily DDT) were first being recognized.

After the banning of DDT in the United States in 1972, wading birds began to recover, and Great Egrets gradually occurred more regularly through the mid-1980s. The species was finally rediscovered breeding in Kentucky in 1986, when two nests were found in the Great Blue Heron nesting colony at Axe Lake (Stamm 1986a). Before the late 1980s Great Egrets were known to have nested only in the vicinity of natural floodplain wetlands of the larger rivers, but on August 19, 1989, a nest containing three young was observed in the Black-crowned Night-Heron colony on Lake Barkley, in Trigg County (Stamm 1990). This discovery not only represented the second confirmed nesting in recent years, but also the first on an impoundment.

Great Egrets winter from coastal regions of the southern United States southward into the tropics (AOU 1983), and locally nesting birds return to western Kentucky during late March and early April. Specific information on breeding in Kentucky is lacking, but nesting activity probably begins before the middle of April. Young have been observed in nests from May 20 (small) to August 19 (ready to fledge), indicating that clutches are completed from mid-April to late June. Kentucky data on clutch size are lacking, but Harrison (1975) gives rangewide clutch size as 3–4, rarely more.

Maslowski Wildlife Productions

Great Egrets in Kentucky have always been found nesting within colonies of more common wading birds, typically Great Blue Herons. Like other large waders, Great Egrets usually place their bulky stick nests high in large trees of a riparian corridor, swamp, or bottomland forest. In contrast, the nest observed on Lake Barkley in 1989 was situated only about 15 feet above the water in a grove of relatively small green ash trees.

Great Egrets were found nesting at two localities during the atlas fieldwork. While the Lake Barkley site was apparently used for only one year, small numbers of egrets continue to nest at Axe Lake. In addition, Great Egret feathers were seen on the ground at the Great Blue Heron nesting colony in Union County in June 1991, and adult birds were observed in the vicinity on several occasions during the atlas period. Nesting could never be confirmed, however. Small numbers also were observed at scattered localities along the Mississippi and lower Ohio River floodplains from Henderson County southwestward, but most or all of these observations probably represent nonbreeders and birds from nesting colonies in adjacent states. For example, small feeding groups observed in western Fulton County certainly originated from the nesting colony at Reelfoot Lake, a few miles south of the Tennessee state line. Breeding at other sites in western Kentucky is possible, especially if the species continues to increase.

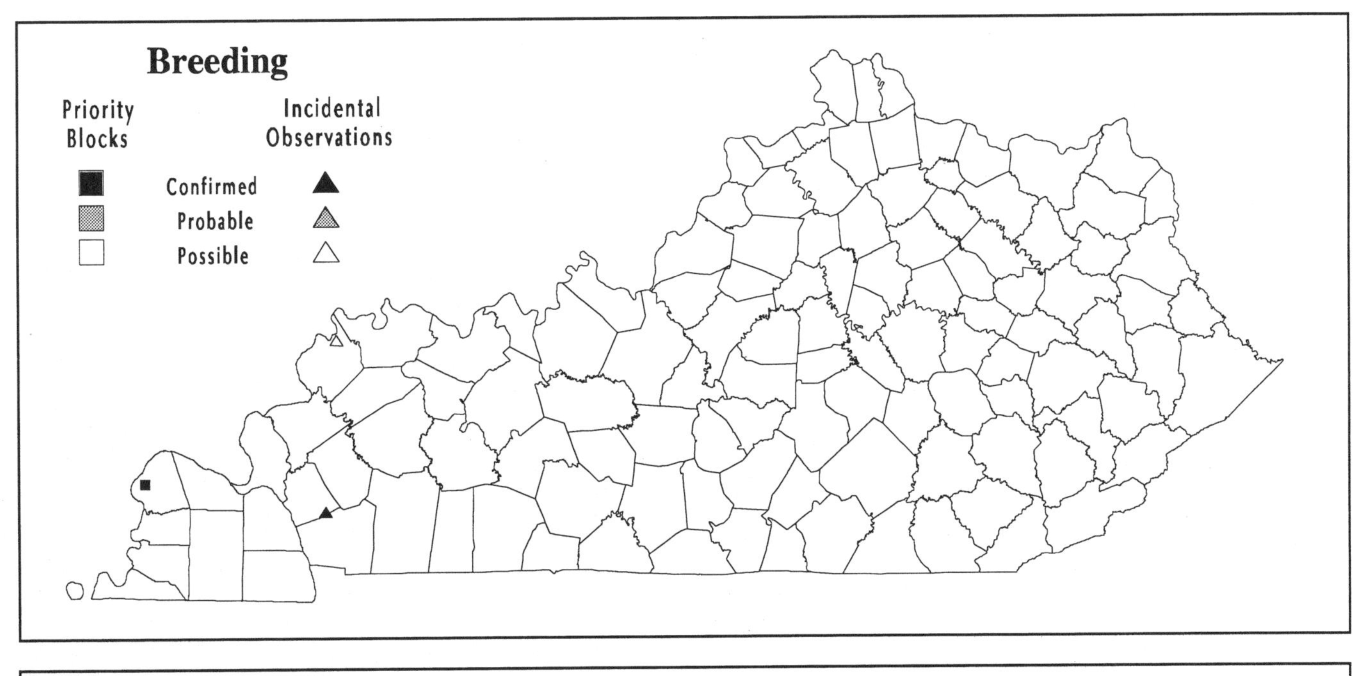

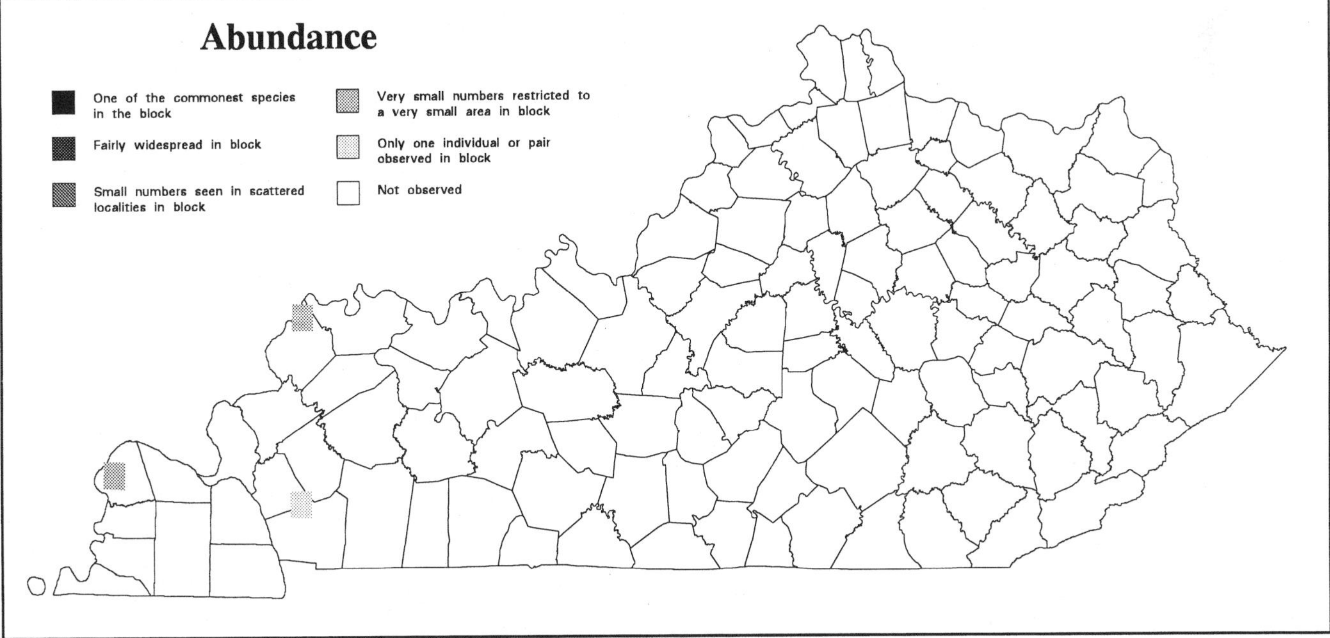

Analysis of Block Data by Physiographic Province Section

Physiographic Province Section	Total Blocks Surveyed	Blocks with Data	% with Data	Section's % for State
Mississippi Alluvial Plain	14	1	7.1	100.0
East Gulf Coastal Plain	36	-	-	-
Highland Rim	139	-	-	-
Shawnee Hills	142	-	-	-
Blue Grass	204	-	-	-
Cumberland Plateau	173	-	-	-
Cumberland Mountains	19	-	-	-

Summary of Breeding Status

Number of Blocks in Which Species Was Recorded		
Total	**1**	**0.1%**
Confirmed	1	100.0%
Probable	-	-
Possible	-	-

Great Egret

Little Blue Heron

Egretta caerulea

The Little Blue Heron is a recent addition to Kentucky's breeding avifauna. Although substantial nesting colonies have been known from along the Mississippi River in western Tennessee since the early 1950s (Ganier 1951) and southeastern Missouri since the mid-1970s (Robbins and Easterla 1992), the species was not discovered nesting in Kentucky until the early 1980s, and then only in very limited numbers.

In the early 1800s the Little Blue Heron was apparently more widespread in the Midwest, and the species at least occasionally nested well north of Kentucky (Butler 1897). During the latter half of the 19th century the species declined substantially, probably in large part because of overhunting for the millinery trade (Mengel 1965). It is possible that Little Blues nested in Kentucky before this decline occurred, but specific references to such are lacking.

Twentieth-century documentation of the Little Blue Heron's occurrence in Kentucky is more substantial (Mengel 1965), although through the 1970s the species was known only as a spring transient and postbreeding summer and fall visitant. A steady decrease in the numbers of postbreeding birds at the Falls of the Ohio was noted during the 1950s (Stamm et al. 1960), and numbers remained low throughout the DDT era of the 1960s and early 1970s (Stamm et al. 1967; Monroe 1976). During the 1980s numbers rebounded to some extent, especially in far western Kentucky (Palmer-Ball and Barron 1982; Stamm 1985) where large nesting colonies have become established nearby in Missouri (Robbins and Easterla 1992).

In 1981 a few Little Blue Herons were discovered in a Black-crowned Night-Heron nesting colony on Lake Barkley, in Trigg County (Thomas 1982). A couple of pairs apparently nested in this colony in 1983, but none could be found in later years (S. Evans, unpub. rpt.; B. Palmer-Ball and J. MacGregor, unpub. rpt.). After the relocation of this heronry about 10 miles north on the lake in the mid-1980s, Little Blues have not been confirmed nesting, but occasional sightings have been made there during the breeding season, indicating that nesting may occur on occasion.

The state's second breeding record was established when a pair of Little Blue Herons was found nesting in a Black-crowned Night-Heron colony on Shippingport Island near the Falls of the Ohio, in Jefferson County, in 1985 (Palmer-Ball and Evans 1986). Although confirmed nesting of Little Blues was not documented at this colony in subsequent years, one or two birds were seen there occasionally through the late 1980s. This site was abandoned in 1992, and Little Blue Herons have not been seen at the relocation site of the Black-crowned Night-Heron colony on the grounds of the Louisville Zoo (G. Michael, pers. comm.).

Little Blues have been found nesting in Kentucky only in association with other wading birds, specifically Black-crowned Night-Herons. In southeastern Missouri the species is presently nesting primarily with Cattle Egrets and a few Snowy Egrets, and hundreds of pairs of Little Blues are in each of two colonies (Robbins and Easterla 1992). Little Blue Herons seem to avoid nesting with larger species like the Great Blue Heron, perhaps because the Little Blue prefers to nest in smaller trees at a lower height above the ground.

Philippe Roca

Little Blue Herons winter from coastal regions of the southern United States southward into the tropics (AOU 1983), and locally nesting birds probably return sometime in April. Specific information on the timing of nesting in Kentucky is virtually absent, but the observation of large young on Shippingport Island on July 12, 1985, indicated that the clutch was likely completed sometime in late May. Kentucky data on clutch size are limited to the Shippingport Island nest, which contained five young. Harrison (1975) gives rangewide clutch size as 3–5, generally 5. As is typical throughout the species's range, a platform nest of sticks is placed in the crotch of a tree less than 30–40 feet above the ground.

During the atlas survey the only confirmed nesting occurred at Shippingport Island in 1985. In addition, two adults and approximately 25 immatures were observed on August 19, 1989, at the relocated Black-crowned Night-Heron nesting colony on Lake Barkley. Although the observation occurred very late in the season, it was included as a possible record based on the presence of breeding birds with this population of Black-crowneds in the early 1980s. Otherwise, Little Blues were observed on scattered blocks across central and western Kentucky, but most were believed or known to represent nonbreeding birds. Such sightings were most common along the Mississippi and lower Ohio Rivers, where birds from the large nesting colonies in southeastern Missouri regularly feed.

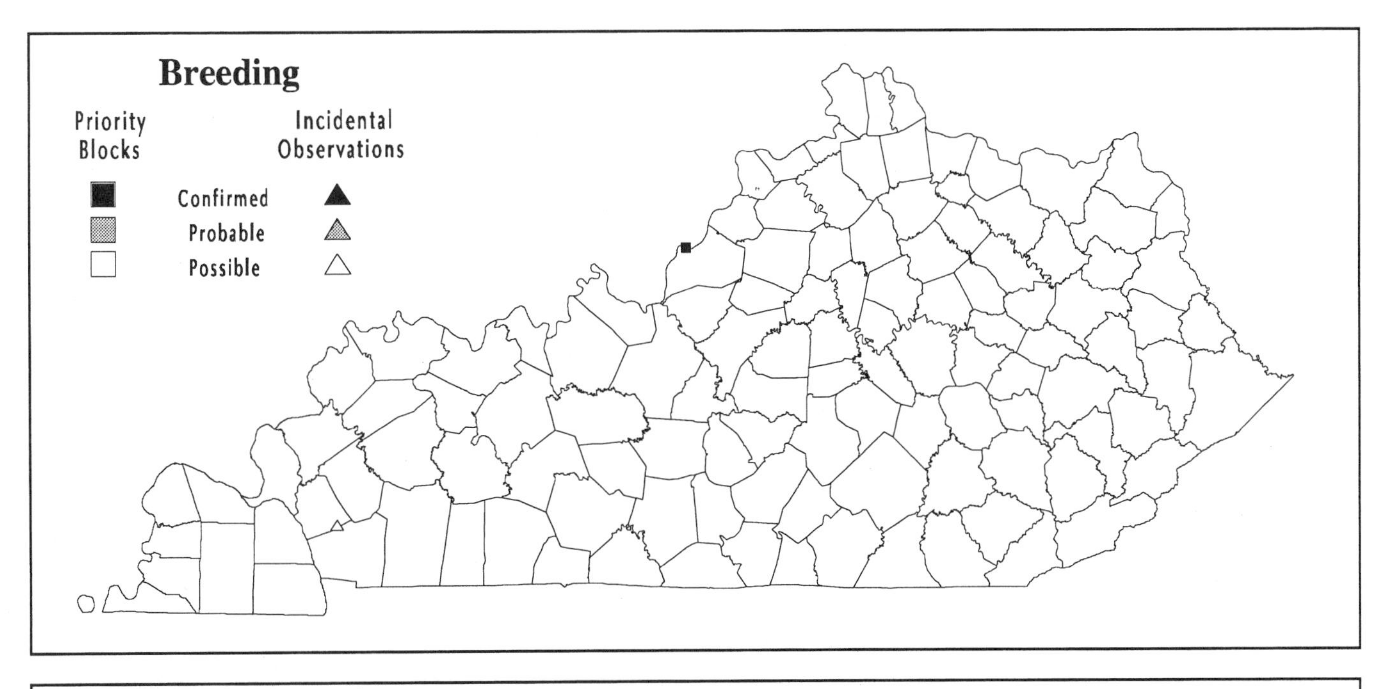

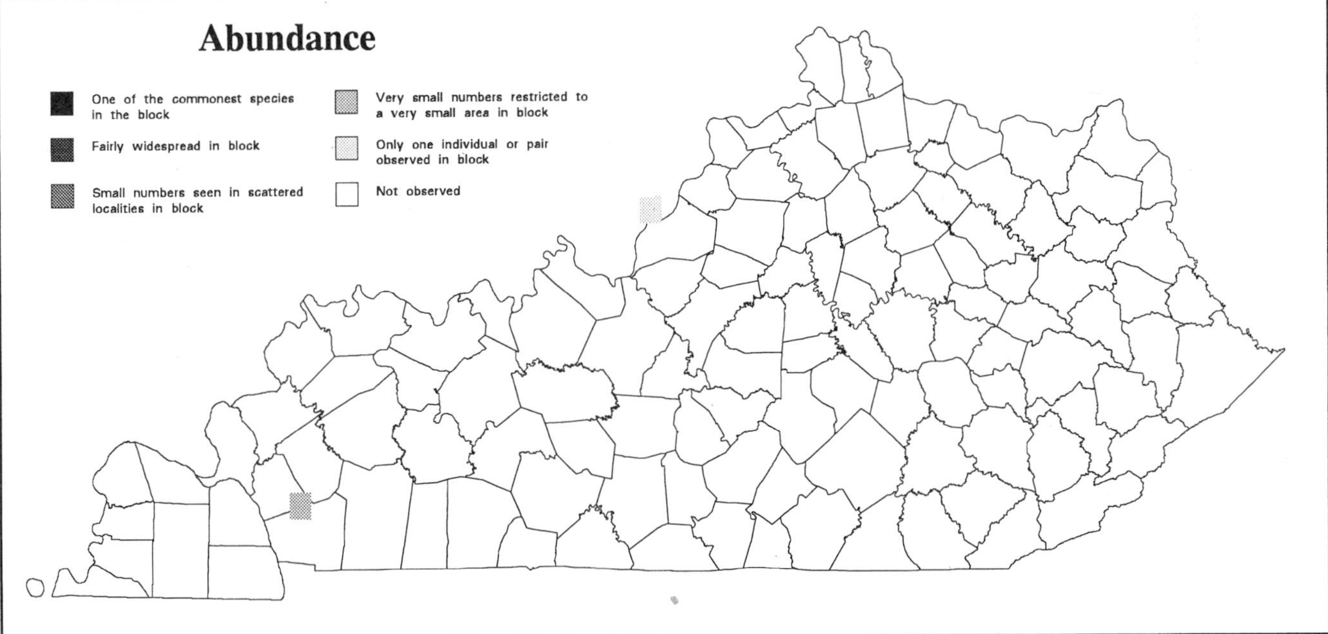

Analysis of Block Data by Physiographic Province Section

Physiographic Province Section	Total Blocks Surveyed	Blocks with Data	% with Data	Section's % for State
Mississippi Alluvial Plain	14	-	-	-
East Gulf Coastal Plain	36	-	-	-
Highland Rim	139	-	-	-
Shawnee Hills	142	-	-	-
Blue Grass	204	1	0.5	100.0
Cumberland Plateau	173	-	-	-
Cumberland Mountains	19	-	-	-

Summary of Breeding Status

Number of Blocks in Which Species Was Recorded		
Total	**1**	**0.1%**
Confirmed	1	100.0%
Probable	-	-
Possible	-	-

Little Blue Heron

Cattle Egret

Bubulcus ibis

The Cattle Egret is a relatively recent addition to the avifauna of Kentucky. The species became established in South America following an apparently natural dispersal across the Atlantic Ocean from Africa sometime between the late 1800s and the early 1930s. It subsequently colonized much of northern South America during the first part of the 20th century (Crosby 1972). The species arrived in the United States in the early 1940s (AOU 1983), expanding rapidly throughout the southern coastal regions. By the late 1970s Cattle Egrets were nesting sporadically throughout much of the eastern United States (R. Peterson 1980). The Cattle Egret was first observed in Kentucky on November 9, 1960, when a single bird was found in Warren County (Wilson 1960). By the mid-1970s it was being seen annually during migration.

Cattle Egrets were first discovered breeding in Kentucky in 1981, when at least one pair nested with Black-crowned Night-Herons on a small island in Lake Barkley, in Trigg County (Thomas 1982). Surveys conducted in later years failed to detect the species, and the birds may have used the site for only one year (S. Evans, unpub. rpt.). Sometime during the late 1980s the Black-crowneds abandoned this island in favor of another about 10 miles to the north. On July 29, 1992, 8–10 Cattle Egret nests were observed at the new site, constituting the first confirmed nesting on Lake Barkley since the 1981 observation. It is not known whether nesting occurred in years when surveys were not done.

On August 1, 1984, a second breeding site was discovered on a sandbar in the Mississippi River near Island No. 9, in Fulton County (S. Evans, unpub. rpt.; Stamm 1984b). Fifty-five nests containing large young were counted within a small portion of a dense stand of young black willows. Fieldwork conducted in the area before 1984 had not revealed the presence of Cattle Egrets, and subsequent surveys have yielded no further evidence of breeding. Thus, it appears that the site was used for only one year.

A third breeding locality was discovered on August 18, 1984, when at least four active nests were found in a Black-crowned Night-Heron nesting colony on Shippingport Island near the Falls of the Ohio, in Jefferson County (Palmer-Ball and Evans 1986). During the atlas fieldwork, Cattle Egret nesting was confirmed only at Shippingport Island from 1985 to 1988 (Stamm 1985, 1986a, 1987b, 1988b; Palmer-Ball and Evans 1986). This site was abandoned in 1992, and Cattle Egrets have not been seen at the relocation site of the Black-crowned Night-Heron colony on the grounds of the Louisville Zoo (G. Michael, pers. comm.).

Cattle Egrets winter in the Caribbean and as far north as the coastal regions of the southern United States (AOU 1983). Little specific data on the timing of nesting activities are available for Kentucky, but it appears that the nesting season may occur relatively late. Birds have been observed at the Shippingport Island heronry as early as May 10, and nest building was observed at the Fulton County nesting area on June 17 (S. Evans, unpub. rpt.). Most clutches are likely completed by the end of June, and young typically fledge during August. Kentucky data on clutch size are scarce, but nests on Shippingport Island have contained mostly three or four young (Palmer-Ball and Evans 1986), and one of the 1992 nests at the Lake Barkley colony contained five eggs.

As is typical of most wading birds, Cattle Egrets construct a shallow platform nest of sticks and twigs in the fork of a tree. Nesting colonies are generally found in thickets or groves of relatively young deciduous trees near or standing in water, and most Kentucky nests have been constructed 10–15 feet above the ground or water (S. Evans, unpub. rpt.; author's notes). Within the Black-crowned Night-Heron colonies, egret nests tend to be clustered in close proximity to one another rather than being scattered singly among those of the night-herons. Cattle Egrets are not normally found nesting with larger wading birds, especially Great Blue Herons, probably because of the Cattle Egret's preference for nesting in smaller trees.

Jeffrey A. Brown

Cattle Egrets were confirmed breeding during the atlas survey only at the Shippingport Island heronry. Possible breeding is indicated for the Lake Barkley heronry based on the occasional observation of birds foraging in the vicinity during the atlas period, although nesting was confirmed there in 1992. Two independent observations of Cattle Egrets in south-central Kentucky near Bowling Green in the summer of 1990 suggested the possibility of nesting somewhere in that area, but in the absence of more substantial evidence these reports were not included in the final data set. The species also was observed at scattered localities throughout western Kentucky during the atlas survey, but most reports were believed or known to represent nonbreeding birds. Such observations were most common along the Mississippi and lower Ohio Rivers, where birds foraging from large nesting colonies near Charleston and Caruthersville in southeastern Missouri accounted for numerous observations in nine priority blocks. During the mid-1980s each of these colonies contained more than 1,500 breeding pairs (Robbins and Easterla 1992). Although these heronries are currently restricted to Missouri, temporary colonies such as the one used in Fulton County in 1984 could appear at any time, and the possibility of nesting should not be overlooked.

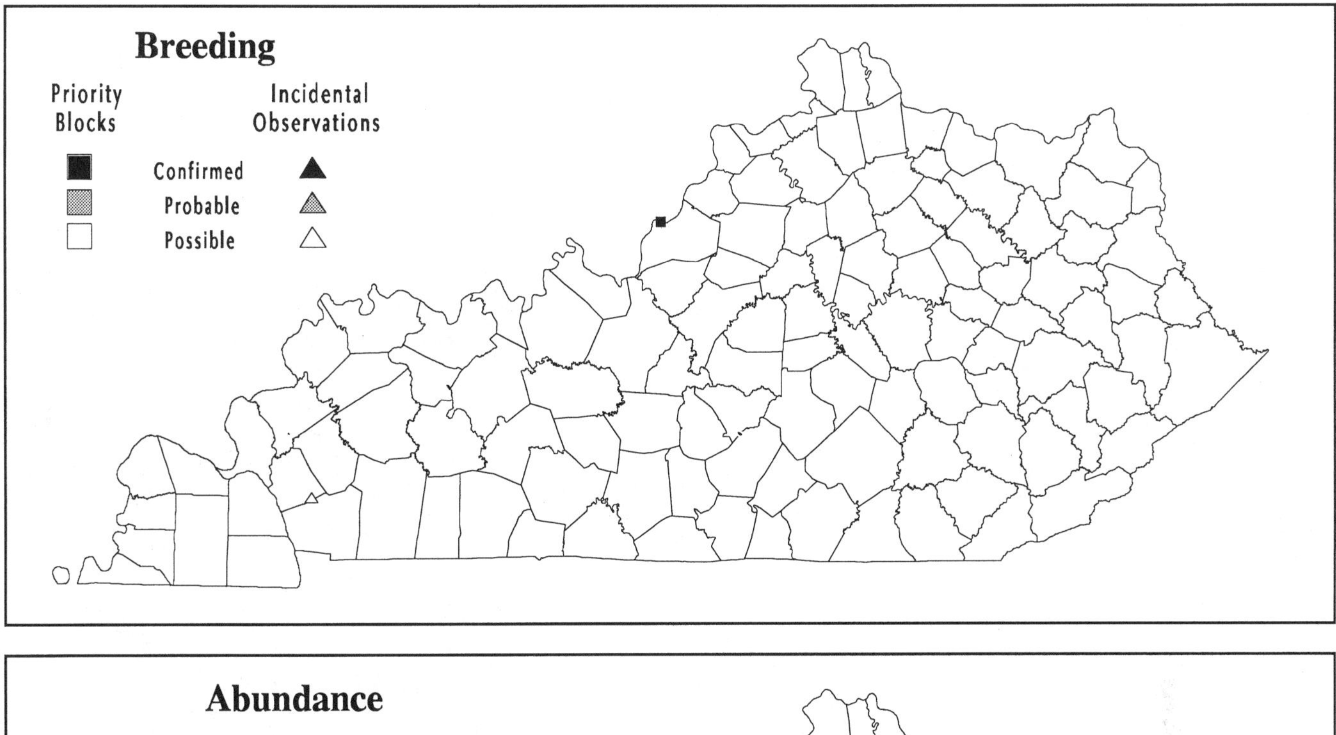

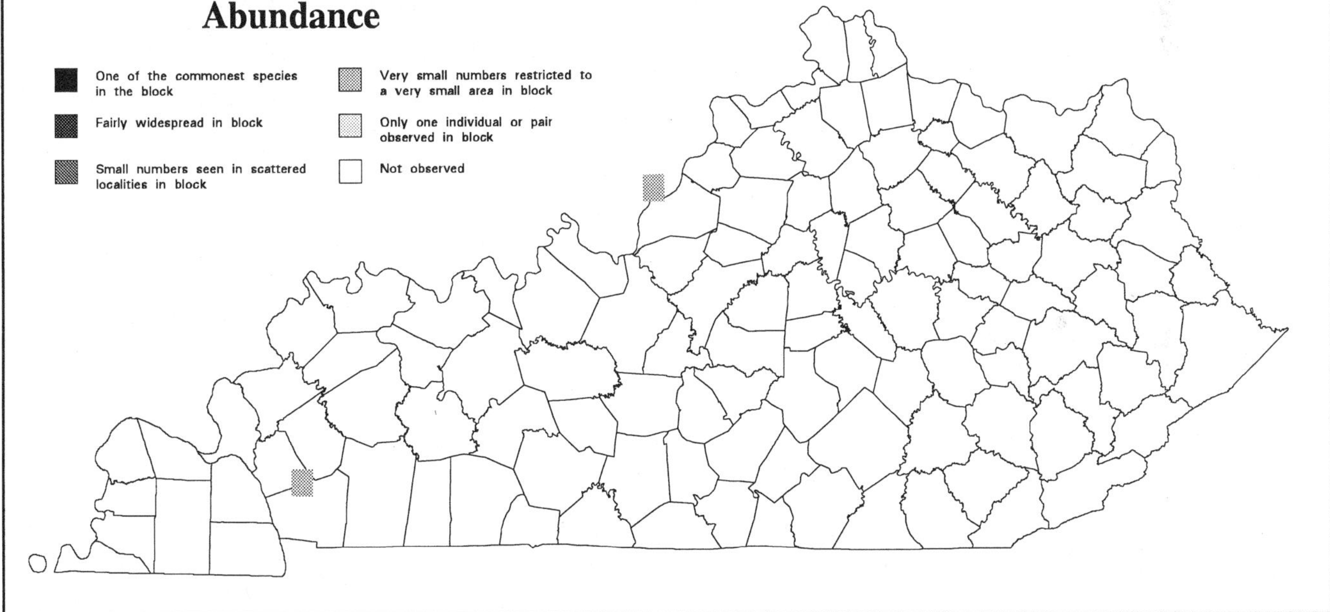

Analysis of Block Data by Physiographic Province Section

Physiographic Province Section	Total Blocks Surveyed	Blocks with Data	% with Data	Section's % for State
Mississippi Alluvial Plain	14	-	-	-
East Gulf Coastal Plain	36	-	-	-
Highland Rim	139	-	-	-
Shawnee Hills	142	-	-	-
Blue Grass	204	1	0.5	100.0
Cumberland Plateau	173	-	-	-
Cumberland Mountains	19	-	-	-

Summary of Breeding Status

Number of Blocks in Which Species Was Recorded		
Total	**1**	**0.1%**
Confirmed	1	100.0%
Probable	-	-
Possible	-	-

Cattle Egret

Green Heron

Butorides virescens

The Green Heron is by far the most widespread nesting heron in Kentucky. The species seems to be well distributed in appropriate habitat throughout the state, although suitable nesting situations are much less abundant in heavily forested regions. Mengel (1965) regarded this small heron as a fairly common to common summer resident statewide, although distributed rather locally in the Cumberland Plateau and the Cumberland Mountains.

Green Herons use a great variety of aquatic habitats, from the largest rivers and reservoirs to small creeks and ponds. The species has adapted quite well to human presence, and it is fairly frequent in settled areas and farmland. While these small waders sometimes use forested creeks and woodland ponds, they are much more often observed in semi-open to open situations.

It is likely that Green Herons are more widely distributed in Kentucky today than at any previous time. Although a substantial amount of wetland habitat has been drained, settlement of the land has included the construction of many lakes and ponds that these small herons use for feeding. Natural lakes and ponds were relatively uncommon throughout most of Kentucky before settlement, and Green Herons may have been largely restricted to river and stream corridors.

Green Herons normally overwinter from the Gulf Coast southward into the tropics of northern South America (AOU 1983), and nesting birds typically return to their Kentucky breeding grounds during the middle of April. Nest building may commence before the end of April, and egg laying has been noted by the first week of May (Stamm and Croft 1968). According to Mengel (1965), a pronounced peak in clutch completion occurs about May 1–10, although clutches have been noted into late June. Late nestings probably represent second attempts to nest following earlier failure because of predation or storms. Young have been observed in the nest most often during June, and fledglings are often on the wing by the middle of July. The average size of 13 clutches or broods reported by Mengel (1965) was 4.3 (range of 3–5).

Forest Cover

Value	% of Blocks	Avg Abund
All	48.4	1.5
1	60.4	1.4
2	63.7	1.6
3	47.4	1.5
4	32.5	1.6
5	21.2	1.3

Nest sites are varied and include shade trees in rural yards, upland woodlots, and woodland edge far from water. More often the nest is situated in a tree or shrub along the shore of a lake or pond. Green Herons usually nest solitarily, but small, loosely associated colonies of up to five pairs have been reported (Harrison 1975; Mengel 1965). It is not unusual, however, to find a few pairs nesting along the margins of Black-crowned Night-Heron colonies.

The nest is typically a flimsy platform of sticks and twigs, placed in the fork of a small tree or shrub. Although deciduous trees are most frequently used, nests have also been found in buttonbush, cedars (Mengel 1965), and Virginia pine (Harm 1973). The average height of 11 nests reported by Mengel (1965) was 14.8 feet (range of 2.5–40.0 feet).

Ron Austing

The atlas fieldwork yielded records of Green Herons in more than 48% of priority blocks statewide, and 46 incidental observations were reported. Occurrence was highest in the East Gulf Coastal Plain and lowest in the Cumberland Plateau and the Cumberland Mountains. Throughout central Kentucky this small heron was found in about half of the priority blocks. Average abundance was relatively low statewide. Occurrence was highest in areas with a good supply of open habitat and considerably lower in primarily forested areas.

Only about 7% of priority block records were for confirmed breeding. Although at least 10 active nests were located during the atlas survey, most confirmed records were based on the observation of recently fledged young.

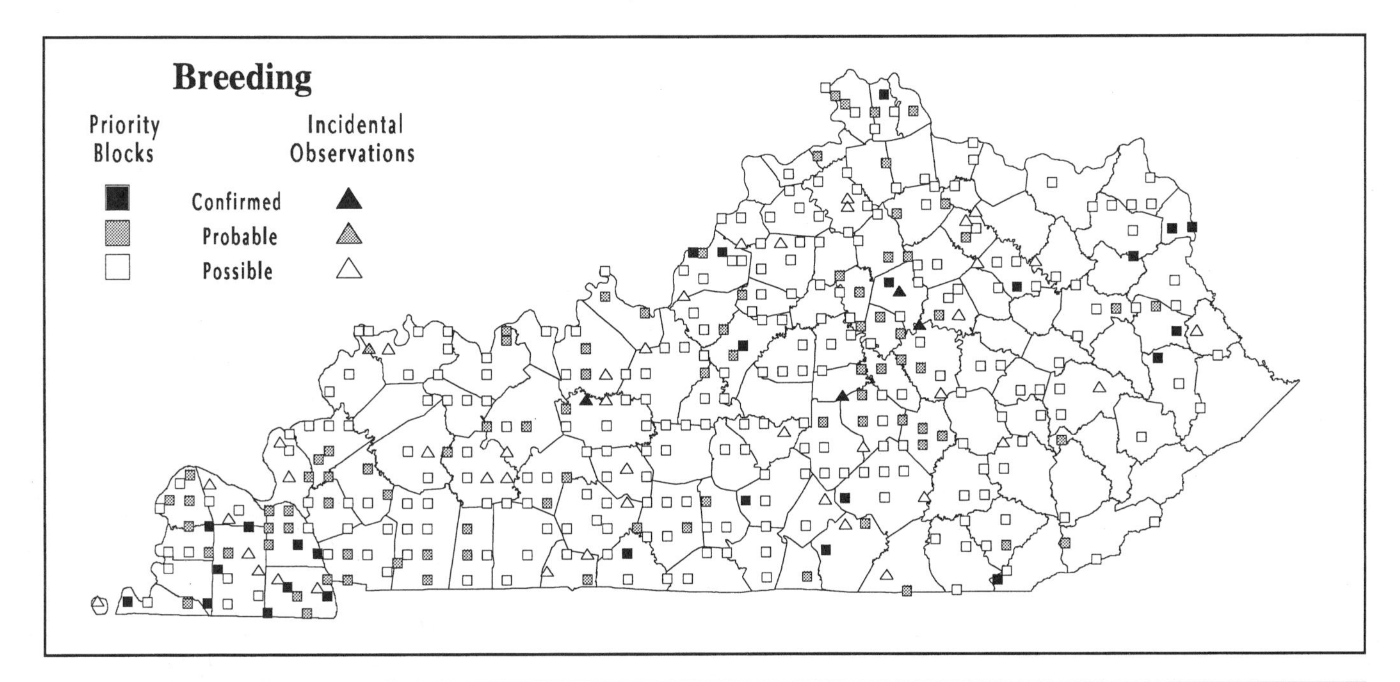

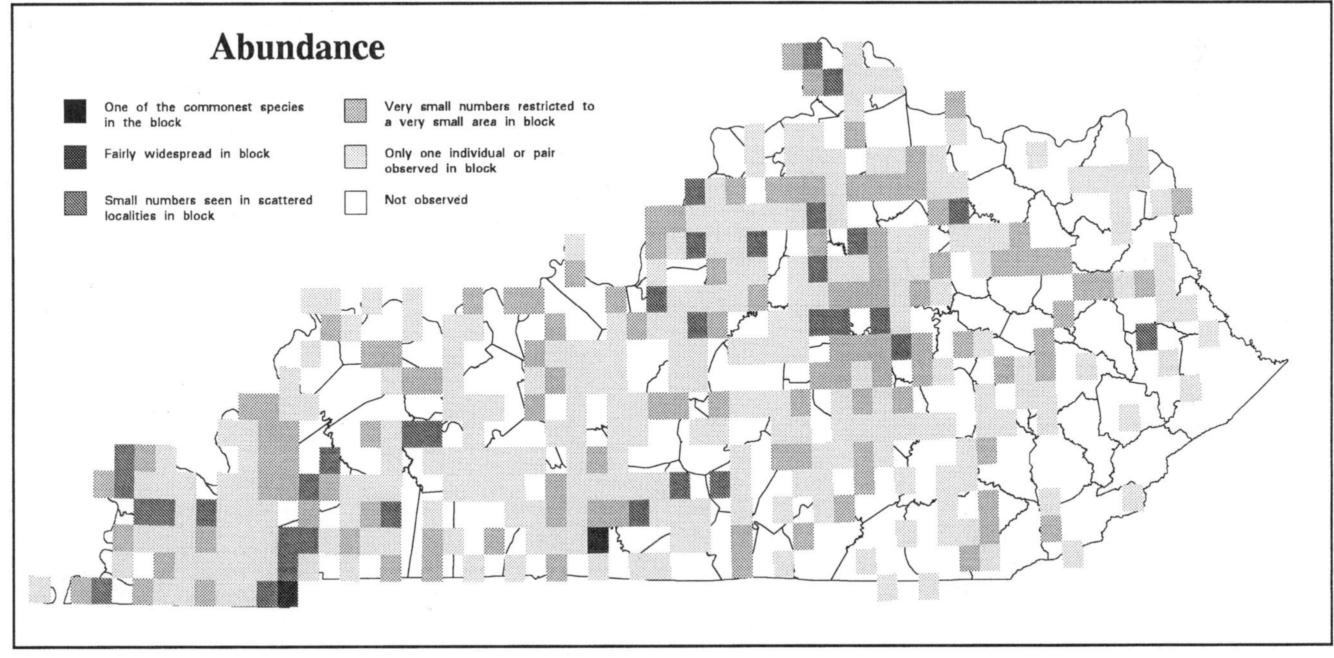

Analysis of Block Data by Physiographic Province Section

Physiographic Province Section	Total Blocks Surveyed	Blocks with Data	Avg Abund	% with Data	Section's % for State
Mississippi Alluvial Plain	14	8	2.4	57.1	2.3
East Gulf Coastal Plain	36	29	1.5	80.6	8.2
Highland Rim	139	80	1.6	57.6	22.7
Shawnee Hills	142	70	1.4	49.3	19.9
Blue Grass	204	112	1.6	54.9	31.8
Cumberland Plateau	173	48	1.3	27.7	13.6
Cumberland Mountains	19	5	1.4	26.3	1.4

Summary of Breeding Status

Number of Blocks in Which Species Was Recorded		
Total	**352**	**48.4%**
Confirmed	25	7.1%
Probable	86	24.4%
Possible	241	68.5%

Green Heron

Black-crowned Night-Heron

Nycticorax nycticorax

The Black-crowned Night-Heron occurs across much of North America as a breeding bird (AOU 1983), but as is the case across most of its range, the species is very locally distributed in Kentucky. Near the few active nesting colonies, Black-crowneds are fairly conspicuous, but throughout the remainder of the state they are known primarily as rare to uncommon transients and post-breeding visitors.

Historically only two nesting colonies were known in the state, both located along major rivers of the Blue Grass in Bourbon and Jefferson Counties. Details of the one in Bourbon County are obscure; it supposedly was situated along the North Fork of the Licking River near Paris, although the river's South Fork, not the North Fork, is much closer to Paris. Nest counts made during the winters of 1949–50 and 1950–51 yielded totals of 283 and 436 nests respectively (Mengel 1965).

Black-crowned Night-Herons have nested in the Louisville area since at least 1930, occupying four different sites along the Ohio River over the years (Mengel 1965; Palmer-Ball and Evans 1986). The colony was originally known from Six-Mile Island, where up to 140 nests were counted in 1937 (Mengel 1965). This site was abandoned in favor of a wooded area below the Falls of the Ohio, where 250 nests were observed in 1949 (Smith 1950). Use of this site declined through the 1950s, and the numbers of birds were greatly reduced by 1959 (Wiley 1964). The heronry was relocated nearby on Sand Island in 1962, when 150 nests were counted (Wiley 1964). This site was used until at least 1966 (Stamm et al. 1967), but it was also abandoned in subsequent years. In 1984 the nesting birds were rediscovered on Shippingport Island (Palmer-Ball and Evans 1986), where up to 300 nests were counted in 1988 (Stamm 1988b). In 1992 this site was abandoned, and in 1993 the colony became reestablished on the grounds of the Louisville Zoo (G. Michael, pers. comm.).

Two additional nesting colonies have been found in recent years. In 1979 Black-crowned Night-Herons were found nesting on an island in Lake Barkley, in Trigg County (Thomas 1982). The heronry then contained about 60 nests, but more than 200 were counted in 1981. Only about 100 nests were in this colony in 1984 (S. Evans, unpub. rpt.), and by 1988 only 6 active nests were observed there (Stamm 1988b). In 1989 it was found that the colony had simply relocated to another island about 10 miles north on the lake and still contained at least 100 nests (Stamm 1990).

A smaller nesting colony was present in western Clark County for a couple of years during the mid-1980s (Stamm 1986a). It was reported to KDFWR personnel in 1986, when 28 active nests were counted, and it was said to have been active the previous year (T. Edwards, pers. comm.). The site was unusual because it was situated in an upland woodlot in rural farmland, more than five miles from the nearest substantial body of water. In 1987 this site was abandoned, probably because of human disturbance (Palmer-Ball 1991a). A relocation site has not been found.

Most atlas survey reports of Black-crowned Night-Herons originated from sites near known breeding colonies, and only records representing locations of nesting colonies that were active during 1985–91 are plotted as confirmed on the maps. One possible record from Lewis County was included in the final data set because the observation may have involved birds from a nesting colony that was discovered along the Ohio River near the mouth of Pond Run, in Greenup County, in May 1993 (P. Morrison, pers. comm.). Otherwise, only a few birds were recorded away from these areas, and all were considered to be nonbreeders.

Philippe Roca

Although a few Black-crowned Night-Herons are seen in Kentucky during the winter months, most locally nesting birds move farther south in winter. At Shippingport Island, birds typically returned during the latter half of March, and courtship and nest building commenced almost immediately (Palmer-Ball and Evans 1986). Incubation has been noted at Louisville as early as mid-April, although later nests are not uncommon, and the nesting season at least occasionally extends into early fall (Stamm 1960). Nesting activity seems to lag several weeks behind at the Lake Barkley heronry, where many birds were still incubating in early June 1986 (Stamm 1986a). Although Kentucky data on clutch size are limited, most nests investigated have contained clutches or broods of three to four eggs or young (Mengel 1965; author's notes).

Older accounts of Black-crowneds nesting in tall cottonwoods on Six-Mile Island and large sycamores and elms on Sand Island certainly do not typify the behavior of the species in Kentucky today. In fact, these birds seem to nest exclusively in relatively low vegetation, usually thickets of young trees less than 30 feet tall. Colonies have been reported over both water and land, but usually in the immediate vicinity of a substantial body of water.

The nest is a platform of sticks, placed in the fork of a tree. Nesting colonies are typically dense, with up to 10 or 12 nests in larger trees. At currently known heronries, nests are situated primarily 10–20 feet above the ground or water.

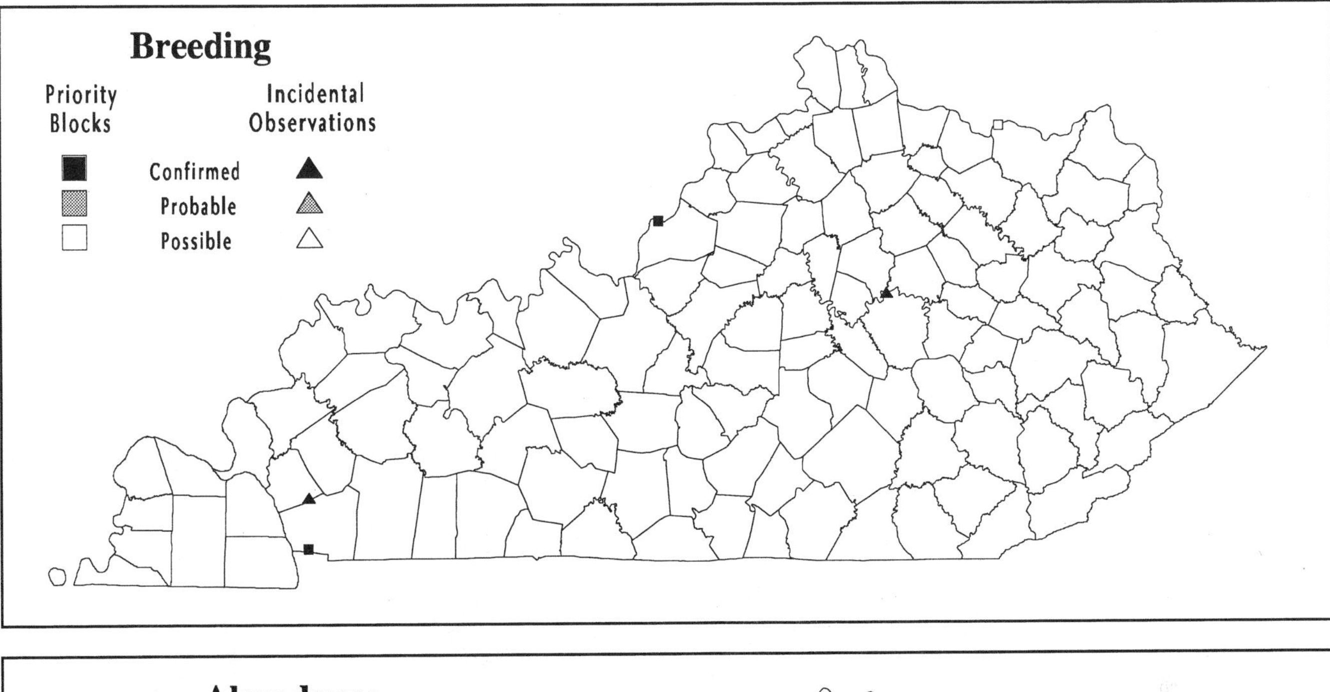

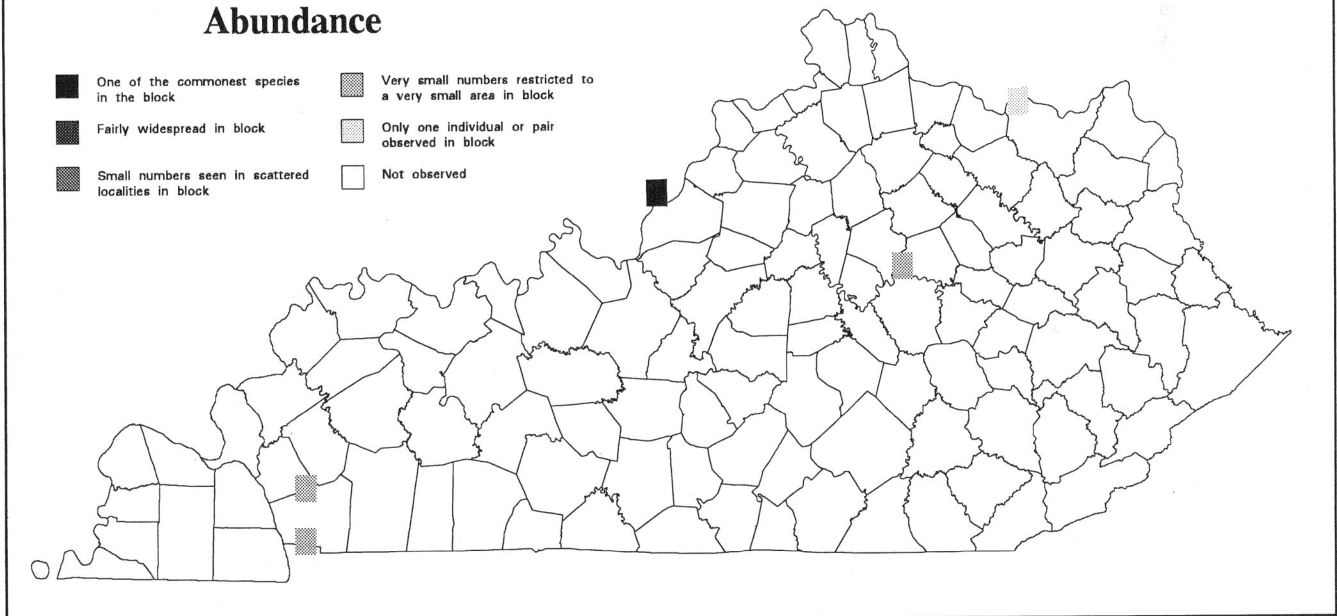

Analysis of Block Data by Physiographic Province Section

Physiographic Province Section	Total Blocks Surveyed	Blocks with Data	% with Data	Section's % for State
Mississippi Alluvial Plain	14	-	-	-
East Gulf Coastal Plain	36	-	-	-
Highland Rim	139	1	0.7	33.3
Shawnee Hills	142	-	-	-
Blue Grass	204	2	1.0	66.7
Cumberland Plateau	173	-	-	-
Cumberland Mountains	19	-	-	-

Summary of Breeding Status

Number of Blocks in Which Species Was Recorded		
Total	**3**	**0.4%**
Confirmed	2	66.7%
Probable	-	-
Possible	1	33.3%

Black-crowned Night-Heron

Yellow-crowned Night-Heron

Nyctanassa violacea

In large part because of its reclusive habits, the Yellow-crowned Night-Heron's breeding status in Kentucky is poorly known. Mengel (1965) considered the species to be an uncommon to fairly common, but locally distributed, summer resident across the central and western parts of the state. More recent evidence indicates that it occurs similarly today.

Little specific information exists on the historical status of this heron in Kentucky. Mengel (1965) included a few references to the species's occurrence in far western Kentucky during the 1800s, but he also noted that its presence elsewhere in the state was not documented until more recently, suggesting that its range had expanded considerably during the first half of the 20th century. The species was first reported in the Louisville area in 1948 (Steilberg 1949) and near Lexington by 1956 (Mayfield 1956). As of 1960, breeding had been documented in only three counties of central Kentucky: at two sites at Louisville, in Jefferson County (Halverson 1955; Fitzhugh 1959, 1961); at a few scattered sites in Fayette County (Mayfield 1956; Webster 1960); and south of Bowling Green, in Warren County (Wilson 1962a). Nonetheless, a more widespread breeding distribution was suggested by additional breeding season records listed by Mengel (1965) for Ballard, Crittenden, Fulton, Hickman, Hopkins, Lyon, Marshall, and Trigg counties. Before the initiation of the atlas project, a few additional nesting localities were added, including two sites at Louisville (Croft and Stamm 1967; Stamm 1977); one site in Lexington (KOS Nest Cards; Stamm 1980, 1983, 1984b); and one site at Danville, in Boyle County (Stamm 1978b).

Yellow-crowned Night-Herons are generally encountered in more closed-in habitats than most other wading birds. Rather than being found along open shorelines and ponds, they are typically seen on woodland pools, forested streams and swamps, and other shallow bodies of water within or near forest or other cover.

These secretive wading birds winter from the coastal regions of the southern United States southward into the tropics (AOU 1983), and locally nesting birds return by early April. Courtship and nest building commence almost immediately, and early clutches likely are completed before the beginning of May (Mengel 1965). Young have been observed in nests most frequently in June, although young from later nestings may not fledge until mid- to late July (Stamm 1977). Mengel (1965) reported the average size of seven broods as 4.1 (range of 3–5).

Unlike most other species of wading birds, Yellow-crowned Night-Herons do not normally nest in dense colonies. More typically, they are found solitarily or in small, loose colonies of only a few pairs. The species usually nests in the midstory of relatively mature forest, typically choosing a site that is open to the ground or water below. Although extensive bottomland forest is often used, the species also nests within small woodlots and narrow riparian corridors, and sometimes in or near residential areas. The nest is a relatively flimsy platform of sticks, typically placed in a horizontal fork well away from the trunk on a large branch below the main canopy. Nests have been reported in a variety of deciduous trees, but most frequently in sycamore, cottonwood, and black walnut. The average height of 13 nests reported by Mengel (1965) was 43 feet (range of 35–70 feet).

Philippe Roca

The atlas survey added a considerable amount of information on the summer occurrence of this wader. Breeding was confirmed at or near traditional sites in Fayette County (Stamm 1985; Stamm and Monroe 1990a) and Jefferson County (Stamm 1985, 1989a), as well as at several other localities: near Axe Lake, in Ballard County; along the Salt River near Shepherdsville, in Bullitt County (Stamm 1985, 1988b); on the Jenny Hole–Highland Creek Unit of Sloughs WMA, in Union County (Stamm 1986a); and at two sites along Obion Creek, in Hickman County (Stamm 1988b).

Reports of Yellow-crowned Night-Herons away from known nesting areas are rare. For this reason, all sightings from May through July were included in the final data set. The presence of more than one bird or repeated observations of single birds in suitable breeding habitat were regarded as probable evidence of breeding. Such reports originated along South Fork Creek, in Barren County; along White Oak Creek, in Greenup County (R. Record, pers. comm.); near Fairdale, in Jefferson County; and at Blizzard Ponds, in McCracken County (Stamm 1986a). In addition, eight priority block records and five incidental observations of single birds in suitable breeding habitat were regarded as indicating possible breeding.

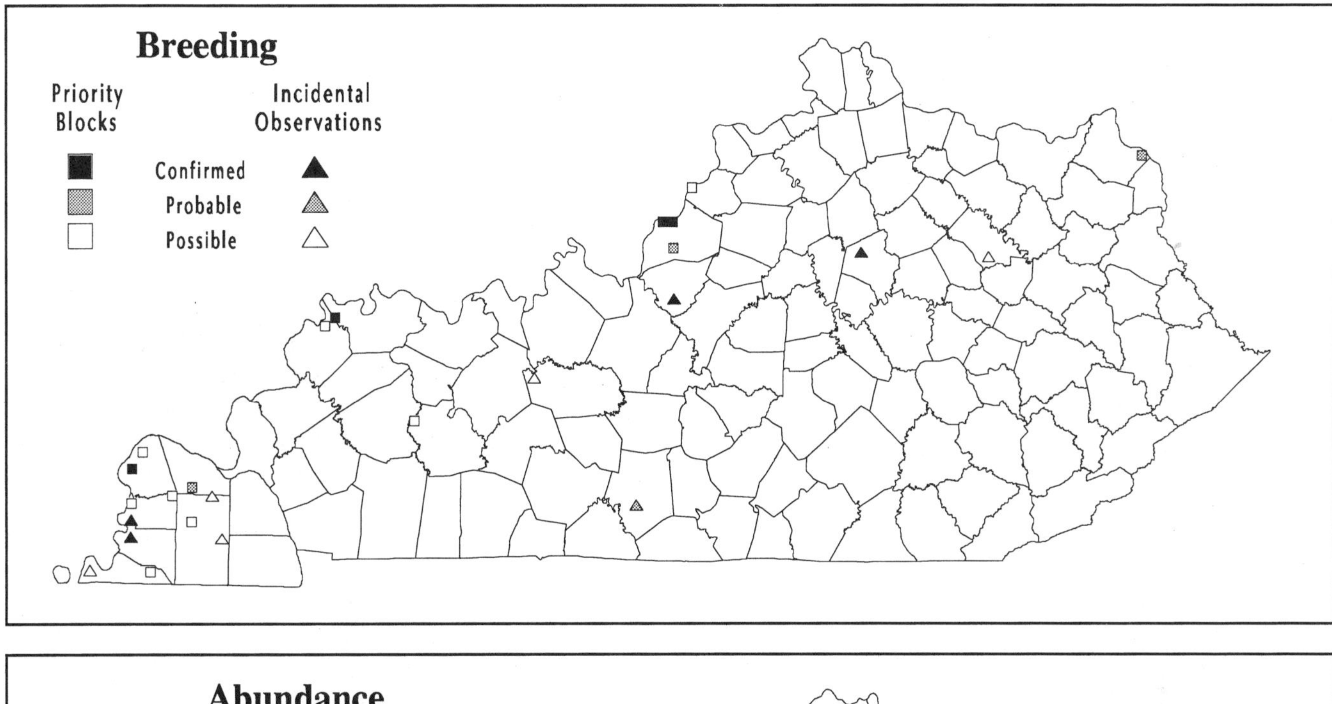

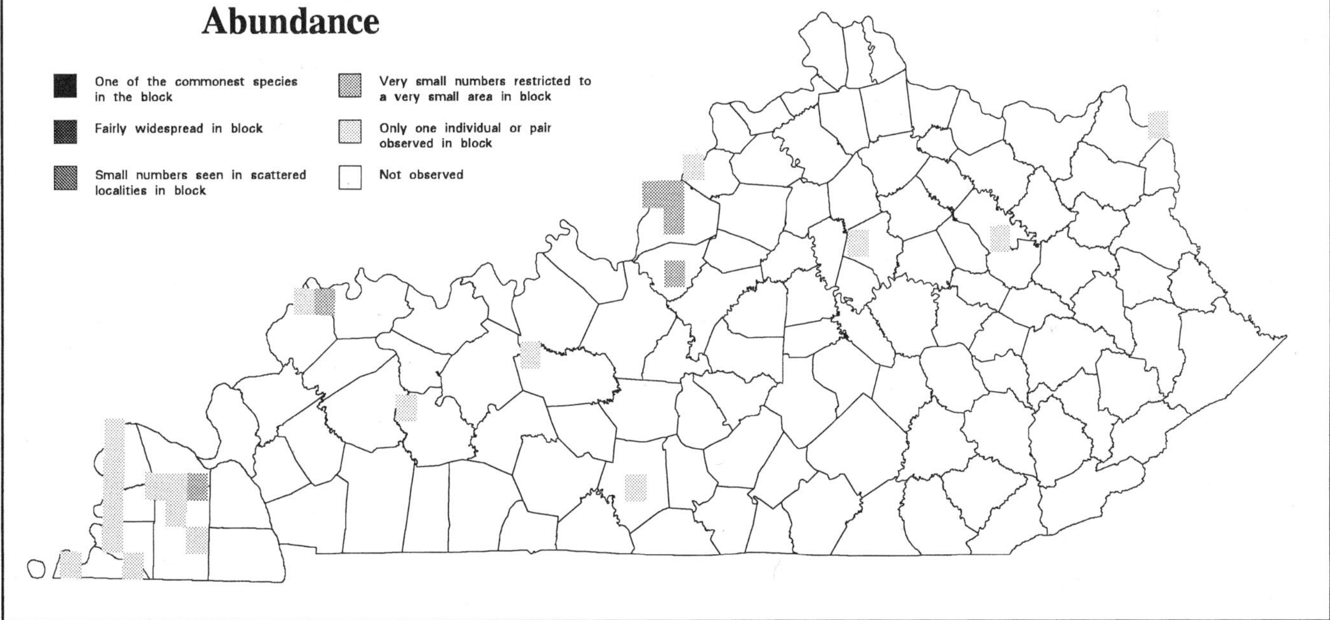

Analysis of Block Data by Physiographic Province Section

Physiographic Province Section	Total Blocks Surveyed	Blocks with Data	% with Data	Section's % for State
Mississippi Alluvial Plain	14	3	21.4	20.0
East Gulf Coastal Plain	36	4	11.1	26.7
Highland Rim	139	-	-	-
Shawnee Hills	142	3	2.1	20.0
Blue Grass	204	4	2.0	26.7
Cumberland Plateau	173	1	0.6	6.6
Cumberland Mountains	19	-	-	-

Summary of Breeding Status

Number of Blocks in Which Species Was Recorded		
Total	**15**	**2.1%**
Confirmed	4	26.7%
Probable	3	20.0%
Possible	8	53.3%

Yellow-crowned Night-Heron

Canada Goose

Branta canadensis

In the late 1950s the Canada Goose was known primarily as a transient and winter resident in Kentucky (Mengel 1965), but historical accounts indicate that at one time the species nested regularly in the state. During the 1800s geese of the now supposedly extinct race *B. c. maxima,* or Giant Canada Goose, apparently nested in parts of western Kentucky. Audubon (1861), for example, noted that the species bred on the Mississippi and lower Ohio Rivers in the early 1800s and specifically mentioned a pair that nested near the mouth of the Green River. Through the mid-1900s nesting also occurred at least occasionally when wild geese bred with crippled or captive individuals (Mengel 1965). Captive flocks have been present since at least the early 1800s, when Audubon himself propagated a flock from pinioned birds while living at Henderson (Audubon 1861).

Beginning in the early 1970s KDFWR initiated an introduction program, placing small numbers of geese in suitable nesting locations. The species was first released at Cave Run Lake, in Rowan County, and subsequently at the State Game Farm in Frankfort. Following successful expansion of these populations, additional releases were conducted at Land Between the Lakes and at several reclaimed surface mines in Muhlenberg and Ohio Counties (R. Pritchert, pers. comm.). Additional birds have originated from similar projects in adjacent states and from efforts by private waterfowl enthusiasts, and a few crippled birds remain each year, often breeding with these locally established birds. In most cases introduced birds have become well established, and the Canada Goose is now a fairly regular, although locally distributed, nesting species throughout much of central and western Kentucky. In addition, populations continue to expand on Cave Run Lake, and birds will likely continue to show up in suitable locales at other sites in eastern Kentucky.

Canada Geese nest in a variety of semi-open to open situations. Although nests are usually restricted to the vicinity of reservoirs and ponds, they occasionally are placed relatively far from water. Canada Geese probably occur most frequently in rural farmland, but some reservoirs and wildlife management areas harbor substantial populations. They are also well established in urban areas and are regulars at parks and cemeteries in Lexington and Louisville.

Pairs initiate territorial behavior during March, and by April many nests have a full clutch of eggs. Most young hatch by early May, and parents attend them through much of the summer. Kentucky data on clutch size are lacking, but Harrison (1975) gives rangewide clutch size as 4–10, commonly 5–6.

Canada Geese form nests in depressions on islands or along or near shores of reservoirs, swamps, ponds, and marshes. Nests are sometimes built on natural sites such as stumps, beaver lodges, and muskrat dens, and old Osprey nests are used on Lake Barkley. Artificial structures are most frequently used, however, and nesting sites made of old tires, cans, and barrels have been placed on many impoundments. Typically with little assistance from the male, the female constructs a crude nest of sticks, grass, and weed stalks and lines it with down (Harrison 1975).

The atlas survey yielded records of Canada Geese in 5% of priority blocks statewide, and there were 10 incidental observations reported. The species was found in all physiographic province sections except the Cumberland Mountains. Occurrence was highest in the Mississippi Alluvial Plain, the East Gulf Coastal Plain, and the Blue Grass.

Mabel Slack

Nearly 53% of priority block records were for confirmed breeding, in large part because of the conspicuousness of family groups. Although a few active nests were reported, most confirmed records were based on the observation of broods of young being attended by adults.

The origin of some birds recorded during the atlas fieldwork was uncertain, and it is likely that a few reports resulted from the observation of captive individuals. Most records involved free-flying birds, and it is believed that the atlas survey produced a fairly accurate picture of the breeding occurrence of Canada Geese in Kentucky. It should be expected that these geese will continue to increase in the future, since suitable habitat is present in abundance and small-scale introductions are continuing.

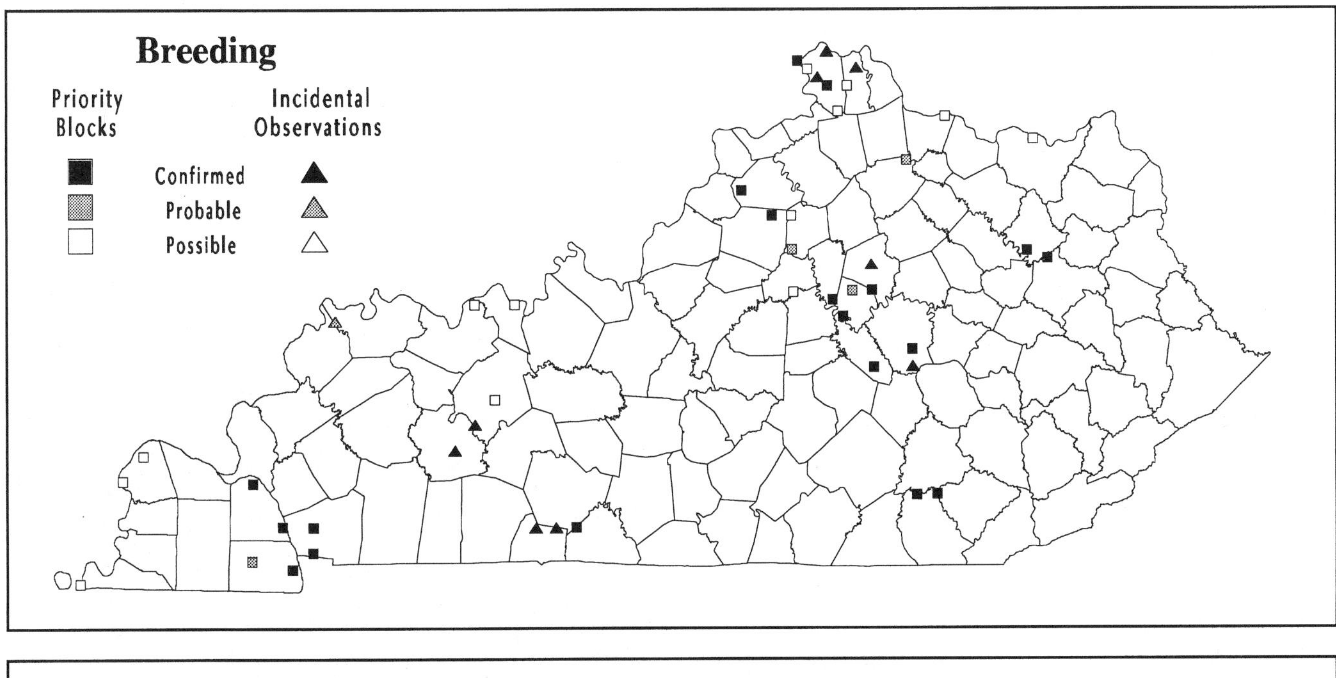

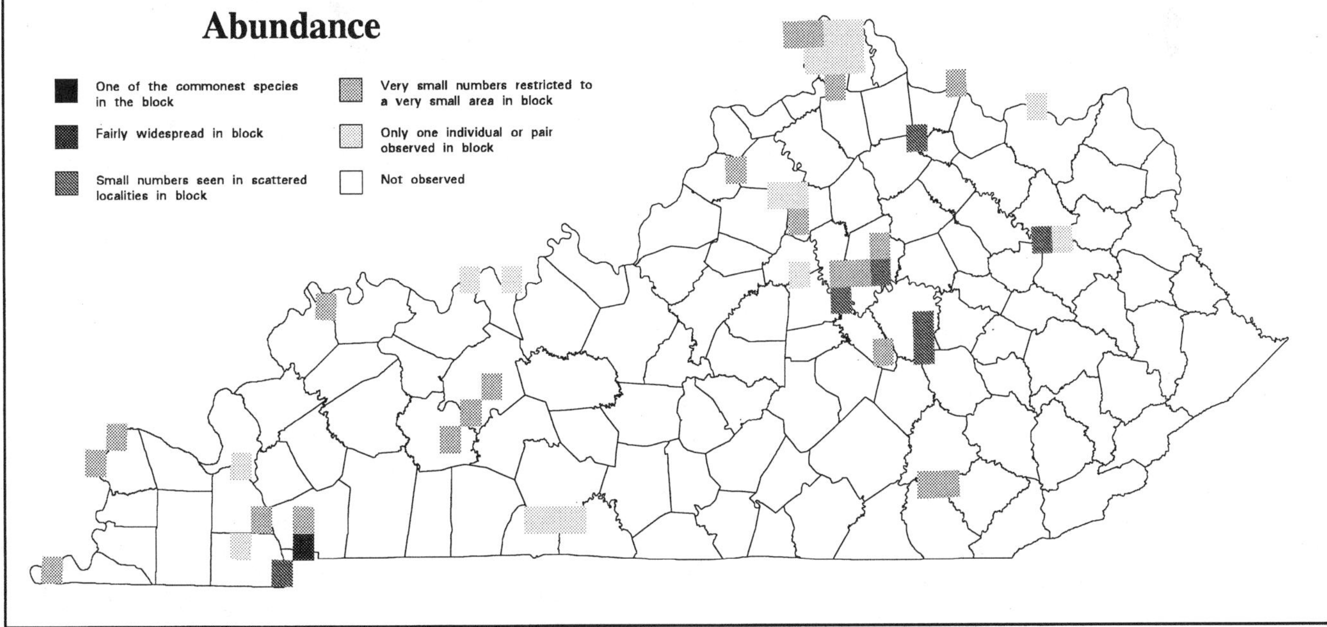

Analysis of Block Data by Physiographic Province Section

Physiographic Province Section	Total Blocks Surveyed	Blocks with Data	% with Data	Section's % for State
Mississippi Alluvial Plain	14	3	21.4	8.3
East Gulf Coastal Plain	36	4	11.1	11.1
Highland Rim	139	3	2.2	8.3
Shawnee Hills	142	3	2.1	8.3
Blue Grass	204	20	9.8	55.6
Cumberland Plateau	173	3	1.7	8.3
Cumberland Mountains	19	-	-	-

Summary of Breeding Status

Number of Blocks in Which Species Was Recorded		
Total	**36**	**5.0%**
Confirmed	19	52.8%
Probable	4	11.1%
Possible	13	36.1%

Wood Duck

Aix sponsa

The Wood Duck is the only species of waterfowl that nests in substantial numbers in Kentucky. As of the early 1960s it had been documented breeding only west of the Cumberland Plateau (Mengel 1965), but today the species is known to nest throughout eastern Kentucky as well. KDFWR annually conducts approximately two dozen summer brood surveys on rivers across the state, and Woodies are recorded throughout all regions and in substantial numbers on most routes (R. Pritchert, unpub. rpt.).

Wood Ducks are encountered in a great variety of aquatic habitats, from the protected bays of large reservoirs to small streams and farm ponds. The species is common along larger rivers and is especially abundant in floodplain sloughs and swamps of western Kentucky. In eastern Kentucky, Woodies inhabit many of the larger streams as well as reservoirs and ponds.

The occurrence of Wood Ducks in Kentucky before settlement was likely somewhat similar to that observed today. Audubon (1861) considered the species abundant in western Kentucky in the early 1800s, and suitable habitat was likely present along the larger streams across the state. The widespread clearing and draining of vast areas of wetland habitat certainly has resulted in declines in some areas, especially in the western third of the state. In contrast, the creation of numerous reservoirs and ponds has probably added suitable habitat in other areas, especially in central and eastern Kentucky.

A few Wood Ducks spend the winter in Kentucky, but most move farther south. Although a few birds regularly show up in late February, the bulk of the breeding population probably returns during early March. According to Mengel (1965), early clutches are completed during mid-March, with a peak in clutch completion during the last ten days of the month. Broods of young are most conspicuous from early May through early June, although there are numerous reports of later nestings, and broods of small young may be observed into July. Although the species raises two broods in the South (Ehrlich et al. 1988), evidence of such in Kentucky is lacking (R. Pritchert, pers. comm.). Most late nestings may be the result of earlier failures. Kentucky data on clutch size are somewhat limited, but there is one report of a nest containing 14 eggs (Mengel 1965), and Allen (1971, 1972) reported clutches of 8–11 eggs at Land Between the Lakes. In addition, "dump nests" involving more than one female have been reported to contain as many as 23 eggs (Allen 1972).

Forest Cover

Value	% of Blocks	Avg Abund
All	25.9	1.7
1	29.7	1.7
2	32.1	1.7
3	25.7	1.6
4	22.1	1.7
5	5.8	1.3

Wood Ducks are cavity nesters, using both natural cavities in trees and nest boxes erected to attract them. Large trees standing along or near lakes, ponds, rivers, and streams are usually used, but nest sites are occasionally found relatively far from the nearest substantial body of water. Both living and dead trees are used. Audubon (1861) perhaps erroneously noted a nest in a rock fissure along the Kentucky River near Frankfort in the early 1800s. Somewhat more credible were his notes concerning the use of abandoned nest cavities of Ivory-billed Woodpeckers near Henderson during the same era. Mengel (1965) gave a range in nest height of 14–25 feet. Although a variety of trees are used, sycamores are perhaps most frequently chosen, in large part because of their abundance along rivers and streams across much of the state.

Herbert Clay Jr. M.D.

Wood Ducks were found on nearly 26% of priority blocks statewide. In addition, 56 incidental reports were obtained during the atlas survey period, most originating from KDFWR brood surveys. The species was found on more than 34% of priority blocks only in the Mississippi Alluvial Plain. Although formerly the species was considered to be absent in eastern Kentucky (Mengel 1965), Wood Ducks were found on nearly 12% of priority blocks on the Cumberland Plateau and on more than 21% of priority blocks in the Cumberland Mountains. It is probable that the species has increased in this region as humans have altered the landscape. In addition to the creation of a few reservoirs and many farm ponds, surface mining reclamation work frequently includes the building of impoundments that are used by Woodies. Average abundance was relatively low statewide. Occurrence was noticeably lower only in extensively forested areas.

Fifty-eight percent of priority block records were for confirmed breeding. The high rate of confirmation can be attributed to the conspicuousness of broods of young. In fact, nearly all of the 109 records of confirmed breeding pertained to the observation of broods of young.

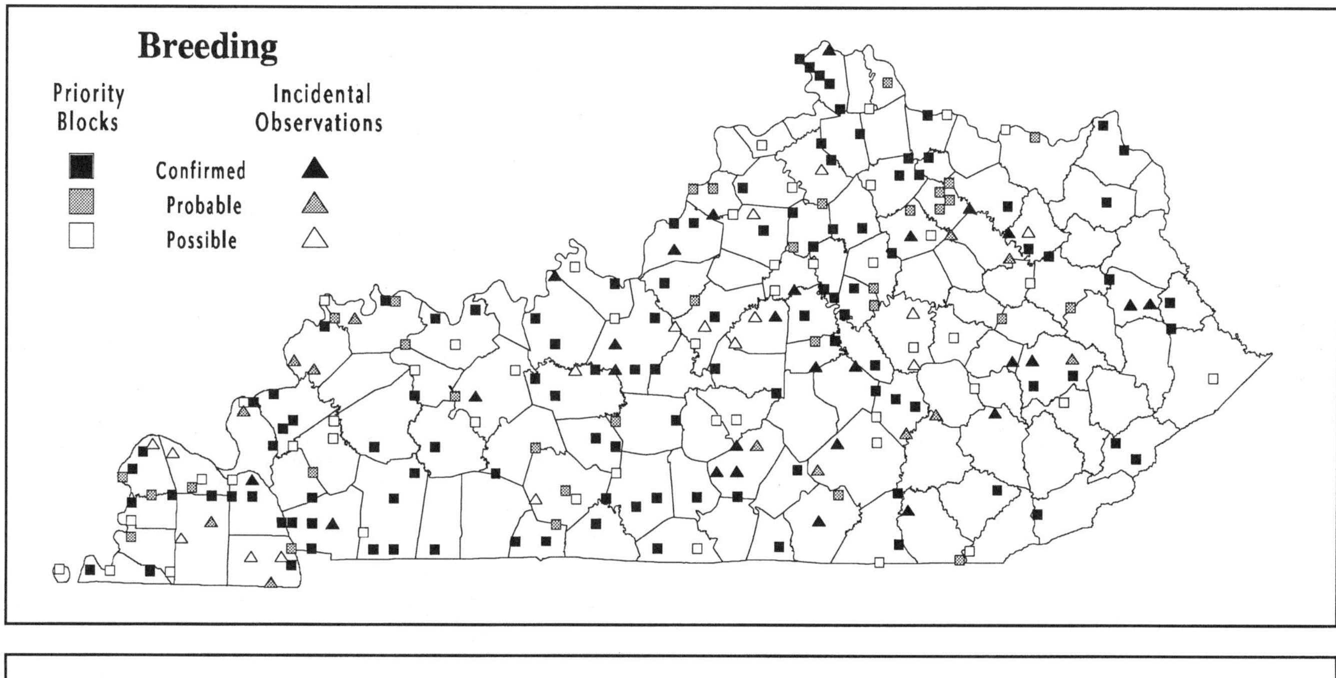

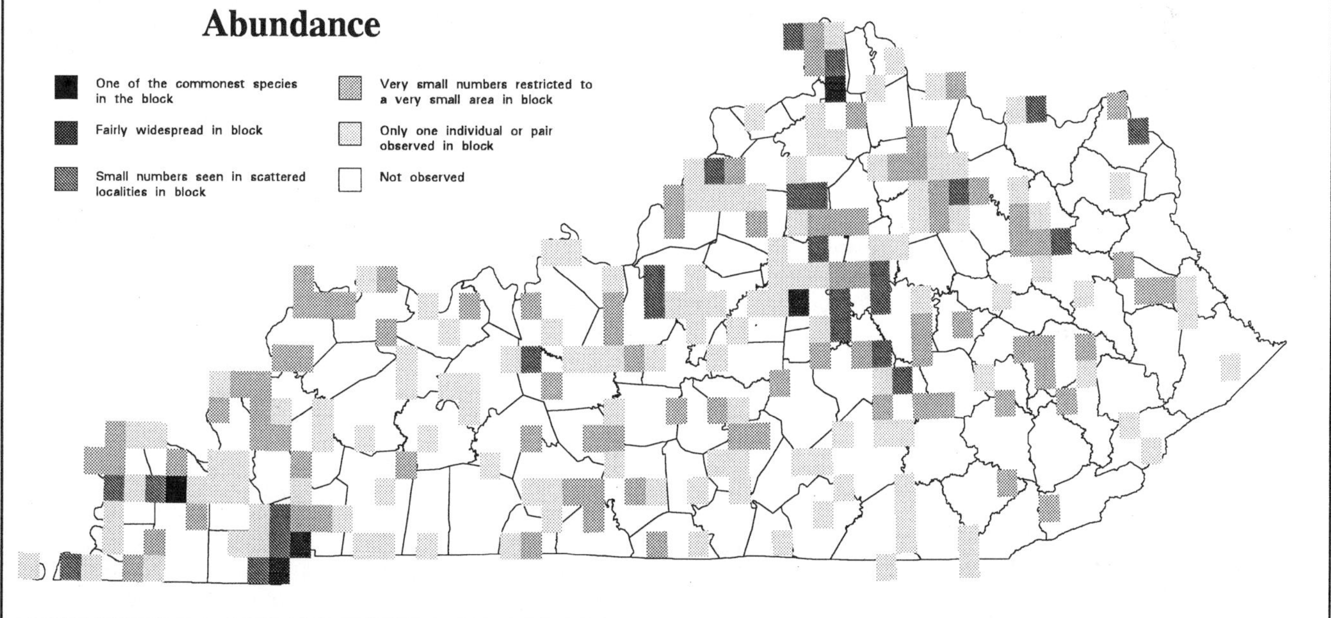

Analysis of Block Data by Physiographic Province Section

Physiographic Province Section	Total Blocks Surveyed	Blocks with Data	Avg Abund	% with Data	Section's % for State
Mississippi Alluvial Plain	14	10	1.7	71.4	5.3
East Gulf Coastal Plain	36	12	2.1	33.3	6.4
Highland Rim	139	38	1.6	27.3	20.2
Shawnee Hills	142	36	1.5	25.4	19.1
Blue Grass	204	68	1.9	33.3	36.2
Cumberland Plateau	173	20	1.5	11.6	10.6
Cumberland Mountains	19	4	1.2	21.1	2.1

Summary of Breeding Status

Number of Blocks in Which Species Was Recorded		
Total	**188**	**25.9%**
Confirmed	109	58.0%
Probable	33	17.6%
Possible	46	24.5%

Mallard

Anas platyrhynchos

The Mallard has increased dramatically as a nesting bird in Kentucky since the late 1950s, when Mengel (1965) stated that it was occasionally reported breeding in central and western Kentucky. Since the mid-1970s breeding Mallards have increased substantially in both range and abundance. The factors responsible for this increase are not known, but a similar trend has been documented throughout the eastern United States (Robbins et al. 1986).

Little is known of the occurrence of nesting Mallards in Kentucky before the 20th century. According to Audubon (1861), many bred in lowland lakes along the Mississippi and lower Ohio Rivers in the early 1800s. Otherwise, the only specific records of nesting before the late 1950s originated from scattered localities in central and western parts of the state, mostly in association with the floodplains of the larger rivers (Mengel 1965). Therefore, it appears that the species is currently more numerous and widespread as a nesting bird than at any previous time.

Mallards can be found on a variety of water bodies, from the largest rivers and reservoirs to the smallest marshes and farm ponds. In contrast to the Wood Duck, this species is found less frequently on unimpounded rivers and streams. Escaped or introduced Mallards quickly become accustomed to the presence of humans, and they are frequently reported nesting at marinas, suburban parks, cemeteries, and rural farm ponds. In general, naturally occurring birds seem to be observed more frequently in less settled areas, although interaction between wild and feral birds probably results in some nestings in settled situations.

The Mallard is most conspicuous in Kentucky during winter, when large numbers can be found about floodplain wetlands, reservoirs, and ponds, especially in western Kentucky. As soon as winter weather breaks, most birds begin moving north, and Mallards become less common by the end of March. A few birds linger later, but most of those present after early April are probably nesting. Birds crippled during the hunting season and unable to migrate constitute an unknown percentage of the nesting population. Kentucky nesting data are scarce, but the appearance of many broods during the last week of April and the first week of May indicates that most clutches are completed during the end of March and early April. At least some Mallards may raise a second brood, as there are a number of observations of broods into June and July, and at least one in September (Stamm 1982a); double-brooding has not been confirmed in Kentucky, however.

Forest Cover

Value	% of Blocks	Avg Abund
All	11.6	1.8
1	23.8	1.8
2	15.3	1.7
3	10.0	1.6
4	5.2	2.7
5	—	—

Mallards typically nest on the ground, usually near a lake or pond but sometimes relatively far from any substantial body of water. The nest is constructed of dead grasses and weed stalks and lined with down. It is usually well concealed among dense herbaceous growth, although sometimes it is situated with little surrounding cover. Kentucky data on clutch size are scarce, but most broods seem to contain 10–12 young. Harrison (1975) gives rangewide clutch size as 8–12, sometimes 6–15.

Bill Schoettler

The atlas survey yielded records of Mallards in nearly 12% of priority blocks statewide, and 18 incidental observations were reported. Mallards were found in all physiographic province sections except the Cumberland Mountains and were observed in only about 2% of the priority blocks in the Cumberland Plateau. In contrast, the species was found on more than one-third of the Mississippi Alluvial Plain priority blocks and more than one-fifth of the priority blocks in the Blue Grass. Occurrence was highest in open areas with limited forest cover and decreased as forestation increased.

Some atlas reports of Mallards probably pertained to domesticated or otherwise nonnaturally occurring individuals, especially in and near cities and towns. In contrast, many reports certainly pertained to naturally occurring birds, especially in the western half of the state. It is possible that some of these birds are descendants of feral birds originally established in artificial situations, but the frequency of occurrence of such individuals is unknown.

Nearly 42% of priority block records were for confirmed breeding. Although a few nests were located, most of the records resulted from the observation of broods of small young.

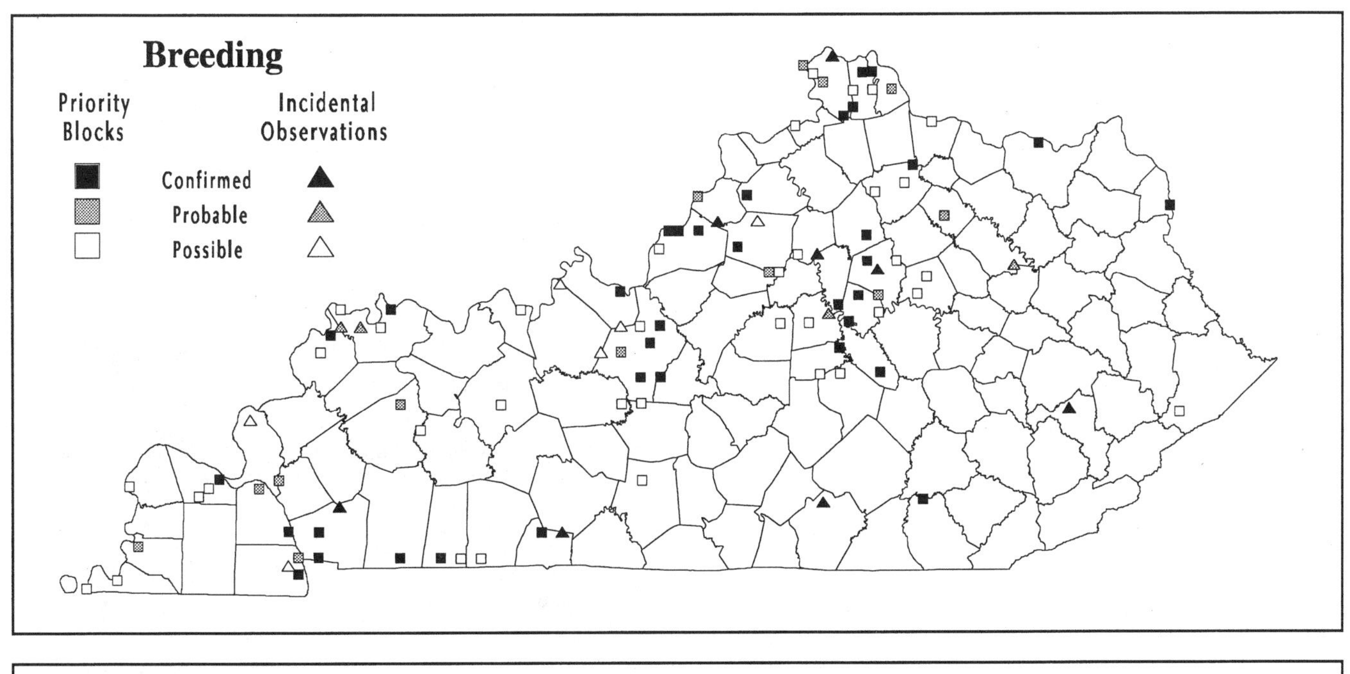

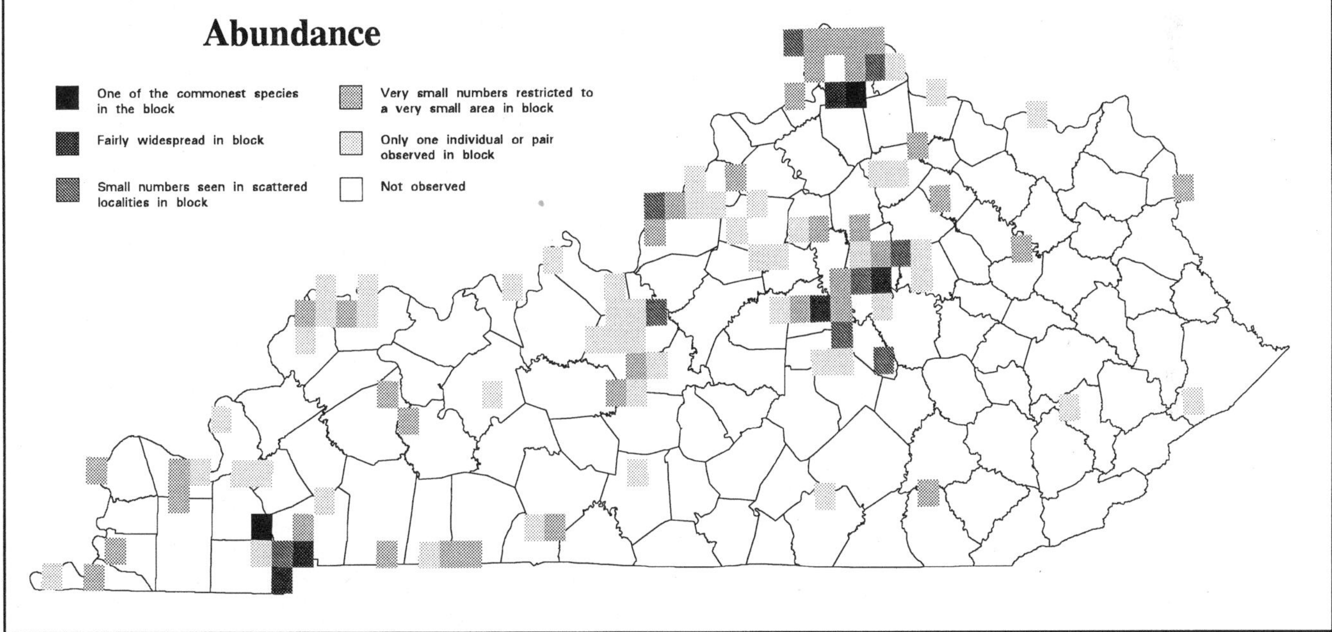

Analysis of Block Data by Physiographic Province Section

Physiographic Province Section	Total Blocks Surveyed	Blocks with Data	Avg Abund	% with Data	Section's % for State
Mississippi Alluvial Plain	14	5	1.6	35.7	6.0
East Gulf Coastal Plain	36	6	2.8	16.7	7.1
Highland Rim	139	16	1.6	11.5	19.0
Shawnee Hills	142	11	1.4	7.8	13.1
Blue Grass	204	43	1.9	21.1	51.2
Cumberland Plateau	173	3	1.7	1.7	3.6
Cumberland Mountains	19	-	-	-	-

Summary of Breeding Status

Number of Blocks in Which Species Was Recorded		
Total	**84**	**11.6%**
Confirmed	35	41.7%
Probable	13	15.5%
Possible	36	42.9%

Blue-winged Teal

Anas discors

The Blue-winged Teal is rare and very local as a breeding bird in Kentucky, apparently nesting only in response to the sporadic presence of suitable habitat. Ornithologists of the 19th century appear to have made no references to nesting of the species (Mengel 1965). As of the early 1960s nesting had been confirmed at only two locations: near Louisville, in Jefferson County, and at the transient lakes in southern Warren County (Mengel 1965; Wilson 1962a).

Before the atlas fieldwork, more recent reports of nesting were limited to additional records from the Louisville area. An adult with a brood of ten young was observed at the Falls of the Ohio on June 13, 1965 (Stamm and Jones 1966; Stamm et al. 1967). Reference was also made to nesting on ponds along the Ohio River in the Louisville area in the mid-1970s (Monroe 1976).

During the atlas fieldwork Blue-winged Teal were recorded on two priority blocks, and five incidental observations were reported. Of these seven records, four were for confirmed breeding. In 1989, an unusually wet year, Blue-winged Teal nested in substantial numbers at McElroy Lake (and probably Chaney Lake) in Warren County (Palmer-Ball and Boggs 1991). In addition, two broods of young were found at another transient lake nearby in Simpson County (also on the Woodburn quad) on June 6, 1989, and a brood of twelve small young accompanied by a female was found at a transient pond in Trigg County, about 1.0 mile northwest of Gracey, on June 21, 1989. In mid-June 1991 at least one brood of young was present at Chaney Lake in Warren County, and no fewer than three broods of young were located on another transient lake near Fort Campbell in southern Christian County. Of the remaining three records, none involved evidence of nesting other than the observation of birds in appropriate nesting habitat during the breeding season.

Blue-winged Teal winter well to the south of Kentucky, occurring regularly as far north as the Gulf Coast (AOU 1983). The first spring transients typically appear in the state by early March, and migrant numbers peak during the latter part of April. Following the departure of most birds in early May, a few initiate nesting if suitable habitat is present. At McElroy Lake in 1989, the first brood of 11 downy young was observed on May 29, indicating completion of early clutches by the first week of May. On June 6 four or five broods were seen, and at least five and probably as many as eight broods were observed there on June 13. By early July many young were indistinguishable in size from the adults; on September 12, however, a late brood of five or six young, seemingly full grown but still incapable of flight, was observed there. Altogether, at least 10–12 broods were probably raised there in 1989, with a peak in clutch completion occurring during the latter half of May (Palmer-Ball and Boggs 1991).

Blue-winged Teal are ground nesters, constructing a nest of dead grasses and down among the dense cover of grasses, weeds, and other herbaceous plants in the immediate vicinity of a body of water (Harrison 1975). Based on the sizes of broods of small young observed during the atlas survey, it appears that most Kentucky clutches contain about 10–12 eggs. Harrison (1975) gives rangewide clutch size as 10–13, sometimes 6–15.

Herbert Clay Jr. M.D.

It is noteworthy that all reports of confirmed nesting during the atlas fieldwork came from transient lakes and ponds in the karst region of the Highland Rim. These areas are used by many migrant teal during March and April, and it appears that if spring rains are sufficient to allow water to remain in these depressions into late May, teal remain to nest. Sporadic nesting is to be expected in other parts of the state, especially in floodplains of the larger rivers, where spring floodwaters sometimes persist long enough to allow nesting.

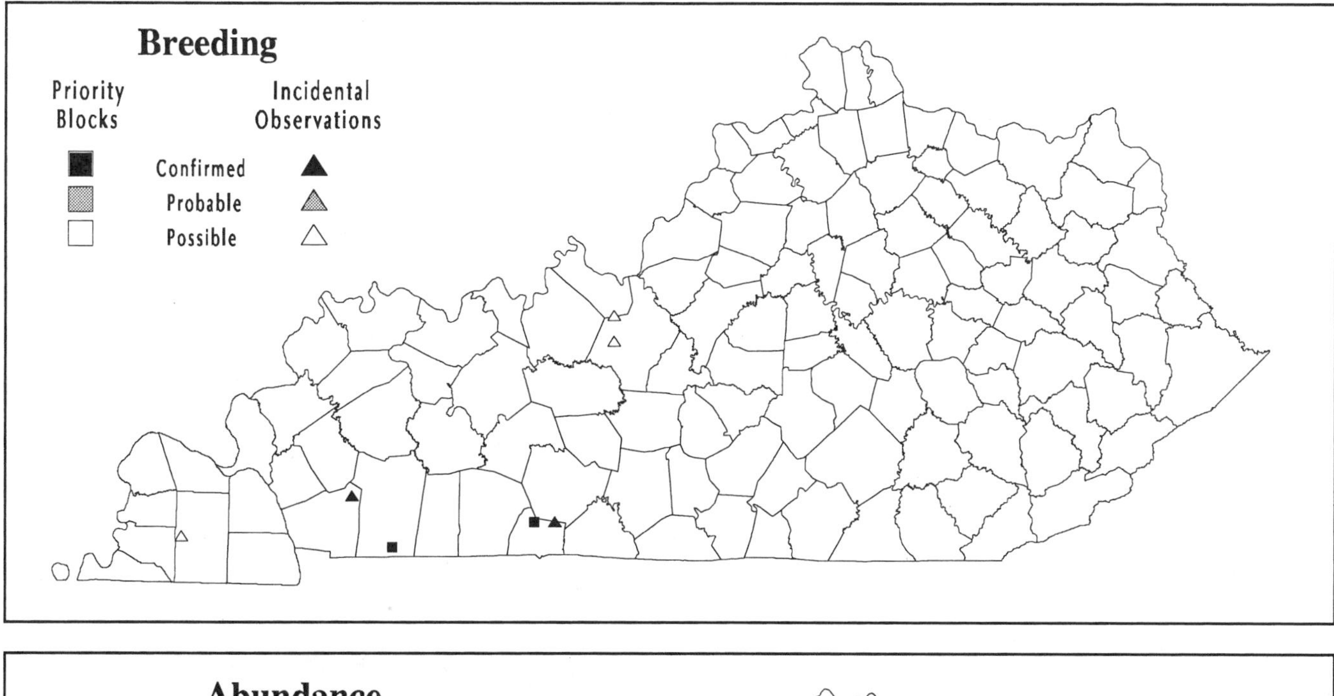

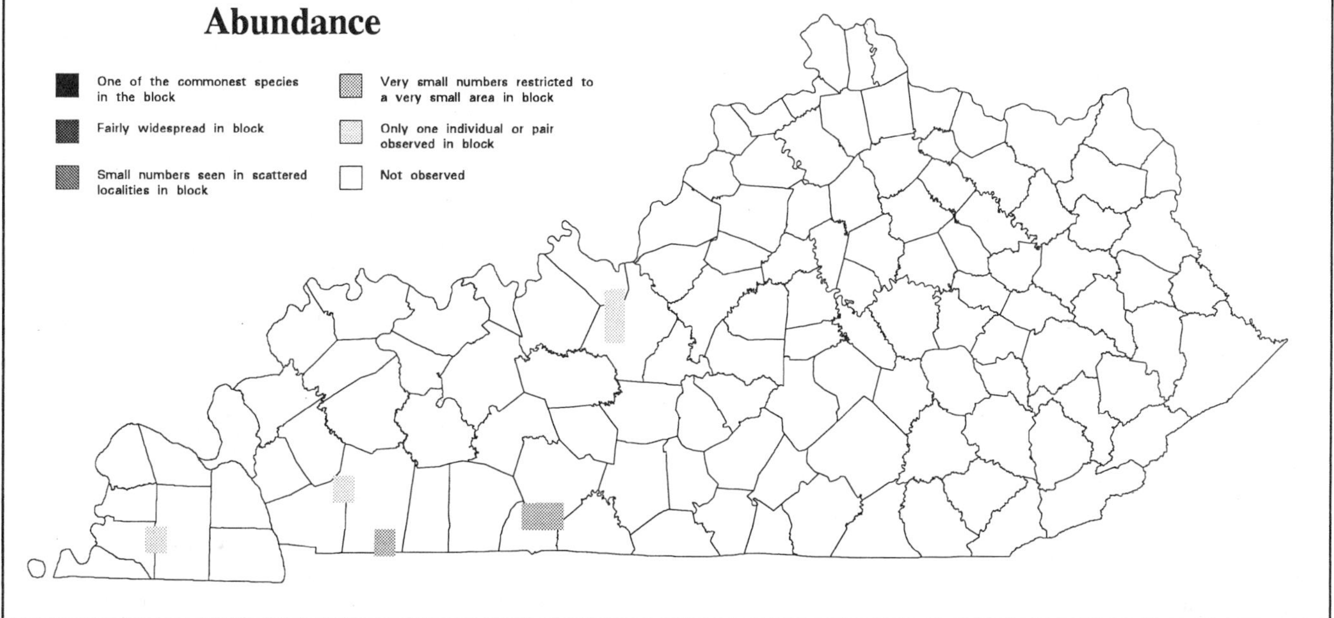

Analysis of Block Data by Physiographic Province Section

Physiographic Province Section	Total Blocks Surveyed	Blocks with Data	% with Data	Section's % for State
Mississippi Alluvial Plain	14	-	-	-
East Gulf Coastal Plain	36	-	-	-
Highland Rim	139	2	1.4	100.0
Shawnee Hills	142	-	-	-
Blue Grass	204	-	-	-
Cumberland Plateau	173	-	-	-
Cumberland Mountains	19	-	-	-

Summary of Breeding Status

Number of Blocks in Which Species Was Recorded		
Total	**2**	**0.3%**
Confirmed	2	100.0%
Probable	-	-
Possible	-	-

Blue-winged Teal

Hooded Merganser

Lophodytes cucullatus

The Hooded Merganser breeds throughout much of the eastern United States, including Kentucky (AOU 1983). As is the case throughout much of its range, however, the species is distributed very locally in the state. Hooded Mergansers have been reported nesting at only a few sites in central and western Kentucky, and nesting has been reported more than once at only two localities. While the female's inconspicuous plumage and the species's reclusive habits during the nesting season may be largely responsible for the lack of information, it is likely that these mergansers have never nested commonly (Mengel 1965). Efforts to protect and restore wetland areas through a variety of public and private efforts may be responsible for a recent increase in observations of nesting.

Mengel (1965) included details of specific nesting records only from Jefferson County, although he believed that small numbers of Hooded Mergansers nested along the Mississippi and lower Ohio Rivers west of Louisville. More recent records of nesting reported before the atlas survey included reports of a female with four young in Jefferson County in May 1966 (Croft and Stamm 1967); a female with four young on a slough along the Mississippi River in Fulton County in June 1966 (Able 1967); a brood of young in the company of a female on a floodplain slough on the West Kentucky WMA, in McCracken County, in May 1980 (C. Nicholson, pers. comm.); and a family group near Lake No. 9, in Fulton County, in June 1980 (Palmer-Ball and Barron 1982).

Hooded Mergansers typically inhabit shallow-water sloughs and ponds in the lowlands along major rivers. The species is more restricted in habitat than the much more common and widespread Wood Duck, seldom being found far from floodplain situations and almost never on fast-flowing creeks and streams. These mergansers are somewhat adaptable, however, and they use a variety of natural and artificial bodies of water, as long as a good stand of fairly mature forest is nearby for nest sites.

Kentucky nesting data are virtually absent, except for the few observations of broods reported in the literature and during the atlas fieldwork. Small young have been observed from late April in Henderson County (Stamm and Monroe 1992) to late May in Jefferson County (Monroe 1947), indicating that most clutches are probably completed by the end of April. By mid-June most young begin approaching the adult female in size, and family groups have been observed into mid-July. Kentucky data on clutch size are lacking, but two broods have contained 12 young (Monroe 1947; author's notes). Harrison (1975) gives rangewide clutch size as 10–12, sometimes 6–18.

The Hooded Merganser is a cavity nester, using both natural cavities in trees and nest boxes erected to attract Wood Ducks. If typical of those used throughout most of its range, nest trees of the Hooded Merganser in Kentucky would be situated in tracts of forest and standing near to or in water of slow-moving streams, sloughs, ponds, or impoundments.

The atlas fieldwork yielded two records of Hooded Mergansers in priority blocks, and two incidental observations were reported. Both incidental records were for confirmed breeding. On May 6, 1986, a brood of 12 downy young was observed in the company of a female on Muddy Slough on the Sauerheber Unit of Sloughs WMA, in Henderson County, and on May 27, 1989, a brood of five half-grown young was observed with a female on Chaney Lake, in Warren County (Palmer-Ball and Boggs 1991). The two priority block reports were both recorded as possible. They involved the observation of a female on a small backwater pond near Hickman, in Fulton County, on June 13, 1990, and three birds on a backwater slough along the Mississippi River at Chalk Bluff, in Hickman County on June 6, 1991 (BBS data).

Gary Meszaros

The atlas data probably demonstrate the rarity of nesting Hooded Mergansers in Kentucky. While the species is certainly overlooked to some extent, small numbers probably nest regularly along the lower Ohio River and Mississippi River floodplains, and sporadically at other suitable locations. A recent record of a female with nine small young on an impoundment near Lake Cumberland, in Pulaski County, on May 11, 1992, indicates the latter (J. Elmore, pers. comm.). If wetland protection measures continue, the population of nesting Hooded Mergansers may increase perceptibly in the future.

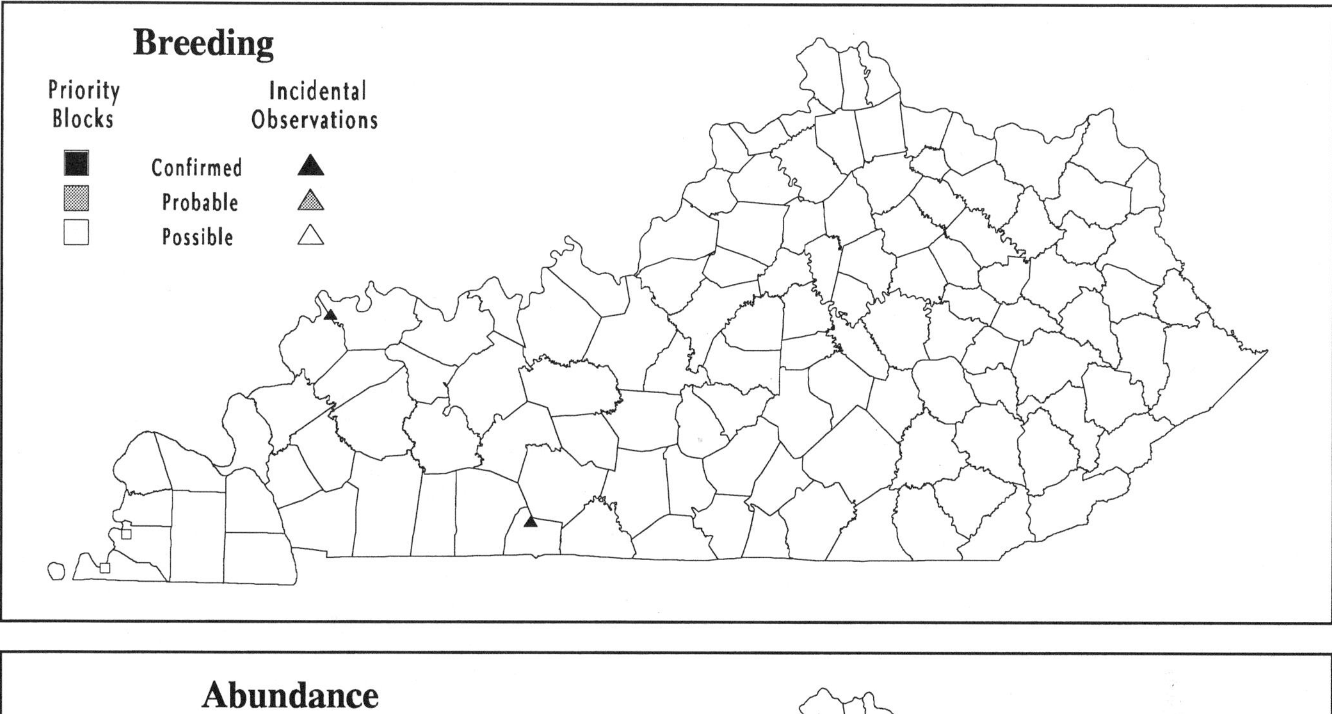

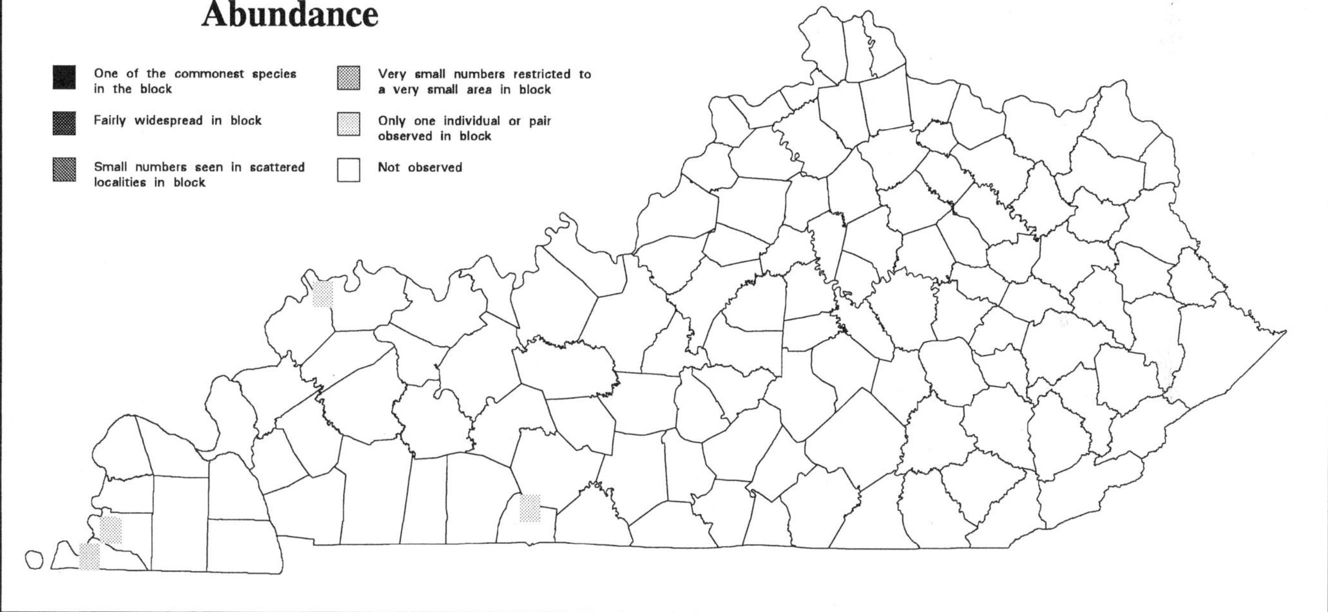

Analysis of Block Data by Physiographic Province Section

Physiographic Province Section	Total Blocks Surveyed	Blocks with Data	% with Data	Section's % for State
Mississippi Alluvial Plain	14	2	14.3	100.0
East Gulf Coastal Plain	36	-	-	-
Highland Rim	139	-	-	-
Shawnee Hills	142	-	-	-
Blue Grass	204	-	-	-
Cumberland Plateau	173	-	-	-
Cumberland Mountains	19	-	-	-

Summary of Breeding Status

Number of Blocks in Which Species Was Recorded		
Total	**2**	**0.3%**
Confirmed	-	-
Probable	-	-
Possible	2	100.0%

Hooded Merganser

Black Vulture

Coragyps atratus

Although the Black Vulture occurs in all of Kentucky's major physiographic regions, it is very locally distributed, being nowhere common and rare or absent in many areas. Mengel (1965) regarded the species as a rare to fairly common resident west of the Cumberland Plateau. Although no convincing evidence of its presence in eastern Kentucky existed at that time, he mentioned the possibility that the species might soon turn up there as a result of the expansion of its range elsewhere in the Appalachians. As he predicted, the species has since been reported at a few points within the Cumberland Mountains (see, e.g., Croft 1969), but it remains scarce to absent throughout most of eastern Kentucky.

Black Vultures occur in a wide variety of semi-open habitats. These scavengers are most frequent in rural situations with a mixture of openings and forest, but they also occur along open clifflines in predominantly forested areas and in isolated tracts of forest in agricultural land. In far western Kentucky the species is most often found in bottomland swamp forest, and the birds are often seen near heronries (Mengel 1965).

Black Vultures were first reported in Kentucky early in the 19th century by Audubon (1861), who observed them throughout the year and noted their occurrence as far east as Cincinnati. The species was probably very locally distributed before settlement, but it was likely present along major river corridors and in bottomland swamps. In central and western Kentucky, the conversion of vast areas of native forest to agricultural use and settlement may have resulted in an increase in the species in some areas (Mengel 1965). Human manipulation of the landscape has resulted in the availability of a substantial food source in the form of farm animals, as well as ubiquitous mammals such as opossums, skunks, and raccoons. Such changes also may explain the species's recent expansion into the Cumberland Mountains and the western Cumberland Plateau.

Unlike the Turkey Vulture, which typically becomes much less common in winter, the numbers of Black Vultures in Kentucky seem to remain fairly consistent throughout the year. It is unclear to what extent local birds move farther south, but it appears that some seasonal movement does take place. Those birds that do winter in the state typically gather in roosts, making their occurrence even more local than in summer. The roosts disperse as soon as winter weather moderates, and pairs begin to search out suitable nest sites during late February and March. According to Mengel (1965), clutches may be completed by early March, although a peak in egg laying probably occurs in early April. The nesting season is rather drawn out, and later clutches have been reported into June. Young typically fledge by mid-August, but sometimes not until mid-September (Pearson and Pearson 1985). According to Mengel (1965), the typical clutch size is two.

Forest Cover

Value	% of Blocks	Avg Abund
All	18.4	1.5
1	16.8	1.8
2	25.8	1.6
3	21.3	1.5
4	9.1	1.2
5	9.6	1.0

Black Vultures nest in a variety of situations, including sheltered crevices and small caves along clifflines, hollow trees and fallen logs in forests, and abandoned houses and barns. The eggs are laid directly on the dirt or debris at the nest site, and the young remain in the nest until they are ready to fly. Nest sites may be used year after year if the birds are not disturbed.

Matthew Patterson

The atlas fieldwork yielded records of Black Vultures in more than 18% of priority blocks statewide, and 20 incidental observations were reported. Occurrence was highest in the Mississippi Alluvial Plain and the Blue Grass and lowest in the Cumberland Mountains and the Cumberland Plateau. Most of the records in the Cumberland Plateau originated from the western Cliff Section, as well as its western transition to the Knobs of the Blue Grass. Average abundance varied somewhat similarly, but values were low across the state. Occurrence was highest in open areas with some forest and decreased as forestation increased.

Vultures may forage far from their nest sites for food, and it is likely that some atlas reports pertained to individuals that were not actually nesting within the block in which they were observed. Despite this likelihood, observations of Black Vultures were recorded as possible, even when the birds were circling high in the air. Probable records were typically based on repeated observations of birds at or near suitable nest sites (e.g., old houses, clifflines), especially when sightings involved more than one bird.

Only three priority block records and one incidental observation were for confirmed breeding. Two confirmed records involved the observation of young in abandoned buildings (Logan and Rowan Counties), one in an unused duck blind (Ballard County), and one along a cliffline (Fayette/Woodford county line).

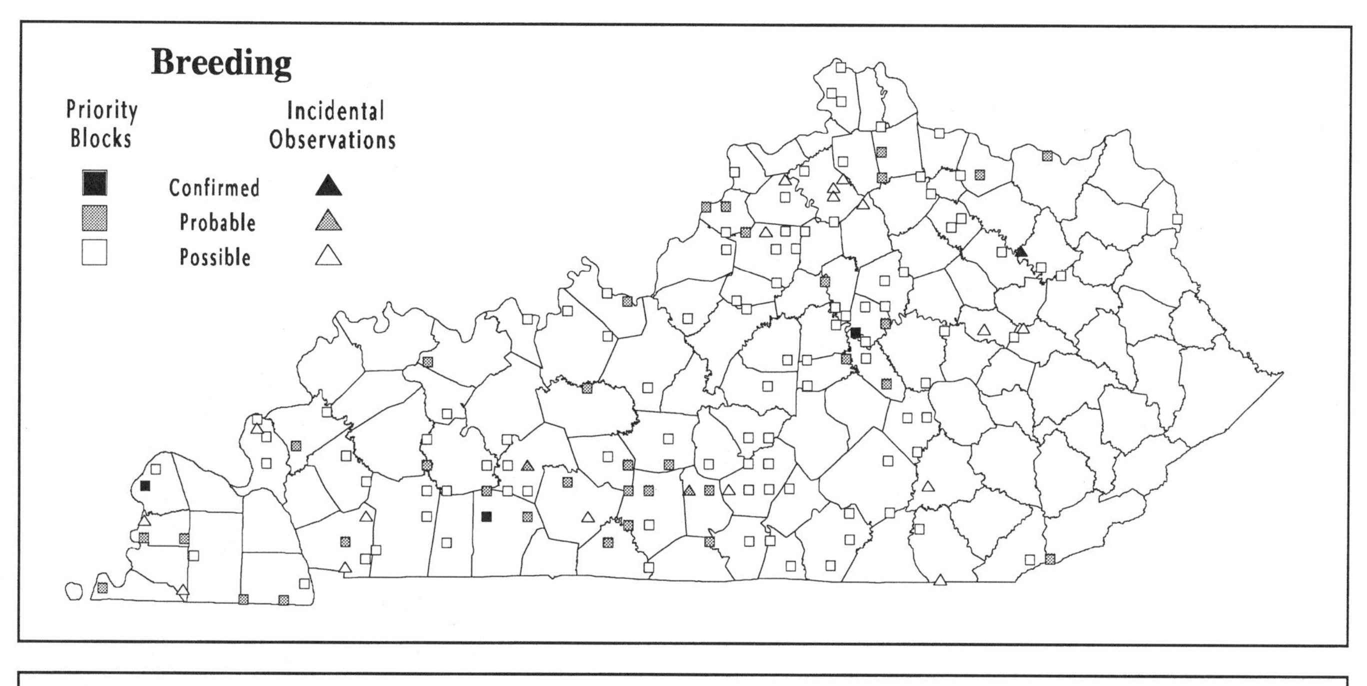

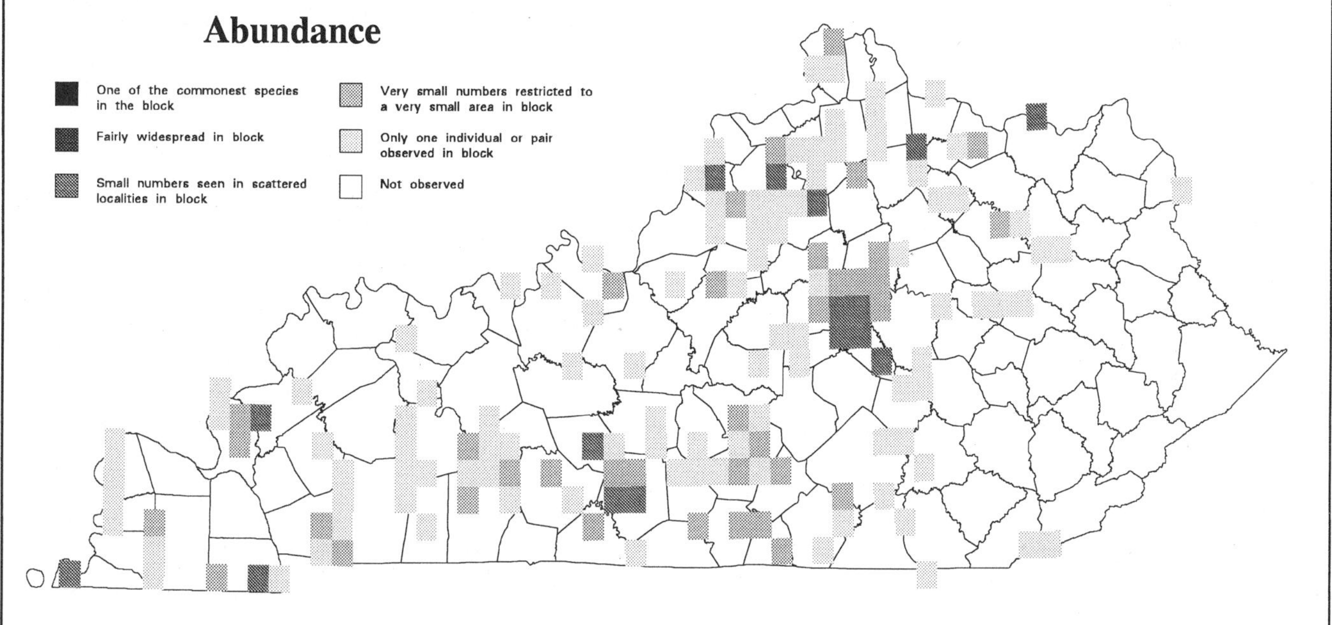

Analysis of Block Data by Physiographic Province Section

Physiographic Province Section	Total Blocks Surveyed	Blocks with Data	Avg Abund	% with Data	Section's % for State
Mississippi Alluvial Plain	14	4	1.5	28.6	3.0
East Gulf Coastal Plain	36	5	1.8	13.9	3.7
Highland Rim	139	31	1.6	22.3	23.1
Shawnee Hills	142	28	1.4	19.7	20.9
Blue Grass	204	55	1.6	27.0	41.0
Cumberland Plateau	173	9	1.0	5.2	6.7
Cumberland Mountains	19	2	1.0	10.5	1.5

Summary of Breeding Status

Number of Blocks in Which Species Was Recorded		
Total	**134**	**18.4%**
Confirmed	3	2.2%
Probable	34	25.4%
Possible	97	72.4%

Turkey Vulture

Cathartes aura

The Turkey Vulture is probably the most common and widespread soaring bird in Kentucky. Conspicuous and well known to most local residents as "buzzards," these vultures occur virtually statewide. Mengel (1965) regarded the species as a common summer resident, breeding locally throughout the state except in the southeastern mountains, where he considered it uncommon.

Turkey Vultures are found in a great variety of habitats, from extensive areas of rugged forest to semi-open and open farmland. The species is rather shy and retiring, however, and its distribution may depend largely on the availability of undisturbed nest sites. This factor also may explain its relative scarcity in extensively cleared areas.

Turkey Vultures may not have been as common and widespread in Kentucky two centuries ago. As has been assumed to be true for the Black Vulture (Mengel 1965), the conversion of vast areas of native forest to agricultural use and settlement may have resulted in an increase in the species. The creation of clearings has likely resulted in increased opportunities for foraging. In addition, these vultures rely heavily on livestock and smaller animals such as opossums, skunks, and raccoons, all of which are present in greater abundance and distribution today because of human activity. Although native animal populations probably sustained numbers of vultures in some areas, food resources may not have been as reliable in others. Therefore, the species was likely more locally distributed in the past than it is today.

Although Turkey Vultures usually can be found in Kentucky throughout the year, most breeding birds migrate southward in winter. Those that remain tend to congregate in large roosts, resulting in the species's absence throughout much of the state from late fall through early spring. Turkey Vultures typically return to Kentucky's skies about the middle of February, and full breeding numbers are probably attained by mid-March. Clutches are completed from late March to mid-May, with an apparent peak during the middle of April (Mengel 1965). Some nesting occurs later into the summer, and large young have been observed at nest sites into early September (Stamm and Croft 1968). Kentucky nests typically contain two eggs (Mengel 1965).

Forest Cover

Value	% of Blocks	Avg Abund
All	79.6	2.4
1	78.2	2.2
2	88.9	2.5
3	86.1	2.5
4	66.9	2.3
5	57.7	1.6

Turkey Vultures nest in a variety of situations, including small shelters and caves in clifflines, hollow trees, stumps, and fallen logs. The eggs are laid directly on the dirt or debris at the nest site. Unlike the Black Vulture, which often uses abandoned houses and barns, the Turkey Vulture seems to use natural sites much more frequently. Observations of active nests in an abandoned barn in Boone County in 1984 and 1985 represent two of only a few reports of the use of artificial sites in the state (L. McNeely, pers. comm.).

The atlas survey yielded records of Turkey Vultures in nearly 80% of priority blocks statewide, and 16 incidental observations were reported. Occurrence was highest in the Blue Grass and the Highland Rim, and lowest in the Mississippi Alluvial Plain and the Cumberland Mountains, but more than half of the blocks in all regions provided data. Average abundance was relatively low statewide, but varied similarly. Occurrence was highest in areas with a good mixture of forested areas and open habitats, and lowest in entirely forested blocks.

Lewis Kornman

Vultures sometimes travel great distances from their nest sites to forage for food, and it is probable that some atlas reports pertained to individuals that were not nesting within the block in which they were observed. Despite this possibility, sightings of Turkey Vultures were recorded as possible even when birds were observed soaring high in the air. Probable records were typically based on repeated observations of birds within a block during the breeding season, especially in the vicinity of suitable nest sites.

Despite the Turkey Vulture's overall abundance, less than 3% of priority block records were for confirmed breeding. Although more than half of the 16 confirmed records were based on the discovery of active nests, several pertained to the observation of recently fledged young. It is believed that the FL code was used correctly in most cases—that is, for young birds not capable of full flight—but a few records may have been based on the observation of full-grown, dark-headed immatures. While it is possible that such birds were fairly far from their nest sites, these observations likely indicated nesting somewhere in the general vicinity, and they were included in the final data set.

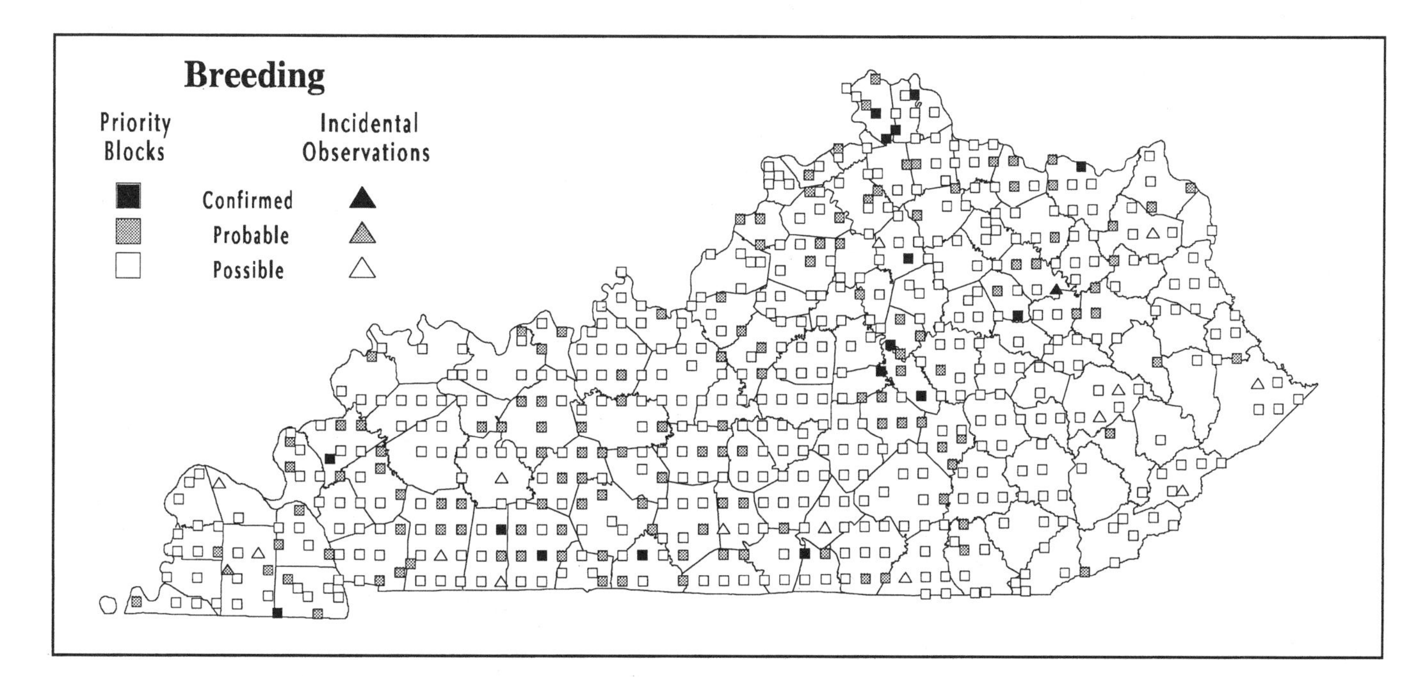

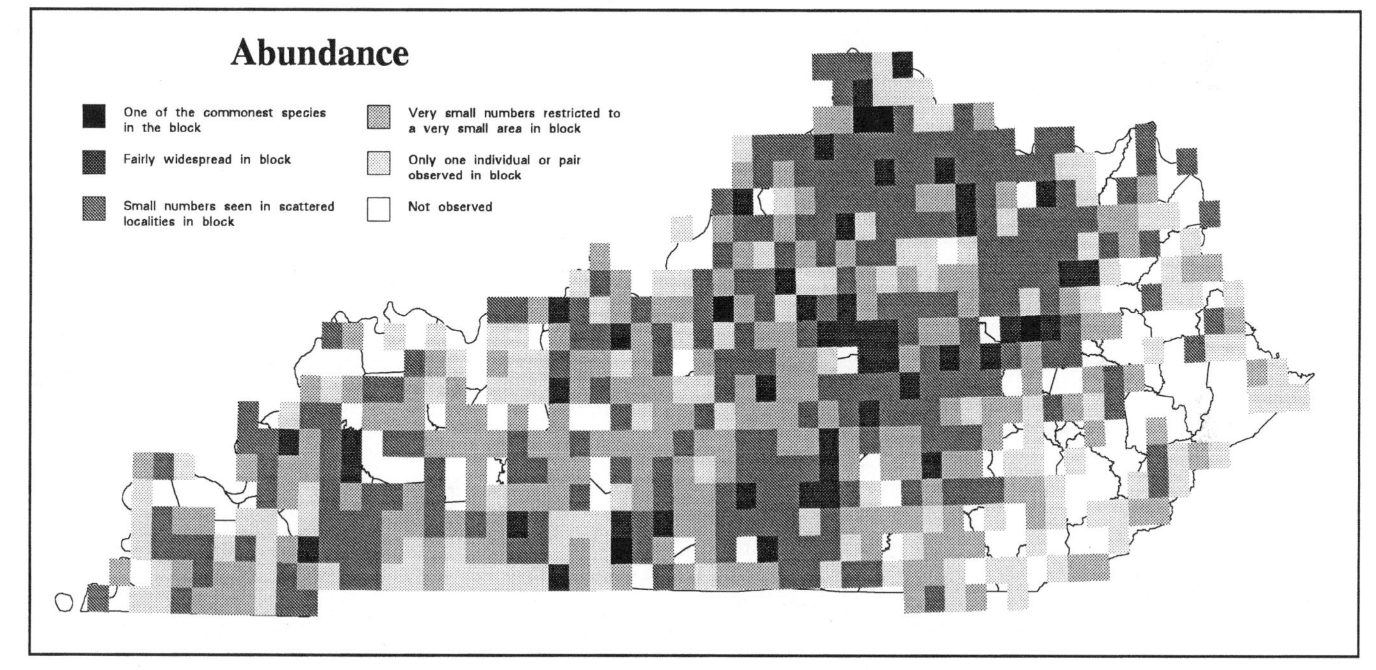

Analysis of Block Data by Physiographic Province Section

Physiographic Province Section	Total Blocks Surveyed	Blocks with Data	Avg Abund	% with Data	Section's % for State
Mississippi Alluvial Plain	14	8	1.8	57.1	1.4
East Gulf Coastal Plain	36	27	2.1	75.0	4.7
Highland Rim	139	128	2.4	92.1	22.1
Shawnee Hills	142	109	2.3	76.8	18.8
Blue Grass	204	190	2.7	93.1	32.8
Cumberland Plateau	173	107	2.0	61.8	18.5
Cumberland Mountains	19	10	1.4	52.6	1.7

Summary of Breeding Status

Number of Blocks in Which Species Was Recorded		
Total	**579**	**79.6%**
Confirmed	16	2.8%
Probable	133	23.0%
Possible	430	74.3%

Osprey

Pandion haliaetus

The breeding occurrence of the Osprey in Kentucky has been poorly documented. Audubon (1861) recorded several pairs nesting in the vicinity of the Falls of the Ohio at Louisville in the early 1800s, but breeding was not reported again until June 1949, when a nest was observed on the Blood River embayment of Kentucky Lake (DeLime 1949). Mengel (1965) regarded the species as an occasional breeder in extreme western Kentucky in the late 1950s, based primarily on DeLime's report and occasional summer observations elsewhere in that part of the state. After the mid-1950s published reports of Ospreys decreased significantly, and there were only two reports of attempted nesting: in the spring of 1968 and 1969 nests were constructed on Lake Barkley, in Trigg and Lyon Counties, respectively, but they were later abandoned in both cases (Stamm 1969, 1970). The Osprey's scarcity during this period corresponded to a widespread decline that apparently occurred because of the accumulation of residues from chlorinated hydrocarbon pesticides (primarily DDT) in natural ecosystems, human disturbance of nest sites, and shooting (Peterson 1969; Terres 1980; Ehrlich et al. 1988).

In response to rangewide declines experienced during the DDT era, state and federal agencies became involved in reintroduction programs during the early 1980s. The Tennessee Valley Authority (TVA) established an Osprey hacking program in the Tennessee portion of Land Between the Lakes in 1981. In 1983 KDFWR began cooperating with TVA to hack birds in the Kentucky portion of Land Between the Lakes, and as a result of this effort 61 Ospreys were released through 1989 (S. Bloemer, pers. comm.). From 1982 to 1984 KDFWR released another 25 birds from four additional sites in central and western Kentucky, mostly within the Blue Grass (S. Evans, unpub. rpt.). Finally, in 1988–89 KDFWR assisted the U.S. Forest Service in hacking Ospreys at Laurel River Lake, in Laurel County, resulting in the release of 10 more young (D. Yancy, pers. comm.).

In part owing to the success of these programs, Ospreys were finally documented breeding again in 1986. That year a nest was found along the Ohio River near Bayou, in Livingston County, and two nests were observed adjacent to Land Between the Lakes on Kentucky Lake and Lake Barkley (Stamm 1986a). One of the individuals of the Livingston County pair was banded, indicating that it had been released at Land Between the Lakes during the early 1980s. Since the mid-1980s the number of breeding birds has slowly increased, mostly on Lake Barkley near the Lyon/Trigg county line. As of 1991 at least six pairs were known to be attempting to nest in this area (S. Bloemer, pers. comm.).

Away from the Land Between the Lakes region, summer observations of Ospreys also have been on the increase on rivers and reservoirs (Stamm 1986a). As of 1992, however, nesting had not been confirmed elsewhere. The species currently seems to be rebounding in most parts of its range, and with suitable nesting habitat now available on numerous impoundments across Kentucky, there is no reason to suspect that Ospreys will not begin nesting at scattered localities in the future.

Although Ospreys are occasionally reported in Kentucky in winter, the species typically winters from the Gulf Coast southward into South America (AOU 1983). During late March and early April breeding birds return northward, and nest building usually begins immediately. Although specific data on nesting in Kentucky are scarce, most birds appear to be incubating by the end of April, and young are typically observed in nests during May and early June. Most young probably fledge by July, and family members often remain loosely associated with one another through late summer. Kentucky data on clutch size are lacking, but Harrison (1975) gives rangewide clutch size as 3, sometimes 2, rarely 4. Recent observations indicate that successful nests have usually fledged one or two young (D. Yancy, pers. comm.).

Gene Boaz

Ospreys construct large stick nests in trees or on artificial structures such as utility poles and navigation markers. Once constructed, nests are often used in succeeding years. Although most nests are situated over water, the Livingston County pair built a nest about half a mile from the Ohio River one year in the late 1980s. Nests constructed above the ground are typically situated rather high, but nests over water may be constructed as low as about 20 feet.

The atlas survey yielded three records of Ospreys in priority blocks, and three incidental observations were reported. All four confirmed records (one priority block and three incidental) involved active nests at previously known sites on the Ohio River in Livingston County, and on Lake Barkley in Lyon and Trigg Counties. Outside of the known nesting areas, the only location where breeding appeared to be possible was Barren River Reservoir, in Barren County, where two Ospreys were observed during the summers of 1985 and 1986 (Stamm 1985; W. Mason, pers. comm.). Although a few additional observations were reported during the atlas project, all involved single, probably nonbreeding, birds, and none were included in the final data set.

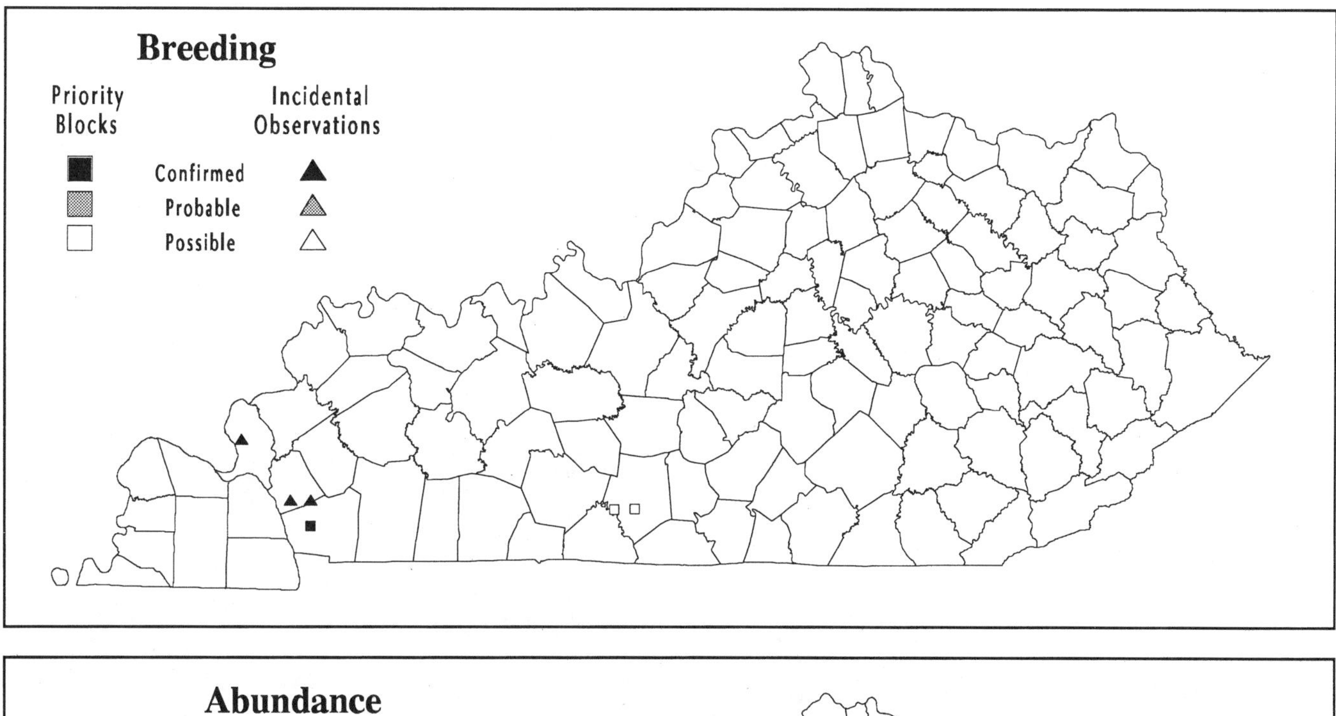

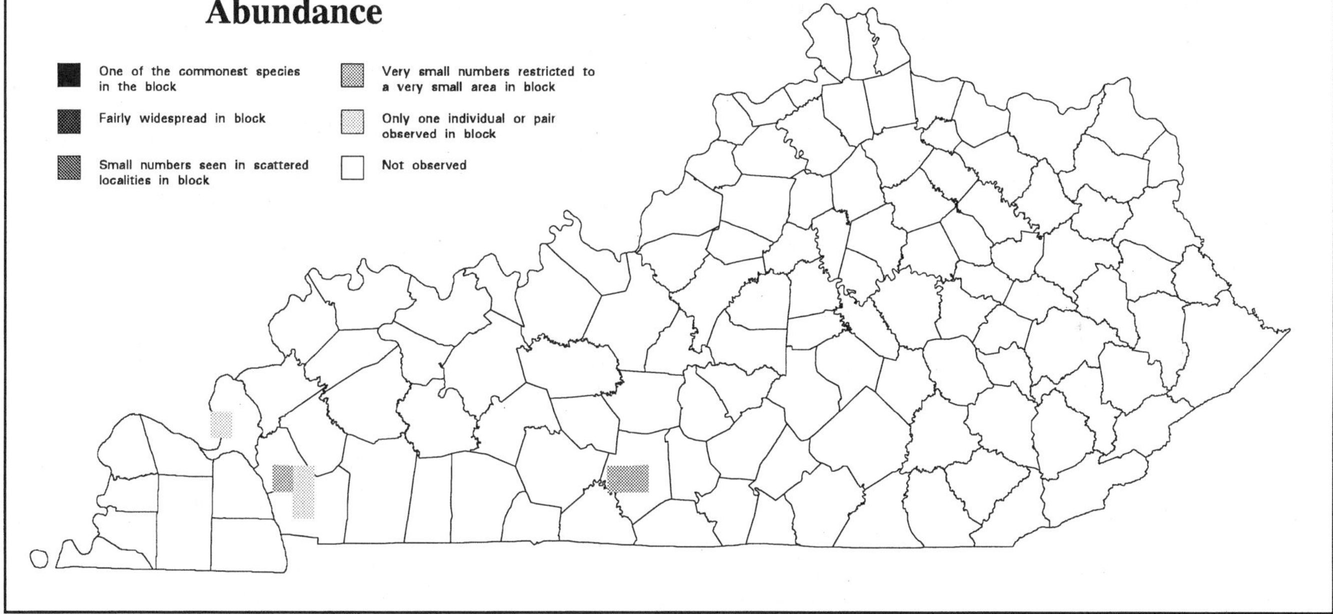

Analysis of Block Data by Physiographic Province Section

Physiographic Province Section	Total Blocks Surveyed	Blocks with Data	% with Data	Section's % for State
Mississippi Alluvial Plain	14	-	-	-
East Gulf Coastal Plain	36	-	-	-
Highland Rim	139	3	2.2	100.0
Shawnee Hills	142	-	-	-
Blue Grass	204	-	-	-
Cumberland Plateau	173	-	-	-
Cumberland Mountains	19	-	-	-

Summary of Breeding Status

Number of Blocks in Which Species Was Recorded		
Total	**3**	**0.4%**
Confirmed	1	33.3%
Probable	-	-
Possible	2	66.7%

Osprey

Mississippi Kite

Ictinia mississippiensis

Although the graceful images of Mississippi Kites coursing through the skies of far western Kentucky are not an uncommon sight today, the species was once known only as an occasional vagrant. Audubon (1861) did not note these kites along the Mississippi River north of Memphis during the early 1800s, but in the latter half of the 19th century they occurred as far north as the prairies of the southern Midwest. Nelson (1877) found the species to be abundant near Cairo, Illinois, adjacent to the confluence of the Mississippi and Ohio Rivers in August 1875, and Pindar (1925) considered the bird a rare (but seemingly regular) summer visitant in Fulton County in the late 1800s. During this era Mississippi Kites were common in the prairies of southern Illinois (Ridgway 1873) and in bottomland habitats of southeastern Missouri (Widmann 1907).

Mengel (1965) assumed that the species still occurred regularly in western Kentucky about the turn of the century. Not long afterward, he believed, it began to disappear, and by the 1920s Mississippi Kites were scarce throughout their former range in the Midwest. The species was absent throughout the mid–Mississippi River Valley north of Memphis until the early 1960s, when small numbers began to reoccupy formerly inhabited areas in and adjacent to the river's floodplain as far north as St. Louis. The Mississippi Kite was first seen again in Kentucky in the mid-1960s, when small numbers were found in Fulton and Hickman Counties (Croft and Rowe 1966).

Since the late 1960s Mississippi Kites have been reported regularly in summer within a limited range in extreme western Kentucky, throughout the Mississippi Alluvial Plain and upstream along the lower Ohio River to near Paducah. Outside this area, the species is reported only occasionally in spring and summer as a vagrant. Although an active nest has not been discovered in Kentucky, breeding has been documented at the Ballard WMA, in Ballard County, based on the observation of family groups, including recently fledged young (Stamm 1984b).

Mississippi Kites winter primarily in central South America (AOU 1983), and summer residents typically reappear in the skies over western Kentucky during the first week of May. Specific information on nesting in Kentucky is lacking, but data obtained in southern Illinois indicate that incubation is likely under way at most nests by early June. Most young fledge by early August (Evans 1981). During the summer, foraging birds may collect in loose groups at favored feeding areas, and such aggregations have numbered up to 30 birds (Bierly 1973). Small groups of adults and young are observed through August, and a few birds may linger into the second week of September.

Throughout much of its range, the Mississippi Kite is found in rolling, upland prairies and farmland. In Kentucky the species primarily occurs in floodplain areas where tracts of bottomland forest are intermixed with or adjacent to farmland. These kites typically nest within tracts of fairly mature to mature forest, although they sometimes build along an isolated corridor of large trees. The nest is constructed of sticks and placed in a crotch of the limbs high up in a tall tree (Evans 1981). Kentucky data on clutch size are lacking, but Evans (1981) found that all nests studied in southern Illinois seemed to contain one or two eggs.

The atlas fieldwork yielded records of Mississippi Kites in 13 priority blocks, and two incidental observations were reported. All reports came from areas considered to be within the species's normal range, although more comprehensive fieldwork along the tributaries of the Mississippi River resulted in observations farther up Obion and Mayfield Creeks than had previously been documented.

Sherri A. Evans

The Mississippi Kite proved to be very difficult to confirm as a nesting bird during the atlas survey, and there was only one report for confirmed breeding. Recently fledged young were observed in the bottomlands near the Reelfoot NWR in southwestern Fulton County and were assumed to have been raised within the immediate vicinity. Most of the probable reports were based on repeated observations of kites in the same area, both during individual breeding seasons and in succeeding years. Adult kites were observed carrying food in several blocks, but because they often forage some distance from the nest, these observations were regarded as only possible evidence of breeding.

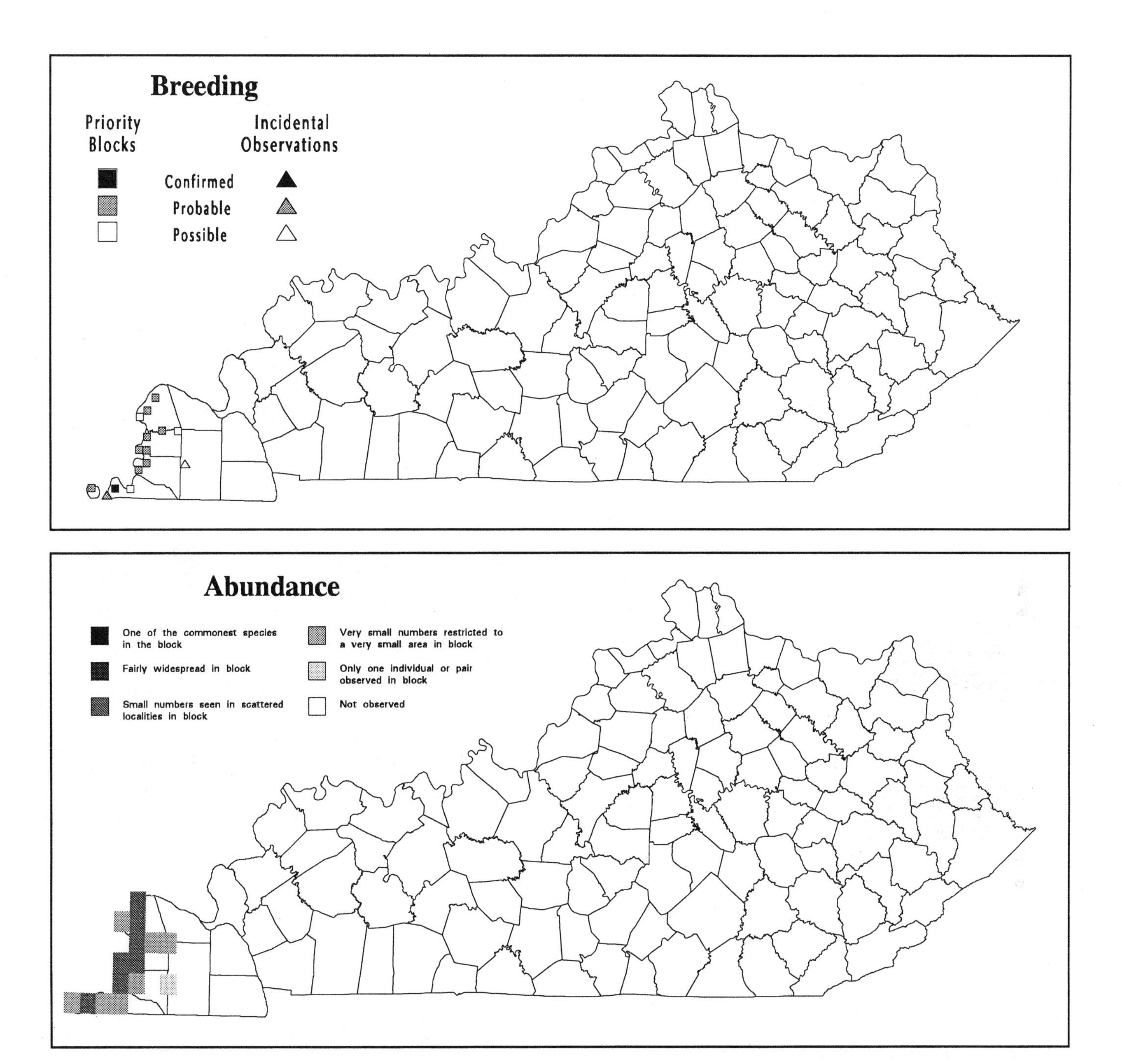

Analysis of Block Data by Physiographic Province Section

Physiographic Province Section	Total Blocks Surveyed	Blocks with Data	% with Data	Section's % for State
Mississippi Alluvial Plain	14	12	85.7	92.3
East Gulf Coastal Plain	36	1	2.8	7.7
Highland Rim	139	-	-	-
Shawnee Hills	142	-	-	-
Blue Grass	204	-	-	-
Cumberland Plateau	173	-	-	-
Cumberland Mountains	19	-	-	-

Summary of Breeding Status

Number of Blocks in Which Species Was Recorded		
Total	**13**	**1.8%**
Confirmed	1	7.7%
Probable	9	69.2%
Possible	3	23.1%

Mississippi Kite

Bald Eagle

Haliaeetus leucocephalus

The breeding status of the Bald Eagle in Kentucky has varied significantly since the time of earliest ornithological accounts. Audubon (1861) described nesting at only one location in western Kentucky during the early 1800s, but based on the abundance of suitable habitat along the floodplains of the Mississippi and Ohio Rivers, it is possible that the species nested in considerable numbers at that time. Pindar (1925) reported that about six pairs regularly nested in the vicinity of Island No. 8, in Fulton County, in the late 1800s. As floodplain wetlands were drained and cleared for agricultural use during the first half of the 20th century, numbers of nesting eagles declined. Mengel (1965) supposed that not more than 5–10 pairs nested in western Kentucky as of about 1960, with most being present along the Mississippi River from the mouth of the Ohio to near Hickman.

During the 1960s the number of Bald Eagles continued to decline. The species disappeared as a breeding bird, and greatly reduced numbers were reported during migration and in winter through the mid-1970s. This decline is widely thought to have occurred at least in part because of the accumulation of residues from chlorinated hydrocarbon pesticides (primarily DDT) in natural ecosystems, continued habitat loss, and shooting (Terres 1980). The banning of DDT in 1972 and recent attempts to restore nesting eagles through hacking programs appear to have aided a recovery in the population. Numbers of wintering birds have recently reached all-time highs (Durell and Yancy 1990), and a small breeding population has become reestablished.

Nesting Bald Eagles reappeared in Kentucky during the atlas period of 1985–91. An initial attempt at the Ballard WMA, in Ballard County, in the spring of 1986 was followed by several failed attempts there and at Land Between the Lakes over the next few years (Stamm 1987a, 1987c; D. Yancy, pers. comm.). By 1990 young were successfully raised at both locations (Stamm and Monroe 1990a), and in 1991 two nests were successful at each location (Stamm and Monroe 1991b; D. Yancy, pers. comm.). A nest constructed in 1991 on the Sauerheber Unit of Sloughs WMA, in Henderson County, contained two young in 1992 (D. Yancy, pers. comm.). In addition, nests were constructed but apparently not used at three locations: just north of Hickman, in Fulton County, in 1987; along the Mississippi River south of Middle Bar, in Hickman County, in 1990; and at Laurel River Lake, in Laurel County, in 1991 (D. Yancy, pers. comm.). All nests constructed but not fledging young are shown on the atlas maps as possible breeding records.

The reappearance of nesting eagles is in part the result of introduction efforts undertaken by conservation agencies in surrounding states. In 1980 the Tennessee Valley Authority initiated a Bald Eagle hacking project on Lake Barkley, just south of the Kentucky/Tennessee state line (Lowe 1980; D. Hammer et al., unpub. rpt.). Subsequently, other agencies have undertaken similar projects—some still ongoing—at additional sites in Tennessee and one site in Indiana. Banding records indicate that a few of these birds have been involved in recent Kentucky nestings.

Bald Eagles typically are conspicuous only during the winter months, when birds from farther north can be found on rivers and reservoirs at scattered localities across the state. Locally nesting birds are involved in territorial defense, courtship, and nest building throughout the late winter. When wintering birds begin departing in late February and early March, the local birds are ready to initiate nesting. In recent years birds usually have been observed incubating by late February, and young appear to hatch by the beginning of April (D. Yancy, pers. comm.). Most young typically fledge by the middle of June, and family members often remain in close proximity to one another throughout the remainder of the summer.

Gene Boaz

Bald Eagles typically nest in the immediate vicinity of a body of water, although sites lying up to about half a mile from water have also been used (S. Bloemer, pers. comm.). All Kentucky nests have been constructed in large, living trees. The nest is a massive structure, composed of large sticks and lined with finer twigs and soft plant material. Nests that persist are often used in succeeding years.

Although Kentucky may never support a substantial nesting population of Bald Eagles, the creation of reservoirs has resulted in an increase in the amount of suitable habitat. It is expected that the nesting population will slowly increase in coming years and that a few pairs may someday nest on reservoirs and large river floodplains throughout the state.

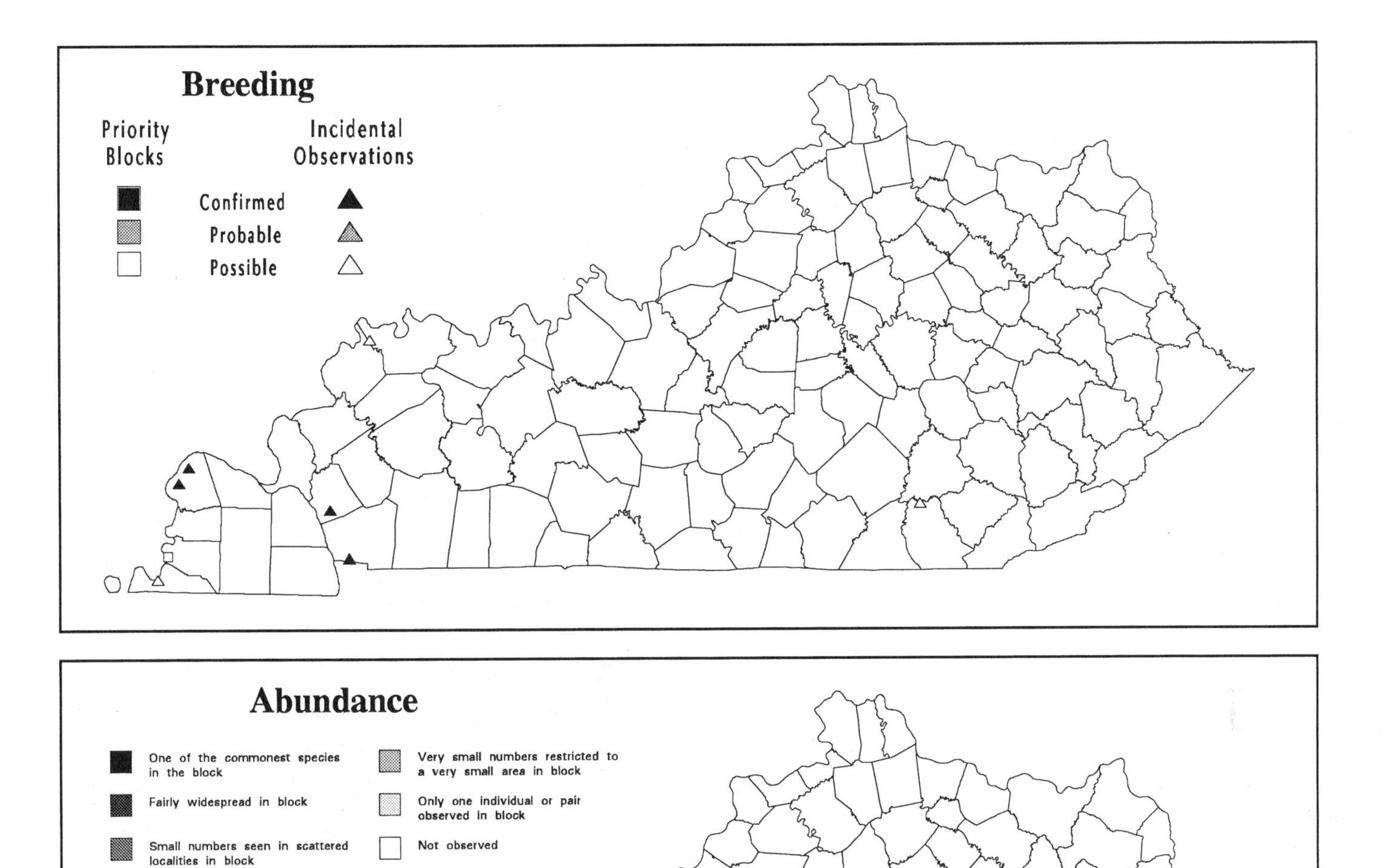

Analysis of Block Data by Physiographic Province Section

Physiographic Province Section	Total Blocks Surveyed	Blocks with Data	% with Data	Section's % for State
Mississippi Alluvial Plain	14	1	7.1	100.0
East Gulf Coastal Plain	36	-	-	-
Highland Rim	139	-	-	-
Shawnee Hills	142	-	-	-
Blue Grass	204	-	-	-
Cumberland Plateau	173	-	-	-
Cumberland Mountains	19	-	-	-

Summary of Breeding Status

Number of Blocks in Which Species Was Recorded		
Total	**1**	**0.1%**
Confirmed	-	-
Probable	-	-
Possible	1	100.0%

Northern Harrier

Circus cyaneus

Until very recently the Northern Harrier was not known to have nested in Kentucky in many years. Audubon (1861) reported that the species occurred along the grassy margins of the Mississippi and lower Ohio Rivers and that he had found its nest in the "barrens of Kentucky," presumably sometime in the early 1800s. The latter observation was probably made in the western Highland Rim, where native prairies were most prevalent. In addition, Hibbs (1927) referred without detail to nesting in Nelson County in 1926. Otherwise, a few more recent summer observations of Northern Harriers in the central and western parts of the state constituted the only evidence of possible nesting before the late 1980s.

During the 1970s harriers were found nesting on extensive reclaimed surface mines in southern Indiana (J. Campbell, pers. comm.). Based on these reports, a specific effort was made during the atlas fieldwork to confirm that harriers nested in similar habitat in Kentucky. Reclaimed mines in the Shawnee Hills were investigated in the spring of 1989, and Northern Harriers were found nesting in small numbers (Palmer-Ball and Barron 1990). On May 26, 1989, a nest containing two eggs was located in open grassland of a reclaimed mine near Paradise, in eastern Muhlenberg County. Later the same day, five downy young were found at a nest site in a thick, grassy field on a reclaimed mine several miles away in southern Ohio County. Further survey work in this region in subsequent years yielded several observations of harriers in similar habitat in adjacent blocks, including seemingly territorial pairs at three localities.

Outside this core nesting area, Northern Harriers known or thought to be nesting were found at two sites. On June 7, 1989, a male was observed defending a large area of grassland in rural farmland of northern Logan County from other raptors, suggesting that a nest or young were present. In June 1990 nesting was confirmed in rural farmland of western Hart County based on the presence of recently fledged young (Clay and Clay 1990).

As indicated by data obtained over the past several years, the bulk of the Northern Harrier nesting population occurs within the Shawnee Hills, specifically in association with the more extensive, recently reclaimed surface mines and other large tracts of idle grassland. Little fieldwork has been done away from a few specific localities, however, and it is possible that the species is more widespread than is currently believed. For example, an incidental report in McCracken County was based on the observation of two birds foraging over extensive grassland in July 1987 (Stamm 1987b). Also, an incidental record from Breathitt County on the Cumberland Plateau involved the observation of at least one bird on extensive, reclaimed mines in late May and early June 1986 (J. Whitaker, pers. comm.). Although further evidence of nesting was not obtained in either case, the potential for nesting in suitable habitat in other regions of the state seems great.

Based on the limited amount of information gathered on Kentucky's nesting Northern Harriers to date, it appears that local nesting biology is similar to that in other parts of the species's range. Nests are situated on the ground amid the dense cover of tall grasses, and they are rather bulky structures built of grass and weed stalks. Kentucky data on clutch size are obviously limited, although the possibly incomplete clutch of two eggs and the brood of five young indicate that local clutches may approach the range-wide average of 5, or frequently 4–6, given by Harrison (1975).

The fact that a clutch was being initiated at the same time that downy young were in a nest several miles away on May 26, 1989, indicates that there is some variation in the commencement of nesting. Based on an incubation period of 24 days (Harrison 1975), the latter clutch must have been completed before the beginning of May. Young on the wing have been observed as early as late June, but mostly during July.

Joseph O. Knight

The extent to which harriers nest on the reclaimed mines of the Shawnee Hills is unclear, but their numbers may depend at least in part on the abundance of prey. Small mammal populations may exhibit dramatic fluctuations, and it is possible that harriers nest on the mines only when prey populations are at or near their greatest densities. Such high population levels may last for only a short time and may occur infrequently (Getz et al. 1987). If the presence of nesting harriers depends on the level of small mammal populations, the birds may not linger to nest in some or many years. It is also possible that nesting birds move about, searching out territories where local mammal populations are especially high.

Northern Harriers inhabit reclaimed mines within the first few years after completion of the initial stages of reclamation. These areas are restored approximately to natural contour and reseeded to grasses and forbs. In most areas, trees are also planted during reclamation, and the harriers probably use the mines only for a limited number of years before woody growth begins to predominate. For this reason their distribution is and may always be somewhat local and temporary, shifting to new areas as prime habitat becomes available.

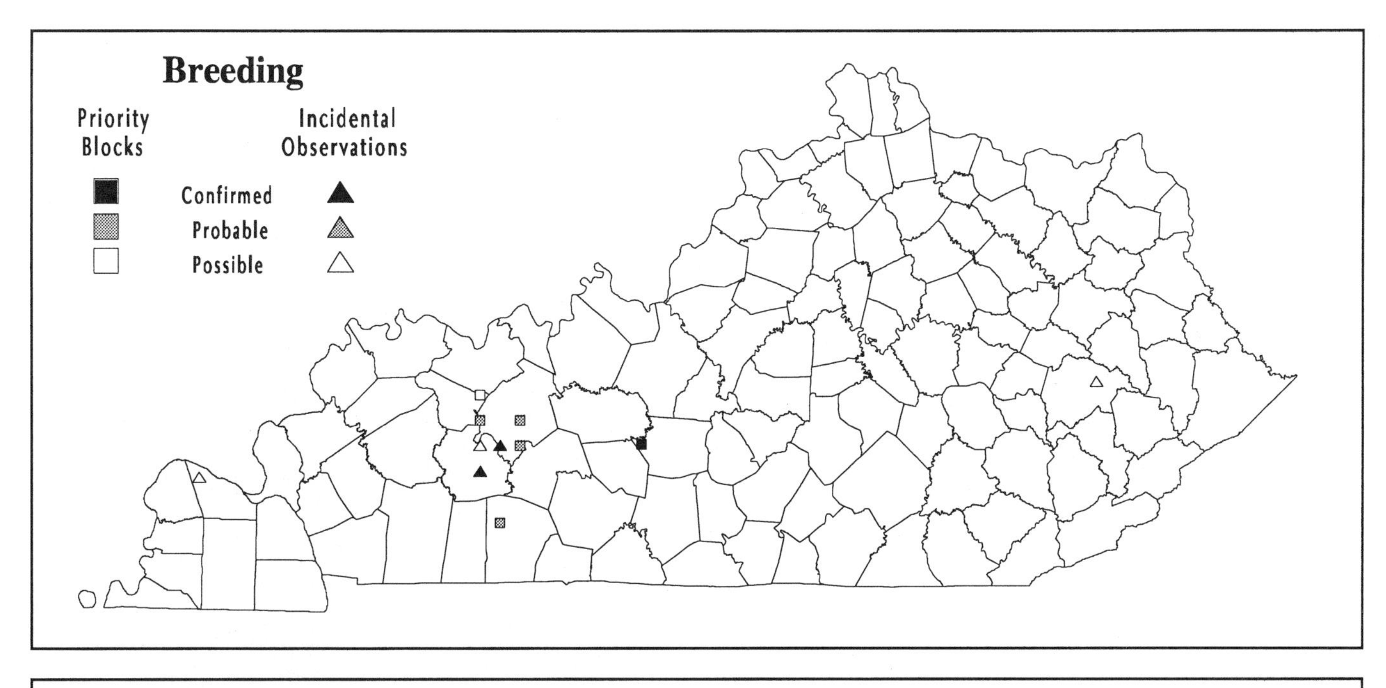

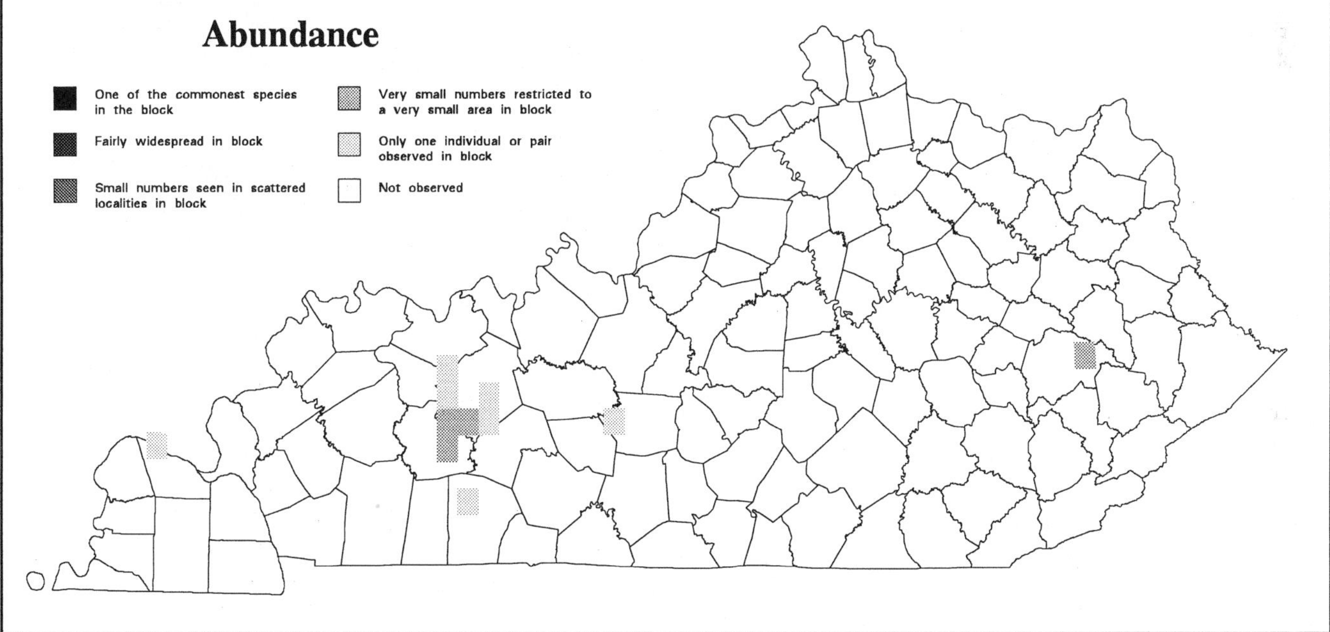

Analysis of Block Data by Physiographic Province Section

Physiographic Province Section	Total Blocks Surveyed	Blocks with Data	% with Data	Section's % for State
Mississippi Alluvial Plain	14	-	-	-
East Gulf Coastal Plain	36	-	-	-
Highland Rim	139	-	-	-
Shawnee Hills	142	6	4.2	100.0
Blue Grass	204	-	-	-
Cumberland Plateau	173	-	-	-
Cumberland Mountains	19	-	-	-

Summary of Breeding Status

Number of Blocks in Which Species Was Recorded		
Total	**6**	**0.8%**
Confirmed	1	16.7%
Probable	4	66.7%
Possible	1	16.7%

Northern Harrier

Sharp-shinned Hawk

Accipiter striatus

Although the breeding status of the Sharp-shinned Hawk in Kentucky has never been well known, the species has been reported locally throughout the state in summer since the early 1800s (Audubon 1861). Mengel (1965) included only a few confirmed records of nesting from central and western Kentucky, but based largely on summer reports of the species, he regarded the Sharp-shinned as a very rare to rare summer resident statewide. In recent years these small accipiters seem to have increased at all seasons, and while confirmed nesting records remain scarce, the species appears to be about as numerous as at any previous time.

Sharp-shinned Hawks are encountered in a variety of semi-open and forested habitats. They are most frequently found in heavily forested areas, but small numbers also occur in semi-open conditions where forest has been fragmented. In general, the Sharp-shinned occurs in relatively greater abundance in forested areas than the Cooper's Hawk, which seems to prefer more open situations. The Cooper's, however, is more common overall, and it may occur in greater abundance, even in regions where Sharp-shinneds attain their highest density.

The Sharp-shinned Hawk may have been more common and widespread in Kentucky before settlement. Being a bird of heavily forested areas, it likely occurred more regularly and in greater numbers throughout parts of the state where native forests have been cleared and highly fragmented for agricultural use and settlement. On the other hand, the introduction of pines into many parts of central and western Kentucky probably has had a positive impact on numbers of Sharp-shinneds in the recent past. Such plantings have created suitable nest sites that have been used by both species of *Accipiter.*

Although Sharp-shinned Hawks can be found in Kentucky throughout the year, birds from farther north pass through during migration and supplement the local population to some degree during winter. Little specific data on breeding in Kentucky are available, but nesting activity likely commences during late March or early April and continues through June. In eastern Calloway County, incubating birds have been observed on two occasions in May (author's notes), and fledged young were recorded during atlas fieldwork most frequently during July. Harrison (1975) gives rangewide clutch size as 4–5, often 3, rarely 6–8.

Forest Cover

Value	% of Blocks	Avg Abund
All	5.4	1.1
1	—	—
2	4.7	1.2
3	3.5	1.0
4	11.0	1.2
5	9.6	1.0

Sharp-shinned Hawks typically nest in extensive tracts of fairly mature forest, but smaller woodlots and corridors of forest are occasionally used. As elsewhere throughout its range, the species in Kentucky seems to use mostly evergreens, especially pines, although there are historical records of nesting in deciduous trees (Mengel 1965). Natural stands of hemlock and pines are probably used in eastern Kentucky, but in the central and western parts of the state nests usually occur in plantings of introduced pines.

The nest is a rather bulky structure of sticks, usually built entirely by the hawks rather than using the old nest of another hawk or crow. Although birds often return to the immediate vicinity of previous nestings, they typically construct a new nest every year. The nest is usually placed quite high in the tree and supported by a horizontal fork next to or near the trunk. Nests for which information is available have ranged from about 35 to 50 feet above the ground.

Ron Austing

The atlas survey yielded 39 records of Sharp-shinned Hawks in priority blocks, and eight incidental observations were reported. Occurrence was highest in the Cumberland Plateau, where the species was observed in about one out of every ten blocks. Sharp-shinneds were scarcely reported throughout the rest of the state and were missed entirely in the Mississippi Alluvial Plain. Most reports from the Blue Grass originated in the more heavily forested northeastern portion.

Although some Sharp-shinned Hawks (especially those in immature plumage) observed in summer may represent nonbreeding birds, atlas reports were recorded as possible unless the season or location indicated a strong possibility that the bird might not be nesting. Some uncertainty also existed concerning whether this species was distinguished from the very similar Cooper's Hawk in all cases. While it is possible that a few misidentifications were made during the atlas fieldwork, enough fully substantiated reports were made to confirm the distribution indicated on the maps.

Only two priority block records and two incidental observations were for confirmed breeding. Both priority block records were based on the observation of recently fledged young in the company of adults. An incidental record from Letcher County was based on the observation of an adult feeding young. An adult incubating on a nest in eastern Calloway County in May 1989 represented the only confirmed report west of the Cumberland Plateau.

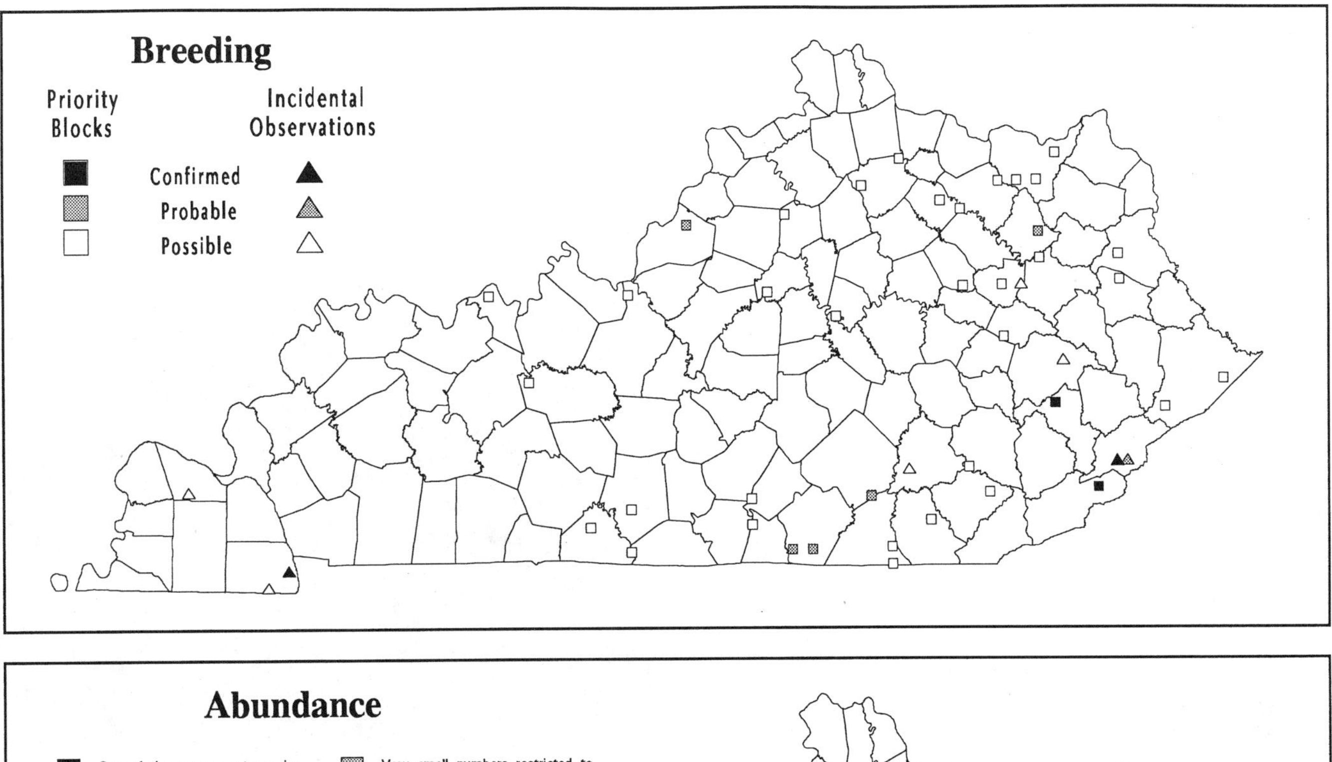

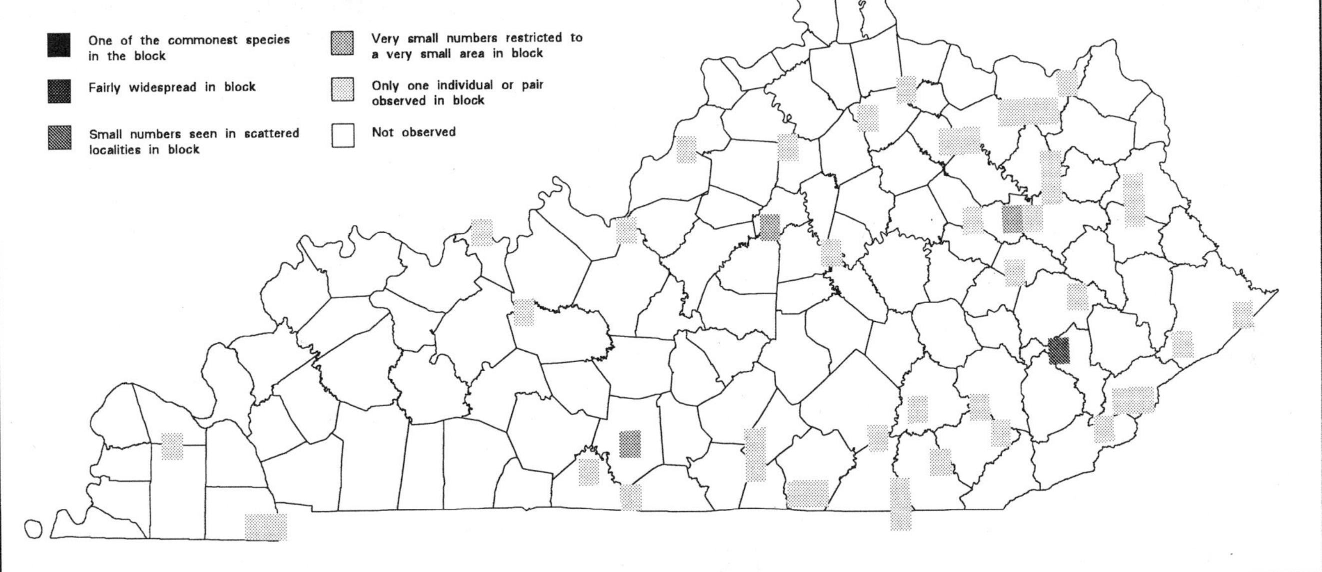

Analysis of Block Data by Physiographic Province Section

Physiographic Province Section	Total Blocks Surveyed	Blocks with Data	% with Data	Section's % for State
Mississippi Alluvial Plain	14	-	-	-
East Gulf Coastal Plain	36	-	-	-
Highland Rim	139	6	4.3	15.4
Shawnee Hills	142	2	1.4	5.1
Blue Grass	204	13	6.4	33.3
Cumberland Plateau	173	17	9.8	43.6
Cumberland Mountains	19	1	5.3	2.6

Summary of Breeding Status

Number of Blocks in Which Species Was Recorded		
Total	**39**	**5.4%**
Confirmed	2	5.1%
Probable	5	12.8%
Possible	32	82.1%

Sharp-shinned Hawk

Cooper's Hawk

Accipiter cooperii

As of the late 1950s Mengel (1965) regarded the Cooper's Hawk as an uncommon to fairly common summer resident statewide. During the 1960s, however, it became much less numerous and was scarcely detected during the summer through the 1970s. Although the accumulation of residues from chlorinated hydrocarbon pesticides like DDT have not been directly linked to this decline, the species has been on the increase since the early 1980s, suggesting that these pesticides may have had some effect. Atlas fieldwork conducted in 1985–91 yielded many more records of Cooper's Hawks than expected, and it now appears that the species may have recovered to the status described by Mengel.

Cooper's Hawks are encountered in a variety of habitats, from large tracts of predominantly forested land to fairly open farmland. Mengel (1965) regarded the species as slightly more numerous in rough, forested country than in more open habitats, but today the species seems to be decidedly more common in the latter. In general this accipiter is relatively more numerous than the Sharp-shinned in semi-open and open habitats, and thus it is more common in settled situations. The species is most conspicuous in rural farmland, where it preys heavily on common species such as the Mourning Dove, the European Starling, and various blackbirds.

The status of the Cooper's Hawk in Kentucky before settlement is not known, but the species probably occurred regularly throughout the state. The extent of forestation may have limited its numbers in some regions, but small numbers likely occurred in the vicinity of natural openings. While the species may have been outnumbered by the Sharp-shinned in heavily forested areas, the Cooper's Hawk likely was more numerous within the native prairies and savannas.

Cooper's Hawks are present in Kentucky throughout the year, but many birds pass through the state during migration, and some from more northerly breeding grounds may supplement the local population in winter. Although little specific information on breeding in Kentucky has been published, it appears that nesting activity commences in April and extends through June. According to Mengel (1965), incubation has been reported as early as April 22, and young in the nest have been reported as late as June 20. A Calloway County nest discovered during the atlas survey, however, contained at least three partially feathered young on June 25, 1989. Kentucky data on clutch size are lacking, but a nest in Jefferson County contained six young on June 14, 1942 (Mengel 1965). Harrison (1975) gives rangewide clutch size as 4–5, sometimes 3, rarely 6.

Forest Cover

Value	% of Blocks	Avg Abund
All	13.6	1.1
1	7.9	1.1
2	19.0	1.1
3	14.4	1.2
4	11.0	1.2
5	9.6	1.0

Although Cooper's Hawks sometimes nest in heavily forested areas, they typically use smaller tracts of fairly mature forest situated in semi-open and open areas. Nests have been reported primarily in deciduous trees (Mengel 1965), but pines are sometimes used (Johnson 1980; KOS Nest Cards), especially in central and western Kentucky.

Ron Austing

The nest is a bulky structure, composed of sticks and small branches and lined with chips of outer bark from oaks or pines (Harrison 1975). Nesting pairs typically construct their own nest, but sometimes the old nest of another hawk or crow is refurbished. Nests are sometimes used in succeeding years. The nest is placed in a crotch of the main trunk or a horizontal fork among the upper branches of the tree. Kentucky nests have been situated from 20 to 60 feet above the ground (Mengel 1965; Johnson 1980; KOS Nest Cards).

The atlas survey yielded records of Cooper's Hawks in nearly 14% of priority blocks, and 31 incidental observations were reported. Occurrence was highest in the Highland Rim and the Blue Grass and lowest in the East Gulf Coastal Plain, although the species was missed entirely in the Cumberland Mountains. Occurrence was highest in areas with a good mixture of open and forested habitats, and lowest in predominantly forested and extensively cleared areas.

Like other raptors, Cooper's Hawks sometimes forage far from their nest sites for food. As a result, some atlas reports may have pertained to individuals that were not nesting in the block in which they were observed. It is also unknown to what extent observations, especially of immature birds, pertained to nonbreeders. Moreover, it was also uncertain whether the two species of *Accipiter* were correctly distinguished in all instances. In general, enough fully substantiated reports of Cooper's Hawks were made to confirm the distribution indicated on the maps.

Only four priority block and five incidental records were for confirmed breeding. Most confirmed records were based on the observation of recently fledged young, although two active nests were observed in the northern Blue Grass, and one was found in Calloway County.

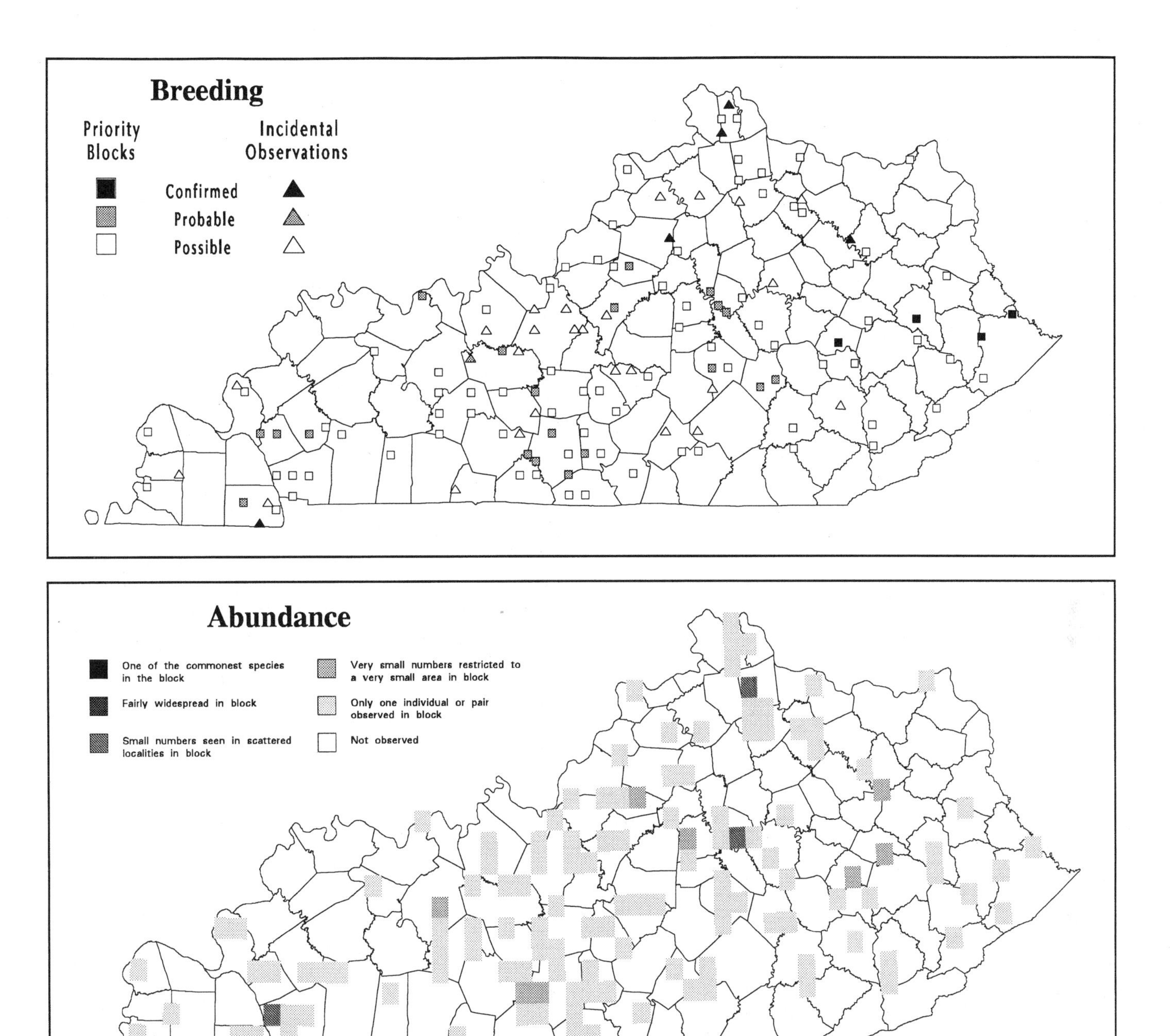

Analysis of Block Data by Physiographic Province Section

Physiographic Province Section	Total Blocks Surveyed	Blocks with Data	% with Data	Section's % for State
Mississippi Alluvial Plain	14	2	14.3	2.0
East Gulf Coastal Plain	36	2	5.6	2.0
Highland Rim	139	30	21.6	30.3
Shawnee Hills	142	18	12.7	18.2
Blue Grass	204	29	14.2	29.3
Cumberland Plateau	173	18	10.4	18.2
Cumberland Mountains	19	-	-	-

Summary of Breeding Status

Number of Blocks in Which Species Was Recorded		
Total	**99**	**13.6%**
Confirmed	4	4.0%
Probable	20	20.2%
Possible	75	75.7%

Cooper's Hawk

Red-shouldered Hawk

Buteo lineatus

Although the Red-shouldered Hawk is a widely distributed nesting bird across much of Kentucky, it occurs very locally in some regions. Throughout the western third of the state the species is conspicuous in lowland forests along major rivers and streams, and it is relatively numerous in mesic forests of the Knobs, the Cumberland Plateau, and the Cumberland Mountains. In contrast, this raptor is much less common in highly cultivated portions of central and western Kentucky, being absent from large parts of the Blue Grass, the Highland Rim, and the Shawnee Hills.

Red-shouldered Hawks are typically found in forested and semi-open habitats, being most numerous in bottomland hardwood forests and floodplain swamps. They also frequently occur in forests of mesic slopes, especially in eastern Kentucky. Although Mengel (1965) thought that the Red-shouldered was largely replaced in eastern Kentucky by Red-tailed and Broad-winged Hawks, these two species occur more frequently in drier, typically upland forests. Although the Red-shouldered inhabits deciduous woodland across most of the state, mixed pine-hardwood forests are frequently used, especially in the Knobs and eastern Kentucky. Unlike the Red-tailed Hawk, which is conspicuous in highly cultivated areas, Red-shouldereds are usually found near more extensive forest, often foraging along woodland borders and in nearby openings.

It is unclear to what extent human alteration of the landscape has affected the distribution and abundance of Red-shouldered Hawks in Kentucky. Audubon (1861) considered the species to be as common in Kentucky as anywhere it was found as of the early 1800s. It is likely that the clearing of vast areas of bottomland forest for agricultural use and development has resulted in an overall decrease in these buteos in central and western Kentucky. In contrast, fragmentation of forests in eastern Kentucky actually may have created more suitable foraging habitat in some areas, since the species seems to hunt mostly along edges.

Red-shouldered Hawks are permanent residents in Kentucky, although birds from farther north pass through during migration and may supplement the local population in winter. During the late winter Red-shouldereds begin noisy territorial calling and courtship. Nest building may be under way by the end of February, and according to Mengel (1965), early clutches are completed by mid-March. A probable peak in clutch completion occurs during the last week or so of March, but egg laying may continue into late April. Most young fledge by mid-June, and family groups remain in close association throughout most of the summer. Mengel (1965) gives the average size of 14 clutches and broods as 2.9 (range of 1–6).

Forest Cover

Value	% of Blocks	Avg Abund
All	23.1	1.4
1	6.9	1.1
2	14.7	1.2
3	25.7	1.3
4	39.0	1.6
5	26.9	1.3

These medium-sized buteos typically nest in fairly mature to mature forest, although sometimes they build along or just inside a forest margin. Nests have been reported only in deciduous trees in Kentucky, but evergreens are sometimes used in other parts of the species's range (Harrison 1975). The nest is constructed of sticks and twigs and lined with finer material, including some down and sprigs of evergreen (Harrison 1975). The nest is usually situated within a substantial crotch, often among the larger branches of the main trunk. The average height of 18 nests reported by Mengel (1965) was 45 feet.

Ron Austing

The atlas survey yielded records of Red-shouldered Hawks in more than 23% of priority blocks statewide, and 21 incidental observations were reported. Occurrence was highest in the Mississippi Alluvial Plain and the East Gulf Coastal Plain, but also higher than the statewide average in the Cumberland Plateau and the Cumberland Mountains. In contrast, occurrence was below average in the Shawnee Hills and the Highland Rim, and lowest in the Blue Grass. Two factors seem to be primarily responsible for the distribution of this species. Most important may be its preference for mesic conditions—especially lowland and swamp forest situations—over drier (especially upland) forest types. Also significant is the amount of mature forest cover. Red-shouldereds occurred much less frequently in extensively cleared areas than in places with a good supply or predominance of forest.

Only about 11% of priority block records were for confirmed breeding. While a few active nests were located, most confirmed records were based on the observation of recently fledged young and adults carrying food. As with other raptors, it is possible that some confirmed records based on the FL and FY codes pertained to birds that were far from nest sites. Despite broad interpretation of these codes, all confirmed records based on FL and FY codes were included as confirmed in the final data set.

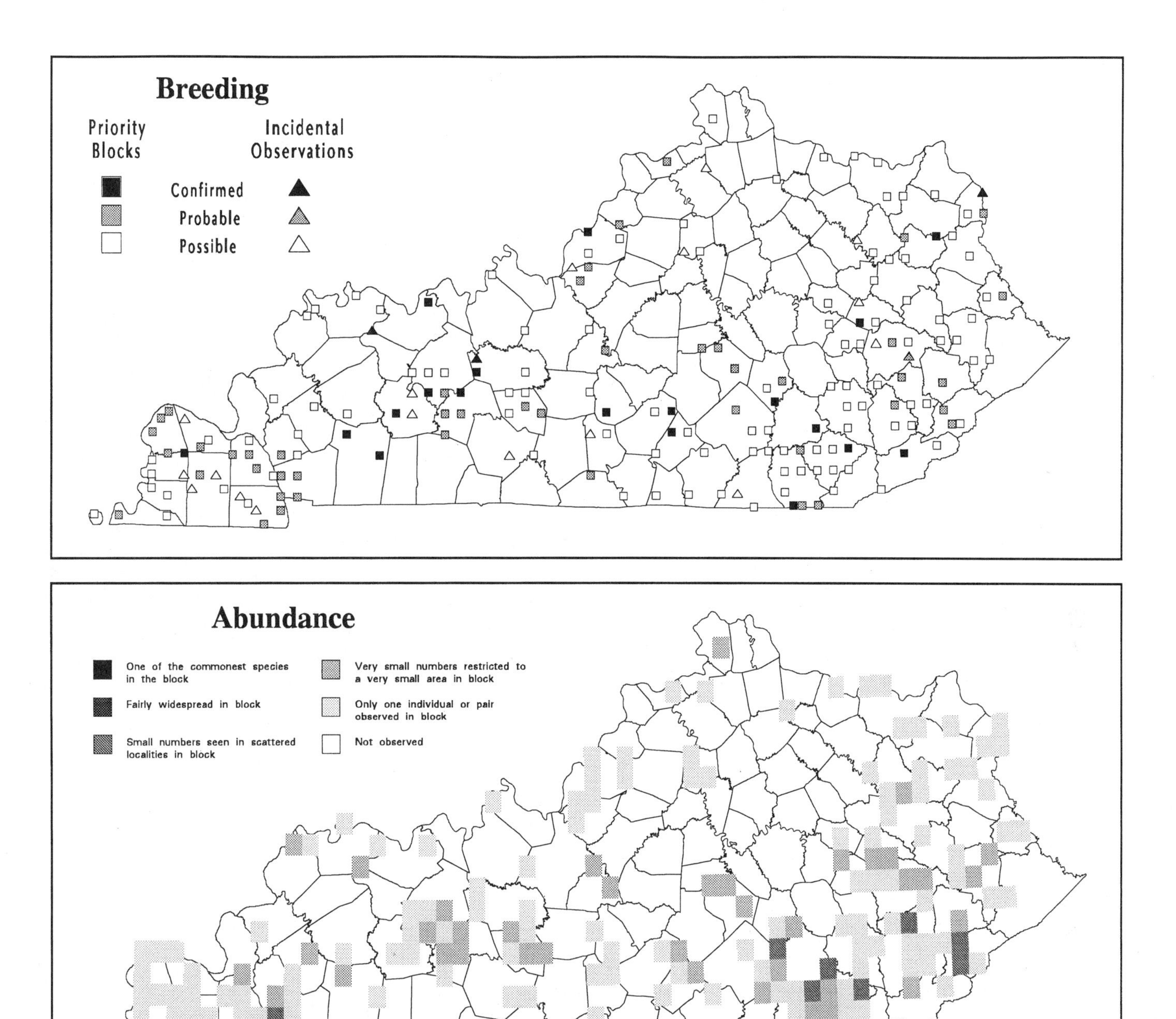

Analysis of Block Data by Physiographic Province Section

Physiographic Province Section	Total Blocks Surveyed	Blocks with Data	Avg Abund	% with Data	Section's % for State
Mississippi Alluvial Plain	14	8	1.0	57.1	4.8
East Gulf Coastal Plain	36	17	1.2	47.2	10.1
Highland Rim	139	22	1.3	15.8	13.1
Shawnee Hills	142	29	1.4	20.4	17.3
Blue Grass	204	22	1.2	10.8	13.1
Cumberland Plateau	173	64	1.5	37.0	38.1
Cumberland Mountains	19	6	1.5	31.6	3.6

Summary of Breeding Status

Number of Blocks in Which Species Was Recorded		
Total	**168**	**23.1%**
Confirmed	19	11.3%
Probable	46	27.4%
Possible	103	61.3%

Red-shouldered Hawk

Broad-winged Hawk

Buteo platypterus

The high-pitched call of the Broad-winged Hawk is a fairly common summer sound throughout much of eastern Kentucky, where the species typically outnumbers all other hawks (Croft et al. 1965). In contrast, across central and western Kentucky this buteo is distributed quite locally, generally in association with more heavily forested regions. Mengel (1965) regarded the species as a rare to fairly common summer resident, encountered chiefly in hilly, forested areas.

Of Kentucky's three nesting buteos, the Broad-winged Hawk is the one most closely associated with forested situations. The species is typically encountered in upland forests of oak-hickory and mixed pine-hardwood, and it is not normally found in the bottomland habitats favored by Red-shouldered Hawks (Mengel 1965). In transitional slope forests the two species may be found nesting near one another, especially on the Cumberland Plateau. Although Broad-wingeds prefer areas of extensive forest, they also use regions of fragmented forest, and the birds are often observed foraging along forest edge of road corridors and other natural and artificial openings.

Broad-winged Hawks have likely declined in Kentucky during the past two centuries. While the clearing and fragmentation of forested areas have benefited some species, this denizen of upland forests has probably disappeared from many parts of central and western Kentucky that have been converted to agricultural land and settlement. In contrast, fragmentation of eastern Kentucky forests may not have affected the species significantly.

Broad-winged Hawks winter from the extreme southern United States southward into the tropics of South America (AOU 1983), and most nesting birds do not return until early April. Breeding activity seems to begin almost immediately, and by late April most birds have initiated courtship and nest building. Kentucky breeding data are very limited, but it appears that most clutches are completed by the middle of May, and young are off most nests by early July. Kentucky data on clutch size are scarce, but Mengel (1965) included a record of two eggs in a nest in Bullitt County in 1937, and a nest observed in Boone County during the atlas fieldwork contained three nearly full-grown young on July 4, 1987 (McNeely 1987). Harrison (1975) gives rangewide clutch size as 2–3, sometimes 1 or 4.

Forest Cover

Value	% of Blocks	Avg Abund
All	25.0	1.4
1	4.0	1.7
2	11.6	1.4
3	20.0	1.4
4	50.6	1.5
5	63.5	1.5

Broad-winged Hawks nest relatively high in large trees, typically building in a crotch of a main branch. Nest trees are usually situated within relatively undisturbed, closed-canopy deciduous or mixed forest. Although evergreens are used in some parts of the species's range, in Kentucky the bird probably primarily uses deciduous trees. The nest is constructed of sticks and lined with finer material, including a few evergreen sprigs, and it is typically not as bulky as that of other buteos (Harrison 1975). Most nests range from 30 to 50 feet above the ground (Ehrlich et al. 1988), although a nest found in Bullitt County in 1937 was estimated to be 70 feet above the ground (Mengel 1965).

Ron Austing

The atlas fieldwork yielded records of Broad-winged Hawks in 25% of priority blocks statewide, and 15 incidental observations were reported. Occurrence in the Cumberland Mountains and the Cumberland Plateau was more than twice that of any physiographic province section in central and western Kentucky. The species was absent from the Mississippi Alluvial Plain (where upland forests are absent) and was found on only one priority block in the East Gulf Coastal Plain. Average abundance was relatively low throughout the state. Occurrence was closely related to percentage of forest cover, increasing significantly as forests became predominant.

As with other raptors, it is possible that some atlas records of Broad-winged Hawks pertained to individuals that were foraging outside of the block in which they were nesting. It is believed, however, that observations of birds in suitable breeding habitat indicated breeding in the general vicinity, and all such reports were included in the final data set. Probable breeding records were based on the observation of more than one bird in suitable breeding habitat, or repeated observations of a single bird.

Only about 8% of priority block records were for confirmed breeding. Although at least two active nests were located, most of the confirmed records were based on the observation of family groups that included recently fledged young. As with other raptors, confirmation of breeding using the FL and FY codes may have pertained in some cases to birds that were not in the immediate vicinity of actual nest sites. Confirmed records based on the FL and FY codes were regarded as indicating nesting in the general vicinity, and all were included in the final data set.

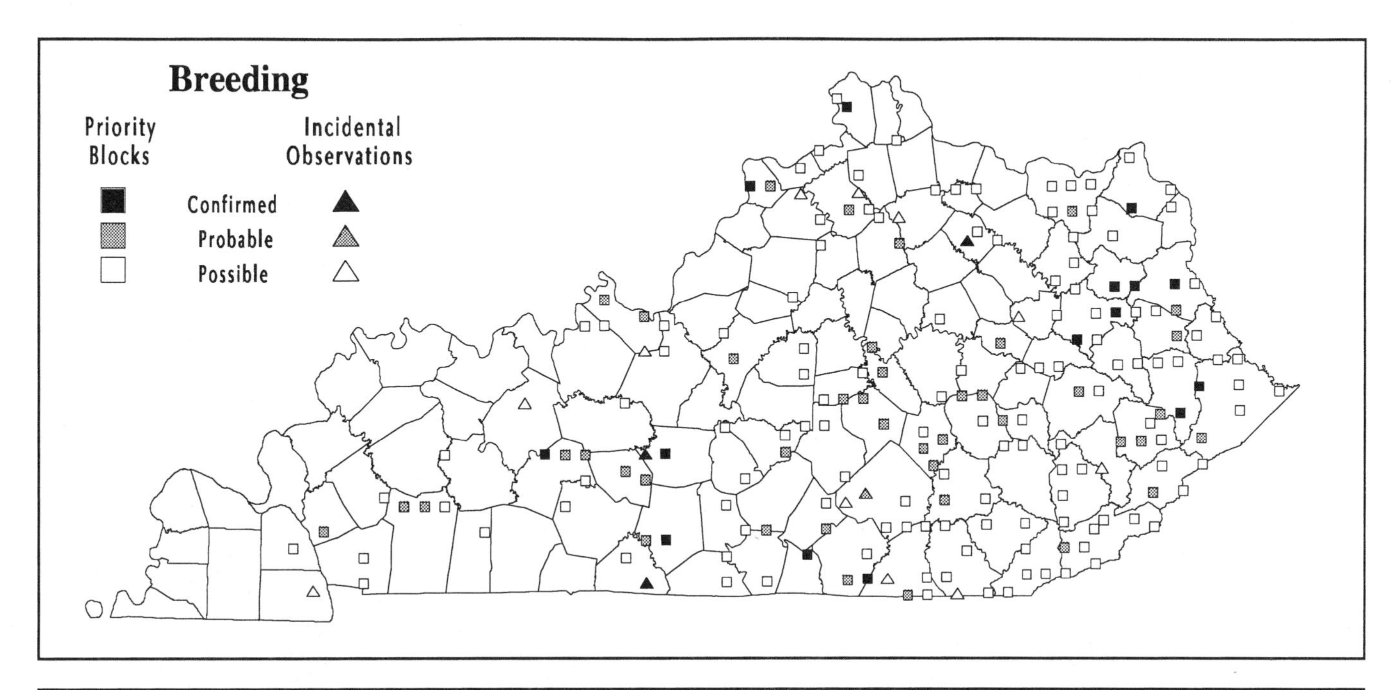

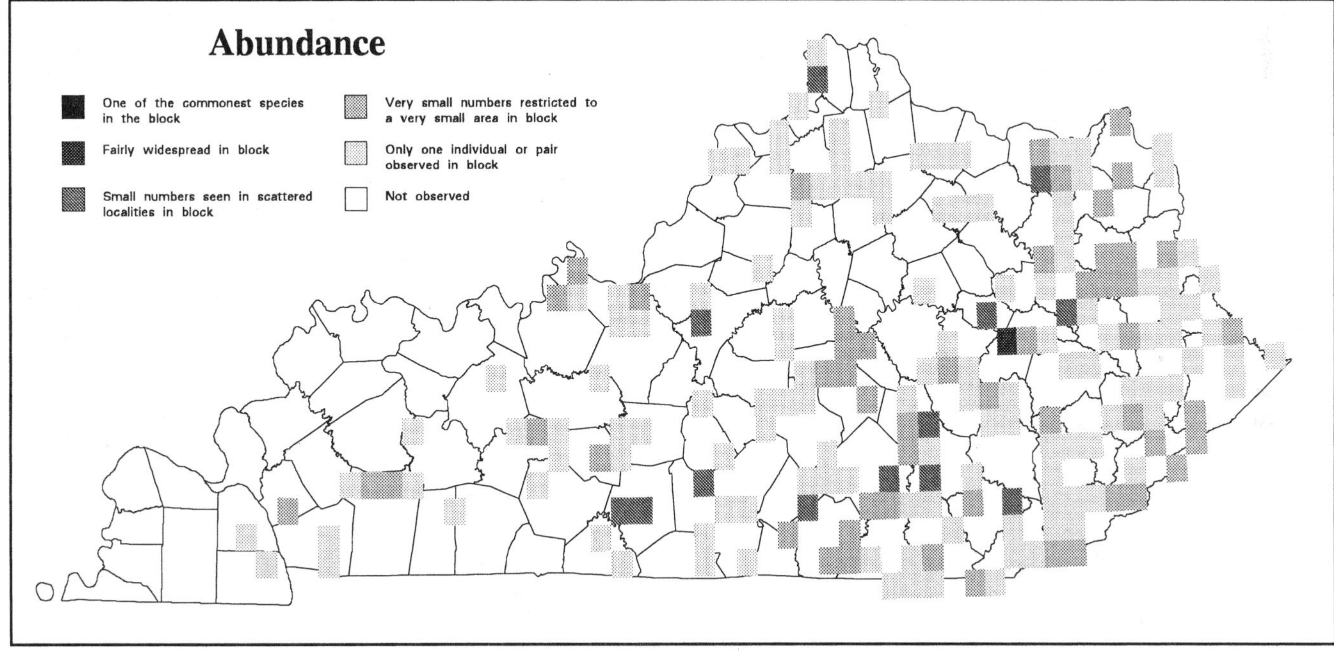

Analysis of Block Data by Physiographic Province Section

Physiographic Province Section	Total Blocks Surveyed	Blocks with Data	Avg Abund	% with Data	Section's % for State
Mississippi Alluvial Plain	14	-	-	-	-
East Gulf Coastal Plain	36	1	1.0	2.8	0.5
Highland Rim	139	25	1.5	18.0	13.7
Shawnee Hills	142	18	1.3	12.7	9.9
Blue Grass	204	46	1.4	22.5	25.3
Cumberland Plateau	173	80	1.5	46.2	44.0
Cumberland Mountains	19	12	1.5	63.2	6.6

Summary of Breeding Status

Number of Blocks in Which Species Was Recorded		
Total	**182**	**25.0%**
Confirmed	15	8.2%
Probable	42	23.1%
Possible	125	68.7%

Broad-winged Hawk

Red-tailed Hawk

Buteo jamaicensis

The Red-tailed Hawk is probably the most widely known raptor in Kentucky. Conspicuous along roadsides throughout much of the state, the species is fairly common in a great variety of natural and altered habitats. Curiously, Red-taileds were apparently not so numerous in the 1950s, when, according to Mengel (1965), they were distributed much differently than today. Mengel considered this buteo to be a summer resident throughout Kentucky, but he regarded it as rather locally distributed and usually found in relatively unsettled country and rough hills typically forested with oak-hickory or mixed pine-hardwood. Moreover, he went on to describe the species as most numerous in the Cumberland Plateau and Mountains but decidedly rare in extensively cultivated areas like the Blue Grass.

Today the Red-tailed Hawk is nowhere more conspicuous than in highly cultivated regions like the Inner Blue Grass, where roadside nests are well distributed. In contrast, regions of rugged forest certainly harbor some birds, but the species is much less conspicuous and does not seem to occur in considerable numbers in such areas. The factors that have resulted in such a change in occurrence and abundance are unclear. It is possible that public education concerning the benefits of raptors has resulted in a dramatic decline in human persecution. This might at least in part explain the species's abundance in farmland and other settled areas, where shooting of "chicken hawks" was formerly much more prevalent. It is also possible that the species's increase in the state is the result of an overall increase documented in other parts of its range (Peterjohn and Rice 1991; Robbins et al. 1986).

Red-tailed Hawks are permanent residents in Kentucky, although birds from farther north and west supplement the local population in winter, especially in the central and western parts of the state (Mengel 1965). Breeding activity typically commences as soon as the weather moderates in late February and early March. Nest building is often under way by the middle of March, and most pairs seem to be incubating by early April. Most clutches probably hatch during the latter part of April, and by mid-June many young are on the wing. Kentucky data on clutch size are scarce, but Mengel (1965) gave the average size of four clutches as 2.25 (range of 2–3).

Forest Cover

Value	% of Blocks	Avg Abund
All	65.5	1.7
1	76.2	1.7
2	79.5	1.8
3	70.9	1.7
4	46.1	1.8
5	26.9	1.4

Red-tailed Hawks usually nest in large trees, but they also use protected sites on cliffs, especially in rugged parts of eastern Kentucky (Mengel 1965). Nest trees are usually located in semi-open situations, including narrow wooded corridors, woodlots, and the margins of larger tracts of forest. The nest is a bulky structure of sticks lined with finer material, including fresh evergreen sprigs (Harrison 1975). Nests are often used in succeeding years. Most are built more than 50 feet above the ground, usually in a large crotch of main branches.

The atlas survey yielded records of Red-tailed Hawks in nearly 66% of priority blocks statewide, and 26 incidental observations were reported. Occurrence was highest in the East Gulf Coastal Plain, the Highland Rim, and the Blue Grass, and lowest in the Cumberland Plateau and the Cumberland Mountains. Average abundance did not vary substantially and was below 2.0 throughout the state. Occurrence was closely related to the amount of forest, being highest in areas with a good supply of open areas and decreasing substantially as forest cover became predominant.

Ron Austing

More than 26% of priority block records were for confirmed breeding. Many active nests were located during atlas fieldwork, but observations of recently fledged young accounted for most confirmed records. Young Red-taileds keep in contact with the parents by a distinctive begging call. This call was used on many occasions to identify young birds, and it usually was interpreted as constituting confirmed evidence of breeding. It is possible that some confirmed records based on the FL code pertained to young birds that were raised some distance from the point of observation, especially since the begging call can be heard for several weeks after fledging. Despite this possibility, such records were interpreted as indicating confirmed breeding somewhere in the vicinity, and all were included in the final data set.

Like most other raptors, Red-tailed Hawks are not reported frequently on Kentucky BBS routes. The average number of individuals per BBS route for the periods 1966–91 and 1982–91 was 0.48 and 0.66, respectively. Trend analysis of these data is obviously limited by sample size, but it yields a significant ($p<.05$) increase of 2.8% per year for the period 1966–91 and a nonsignificant increase of 2.8% per year for the period 1982–91.

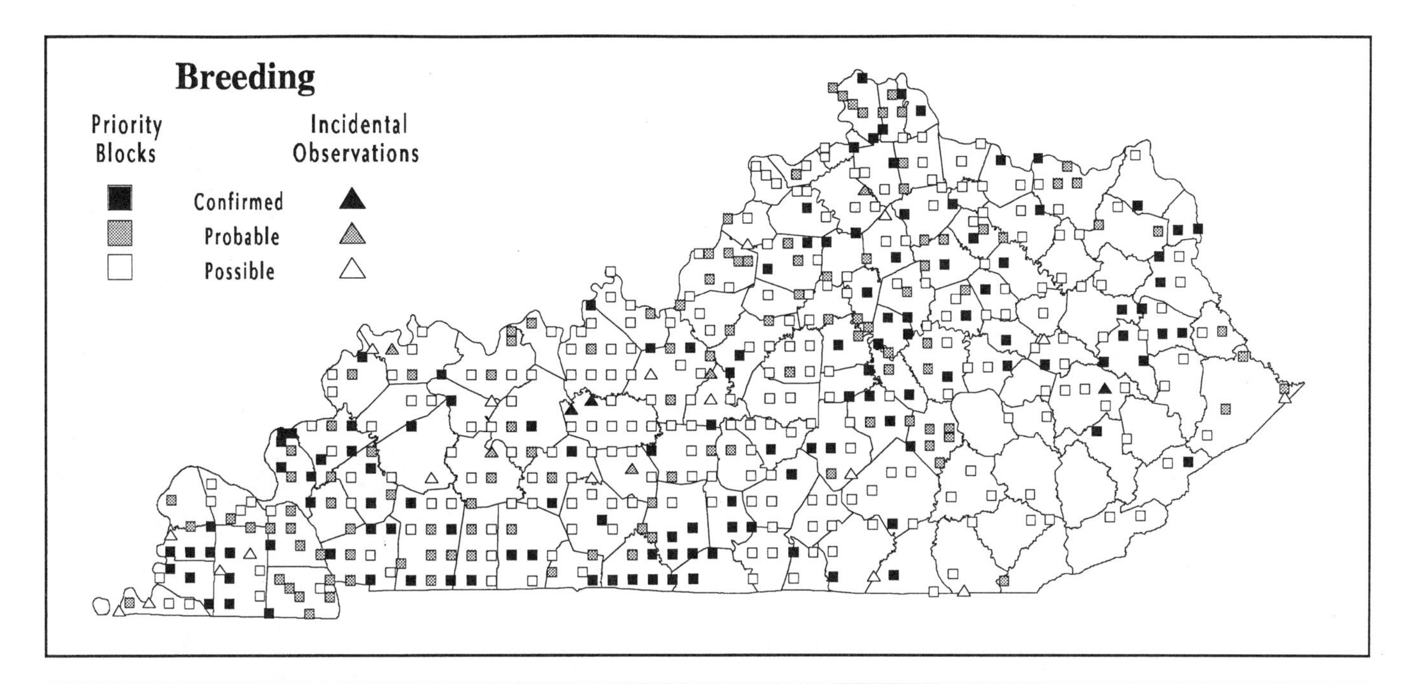

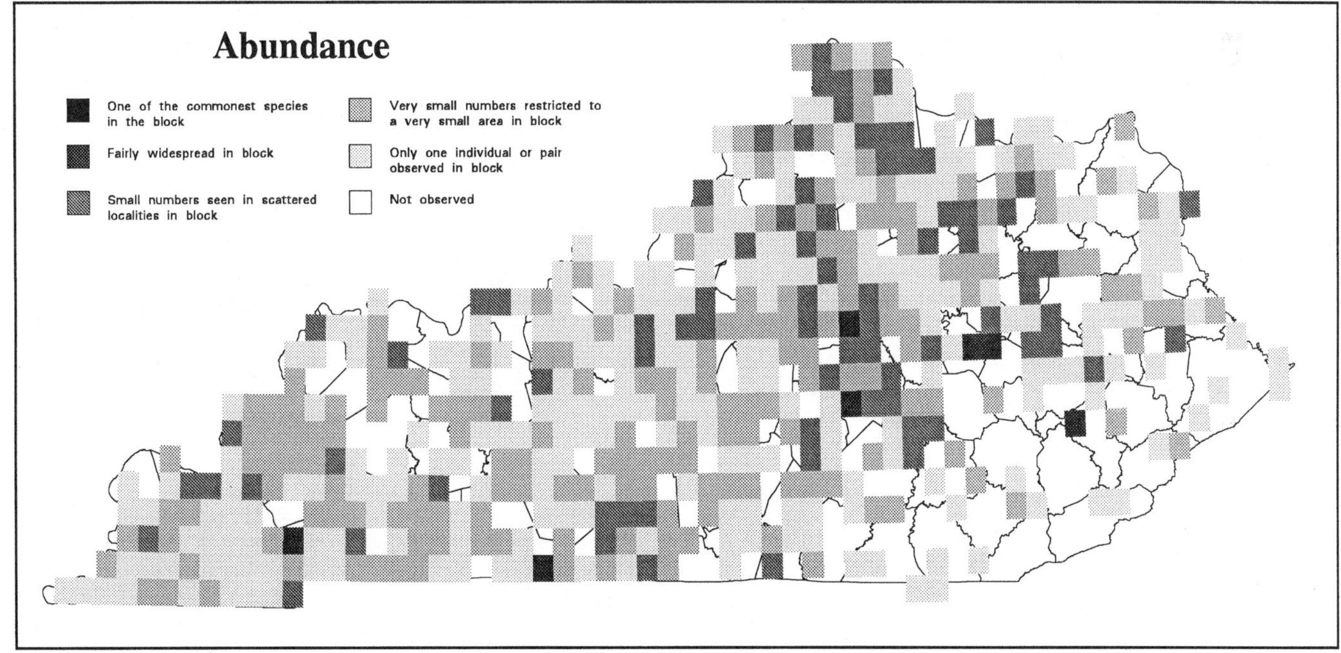

Analysis of Block Data by Physiographic Province Section

Physiographic Province Section	Total Blocks Surveyed	Blocks with Data	Avg Abund	% with Data	Section's % for State
Mississippi Alluvial Plain	14	8	1.5	57.1	1.7
East Gulf Coastal Plain	36	31	1.6	86.1	6.5
Highland Rim	139	112	1.7	80.6	23.5
Shawnee Hills	142	96	1.6	67.6	20.2
Blue Grass	204	164	1.9	80.4	34.5
Cumberland Plateau	173	63	1.6	36.4	13.2
Cumberland Mountains	19	2	1.0	10.5	0.4

Summary of Breeding Status

Number of Blocks in Which Species Was Recorded		
Total	**476**	**65.5%**
Confirmed	125	26.3%
Probable	117	24.6%
Possible	234	49.2%

American Kestrel

Falco sparverius

The distinctive silhouettes of American Kestrels perched on electrical wires or hovering over potential prey are a conspicuous sight throughout much of Kentucky. Mengel (1965) regarded the species as a fairly common to common resident statewide. Only in extensively forested portions of eastern Kentucky do these small falcons become less numerous and distinctly harder to find. Throughout the remainder of the state kestrels are regularly encountered and distributed rather uniformly in suitable habitat.

American Kestrels are primarily birds of semi-open and open habitats, although an occasional pair may be seen near openings and cliff edges in extensively forested terrain (Mengel 1965). These diminutive raptors are most abundant in rural farmland, where they hunt over fields and pastures. Substantial numbers also occur in other altered habitats, including urban areas, city parks, golf courses, industrial parks, rural road and utility corridors, and reclaimed surface mines.

It is likely that these small falcons have increased in occurrence and abundance since Kentucky was first settled. The clearing and fragmentation of vast areas of unbroken forest have resulted in the creation of an abundance of suitable habitat where the species must have been rare or absent formerly. It is also likely that the open farmland that has largely replaced the prairies and savannas of central and western Kentucky supports as many birds as the native habitats did prior to conversion.

Although American Kestrels are permanent residents in Kentucky, some seasonal movement away from breeding areas seems to occur, and birds from farther north may supplement the local population in winter (Mengel 1965). During March resident birds begin setting up nesting territories. According to Mengel, early clutches are completed during the latter half of March, but a peak in clutch completion probably does not occur until the middle of April. Young are typically found in nests during the latter half of May, but at least occasionally into June (e.g., Stamm and Croft 1968; KOS Nest Cards). Recently fledged young are most conspicuous during June, and family groups often remain together well into the summer. Kentucky data on clutch size are scarce, but most nests have contained five eggs (Mengel 1965; Phillips 1973; Stamm and Croft 1968). Harrison (1975) gives rangewide clutch size as 3–5, commonly 4–5.

Forest Cover

Value	% of Blocks	Avg Abund
All	57.8	1.8
1	82.2	1.9
2	80.5	2.0
3	66.5	1.7
4	17.5	1.4
5	7.7	1.5

American Kestrels typically nest in cavities. A great variety of natural and artificial sites are used, including natural cavities and old woodpecker nests in trees, nooks and crannies in barns, homes, buildings, and bridges, nest boxes, and occasionally small recesses along clifflines and artificial rock faces of road cuts and quarries. Nest sites typically lie within fairly open settings, although occasionally along forest edge. Nest cavities are typically located 20 to 50 feet above the ground.

The atlas fieldwork yielded records of American Kestrels in nearly 58% of priority blocks statewide, and 27 incidental observations were reported. Occurrence was highest in the Highland Rim, the Blue Grass, and the Mississippi Alluvial Plain, and lowest in the Cumberland Plateau and the Cumberland Mountains. Average abundance was relatively low throughout the state and not substantially variable. Occurrence was closely related to percentage of forest cover, being highest in areas with substantial open areas and decreasing significantly as forest cover became predominant.

Ron Austing

Nearly 17% of priority block records were for confirmed breeding. Although a number of active nests were discovered, most confirmed records were based on the observation of family groups including recently fledged young, and a few were based on the observation of adults carrying food. As with other raptors, it is possible that some confirmed records based on the FL and FY codes pertained to birds that were some distance from nest sites. Kestrels, however, do not forage as far away from nest sites as larger raptors, and all confirmed records based on FL and FY codes were included in the final data set.

Like most other raptors, American Kestrels are reported in very small numbers on Kentucky BBS routes. The average number of individuals per BBS route for the periods 1966–91 and 1982–91 was 1.40 and 1.43, respectively. Trend analysis of these data is obviously limited by sample size, but it yields a nonsignificant decrease of 0.6% per year for the period 1966–91 and a significant ($p<0.1$) increase of 6.5% per year for the period 1982–91.

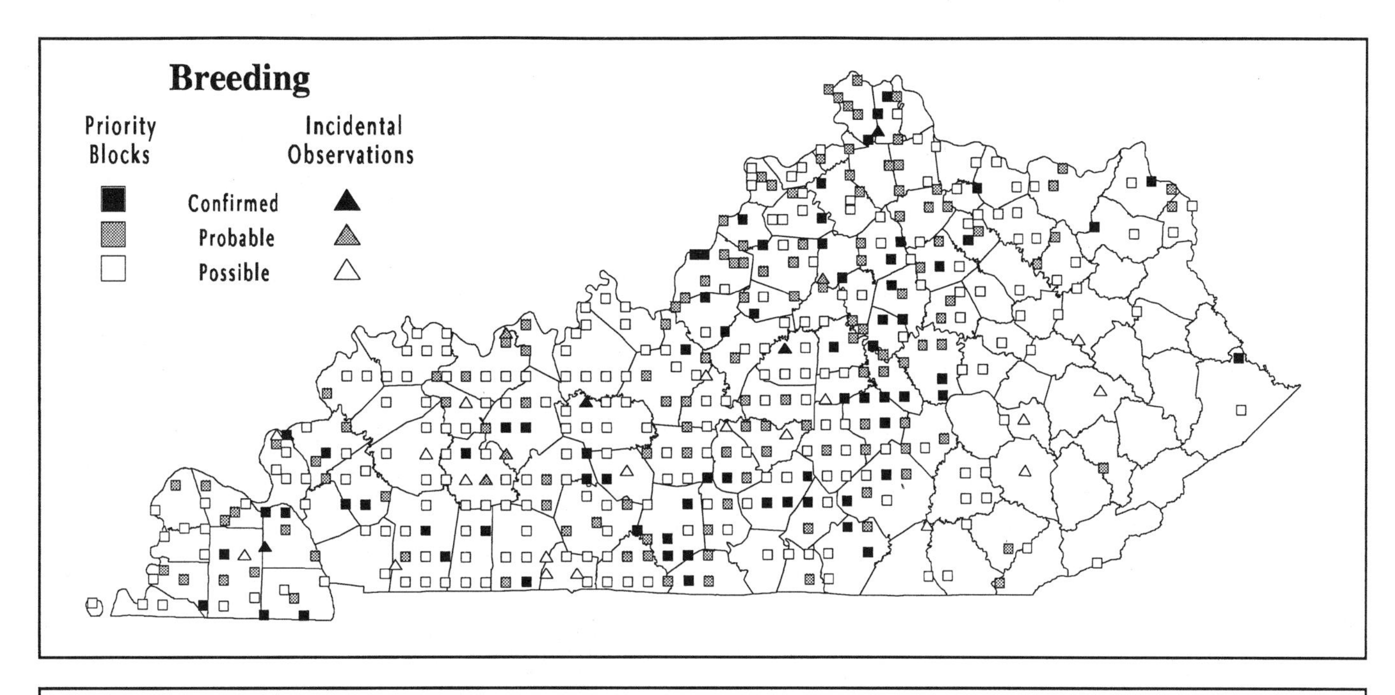

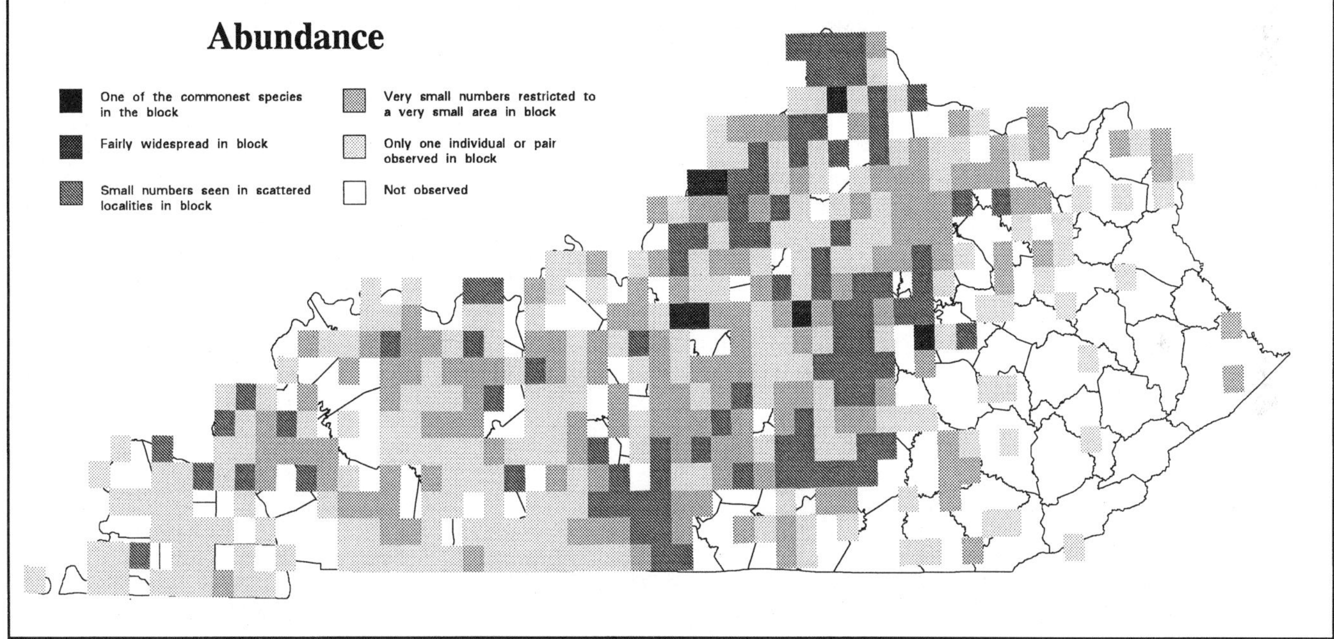

Analysis of Block Data by Physiographic Province Section

Physiographic Province Section	Total Blocks Surveyed	Blocks with Data	Avg Abund	% with Data	Section's % for State
Mississippi Alluvial Plain	14	10	1.2	71.4	2.4
East Gulf Coastal Plain	36	22	1.3	61.1	5.2
Highland Rim	139	108	1.9	77.7	25.7
Shawnee Hills	142	91	1.6	64.1	21.7
Blue Grass	204	156	2.1	76.5	37.1
Cumberland Plateau	173	31	1.3	17.9	7.4
Cumberland Mountains	19	2	1.5	10.5	0.5

Summary of Breeding Status

Number of Blocks in Which Species Was Recorded		
Total	**420**	**57.8%**
Confirmed	71	16.9%
Probable	129	30.7%
Possible	220	52.4%

American Kestrel

Ruffed Grouse

Bonasa umbellus

Although poorly known owing to its cryptic plumage and secretive habits, the Ruffed Grouse is well established over a large part of Kentucky. The species is broadly distributed throughout the Cumberland Plateau and the Cumberland Mountains. It is also scattered locally through the central and western portions of the state, largely as a result of a restoration program initiated by KDFWR in 1984 (J. Sole, unpub. rpt.).

At one time grouse occurred in suitable habitat throughout Kentucky. Presettlement conditions provided these birds with an abundance of woodland habitat, and the earliest reports indicated a plentiful population in most regions. Audubon (1861) reported encountering the species in large numbers along the Ohio River around 1820 and noted that the bird frequented the "groves," that is, scattered woodlands, in the barrens. Its abundance in many other parts of the state can be inferred from local accounts of its disappearance (Mengel 1965). As vast areas of forest were cleared for farmland, numbers of grouse declined. By the early 1900s forests throughout most of central and western Kentucky had become so fragmented and the species sought so vigorously by unregulated market hunters that the bird disappeared from all but the largest tracts of woodland.

As of the 1950s Mengel (1965) regarded the Ruffed Grouse as a rare to fairly common bird of eastern Kentucky, occurring west of the Cumberland Plateau only in the immediately adjacent eastern Knobs. He found the species to be most common in the more remote and rugged regions and more locally distributed in the settled, less hilly portions of the Plateau. In western Kentucky, there was evidence of a small population persisting in Lyon County, in what is now Land Between the Lakes, through the mid-1940s (Mengel 1965).

The restoration program undertaken by KDFWR has concentrated on returning grouse to the forests of the Knobs, the northern Blue Grass, the Shawnee Hills, and the western Highland Rim in Land Between the Lakes (J. Sole, unpub. rpt.). Overall the birds appear to be doing well, and most populations are increasing. If this program continues to be successful, grouse may once again be encountered in substantial tracts of forest throughout the state.

Ruffed Grouse typically inhabit early successional stages of forest, although they often venture into brushy woodland openings and edge to forage. In eastern Kentucky the species moves about freely from the narrow, cliff-bordered ridges and xeric slopes to the more protected ravines dominated by rhododendron and hemlock. In the Knobs grouse are more likely to be found on slopes in remote tracts of large forests, often along major streams.

Ruffed Grouse are permanent residents in the state, moving about a surprisingly small home range throughout the year. The males' unique drumming can be heard occasionally throughout the year but becomes more frequent during March and peaks in April (Mengel 1965; Triquet et al. 1988). Egg laying begins in late March and extends to late May, with a peak in the latter part of April (Mengel 1965). Broods of young are most conspicuous during June, although family groups often remain together into late summer. Brood dispersal typically occurs from mid-September through October (J. Sole, pers. comm.). Mengel (1965) gave the average size of 13 clutches as 8.2 (range of 4–13).

Ruffed Grouse usually nest in forested situations, close to early successional habitat. Eight of nine nest sites studied by Hardy (1950) occurred in young second-growth forest that had been previously disturbed by cultivation or logging. Nests occur mostly on gentle slopes, although sometimes on flat ground or steep slopes, and they have been found in a variety of forest types.

Alvin E. Staffan

Grouse nest on the ground, typically in a sheltered situation such as at the base of a tree, against a rock ledge, or under an uprooted stump or the branches of a fallen tree (Hardy 1950; Croft and Stamm 1967). The nest itself is a shallow, cup-shaped depression in the litter of the forest floor. Grouse apparently rely on their plumage to avoid nest predation, as nests are typically not significantly concealed by surrounding or overhanging vegetation, although the female may scatter leaves on the eggs when she is off the nest (Hardy 1950).

The atlas survey was highly deficient in gathering information on the Ruffed Grouse, and the species was found in only about 6% of priority blocks. Fifty-two incidental reports were obtained for the atlas period, predominantly from KDFWR Drumming Surveys and additional contributions from within the department. Although these data yield an approximate picture of the species's current range, they do not adequately depict either the grouse's abundance or the breadth of its distribution, especially in eastern Kentucky.

Eleven confirmed reports in priority blocks were generated directly by the atlas fieldwork. Most of these involved the observation of family groups including small young, but a nest with eggs was located on Black Mountain on June 23, 1989, a relatively late date that likely represented an attempt to renest following earlier failure.

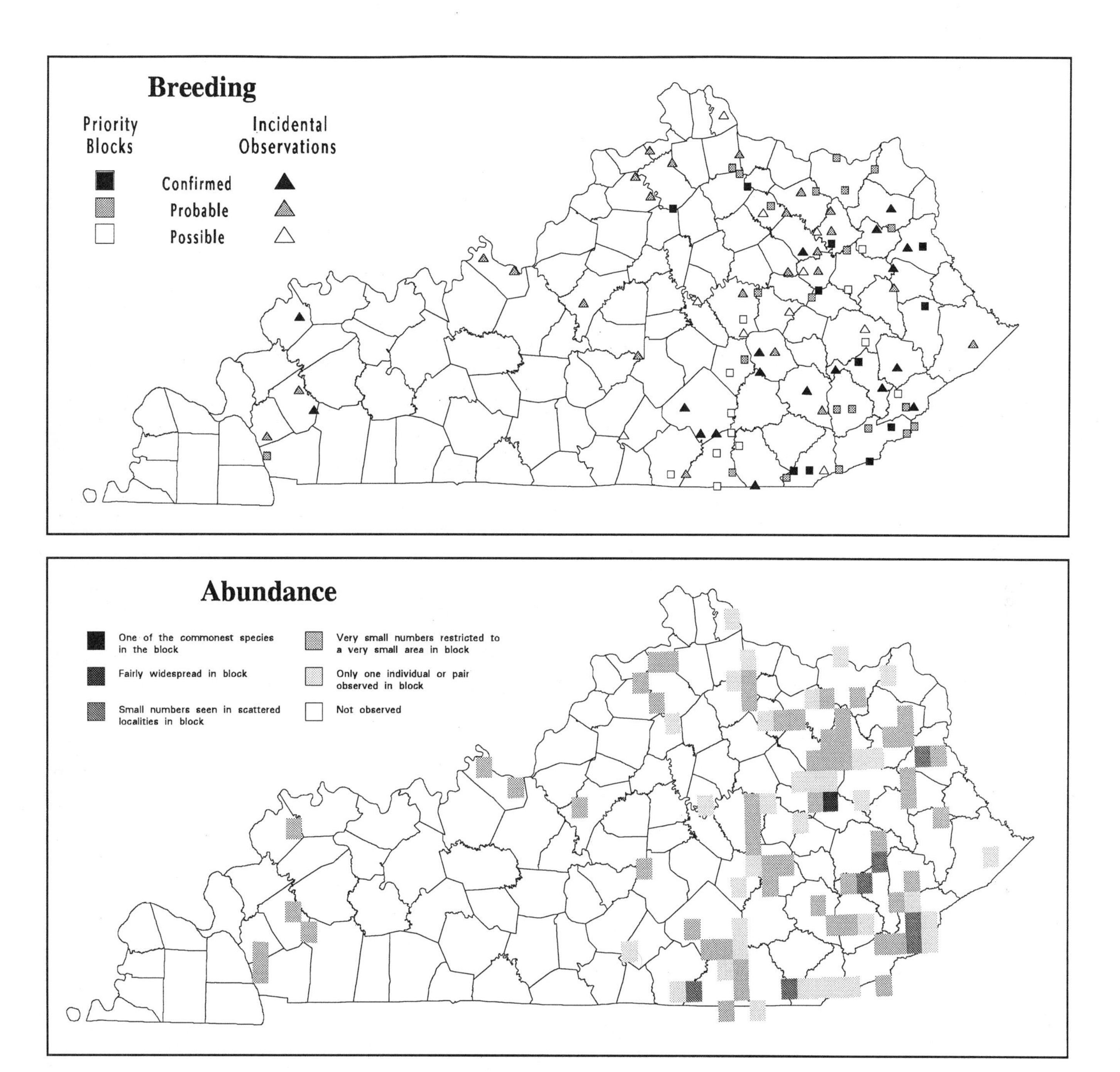

Analysis of Block Data by Physiographic Province Section

Physiographic Province Section	Total Blocks Surveyed	Blocks with Data	% with Data	Section's % for State
Mississippi Alluvial Plain	14	-	-	-
East Gulf Coastal Plain	36	-	-	-
Highland Rim	139	1	0.7	2.2
Shawnee Hills	142	-	-	-
Blue Grass	204	12	5.9	26.7
Cumberland Plateau	173	24	13.9	53.3
Cumberland Mountains	19	8	42.1	17.8

Summary of Breeding Status

Number of Blocks in Which Species Was Recorded		
Total	**45**	**6.2%**
Confirmed	11	24.4%
Probable	22	48.9%
Possible	12	26.7%

Ruffed Grouse

Wild Turkey

Meleagris gallopavo

Although Wild Turkeys once inhabited virtually all of Kentucky, they disappeared from most or all of the state by the early 1900s (Mengel 1965). It is unclear how many birds from native stock survived, but since the early 1930s there has been an ongoing effort to reestablish the species in various regions. Wild Turkeys are now increasing dramatically in many areas, in large part because of more recent restocking efforts by KDFWR.

Before settlement, substantial numbers of Wild Turkeys occurred in wooded and semi-open habitats throughout the state. Early accounts of the bird's abundance are found in many sources (Audubon 1861; Wright 1915), but as vast areas of forest were cleared for farming during the late 1800s and early 1900s Wild Turkeys declined significantly. By about 1930 they were virtually extirpated (Mengel 1965). A small population of native birds apparently survived in what was formerly the Kentucky Woodlands NWR in Lyon and Trigg Counties, now Land Between the Lakes (Baker 1943; Mengel 1965). Conclusive evidence of the persistence of native birds in other parts of Kentucky is absent, but Mengel was convinced of the likelihood that small numbers of birds survived in several areas.

By the late 1950s turkeys had become reestablished at scattered localities, from the more remote mountains of southeastern Kentucky to large tracts of public land in the central and western parts of the state (Mengel 1965). Much of this effort was accomplished using birds from the Kentucky Woodlands NWR population (Nelson 1959). Since that time Wild Turkey restocking efforts have continued at additional locations (e.g., see Davis et al. 1980). The population has increased dramatically during the past decade, and observations are reported from new localities every year. If the restocking effort meets with further success, numbers of Wild Turkeys should continue to increase in many parts of Kentucky.

Wild Turkeys use a variety of habitats, from forests and woodland openings to forest edge and the margins of farm fields. They are most abundant in forests dominated by oaks and other nut-bearing species, which provide a staple of their diet (Baker 1943). The bird seems to do best in semi-open habitats affording a mixture of woodlands, fields, and edges, while areas of highly fragmented forest support only minimal numbers.

Wild Turkeys are permanent residents in the state, wandering about over fairly large home ranges. The birds spend the winter in flocks that sometimes number 100 or more individuals. During late March and early April the flocks break up, and the males initiate their courtship calling. Clutches are likely completed during the latter half of April and early May, and broods of young have been observed as early as the middle of May (Baker 1943). Most broods of small young are seen in June and July, and juveniles remain with their parents through the remainder of the summer. Kentucky data on clutch size are limited, but Baker (1943) gave a range in clutch size at Kentucky Woodlands NWR of 5–14 eggs, and F. Hardy observed a nest containing 12 eggs in McCreary County on May 15, 1951 (Lovell 1951).

Nests are placed on the ground, usually in forest but sometimes along forest edge. The nest is typically a shallow depression formed in the litter of the forest floor and is generally situated on flat ground or a gentle slope. It is often placed near the base of a tree or fallen log, but occasionally it lies some distance from any such object. Wild Turkeys apparently depend on their plumage for protection of the nest, as concealing vegetation is rarely present.

Gene Boaz

The atlas survey did not yield a substantial amount of information on nesting Wild Turkeys. The species was recorded in only about 6% of priority blocks statewide. Most of the 33 incidental observations came from KDFWR Gobbler Surveys and additional contributions from department personnel. Although these data suggest that the turkey is distributed statewide, they do not adequately depict either the species's true occurrence or abundance. All confirmed reports of nesting resulted from the observation of young birds in the company of adults.

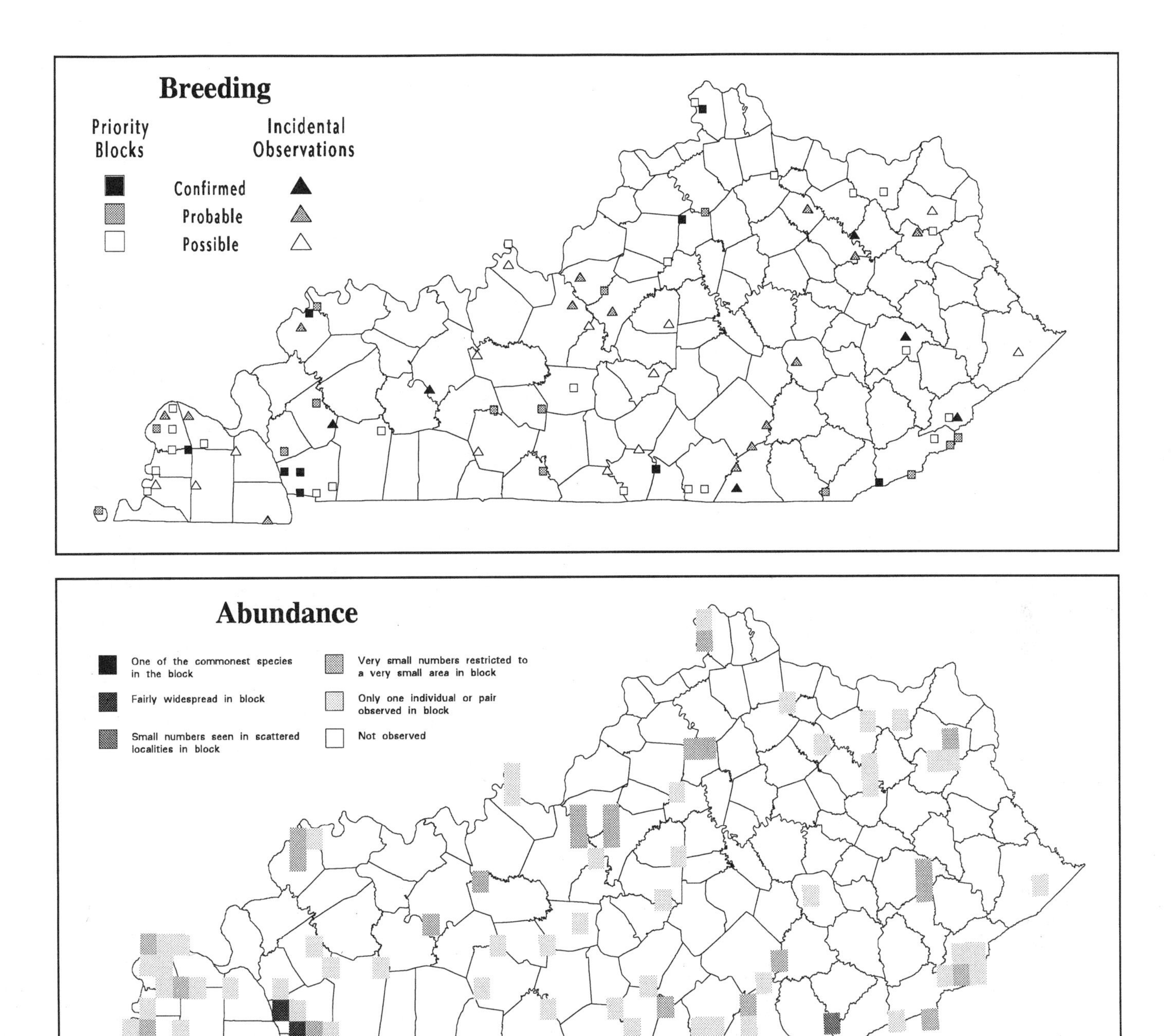

Analysis of Block Data by Physiographic Province Section

Physiographic Province Section	Total Blocks Surveyed	Blocks with Data	% with Data	Section's % for State
Mississippi Alluvial Plain	14	5	35.7	10.9
East Gulf Coastal Plain	36	4	11.1	8.7
Highland Rim	139	10	7.2	21.7
Shawnee Hills	142	7	4.9	15.2
Blue Grass	204	9	4.4	19.6
Cumberland Plateau	173	5	2.9	10.9
Cumberland Mountains	19	6	31.6	13.0

Summary of Breeding Status

Number of Blocks in Which Species Was Recorded		
Total	**46**	**6.3%**
Confirmed	9	19.6%
Probable	14	30.4%
Possible	23	50.0%

Wild Turkey

Northern Bobwhite

Colinus virginianus

The Northern Bobwhite is certainly the most widespread gallinaceous bird in Kentucky. The species has adapted well to most of the changes humans have made to the landscape, and it has been stocked at many localities through a number of efforts (see Mengel 1965; Alsop 1971).

Northern Bobwhites use a great variety of semi-open and open habitats. Although the species is found in a few naturally occurring situations, such as patches of remnant prairie, today it is primarily a bird of altered habitats. These quail are most frequent in rural farmland that has a good supply of fencerows, brushy borders, and other patches of dense cover. In contrast, numbers are much lower in intensively managed farmland, especially where fescue is the predominant grass. Substantial numbers also occur in a variety of other habitats, including reclaimed surface mines, abandoned homesites, and young pine plantations.

Before altered habitats were available, Northern Bobwhites were probably restricted to the native prairies and other naturally open and semi-open situations with brushy cover. As vast areas of forest were converted to agricultural land and settlement, these quail likely moved into new areas, soon occupying suitable habitat throughout much of the state. For example, Audubon (1861) considered them to be abundant in those parts of Kentucky with which he was familiar in the early 1800s. In contrast, the species has likely decreased in areas now intensively managed for agricultural use and heavily settled.

The Northern Bobwhite is a permanent resident in Kentucky. Family groups typically join together during the winter to form coveys of up to a few dozen birds. During the end of March and early April these flocks break up, and males typically begin their territorial calling by mid-April. Nesting commences during the middle of May, and Mengel (1965) gave a peak in clutch completion during early June. The nesting season is rather prolonged, however, and more recent data suggest that nesting peaks in July (J. Sole, unpub. rpt.). Later nests are not uncommon, and both nests containing eggs and small broods of young have been noted well into September (Croft and Stamm 1967; Stamm 1990; J. Sole, unpub. rpt.). Recent data indicate that both males and females are polygamous and that some females may lay three or more clutches during the breeding season (J. Sole, pers. comm.). The rate of nest failure is relatively high, and although some birds may raise more than one brood, most late nests likely represent renesting attempts owing to failure from predation or inclement weather. Kentucky data on clutch size are somewhat limited, but Mengel (1965) gave the average size of five clutches as 15.0 (range of 12–19). Large clutches may include eggs from more than one female.

Forest Cover

Value	% of Blocks	Avg Abund
All	78.1	2.5
1	83.2	2.7
2	91.1	2.7
3	86.1	2.5
4	62.3	2.0
5	32.7	1.4

Bobwhites typically nest in fairly open situations, including fallow fields, hayfields, and cedar thickets (Mengel 1965). The nest is situated on the ground, usually near a forest margin, fencerow, or other feature that affords some degree of cover. The nest site is formed in a shallow depression, usually concealed by overhanging vegetation, and lined with grasses or other plant material from nearby (Harrison 1975).

Herbert Clay Jr. M.D.

The atlas fieldwork yielded records of Northern Bobwhites in more than 78% of priority blocks, and 14 incidental observations were reported. Occurrence was higher than 90% in the East Gulf Coastal Plain, the Shawnee Hills, the Highland Rim, and the Mississippi Alluvial Plain, and lowest in the Cumberland Plateau and the Cumberland Mountains. Average abundance varied somewhat similarly. Occurrence was lower in the Blue Grass section because of the extent of clearing and lack of suitable cover in the farmland of the Inner Blue Grass. Occurrence and average abundance were highest in cleared areas with some forest cover and lowest in predominantly forested areas.

Despite the abundance of Northern Bobwhites, only about 11% of priority block reports were for confirmed breeding. Nearly all of the 62 confirmed reports were for young birds in the company of adults, but a few active nests were located.

Northern Bobwhites are frequently reported on most Kentucky BBS routes. The average number of individuals per BBS route for the periods 1966–91 and 1982–91 was 26.4 and 24.7, respectively. According to these data, the species ranked 13th in abundance on BBS routes during the period 1982–91. Trend analysis of the data reveals a significant ($p<.01$) decrease of 2.2% per year for the period 1966–91 and a significant ($p<.05$) decrease of 2.4% per year for the period 1982–91. These decreases partially result from the severe winters of the late 1970s (Stamm 1980; Robbins et al. 1986; J. Sole, unpub. rpt.), but an overall trend in the loss of optimal habitat is likely involved also (J. Sole, unpub. rpt.).

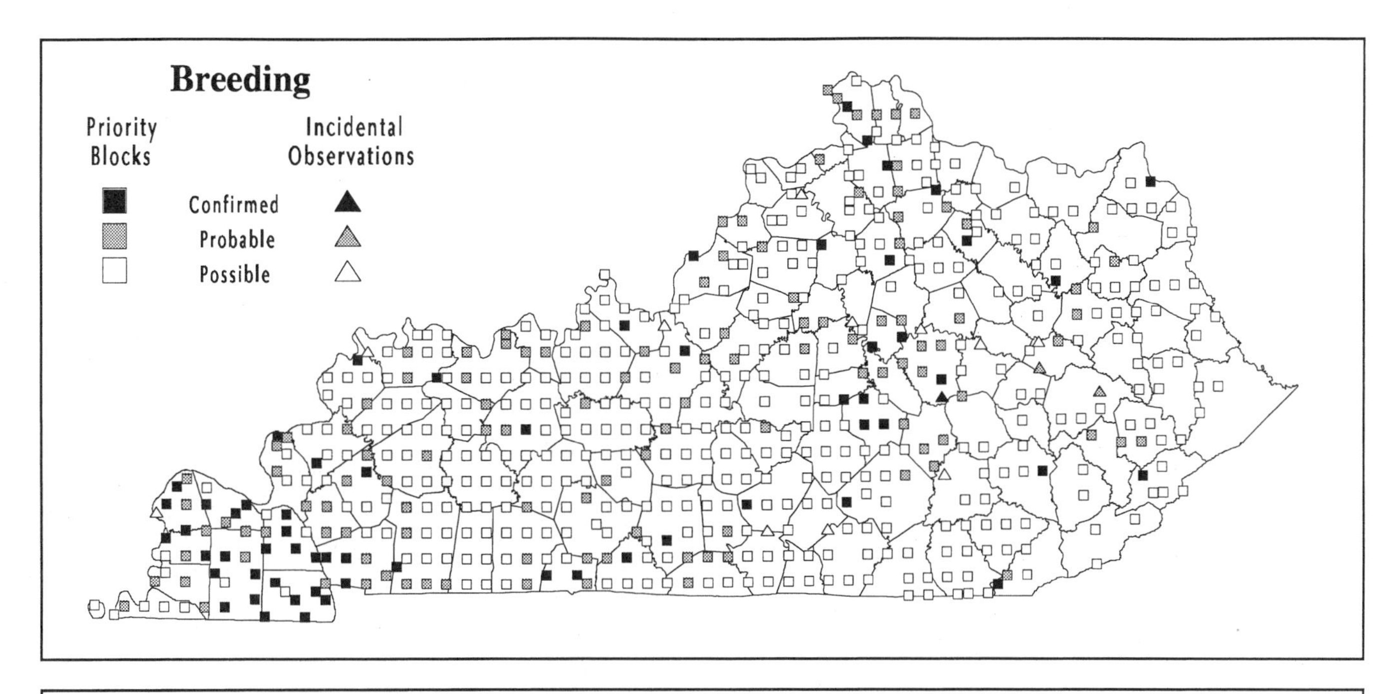

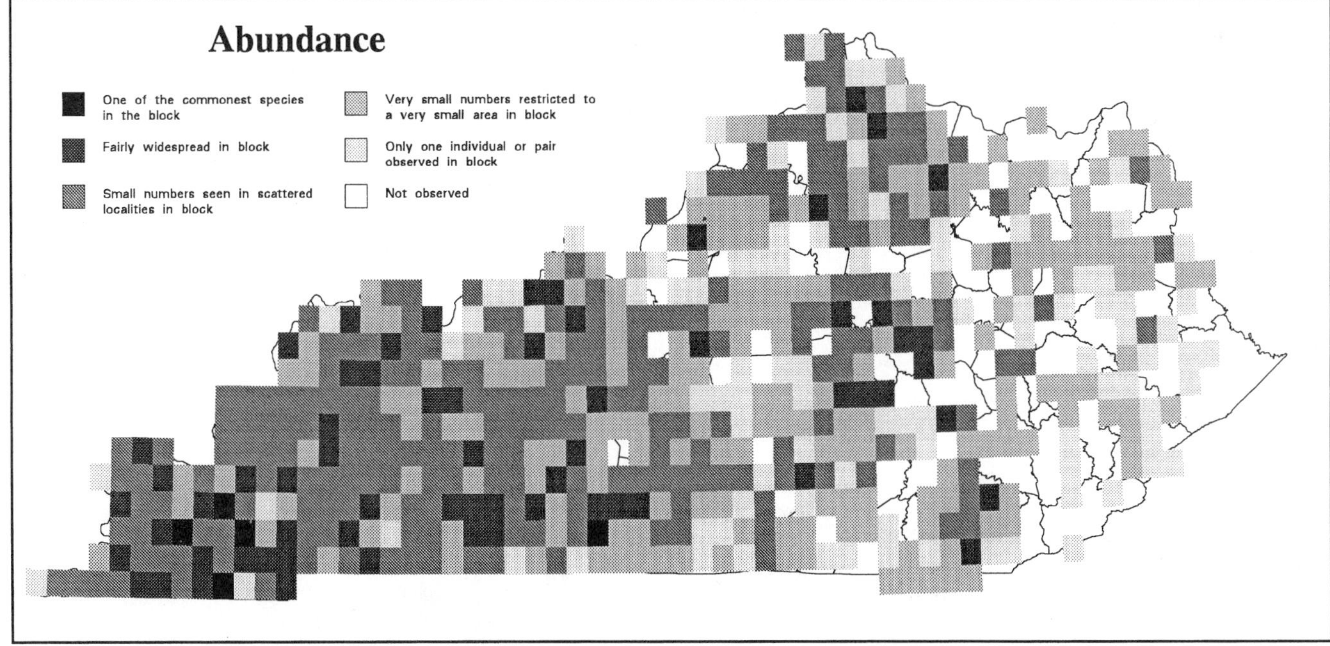

Analysis of Block Data by Physiographic Province Section

Physiographic Province Section	Total Blocks Surveyed	Blocks with Data	Avg Abund	% with Data	Section's % for State
Mississippi Alluvial Plain	14	13	2.9	92.9	2.3
East Gulf Coastal Plain	36	36	3.4	100.0	6.3
Highland Rim	139	132	2.5	95.0	23.2
Shawnee Hills	142	137	2.8	96.5	24.1
Blue Grass	204	145	2.3	71.1	25.5
Cumberland Plateau	173	96	1.9	55.5	16.9
Cumberland Mountains	19	9	1.4	47.4	1.6

Summary of Breeding Status

Number of Blocks in Which Species Was Recorded		
Total	**568**	**78.1%**
Confirmed	62	10.9%
Probable	119	21.0%
Possible	387	68.1%

King Rail

Rallus elegans

Historical documentation of the presence of the King Rail in Kentucky is scarce, but Mengel (1965) regarded the species as a locally uncommon summer resident in the late 1950s. Although this secretive bird went largely undetected during early settlement, the difficulty in locating nesting birds may be partially responsible for the paucity of reports. Mengel listed breeding records for Clinton, Fulton, Henderson, and Jefferson Counties and gave only a few additional references to its occurrence in other areas. Since that time, there has been only one documented nesting record from the Reelfoot NWR in Fulton County (Stamm 1984b), and today the species must be considered among the state's rarest nesting birds.

Although this large rail may have always been locally distributed in Kentucky, floodplain sloughs and marshes along the larger rivers must have provided prime nesting habitat. Loss of such habitat is primarily responsible for the species's decline. During the last two centuries vast areas of floodplain wetlands have been drained, cleared, and converted to row-crop agriculture, resulting in the loss of most suitable nesting habitat. Wetland protection efforts have begun to slow this trend, and the recent increase in the use of "moist soils" techniques for waterfowl management holds promise for returning a large amount of acreage to conditions more suitable for nesting.

King Rails inhabit marshes and the marshy borders of lakes and ponds. In Kentucky this type of habitat is most frequently encountered in the floodplains of the larger rivers, but similar areas are scattered across much of the state. Favored habitat is typically dominated by a thick growth of herbaceous emergent vegetation such as cattails, bulrushes, and sedges.

Although King Rails have been known to winter to the north of Kentucky, observations suggest that most or all of the state's small breeding population moves south in winter. The few records of spring migrants have occurred from late March through mid-April, and summer residents probably return about the same time (Mengel 1965). Mengel indicated that nesting activity commences during the middle of April and extends through May, with a probable peak in clutch completion in mid-May. The Reelfoot NWR nest contained 13 eggs on May 27, 1984. Small young have been observed from early May through mid-June, although family groups have been seen well into July. The average size of four clutches described by Mengel was 12.3 (range of 11–13).

King Rails sometimes nest in small marshes, but they are more often found in larger tracts of suitable habitat. The nest is woven into the dense cover of emergent wetland plants and is typically situated several inches above shallow water or saturated ground. It is constructed primarily of the dead stems of whatever plants are abundant and is formed into a thick platform with a shallow, saucerlike depression. The King Rail is somewhat dependent upon a relatively stable water level during the breeding season, and it is likely that abnormally wet or dry periods cause abandonment of nests owing to inundation or lack of food. Late in the season, family groups have been observed in sloughs that have dried up but still offer some foraging opportunity (Mengel 1965).

During the atlas survey there was only one incidental report of the King Rail. On two occasions in June 1990, a single bird was observed feeding along the margin of a roadside slough on the Reelfoot NWR, several hundred yards from the 1984 nesting location (Hendricks et al. 1991).

Ron Austing

While King Rails probably nest at a few locations that are currently unknown, the scarcity of suitable habitat certainly limits their numbers. It can only be hoped that continuing efforts to preserve wetland habitats will assist this species in recovering as a nesting bird in the state.

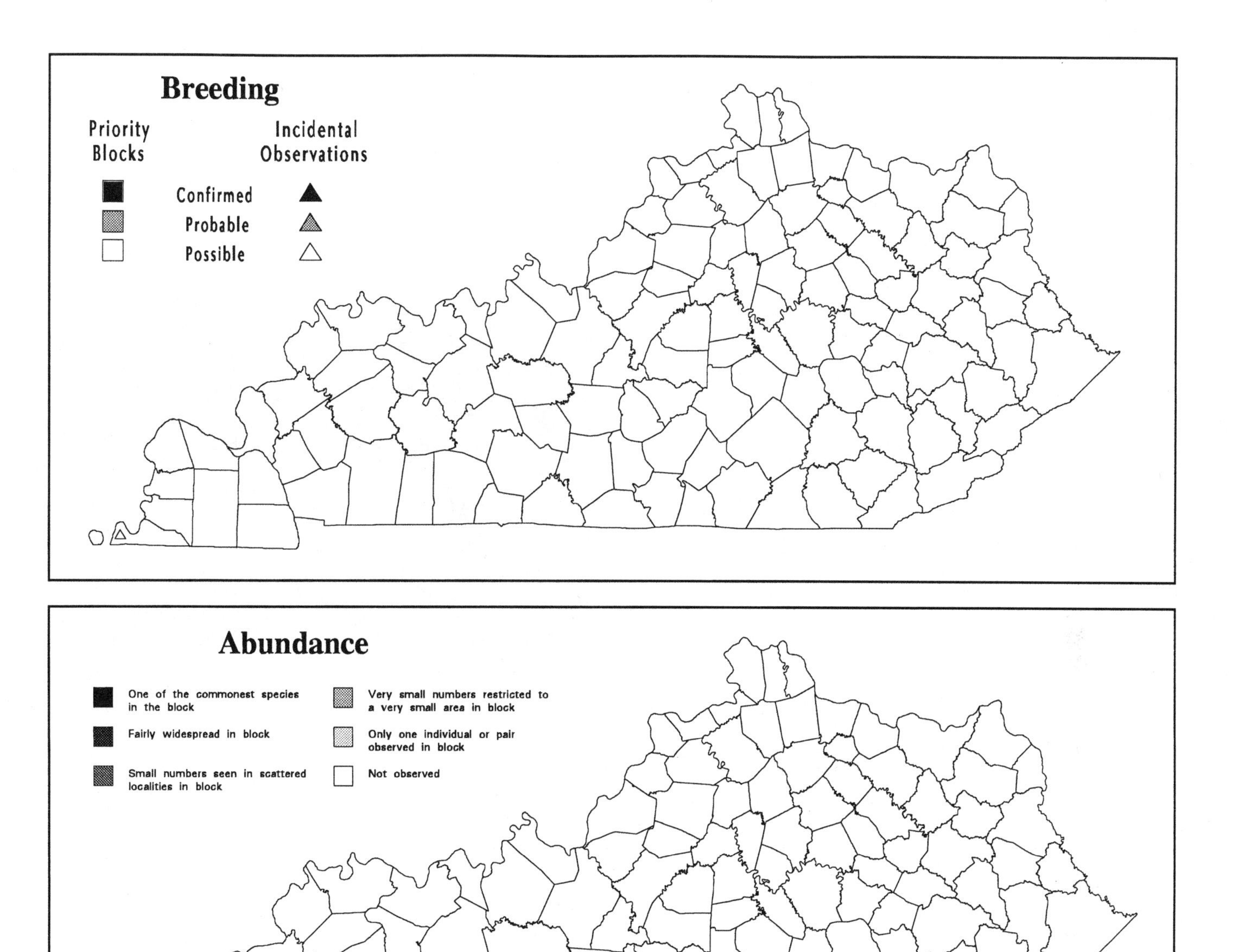

Analysis of Block Data by Physiographic Province Section

Physiographic Province Section	Total Blocks Surveyed	Blocks with Data	% with Data	Section's % for State
Mississippi Alluvial Plain	14	-	-	-
East Gulf Coastal Plain	36	-	-	-
Highland Rim	139	-	-	-
Shawnee Hills	142	-	-	-
Blue Grass	204	-	-	-
Cumberland Plateau	173	-	-	-
Cumberland Mountains	19	-	-	-

Summary of Breeding Status

Number of Blocks in Which Species Was Recorded		
Total	-	-
Confirmed	-	-
Probable	-	-
Possible	-	-

Common Moorhen

Gallinula chloropus

The Common Moorhen is a very rare breeding bird in Kentucky. Although suitable nesting habitat occurs at scattered localities across the state, the species is only occasionally reported in summer. Before the early 1980s the only confirmed breeding record had come from McElroy Lake, in Warren County, in 1935 (Wilson 1940). Moorhens were also reported in summer in Fulton County in the 1890s, without specific detail (Pindar 1925). Suitable habitat must have certainly existed in Fulton County at the time, and summering birds were likely part of the substantial nesting population at Reelfoot Lake, Tennessee (Mengel 1965).

Common Moorhens typically inhabit marshes and the marshy borders of lakes and ponds. Although they must have occurred in marshy wetlands before settlement, such natural sites have been all but eliminated during the present century. In the absence of suitable natural sites, moorhens are now encountered in artificially created situations, including the margins of reservoirs, fish hatchery ponds, and other types of shallow-water impoundments where cattails and other emergent aquatic vegetation thrive.

Moorhens winter from the coastal regions of the southern United States southward into Central America (AOU 1983). During the middle of April small numbers of transients pass through on their way back to breeding areas that are scattered throughout the northern United States and southern Canada. While migrants may linger well into May, it appears that birds still present late in the month are likely breeding. Specific Kentucky data on nesting are scarce, but most clutches are likely completed during June, and most young are probably flying by late July. Wilson's observation of immatures in Warren County in early August 1935 (Wilson 1940) indicates that nesting activity may continue well into summer. Although Kentucky data on clutch size are lacking, Harrison (1975) gives rangewide clutch size as 6–17, typically 10–12.

According to Harrison (1975), the nest is usually placed on or just above the water or marshy ground. It consists of a thick platform of herbaceous plant material, interwoven with and concealed to some degree by the existing vegetation. The birds usually build an angled ramp of vegetation that leads down to the water from the top of the platform.

Recent summer records of Common Moorhens were all reported during the atlas survey period. The only successful nesting was reported from a cattail pond on a reclaimed surface mine in southern Ohio County, where four half-grown young were observed in the company of both parents on July 16, 1989 (Palmer-Ball and Barron 1990). An unsuccessful nesting attempt was documented at the Minor Clark Fish Hatchery, in Rowan County, where a nest containing two eggs on June 7, 1985, was later found destroyed (Stamm 1985). In addition, birds were observed on more than one occasion in suitable nesting habitat at the Frankfort Fish Hatchery, in Franklin County, in 1985 (Stamm 1985), and at the Sauerheber Unit of Sloughs WMA, in Henderson County, in 1986 (Stamm 1986a). Finally, a single moorhen was found in suitable nesting habitat along Mayfield Creek, in Carlisle County, on June 27, 1990, but further evidence of nesting could not be obtained (Stamm and Monroe 1990a).

Maslowski Wildlife Productions

Although the Common Moorhen remains a scarce summer resident, wetland protection measures now being undertaken hold promise for providing additional nesting habitat. With the species scattered locally across the state, the potential for use of newly created habitat seems great.

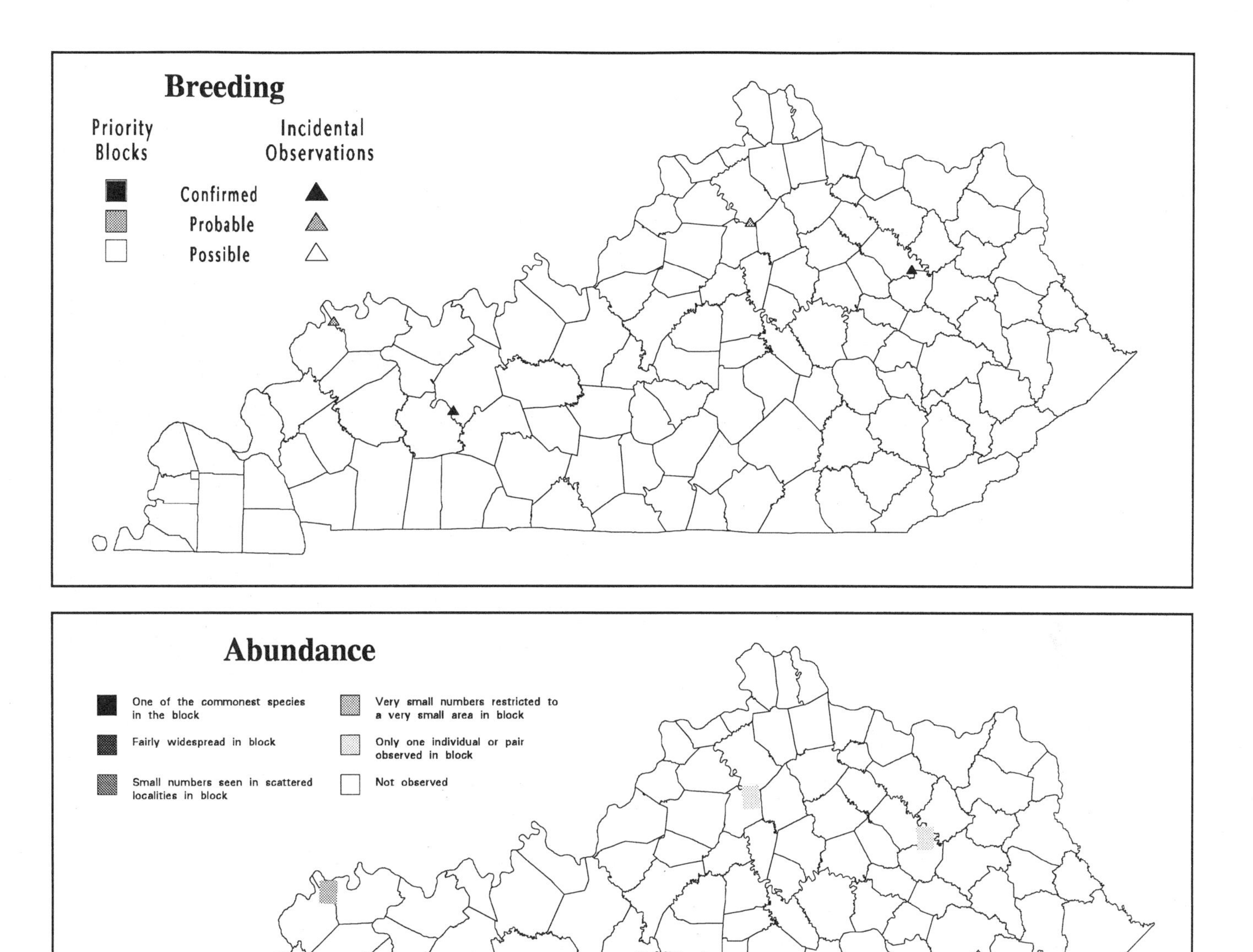

Analysis of Block Data by Physiographic Province Section

Physiographic Province Section	Total Blocks Surveyed	Blocks with Data	% with Data	Section's % for State
Mississippi Alluvial Plain	14	-	-	-
East Gulf Coastal Plain	36	1	2.8	100.0
Highland Rim	139	-	-	-
Shawnee Hills	142	-	-	-
Blue Grass	204	-	-	-
Cumberland Plateau	173	-	-	-
Cumberland Mountains	19	-	-	-

Summary of Breeding Status

Number of Blocks in Which Species Was Recorded		
Total	**1**	**0.1%**
Confirmed	-	-
Probable	-	-
Possible	1	100.0%

Common Moorhen

Killdeer

Charadrius vociferus

Primarily owing to its abundance in settled areas and the conspicuousness of its calls, the Killdeer is certainly the most widely recognized member of the shorebird family occurring in Kentucky. Killdeers are well distributed in suitable habitat, and they are absent only in extensively forested regions. Mengel (1965) regarded the species as a common summer resident, breeding throughout the state.

Killdeers are birds of semi-open and open habitats. They are most frequent in areas that offer an abundance of bare or sparsely vegetated ground, especially near lakes and ponds that have an exposed shoreline. In summer these plovers are most often found in rural farmland and settlement, where they forage about lawns, pastures, and crop fields. They also occur in a great variety of rural and urban situations, including seldom-used road corridors, railroad rights-of-way, industrial parks, airports, and parking lots.

The Killdeer must have been much less common and widespread in Kentucky before altered habitats were created. The conversion of vast areas of forest and prairie to agricultural use and settlement has resulted in the spread and increase of the species throughout most of the state. It is likely that these shorebirds were formerly restricted to the sand and gravel bars of the larger rivers, and perhaps those portions of the prairies and barrens of western and central Kentucky that were more heavily affected by herds of large grazing mammals. Audubon (1861) reported breeding in the region in the early 1800s, and by this time the species may have already begun to use agricultural land.

Although a few Killdeers can be found throughout the winter, most birds move farther south in late fall. Many return in early March, and it is not uncommon for early clutches to be completed by the middle of the month, especially in western Kentucky. According to Mengel (1965), a peak in early clutch completion occurs during the last week or so of March. Many pairs apparently raise a second brood, and later nestings are initiated soon after the first young are on the wing. Late nests have been reported into July (Croft and Stamm 1967; Stamm and Croft 1968; Larson 1973). By the middle of July large aggregations, up to more than 100 birds, often collect at favored feeding areas. Although the typical clutch consists of four eggs, Mengel (1965) notes that a few containing three eggs may be complete.

Forest Cover

Value	% of Blocks	Avg Abund
All	70.0	2.2 *
1	97.0	2.4
2	91.1	2.5
3	77.8	2.1
4	35.1	1.8
5	7.7	1.2

Killdeers nest on the ground, typically in open situations such as gravel road shoulders and rooftops, margins of parking lots, well-grazed pastures, and recently planted crop fields. Nests in natural sites are seldom observed today, but birds do nest on the few sand, gravel, and rock bars that remain along the larger rivers. The nest is a shallow scrape formed among small rocks, pebbles, or other camouflaging debris, although sometimes it is situated on bare dirt. The adults are well known for their broken wing displays, which lure intruders away from eggs or young.

Ron Austing

The atlas fieldwork yielded reports of Killdeers in 70% of priority blocks statewide, and 24 incidental observations were reported. Occurrence fell between 83% and 95% in all physiographic province sections in central and western Kentucky, but the species was found on only about one-fourth of the priority blocks in the Cumberland Plateau and the Cumberland Mountains. Average abundance was also lower in eastern Kentucky. Occurrence and average abundance were closely related to amount of open habitat, which approximated the amount of suitable nesting habitat.

Slightly more than 20% of priority block reports were for confirmed breeding. Killdeers are relatively conspicuous in their breeding behavior and choice of nest sites. A number of active nests were located during the atlas work, but most records of confirmed breeding involved the observation of juvenile birds in the company of adults and the conspicuous broken-wing distraction displays.

Killdeers are reported in small numbers on most Kentucky BBS routes. The average number of individuals per BBS route for the periods 1966–91 and 1982–91 was 4.8 and 6.7, respectively. Trend analysis of these data reveals a significant ($p<.01$) increase of 3.1% per year for the period 1966–91 and a nonsignificant increase of 1.4% per year for the period 1982–91.

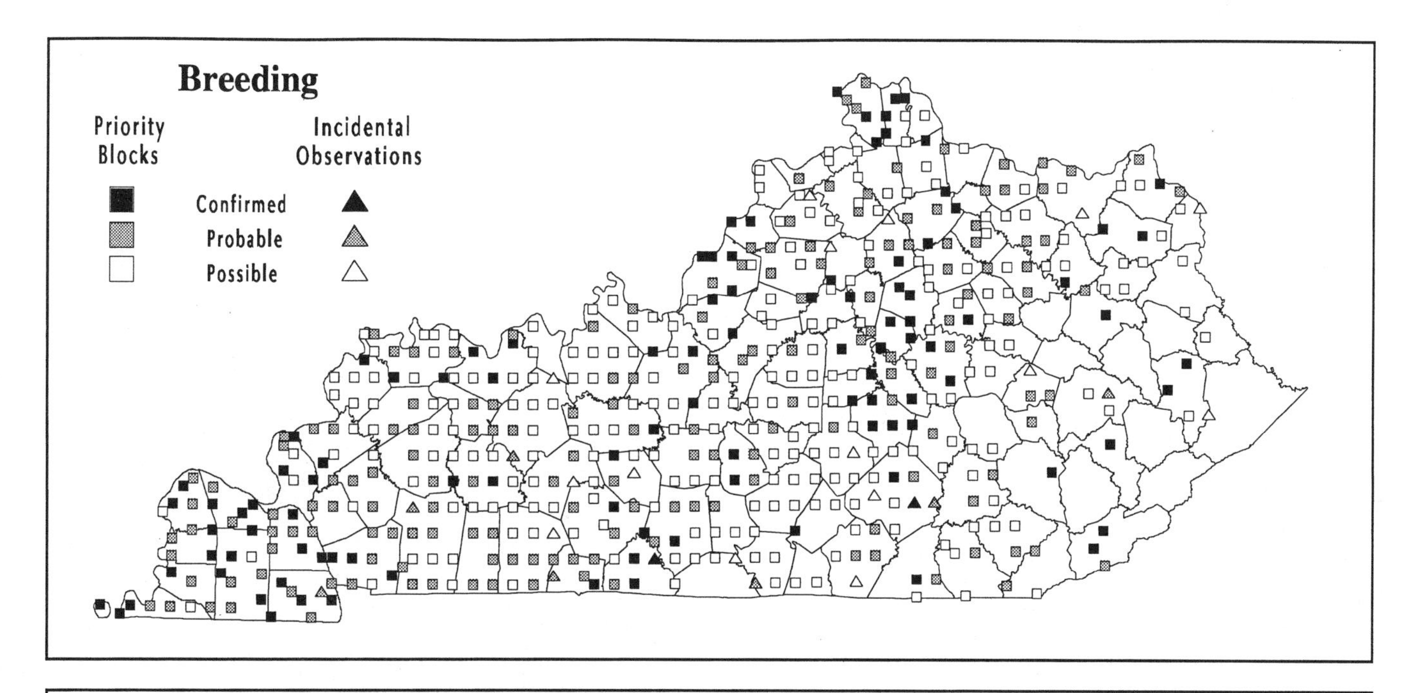

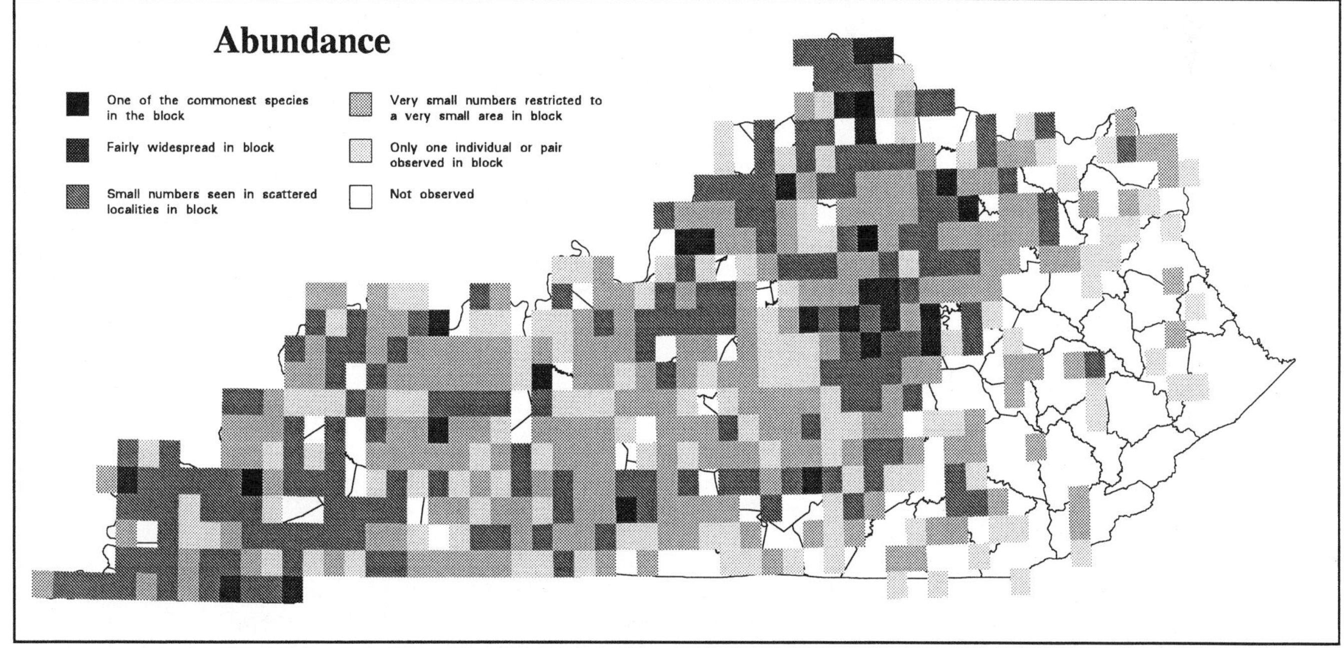

Analysis of Block Data by Physiographic Province Section

Physiographic Province Section	Total Blocks Surveyed	Blocks with Data	Avg Abund	% with Data	Section's % for State
Mississippi Alluvial Plain	14	13	2.8	92.9	2.6
East Gulf Coastal Plain	36	34	2.7	94.4	6.7
Highland Rim	139	118	2.2	84.9	23.2
Shawnee Hills	142	119	2.1	83.8	23.4
Blue Grass	204	173	2.4	84.8	34.0
Cumberland Plateau	173	47	1.7	27.2	9.2
Cumberland Mountains	19	5	1.4	26.3	1.0

Summary of Breeding Status

Number of Blocks in Which Species Was Recorded		
Total	**509**	**70.0%**
Confirmed	104	20.4%
Probable	171	33.6%
Possible	234	46.0%

Killdeer

Spotted Sandpiper

Actitis macularia

The Spotted Sandpiper is a rare and sporadic nesting bird in Kentucky. Its occurrence apparently depends largely on the availability of suitable habitat. Breeding has been documented from two sites in the north central part of the state, but elsewhere scattered accounts of summering birds have been reported without evidence of nesting. Mengel (1965) correctly noted that many early reports of breeding (e.g., Wilson 1942) may have been based largely on the conclusion that "summer" sightings indicated nesting.

At the Falls of the Ohio, in Jefferson County, nesting has been reported in various years when summer water levels have left the riverbed exposed for a sufficient time. Young too small to fly have been observed there on at least a half dozen occasions since the first report sometime before 1960 (Mengel 1965). During the mid-1960s the nesting population was quite large: 10–12 pairs that seemed to be territorial were observed in June 1964 (Stamm 1966). From 1965 to 1970 active nests were located on three occasions (Stamm 1966; Stamm et al. 1967; KOS Nest Cards). Since the mid-1960s little effort has been made to monitor the status of the nesting population at the Falls of the Ohio. Although more recent fieldwork seems to indicate that the species has not summered in such substantial numbers in subsequent years, breeding was confirmed at the site in 1985 (Stamm 1985; Palmer-Ball and Hannan 1986).

Away from the Falls of the Ohio, there is only one additional record of confirmed breeding. Two downy young were observed in the company of an adult, foraging along an impoundment at the State Game Farm, in Franklin County, on June 18, 1954 (Despard 1954). A number of other summer sightings of birds in suitable breeding habitat have been reported from scattered localities throughout Kentucky (Mengel 1965), but none have been accompanied by convincing evidence of nesting. Furthermore, because spring migrants may linger into early June and returning fall birds typically arrive by early July, many "summer" records in the literature may pertain to transients. Nonbreeding Spotted Sandpipers may also remain throughout the summer.

The status of nesting Spotted Sandpipers in Kentucky before 1900 is not clear. Audubon (1861) commonly noted migrants along the Ohio River but apparently did not observe breeding in the early 1800s. Before the damming of the Ohio River and most of its tributaries, it is likely that suitable nesting habitat was present in much greater abundance. Many gravel and sand bars have been inundated, and suitable habitat is now restricted to a few locations immediately downstream from dams on the Ohio River. Even in these areas, human disturbance and widely fluctuating water levels probably preclude nesting at most sites. Only at the Falls of the Ohio is habitat sufficient and human disturbance limited enough to allow successful nesting on a somewhat regular basis. While some impoundments and smaller rivers seem to offer marginal nesting opportunities, the species does not appear to have used these situations to any great extent.

Spotted Sandpipers winter primarily along the Gulf Coast and southward into the tropics (AOU 1983), and numbers of spring migrants arrive during early May. By the end of the month only a few birds remain, and if water levels are suitable at this time, nesting may be attempted. Based on the limited data, it appears that most clutches are completed from the latter half of June to early July. Only three nests have ever been found in Kentucky, all at the Falls of the Ohio. A nest containing four eggs was found there on July 4, 1965 (Stamm 1966); one containing a chick, still moist, was found on July 16, 1966 (Croft and Stamm 1967); and the third, containing three eggs, was studied on July 10–13, 1970 (KOS Nest Cards). According to Harrison (1975), the typical clutch consists of four eggs.

Ron Austing

The nest is placed on the ground and is typically concealed in moderately thick vegetation. It is often situated on the edge of an exposed area of sand or gravel along the margin of a body of water, but Spotted Sandpipers also regularly nest far from water, in dry fields and pastures (Harrison 1975). The nest itself usually consists of a shallow depression loosely enclosed with dry grass, fine weed stalks, and other dead plant material (Stamm 1966). The Falls of the Ohio nests have been situated on small deposits of sand among the exposed rock ledges, and the one discovered in 1965 was sheltered by a small sapling, young tree sprouts, and various forbs.

During the atlas fieldwork, nesting Spotted Sandpipers were found only at the Falls of the Ohio, where three downy young were observed in the company of an adult on July 12, 1985 (Stamm 1985). Territorial behavior and copulation were observed in suitable habitat surrounding McElroy Lake, in Warren County, in early June 1989. Although a nest could not be located, a sudden rise in the water level during the first week of July may have inundated any active nests (Palmer-Ball and Boggs 1991). A second probable record involved a pair of birds that was observed on a small impoundment along the Green River about three miles southwest of Poverty on June 20, 1990. The only other report involved the observation of a bird on a backwater area near the East Bend Power Plant, in Boone County, on May 30–June 18, 1988. Several other records were omitted owing to the date of observation or unsuitability of habitat for nesting.

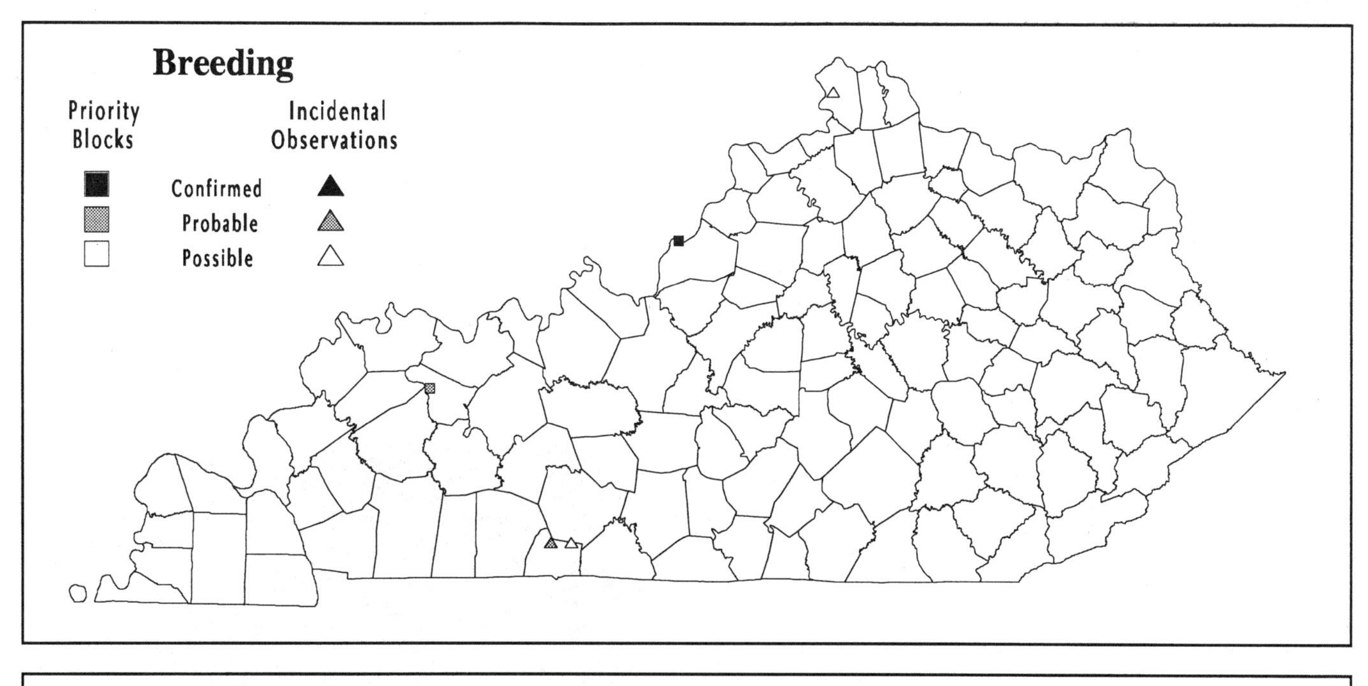

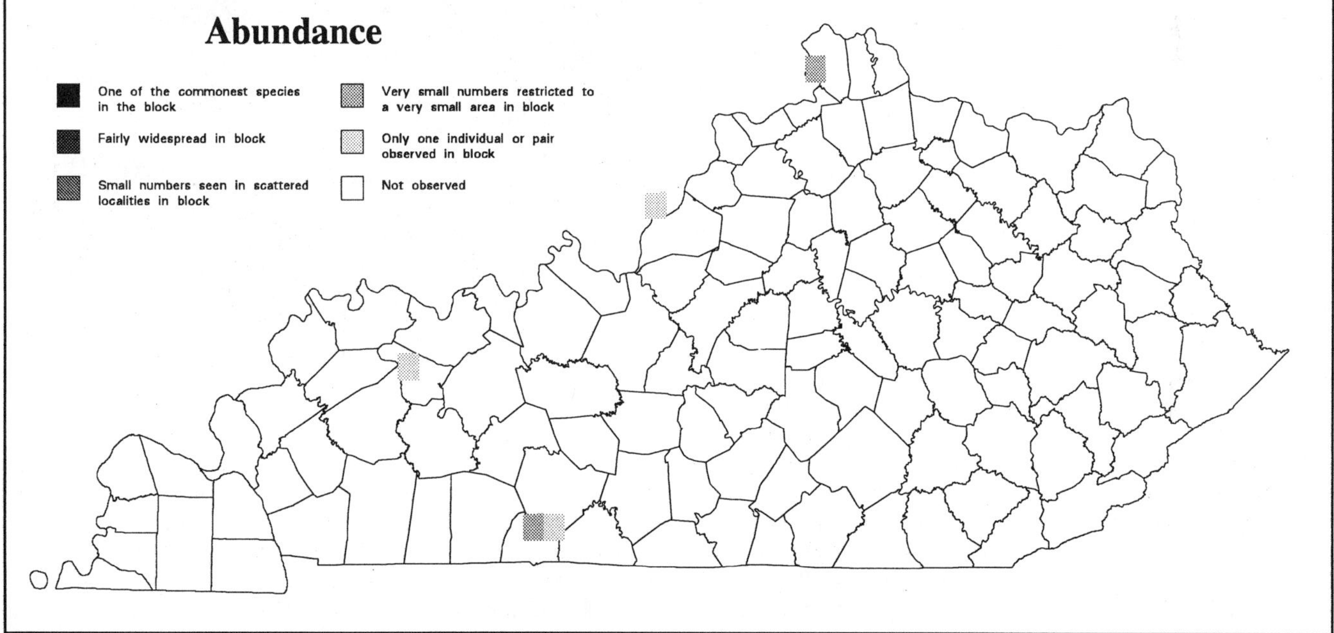

Analysis of Block Data by Physiographic Province Section

Physiographic Province Section	Total Blocks Surveyed	Blocks with Data	% with Data	Section's % for State
Mississippi Alluvial Plain	14	-	-	-
East Gulf Coastal Plain	36	-	-	-
Highland Rim	139	-	-	-
Shawnee Hills	142	1	0.7	50.0
Blue Grass	204	1	0.5	50.0
Cumberland Plateau	173	-	-	-
Cumberland Mountains	19	-	-	-

Summary of Breeding Status

Number of Blocks in Which Species Was Recorded		
Total	**2**	**0.3%**
Confirmed	1	50.0%
Probable	1	50.0%
Possible	-	-

American Woodcock

Scolopax minor

The breeding status of the American Woodcock is not well documented in Kentucky, in large part because of its crepuscular habits. From the limited amount of information available, the species seems to be a fairly common and widespread nesting bird in suitable habitat throughout the state. Mengel (1965) regarded the woodcock as an uncommon summer resident, being most abundant in the Cumberland Plateau and the Shawnee Hills and least numerous in the Inner Blue Grass and the Mississippi Alluvial Plain.

American Woodcock occur in semi-open, often successional habitats, being most numerous in moist, lowland situations. While these odd shorebirds are largely absent from extensive areas of mature forest, they are frequently encountered in woodland edge. A few birds are found along the margins of remnant prairie patches, wetlands, and naturally disturbed areas, but woodcock are primarily birds of altered habitats today. The species is most common in rural areas where habitats are highly fragmented and early successional vegetation is intermixed with openings and forest.

The status of the American Woodcock before settlement is unclear, but the species was likely not as broadly distributed as it is today, being largely absent from extensive areas of mature forest. Woodcock probably occurred in and around natural woodland openings, including bogs, wetlands, and areas disturbed by storms and fire. In the native prairies and savannas, they also probably found suitable habitat in relative abundance, especially in areas of transition to forest. Audubon (1861) indicated that woodcock bred abundantly in Kentucky in the early 1800s, by which time they were likely using artificial clearings in addition to naturally occurring habitat.

A few American Woodcock probably winter in southern and western Kentucky, especially in milder years, but most birds move farther south in late fall. At the onset of the first warm days of late winter, males return northward, and their unique courtship displays become common by early February. It is not known to what extent transient individuals participate in such behavior along the migratory route, but it is possible that not all courtship behavior observed pertains to locally breeding birds. These displays typically peak during the first half of March, and clutches have been observed as early as March 17 (Morse 1948). According to Mengel (1965), clutch completion probably peaks during the last week or so of March but continues into mid-April. The average size of 11 clutches reported by Mengel (1965) was 4.1 (range of 4–5), and Nelson (1973; KOS Nest Cards) reported four eggs in each of three nests located in western Kentucky. Following hatching, the family forages together for some time, and a number of broods of young have been reported in the literature.

Woodcock are ground nesters, choosing a variety of sites, including open woods, woodland edge, and old fields overgrown with weeds or thickets of young trees (Russell 1954). The eggs are laid in a shallow saucer or platform of dead leaves among sparse ground cover and are typically not concealed by overhanging vegetation (Morse 1948; Nelson 1973). Incubating females rely so heavily on their camouflaged plumage that they can sometimes be touched without flushing (Harrison 1975).

Unfortunately, the atlas survey generated only a minimal amount of information on the breeding occurrence of the American Woodcock. The species was reported on less than 10% of priority blocks statewide, and 30 incidental observations were reported. A few of the priority block records and most of the incidental reports were obtained through correspondence with KDFWR personnel. In particular, results of woodcock singing routes for the atlas period were incorporated into the final data set, as were miscellaneous observations from the department's small game program (J. Sole, pers. comm.).

Alvin E. Staffan

Because of the limited data obtained, comparisons of occurrence by physiographic province section are not meaningful. Occurrence was probably highest in the Blue Grass because a higher percentage of priority blocks were completed by atlas volunteers, who were typically able to devote a great deal more time to surveying blocks than seasonal atlasers. Seasonal atlasers also did not start surveys until June, when most courtship activity had ceased.

Although obviously deficient, the atlas data do seem to show a relatively uniform statewide distribution. Woodcock are probably less common in the highly cleared regions, but even in such areas as the Inner Blue Grass and the southern Highland Rim, at least a few birds are present. Likewise, the species is not especially widespread in heavily forested regions, but edges and small openings in such areas provide some suitable habitat.

Despite the limited number of reports, nearly 30% of priority block records were for confirmed breeding. Most of the 19 confirmed records pertained to the discovery of active nests, although some involved the observation of young birds in the company of adults.

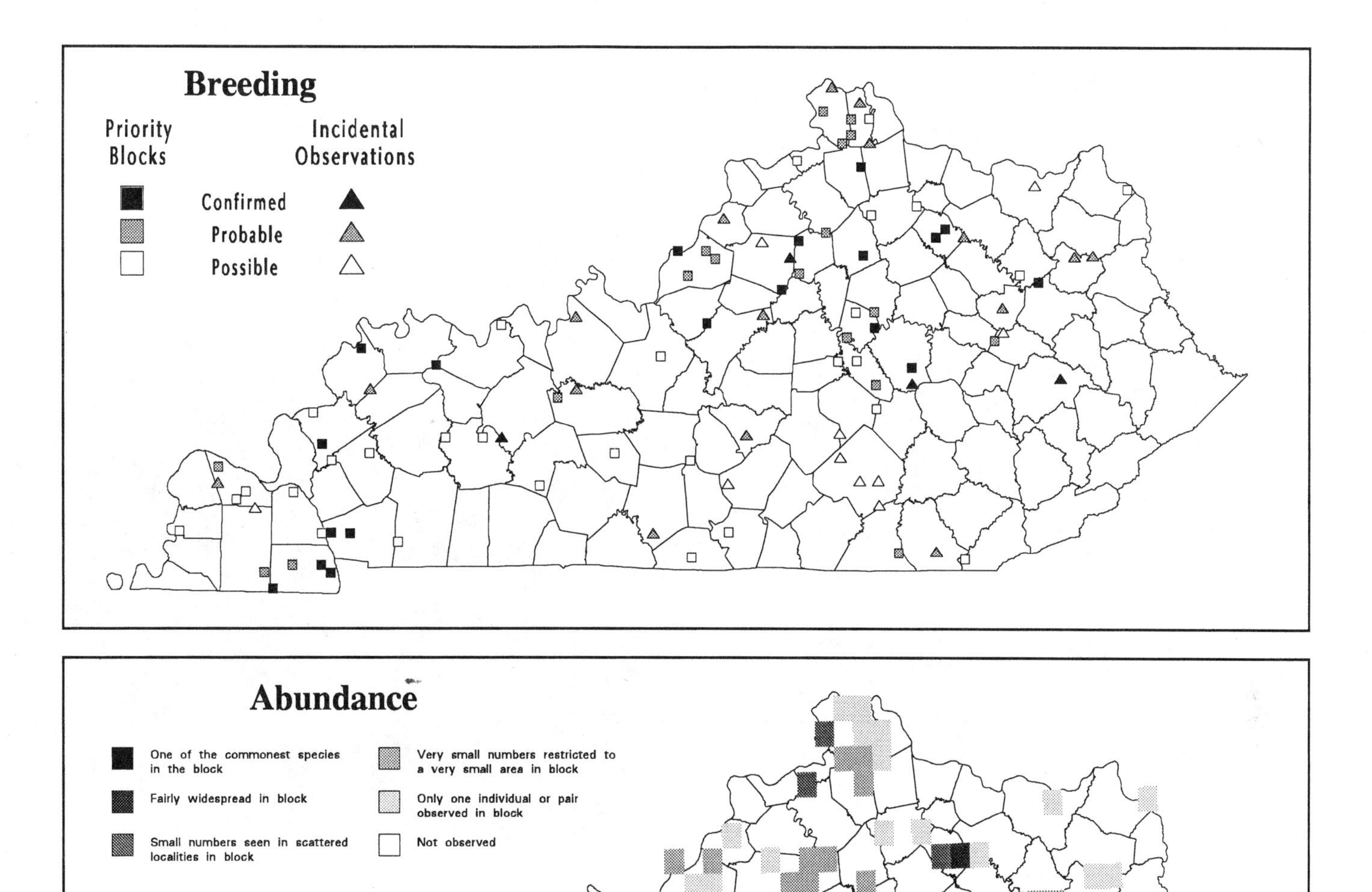

Analysis of Block Data by Physiographic Province Section

Physiographic Province Section	Total Blocks Surveyed	Blocks with Data	% with Data	Section's % for State
Mississippi Alluvial Plain	14	1	7.1	1.5
East Gulf Coastal Plain	36	10	27.8	15.1
Highland Rim	139	8	5.8	12.1
Shawnee Hills	142	11	7.7	16.7
Blue Grass	204	31	15.2	47.0
Cumberland Plateau	173	4	2.3	6.1
Cumberland Mountains	19	1	5.3	1.5

Summary of Breeding Status

Number of Blocks in Which Species Was Recorded		
Total	**66**	**9.1%**
Confirmed	19	28.8%
Probable	18	27.3%
Possible	29	43.9%

American Woodcock

Least Tern

Sterna antillarum

Although the Least Tern is widely distributed along the Atlantic and Gulf Coasts in summer, the interior race (*S. a. athalassos*) nests very locally throughout the central United States, occurring primarily at scattered points along the lower Mississippi River and some of its larger western tributary systems—the Arkansas, the Missouri, and the Red Rivers (USFWS 1990). In Kentucky it has been found nesting regularly at only a few sites along the Mississippi River, although small numbers have been reported sporadically on the lower Ohio River upstream as far as Louisville (Mengel 1965; Stamm 1968; S. Evans, unpub. rpt.). In 1985 the interior race was listed as Federally Endangered, in large part because of the dramatic loss of nesting habitat (USFWS 1990).

The historical status of the Least Tern in Kentucky was not well documented. In the early 1800s Audubon (1861) regularly noted birds on the Ohio River, but he did not observe breeding. Before the Ohio River was altered by the construction of navigation dams, suitable habitat was present in abundance upstream to about Henderson County and locally at least to Louisville. Dams raised the water level substantially, altering formerly used nesting sites such as Bell Island, in Union County, where up to 30 pairs nested as recently as 1953 (Hardy 1957). In July 1967 a single pair of Least Terns attempted to nest at the Falls of the Ohio at Louisville, but human disturbance caused abandonment (Stamm 1968).

Since at least 1984 most of Kentucky's Least Terns have nested in three distinct colonies on the Mississippi River in Carlisle, Hickman, and Fulton Counties. A few other significant nesting colonies occur within this same stretch of the river, but they lie within Missouri. Most years the nesting population at the three Kentucky sites numbers 100–200 nesting pairs (R. Renken, pers. comm.). In addition, small numbers of birds have nested at several other places in this same stretch of the river. Such sites are usually used only when water levels are especially low, exposing suitable habitat that is normally inundated. On the lower Ohio River, small numbers of birds are observed each summer on scattered sandbars and dredge piles as far upstream as Union County. Although nesting was documented at only one locality on the lower Ohio River during the atlas survey period, a 1994 survey documented nesting at three sites, two in Livingston County and one in Union County (Palmer-Ball 1995).

Least Terns apparently winter along the Gulf Coast from Texas to northern South America (USFWS 1990). The first spring arrivals often do not appear until the second or third week of May, and it may be the end of the month before the entire nesting population has returned. Studies on Kentucky's Mississippi River population have been conducted by the Missouri Department of Conservation (MDOC) since 1986, and they provide the most up-to-date information available on the state's nesting terns (Smith and Renken 1990).

River levels often dictate the commencement of nesting, and although nests have been found as early as May 13, high water may delay the onset of egg laying into June (R. Renken, pers. comm.). An early peak in clutch completion typically occurs during the first part of June, and a second peak often occurs during July (Smith and Renken 1990). While some pairs may raise two broods, MDOC survey results suggest that most late nesting involves one-year-old birds and birds that have experienced nesting failure during the early cycle (R. Renken, pers. comm.). According to MDOC data, average clutch size for sites in and near Kentucky from 1986 to 1989 was 2.4 (range of 1–4). Early nests averaged 2.5 eggs, while later nests averaged 1.8 eggs (Smith and Renken 1990).

Missouri Department of Conservation

Least Terns typically nest on exposed sand and gravel bars within or along the main river channel, although the species is opportunistic and sometimes uses alternative sites. For example, a few birds were suspected of nesting on a deposit of sand in agricultural fields near the Mississippi River in Hickman County in 1984 (S. Evans, unpub. rpt.), and the species attempted to use sandy agricultural fields in Fulton County during flooding in June 1990 (Stamm and Monroe 1990a). The nest itself is nothing more than a shallow scrape in the sand or gravel. It is often situated among the clutter of a high-water line of driftwood and other debris, if present.

During the atlas survey Least Terns were found at all three traditionally known nesting colonies on the Mississippi River. In addition, nesting was confirmed at two sites that were used only once. In farmland of Kentucky Bend, in Fulton County, at least six nests were located during flood conditions in early June 1990. At that time the traditionally used sandbars were completely inundated, and nesting was attempted on high, sandy points in large bare fields. In 1986, 14 nests were found at the head of Island No. 8, in Fulton County (Smith 1986), where the species has nested previously (S. Evans, unpub. rpt.).

Small numbers of birds were also found on the lower Ohio River between Paducah and Smithland Dam during the atlas survey period. At one site near the mouth of the Tennessee River, in Livingston County, two nests were found in 1986 by the U.S. Army Corps of Engineers, but a sudden rise in the water level inundated them (T. Siemsen, pers. comm.). At a few other sites on the Mississippi and lower Ohio Rivers, terns were present during the breeding season, but neither nests nor chicks could be found.

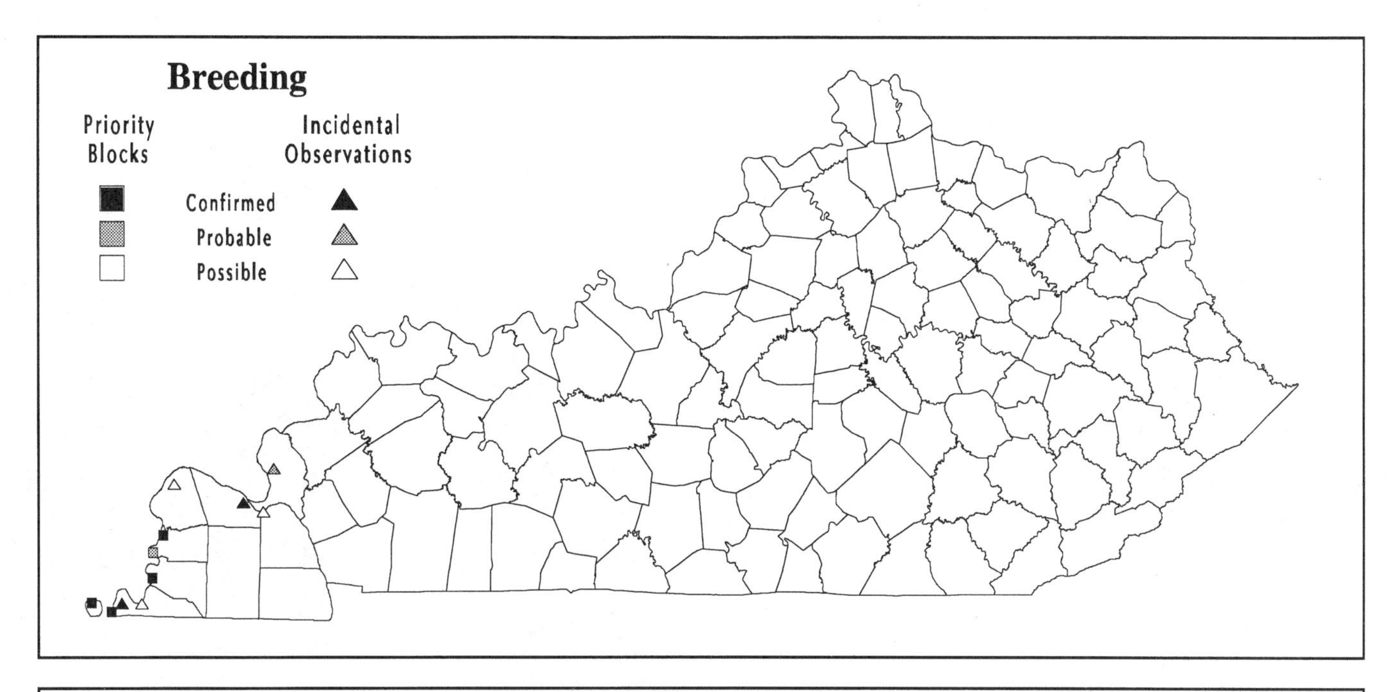

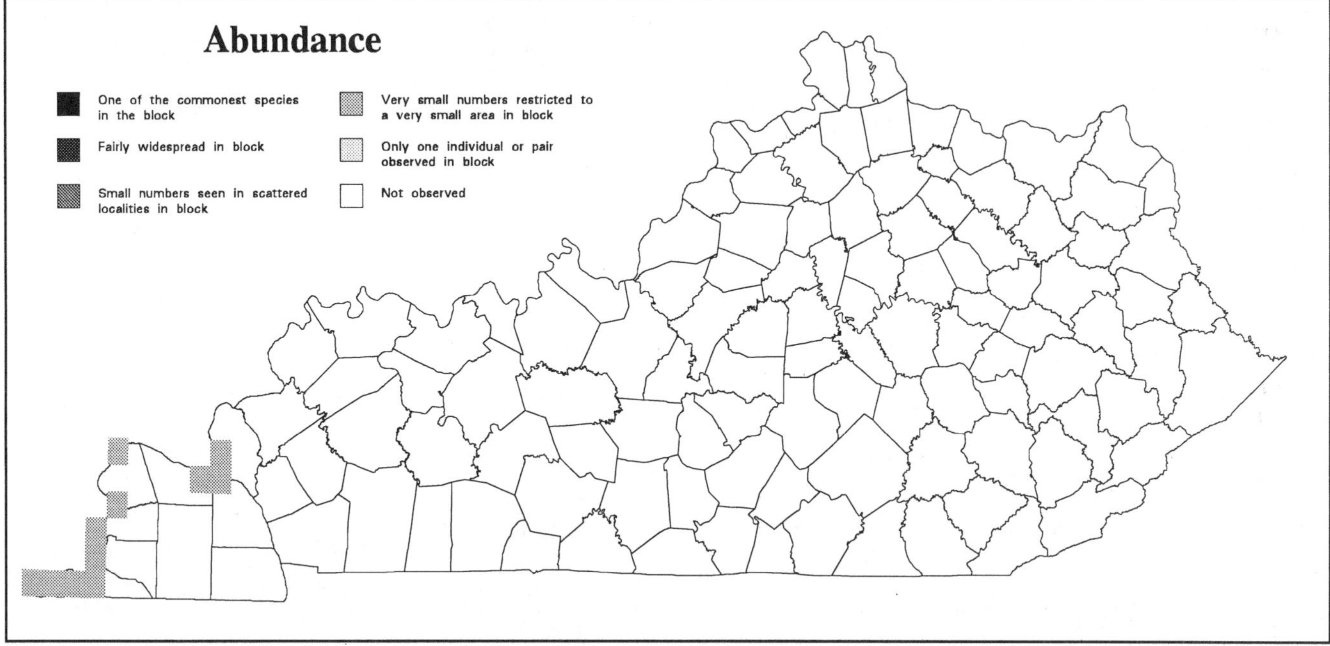

Analysis of Block Data by Physiographic Province Section

Physiographic Province Section	Total Blocks Surveyed	Blocks with Data	% with Data	Section's % for State
Mississippi Alluvial Plain	14	5	35.7	100.0
East Gulf Coastal Plain	36	-	-	-
Highland Rim	139	-	-	-
Shawnee Hills	142	-	-	-
Blue Grass	204	-	-	-
Cumberland Plateau	173	-	-	-
Cumberland Mountains	19	-	-	-

Summary of Breeding Status

Number of Blocks in Which Species Was Recorded		
Total	**5**	**0.7%**
Confirmed	4	80.0%
Probable	1	20.0%
Possible	-	-

Least Tern

Rock Dove

Columba livia

The Rock Dove became established in North America following introductions into the eastern United States and Canada in the early 1600s (Terres 1980). The species has apparently been present in Kentucky since the time of settlement. Common about cities and towns, the domestic pigeon is a well-known part of the breeding avifauna today.

Maslowski Wildlife Productions

Rock Doves are typically found in cities and towns, although substantial numbers are also encountered in association with rural settlement, especially farmland. Away from such situations the species is rare or absent, being found regularly only around artificially created features such as bridges, quarries, and road cuts. Although these doves sometimes use natural cliffs, this behavior has been observed in Kentucky only rarely (e.g., Croft 1969).

Little detail is known concerning the breeding biology of Rock Doves in Kentucky. They are permanent residents, and breeding is believed to continue nearly throughout the entire year, especially during milder winters. In most years nesting activity seems to be initiated at the onset of the first warm weather of late winter and continues at least through early fall.

Rock Doves build a flimsy nest of weed stalks, straw, and debris in a nook or cranny of a building, barn, bridge, rock wall, or cliff. According to Harrison (1975), two eggs are normally laid.

The atlas survey yielded reports of Rock Doves in 41% of priority blocks statewide, and 38 incidental observations were reported. Occurrence was highest in the Mississippi Alluvial Plain, the Blue Grass, and the Highland Rim, although average abundance was relatively low throughout the state. Occurrence and average abundance were highest in open areas and decreased as forest cover increased. Away from urban areas Rock Doves were not reported regularly except in open farmland, and elsewhere the species was only infrequently encountered.

Forest Cover

Value	% of Blocks	Avg Abund
All	41.0	2.1
1	59.4	2.3
2	52.1	2.3
3	39.1	2.0
4	26.0	1.8
5	17.3	1.6

More than 15% of records in priority blocks were for confirmed breeding. Although a number of active nests were located during the atlas work, many confirmed records also resulted from the observation of nest building and recently fledged young.

Rock Doves are not recorded in large numbers on most Kentucky BBS routes. The average number of individuals per BBS route for the periods 1966–91 and 1982–91 was 4.70 and 6.35, respectively. Trend analysis of these data shows slight, nonsignificant decreases of 0.9% per year for the period 1966–91 and 5.7% per year for the period 1982–91.

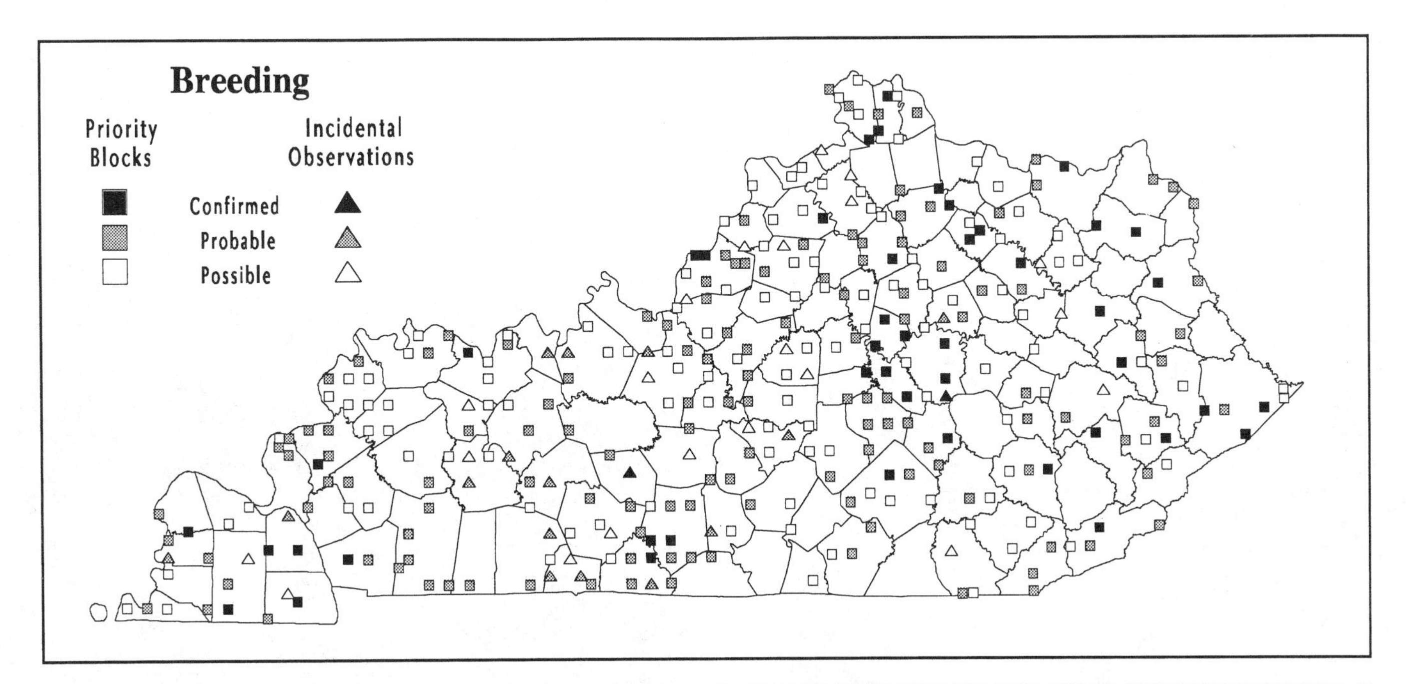

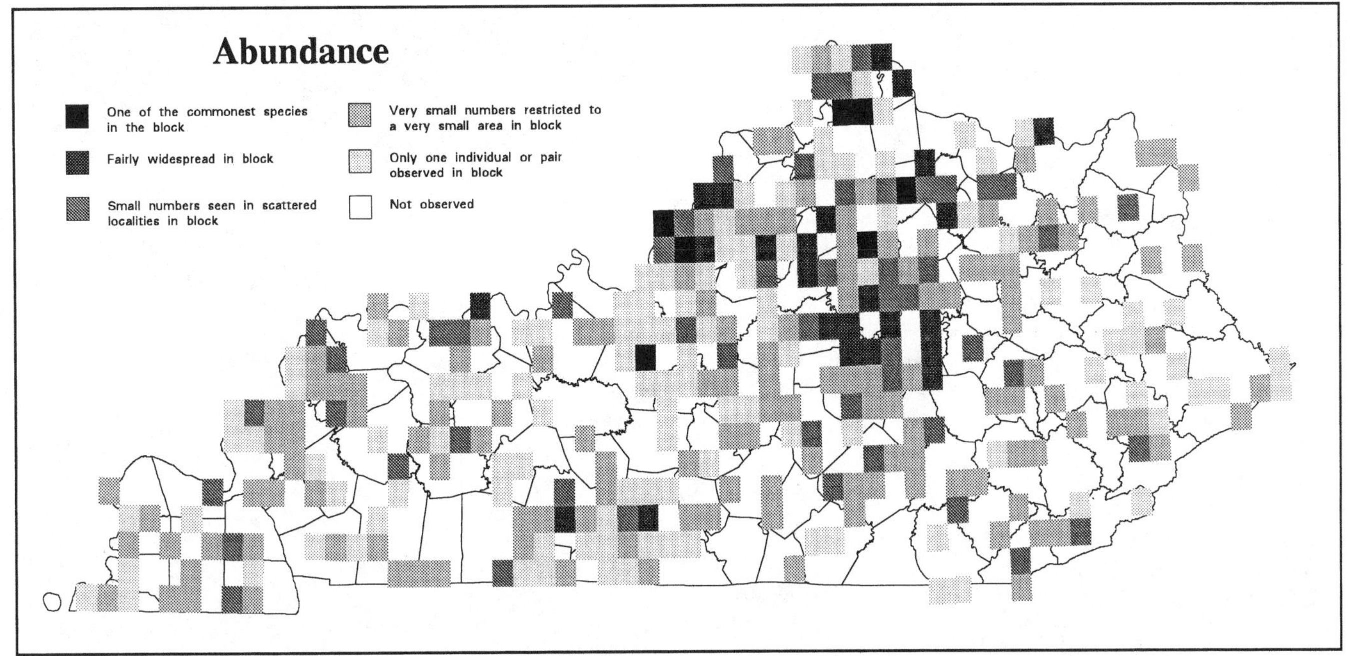

Analysis of Block Data by Physiographic Province Section

Physiographic Province Section	Total Blocks Surveyed	Blocks with Data	Avg Abund	% with Data	Section's % for State
Mississippi Alluvial Plain	14	8	1.6	57.1	2.7
East Gulf Coastal Plain	36	9	2.1	25.0	3.0
Highland Rim	139	66	1.8	47.5	22.1
Shawnee Hills	142	47	1.9	33.1	15.8
Blue Grass	204	112	2.5	54.9	37.6
Cumberland Plateau	173	50	1.8	28.9	16.8
Cumberland Mountains	19	6	2.0	31.6	2.0

Summary of Breeding Status

Number of Blocks in Which Species Was Recorded		
Total	**298**	**41.0%**
Confirmed	46	15.4%
Probable	133	44.6%
Possible	119	39.9%

Rock Dove

Mourning Dove

Zenaida macroura

The Mourning Dove is undoubtedly one of the most common and widely recognized members of Kentucky's avifauna. The species is frequent wherever humans have altered the land, especially for agricultural use and settlement. Mengel (1965) regarded the Mourning Dove as a common to abundant permanent resident, although somewhat less numerous in mountainous portions of eastern Kentucky.

Mourning Doves are typically found in semi-open to open habitats, characterized by an abundance of open ground with short or sparse vegetation. Such situations are most abundant in farmland, but city parks, suburban yards, and a great variety of altered habitats also offer favored nesting areas. Although occasionally found in natural situations such as open woodlands of ridgetops, where fire or xeric conditions maintain an open canopy, this dove is primarily a bird of artificial habitats today.

The Mourning Dove has likely benefited immensely from the changes humans have made upon Kentucky's landscape. Although substantial numbers of doves probably inhabited the grassy barrens in presettlement times, the species must have been virtually absent from the unbroken forests that once covered much of the state. The clearing of vast areas and subsequent conversion to agricultural land and settlement surely resulted in a dramatic increase in the number of Mourning Doves in most areas.

Mourning Doves are permanent residents in Kentucky, but in many areas they are decidedly less common in winter (Mengel 1965). Sporadic periods of unusually warm weather often elicit courtship behavior, and breeding commences in earnest once the first warm days of late winter arrive. Eggs have been observed by mid-February (Mengel 1965), and fledglings have been reported by the last week of March (Young 1973). An early peak in clutch completion, however, occurs during the last ten days of March (Mengel 1965). Later peaks are not evident, but the species apparently raises several broods, and active nests have been observed into October (Nelson 1981). In August, large flocks often gather in prime feeding areas such as freshly harvested agricultural fields. Although clutch size has been reported to vary from one to three, most clutches contain two eggs (Mengel 1965; Nelson 1981).

Forest Cover

Value	% of Blocks	Avg Abund
All	95.7	3.3
1	99.0	3.6
2	100.0	3.7
3	99.6	3.4
4	91.6	2.8
5	69.2	1.8

Mourning Doves nest in a great variety of situations. Most frequently the birds constructs their flimsy stick nest in the branches of a conifer or densely leaved deciduous tree, shrub, or thick growth of vines, but they will also nest on ledges on homes and buildings. These doves also build atop or use the abandoned nests of other birds (Mengel 1965). Nelson (1981) found that the species sometimes reused the same nest during a single breeding season. In open areas, especially farmland, the species will also nest on the ground. Mengel (1965) gave the average height of 33 nests situated above the ground as 10.6 feet (range of 2–40 feet), although Stamm (1963a) observed a nest 50 feet high.

Alvin E. Staffan

The atlas fieldwork yielded records of Mourning Doves in nearly 96% of priority blocks statewide, and three incidental observations were reported. The species ranked 11th according to the number of priority block records, but 4th by both total and average abundance. Occurrence and average abundance were highest in central and western Kentucky and noticeably lower in the Cumberland Mountains and the Cumberland Plateau. Occurrence and average abundance were closely related to the amount of open habitat, being highest in predominantly open areas and decreasing as forest cover became predominant.

Despite the Mourning Dove's abundance, less than 22% of priority block records were for confirmed breeding. A number of active nests with eggs or young were observed, especially in residential areas, but most confirmed records were based on the observation of recently fledged young. A few others involved the observation of birds engaged in nest building and of used nests.

Mourning Doves are typically reported in large numbers on Kentucky BBS routes. The average number of individuals per BBS route for the periods 1966–91 and 1982–91 was 36.34 and 43.82, respectively. According to these data, the Mourning Dove ranked 8th in abundance on BBS routes during the period 1982–91. Trend analysis of these data reveals a nonsignificant increase of 0.8% per year for the period 1966–91 and a nonsignificant decrease of 0.1% per year for the period 1982–91.

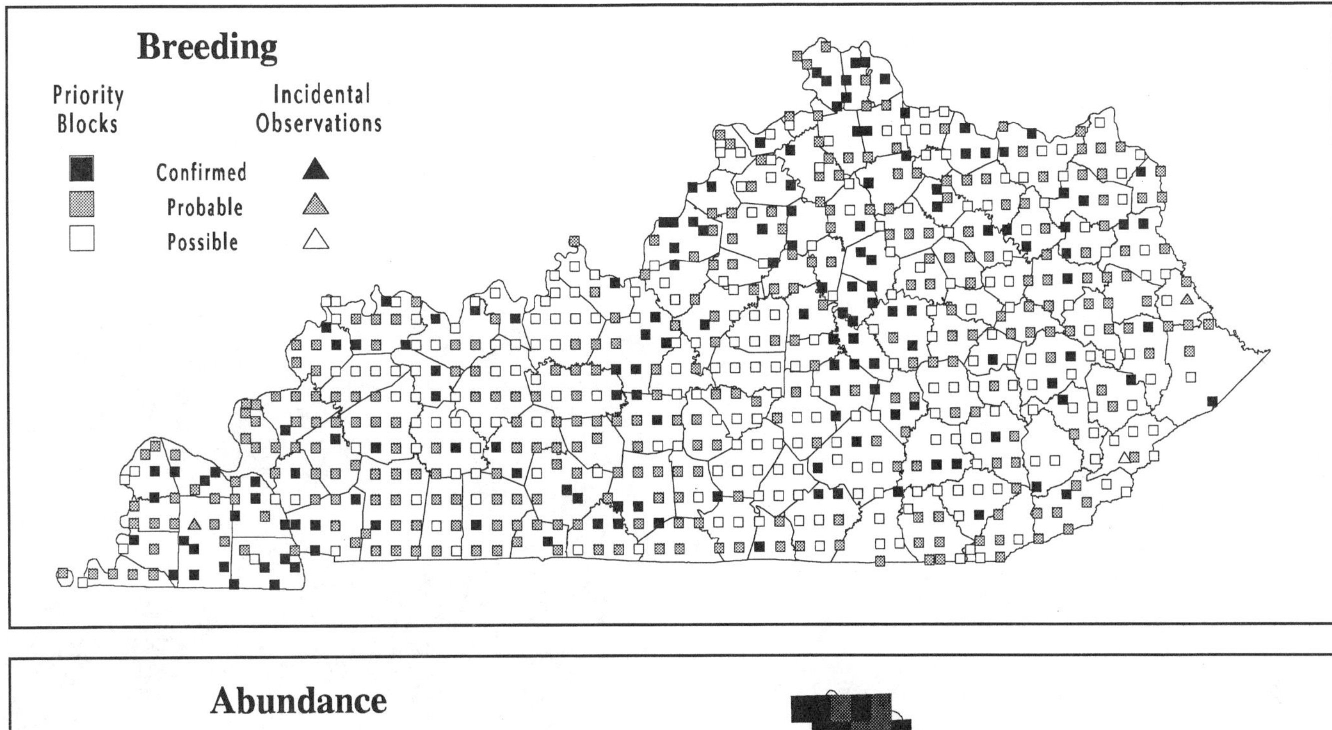

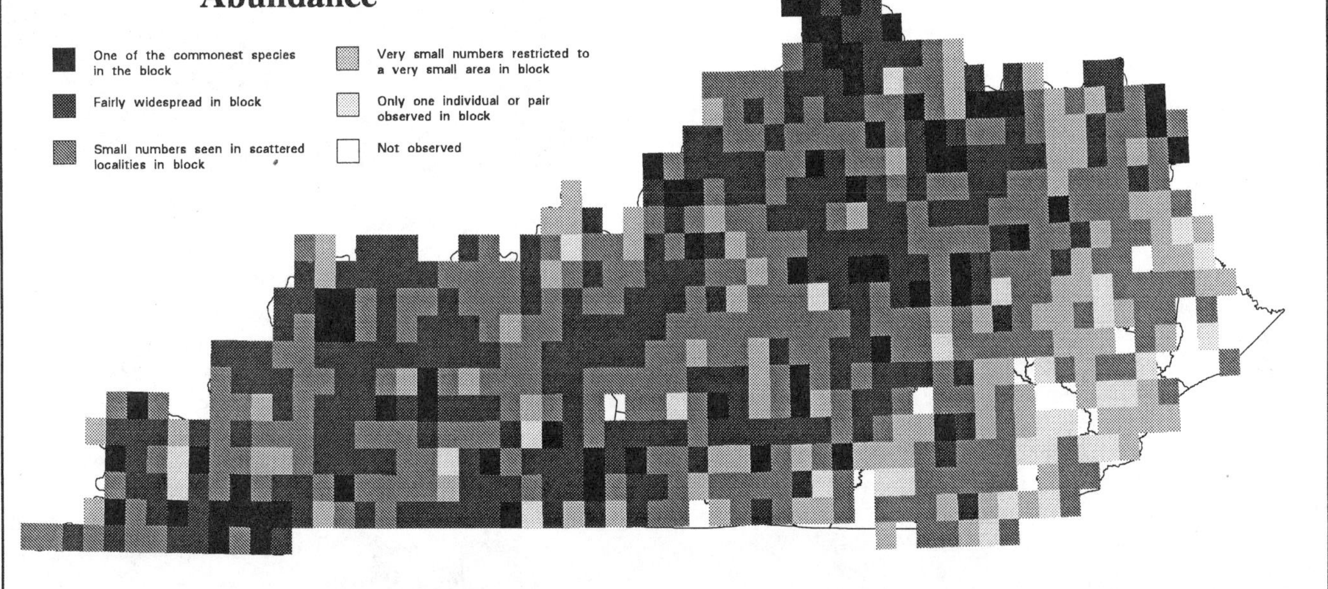

Analysis of Block Data by Physiographic Province Section

Physiographic Province Section	Total Blocks Surveyed	Blocks with Data	Avg Abund	% with Data	Section's % for State
Mississippi Alluvial Plain	14	14	3.5	100.0	2.0
East Gulf Coastal Plain	36	35	3.8	97.2	5.0
Highland Rim	139	138	3.4	99.3	19.8
Shawnee Hills	142	140	3.5	98.6	20.1
Blue Grass	204	204	3.6	100.0	29.3
Cumberland Plateau	173	148	2.7	85.5	21.3
Cumberland Mountains	19	17	2.2	89.5	2.4

Summary of Breeding Status

Number of Blocks in Which Species Was Recorded		
Total	**696**	**95.7%**
Confirmed	150	21.6%
Probable	315	45.3%
Possible	231	33.2%

Mourning Dove

Black-billed Cuckoo

Coccyzus erythropthalmus

Very little is known about the breeding status of the Black-billed Cuckoo in Kentucky. Mengel (1965) regarded the species as an uncommon summer resident throughout the Cumberland Plateau and the Cumberland Mountains, but this belief was based solely on breeding season sightings from approximately a half dozen counties of eastern Kentucky. For example, Barbour (1951) regarded this cuckoo as a common summer resident near Morehead, in Rowan County, but he included no detailed records of nesting. From areas west of the Cumberland Plateau, Mengel (1965) included only two summer records from Hopkins County.

Since the late 1950s two records of confirmed breeding have been published for Franklin County (Stamm and Croft 1968) and Lyon County (Stamm 1977; W.H. Brown, pers. comm.), and occasional summer sightings have been reported from scattered localities across the eastern two-thirds of the state. Most recently, the atlas survey yielded three more records of confirmed breeding and more than two dozen summer sightings. While the Black-billed Cuckoo's breeding status remains unclear, the body of knowledge now at hand suggests that the species occurs sporadically throughout nearly the entire state, and that a shift in the breeding population may have occurred since the 1950s.

Black-billed Cuckoos are typically encountered in successional habitats, most often along woodland borders and in old fields reverting to forest. The species also inhabits forested areas, although some degree of disturbance (i.e., natural or artificial openings) appears to be required. These cuckoos also have been reported in fairly open situations where thickets of young trees are scattered about, typically along moist drainages.

Black-billed Cuckoos winter primarily in South America (AOU 1983), and birds begin to return to Kentucky during the last week of April. Some individuals appear to arrive much later, however, and it may be sometime in June before the full complement of breeding birds has returned. Specific data on nesting activity in Kentucky are scarce, but the limited information suggests an extended breeding season. During the atlas survey, two fledglings were observed being attended by a pair of adults in Jefferson County on the surprisingly early date of May 24, 1989 (Stamm 1989b), indicating probable clutch completion immediately upon arrival in late April. Probably indicative of nesting at a more normal time are three reports, the latter two a result of the atlas fieldwork: a nest containing three young in Franklin County on May 28, 1967 (Stamm and Croft 1968); an adult observed incubating three eggs in a nest in northern Madison County on June 21, 1990; and an adult observed carrying food in Martin County on June 6, 1991. In contrast, a record of four eggs being incubated in a nest in Lyon County on September 6, 1976 (Stamm 1977), indicates that nesting at least occasionally occurs much later. Two of the three clutches reported in Kentucky contained three eggs, while the third contained four. Harrison (1975) gives rangewide clutch size as 2–4, occasionally 5.

The nest is a rather flimsy structure built of sticks and lined with dead leaves, cottony plant material, and pine needles (Harrison 1975). It is typically placed on a horizontal branch among or just inside the cover of the outer crown of foliage of a tree, shrub, or tangle of vines. While one of Kentucky's reported nests was found in a red cedar, the other two were found in relatively small deciduous trees. These nests were placed five, six to seven, and eight feet above the ground, respectively. This species occasionally lays its eggs in the nests of other birds, including the Yellow-billed Cuckoo (Harrison 1975), although such behavior has not been reported in Kentucky.

Hal H. Harrison

The atlas survey yielded records of Black-billed Cuckoos in 3% of priority blocks, and nine incidental observations were reported. Interestingly, nearly two-thirds of the priority block records came from the Blue Grass, a region from which Mengel (1965) had included only one obscure record. Moreover, occurrence in the Blue Grass was more than twice that in the Cumberland Plateau, and the species was not recorded in the Cumberland Mountains. Seasonal atlasers who covered most of eastern Kentucky were familiar with the species's call, and it is unclear why so few reports came from the only region where Mengel regarded the species as occurring regularly less than forty years ago. A possible answer may be found in results of trend analysis of regional BBS data for the period 1965–79. This analysis shows a significant increase in the number of Black-billed Cuckoos in the Highland Rim and the Blue Grass across portions of three states (Robbins et al. 1986), suggesting an expansion in nesting range may account for the results of Kentucky's atlas survey.

As noted previously, spring migrant Black-billed Cuckoos may sometimes occur well into June. In addition, the occurrence of wandering, nonbreeding birds may account for some midsummer observations within the region (B. Peterjohn, pers. comm.). All but 5 of the 31 atlas records were based on single-date observations, so some possible records may have pertained to birds that were not nesting locally. Nonetheless, all records based on the observation of a bird in suitable breeding habitat were included in the final data set.

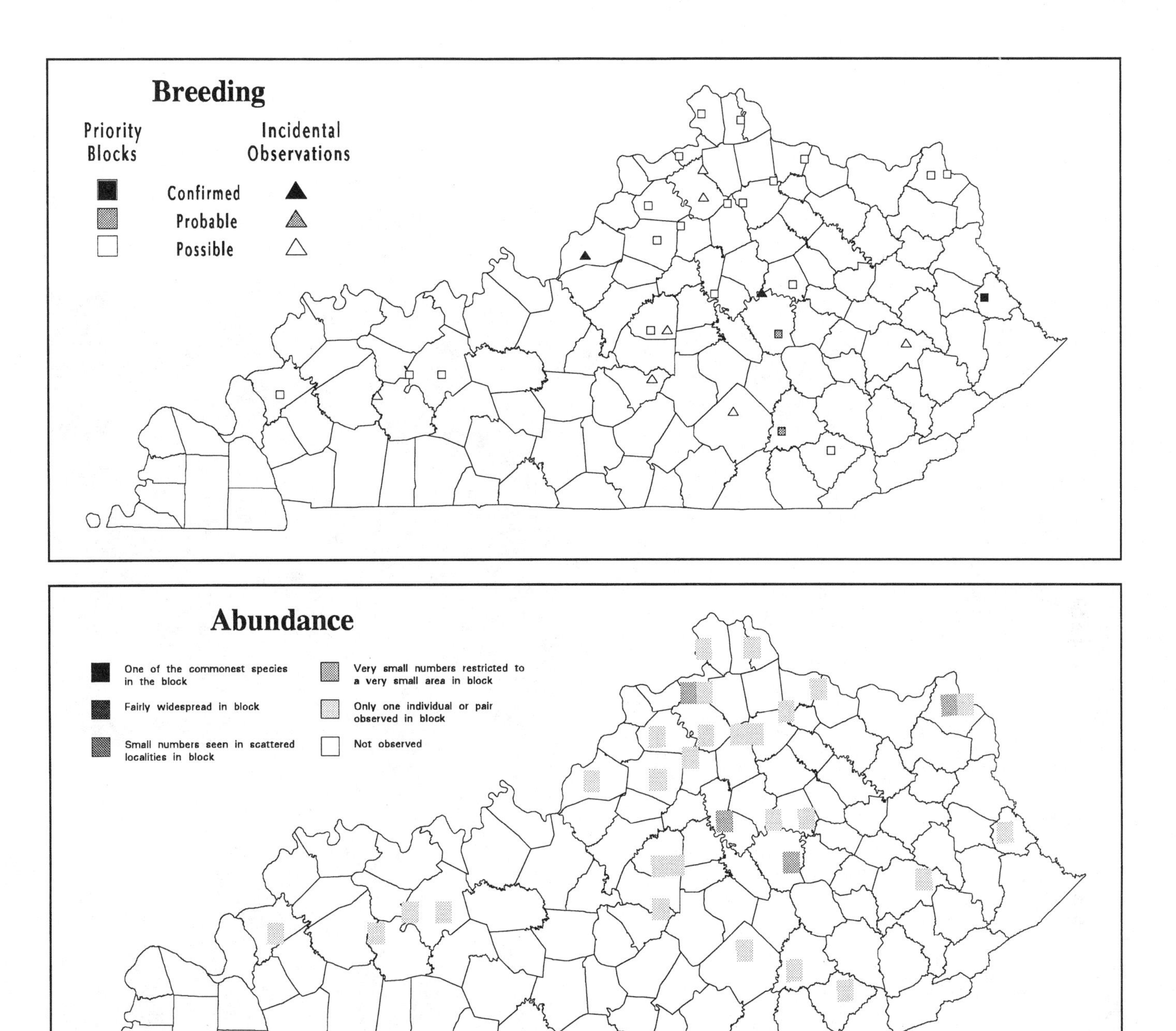

Analysis of Block Data by Physiographic Province Section

Physiographic Province Section	Total Blocks Surveyed	Blocks with Data	% with Data	Section's % for State
Mississippi Alluvial Plain	14	-	-	-
East Gulf Coastal Plain	36	-	-	-
Highland Rim	139	-	-	-
Shawnee Hills	142	3	2.1	13.6
Blue Grass	204	14	6.9	63.6
Cumberland Plateau	173	5	2.9	22.7
Cumberland Mountains	19	-	-	-

Summary of Breeding Status

Number of Blocks in Which Species Was Recorded		
Total	**22**	**3.0%**
Confirmed	1	4.5%
Probable	2	9.1%
Possible	19	86.4%

Black-billed Cuckoo

Yellow-billed Cuckoo

Coccyzus americanus

The distinctive notes of the Yellow-billed Cuckoo are a fairly common summer sound throughout much of Kentucky. Mengel (1965) regarded the species as a fairly common to common summer resident of forested areas statewide.

Yellow-billed Cuckoos are encountered in a variety of habitats, from open woodland and forest edge to extensively cleared areas with only scattered patches of trees. The species seems to be most numerous in semi-open to open situations where thickets of young trees are intermixed with small patches or corridors of woodland. In contrast, these cuckoos seem to avoid extensive tracts of closed-canopy forest. Although Yellow-billeds can be found in naturally open woodland and along natural forest margins, they are typically observed in artificially created habitats today.

Human alteration of the landscape has likely affected the Yellow-billed Cuckoo in a variety of ways. The clearing of native open-canopy forest types for agricultural use and settlement has probably eliminated some optimal habitat, especially in central and western Kentucky. In contrast, the fragmentation of extensive tracts of closed-canopy forest has likely created suitable habitat in many areas that were formerly unoccupied.

Yellow-billed Cuckoos winter in South America (AOU 1983), and most birds reappear in Kentucky during the first half of May. The species's migratory season is relatively prolonged, and it may be mid-June before the entire nesting population has returned. Although nests containing eggs have been observed as early as May 20 (Hays 1957), a peak in clutch completion probably occurs in early June (Mengel 1965). Nesting activity continues through late summer, and nests containing eggs have been observed into mid-September. These later nests suggest that a second brood is sometimes raised (Mengel 1965), but they may also represent renesting after earlier failure, or even first attempts of pairs arriving on the breeding grounds late in the season (Peterjohn and Rice 1991). According to Mengel (1965), the average size of 20 clutches or broods thought to be complete was 2.5 (range of 2–4); however, a clutch of 5 eggs was reported from Hopkins County in 1971 (KOS Nest Cards).

Forest Cover

Value	% of Blocks	Avg Abund
All	72.6	2.3
1	85.1	2.2
2	86.8	2.4
3	78.7	2.3
4	50.7	2.0
5	34.6	2.2

These cuckoos usually nest along a forest margin or within a corridor of trees, but not typically in canopied forest. The nest is usually placed in the dense cover of a tree, shrub, or tangle of vines. Both deciduous and coniferous vegetation are used. The nest is a flimsy, rather shallow saucer of sticks, typically constructed in a fork of small limbs or a tangle of vines. The average height of 14 nests reported by Mengel (1965) was 8.2 feet (range of 2.5–20.0 feet). This species sometimes lays its eggs in the nests of other birds, and Nelson (1981) reported the use of an active Mourning Dove nest by a Yellow-billed Cuckoo in Henderson County.

Ron Austing

The atlas fieldwork yielded records of Yellow-billed Cuckoos in nearly 73% of priority blocks statewide, and 15 incidental observations were reported. Occurrence was greater than 79% in all physiographic province sections except the Cumberland Plateau and the Cumberland Mountains. Likewise, average abundance was greater west of the Cumberland Plateau. Occurrence was highest in relatively open areas and decreased significantly as forestation increased. Apparently an important contributor to this trend is the species's scarcity in unbroken tracts of fairly mature forest, which occur much more frequently in eastern Kentucky.

Despite the Yellow-billed Cuckoo's relative abundance, less than 6% of priority block records were for confirmed breeding. Although a few active nests were located, most confirmed records were based on the observation of adults carrying nesting material or food for young, and several involved the observation of recently fledged young in the company of adults. The species also engages in a vocal distraction display when young are in danger, and this behavior was used to confirm nesting on at least three occasions.

Yellow-billed Cuckoos are reported fairly frequently on most Kentucky BBS routes. The average number of individuals reported per BBS route for the periods 1966–91 and 1982–91 was 9.20 and 8.53, respectively. According to these data, the species ranked 28th in abundance on BBS routes during the period 1982–91. Trend analysis of these data shows a significant ($p<0.1$) decrease of 1.8% per year for the period 1966–91, but a highly significant ($p<.01$) decrease of 6.5% per year for the period 1982–91. Reasons for this decline are not understood, but they may involve natural fluctuation in the breeding population.

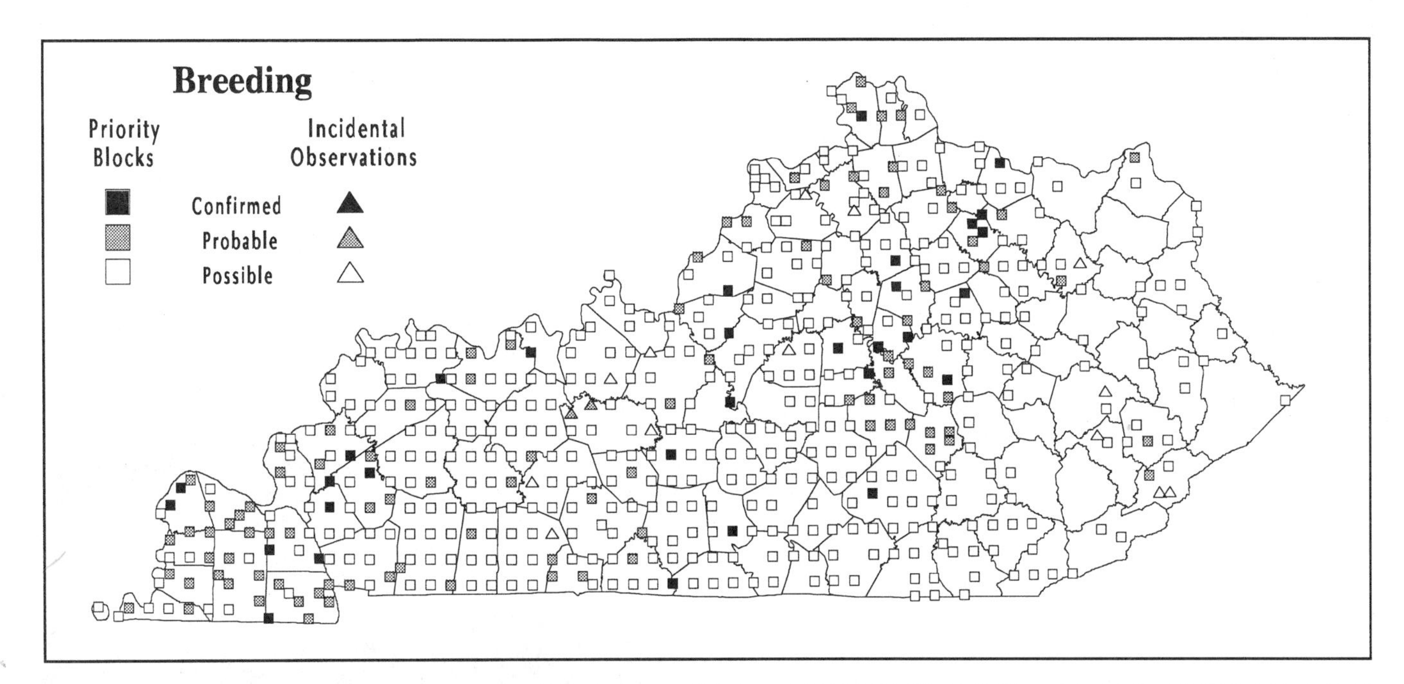

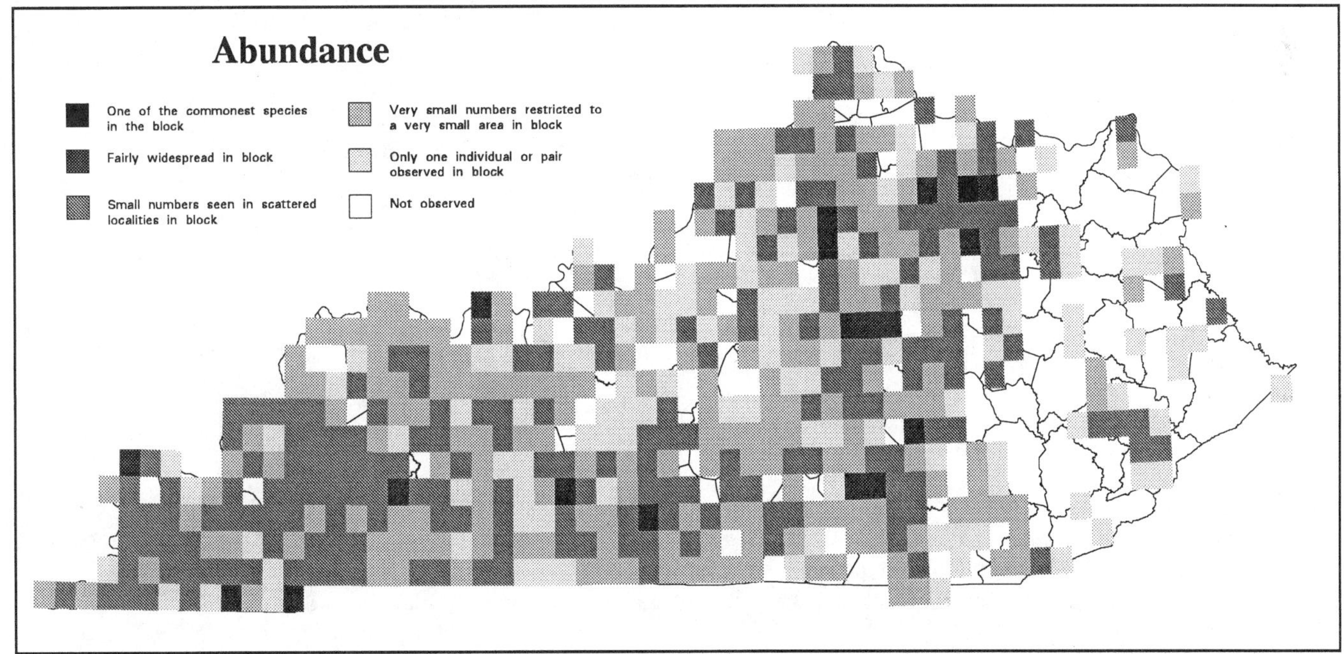

Analysis of Block Data by Physiographic Province Section

Physiographic Province Section	Total Blocks Surveyed	Blocks with Data	Avg Abund	% with Data	Section's % for State
Mississippi Alluvial Plain	14	14	2.6	100.0	2.7
East Gulf Coastal Plain	36	34	2.5	94.4	6.4
Highland Rim	139	126	2.3	90.6	23.9
Shawnee Hills	142	120	2.3	84.5	22.7
Blue Grass	204	162	2.2	79.4	30.7
Cumberland Plateau	173	66	1.9	38.2	12.5
Cumberland Mountains	19	6	1.7	31.6	1.1

Summary of Breeding Status

Number of Blocks in Which Species Was Recorded		
Total	**528**	**72.6%**
Confirmed	31	5.9%
Probable	96	18.2%
Possible	401	75.9%

Yellow-billed Cuckoo

Barn Owl

Tyto alba

The breeding status of the Barn Owl in Kentucky is poorly known. Despite an abundance of seemingly suitable habitat, the species is observed only sporadically across the state. Most recent reports have resulted from young being found accidentally, most often as a result of cutting down a large tree during the nesting season. Despite the species's present scarcity, at one time Barn Owls were regarded as much more numerous and widespread. Mengel (1965) noted the paucity of published records but stated that Barn Owls likely occurred "widely and regularly, and probably in unsuspected numbers." While this may still be true today, these secretive owls are certainly not among the state's most numerous raptors.

Barn Owls inhabit a variety of semi-open and open habitats. Mengel (1965) considered them to be most frequently found in farm country, and it is likely that rural farmland continues to harbor a persistent breeding population. Today these owls are also reported regularly from older residential areas of cities and towns, where large shade trees provide nest sites. Small numbers are even reported occasionally from larger cities.

Before the 19th century Barn Owls likely were absent or very scarce. Audubon (1861) did not note seeing the species in the state, although it is possible that small numbers occurred in the open savannas of the Blue Grass and in transitional woodlands surrounding the native prairies, where cavities in large trees would have provided nesting sites and small mammal prey was likely abundant. The species was not observed in Ohio until the late 1800s (Langdon 1879), nor in West Virginia until 1909 (Hall 1983). As humans cleared the native forests, however, these owls likely invaded areas that were formerly unsuitable, becoming fairly numerous and widespread. By the mid-1900s substantial populations were known from some parts of the Midwest (Peterjohn and Rice 1991), and significant numbers may have been present in Kentucky (Mengel 1965). More recently the species has shown declines in many areas, probably at least in part as a result of changing land use. The conversion of hayfields and pastureland to row-crop production has resulted in the loss of much suitable foraging habitat (Colvin 1985). Additional factors may include competition from other owls, an accumulation of pesticide residues in prey, and a reduction in the number of suitable nest sites.

Barn Owls are present throughout the year, although there may be some seasonal movement and migrants from farther north may sometimes be present (Mengel 1965). Some birds apparently remain in the same area throughout the year, especially if prey is abundant (Mengel 1965). Little specific information on nesting in Kentucky is available, but regional accounts illustrate that the timing of nesting may be closely related to the availability of prey (Peterjohn and Rice 1991). From the limited data available, it appears that nest sites are typically chosen in early spring and that most clutches are completed from mid-April through May. The observation of a nest containing eggs in the attic of a cabin in Trigg County on February 24, 1973, indicates that clutches are at least occasionally initiated relatively early (KOS Nest Cards). Wilson (1969) reported young in a church tower nest in Warren County on May 16, 1967, and a nest in the attic of a house in Richmond, in Madison County, contained five young on May 19, 1982 (Stamm 1982b). Evidence of later nestings is provided by a report of young in a tree cavity nest in Jefferson County in early July 1987 (Stamm 1987b) and a record of nearly fledged young in a tree cavity nest near Burlington, in Boone County, on July 16, 1987 (McNeely 1988). Family groups including recently fledged young have been found mostly during late June and July (Stevenson 1985; R. Brown, pers. comm.; author's notes). Kentucky data on clutch size are lacking, but Harrison (1975) reports that range-wide clutch size is 3–11, generally 5–7.

Barn Owls choose a variety of natural and artificial sites in which to nest. Although the species frequently uses natural cavities in large trees, many reports have come from old houses, barns, or buildings. The eggs are laid directly on the floor or on a shelf of some sort, often among remnants of cast pellets at traditional sites (Harrison 1975). The birds typically choose nest sites that lie well above ground level.

Alvin E. Staffan

The atlas survey yielded records of Barn Owls in less than 2% of priority blocks statewide, and eight incidental observations were reported. The species was not reported from the Mississippi Alluvial Plain, and the only Highland Rim records were incidental reports. It was not surprising that more records of this rather secretive, nocturnal species were not obtained during the atlas fieldwork. Only a few sightings are reported each year, and most atlasing was not accomplished in a manner that would yield new information. In fact, fewer than half of atlas reports were directly attributable to atlasing activities, most arising from miscellaneous reports generated during the atlas period.

Two priority block records and five incidental reports were for confirmed breeding. Records from Boone, Daviess, Jefferson, and McCracken Counties were based on the observation of active nests, while the Boyle, Livingston, and Marshall County records involved recently fledged young. Probable records were based on the observation of more than one bird on one occasion or on multiple observations of single birds.

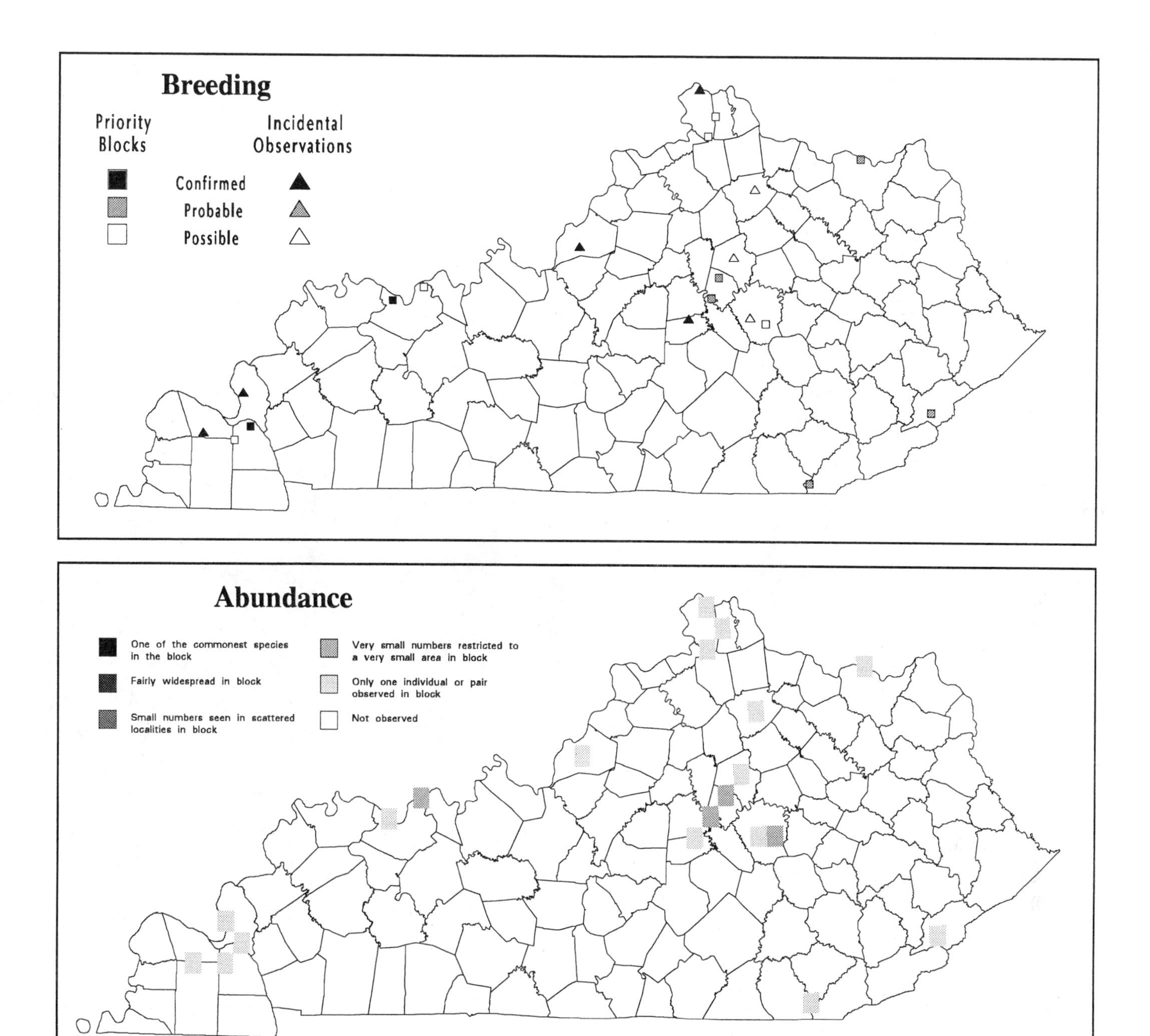

Analysis of Block Data by Physiographic Province Section

Physiographic Province Section	Total Blocks Surveyed	Blocks with Data	% with Data	Section's % for State
Mississippi Alluvial Plain	14	-	-	-
East Gulf Coastal Plain	36	2	5.6	16.7
Highland Rim	139	-	-	-
Shawnee Hills	142	2	1.4	16.7
Blue Grass	204	6	2.9	50.0
Cumberland Plateau	173	1	0.6	8.3
Cumberland Mountains	19	1	5.3	8.3

Summary of Breeding Status

Number of Blocks in Which Species Was Recorded		
Total	**12**	**1.7%**
Confirmed	2	16.7%
Probable	5	41.7%
Possible	5	41.7%

Barn Owl

Eastern Screech-Owl

Otus asio

The mellow, trilling calls of the Eastern Screech-Owl are relatively common sounds throughout most of Kentucky. Although Mengel (1965) regarded the species as a rare to uncommon resident that had been on the decrease, evidence indicates that these small owls are present in relative abundance today. It is unclear why there was a perceived decrease in screech-owls during the early to mid-1900s. As Mengel noted, most earlier reports described a rather common and widely distributed species.

Eastern Screech-Owls are encountered in a great variety of natural and altered habitats, from large tracts of mature forest to small woodlots in extensively cleared areas. They are especially common in semi-open areas of fragmented forest, intermixed with an abundance of small openings and edge. Such situations are most common in rural farmland and settlement, where screech-owls are well known by local residents. In addition, these small owls are often found in suburban parks and yards of cities and towns.

The status of screech-owls in Kentucky before settlement is not known, but they likely occurred in substantial numbers throughout a large part of the state. Although the species seems to be somewhat less numerous in extensive areas of fairly mature forest today, native forests may have offered better opportunity for nesting. At that time natural forest openings and open woodland in transition to grasslands likely provided optimum habitat. While clearing of vast areas of forest has likely caused the loss of some suitable habitat, the fragmentation of forest and consequent creation of edges and small openings have improved habitat in many areas.

Eastern Screech-Owls are permanent residents in Kentucky, although young birds often disperse widely (G. Ritchison, pers. comm.). Nesting pairs set up or reestablish territories during early spring, and most clutches are completed from late March to early April (Mengel 1965; Phillips 1973). Nesting activity typically peaks by the middle of May, and most young are probably on the wing by early June. Phillips (1973) alluded to some pairs nesting a second time, but documentation of this is lacking. Although screech-owls may be heard all year long, calling is most frequent in late summer and early fall, when young birds disperse into new areas and winter territories are established. The average size of four clutches described by Mengel (1965), was 5.0 (range of 4–6). Phillips (1973) similarly recorded an average of five eggs for six clutches found in Wood Duck boxes. According to Harrison (1975), rangewide clutch size is 2–7, generally 4–5.

These small owls usually nest in natural tree cavities, although abandoned woodpecker holes are also frequently used. In addition, artificial nest boxes are readily accepted if placed in suitable habitat, and the species has even been reported using the attic of a home (Alsop 1971). Although nest trees have been found in dense forest, they are more typically located in fairly open, parklike situations or along woodland borders. No nest is built, so the eggs are laid directly on wood chips or other debris on the floor of the cavity. Harrison (1975) gives a range in nest height of 5–30 feet.

Despite only a limited amount of survey work for nocturnal species, Eastern Screech-Owls were recorded on nearly 26% of priority blocks statewide, and 23 incidental observations were reported. Occurrence was highest in the Blue Grass and the East Gulf Coastal Plain, where volunteer atlasers surveyed a significant proportion of priority blocks. Volunteers usually devoted at least some time to surveying for nocturnal species, probably accounting in large part for the greater percentages. In contrast, screech-owls were not recorded in more than 23% of priority blocks within any region surveyed primarily by seasonal atlasers. Overall, the species seems to be less common in both extensively cleared and heavily forested areas, and more common in semi-open habitats with a good supply of mature forest.

Alvin E. Staffan

Although occurrence figures are relatively low, values for this species are much higher than those for other regularly occurring owls. In large part this is owing to the fact that individuals will respond to imitations of their calls during the day. Thus, unlike other owl species, screech-owls were regularly recorded during daytime atlasing.

Since most atlas fieldwork was conducted after young had fledged, it is possible that some records pertained to birds that had already dispersed from nesting areas. Despite this possibility, calling birds in suitable breeding habitat were believed to indicate breeding in the vicinity, and all records were included in the atlas data set. Based on the limited amount of data generated by the atlas fieldwork and general observations, it seems that screech-owls are well distributed across the state.

Nearly 10% of priority block records were for confirmed breeding. Although most of the 18 confirmed records were based on the observation of recently fledged young, at least six active nests were reported during the survey.

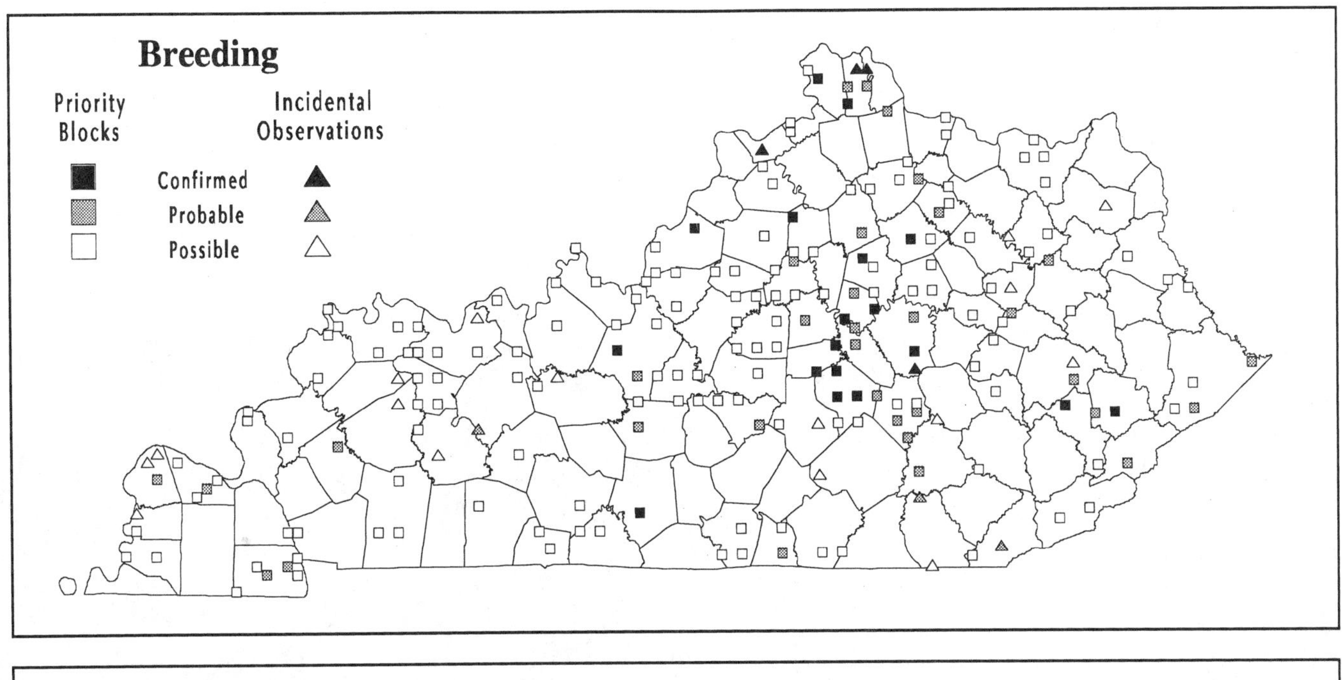

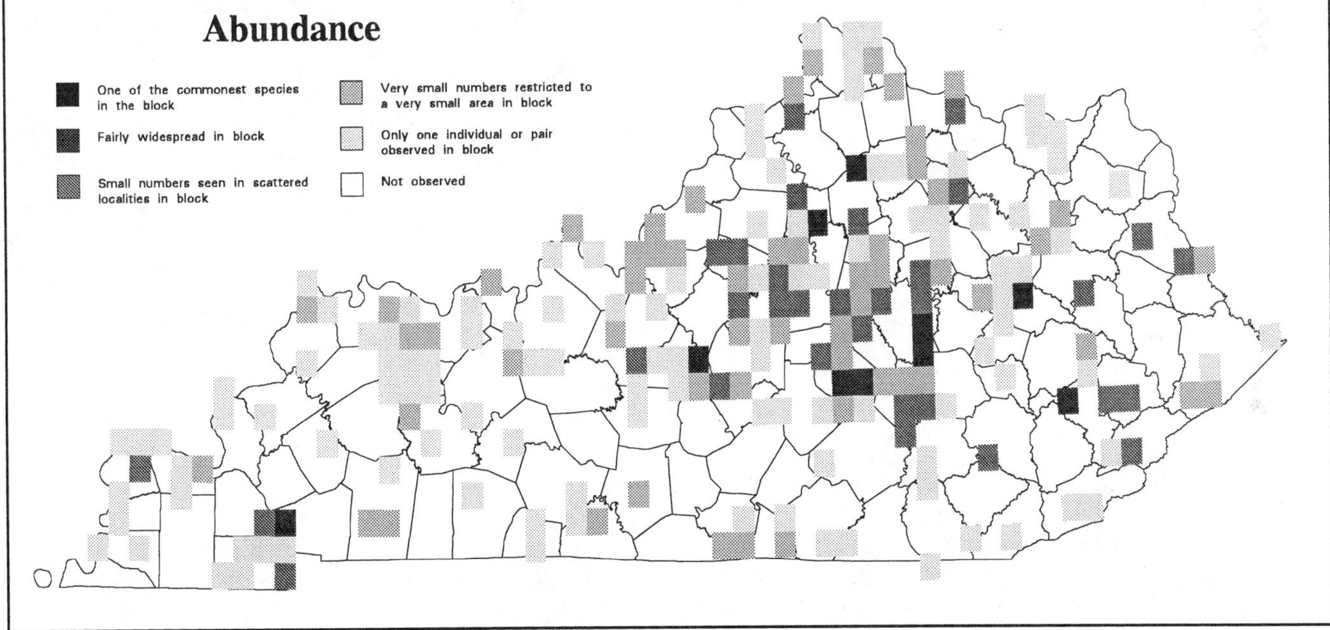

Analysis of Block Data by Physiographic Province Section

Physiographic Province Section	Total Blocks Surveyed	Blocks with Data	% with Data	Section's % for State
Mississippi Alluvial Plain	14	3	21.4	1.6
East Gulf Coastal Plain	36	12	33.3	6.4
Highland Rim	139	32	23.0	17.1
Shawnee Hills	142	32	22.5	17.1
Blue Grass	204	75	36.8	40.1
Cumberland Plateau	173	30	17.3	16.0
Cumberland Mountains	19	3	15.8	1.6

Summary of Breeding Status

Number of Blocks in Which Species Was Recorded		
Total	**187**	**25.7%**
Confirmed	18	9.6%
Probable	33	17.6%
Possible	136	72.7%

Eastern Screech-Owl

Great Horned Owl

Bubo virginianus

The Great Horned Owl is a locally uncommon to fairly common breeding bird, occurring across the entire state. Compared with other large raptors, the species is one of the most numerous, and most regions appear to harbor substantial populations.

Great Horned Owls are encountered in a wide variety of natural and altered habitats, from extensive tracts of mature forest to woodlots in predominantly cultivated areas. The species usually avoids heavily forested bottomland situations in favor of more xeric, typically upland, forest types. Mengel (1965) described the species as a bird of forested regions, being rare or absent in highly cultivated or settled areas. In contrast, current observations tend to suggest a much different pattern of abundance. Today Great Horned Owls are conspicuous in regions of open farmland, and the species occurs regularly in suburban parks and yards of cities and towns. In contrast, this owl no longer appears to be as numerous in heavily forested regions, although it certainly must be somewhat less conspicuous in such areas.

Great Horned Owls likely occurred in Kentucky in considerable numbers before altered habitats were available. Based on his assessment that these large owls were most common in forested regions, Mengel (1965) suspected that Great Horneds must have been present in considerable numbers before European settlers arrived. Audubon (1861) noted that the species was one of the most common along the shores of the Ohio and Mississippi Rivers in the early 1800s. While it is likely that the spread of agriculture and settlement has resulted in a decrease in numbers in some areas, the species's abundance in most rural situations suggests that human alteration of the landscape has not had too negative an effect overall.

Great Horned Owls are permanent residents in Kentucky, and they begin nesting earlier than all other birds. Pairs begin courtship in early winter, when their deep, resonant hooting is most frequently heard. It is not uncommon for eggs to be laid before the end of January, and a peak in clutch completion probably occurs by mid-February (Mengel 1965). By the middle of March young have hatched in most nests, and many fledglings are on the wing by the end of April. The young birds remain in contact with the adults using a loud, screechlike call that may be heard well into summer. Although Kentucky data on clutch size are limited, most pairs probably lay two eggs.

Great Horned Owls nest in many situations, from mature forest to isolated woodlots and wooded corridors in otherwise predominantly cleared areas. While they often use the abandoned nests of hawks or crows, these large owls also frequently use natural tree cavities. Protected recesses of structures including barns and bridges are also occasionally used, as are ledges of natural or artificial cliffs. Most nest sites are relatively high, ranging from 15 to 60 feet or more above the ground.

The atlas survey yielded records of Great Horned Owls in about 15% of priority blocks statewide, and 42 incidental observations were reported. Occurrence was highest in the East Gulf Coastal Plain and the Blue Grass, where volunteer atlasers covered a much higher proportion of priority blocks. Volunteers typically were able to devote at least some time to surveying for nocturnal species. In contrast, the species was largely missed in regions surveyed primarily by seasonal atlasers, whose schedules did not typically allow for nocturnal survey work. Although most trends in occurrence are masked by this bias, the species's abundance in the Blue Grass contrasts sharply with Mengel's (1965) statement that as of the mid- to late 1950s this owl was nearly absent in highly cultivated portions of the region. The reasons for this change in distribution are unclear, but they may include a shift in habitat preference, a reduction in human persecution, or an adaptation to more open habitat types.

Ron Austing

Almost 23% of priority block records were for confirmed breeding. About half of the 25 confirmed records were based on the observation of recently fledged young, but at least six active nests were found in priority blocks. A number of incidental records were based on the observations of nests as well. In several cases, begging calls were used as the basis for FL codes, even as late as mid-summer. While some of these records may have involved young birds that were very close to nest sites, others (especially late in the season) may have involved individuals that were quite a distance from where they were hatched. Despite this fact, all records of confirmed breeding based on the FL code were believed to indicate breeding somewhere in the vicinity, and all were included in the atlas data set. Probable records were typically based on the presence of calling birds over an extended period, but some involved the observation of a pair.

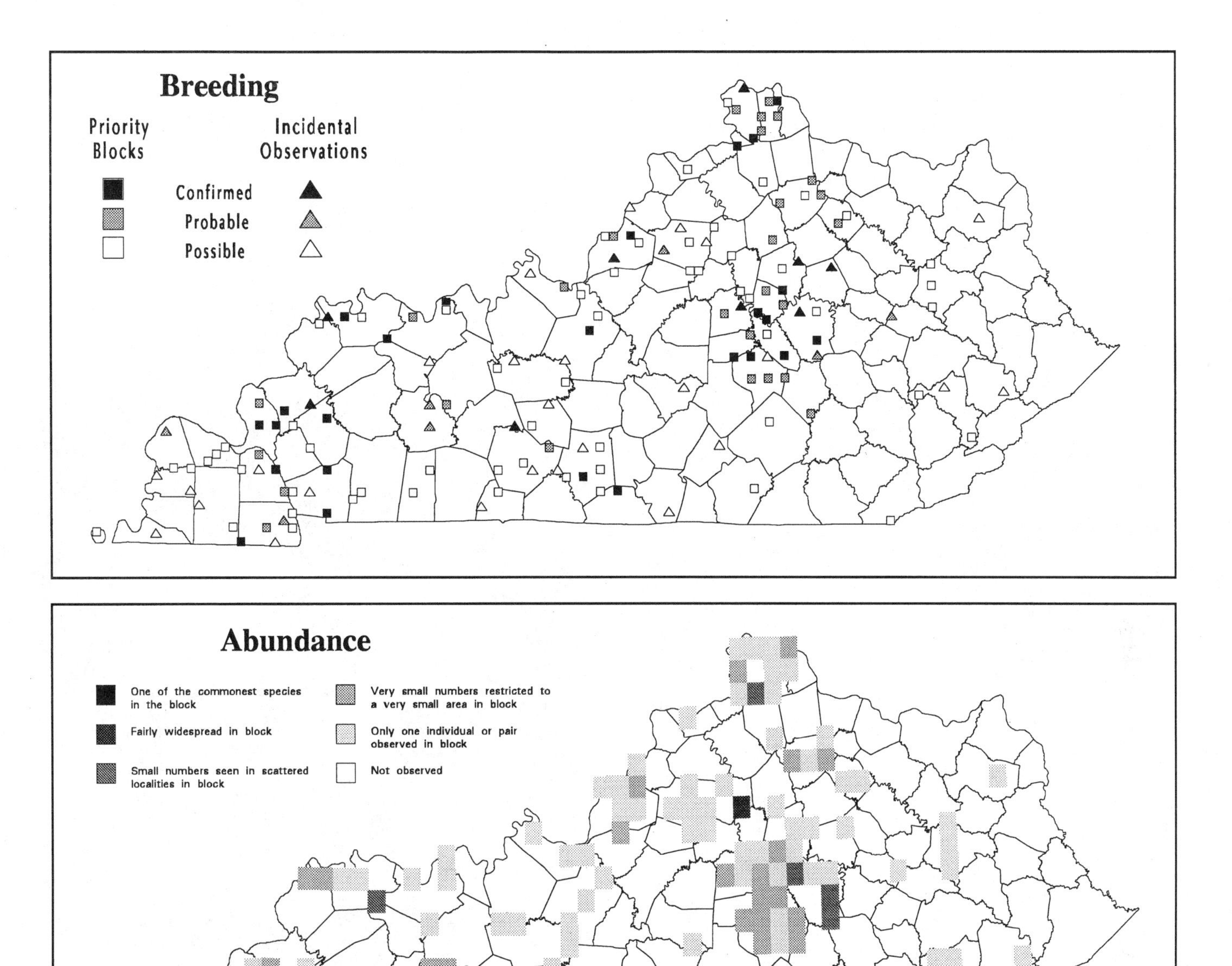

Analysis of Block Data by Physiographic Province Section

Physiographic Province Section	Total Blocks Surveyed	Blocks with Data	% with Data	Section's % for State
Mississippi Alluvial Plain	14	3	21.4	2.7
East Gulf Coastal Plain	36	12	33.3	10.9
Highland Rim	139	26	18.7	23.6
Shawnee Hills	142	16	11.3	14.5
Blue Grass	204	45	22.1	40.9
Cumberland Plateau	173	7	4.0	6.4
Cumberland Mountains	19	1	5.3	0.9

Summary of Breeding Status

Number of Blocks in Which Species Was Recorded		
Total	**110**	**15.1%**
Confirmed	25	22.7%
Probable	27	24.5%
Possible	58	52.7%

Great Horned Owl

Barred Owl

Strix varia

The distinctive hoots of the Barred Owl are a fairly common sound throughout much of Kentucky. The species is somewhat locally distributed in the forests of the Cumberland Plateau and the Cumberland Mountains, and relatively infrequent in cleared portions of central Kentucky. In contrast, it is common and conspicuous throughout the western third of the state, especially in lowland situations (Mengel 1965).

Barred Owls occur in a variety of forested and semi-open habitats, preferring mesic situations, including moist ravines, riparian corridors, bottomland hardwood forests, and the margins of floodplain swamps and sloughs. These owls probably reach their maximum breeding density in lowland forests of the western third of Kentucky. In contrast, they typically avoid drier forests of uplands, where Great Horned Owls seem to predominate.

It is likely that Barred Owls were more numerous and widespread in Kentucky two centuries ago. The fragmentation of vast forested areas, however, may have benefited Barred Owls to some extent by creating edge habitat for foraging. Mengel (1965), for example, suggested that the species had moved into areas abandoned by Great Horned Owls as settlement increased. While Barred Owls do inhabit areas of dissected forest and moderate settlement, clearing of native forests probably has been extensive enough in many areas to reduce numbers substantially. Lowland habitats typically have been affected to a greater degree than uplands and slopes, and most of the river and stream floodplains throughout the state have been cleared for agricultural purposes. Fortunately, it appears that Barred Owls have been able to sustain their numbers in remaining forest, inhabiting fragmented tracts and corridors of mesic forest.

Barred Owls are permanent residents in Kentucky. Although they do not initiate nesting as early as do Great Horned Owls, many territories are probably held throughout the year, and calling becomes more frequent in late winter. According to Mengel (1965), some clutches may be laid in February, although most are probably completed in March. By late April or early May young are on the wing, and both juveniles and family groups are regularly observed through June. Mengel (1965) gave a range in typical clutch size of two to four.

Barred Owls nest most frequently in fairly mature to mature tracts of forest, although they sometimes choose a nest site within a settled area or along woodland edge. Although natural tree cavities are typically chosen, these owls occasionally use abandoned hawk or crow nests. The eggs are laid directly on the debris of the nest cavity or old nest, although a few sprigs of pine may be added to open nests (Harrison 1975).

The atlas fieldwork yielded records of Barred Owls in 15% of priority blocks, and 33 incidental observations were reported. Occurrence was highest in the Mississippi Alluvial Plain and the East Gulf Coastal Plain, and decreased eastward. The species was not recorded in the Cumberland Mountains, although it likely occurs there in suitable habitat. Like other nocturnal species, Barred Owls were not surveyed adequately by the atlas fieldwork, and the distribution of records was influenced to some degree by observer coverage. Volunteer atlasers were able to devote greater effort to surveying for nocturnal species, yielding comparatively more records than seasonal atlasers. From the records accumulated, it is clear that the species is distributed throughout the state, although it occurs decidedly more often in the western third. In addition, results tend to suggest a more widespread occurrence on the Cumberland Plateau than indicated by Mengel (1965). Croft (1969) found the species to be more common than expected in southeastern Kentucky, and it is likely that the atlas results are further indication of the presence of a substantial nesting population there.

Ron Austing

Less than 14% of priority block records were for confirmed breeding. No active nests were located, and most of the 15 confirmed records were based on the observation of recently fledged young. Some of these reports were based on young identified by their distinctive begging calls. While some of these young may have been relatively far from the actual nest site, it was thought that observation of begging young indicated breeding somewhere in the vicinity, and the reports were included as confirmed in the final data set. Probable records were based on the observation or calls of evidently paired birds, or detection of the species in a certain area on more than one occasion.

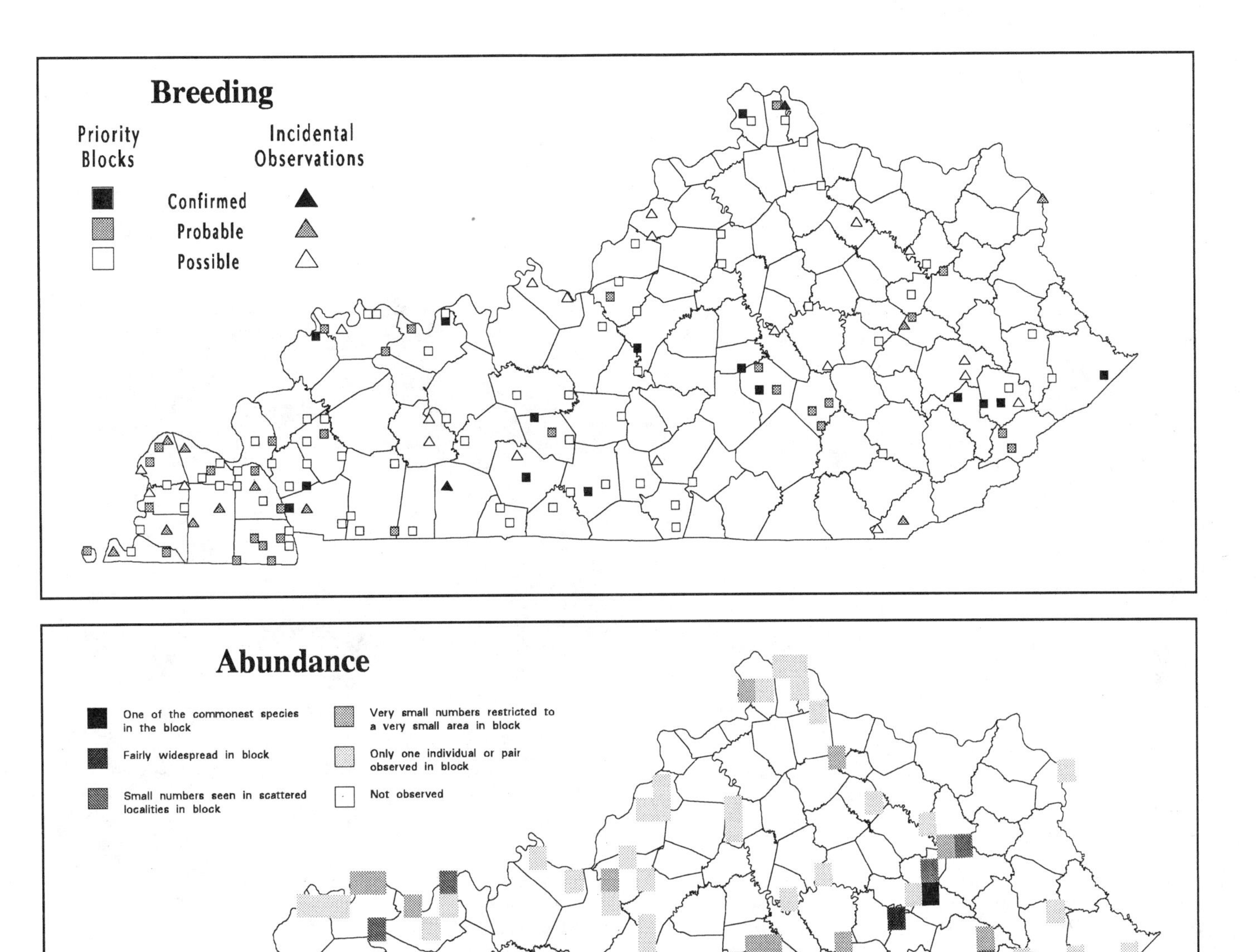

Analysis of Block Data by Physiographic Province Section

Physiographic Province Section	Total Blocks Surveyed	Blocks with Data	% with Data	Section's % for State
Mississippi Alluvial Plain	14	8	57.1	7.3
East Gulf Coastal Plain	36	17	47.2	15.6
Highland Rim	139	26	18.7	23.9
Shawnee Hills	142	23	16.2	21.1
Blue Grass	204	19	9.3	17.4
Cumberland Plateau	173	16	9.2	14.7
Cumberland Mountains	19	-	-	-

Summary of Breeding Status

Number of Blocks in Which Species Was Recorded		
Total	**109**	**15.0%**
Confirmed	15	13.8%
Probable	31	28.4%
Possible	63	57.8%

Barred Owl

Short-eared Owl

Asio flammeus

Until very recently the Short-eared Owl was known only as a rare to uncommon transient and winter resident in Kentucky. Although the species turned up in some part of the state nearly every winter, nesting was not considered a possibility until the late 1980s, when a substantial wintering population was discovered in Ohio County (Eaden and Eaden 1988).

During the mid-1970s small numbers of Short-eared Owls were found nesting on grassy, reclaimed surface mines in southwestern Indiana (Keller et al. 1986; J. Campbell, pers. comm.). When similar habitat in the Shawnee Hills of Kentucky was investigated in the spring of 1989, Short-eared Owls were discovered nesting in small numbers there as well. On May 18, 1989, a nest containing five young was located in southern Ohio County (Stamm and Clay 1989), and on May 25 a nest containing six young was found in eastern Muhlenberg County (Palmer-Ball and Barron 1990). In addition, a family group of at least four or five young was observed at another location in southern Ohio County on June 14 (Palmer-Ball and Barron 1990).

Young in the Ohio County nest were judged to range from two or three days old to about six days old when they were first discovered on May 18 (Stamm and Clay 1989), and the young in Muhlenberg County were slightly larger when observed on May 25 (Palmer-Ball and Barron 1990). Based on a duration of incubation of about 26–28 days (Ehrlich et al. 1988), the first two clutches were apparently completed in the latter half of April. In contrast, the group of young observed in Ohio County were full grown and in full flight on June 14 (Palmer-Ball and Barron 1990). Based on a nestling period of 31–36 days (Ehrlich et al. 1988), these young probably hatched by early May, meaning that their clutch was likely completed a week or two earlier than the other two. The size of these broods lies within the normal range of 4–7, given by Ehrlich et al. (1988) for the species.

As far as is known, Kentucky's nesting population of Short-eared Owls is restricted to a few of the more extensive, recently reclaimed surface mines of Ohio and Muhlenberg Counties. Little fieldwork has been undertaken away from a few specific localities, however, and it is possible that the species is more widespread than currently believed. Although the extent to which Short-eareds nest on the mines is unclear, their numbers probably depend on the abundance of small mammals. Because small mammal populations often exhibit dramatic cyclical fluctuations (Getz et al. 1987), it is possible that the owls only nest on the mines when prey populations are at or near their greatest densities. If this is true, the birds may not linger to nest in some years. It is also possible that nesting owls move about from year to year, searching out nest sites where local mammal populations are especially high.

Short-eared Owls inhabit reclaimed mines that have been restored to natural contour and reseeded to grasses and forbs. In most areas, trees are also planted during reclamation, and the owls probably use the mines for a limited number of years before woody growth begins to predominate. For this reason, their distribution is (and may always be) somewhat local, shifting as prime habitat becomes available.

All of the observations of nesting Short-eared Owls described above were made during the atlas period, and most were generated by atlas fieldwork. It is unclear how long the species has been breeding on reclaimed mines in the area, but a small nesting population possibly has been established for a number of years.

Ron Austing

Although these recent observations represent the first documented nesting records for Kentucky, Wilson (1923) reported without detail that the Short-eared Owl was resident in Calloway County. Even at the time of his account, remnant prairie had been greatly reduced in western Kentucky, but much of the East Gulf Coastal Plain, including a large portion of Calloway County, was covered by native prairie before settlement. Because of the species's association with grassland habitats, it must be presumed that Wilson's account, if accurate, pertained to Short-eared Owls using such habitat. Moreover, a relatively recent report of Short-eared Owls on a reclaimed mine in Breathitt County on the Cumberland Plateau (Allaire et al. 1982) suggests breeding, especially since the date of observation was April 8.

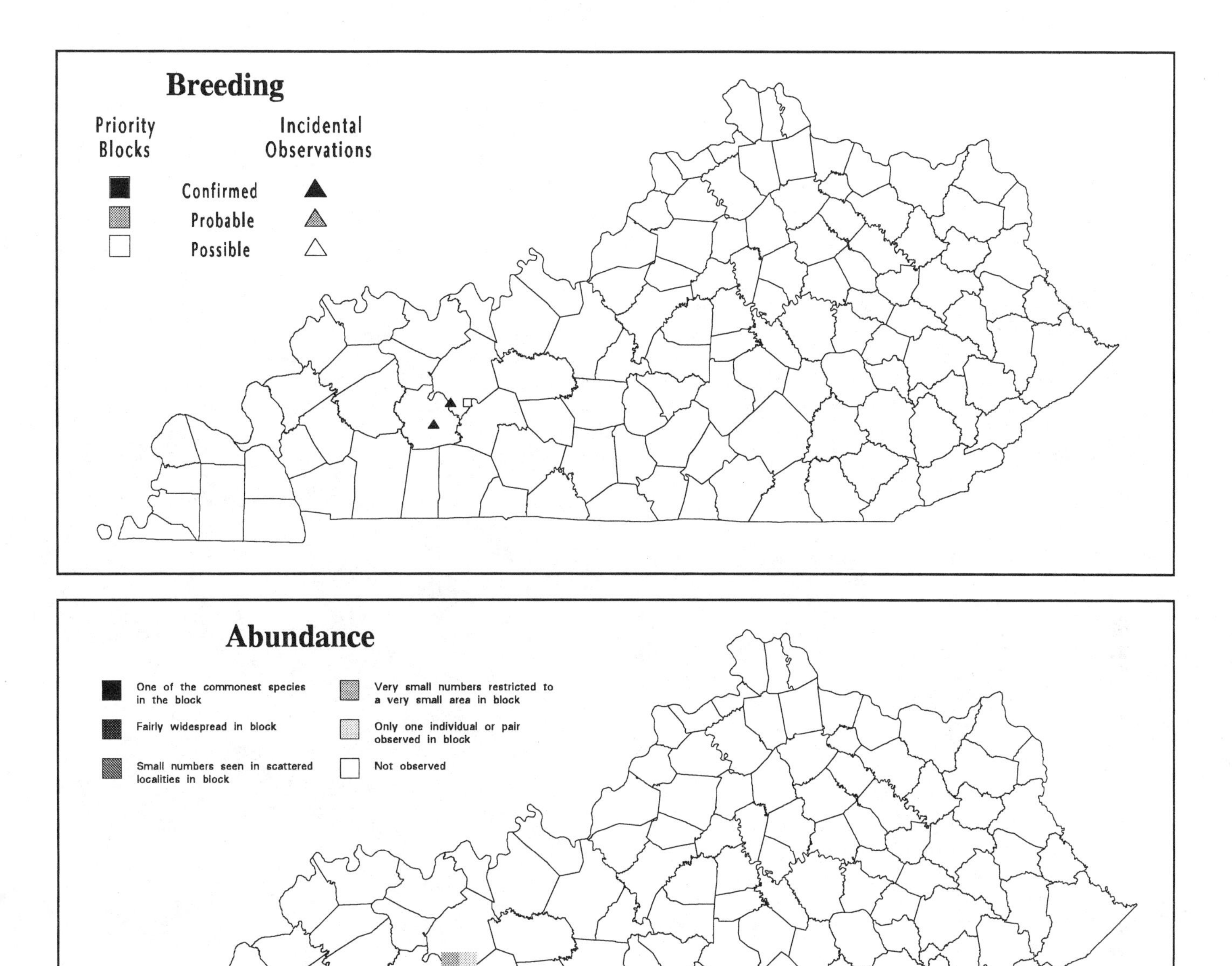

Analysis of Block Data by Physiographic Province Section

Physiographic Province Section	Total Blocks Surveyed	Blocks with Data	% with Data	Section's % for State
Mississippi Alluvial Plain	14	-	-	-
East Gulf Coastal Plain	36	-	-	-
Highland Rim	139	-	-	-
Shawnee Hills	142	1	0.7	100.0
Blue Grass	204	-	-	-
Cumberland Plateau	173	-	-	-
Cumberland Mountains	19	-	-	-

Summary of Breeding Status

Number of Blocks in Which Species Was Recorded		
Total	**1**	**0.1%**
Confirmed	-	-
Probable	-	-
Possible	1	100.0%

Short-eared Owl

Common Nighthawk

Chordeiles minor

Although widely known by both sight and call notes, the Common Nighthawk is very locally distributed across Kentucky as a breeding bird. In fact, while it is numerous in most urban situations, the nighthawk is rare to absent elsewhere. Mengel (1965) regarded the species as an uncommon to common summer resident in suitable habitat statewide.

Although nighthawks use natural habitats extensively in some parts of their nesting range, they are birds of altered situations in Kentucky. Most breeding birds are found in and about cities and towns, but nighthawks are also occasionally reported in other artificial habitats, including abandoned roadbeds, railroad rights-of-way, and reclaimed surface mines. Mengel (1965) suggested that small numbers might nest in open farmland. It is possible that a few individuals nest in naturally occurring openings, but such cases have gone unreported.

Even though nighthawks are locally distributed in Kentucky today, they were certainly much less common and even more locally distributed before the creation of altered habitats. Before the arrival of European settlers, the species would have been restricted to nesting in rocky glades and sparsely vegetated grasslands. Such habitats were probably very scarce and limited to parts of central and western Kentucky. The conversion of vast areas of forest to settlement has resulted in the creation of a relative abundance of suitable habitat, especially where flat rooftops are available as nest sites.

Common Nighthawks winter throughout South America (AOU 1983), and they return to Kentucky in large numbers by the first week of May. Males begin their unique courtship displays immediately upon arrival, but the full complement of nesting birds may not return until the middle of May, especially in the northern part of the state. Based on a limited number of records, it appears that egg laying may commence by the first week of May in western Kentucky (Larson 1970), although clutches are regularly completed into mid-July (Mengel 1965). Nighthawks studied by Larson (1970) on the Murray State University campus regularly raised two broods, with peaks in clutch completion occurring from early to mid-May and again in mid- to late June. Some late nestings must represent renesting following earlier failure because of severe weather or predation. By late July most young from second clutches have taken to the wing, and nesting activities have been completed by mid-August. During late August and early September large numbers of fall migrants pass through the region on their way south.

Like other goatsuckers, nighthawks do not build a nest but rather lay eggs directly on an open, flat surface of gravel, rock, or dirt. Throughout Kentucky the most frequently used substrates are the gravel rooftops of buildings. The typical clutch contains two eggs, although among 11 nests studied at Murray in 1968, Larson (1970) found one clutch containing only one egg. Larson also noted that the normal period of incubation was 17–18 days and that the typical period from hatching to fledging was 21–23 days.

The atlas survey yielded records of Common Nighthawks in less than 11% of priority blocks statewide, and 19 incidental observations were reported. Occurrence was highest in the Blue Grass and the East Gulf Coastal Plain, and lowest in the Cumberland Mountains and the Cumberland Plateau. In general, distribution was closely tied to the presence of urban situations, although a number of reports originated from industrial buildings and schools in less heavily settled areas. The species was likely missed in some blocks, especially in those regions surveyed primarily by seasonal atlasers. Occurrence was highest in the Blue Grass and East Gulf Coastal Plain in part because a higher proportion of blocks in those regions were completed by volunteers. A small number of sightings recorded as observed—such as observations of flyover birds in seemingly unsuitable breeding habitat—were omitted from the data set and are not included on the maps.

Mabel Slack

Only about 10% of priority block records were for confirmed breeding. One active nest was located near Danville, in Boyle County, but most of the eight confirmed records were based on the observation of recently fledged young. While it is possible that some confirmed records based on the FL code pertained to individuals that were some distance from the actual nest site, it was believed that such observations indicated nesting somewhere in the vicinity, and all were included in the final data set.

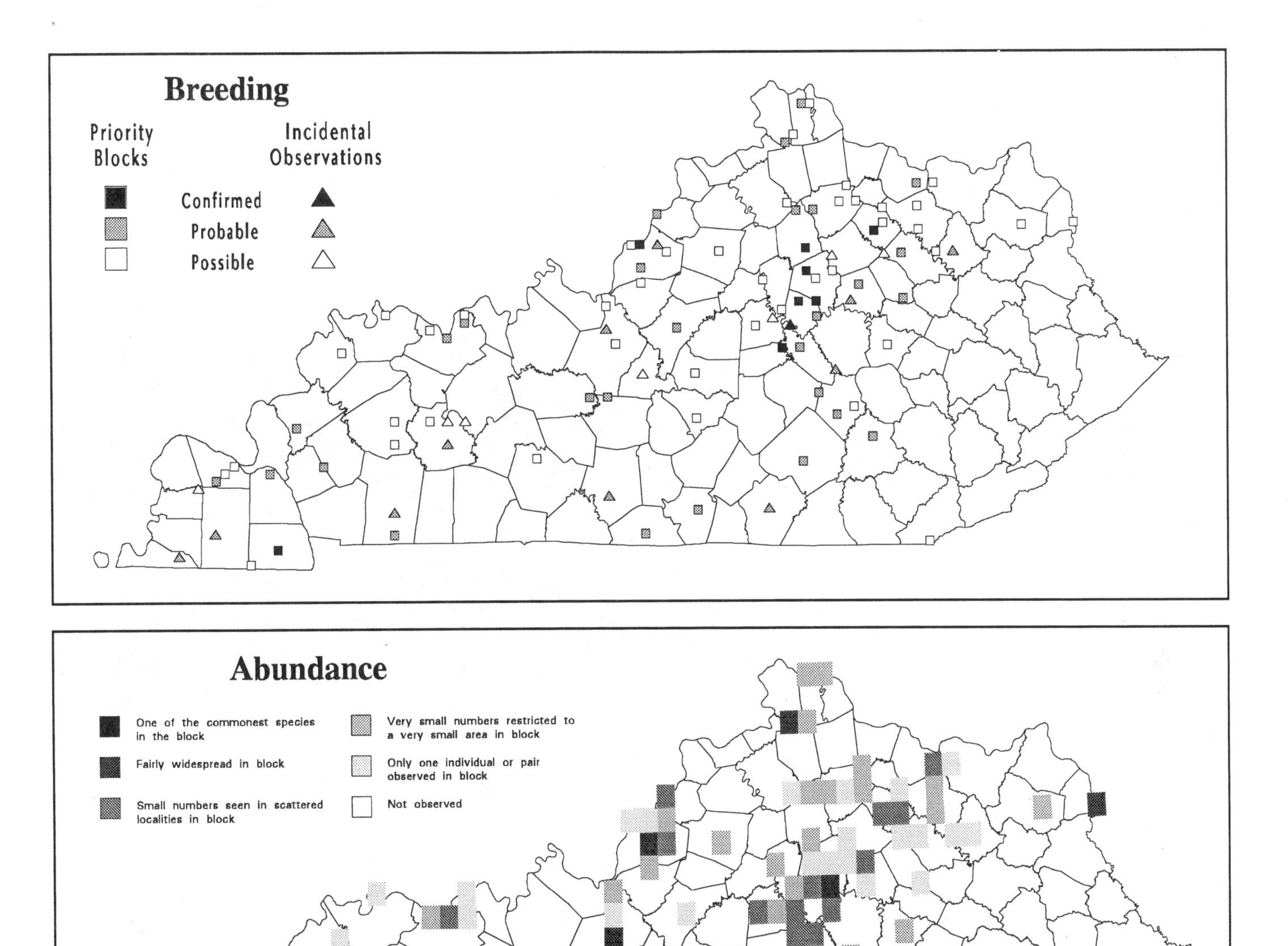

Analysis of Block Data by Physiographic Province Section

Physiographic Province Section	Total Blocks Surveyed	Blocks with Data	% with Data	Section's % for State
Mississippi Alluvial Plain	14	1	7.1	1.3
East Gulf Coastal Plain	36	5	13.9	6.5
Highland Rim	139	8	5.8	10.4
Shawnee Hills	142	13	9.2	16.9
Blue Grass	204	44	21.6	57.1
Cumberland Plateau	173	5	2.9	6.5
Cumberland Mountains	19	1	5.3	1.3

Summary of Breeding Status

Number of Blocks in Which Species Was Recorded		
Total	**77**	**10.6%**
Confirmed	8	10.4%
Probable	28	36.4%
Possible	41	53.2%

Common Nighthawk

Chuck-will's-widow

Caprimulgus carolinensis

The Chuck-will's-widow is a variably rare to common summer resident west of the Cumberland Plateau (Mengel 1965). This large nightjar is nowhere common, but in the East Gulf Coastal Plain and portions of the Highland Rim and Shawnee Hills it usually can be found in suitable habitat. Throughout the rest of central and western Kentucky, Chuck-will's-widows occur locally, often being absent from seemingly suitable habitat. The species becomes even more scarce to the north and east, and it is absent across most of the Cumberland Plateau and the Cumberland Mountains.

Chuck-will's-widows are encountered in semi-open and open habitats with scattered tracts of forest. In contrast to the Whip-poor-will, which is generally more common in areas with greater forest cover, this species is usually absent in extensively forested areas. This trend is especially apparent in the vicinity of Kentucky Lake and Lake Barkley in western Kentucky. Whip-poor-wills are very common and Chuck-will's-widows are relatively rare within the Land Between the Lakes, where fairly mature forest covers much of the rolling terrain. In contrast, in the more highly fragmented forests of eastern Calloway County, the Whip-poor-will is outnumbered by the Chuck. The Chuck-will's-widow is also much more common in drier forests where the understory and midstory levels are relatively open. Optimal sites in southern and western Kentucky are dominated by oaks and hickories, with scattered cedars or introduced pines.

It is unknown whether Chuck-will's-widows occurred in Kentucky before European settlement. There is some evidence that the species has extended its range northward in the state since the early 1900s (Schneider 1944; Croft and Stamm 1964). Audubon (1861) made no mention of the Chuck-will's-widow as of the early 1800s, although it was reported from Fulton County during the 1890s (Pindar 1925). Likewise, it was unknown in the state of Ohio until the early 1930s (Thomas 1932). The species may have inhabited the open woodlands that occurred in association with the native prairies of southern and western Kentucky, but it was likely absent from the rest of the state, especially those parts covered by more extensive blocks of mature forest. It is likely that fragmentation of forests has resulted in the creation of suitable habitat and an overall increase in the species.

Chuck-will's-widows winter primarily in the tropics of Middle America (AOU 1983), and summer residents begin returning to their Kentucky breeding grounds during the latter half of April. Kentucky data on nesting activity are quite limited, but territorial behavior commences soon after arrival, and egg laying has been documented by late April (Stamm 1969). According to Mengel (1965), most clutches are probably completed during the first half of May. Song continues regularly into mid-July, but little is known about the species's average date of departure. Mengel (1965) reported that all known Kentucky clutches have contained two eggs.

The Chuck-will's-widow typically nests in open woodland, often near an opening or along an edge. Although both deciduous and mixed forest are used, the species seems to choose a site among or near cedars or pines if these trees are present. A nest is not constructed, but the eggs are laid directly on the leaves of the forest floor. The incubating bird relies heavily on its camouflaged plumage, and the nest site is typically unprotected by overhanging vegetation.

Ron Austing

Despite the limited amount of coverage of nocturnal species, Chuck-will's-widows were reported on more than 11% of priority blocks, and 19 incidental observations were recorded. Occurrence was highest in the East Gulf Coastal Plain, where the species is considered to be as numerous as anywhere in the state. Volunteer participation was also high in this region, and the higher percentage in part reflects the greater effort at surveying nocturnal species. Occurrence in the Highland Rim and the Shawnee Hills was noticeably higher than in the Blue Grass, but still less than half the occurrence in the East Gulf Coastal Plain. Bottomland forests of the Mississippi Alluvial Plain are typically not used by the Chuck-will's-widow, explaining its rarity in that section. Not surprisingly, the species was almost entirely absent on the Cumberland Plateau, and despite its presence nearby in Tennessee and Virginia (Croft and Stamm 1964; Croft 1969; A. Barron, pers. comm.), it went unrecorded in the Cumberland Mountains.

Only three priority block records were for confirmed breeding. One involved the observation of an active nest, and the other two were based on the observation of recently fledged young and distraction displays by adults near young.

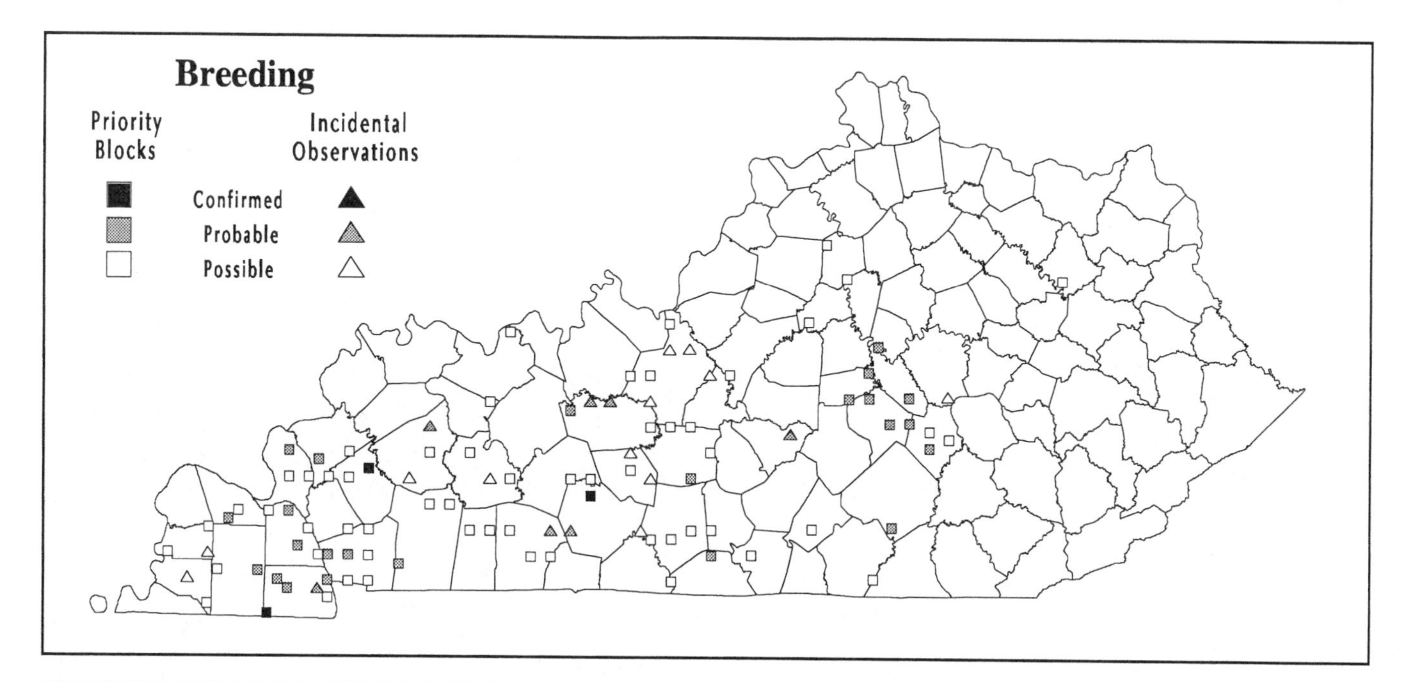

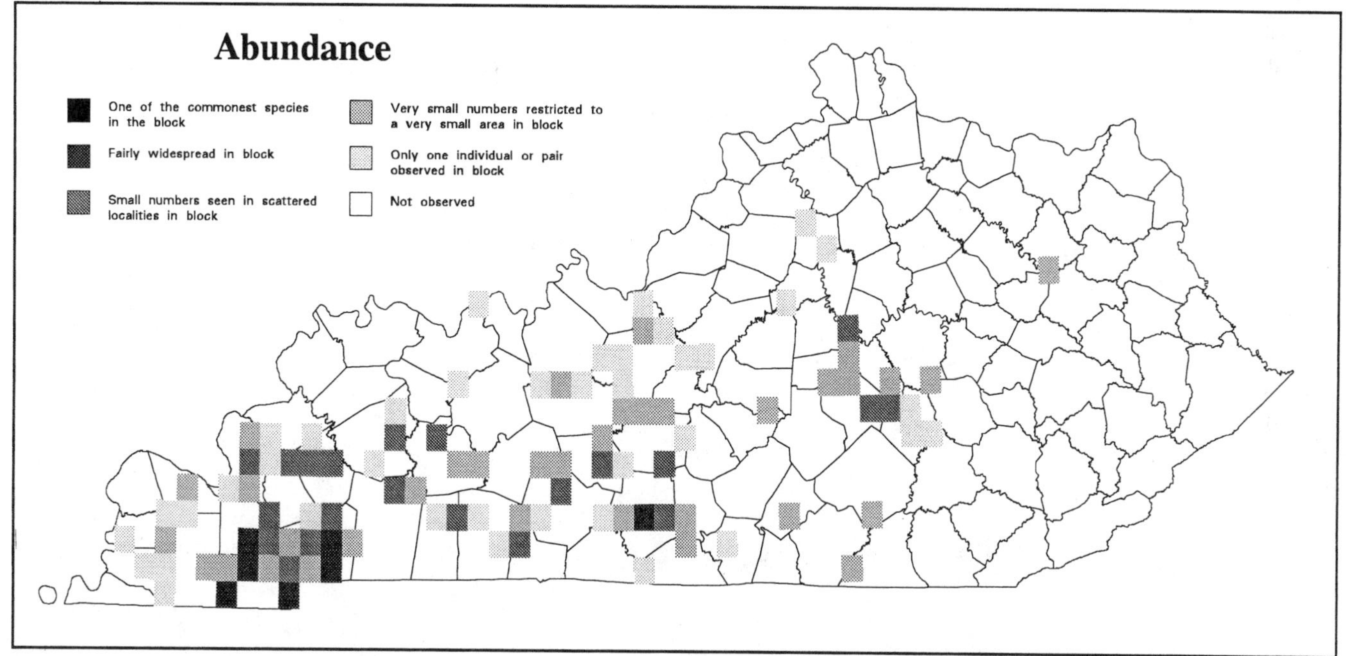

Analysis of Block Data by Physiographic Province Section

Physiographic Province Section	Total Blocks Surveyed	Blocks with Data	% with Data	Section's % for State
Mississippi Alluvial Plain	14	1	7.1	1.2
East Gulf Coastal Plain	36	16	44.4	19.3
Highland Rim	139	27	19.4	32.5
Shawnee Hills	142	24	16.9	28.9
Blue Grass	204	12	5.9	14.5
Cumberland Plateau	173	3	1.7	3.6
Cumberland Mountains	19	-	-	-

Summary of Breeding Status

Number of Blocks in Which Species Was Recorded		
Total	**83**	**11.4%**
Confirmed	3	3.6%
Probable	24	28.9%
Possible	56	67.5%

Chuck-will's-widow

Whip-poor-will

Caprimulgus vociferus

The Whip-poor-will's well-known song is a fairly common summer sound throughout a large part of Kentucky. While the species is absent from many areas that have been extensively altered by humans, it remains quite numerous across much of the state where forests remain intact or only moderately fragmented. Mengel (1965) regarded the Whip-poor-will as a variably rare to common summer resident, being most common on the Cumberland Plateau and rarest in the Jackson Purchase.

Although Whip-poor-wills are generally regarded as woodland birds, they usually do not occur in extensive areas of unbroken forest. More typically they are found in disturbed forest and forest edge habitats, where they forage in openings for insect prey. Formerly the species must have occurred in and along natural forest openings and borders, but it is primarily a bird of altered habitats today. These nightjars are found in or adjacent to a great variety of semi-open situations, including rural farmland, power line and roadway corridors, clear-cut and selectively logged forest tracts, old fields, and reclaimed surface mines. In general, they are more common in areas of greater forest cover than Chuck-will's-widows. Although the two species are sometimes heard calling side by side, Whip-poor-wills tend to occupy a greater range of habitats, from mesic slope forests to subxeric, upland forests.

Although it is not known how widespread and abundant Whip-poor-wills were in Kentucky before settlement, they may have been fairly common locally. As of the early 1800s Audubon (1861) considered the species to be plentiful in the barrens, occurring in the open grasslands rather than in areas of extensive forest. Because the Whip-poor-will is now distributed across much of the state, Audubon's observations would suggest that fragmentation of native forests has resulted in an increase in the amount of suitable nesting habitat. While this may be true, it is also likely that conversion of open woodlands to intensive agricultural use and settlement has reduced the amount of suitable habitat in some areas as well.

Whip-poor-wills winter primarily from the extreme southern United States south through Middle America (AOU 1983). Although the species has been reported occasionally before the end of March, most birds arrive on their Kentucky breeding grounds during mid-April. Territorial behavior commences soon thereafter, and according to Mengel (1965), clutches may be completed from late April to mid-June. There is no evidence that the species is double-brooded, and late nestings may represent renesting following earlier failure resulting from predation or inclement weather. Most eggs likely hatch by mid-June, and most young are probably on the wing by early July. Although song is occasionally heard into late summer, little is known about the average date of departure of nesting birds. As is typical rangewide, Kentucky clutches have all contained two eggs (Mengel 1965).

Whip-poor-wills nest on the ground, typically choosing a site near a forest margin. Like other ground nesters, they usually avoid selecting sites in lowland forest that are prone to flooding. A nest is not constructed, but rather the eggs are laid directly on the leaves of the forest floor. The incubating bird relies heavily upon its camouflaged plumage, and the nest site may be unprotected by overhanging vegetation.

Gene Boaz

Like other nocturnal species, the Whip-poor-will was poorly sampled by the atlas survey. The species was reported on more than 17% of priority blocks, and 23 incidental observations were recorded. Occurrence was highest in the East Gulf Coastal Plain, where volunteer atlasers were able to accomplish more survey work for night birds. Otherwise, analysis of the data is not possible because of the paucity of information. The species was reported in all physiographic province sections, and distribution was relatively even across the state.

Seven priority block records were for confirmed breeding. All seven reports pertained to the observation of active nests with eggs or small young.

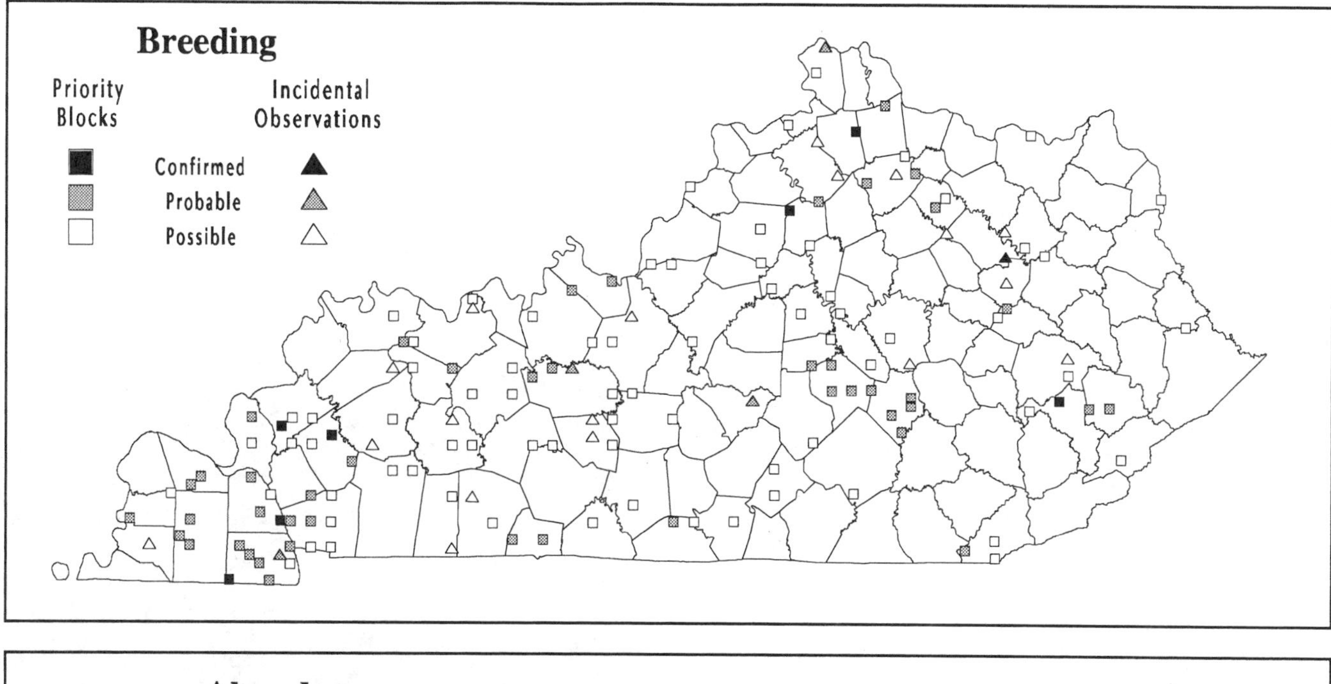

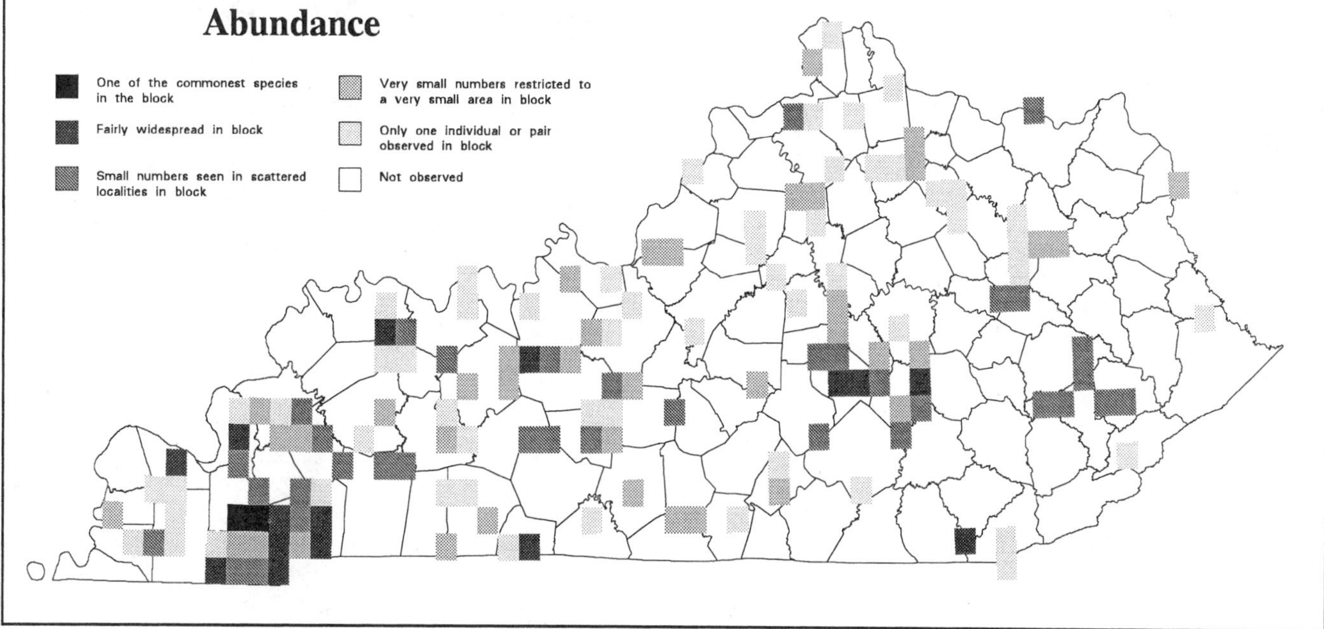

Analysis of Block Data by Physiographic Province Section

Physiographic Province Section	Total Blocks Surveyed	Blocks with Data	% with Data	Section's % for State
Mississippi Alluvial Plain	14	1	7.1	0.8
East Gulf Coastal Plain	36	17	47.2	13.6
Highland Rim	139	25	18.0	20.0
Shawnee Hills	142	34	23.9	27.2
Blue Grass	204	30	14.7	24.0
Cumberland Plateau	173	15	8.7	12.0
Cumberland Mountains	19	3	15.8	2.4

Summary of Breeding Status

Number of Blocks in Which Species Was Recorded		
Total	**125**	**17.2%**
Confirmed	7	5.6%
Probable	45	36.0%
Possible	73	58.4%

Whip-poor-will

Chimney Swift

Chaetura pelagica

Common throughout most of Kentucky in the summer, Chimney Swifts course through the sky, twittering conspicuously as they forage for flying insects. Mengel (1965) regarded the species as a common summer resident in cleared and settled areas throughout the state, but less numerous in extensively forested areas. Although the swift is broadly distributed in association with human settlement, small numbers can be encountered just about anywhere.

Swifts occur in a great variety of semi-open and open habitats. These small aerialists are common wherever suitable nest sites are available, and they inhabit both urban and suburban situations as well as rural farmland and countryside. Within the large river floodplains of far western Kentucky, swifts are regularly observed foraging far from the nearest buildings over swamps and sloughs, where they probably still nest in hollow trees.

Chimney Swifts must have been less widespread in Kentucky before settlement, when only natural nesting sites were available. By the early 1800s Audubon (1861) had already noted the widespread use of chimneys for nesting. Furthermore, he even stated that the species "earlier nested in trees in [western] Kentucky," suggesting that the use of natural sites was something of the past even then. Today, the availability of artificial nesting sites is certainly responsible for the presence of swifts in many areas where they were formerly absent or occurred very locally.

Chimney Swifts typically return to Kentucky from their wintering grounds in South America during the second and third weeks of April. They soon become numerous in the skies overhead, but the full complement of breeding birds may not return until the beginning of May. Commencement of the nesting season depends somewhat on the weather; periods of cool, rainy conditions regularly delay nesting by some weeks. Courtship and nest building may require several weeks, and it is probable that egg laying peaks during the first half of June. There is no specific evidence of the raising of two broods, and because incubation and nestling periods are relatively long, it is believed that swifts raise only one brood. The average size of six clutches reported by Mengel (1965) was 3.3 (range of 2–5), although the average size of five clutches summarized by Croft and Stamm (1967) and Stamm and Croft (1968) was 4.6 (range of 4–6).

Forest Cover

Value	% of Blocks	Avg Abund
All	91.1	2.7
1	95.0	2.9
2	95.8	2.8
3	91.7	2.7
4	86.4	2.5
5	75.0	1.9

Chimney Swifts choose a variety of sites in which to nest. Although they most frequently use chimneys, other sites are utilized, including the inside walls of silos, abandoned homes and barns, and old wells. Artificial sites are used primarily, but the species likely nests in hollow trees in some unsettled areas. Birds nesting in rugged parts of eastern Kentucky may also use crevices in clifflines. The adults snap small twigs from the ends of dead branches while in flight and carry them to the nest site. The nest itself is a saucer-shaped structure attached to a vertical surface. It is composed entirely of twigs cemented together with sticky saliva that hardens, holding the nest together. Most nests are situated well down into the cavity, where they are protected from heavy rains.

Ron Austing

The atlas fieldwork yielded records of Chimney Swifts in more than 91% of priority blocks statewide, and six incidental observations were reported. The species ranked 16th according to the number of priority block records, 23rd by total abundance, and 27th by average abundance. Occurrence was highest in the Highland Rim and the Blue Grass, although it was greater than 85% throughout all regions except the Cumberland Mountains. Average abundance varied only slightly statewide. As expected, occurrence and average abundance were highest in areas with little forest cover and lowest in areas predominantly covered in forest. Swifts range widely to feed, and some priority block observations may have pertained to birds foraging outside the block in which they actually nested.

Nearly 12% of priority block records were for confirmed breeding. Although many of the 79 confirmed records involved the observation of family groups, most were based on the discovery of active nests, including one in an old well in eastern Calloway County.

Chimney Swifts are reported in fairly substantial numbers on most Kentucky BBS routes. The average number of individuals per BBS route for the periods 1966–91 and 1982–91 was 21.02 and 28.28, respectively. According to these data, the species ranked 12th in abundance on BBS routes during the period 1982–91. Trend analysis of these data shows significant decreases of 1.3% per year ($p<.05$) for the period 1966–91 and 2.4% per year ($p<0.1$) for the period 1982–91.

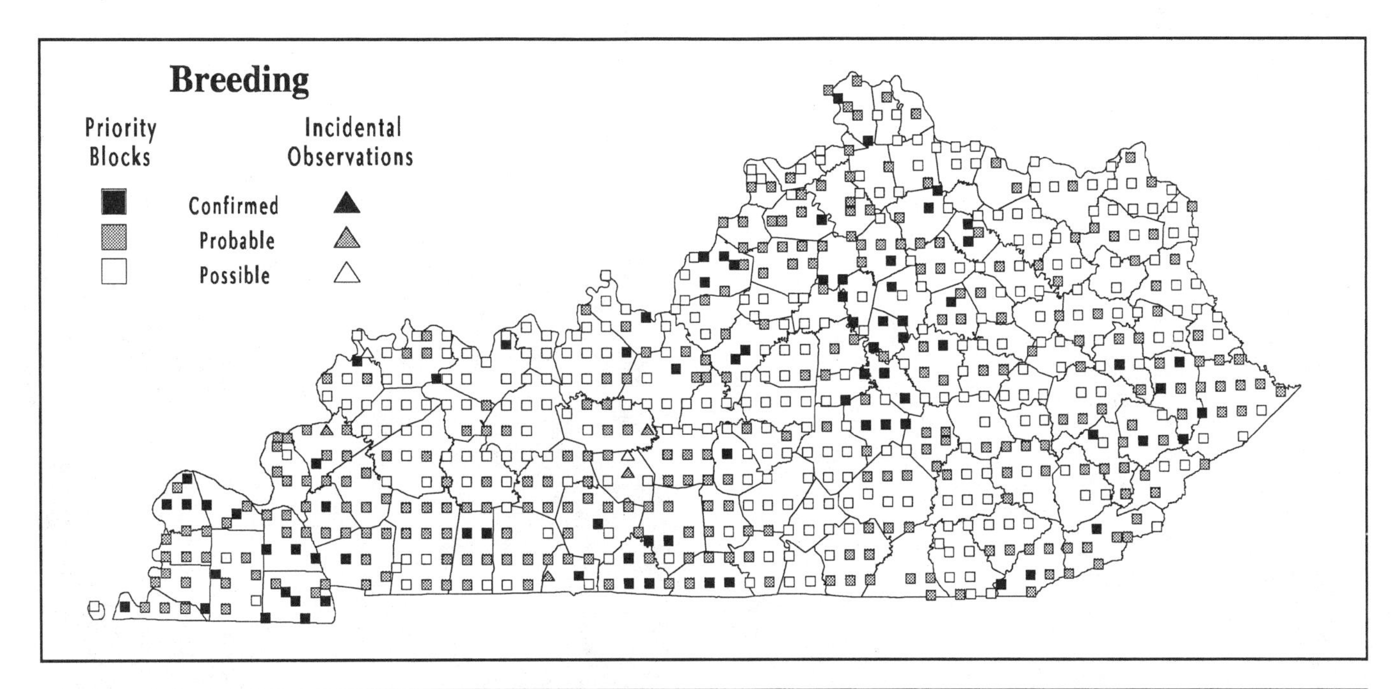

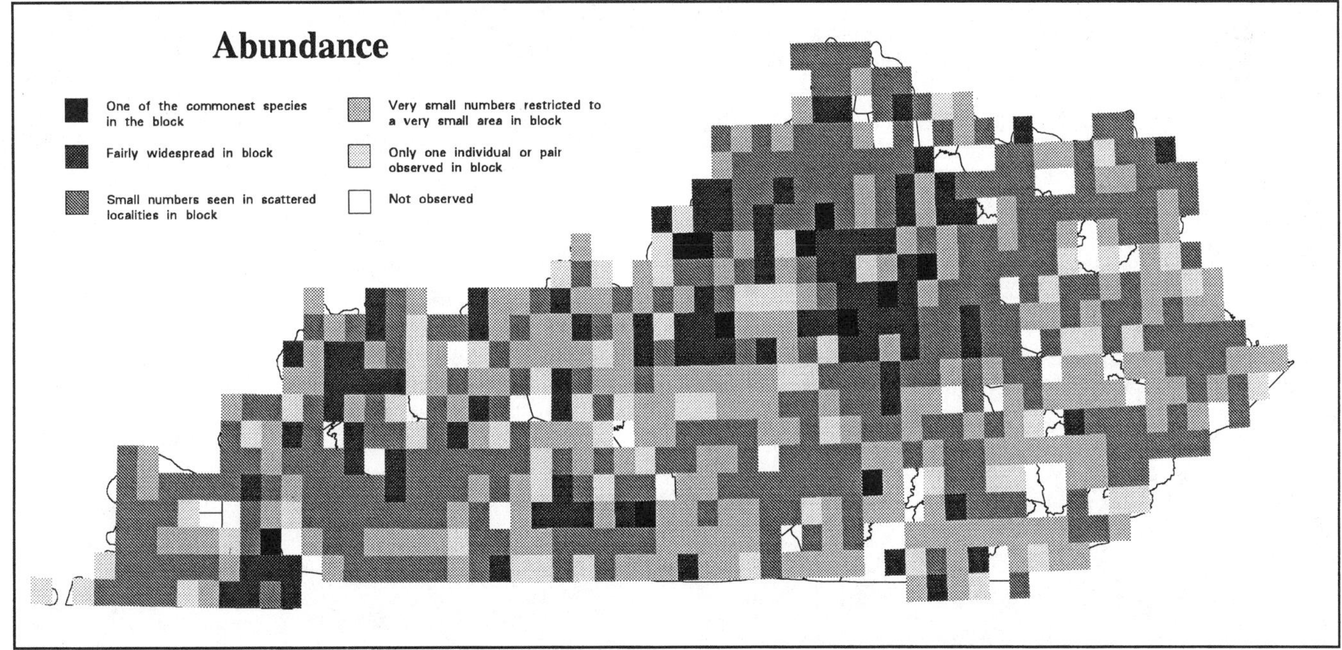

Analysis of Block Data by Physiographic Province Section

Physiographic Province Section	Total Blocks Surveyed	Blocks with Data	Avg Abund	% with Data	Section's % for State
Mississippi Alluvial Plain	14	12	2.5	85.7	1.8
East Gulf Coastal Plain	36	33	2.8	91.7	5.0
Highland Rim	139	133	2.6	95.7	20.1
Shawnee Hills	142	128	2.6	90.1	19.3
Blue Grass	204	193	3.0	94.6	29.2
Cumberland Plateau	173	149	2.4	86.1	22.5
Cumberland Mountains	19	14	2.1	73.7	2.1

Summary of Breeding Status

Number of Blocks in Which Species Was Recorded		
Total	**662**	**91.1%**
Confirmed	79	11.9%
Probable	297	44.9%
Possible	286	43.2%

Ruby-throated Hummingbird

Archilochus colubris

Although relatively inconspicuous during the nesting season, the Ruby-throated Hummingbird is a surprisingly numerous summer resident throughout much of Kentucky. Mengel (1965) regarded the species as a common summer resident statewide, occurring in a great variety of habitat types. Although these hummingbirds are frequently observed during migration and sometimes breed in settled situations, they usually nest in relatively undisturbed areas. The species is most often found nesting in fairly mature to mature forest, but woodland borders, forested riparian corridors, orchards, rural yards, and suburban parks are also used.

It is possible that Ruby-throated Hummingbirds are more widespread in Kentucky today than before European settlement. Although the clearing of native forests has resulted in the loss of much suitable nesting habitat, the fragmentation of remaining forest may have resulted in an increase in numbers across much of the state because of the creation of edge habitats. Moreover, the introduction of feeders and nonnative flowers and shrubs that bloom prolifically may have benefited the species in settled areas.

Ruby-throated Hummingbirds winter primarily in the tropics of Middle America (AOU 1983), and the first males typically return to Kentucky during the third week of April. Migrants are conspicuous through the end of May, but by the beginning of June most birds have probably returned to their local breeding grounds. Males establish territories as soon as they return, and the unique courtship displays commence as soon as the females arrive. Not long after mating, the males typically disperse to favored feeding areas, while the females engage in nest building. According to Mengel (1965), early clutches may be completed before the middle of May, but most egg laying probably does not commence until sometime in June. The presence of an adult on a nest in Franklin County on August 2, 1966, and young in a nest in Daviess County until August 18, 1968, indicate that nesting occasionally occurs much later (Croft and Stamm 1967; KOS Nest Cards). Clutches normally contain two eggs (Mengel 1965). Fledglings begin to appear during the latter part of June, and by late July many birds have dispersed from nesting areas.

Forest Cover

Value	% of Blocks	Avg Abund
All	69.6	1.8
1	54.5	1.6
2	66.3	1.7
3	70.9	1.8
4	79.9	2.0
5	78.8	1.8

The nest is typically constructed at the base of a fork in the outer branches of a tree or shrub. Ruby-throated Hummingbirds often choose a branch that angles downward toward the tip, but horizontal limbs are also used. In Kentucky all reported nests have been in deciduous trees, most often in sycamores and elms. Most nests seem to be situated over a stream or some other natural or artificial opening. A unique nest observed in Henderson County was built on top of a ripening peach (Cooper 1952).

The nest is intricately constructed of fine plant down, held together and attached to the branch with spider silk, and invariably covered with lichens. The average height of eight nests reported by Mengel (1965) was 9.9 feet (range of 3–15 feet), although nests have been observed as high as 30–40 feet (M. Stinson, pers. comm.).

Ron Austing

Ruby-throated Hummingbirds were recorded on nearly 70% of priority blocks statewide, and 20 incidental observations were reported. Occurrence was highest in the Mississippi Alluvial Plain and lowest in the Highland Rim, but relatively uniform across the state. Average abundance was also not highly variable. These hummingbirds were recorded least often in extensively cleared areas, and occurrence was highest in predominantly forested areas with some openings. Owing to early dispersal of males, it is likely that some atlas records pertained to individuals that were not breeding in the block in which they were observed. Despite this possibility, it is believed that most atlas observations indicated breeding somewhere in the vicinity, and all records were included in the final data set.

Nearly 10% of priority block records were for confirmed breeding. A majority of the 50 confirmed reports involved the observation of recently fledged young. A few atlasers broadly applied this classification, to include birds observed in juvenile plumage. While the FL code was intended to be interpreted more conservatively, most use of the FL code to confirm observations was probably applied in situations that afforded good nesting habitat, and all such reports were included as confirmed in the final data set. Other confirmed reports were based on the observation of nest building, active nests, used nests, and females with visible brood patches.

Hummingbirds are typically reported in very small numbers on Kentucky BBS routes. The average number of individuals per BBS route for the periods 1966–91 and 1982–91 was 0.59 and 0.85, respectively. Trend analysis of these data reveals slight, nonsignificant increases of 0.9% per year for the period 1966–91 and 1.4% per year for the period 1982–91.

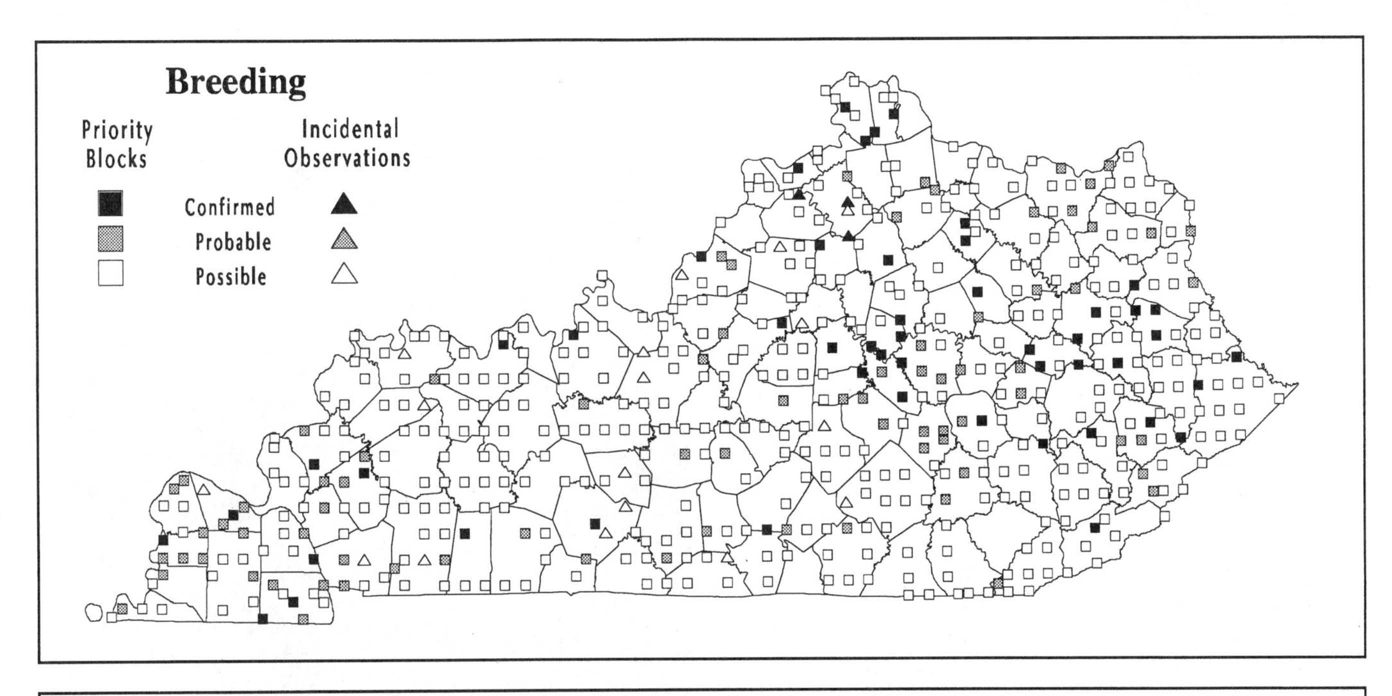

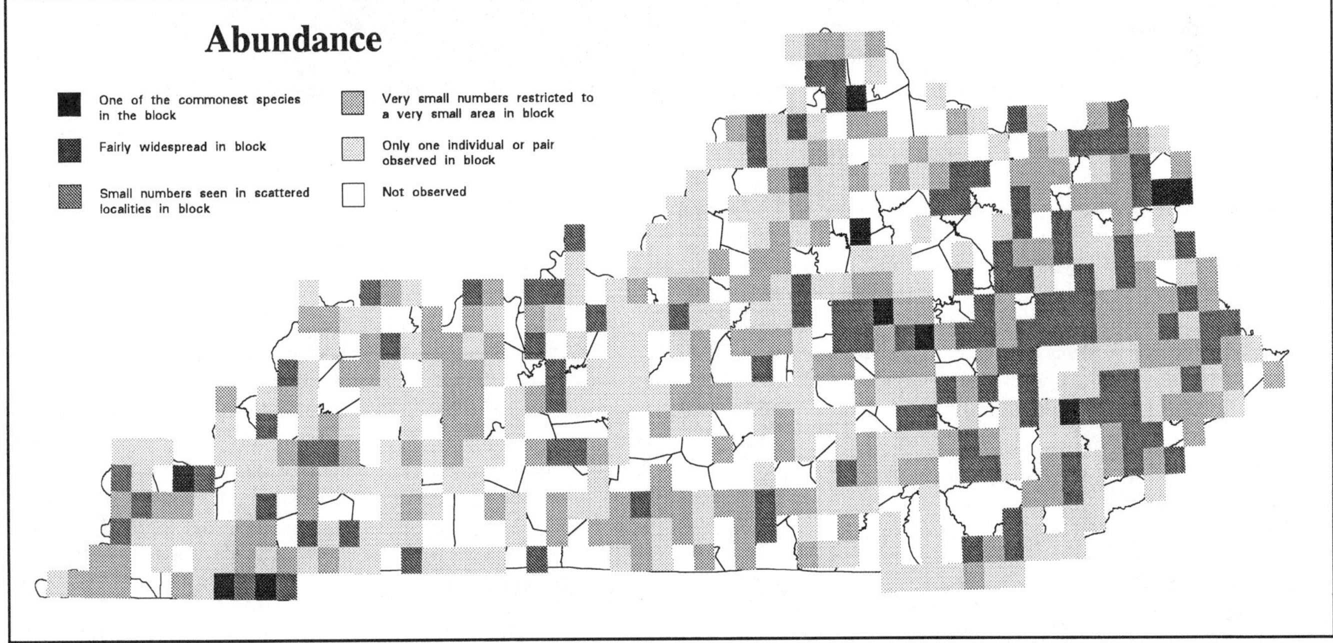

Analysis of Block Data by Physiographic Province Section

Physiographic Province Section	Total Blocks Surveyed	Blocks with Data	Avg Abund	% with Data	Section's % for State
Mississippi Alluvial Plain	14	12	2.2	85.7	2.4
East Gulf Coastal Plain	36	28	1.8	77.8	5.5
Highland Rim	139	84	1.6	60.4	16.6
Shawnee Hills	142	93	1.6	65.5	18.4
Blue Grass	204	136	1.9	66.7	26.9
Cumberland Plateau	173	138	2.0	79.8	27.3
Cumberland Mountains	19	15	1.7	78.9	3.0

Summary of Breeding Status

Number of Blocks in Which Species Was Recorded		
Total	**506**	**69.6%**
Confirmed	50	9.9%
Probable	78	15.4%
Possible	378	74.7%

Ruby-throated Hummingbird

Belted Kingfisher

Ceryle alcyon

The rattling notes of the Belted Kingfisher are a familiar sound throughout most of Kentucky. This conspicuous bird probably occurs at one time or another on just about every sizable body of water across the state. Mengel (1965) regarded the species as common in central and western Kentucky, but decreasing eastward.

Ron Austing

Belted Kingfishers are encountered in a great variety of habitats, from small creeks and farm ponds to the margins of larger rivers and reservoirs. Their occurrence is limited by the availability of suitable nest sites during the breeding season, and for that reason they are sometimes absent from areas of extensive forest as well as the larger river floodplains in summer.

It is possible that kingfishers were not as widespread in Kentucky before artificial habitats were available. The species certainly must have been distributed widely, nesting in natural banks along rivers and streams across the state. Audubon (1861) noted kingfishers using ponds for foraging by the early 1800s, and the creation of an abundance of reservoirs has certainly provided much additional foraging habitat in more recent times. Furthermore, road cuts and other vertical banks created by humans now provide opportunities for nesting in areas where formerly the species was likely excluded because of the absence of suitable nest sites.

Although Belted Kingfishers are permanent residents in Kentucky, many birds move farther south in winter, especially during periods of severe weather. Typically in late March and early April, noisy altercations between territorial males signify the commencement of nesting activity. The reuse of traditional burrows allows some pairs to initiate nesting relatively early, and according to Mengel (1965), clutches are sometimes completed before the beginning of April. If nest building must be undertaken, clutches may be completed later in the season, and Mengel notes that egg laying continues into mid-May. Most young appear in June, although a few family groups are usually apparent into July. Mengel (1965) gave the average size of six clutches and broods as 6.2 (range of 5–7).

Forest Cover

Value	% of Blocks	Avg Abund
All	49.9	1.5
1	42.6	1.7
2	54.2	1.4
3	46.1	1.5
4	56.5	1.6
5	46.2	1.5

Kingfishers excavate a nesting burrow in a vertical bank of dirt or sand, often digging a tunnel more than six feet in length that ends in an enlarged chamber (Harrison 1975). A variety of natural and artificial sites are chosen, including river and stream banks, road cuts, and the walls of sand and gravel pits. The burrow is seldom built in a bank that is less than 8–10 feet high.

Belted Kingfishers were recorded on nearly 50% of priority blocks statewide, and 45 incidental observations were reported. Occurrence was highest in the Mississippi Alluvial Plain and lowest in the Shawnee Hills, although it was relatively consistent throughout the rest of the state (47–58%). The high occurrence in the Mississippi Alluvial Plain is somewhat surprising, because nest sites are scarce within the river's floodplain. Nesting kingfishers may travel a considerable distance to forage for food, and it is likely that birds observed on floodplain sloughs and creeks were actually nesting in adjacent uplands. Average abundance was low across the state. Both occurrence and average abundance were not highly variable relative to the percentage of forest cover, indicating that the species's presence was more closely related to the abundance of suitable foraging areas and nest sites. As noted, because kingfishers sometimes range widely from nest sites for food, some priority block records may have pertained to individuals not nesting in the immediate vicinity. Unfortunately, it was impossible to determine the origin of most birds, and all reports of kingfishers were incorporated into the final data set.

Less than 7% of priority block reports were for confirmed breeding. Most of the 25 confirmed records involved the observation of recently fledged young and adults carrying food for young, although a few occupied nests were recorded. In addition, formerly used nesting burrows were observed on a few blocks.

Kingfishers are typically reported in very small numbers on Kentucky BBS routes. The average number of individuals per BBS route for the periods 1966–91 and 1982–91 was 0.61 and 0.54, respectively. Trend analysis of these data reveals a significant ($p<.01$) decrease of 2.1% per year for the period 1966–91, but a nonsignificant decrease of 1.9% per year for the period 1982–91.

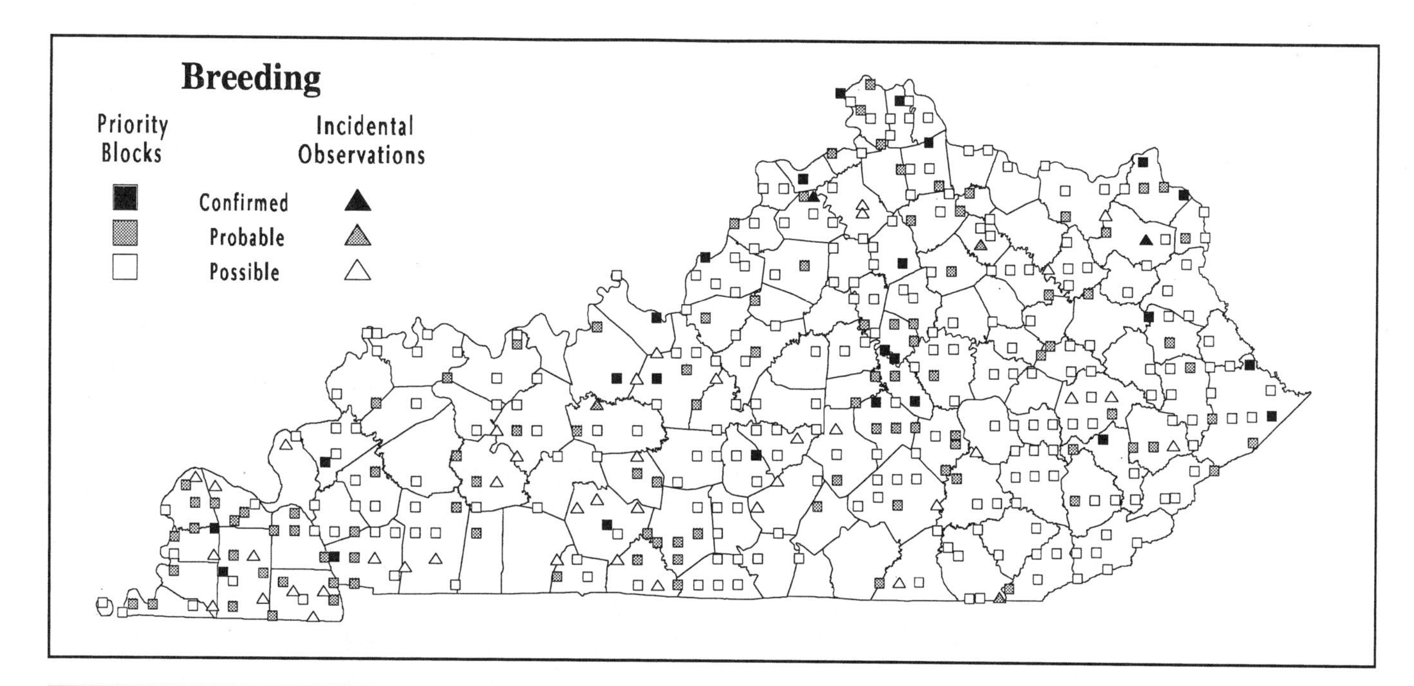

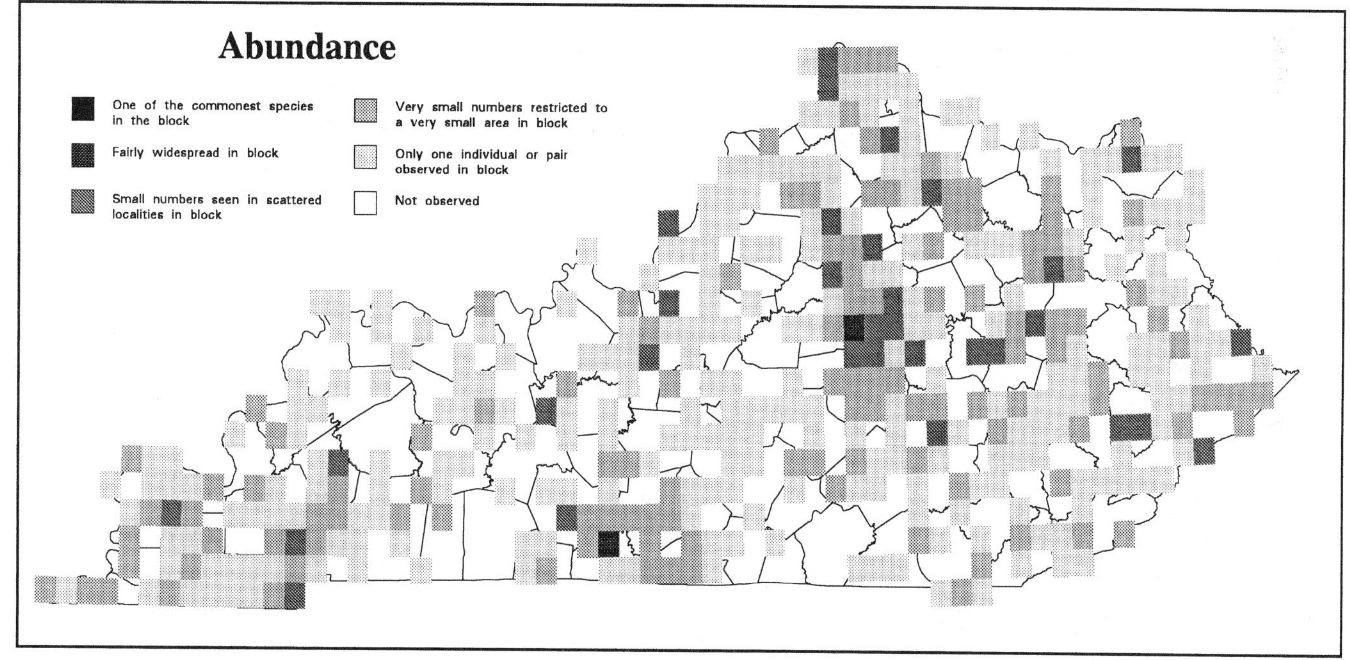

Analysis of Block Data by Physiographic Province Section

Physiographic Province Section	Total Blocks Surveyed	Blocks with Data	Avg Abund	% with Data	Section's % for State
Mississippi Alluvial Plain	14	11	1.7	78.6	3.0
East Gulf Coastal Plain	36	21	1.4	58.3	5.8
Highland Rim	139	66	1.5	47.5	18.2
Shawnee Hills	142	48	1.3	33.8	13.2
Blue Grass	204	118	1.6	57.8	32.5
Cumberland Plateau	173	90	1.5	52.0	24.8
Cumberland Mountains	19	9	1.3	47.4	2.5

Summary of Breeding Status

Number of Blocks in Which Species Was Recorded		
Total	**363**	**49.9%**
Confirmed	25	6.9%
Probable	102	28.1%
Possible	236	65.0%

Belted Kingfisher

Red-headed Woodpecker

Melanerpes erythrocephalus

The Red-headed Woodpecker is a variably uncommon to common summer resident across central and western Kentucky. In contrast, it is locally distributed and relatively rare throughout the Cumberland Plateau and the Cumberland Mountains. The species was once considered to be the commonest of woodpeckers, but a notable decrease was noted during the first half of the 20th century, perhaps in response to the increase in European Starlings (Mengel 1965). More recent reports (e.g., Jones 1975; Robbins et al. 1986) indicate that the species may have partially recovered in some areas.

Red-headed Woodpeckers are found in a great variety of habitats, but they occur most frequently in semi-open to open areas with some large trees. The species is most conspicuous in western Kentucky, where it inhabits bottomland forests, swamps, and the margins of floodplain sloughs. Elsewhere, this woodpecker most frequently inhabits rural farmland with scattered trees or small woodlots, but it is also regularly found in parkland, riparian corridors, and the margins of reservoirs. The species generally avoids mature, closed-canopy forest during the breeding season, probably because of its active, flycatching habits. The exception to this trend is mature bottomland forest, where the midstory is typically open. In eastern Kentucky Red-headed Woodpeckers seem to be restricted to altered habitats, including roadway and utility corridors, forest clear-cuts, golf courses, and the margins of reclaimed strip mines and reservoirs.

It is likely that human alteration of the landscape has affected the status of the Red-headed Woodpecker in Kentucky in a variety of ways. Clearing and dissection of the extensive upland forests that once covered much of the state have certainly created much suitable habitat and resulted in an expansion of birds into areas where they were formerly excluded. In contrast, the clearing of bottomland forests of the western half of the state has resulted in the loss of a substantial amount of suitable habitat in lowland areas. Before settlement the species may have been numerous in and about the native prairies and savannas of western and central Kentucky, but it appears that the open farmland that has replaced most of these natural habitats also supports a substantial population.

Forest Cover

Value	% of Blocks	Avg Abund
All	40.9	1.9
1	73.3	1.9
2	66.3	2.0
3	35.2	1.9
4	9.1	1.7
5	3.8	1.5

Red-headed Woodpeckers are present in Kentucky throughout the year, but many birds move about seasonally, seeking out territories with a sufficient winter food supply of acorns and other nuts (Mengel 1965). In addition, birds from farther north pass through during migration, and some may overwinter, especially in western Kentucky. By late April or early May, most of the nonresidents have departed, and summering birds have reestablished their nesting territories. Mengel (1965) indicated that early clutches were completed during the first ten days of May, with an early peak in clutch completion of May 20. A later peak during the latter part of July may indicate the raising of a second brood by some pairs, and late clutches may be completed in early August (Mengel 1965; Stamm and Croft 1968). The average size of five clutches described by Mengel (1965) was 6.2 (range of 5–8).

Alvin E. Staffan

The nest typically is excavated in dead wood of a branch or the upper trunk of a tree, although utility poles are also used. Most nest sites lie along or just inside forest margins, among small groves of trees, or in trees standing alone in otherwise cleared areas. The average height of 14 nest cavities reported by Mengel (1965) was 29 feet (range of 10–60 feet).

The atlas survey yielded records of Red-headed Woodpeckers in nearly 41% of priority blocks, and 28 incidental observations were reported. As expected, occurrence and abundance were highest in the western part of the state and decreased eastward. The species was much less widespread in the Cumberland Plateau and was not recorded in the Cumberland Mountains. Occurrence of these woodpeckers was closely related to the amount of open habitat, being greatest in areas with limited forest cover and nearly absent in areas that are predominantly forested.

Almost 21% of priority block records were for confirmed breeding. Most of these records involved the observation of active nest cavities, although some were based on the observation of recently fledged young, and a few involved adults carrying food.

Red-headed Woodpeckers are scarce or absent on most Kentucky BBS routes. The average number of individuals per BBS route for the periods 1966–91 and 1982–91 was 0.93 and 1.73, respectively. Although the sample size is small, trend analysis of these data indicates a significant ($p<.05$) increase of 5.3% per year for the period 1966–91, but a nonsignificant increase of 1.4% per year for the period 1982–91. The long-term increase may reflect the recovery from the decline noted previously.

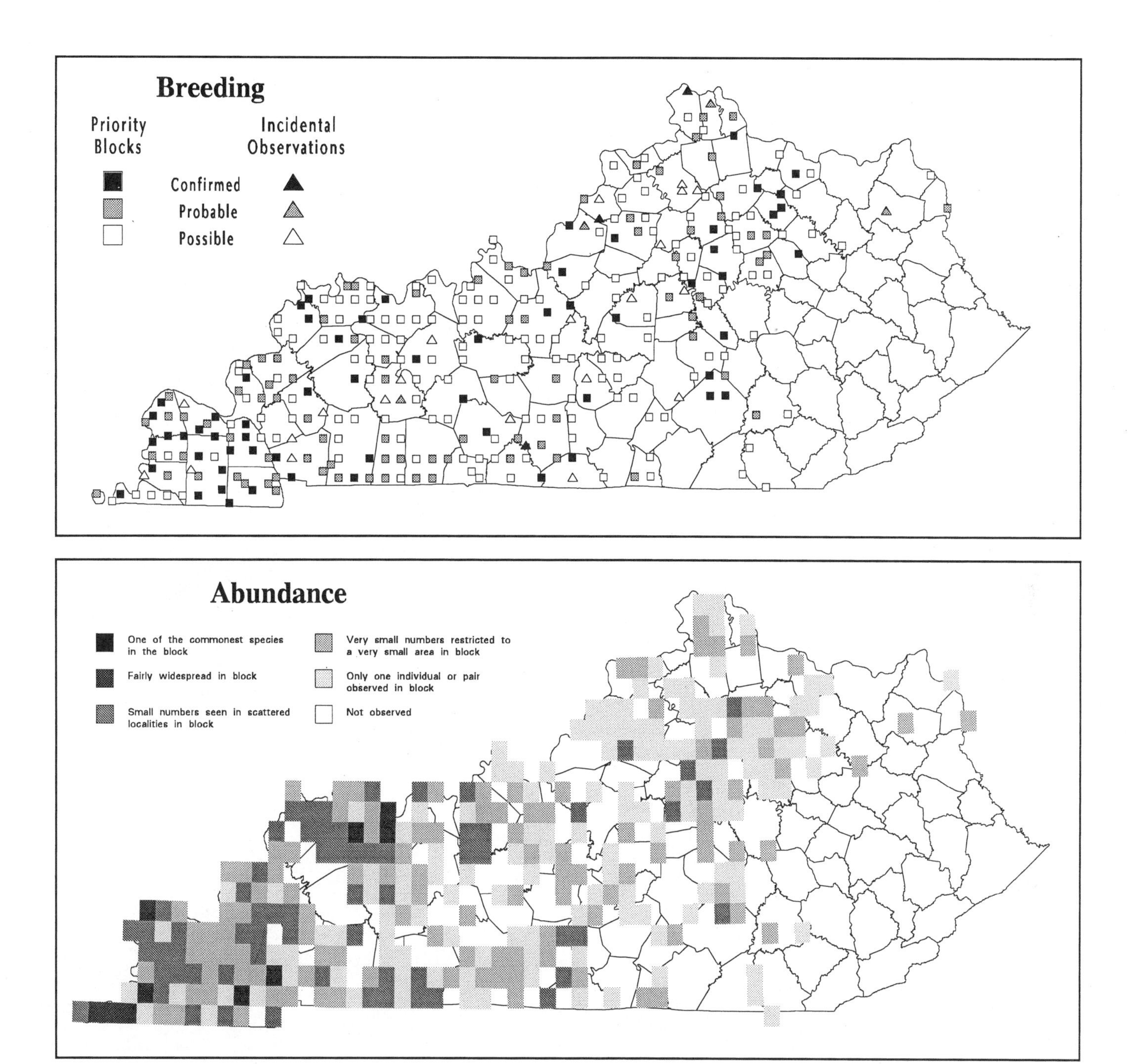

Analysis of Block Data by Physiographic Province Section

Physiographic Province Section	Total Blocks Surveyed	Blocks with Data	Avg Abund	% with Data	Section's % for State
Mississippi Alluvial Plain	14	13	3.1	92.9	4.4
East Gulf Coastal Plain	36	33	2.3	91.7	11.1
Highland Rim	139	78	1.9	56.1	26.3
Shawnee Hills	142	85	2.1	59.9	28.6
Blue Grass	204	81	1.5	39.7	27.3
Cumberland Plateau	173	7	1.3	4.0	2.4
Cumberland Mountains	19	-	-	-	-

Summary of Breeding Status

Number of Blocks in Which Species Was Recorded		
Total	**297**	**40.9%**
Confirmed	61	20.5%
Probable	86	29.0%
Possible	150	50.5%

Red-headed Woodpecker

Red-bellied Woodpecker

Melanerpes carolinus

The Red-bellied Woodpecker is a conspicuous permanent resident throughout all of Kentucky except the heavily forested portions of the Cumberland Plateau and the Cumberland Mountains, where it is decidedly less numerous and more locally distributed (e.g., Hudson 1971). Mengel (1965) summarized the species's occurrence as uncommon to fairly common locally in eastern Kentucky, but common westward.

The Red-bellied Woodpecker is found in a great variety of habitats, preferring semi-open and open situations with at least a few large trees rather than closed-canopy forest. The species is most frequent in and about settled areas, especially rural farmland where extensive forest has been dissected into scattered woodlots and corridors. Substantial numbers are also encountered in suburban yards, urban parks, and riparian corridors. Although the birds appear to avoid extensive forest with a closed canopy in summer, they frequently use forest edge, semi-open forest disturbed by selective logging, and both natural and artificial openings in forest.

It is likely that changes associated with human settlement have affected the Red-bellied Woodpecker's occurrence in Kentucky in several ways. The removal of open-canopy forests that formerly occurred in association with the prairies and savannas of the western and central parts of the state has likely reduced the species's numbers in some regions. In contrast, the fragmentation of extensive areas of unbroken, closed-canopy forest has certainly resulted in an increase in the species in many other areas, especially in eastern Kentucky. Overall, this woodpecker seems to have adapted well to alterations in the landscape, and it may be as widespread and common today as ever before.

Red-bellied Woodpeckers are permanent residents in Kentucky. The territorial calling of males commences with the first warm spells of spring, and early clutches may be completed by mid-April (Mengel 1965). Mengel gave a probable peak in early clutch completion of late April. Although it is possible that some pairs raise two broods, there seem to be no records of active nests later than early June. Surprisingly, Kentucky data on clutch size are limited to one record of a nest containing four eggs (Mengel 1965), but Harrison (1975) gives rangewide clutch size as 3–8, generally 4–5.

Forest Cover

Value	% of Blocks	Avg Abund
All	81.8	2.4
1	87.1	2.4
2	94.7	2.6
3	90.0	2.5
4	64.3	2.1
5	40.4	1.7

The nest cavity is usually excavated in a large branch or the trunk of a dead tree, but dead snags in otherwise healthy trees are also used. Nest sites that lie along or just inside forest edge are usually chosen, although sometimes a solitary tree surrounded by cleared land is used. The average height of four nests listed by Mengel (1965) was 44 feet (range of 15–90 feet). A more recent nest was reported only 12 feet above the ground in Madison County in 1968 (KOS Nest Cards).

The atlas survey yielded records of Red-bellied Woodpeckers in nearly 82% of priority blocks statewide, and 13 incidental observations were reported. Occurrence was highest throughout central and western Kentucky, with an obvious decrease in the Cumberland Plateau and the Cumberland Mountains. Red-bellied Woodpeckers were most common in areas of dissected forest and decreased as the percentage of forestation increased.

Ron Austing

Just over 19% of priority block records were for confirmed breeding. Most reports involved the observation of adults leaving or entering active nest cavities, although some were based on the observation of recently fledged young. Young Red-bellied Woodpeckers have a raspy begging call that is conspicuous; atlasers often used this call to locate dependent young, and such records were interpreted as confirmed.

Red-bellied Woodpeckers are recorded in moderate numbers on most Kentucky BBS routes. The average number of individuals per BBS route for the periods 1966–91 and 1982–91 was 7.06 and 8.34, respectively. According to these data, Red-bellieds ranked 30th in abundance on BBS routes during the period 1982–91. Trend analysis of these data shows that the average number of individuals per BBS route has not changed over the period 1966-91 and that there has been only a nonsignificant increase of 1.5% per year for the period 1982–91.

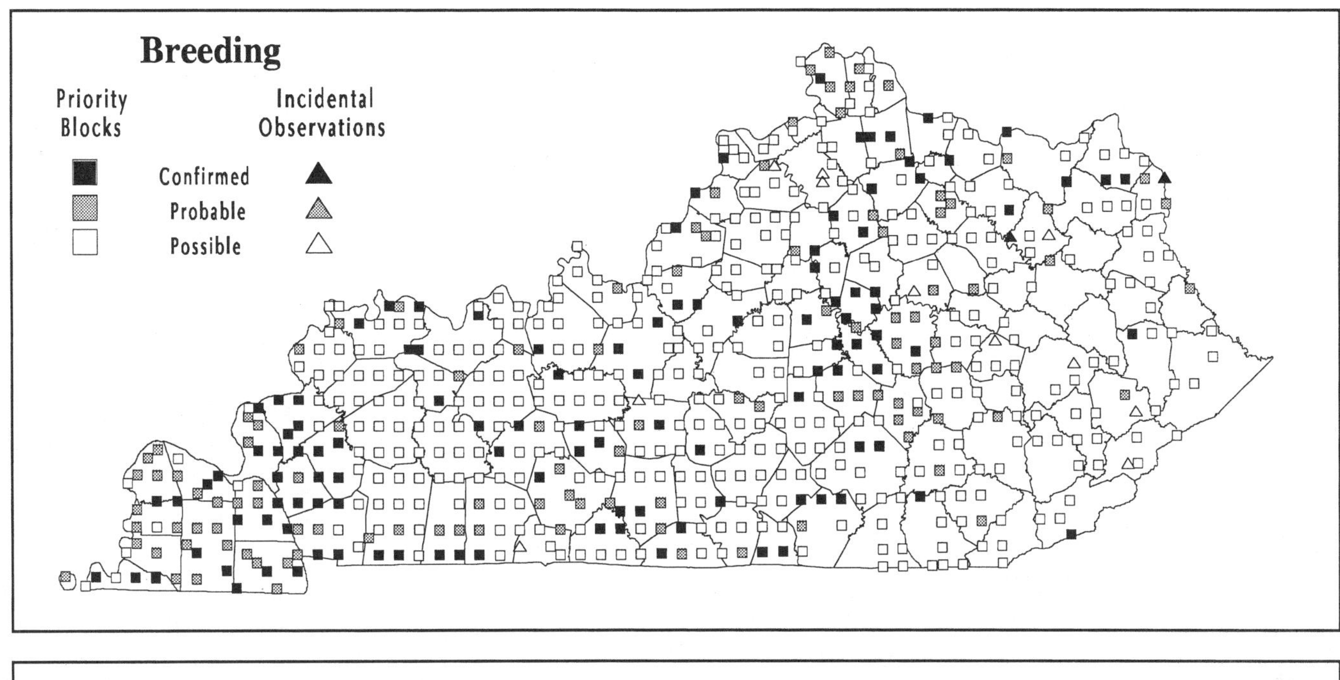

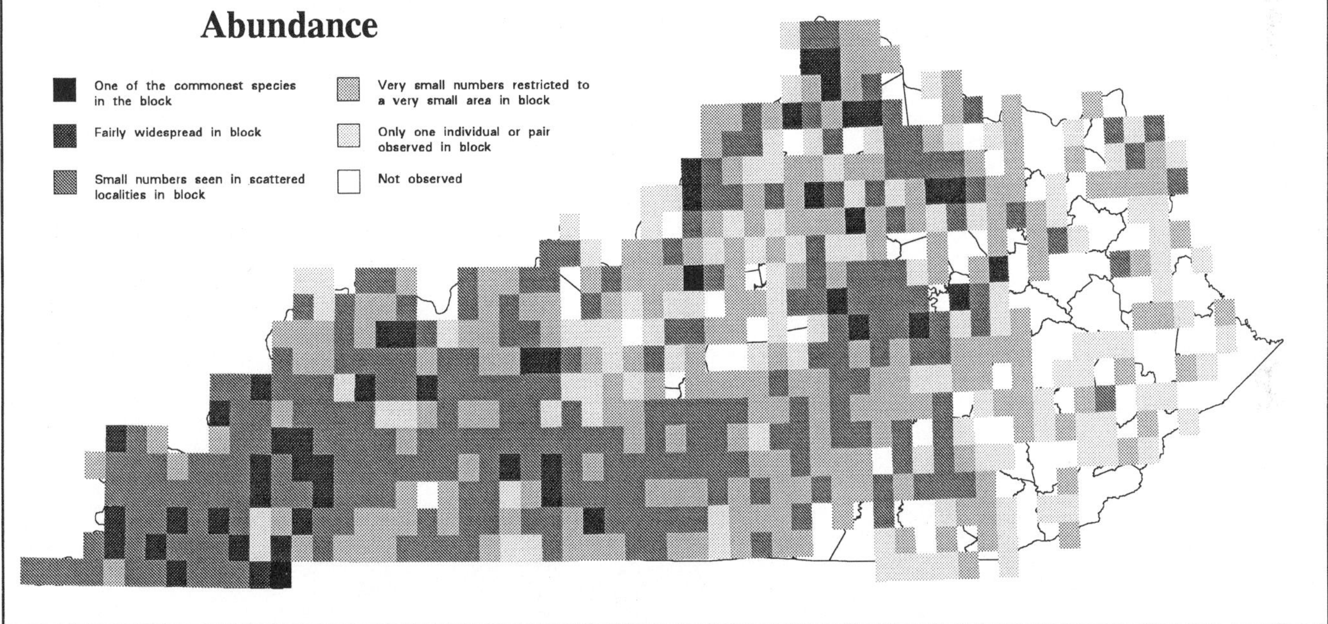

Analysis of Block Data by Physiographic Province Section

Physiographic Province Section	Total Blocks Surveyed	Blocks with Data	Avg Abund	% with Data	Section's % for State
Mississippi Alluvial Plain	14	14	3.1	100.0	2.4
East Gulf Coastal Plain	36	36	3.1	100.0	6.1
Highland Rim	139	135	2.5	97.1	22.7
Shawnee Hills	142	136	2.6	95.8	22.9
Blue Grass	204	169	2.3	82.8	28.4
Cumberland Plateau	173	96	1.8	55.5	16.1
Cumberland Mountains	19	9	1.4	47.4	1.5

Summary of Breeding Status

Number of Blocks in Which Species Was Recorded		
Total	**595**	**81.8%**
Confirmed	115	19.3%
Probable	112	18.8%
Possible	368	61.8%

Red-bellied Woodpecker

Downy Woodpecker

Picoides pubescens

The Downy Woodpecker is undoubtedly the most numerous and widespread woodpecker in Kentucky. Although perhaps locally outnumbered by other species in some areas, and definitely less conspicuous than some of the larger woodpeckers, it is fairly common to common statewide and occupies a much greater variety of habitats than any other woodpecker. Mengel (1965) regarded the Downy Woodpecker as a fairly common to common permanent resident throughout the state.

Downy Woodpeckers are found in a great variety of habitats, from mature, closed-canopy forest to small woodlots and narrow wooded corridors in extensively cleared areas. The species seems to be most abundant in semi-open areas with a fairly good supply of forest, including rural farmland and settlement, wooded suburban parks and yards, and regenerating forest disturbed by logging. In contrast, this woodpecker is not as conspicuous in extensive forest and may not be as abundant where no natural or artificial openings occur. Both deciduous and mixed forest types are used.

Although the Downy Woodpecker seems to have adapted relatively well to the changes that humans have brought upon the landscape, the species may have been more numerous in Kentucky two centuries ago. The clearing of vast expanses of forested land for agricultural use and settlement has eliminated much suitable nesting habitat. Fortunately, this small woodpecker's use of such a wide variety of habitats has allowed for its continued abundance today.

Downy Woodpeckers are permanent residents in Kentucky, and numbers appear to vary little throughout the year. The first territorial drumming can be heard at the onset of warmer weather in early spring, and it is likely that nesting territories are established by mid-April. According to Mengel (1965), early clutches are completed by the end of April, and a peak in early clutch completion occurs during the first ten days of May. Although some pairs may raise a second brood, evidence of such in Kentucky is lacking. The average size of four clutches or broods listed by Mengel (1965) was 4.5 (range of 4–5).

Forest Cover

Value	% of Blocks	Avg Abund
All	90.5	2.4
1	89.1	2.4
2	88.9	2.3
3	93.5	2.4
4	92.2	2.5
5	80.8	2.3

The nest cavity is typically excavated in a dead limb or snag, but fence posts are also regularly used. This woodpecker's small size allows it to use smaller limbs for its nest cavity. For this reason, it may be found nesting far from mature trees. The nest tree may be situated in the midstory of mature forest, but more frequently it is near or along a woodland edge or opening. The average height of six nests given by Mengel (1965) was 21 feet (range of 9–40 feet).

The Downy Woodpecker was reported more frequently than any other woodpecker during the atlas survey. It was found in nearly 91% of priority blocks statewide, and 11 incidental observations were reported. The species ranked 18th by the number of priority block records, but only 13th by total abundance and 33rd by average abundance. These figures indicate that while the Downy Woodpecker is widespread, it is not as abundant or as conspicuous as many other common species of nesting birds. This woodpecker was found in more than 85% of priority blocks in all physiographic province sections except the Cumberland Mountains. Average abundance was not highly variable, but the species was found slightly less often in extensively forested areas.

Alvin E. Staffan

Despite the Downy Woodpecker's abundance, only about 12% of observations were for confirmed breeding. Most confirmed records were based on the observation of recently fledged young following adults, although at least a dozen active nests were found, and adults carrying food for young were observed twice.

Small numbers of Downy Woodpeckers are reported on most Kentucky BBS routes. The average number of individuals per BBS route for the periods 1966–91 and 1982–91 was 3.47 and 3.76, respectively. Trend analysis of these data yields slight, non-significant decreases of 0.9% per year for the period 1966–91 and 0.2% per year for the period 1982–91.

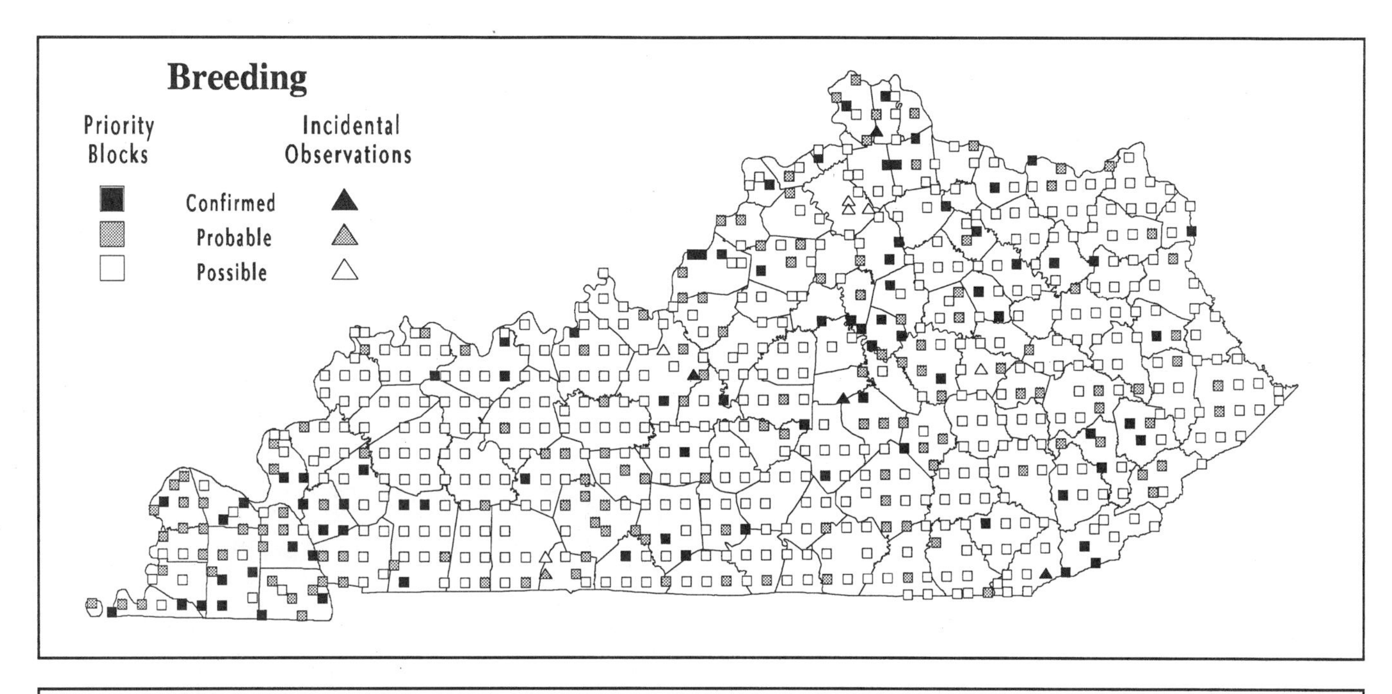

Analysis of Block Data by Physiographic Province Section

Physiographic Province Section	Total Blocks Surveyed	Blocks with Data	Avg Abund	% with Data	Section's % for State
Mississippi Alluvial Plain	14	14	2.9	100.0	2.1
East Gulf Coastal Plain	36	36	2.5	100.0	5.5
Highland Rim	139	130	2.2	93.5	19.8
Shawnee Hills	142	135	2.4	95.1	20.5
Blue Grass	204	174	2.6	85.3	26.4
Cumberland Plateau	173	155	2.5	89.6	23.6
Cumberland Mountains	19	14	2.3	73.7	2.1

Summary of Breeding Status

Number of Blocks in Which Species Was Recorded		
Total	**658**	**90.5%**
Confirmed	80	12.2%
Probable	141	21.4%
Possible	437	66.4%

Downy Woodpecker

Hairy Woodpecker

Picoides villosus

The Hairy Woodpecker's piercing call notes are an uncommon to fairly common sound in forested habitats throughout Kentucky. For the most part the species's abundance is related directly to the amount of fairly mature to mature forest present. Mengel (1965) regarded the bird as variably rare and local in portions of central and western Kentucky, and common in extensively forested areas across the state. In general the species is nowhere more numerous than its smaller relative, the Downy Woodpecker, but it may approach the Downy in abundance in extensively forested portions of eastern Kentucky.

Hairy Woodpeckers can be found in a great variety of forested and semi-open habitats, although they are seldom seen far from a tract of fairly mature woodland. The species is usually associated with deciduous forest, but mixed pine-hardwood communities are used, especially in eastern Kentucky. While these woodpeckers occur most frequently in extensively forested areas, they also can be found in rural farmland, wooded suburban parks, the margins of floodplain swamps and sloughs, and forest slightly to moderately disturbed by logging.

It is likely that Hairy Woodpeckers were more common and widespread in Kentucky before European settlement. While the Downy Woodpecker probably has not been affected so severely by recent changes in the landscape, this denizen of larger tracts of more mature woodlands has certainly decreased as a result of the clearing and fragmentation of native forests.

Hairy Woodpeckers are permanent residents, and territorial behavior is initiated not long after the arrival of the first warm spells of early spring. Specific breeding records from Kentucky are relatively scarce, but young have been detected in nests as early as mid-April (Stamm and Croft 1968; Mengel 1965). According to Mengel, egg laying continues into early June, but most clutches are probably completed in April and May. Young are heard calling from nest cavities most frequently in the latter half of May, and family groups are encountered most often in June. Like most other woodpeckers, the Hairy Woodpecker likely raises only one brood. Specific Kentucky data on clutch size are limited to two nests containing four eggs (Mengel 1965). Harrison (1975) gives rangewide clutch size as 3–6, commonly 4.

Forest Cover

Value	% of Blocks	Avg Abund
All	53.9	1.7
1	33.7	1.4
2	54.2	1.6
3	57.8	1.6
4	55.2	1.9
5	71.2	2.0

The birds usually excavates their nest cavity in a dead limb of a large tree. The nest tree is typically situated in the midstory of fairly mature or mature forest, although it may lie just inside or occasionally along a forest margin. In Kentucky this woodpecker prefers deciduous trees, although Harrison (1975) mentions the use of conifers in other parts of its range. Kentucky data on nest height are scarce, but records have indicated a range of 7 to 35 feet (Mengel 1965; KOS Nest Cards; K. Caminiti, pers. comm.).

The atlas survey yielded records of Hairy Woodpeckers in nearly 54% of priority blocks statewide, and 24 incidental observations were reported. Somewhat unexpectedly, occurrence by physiographic province section was not highly variable, although occurrence was highest in the Cumberland Mountains and lowest in the Blue Grass. Average abundance by physiographic province section showed a more pronounced trend, which was related to the percentage of forest cover in each region. Occurrence and average abundance were lowest in extensively cleared areas and generally increased with the degree of forestation.

Alvin E. Staffan

Less than 10% of priority block records were for confirmed breeding. Of the 38 confirmed records, almost all were based on the observation of young in the company of adults, although at least one active nest was located.

Hairy Woodpeckers are not well represented in Kentucky BBS data. The average number of individuals per BBS route for the periods 1966–91 and 1982–91 was 0.45 and 0.52, respectively. Although the sample size is small, trend analysis of these data shows a significant ($p<.05$) increase of 1.7% per year for the period 1966–91 and a nonsignificant decrease of 0.9% per year for the period 1982–91.

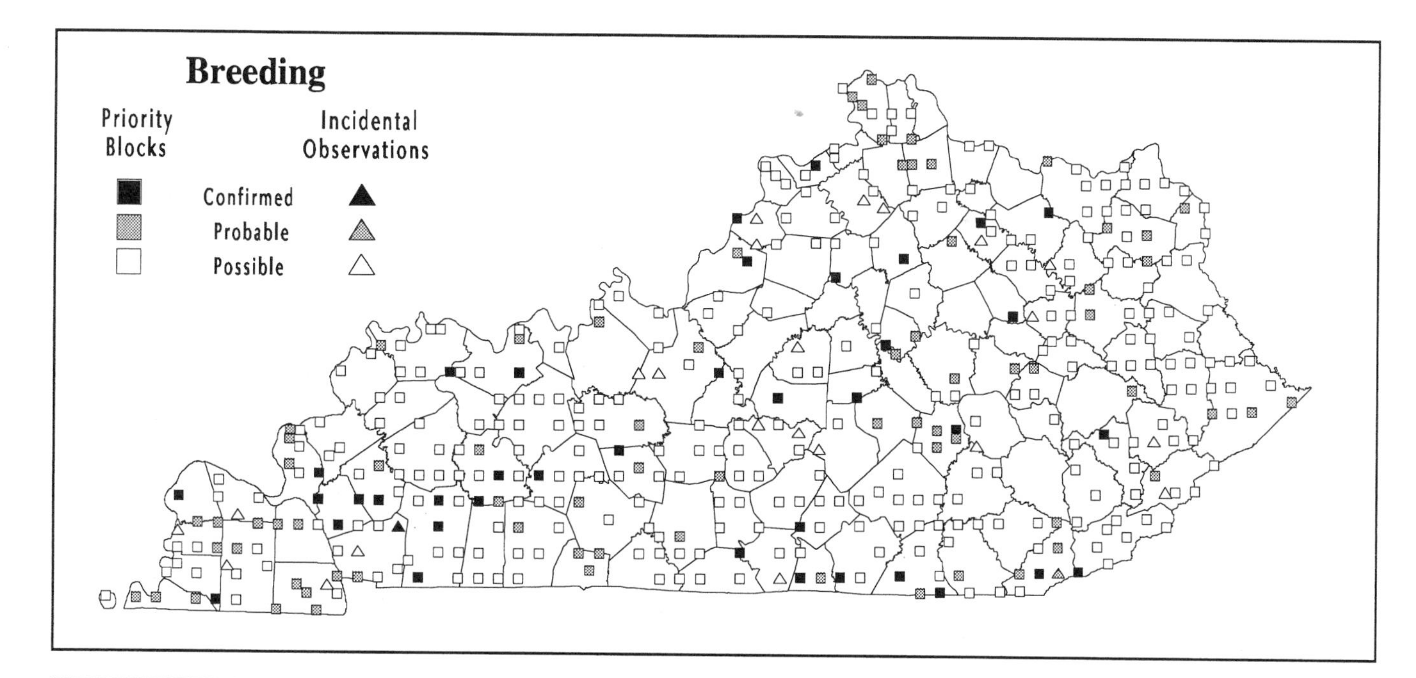

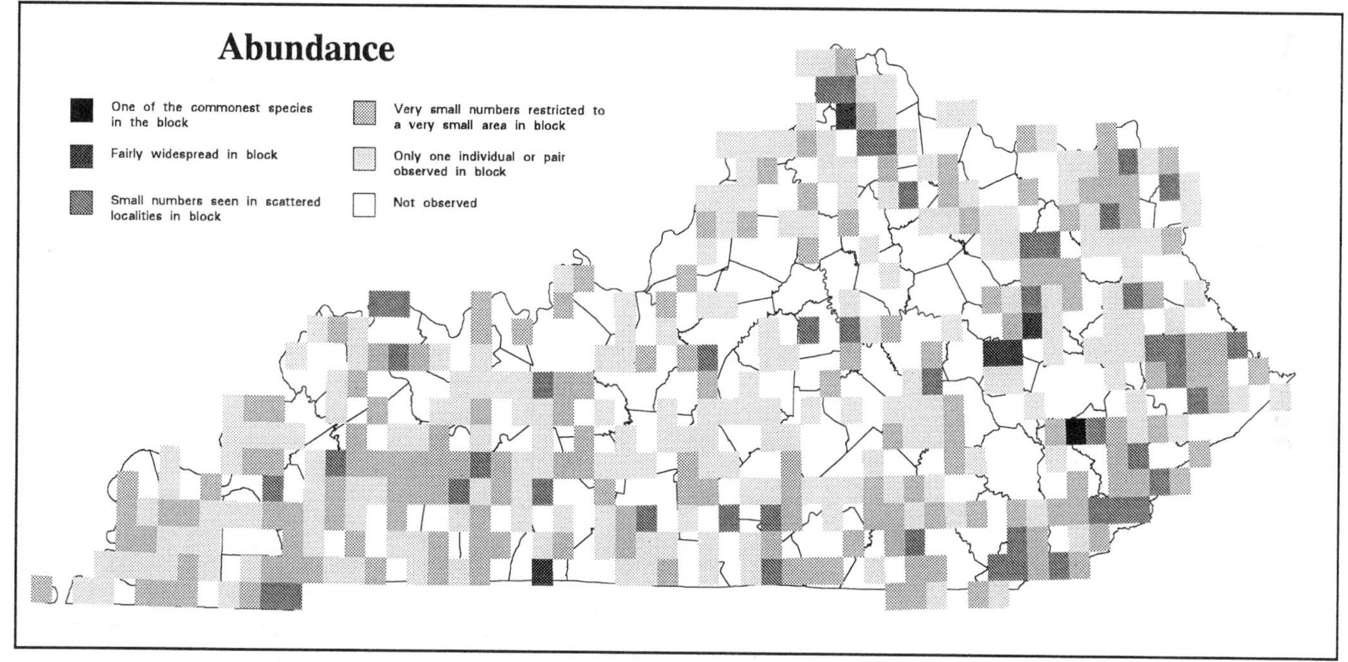

Analysis of Block Data by Physiographic Province Section

Physiographic Province Section	Total Blocks Surveyed	Blocks with Data	Avg Abund	% with Data	Section's % for State
Mississippi Alluvial Plain	14	9	1.4	64.3	2.3
East Gulf Coastal Plain	36	23	1.6	63.9	5.9
Highland Rim	139	77	1.6	55.4	19.6
Shawnee Hills	142	84	1.6	59.2	21.4
Blue Grass	204	88	1.6	43.1	22.4
Cumberland Plateau	173	98	1.8	56.6	25.0
Cumberland Mountains	19	13	2.3	68.4	3.3

Summary of Breeding Status

Number of Blocks in Which Species Was Recorded		
Total	**392**	**53.9%**
Confirmed	38	9.7%
Probable	73	18.6%
Possible	281	71.7%

Hairy Woodpecker

Red-cockaded Woodpecker

Picoides borealis

Most of the Red-cockaded Woodpecker's nesting range lies well to the south and east of Kentucky, in the Coastal Plain from Virginia to Texas (AOU 1983). There the species is found in several mature pine forest types (USFWS 1993). Outside this region, small pockets of nesting birds occur northward into the southern Appalachians, occupying mixed pine-hardwood forest communities as far north as eastern Kentucky. Throughout most of its range, this woodpecker has declined dramatically in the 20th century, and it is listed as Federally Endangered (USFWS 1993).

The habitat used by Red-cockaded Woodpeckers in Kentucky is quite different from those in which the species is most common farther south. The latter are typically very open pine forest communities with a ground cover dominated by grasses. While it is possible that small patches of habitat approaching such a condition may have occurred in Kentucky before European settlement, it is likely that the habitat used more typically has been a community based on mixed pitch/shortleaf pine-oak-hickory dominance with a nearly closed or only partially open canopy. Historical descriptions and vegetative surveys suggest that such habitat was locally abundant and may have even predominated on xeric and subxeric uplands of a narrow region of the western Cumberland Plateau known as the Cliff Section, especially in the southern part of the state (Braun 1950; Campbell et al. 1991). While these forests may never have approached the openness of some southern pine forests, it is likely that they were maintained by natural fires. It is also possible that fires set by Native Americans for habitat manipulation enhanced such areas and may have been responsible for local increases in population levels, although such speculation will never be proved (Martin 1989; Campbell et al. 1991).

It appears that the Red-cockaded Woodpecker only recently has declined substantially in Kentucky. Mengel (1965) regarded the species as uncommon to fairly common locally in pine-oak communities of the Cumberland Plateau. He further noted that "small bands composed of single family groups, usually 4 to 7 birds, are frequently seen from late June onward. Occasional groups of 12 to 15 birds evidently represent aggregations of several families; I have seen bands of this size on various occasions in Laurel and Whitley counties." These observations are remarkable and indicate a population as of the early 1950s that was much more abundant and widespread than indicated thirty years later (see Jackson et al. 1976; Schmaltz 1981; Murphy 1982). The reasons for this decrease appear to be the loss of old-growth pines for nesting and a shift in forest composition toward a more deciduous-dominated, closed-in midstory situation (Whitt 1974). It is apparent that the suppression of fire has played at least a partial role in the decline, since periodic fire probably maintained the pine component of the forest, as well as the open nature of the forest structure (Martin 1989; Campbell et al. 1991). In response to the species's dramatic decline, the U.S. Forest Service has instigated an aggressive management plan for large portions of the southern Daniel Boone National Forest to provide nesting habitat for the Red-cockaded Woodpecker. Part of this management involves the setting aside of large areas for expansion of the nesting population as currently used nest trees die and foraging territories are filled.

Mengel (1965; pers. comm.) documented the presence of Red-cockaded Woodpeckers from as far north in the Cliff Section as the Red River Gorge area of Powell and Wolfe counties. Outside the Cliff Section, the bird's historical status is somewhat unclear. While suitable nesting habitat appears to have been present in other parts of Kentucky, firm evidence of the species's presence is lacking (Mengel 1965). Early records were published for several localities, but Mengel (1965) regarded most as erroneous and seemed to put credence only in reports from the Mammoth Cave area (Wilson 1961; Mengel 1965).

James Parnell

Red-cockaded Woodpeckers are permanent residents in Kentucky, and territories seem to be maintained throughout the year. Permanent roosting and nesting cavities are excavated in living old-growth pines. Such cavities typically are used repeatedly until the tree dies. Specific data on breeding in Kentucky are scarce, but nesting activity apparently commences by mid-April. Incubation has been noted by early May (Murphy 1982), and young have been heard calling from nest cavities from mid-May (Murphy 1982) to early July (Larson 1979). Most young probably fledge during the first half of June (Murphy 1982; KOS Nest Cards). Kentucky data on clutch size are lacking, but Harrison (1975) gives rangewide clutch size as 3–5.

The atlas fieldwork resulted in only one incidental observation of a Red-cockaded Woodpecker in McCreary County; however, 1985–91 data supplied by the U.S. Forest Service (S. Phillips, pers. comm.) yielded about a dozen additional locations for Red-cockaded Woodpecker activity on seven quadrangles. As of 1991 all active colonies were within the Daniel Boone National Forest in Laurel and Whitley counties. In 1991 there were two active nests, one in each county. It only can be hoped that U.S. Forest Service management of Kentucky's Red-cockaded Woodpecker population will reverse the downward trend of the last 40 years.

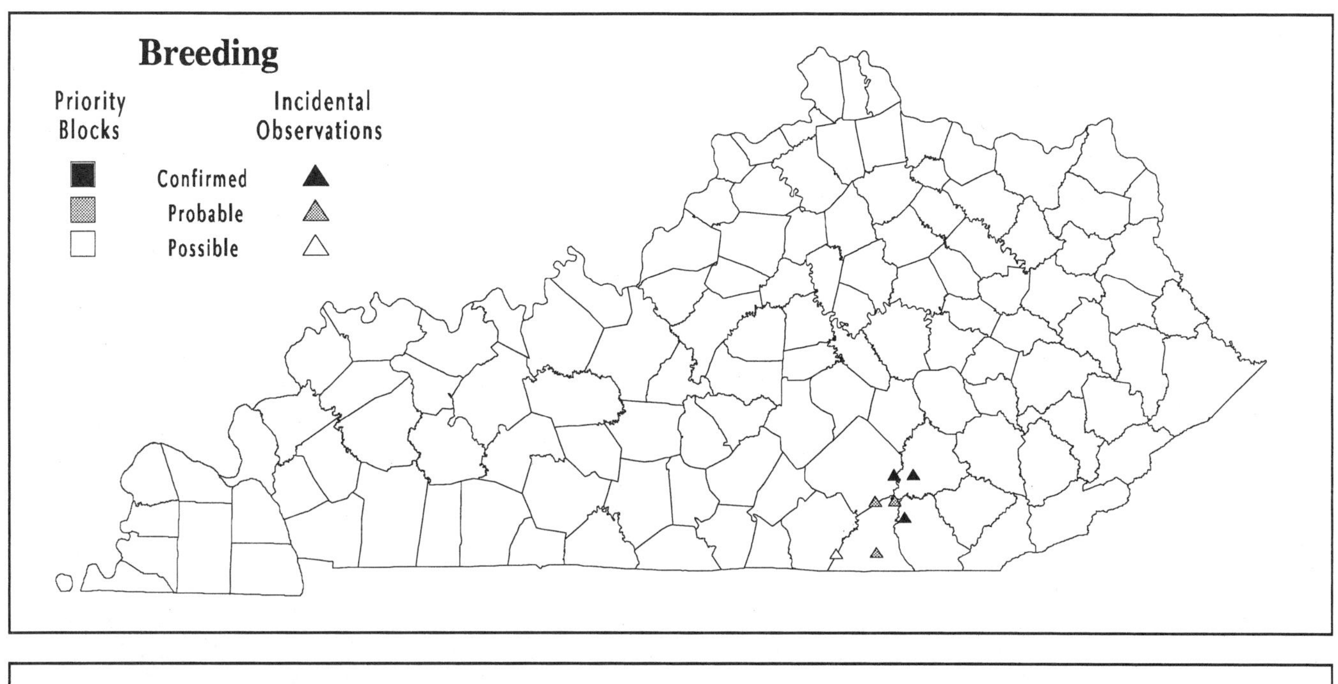

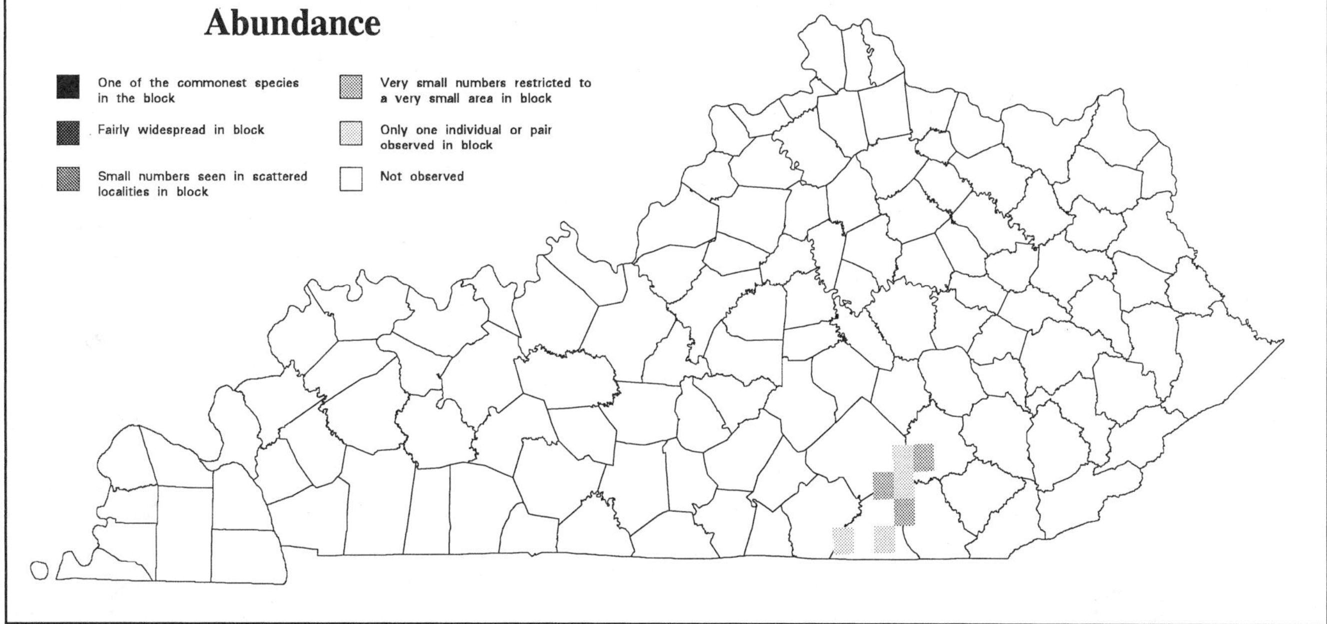

Analysis of Block Data by Physiographic Province Section

Physiographic Province Section	Total Blocks Surveyed	Blocks with Data	% with Data	Section's % for State
Mississippi Alluvial Plain	14	-	-	-
East Gulf Coastal Plain	36	-	-	-
Highland Rim	139	-	-	-
Shawnee Hills	142	-	-	-
Blue Grass	204	-	-	-
Cumberland Plateau	173	-	-	-
Cumberland Mountains	19	-	-	-

Summary of Breeding Status

Number of Blocks in Which Species Was Recorded		
Total	-	-
Confirmed	-	-
Probable	-	-
Possible	-	-

Red-cockaded Woodpecker

Northern Flicker

Colaptes auratus

The distinctive calls of the Northern Flicker are a fairly common year-round sound throughout much of Kentucky. Although the species is decidedly less numerous in the extensively forested portions of eastern Kentucky, it is a fairly conspicuous summer resident of suitable habitat throughout the state. Mengel (1965) regarded the flicker as a fairly common to common summer resident statewide.

The Northern Flicker is found in a great variety of semi-open and open habitats with at least some large trees nearby. It occurs most frequently in rural farmland and settlement, wooded suburban parks and yards, and other areas that offer a combination of dissected woodland tracts and open ground. In contrast, extensive tracts of fairly mature to mature forest are usually avoided, although natural or artificial openings in extensive forest are often used.

It is unclear to what extent the status of the Northern Flicker has changed in Kentucky since settlement. The widespread clearing and dissection of forest land for agricultural use and settlement likely have had both beneficial and detrimental effects on the species. The clearing of natural semi-open woodlands in central and western Kentucky has probably resulted in a loss of optimal habitat. Moreover, the suppression of fire has likely contributed to the general decline of more open forest types throughout much of the state. Such woodlands have become less suitable for flickers as canopy closure has increased. In contrast, the dissection of vast regions of unbroken, closed-canopy forest has certainly resulted in the creation of an abundance of suitable nesting habitat in areas that were formerly unoccupied.

Although Northern Flickers are present in the state throughout the year, many birds appear to move about seasonally. In addition, transients from more northerly breeding areas pass through in migration, and some may overwinter (Mengel 1965). Flickers are among the most vocal of nesting woodpeckers. At the onset of the first warm spells of early spring, their territorial calling and bantering become evident, and drumming and nest construction soon follow. According to Mengel (1965), early clutches are completed during mid-April, and there is likely a peak of clutch completion from late April to early May. Like other woodpeckers, the flicker appears to raise only one brood, although some clutches may not be completed until early June. Kentucky data on clutch size are scarce, but Mengel gives the average size of seven clutches or broods believed to be complete as 6.3 (range of 4–9), and Harrison (1975) gives rangewide clutch size as 3–10, typically 6–8.

Forest Cover

Value	% of Blocks	Avg Abund
All	79.9	2.2
1	85.1	2.2
2	89.5	2.3
3	83.5	2.2
4	66.9	2.0
5	57.7	1.8

The nest cavity is typically excavated from a dead limb or snag of relatively large diameter. Dead trees are most often used, although the bird also chooses dead limbs in living trees and artificial sites such as utility poles. The nest tree is often situated alone, among a small cluster of trees, or along a fencerow in otherwise open areas, although trees along a riparian corridor or just inside a woodland border also are used. The average height of 15 nests given by Mengel was 16.9 feet (range of 5–60 feet).

Alvin E. Staffan

The atlas survey yielded reports of Northern Flickers in nearly 80% of priority blocks statewide, and 19 incidental observations were reported. Occurrence was highest in central and western Kentucky, and somewhat lower in the east. Average abundance was fairly uniform across the state. Occurrence was highest in areas with an abundance of open habitat and some forest, and decreased as forestation increased.

Although flickers are relatively numerous, only about 10% of priority block records were confirmed. Most confirmed reports involved the observation of recently fledged young, although several were based on adults seen entering nest cavities and young that were seen or heard in nests. One interesting nest containing young was located about six miles southwest of Paris, in Bourbon County. The nest was in open farmland devoid of suitable nest trees and had been excavated about three feet above the ground in a wooden fence post.

The Northern Flicker is reported in small to moderate numbers on most Kentucky BBS routes. The average number of individuals reported per BBS route for the periods 1966–91 and 1982–91 was 4.56 and 5.01, respectively. Trend analysis of these data reveals a significant ($p<.01$) decrease of 2.1% per year for the period 1966–91, but a nonsignificant decrease of 3.9% per year for the period 1982–91. Competition with European Starlings for nest cavities and the severe winters of the late 1970s may be largely responsible for the long-term decrease (Robbins et al. 1986).

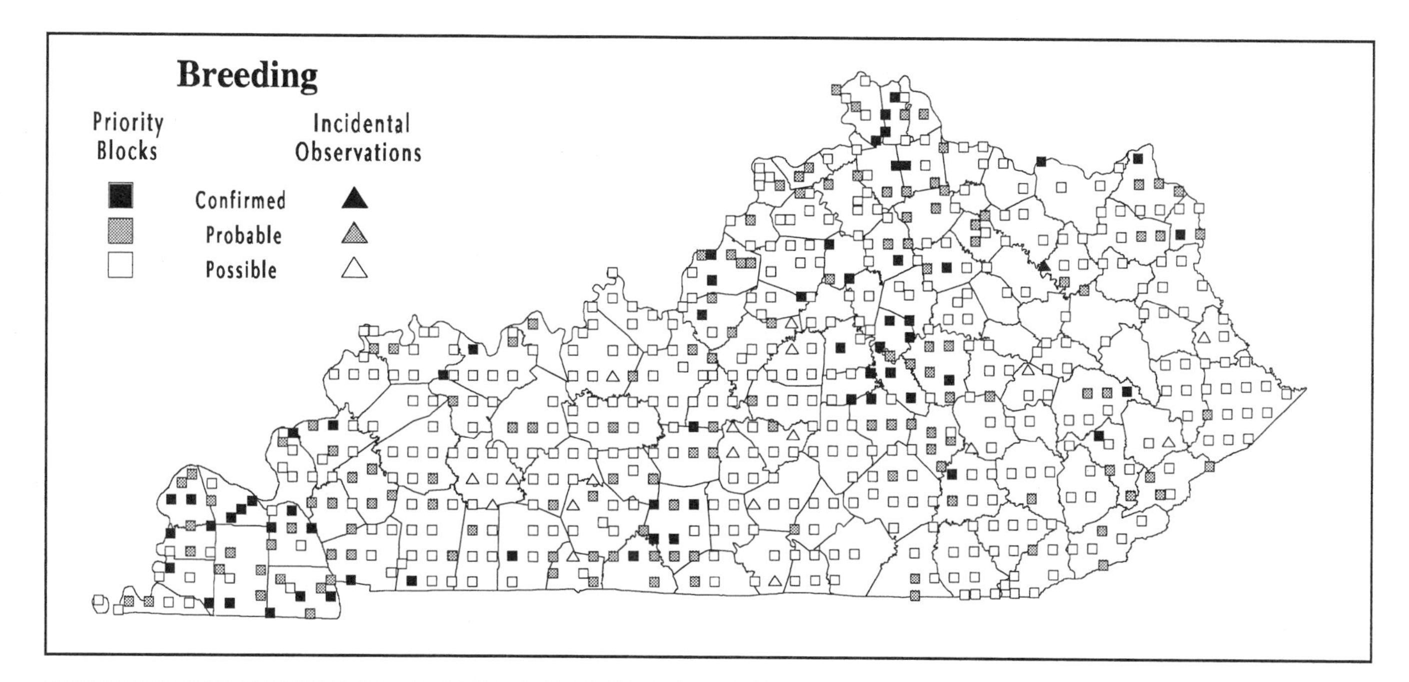

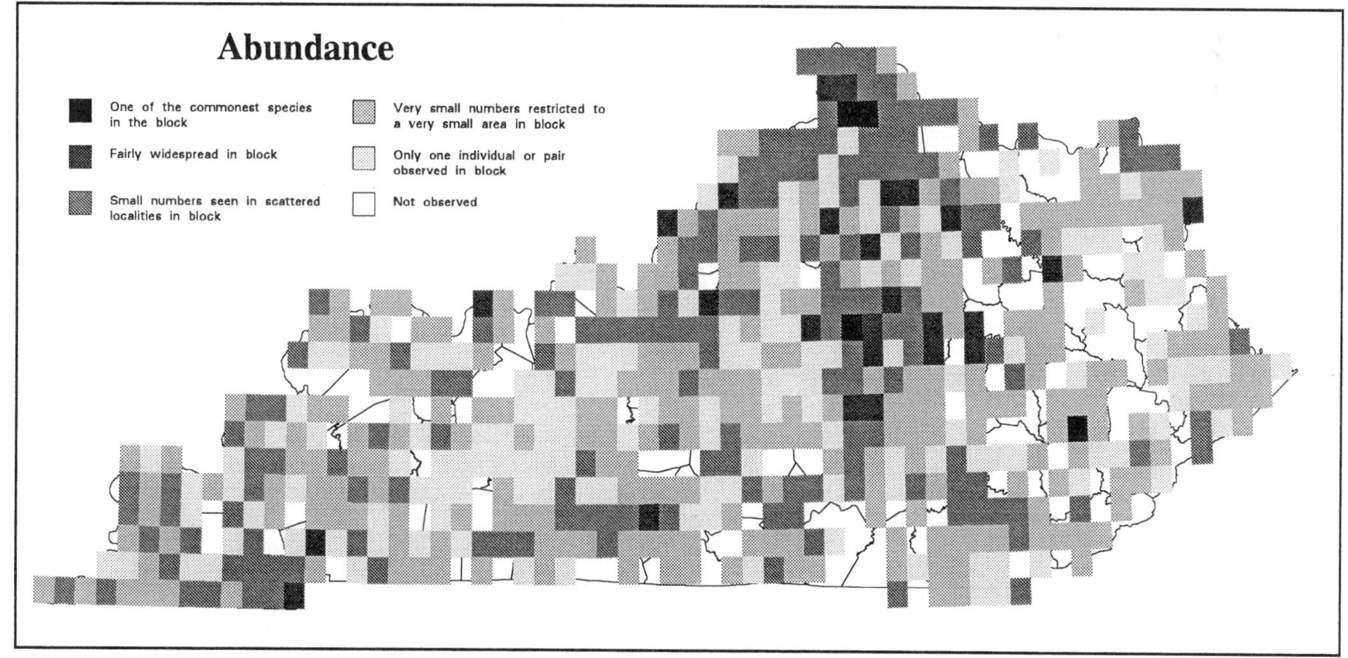

Analysis of Block Data by Physiographic Province Section

Physiographic Province Section	Total Blocks Surveyed	Blocks with Data	Avg Abund	% with Data	Section's % for State
Mississippi Alluvial Plain	14	13	2.2	92.9	2.2
East Gulf Coastal Plain	36	33	2.3	91.7	5.7
Highland Rim	139	114	2.2	82.0	19.6
Shawnee Hills	142	110	1.9	77.5	18.9
Blue Grass	204	178	2.5	87.3	30.6
Cumberland Plateau	173	120	2.0	69.4	20.7
Cumberland Mountains	19	13	1.7	68.4	2.2

Summary of Breeding Status

Number of Blocks in Which Species Was Recorded		
Total	**581**	**79.9%**
Confirmed	60	10.3%
Probable	136	23.4%
Possible	385	66.3%

Northern Flicker

Pileated Woodpecker

Dryocopus pileatus

The Pileated Woodpecker's resounding call is a familiar sound in forested areas throughout Kentucky. Mengel (1965) summarized the species's status as uncommon to fairly common in all parts of the state except the Blue Grass, where he considered it rare and restricted to parts of the Outer Blue Grass and Knobs subsections.

Pileated Woodpeckers typically inhabit fairly mature to mature forest, and their abundance is generally proportional to the amount of such habitat present. Although noticeably less abundant away from extensive forest, the species is still fairly frequent in semi-open areas with at least some forest and large trees for nesting. These include wooded urban parks, suburban yards, rural farmland, and riparian corridors.

The number of these large woodpeckers in Kentucky undoubtedly has decreased since the beginning of the 19th century. The species seems to have adapted fairly well to human alteration of the landscape, and a considerable population continues to reside in the mature forest that remains. It is clear, however, that the widespread clearing of virgin forests for conversion to agricultural use and settlement has eliminated a substantial amount of optimal habitat.

Pileated Woodpeckers are permanent residents, and nesting pairs typically require large home ranges. Territorial behavior in the form of calling and drumming becomes evident during early spring, and nesting activity usually commences in late March. According to Mengel (1965), clutches may be completed by the first week of April, with a probable peak in clutch completion during mid-April. Like most other woodpeckers, Pileateds apparently raise only one brood, and late clutches seem to be completed by mid-May. The average size of 11 clutches or broods thought to be complete was given by Mengel (1965) as 3.2 (range of 2–4).

Nest cavities are usually excavated in dead trees, although living trees or dead snags in living trees are also used. Sometimes Pileateds take over and enlarge cavities started by other woodpeckers. Choice cavities may be used for several years if they remain intact. In Kentucky, deciduous trees are used primarily, although Red-cockaded Woodpecker cavities in pines are sometimes enlarged and used. Nest trees are usually located within forest interior, but often along the edge of an open stream or road corridor, and more rarely in smaller woodlots or groves of trees in otherwise open situations. The average height of 13 active nests reported by Mengel (1965) was 38 feet (range of 18–70 feet), although more recent nests have been reported as low as 12 feet (KOS Nest Cards).

Forest Cover

Value	% of Blocks	Avg Abund
All	58.3	1.8
1	30.7	1.7
2	48.4	1.5
3	61.7	1.7
4	77.2	2.0
5	76.9	2.0

The atlas fieldwork yielded reports of Pileated Woodpeckers in more than 58% of priority blocks statewide, and 34 incidental

Ron Austing

observations were reported. As expected, occurrence by physiographic province section was closely related to the degree of forestation. Occurrence was highest in the Cumberland Plateau, the Cumberland Mountains, and the Mississippi Alluvial Plain, and lowest in the Shawnee Hills, the East Gulf Coastal Plain, and the Blue Grass. The high occurrence in the Mississippi Alluvial Plain probably can be attributed to the distribution and type of forest in this region. Most of the remaining forest is situated in fairly contiguous blocks, making it attractive to Pileateds. In addition, the forests of the Mississippi Alluvial Plain are floodplain types, seemingly among the most favored by the species. Average abundance was relatively low and unvaried across the state.

Less than 6% of priority block records were for confirmed breeding. Most of the 24 confirmed reports were based on the observation of recently fledged young in the company of adults. Other confirmed reports were based on birds entering nest cavities and nests containing young.

Pileated Woodpeckers are recorded in very small numbers on most Kentucky BBS routes. The average number of individuals reported per BBS route for the periods 1966–91 and 1982–91 was 1.30 and 2.10, respectively. Trend analysis of these data yields small, nonsignificant decreases of 0.3% per year for the period 1966–91 and 3.0% per year for the period 1982–91.

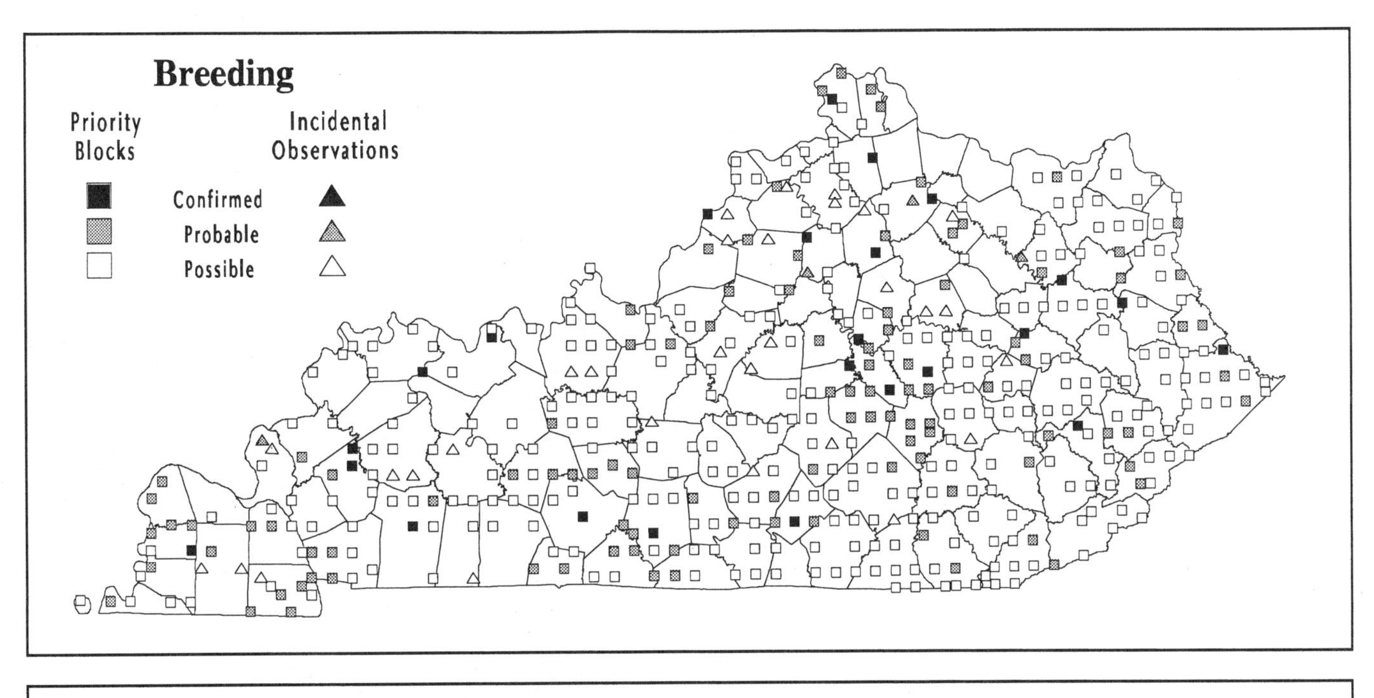

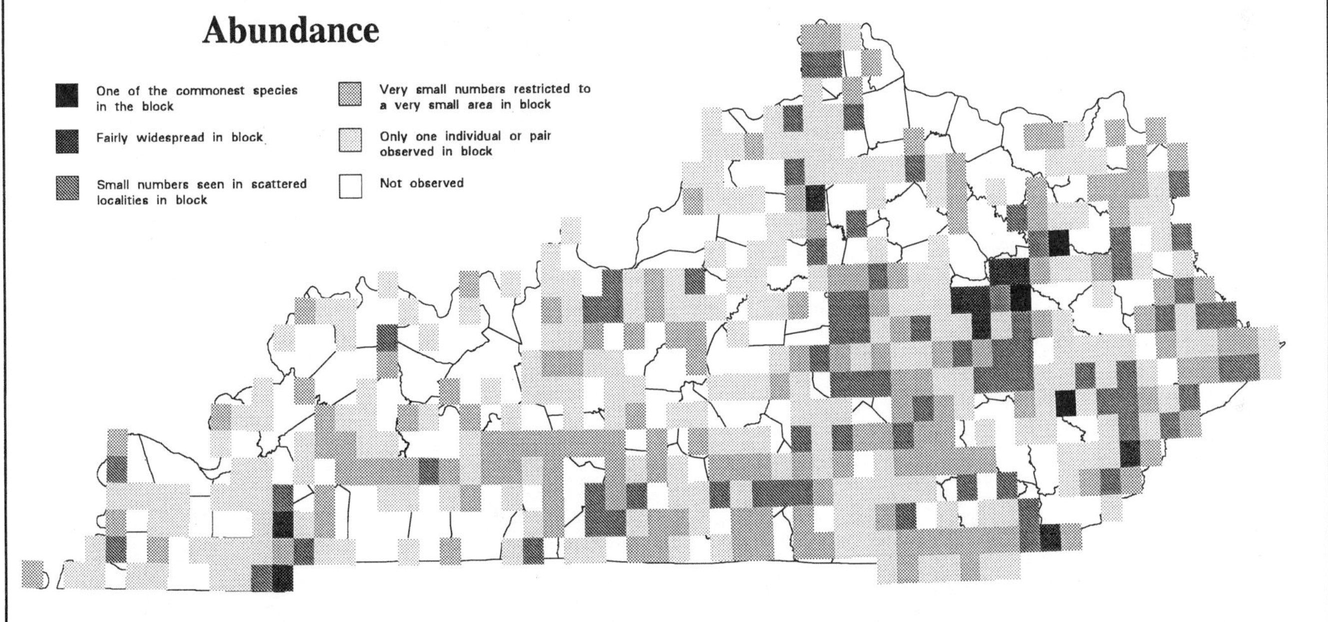

Analysis of Block Data by Physiographic Province Section

Physiographic Province Section	Total Blocks Surveyed	Blocks with Data	Avg Abund	% with Data	Section's % for State
Mississippi Alluvial Plain	14	10	1.7	71.4	2.4
East Gulf Coastal Plain	36	18	1.3	50.0	4.2
Highland Rim	139	83	1.8	59.7	19.6
Shawnee Hills	142	72	1.5	50.7	17.0
Blue Grass	204	95	1.8	46.6	22.4
Cumberland Plateau	173	132	1.9	76.3	31.1
Cumberland Mountains	19	14	2.0	73.7	3.3

Summary of Breeding Status

Number of Blocks in Which Species Was Recorded		
Total	**424**	**58.3%**
Confirmed	24	5.7%
Probable	106	25.0%
Possible	294	69.3%

Pileated Woodpecker

Eastern Wood-Pewee

Contopus virens

The distinctive, whistled song of the Eastern Wood-Pewee is a relatively common summer sound across much of Kentucky. Mengel (1965) regarded the species as a common summer resident throughout the state, although he stated that it was not as numerous in the more mesic forest communities of the Cumberland Plateau and the Cumberland Mountains.

Wood-pewees are found in a great variety of semi-open habitats, from bottomland hardwood forests of the large river floodplains in western Kentucky to mixed pine-hardwood associations of the Cumberland Plateau and the Cumberland Mountains. They are most frequent in habitats with some degree of openness, whether it is the result of forest structure, natural disturbance, or human alteration. Their primary feeding habit involves watching for passing insect prey from an exposed perch, and more closed-in situations may thus be less favorable for foraging. For this reason, the species may be absent from younger, second-growth forest where an open midstory has not developed. In such habitat, wood-pewees often frequent edges and road or stream corridors. Understory removal, such as is typically found in parks and rural yards, often creates an open aspect to forested areas, and wood-pewees occur frequently in such situations.

The Eastern Wood-Pewee seems to have adapted fairly well to changes in the landscape that have occurred since European settlement. Fragmentation of forested areas has created an abundance of edge habitat, and the species even occurs in small woodland tracts in open areas. In many cases these artificial edges have replaced open-canopy woodland and old-growth forest that have been cleared or altered by logging. Although there have certainly been local declines, the species may be about as common statewide now as at any previous time.

Eastern Wood-Pewees winter primarily in South America (AOU 1983), and they usually do not become conspicuous until after the first week of May. In fact, it may be the last week in the month before the full complement of breeding birds has returned. Nest building takes place during late May and early June, with a peak of clutch completion about June 10 (Mengel 1965). Although a later, less noticeable peak in clutch completion (August 1–10) may involve renestings of birds that have experienced failure because of storms or predation, it is apparent that some pairs raise two broods (Mengel 1965). The average size of five clutches and one brood reported by Mengel was 2.7 (range of 2–4). Fuller (1951) documented a pair that used the same nest twice in one season in Marshall County.

Forest Cover

Value	% of Blocks	Avg Abund
All	93.0	2.7
1	93.1	2.4
2	96.3	2.9
3	93.0	2.8
4	90.3	2.7
5	88.5	2.3

Wood-pewees typically choose a large deciduous tree for the nest site, but they may use conifers in mixed forest types. The nest is usually placed at the base of a horizontal fork in the outer portion of a relatively large, horizontal branch, often open to the ground below. The average height of 13 nests reported by Mengel (1965) was 30.8 feet (range of 15–50 feet). The nest is a shallow, compact cup constructed of soft plant material and covered with lichens (Harrison 1975).

Ron Austing

Eastern Wood-Pewees were recorded in 93% of the priority blocks statewide, and eight incidental observations were reported. The species ranked 13th according to the number of priority blocks, 21st by total abundance, and 25th by average abundance. Occurrence was greater than 90% in all regions except the Cumberland Plateau. Average abundance was relatively uniform across most of the state, but it was lowest in the Cumberland Mountains. Occurrence and average abundance varied only slightly by percentage of forest cover, being lowest in predominantly forested areas. Slightly lower figures in eastern Kentucky may have resulted in part from the average age of the forest, which typically has not matured enough to have developed an open midstory.

Despite its abundance, the Eastern Wood-Pewee was relatively hard to confirm as a breeder, and only about 7% of priority block records were confirmed. Most confirmed reports were based on the observation of adults carrying food, although a few active nests and recently fledged young were observed. Parasitism by Brown-headed Cowbirds was indicated by several records of wood-pewees feeding recently fledged cowbirds.

Eastern Wood-Pewees are reported on most Kentucky BBS routes in moderate numbers. The average number of individuals per BBS route for the periods 1966–91 and 1982–91 was 7.65 and 8.24, respectively. According to these data, the wood-pewee ranked 31st in abundance during the 1982–91 period. Trend analysis of these data shows a significant ($p<.05$) decrease of 1.7% per year for the period 1966–91 and a nonsignificant decrease of 1.3% per year for the period 1982–91.

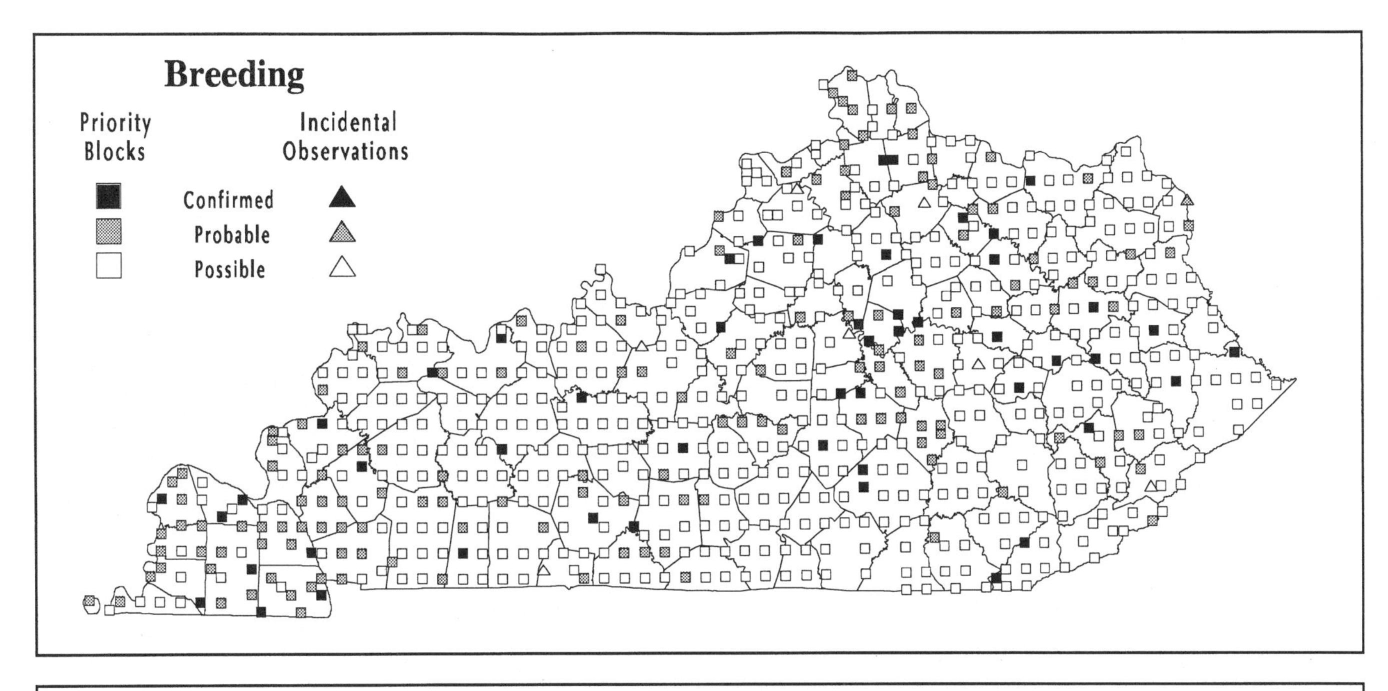

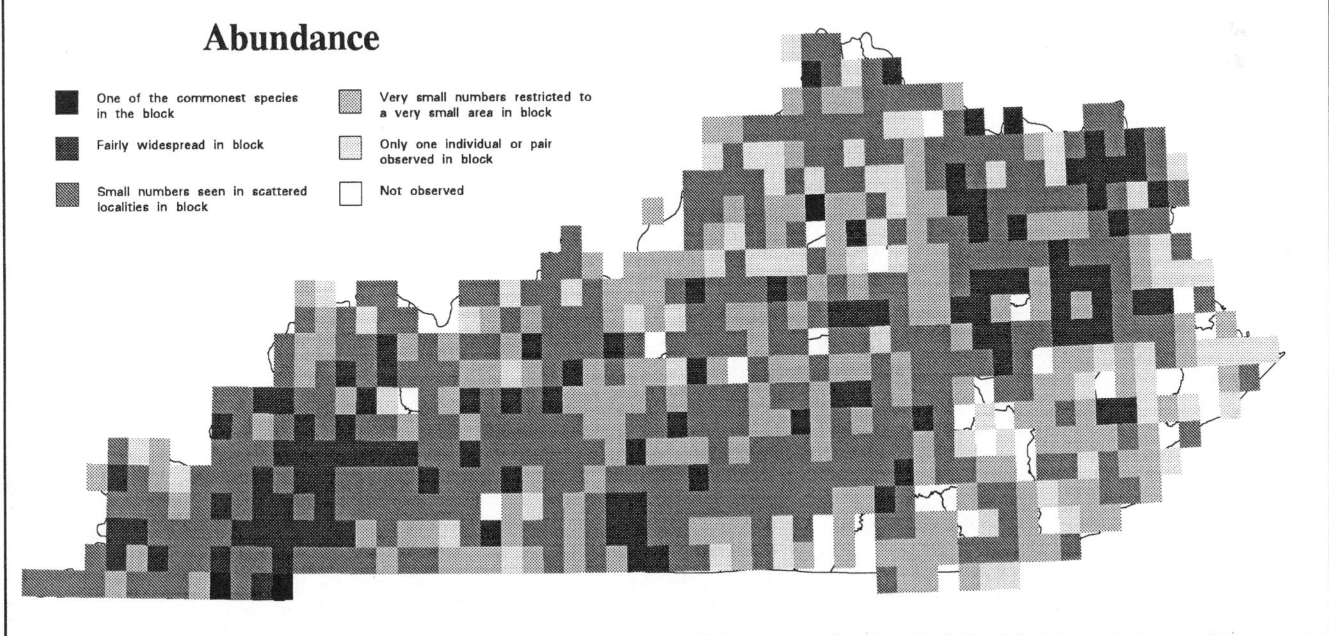

Analysis of Block Data by Physiographic Province Section

Physiographic Province Section	Total Blocks Surveyed	Blocks with Data	Avg Abund	% with Data	Section's % for State
Mississippi Alluvial Plain	14	14	3.1	100.0	2.1
East Gulf Coastal Plain	36	36	3.1	100.0	5.3
Highland Rim	139	135	2.8	97.1	20.0
Shawnee Hills	142	137	2.9	96.5	20.3
Blue Grass	204	189	2.6	92.6	28.0
Cumberland Plateau	173	147	2.6	85.0	21.7
Cumberland Mountains	19	18	2.1	94.7	2.7

Summary of Breeding Status

Number of Blocks in Which Species Was Recorded		
Total	**676**	**93.0%**
Confirmed	50	7.4%
Probable	144	21.3%
Possible	482	71.3%

Eastern Wood-Pewee

Acadian Flycatcher

Empidonax virescens

The Acadian Flycatcher is the only member of the genus *Empidonax* that is widespread in Kentucky in summer. Although the species is not typically considered abundant, its occurrence is relatively predictable in appropriate habitat throughout the state. Mengel (1965) regarded the Acadian Flycatcher as a fairly common to common summer resident statewide.

This small flycatcher is a bird of forested areas, and it usually occurs near water. The species is most frequently found along creeks and streams, but substantial numbers also occur along larger rivers, in swamps, and in forested reservoir backwaters. The only habitats where Acadians commonly nest relatively far from water are bottomland hardwood forests in river floodplains and in mesophytic forests of upland ravines. Otherwise, it is rare to hear the male's abrupt, two-syllable song out of sight of a stream or swamp. Although the Acadian Flycatcher sometimes occurs in relatively small tracts of woodland and along narrow riparian corridors, it is much more abundant in extensively forested areas. Thus the species may be rare or absent where clearing for settlement or agriculture has been extensive.

It is likely that the Acadian Flycatcher is less numerous in Kentucky today than it was two centuries ago. Extensive clearing and fragmentation of forested areas, especially in bottomlands, have resulted in a substantial reduction in the amount of suitable habitat. The species does not require mature forest, however, so it remains fairly abundant in areas that still offer moderately large blocks of woodland.

Acadian Flycatchers winter in the tropics as far south as northern South America (AOU 1983), but they return to Kentucky during the first week of May. Males quickly set up territories, and nest building is often under way by the middle of the month. Early clutches are at least occasionally laid by the last week of May (Altsheler 1962; Stamm and Monroe 1990a). A peak in clutch completion occurs June 1–10 (Mengel 1965), but at least some pairs raise two broods (Altsheler 1962). Late nestings also may represent renesting following failure because of severe storms, flooding, or predation. Two broods were raised in the same nest in Bullitt County in 1961 (Altsheler 1962). The average size of 15 clutches known or thought to be complete was given by Mengel (1965) as 2.6 (range of 2–3), although Croft (1969) reported a nest containing four eggs.

Forest Cover

Value	% of Blocks	Avg Abund
All	67.5	2.2
1	31.7	1.6
2	55.8	2.1
3	74.3	2.1
4	86.4	2.4
5	94.2	2.4

The nest is a shallow saucer composed of grass and other plant material that is woven together and attached to a fork in a small branch. Its construction is often fragile enough that the eggs are visible from below. Nests are often placed among clusters of leaves, typically near the end of a pendant branch hanging over a stream, a wet weather creekbed, or occasionally a road. Numerous long pieces of grass or other material are usually left hanging off the main structure, probably to mimic flood debris left on lower branches by high water. Thus the nest is less conspicuous to predators. The average height of 23 nests given by Mengel (1965) was 11.7 feet (range of 2–28 feet), and Stamm (1961b) reported a nest 30 feet above the ground.

Ron Austing

The atlas survey yielded records of Acadian Flycatchers in nearly 68% of priority blocks statewide, and 23 incidental observations were reported. Occurrence was highest in the Mississippi Alluvial Plain and the Cumberland Plateau, and lowest in the Blue Grass. The high value in the Mississippi Alluvial Plain can be attributed to the predominance of floodplain forests. Average abundance was highest in the Cumberland Mountains and lowest in the Shawnee Hills, but it was not highly variable. Occurrence and average abundance were closely related to the percentage of forest cover, increasing substantially as forestation increased.

The Acadian Flycatcher's song is unmistakable, but its inconspicuous plumage and secretive forest habits made this species difficult for atlasers to confirm. As a result, only about 7% of the priority block records were for confirmed breeding. Most confirmed records were based on the observation of active nests, although observation of recently fledged young, adults carrying food, and nest building were also reported.

Acadian Flycatchers are reported in small numbers on most Kentucky BBS routes. The average number of individuals per BBS route for the periods 1966–91 and 1982–91 was 3.07 and 2.09, respectively. Trend analysis of these data shows slight, nonsignificant increases of 2.5% per year for the period 1966–91 and 1.3% per year for 1982–91.

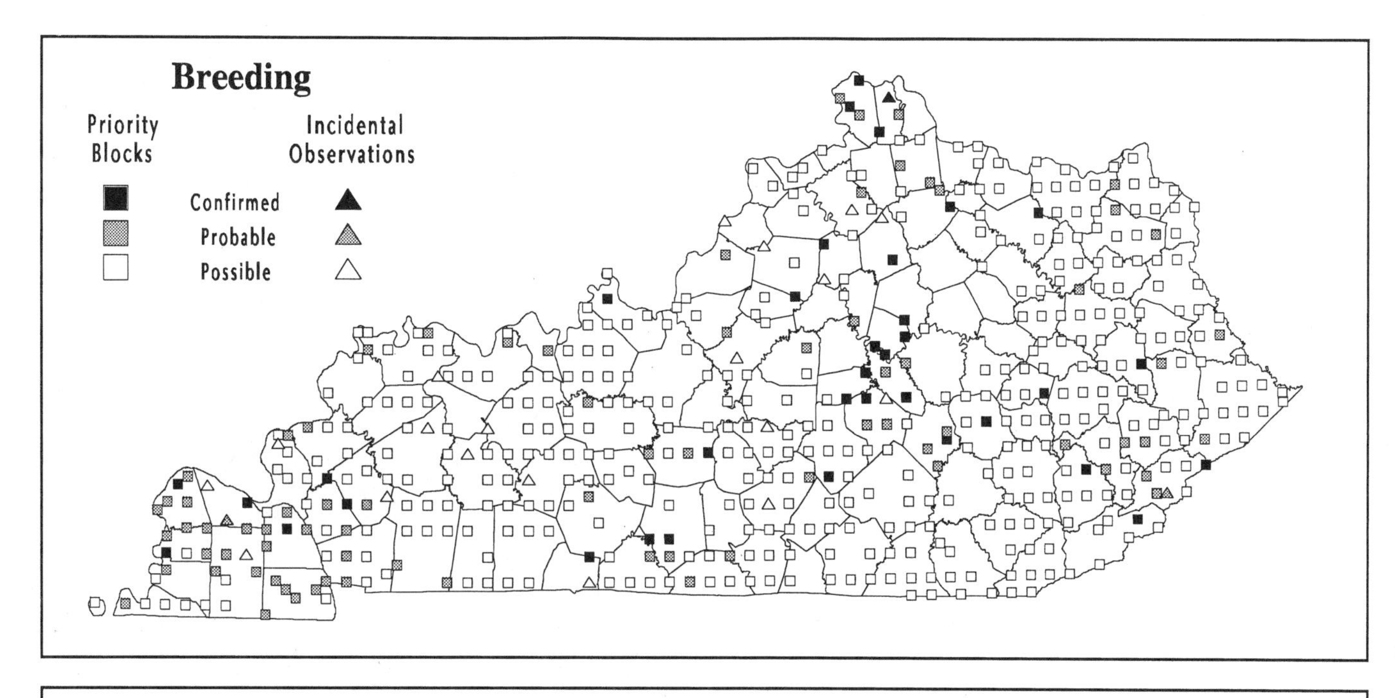

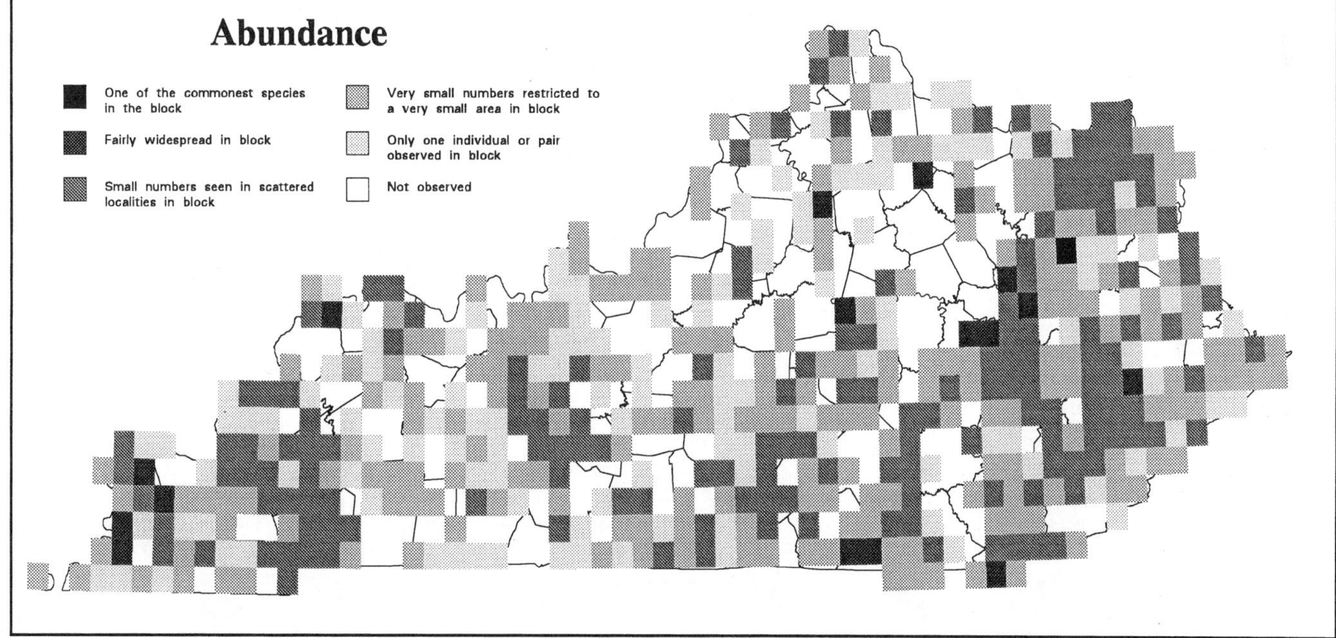

Analysis of Block Data by Physiographic Province Section

Physiographic Province Section	Total Blocks Surveyed	Blocks with Data	Avg Abund	% with Data	Section's % for State
Mississippi Alluvial Plain	14	13	2.3	92.9	2.6
East Gulf Coastal Plain	36	26	2.3	72.2	5.3
Highland Rim	139	90	2.1	64.7	18.3
Shawnee Hills	142	101	1.9	71.1	20.6
Blue Grass	204	97	2.1	47.5	19.8
Cumberland Plateau	173	151	2.3	87.3	30.8
Cumberland Mountains	19	13	2.5	68.4	2.6

Summary of Breeding Status

Number of Blocks in Which Species Was Recorded		
Total	**491**	**67.5%**
Confirmed	35	7.1%
Probable	77	15.7%
Possible	379	77.2%

Acadian Flycatcher

Willow Flycatcher

Empidonax traillii

The Willow Flycatcher was formerly regarded as conspecific with the nearly indistinguishable Alder Flycatcher (*E. alnorum*). The two were collectively known as the Traill's Flycatcher (*E. traillii*) until 1973, when they were split taxonomically and new common names were assigned to each (AOU 1973). The nesting ranges of both species lie predominantly to the north of Kentucky, but the Willow is the more southern of the two. Its range extends southward into the northern part of the state.

Before about 1960 very little was known about this small flycatcher in Kentucky. Mengel (1965) listed only one breeding record from Jefferson County, and only a handful of records of migrants had been reported as of the late 1950s. Based on information from surrounding states, this paucity of records likely resulted from the species's absence rather than its being overlooked. For example, Willow Flycatchers were virtually unknown in the Cincinnati area until the 1940s, when they became widespread (Kemsies and Randle 1953). About this same time a southward expansion of the Traill's Flycatcher's nesting range was being documented elsewhere in the Ohio Valley and the southern Appalachian Mountains (Croft 1961; Peterjohn and Rice 1991). During the 1960s reports began to accumulate in north central Kentucky, and by the mid-1970s the bird was scattered about the northern half of the state. Today the species occurs nowhere commonly, but it is reported regularly at scattered localities throughout the northern two-thirds of the state.

The Willow Flycatcher occurs in a variety of early successional habitats. The species is encountered most often in patches of young trees along open stream corridors or in marshy areas, but it is also found occasionally in drier areas, especially in old fields and pastures regenerating from past agricultural use. As its name implies, this small flycatcher is often found in willows, but it also can be seen in thickets of other species, including alder, mulberry, black locust, indigo bush, and maple. In general, Willow Flycatchers are found much more frequently in patches of young trees situated in fairly open areas than in those surrounded predominantly by forest.

It is likely that the Willow Flycatcher's southward expansion into Kentucky has at least partly resulted from habitat alteration. Appropriate nesting habitat likely was extremely limited until relatively recent times. All presently occupied summer habitat has been created or altered by humans. It is possible that the only naturally occurring nesting habitat was provided by thickets of young trees recolonizing wet prairies after fires and patches of early successional woody vegetation in wetlands or riparian zones along major rivers.

Willow Flycatchers winter in Middle America (AOU 1983), and they return to their Kentucky breeding grounds relatively late. Most birds do not appear until the middle of May, and the full complement of breeding birds may not return until early June. The species is infrequently encountered as a transient, and most individuals heard singing in appropriate nesting habitat, even during the latter half of May, are probably on territory. Details of nesting were reported primarily during the 1960s, when the species first appeared in significant numbers (Croft 1961, 1962, 1964; Mengel 1965; Dubke 1966b; Stamm and Jones 1966; Croft and Stamm 1967; Stamm and Croft 1968; Croft and Lawrence 1970). Nest building commences by mid-June, and egg laying likely extends from mid-June to mid-July, with a peak in clutch completion occurring from late June to early July. Most young probably fledge during the latter half of July. It appears that only one brood is raised. The average size of eight clutches or broods reported in the sources noted above is 3.2 (range of 2–4).

Alvin E. Staffan

The nest is a sturdy cup, typically constructed of cottony or silky plant material and grass (Croft 1961). It is usually found sheltered among the leaves in an outer, often upright fork of the branch of a small tree or shrub. The average height of 17 nests reported in the sources noted above is 6.2 feet (range of 3–15 feet).

Willow Flycatchers were recorded in about 6% of priority blocks, and 12 incidental observations were reported. The species was found in all physiographic province sections except the Mississippi Alluvial Plain and the Cumberland Mountains. Occurrence was highest in the Blue Grass, where these flycatchers were found in nearly one of every ten blocks. A number of atlas reports represented new county breeding and breeding season records.

Only 15% of priority block reports were for confirmed breeding. It is likely, however, that most atlas records represented breeding birds, since migrants are so uncommon. Several active nests were found in Jefferson County near the Falls of the Ohio, but most of the six confirmed records were based on the observation of recently fledged young.

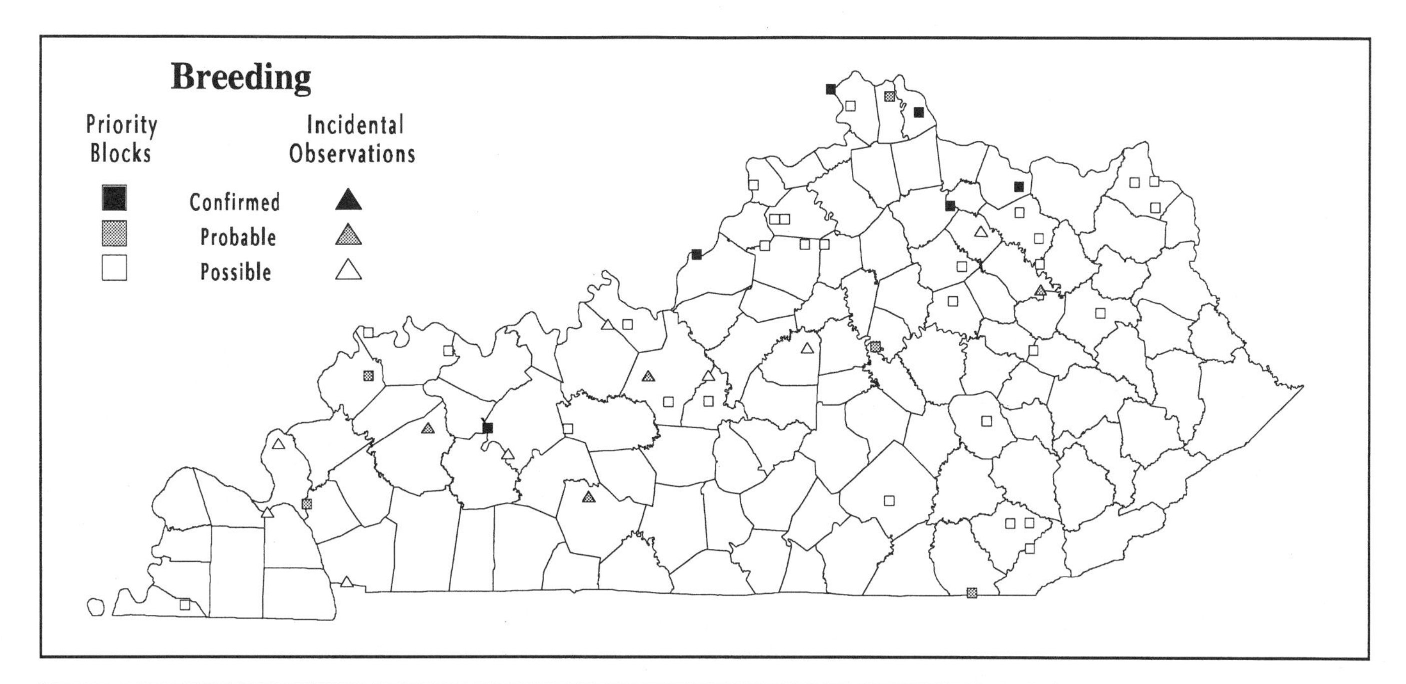

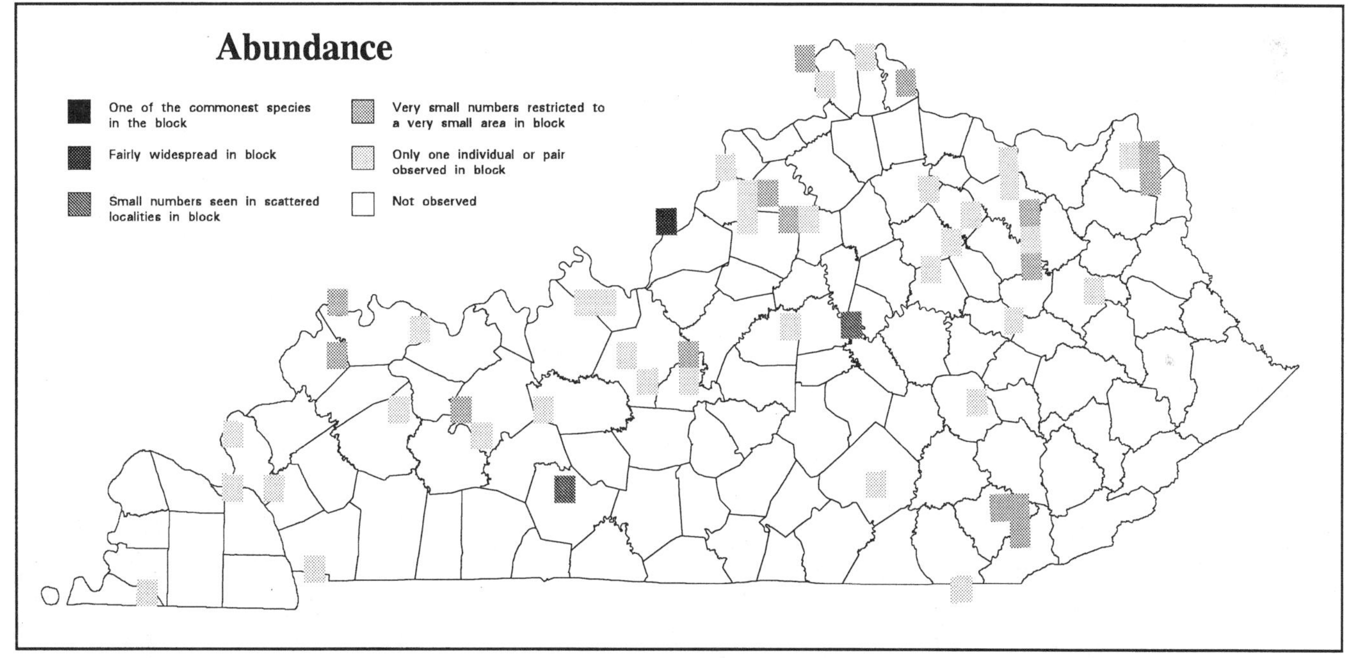

Analysis of Block Data by Physiographic Province Section

Physiographic Province Section	Total Blocks Surveyed	Blocks with Data	% with Data	Section's % for State
Mississippi Alluvial Plain	14	-	-	-
East Gulf Coastal Plain	36	1	2.8	2.5
Highland Rim	139	5	3.6	12.5
Shawnee Hills	142	5	3.5	12.5
Blue Grass	204	19	9.3	47.5
Cumberland Plateau	173	10	5.8	25.0
Cumberland Mountains	19	-	-	-

Summary of Breeding Status

Number of Blocks in Which Species Was Recorded		
Total	**40**	**5.5%**
Confirmed	6	15.0%
Probable	5	12.5%
Possible	29	72.5%

Willow Flycatcher

Least Flycatcher

Empidonax minimus

For many years the Least Flycatcher has eluded confirmation as a nesting bird in Kentucky. The species's breeding range lies predominantly to the north of the state, but it also extends southward through the Appalachian Mountains as far as northern Georgia (AOU 1983). Least Flycatchers occur regularly in the mountains of western Virginia (Kain 1987) and eastern Tennessee (Robinson 1990), and occasional reports of territorial birds have come from the summit of Black Mountain, in Harlan County. Unfortunately, however, efforts to confirm nesting of these birds have not been successful.

Breiding (1947) was the first ornithologist to report this small flycatcher on Black Mountain in summer. During two days of fieldwork in early July 1944, a single bird was recorded. The species went unreported there through the 1970s, but small numbers were reported in summer on several occasions during the early 1980s (Stamm 1983, 1984b, 1985).

In addition to observations from Black Mountain, three reports of Least Flycatchers came from other parts of eastern Kentucky in summer before the atlas fieldwork. Croft (1969) reported a territorial bird in woodland at the base of Pine Mountain near Whitesburg in 1967; D. Coskren reported a single territorial bird in Wolfe County in 1979 and 1980 (Stamm 1979a, 1980); and W.C. Greene Jr. reported another territorial bird in Elliott County in June 1979 (Stamm 1979a).

During the atlas survey there were several reports of territorial birds at the summit of Black Mountain. On May 27, 1985, a bird was observed carrying food or nesting material (Stamm 1985; D. Noonan, pers. comm.), but further evidence of nesting was not obtained. In late June 1986 and 1989 several territorial birds were seen and heard there (Stamm 1986a, 1989a). Away from Black Mountain, the species was found at only one location during the atlas fieldwork. At least two singing males were present over a two-week period in early successional habitat along Clear Creek near Chenoa, in Bell County. Other than the continued presence of a singing bird at the same locality for more than one week, no further evidence of nesting was obtained at either of these sites.

The Least Flycatcher is a bird of early successional forest and forest edge. At the summit of Black Mountain, it has been found in stands of young deciduous trees (mainly oaks) growing in areas disturbed by fire or logging, as well as early successional woody vegetation and woodland edge on an abandoned ski slope.

It is possible that habitat changes have caused the species's sporadic appearance on Black Mountain. Optimal nesting habitat was widespread at the summit of the mountain during the early 1980s, after a fire had killed many of the larger trees. During this time a thick growth of young trees colonized much of the area, and Least Flycatchers were conspicuous. In subsequent years the forest continued to grow, and optimal habitat is no longer abundant. It is interesting that suitable habitat along the crest of Cumberland Mountain has never yielded observations of birds in summer.

Gary Meszaros

An active nest has never been found in Kentucky, but according to Harrison (1975), the nest is a compact cup constructed of a variety of soft plant materials. Nests are commonly situated 10–20 feet above the ground (range of 2–60 feet), and rangewide clutch size is 3–6, commonly 4 (Harrison 1975).

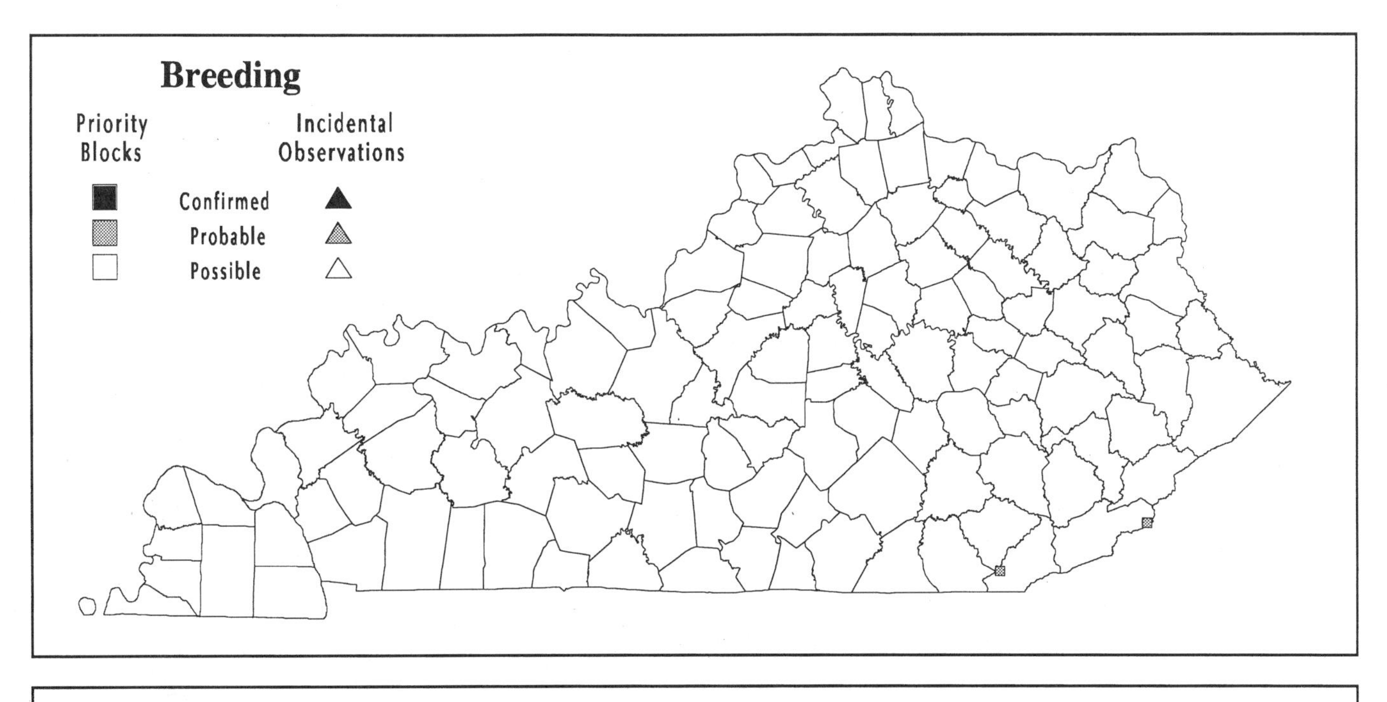

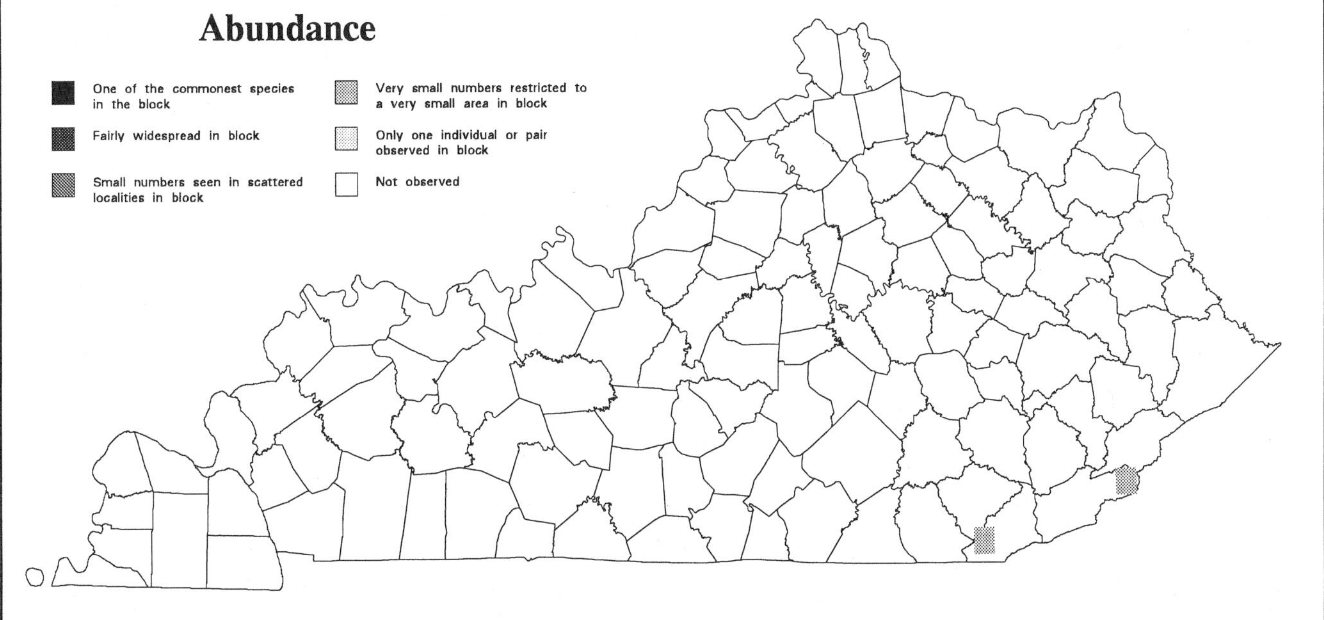

Analysis of Block Data by Physiographic Province Section

Physiographic Province Section	Total Blocks Surveyed	Blocks with Data	% with Data	Section's % for State
Mississippi Alluvial Plain	14	-	-	-
East Gulf Coastal Plain	36	-	-	-
Highland Rim	139	-	-	-
Shawnee Hills	142	-	-	-
Blue Grass	204	-	-	-
Cumberland Plateau	173	-	-	-
Cumberland Mountains	19	2	10.5	100.0

Summary of Breeding Status

Number of Blocks in Which Species Was Recorded		
Total	**2**	**0.3%**
Confirmed	-	-
Probable	2	100.0%
Possible	-	-

Least Flycatcher

Eastern Phoebe

Sayornis phoebe

The Eastern Phoebe is one of the most widespread nesting flycatchers in Kentucky. Even though its plumage is relatively inconspicuous, this flycatcher's easily recognized song is frequently heard throughout much of the state. Mengel (1965) regarded the species as a fairly common to common summer resident statewide.

Eastern Phoebes occur in a great variety of forested and semi-open habitats. In general, their abundance seems most directly related to the availability of suitable nest sites. Although the species occurs regularly along natural rock outcroppings, it is much more a bird of altered habitats today, being especially conspicuous in rural countryside.

The Eastern Phoebe must have occurred much less frequently in Kentucky before settlement. The species's widespread distribution results in large part from its adaptation to artificial nesting situations. Before artificial nest sites became available, phoebes were restricted to areas with natural outcroppings of rock. Such features are unevenly distributed across the state, and their scarcity or absence would have been a limiting factor in many areas.

Phoebes are among the earliest of Kentucky's songbirds to arrive on their breeding grounds. A few birds may overwinter during milder years, but the first significant warm spells of late winter bring with them the spirited songs of returning males. Many birds are on territory by early March, and most breeding pairs are established by April 1. According to Mengel (1965), an early peak in clutch completion occurs during the first ten days of April. The species is regularly double-brooded, although a later peak in clutch completion is poorly defined. Mengel (1965) reported that the average size of 18 clutches thought or known to be complete was 4.1 (range of 2–5), and Croft and Stamm (1967) reported an average size of nine clutches or broods as 4.6 (range of 3–5). A nest observed in Carroll County on May 31, 1987, contained six eggs.

Forest Cover

Value	% of Blocks	Avg Abund
All	84.6	2.3
1	62.4	2.0
2	83.7	2.2
3	86.5	2.4
4	95.5	2.5
5	90.4	2.6

The nest is typically situated beneath the cover of a natural or artificial overhang. Clifflines, cave entrances, and other rock outcroppings serve as natural nest sites. Audubon (1861) noted that the species nested in the sinkholes of the barrens, sometimes as much as 20 feet below the surface of the ground. A variety of artificial sites are used, including bridges, culverts, road cuts, porches of rural homes, and barns.

The nest itself is usually constructed on a small, horizontal shelf or is at least begun on some sort of projection if only a vertical surface is available. New nests are sometimes constructed atop old phoebe or swallow nests. The nest is a neatly cupped mound composed primarily of plant material, although some mud may be used to strengthen the structure. It is invariably covered by an outer layer of moss and lined with fine grass or animal hair. The average height of 27 nests summarized by Croft and Stamm (1967) and Stamm and Croft (1968) was 6.5 feet (range of 3–18 feet).

Alvin E. Staffan

The Eastern Phoebe was the second most frequently recorded flycatcher during the atlas fieldwork. It was found in nearly 85% of priority blocks statewide, and 18 incidental observations were reported. The species ranked 24th according to the number of priority blocks, 33rd by total abundance, and 38th by average abundance. Only in the Mississippi Alluvial Plain, where suitable nest sites are relatively scarce, was the species reported in less than 75% of priority blocks. Average abundance was fairly consistent across the state. Occurrence and average abundance were higher in forested areas than in predominantly open areas. Phoebes were especially conspicuous in the extensively forested areas of eastern Kentucky, where they were more common as a "yard bird" than American Robins.

The phoebe was one of the least difficult species to confirm as breeding, and 38% of priority block records were confirmed. Nests are often placed conspicuously, especially under bridges, and both active and recently used nests were usually easy to find if a few bridges and culverts were present within a block. Other confirmed records were based on the observation of recently fledged young, nest building, and adults carrying food for young.

Eastern Phoebes are reported on most Kentucky BBS routes in small numbers. The average number of individuals per BBS route for the periods 1966–91 and 1982–91 was 3.70 and 4.83, respectively. Trend analysis of these data shows a significant ($p<.05$) increase of 1.8% per year for the period 1966–91 and a nonsignificant increase of 2.2% per year for the period 1982–91. These data suggest that short-term declines apparently caused by severe winter weather (Wilson and Stamm 1960; Robbins et al. 1986) have not affected the nesting population overall.

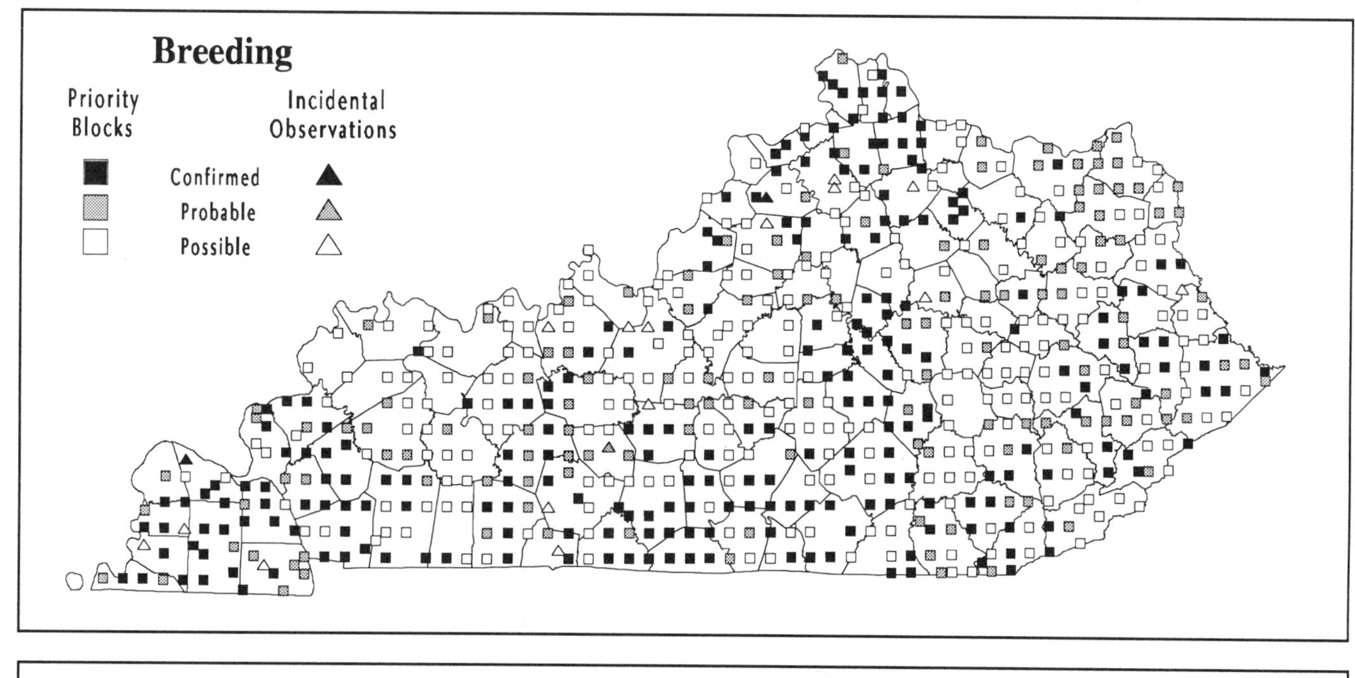

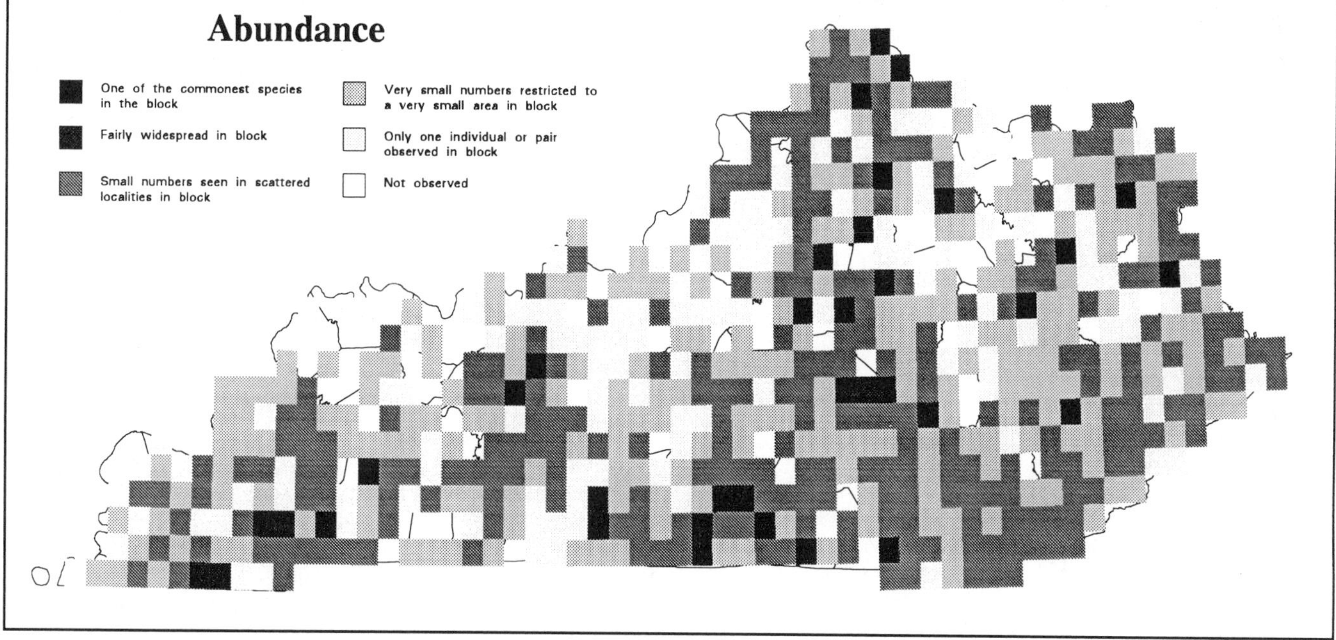

Analysis of Block Data by Physiographic Province Section

Physiographic Province Section	Total Blocks Surveyed	Blocks with Data	Avg Abund	% with Data	Section's % for State
Mississippi Alluvial Plain	14	7	2.0	50.0	1.1
East Gulf Coastal Plain	36	32	2.3	88.9	5.2
Highland Rim	139	123	2.4	88.5	20.0
Shawnee Hills	142	108	2.1	76.1	17.6
Blue Grass	204	163	2.3	79.9	26.5
Cumberland Plateau	173	165	2.4	95.4	26.8
Cumberland Mountains	19	17	2.6	89.5	2.8

Summary of Breeding Status

Number of Blocks in Which Species Was Recorded		
Total	**615**	**84.6%**
Confirmed	234	38.0%
Probable	136	22.1%
Possible	245	39.8%

Eastern Phoebe

Great Crested Flycatcher

Myiarchus crinitus

The Great Crested Flycatcher's distinctive, whistled call notes are a frequent summer sound across most of Kentucky. Mengel (1965) regarded the species as a common summer resident throughout the state except on the higher parts of Black Mountain, but Davis et al. (1980) found it regularly in dry oak and northern hardwood forests along the ridge of Cumberland Mountain.

These large flycatchers are found in a great variety of forested habitats, although they prefer open woodland, where they forage primarily high in the forest canopy and upper midstory levels. The species also occurs frequently along forest edge and in semi-open areas adjacent to forest, inhabiting altered situations such as suburban parks as well as naturally occurring openings associated with wetlands, riparian corridors, and clifflines.

Whether the Great Crested Flycatcher's status in Kentucky has changed appreciably in the last two centuries is unclear. The species has seemingly adapted fairly well to most alteration of the landscape that has resulted from human activity. Dissection of extensive areas of forest has resulted in the creation of an abundance of edge that is suitable for nesting, replacing to some degree naturally open woodlands and old-growth forests that likely afforded excellent nesting habitat. In contrast, logging of old-growth forests has typically resulted in a predominance of younger forests with more closed-in canopy and midstory conditions. This change in forest structure may have eliminated the species from large areas of remaining forest.

Great Crested Flycatchers winter primarily in Middle America (AOU 1983), and birds typically return to Kentucky during the last two weeks of April. Nesting is a prolonged activity for this species, and apparently only one brood is raised (Mengel 1965). Egg laying commences by mid-May, and clutch completion apparently peaks during the first ten days of June (Mengel 1965). Mengel gave the average size of 10 clutches or broods as 4.5 (range of 3–5), although a nest in Daviess County contained six young on June 21, 1972 (KOS Nest Cards).

Forest Cover

Value	% of Blocks	Avg Abund
All	67.3	2.1
1	79.2	2.0
2	84.2	2.1
3	77.0	2.1
4	40.3	1.9
5	19.2	1.7

Unlike other flycatchers that breed in Kentucky, this species nests in cavities, most frequently using old woodpecker nests and natural cavities in trees but sometimes nesting in artificial structures such as nest boxes, mailboxes, or utility poles. The lower portion of the cavity is filled with plant material of various types and usually some snakeskin or clear plastic wrapping (Harrison 1975). The eggs are laid in a well-formed cup lined with finer grasses. The average height of 10 nests reported by Mengel (1965) was 8.6 feet (range of 4–15 feet), while the average height of 10 nests reported by Croft and Stamm (1967), Stamm and Croft (1968), and KOS Nest Cards was 17.2 feet (range of 7–35 feet).

Bill Schoettler

The atlas survey yielded records of Great Crested Flycatchers in more than 67% of priority blocks statewide, and 15 incidental observations were reported. Occurrence was higher than 85% in all regions in the western half of the state, but only about 71% in the Blue Grass and less than 30% in the Cumberland Plateau and the Cumberland Mountains. While the lower occurrence in the Blue Grass probably can be attributed to the extent of deforestation in the Inner Blue Grass subsection, the even lower totals in eastern Kentucky apparently indicate the species's preference for open woodland habitats and edge. It is possible that the average forest over much of eastern Kentucky is too young to provide optimal habitat because of the lack of mature forest structure, including more open midstory levels. Appropriate situations may be provided in greater abundance under artificial conditions in the more highly fragmented forests of central and western Kentucky. Average abundance was highest in the Mississippi Alluvial Plain and lowest in the Cumberland Mountains and the Cumberland Plateau. These trends are probably most closely related to the degree of forestation, as occurrence decreased substantially in predominantly forested areas.

Despite the Great Crested Flycatcher's overall abundance and conspicuous calls, only about 8% of priority block records were for confirmed breeding. At least six active nests were located, but most confirmed records were based on the observation of recently fledged young. Nest building and adults carrying food for young were also used to confirm breeding.

Great Crested Flycatchers are reported in small numbers on most Kentucky BBS routes. The average number of individuals per BBS route for the periods 1966–91 and 1982–91 was 2.84 and 3.02, respectively. Trend analysis of these data yields only small, non-significant decreases of 0.4% per year for the period 1966–91 and 2.1% per year for 1982–91.

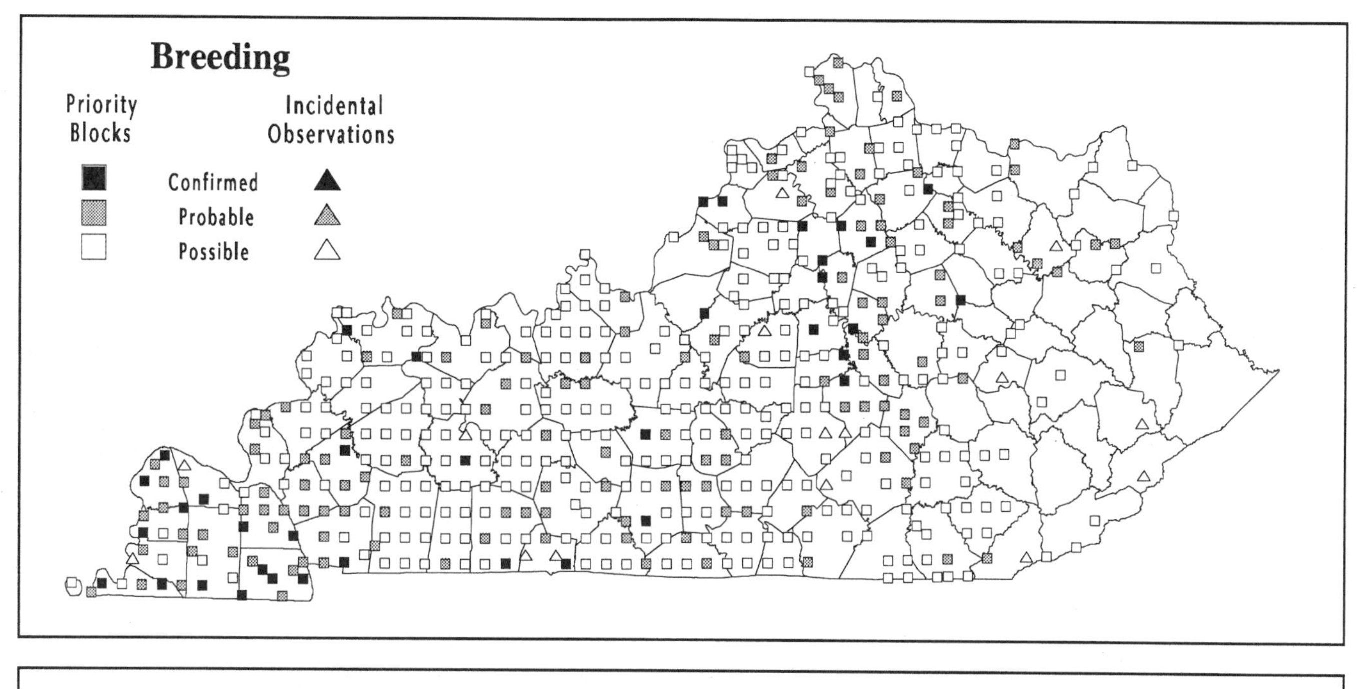

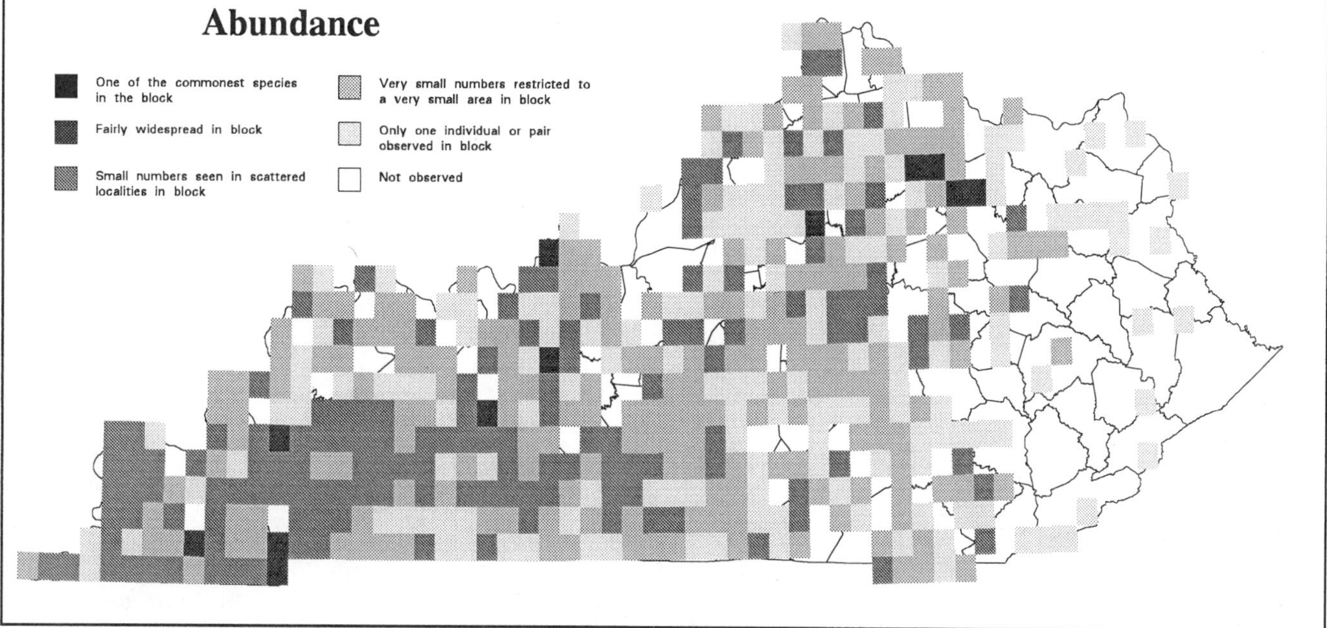

Analysis of Block Data by Physiographic Province Section

Physiographic Province Section	Total Blocks Surveyed	Blocks with Data	Avg Abund	% with Data	Section's % for State
Mississippi Alluvial Plain	14	12	2.8	85.7	2.5
East Gulf Coastal Plain	36	33	2.6	91.7	6.7
Highland Rim	139	123	2.1	88.5	25.2
Shawnee Hills	142	122	2.2	85.9	24.9
Blue Grass	204	144	2.0	70.6	29.4
Cumberland Plateau	173	50	1.5	28.9	10.2
Cumberland Mountains	19	5	1.6	26.3	1.0

Summary of Breeding Status

Number of Blocks in Which Species Was Recorded		
Total	**489**	**67.3%**
Confirmed	37	7.6%
Probable	134	27.4%
Possible	318	65.0%

Great Crested Flycatcher

Eastern Kingbird

Tyrannus tyrannus

The Eastern Kingbird is one of the most conspicuous summer birds throughout much of central and western Kentucky. Pairs noisily defend their territories with an intensity and persistence unrivaled among the state's summer avifauna. Only in the more heavily forested portions of eastern Kentucky does the species become much less common. There it is restricted primarily to river floodplains and other open areas where the forests have been cleared. Mengel (1965) regarded the kingbird as a fairly common to common summer resident wherever suitable habitat was available.

Eastern Kingbirds are birds of semi-open to open habitats with scattered trees. They are most conspicuous in rural farmland and settlement, although substantial numbers inhabit a variety of other situations, including suburban parks, riparian corridors, open wetlands, reservoir margins, and reclaimed surface mines. They forage predominantly from exposed perches, and the utility wires that crisscross the rural countryside provide optimal opportunity to capture passing winged prey.

Kingbirds were likely present in substantial numbers in some parts of Kentucky two centuries ago, especially in the barrens of southern and western Kentucky and the savannas of the Blue Grass. Overall, however, these flycatchers are certainly more widespread and common across the state today. The clearing and fragmentation of extensive areas of unbroken forest have no doubt created much suitable habitat where the species was formerly absent. In fact, in highly altered regions like the Inner Blue Grass, the kingbird is as conspicuous as any other breeding species.

Eastern Kingbirds winter in South America (AOU 1983), and most birds return during the latter half of April. Territory establishment and nest building may be prolonged, and although egg laying may commence by the second week of May, a peak in clutch completion does not occur until around June 1 (Mengel 1965). Although there are some records of late nestings, confirmed cases of double-brooding are lacking, and most late attempts are likely the result of renesting efforts following failure because of severe weather or predation. Kingbirds leave the breeding grounds quite early, and very few birds remain after the first week of September. The average size of three clutches listed by Mengel (1965) was 3.3 (range of 3–4), and Harrison (1975) gives rangewide clutch size as 3–5.

Forest Cover

Value	% of Blocks	Avg Abund
All	82.7	2.8
1	97.0	3.1
2	99.5	3.1
3	95.7	2.7
4	55.8	2.3
5	15.4	1.6

Eastern Kingbirds usually construct their nests high in large trees, often choosing one that is standing alone or along a fencerow. In much of Kentucky, sycamores seem to be favored, although the birds use a variety of deciduous trees. Nests also are occasionally observed in pines and on dead snags, the latter especially over water. The nest is usually placed near the end of a relatively sturdy branch, often in a fork among the leaves of the outer crown. Mengel (1965) gave the average height of 11 nests as 33 feet, with the highest being 75 feet above the ground. Over water, nests may be placed much lower. An atypical nest located during atlas work in a rural yard in Bourbon County was placed only five feet above the ground in a ten-foot-tall crabapple tree. The nest is relatively bulky and composed of a variety of coarse plant material and lined with finer grasses.

Ron Austing

Eastern Kingbirds were recorded on nearly 83% of priority blocks statewide, and eight incidental observations were reported. The species ranked 29th according to the number of priority blocks, 26th by total abundance, and 22nd by average abundance. Occurrence was greater than 95% in all physiographic province sections west of the Cumberland Plateau. Average abundance varied similarly, also being lowest in eastern Kentucky. Both occurrence and average abundance were closely related to the percentage of forest cover, being highest in relatively open areas and decreasing substantially as forestation increased.

Nearly 24% of priority block records were for confirmed breeding. Although a number of active nests were located, most confirmed reports were based on the observation of family groups of recently fledged young. Bobtailed young typically are conspicuous in July, as they perch on telephone lines and bare branches, begging persistently for food. Nest building and adults carrying food for young also accounted for a substantial number of confirmed records.

Eastern Kingbirds are reported in small to moderate numbers on most Kentucky BBS routes. The average number of individuals per BBS route for the periods 1966–91 and 1982–91 was 5.88 and 7.05, respectively. Trend analysis of these data shows a slight, nonsignificant increase of 0.4% per year for the period 1966–91 and a slight, nonsignificant decrease of 0.4% per year for the period 1982–91.

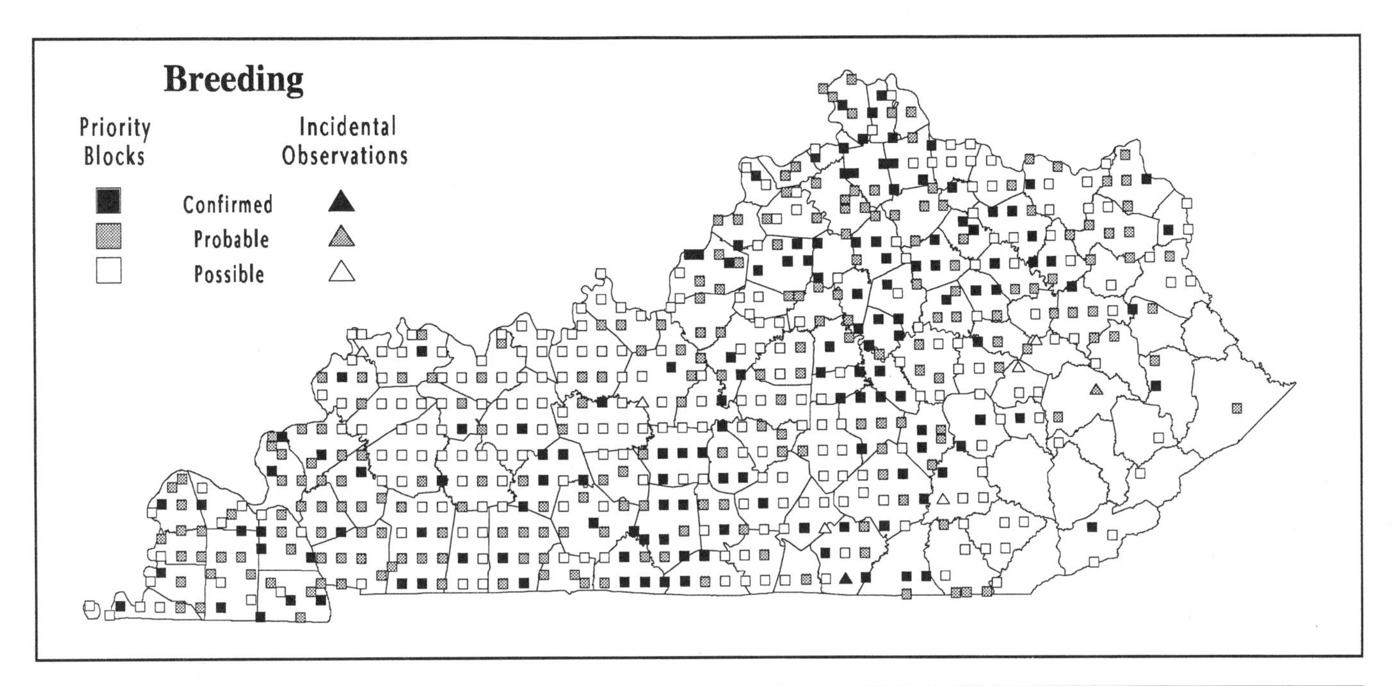

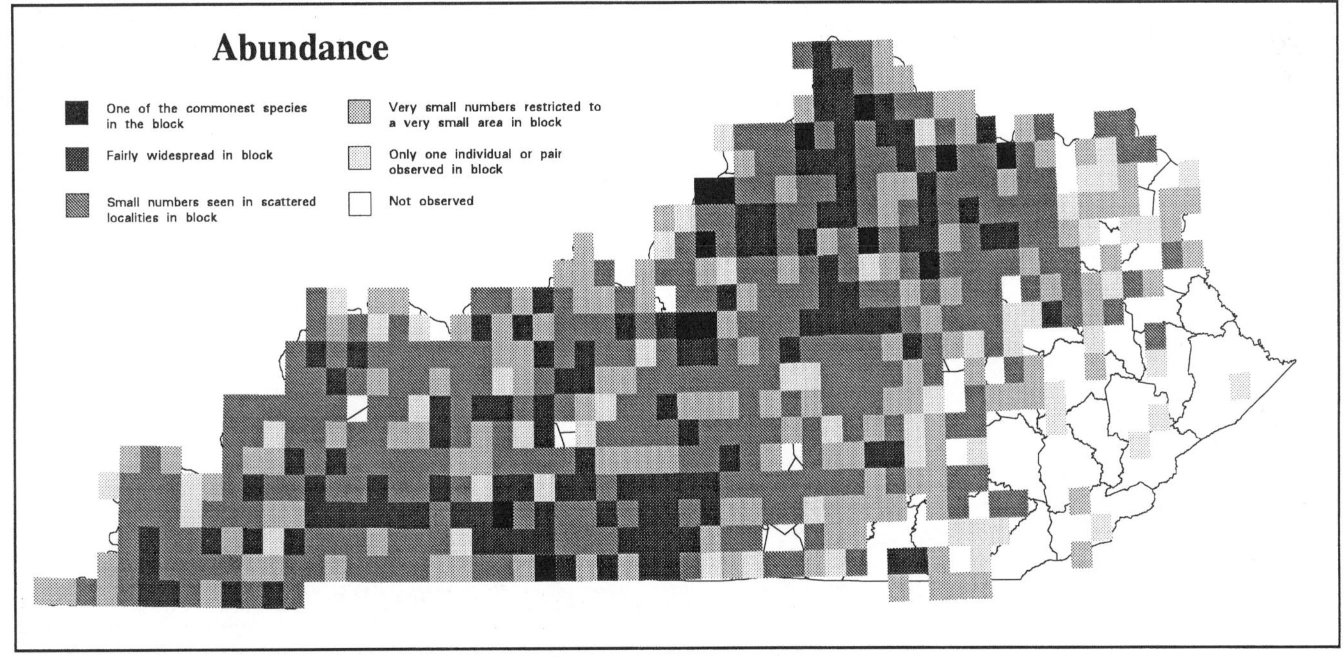

Analysis of Block Data by Physiographic Province Section

Physiographic Province Section	Total Blocks Surveyed	Blocks with Data	Avg Abund	% with Data	Section's % for State
Mississippi Alluvial Plain	14	14	2.4	100.0	2.3
East Gulf Coastal Plain	36	36	2.9	100.0	6.0
Highland Rim	139	135	3.0	97.1	22.5
Shawnee Hills	142	136	2.8	95.8	22.6
Blue Grass	204	200	3.0	98.0	33.3
Cumberland Plateau	173	75	2.1	43.4	12.5
Cumberland Mountains	19	5	1.6	26.3	0.8

Summary of Breeding Status

Number of Blocks in Which Species Was Recorded		
Total	**601**	**82.7%**
Confirmed	142	23.6%
Probable	219	36.4%
Possible	240	39.9%

Eastern Kingbird

Horned Lark

Eremophila alpestris

The Horned Lark is a locally distributed summer resident throughout Kentucky, its occurrence being closely tied to the availability of suitable open habitat. The species is often overlooked because of its preference for large expanses of open ground and its cryptic plumage. Mengel (1965) regarded the Horned Lark as an uncommon to fairly common, but local, summer resident.

Horned Larks are birds of open habitats with a predominance of short or sparse ground cover and patches of bare ground. In Kentucky the species is restricted to altered habitats, being most frequent in tilled agricultural land. These larks also can be found in a variety of other open situations, including overgrazed pastures, airports, and reclaimed surface mines.

Before altered habitats became available, the Horned Lark was either absent or very locally distributed in Kentucky. Suitable habitat may have been present locally in the intensively grazed savannas of the Blue Grass and the native prairies of western Kentucky, but even in these regions suitable habitat was likely scarce. It is possible that cultivation by Native Americans provided the first opportunity for nesting, but Mengel (1965) and several other authors also have suggested that the species may not have nested in Kentucky until the beginning of the 20th century. It is certain that more recent clearing of land for settlement and agricultural purposes has resulted in a dramatic increase in the amount of available nesting habitat. Although present-day agricultural practices result in the destruction of many nests, the species persists in most areas where a moderate amount of land is tilled or grazed.

Kentucky's nesting Horned Larks are resident throughout the year, although there may be some seasonal movement of local birds, and individuals from farther north supplement the resident population in winter (Mengel 1965). As soon as winter weather begins to moderate, males start their aerial singing, and pairs may begin nest building by mid-February. A peak in the completion of early clutches occurs during the last ten days of March (Mengel 1965). Because nests are often placed along the margins of tilled fields, many are destroyed by spring planting, and a large number of pairs probably renest. Although later nestings often reflect these renesting efforts, a second brood is often raised, and there is another, smaller peak of clutch completion in early June (Mengel 1965). Mengel gave the average size of nine clutches and broods considered complete as 3.5 (range of 3–5).

Forest Cover

Value	% of Blocks	Avg Abund
All	16.5	1.7
1	37.6	1.8
2	27.9	1.7
3	10.9	1.7
4	2.6	1.2
5	—	—

Horned Larks nest on the ground and well away from woody vegetation. The nest is formed within a shallow scrape or natural depression in the soil. It is constructed of grasses and other plant material and lined with finer grasses. Nests are usually placed next to a clump of grass or crop stubble.

Ron Austing

The atlas survey yielded records of Horned Larks in about 17% of priority blocks statewide, and 27 incidental observations were reported. Occurrence was highest in physiographic province sections with an abundance of open farmland and generally decreased from western Kentucky eastward. It is possible that the atlas fieldwork did not accurately survey the occurrence of nesting Horned Larks. The species is often transitory in breeding areas, remaining only as long as habitat is suitable for nesting. For example, bare fields used for nesting in spring are abandoned once the crops become tall, and the species may be missed if breeding areas are visited late in the season. Since most of the atlas fieldwork was done in June and July, nesting birds may have been missed in many areas. Average abundance was relatively low statewide, but varied similarly. Both occurrence and average abundance were inversely related to the percentage of forest cover, being highest in very open areas.

Less than 12% of priority block records were for confirmed breeding. Most of the 14 confirmed records were based on the observation of recently fledged young in their distinctive juvenile plumage. Two confirmed reports were based on adults carrying food, and the only observations of active nests were incidental reports at two locations.

Horned Larks are not reported on most Kentucky BBS routes, and on those where they do occur, they are only rarely observed. The average number of individuals reported per BBS route for the periods 1966–91 and 1982–91 was 0.83 and 0.53, respectively. Trend analysis of these data yields a significant (p<.01) decrease of 5.6% per year for the period 1966–91, but a nonsignificant decrease of 2.2% per year for the period 1982–91.

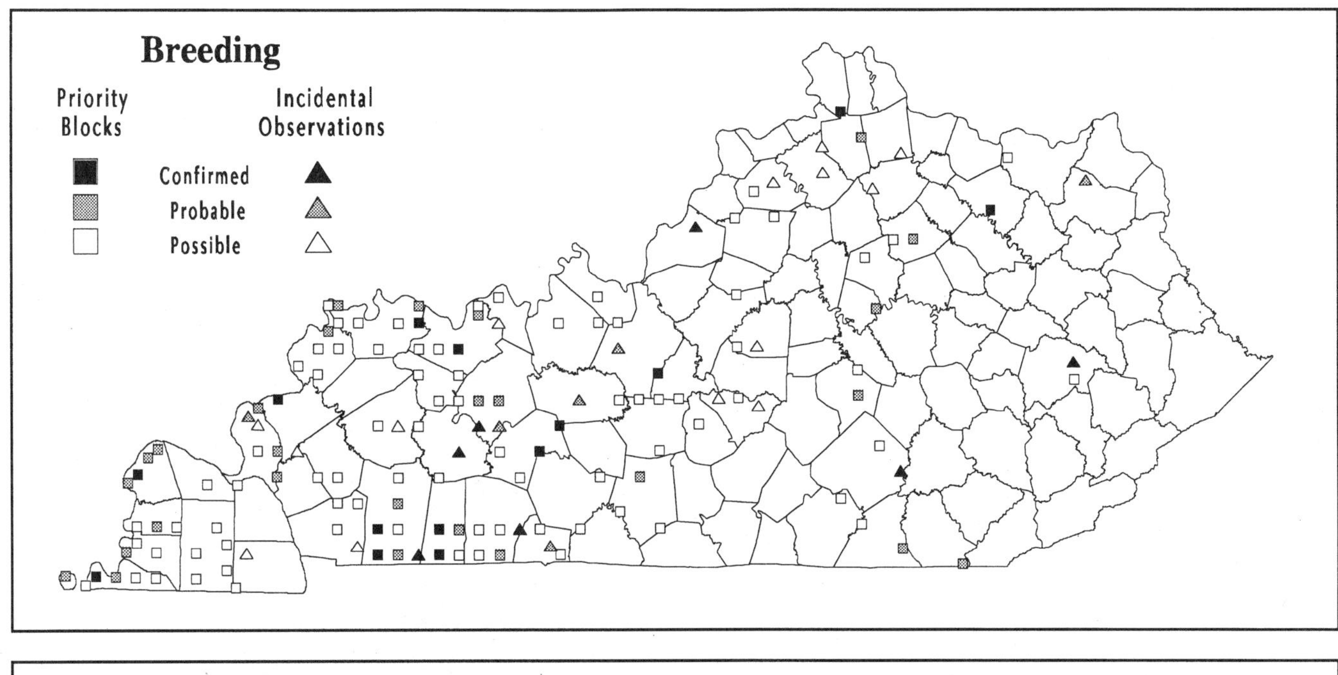

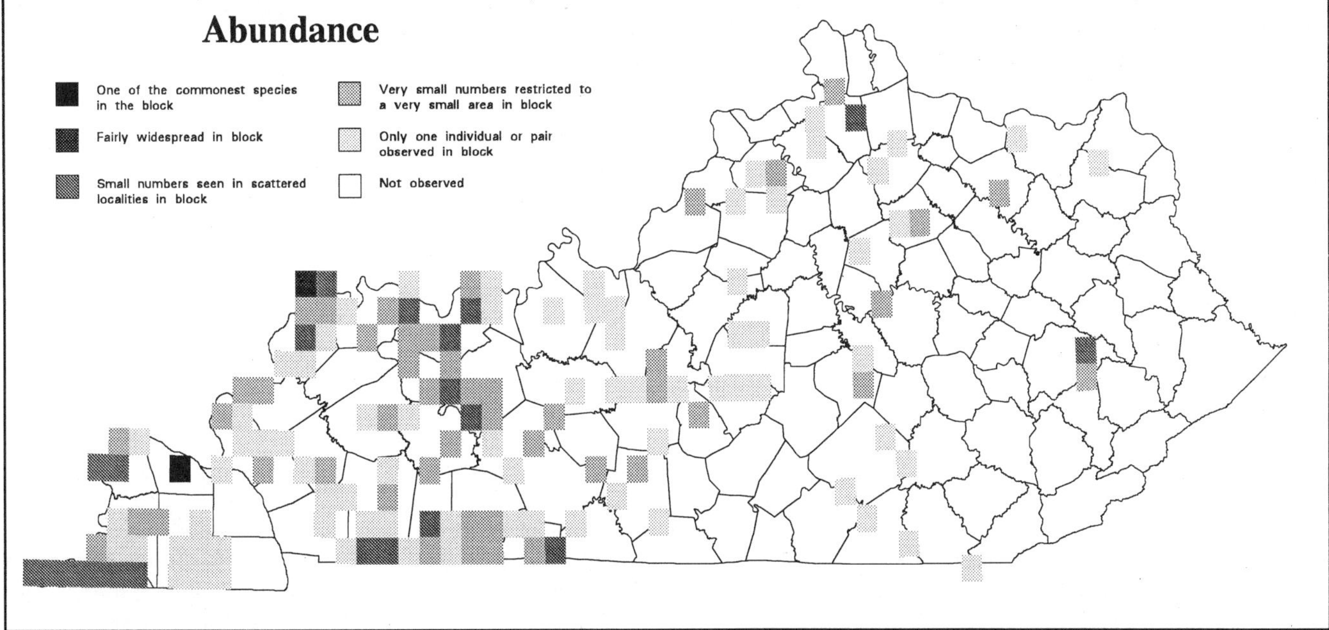

Analysis of Block Data by Physiographic Province Section

Physiographic Province Section	Total Blocks Surveyed	Blocks with Data	Avg Abund	% with Data	Section's % for State
Mississippi Alluvial Plain	14	11	2.4	78.6	9.2
East Gulf Coastal Plain	36	13	1.6	36.1	10.8
Highland Rim	139	37	1.6	26.6	30.8
Shawnee Hills	142	41	1.8	28.9	34.2
Blue Grass	204	14	1.4	6.9	11.7
Cumberland Plateau	173	3	1.3	1.7	2.5
Cumberland Mountains	19	1	1.0	5.3	0.8

Summary of Breeding Status

Number of Blocks in Which Species Was Recorded		
Total	**120**	**16.5%**
Confirmed	14	11.7%
Probable	27	22.5%
Possible	79	65.8%

Purple Martin

Progne subis

The Purple Martin is a fairly conspicuous summer resident throughout much of Kentucky, being broadly distributed in association with human settlement. Mengel (1965) regarded the species as a fairly common to common summer resident wherever suitable nest sites were found.

Purple Martins occur in a great variety of habitats during the nesting season. The species is especially common near larger bodies of water, but it can be found in many situations, from urban parks and suburban yards to rural farmland and settlement. While these large swallows are occasionally found in isolated clearings in predominantly forested areas, they are much more common in open habitats.

If Purple Martins occurred in Kentucky before European settlement, they must have been very locally distributed. Although they probably used natural sites to some extent, Mengel (1965) could find no evidence of this. As of the early 1800s, Audubon (1861) noted that people commonly erected houses for the birds' use, and it is likely that, as in other regions of North America, Native Americans were the first to provide artificial nest cavities (Terres 1980). In addition, the clearing of land for crops also provided the first artificially open habitats and probably assisted the species's spread into new areas. As human settlement progressed, martins continued to spread throughout the countryside wherever nest boxes were provided.

Purple Martins are among the first of Kentucky's summer residents to return in early spring. A few males often return by early March, and nesting pairs become established during the first half of April. According to Mengel (1965), early clutches are completed by the end of April, and the latest ones are probably completed by the end of June. The species is believed to be single-brooded, and occasional late nestings are likely the result of earlier failure because of inclement weather. Robbins et al. (1986) noted that extended periods of rainy weather often result in high mortality of nestlings and adult birds, especially if the rain occurs during the time hatchlings are being fed in the nest. Most young fledge during July, after which both young and adults begin wandering widely. During the latter half of July and August large concentrations of martins roost communally. By the beginning of September most birds have departed for their wintering grounds in South America. Surprisingly, published Kentucky data on clutch size are limited to the observation of a brood of five young in a nest in Pulaski County (Mengel 1965), but Harrison (1975) gives rangewide clutch size as 3–8, commonly 4 or 5.

Forest Cover

Value	% of Blocks	Avg Abund
All	60.8	2.2
1	76.2	2.2
2	73.7	2.3
3	69.1	2.2
4	40.3	2.2
5	7.7	1.2

Purple Martins likely use artificial cavities exclusively in Kentucky. Throughout parts of its range the species has been known to use natural cavities, and it still does so on occasion (Mengel 1965; Terres 1980), but today martins are rarely found using anything except nest boxes or gourds erected to attract them. Inside the nest cavity, a relatively loose layer of grasses, twigs, bark, and other natural or artificial materials is constructed (Harrison 1975). A sprinkling of green leaves is usually added to the nest, and the eggs are laid in a shallow cup in a back corner of the cavity.

Ron Austing

The atlas survey yielded records of Purple Martins in nearly 61% of priority blocks statewide, and 34 incidental observations were reported. Occurrence was highest in the western two-thirds of the state and lowest in the Cumberland Plateau and the Cumberland Mountains. Average abundance was relatively uniform statewide. Occurrence was inversely related to the degree of forestation, which probably is also inversely related to the degree of settlement. Martins may range widely to feed, and they disperse soon after the young have fledged. Thus, some atlas records may have pertained to birds that were far from their nest sites, especially after late June. For this reason, some reports of martins foraging over unsuitable nesting habitat were not included in the final data set.

The Purple Martin was one of only a few species for which more than 50% of priority block records were for confirmed breeding. This can be attributed to the conspicuousness of nest sites. While most confirmed reports were based on the observation of active nests, a considerable number were based on the observation of recently fledged young, and a few involved nest building.

Purple Martins are not reported in substantial numbers on most Kentucky BBS routes. The average number of individuals recorded per BBS route for the periods 1966–91 and 1982–91 was 5.67 and 6.03, respectively. Trend analysis of these data shows a significant ($p<.01$) increase of 4.2% per year for the period 1966–91, but a nonsignificant increase of 5.8% per year for the period 1982–91. The long-term increase probably reflects an increase in settlement along rural BBS routes.

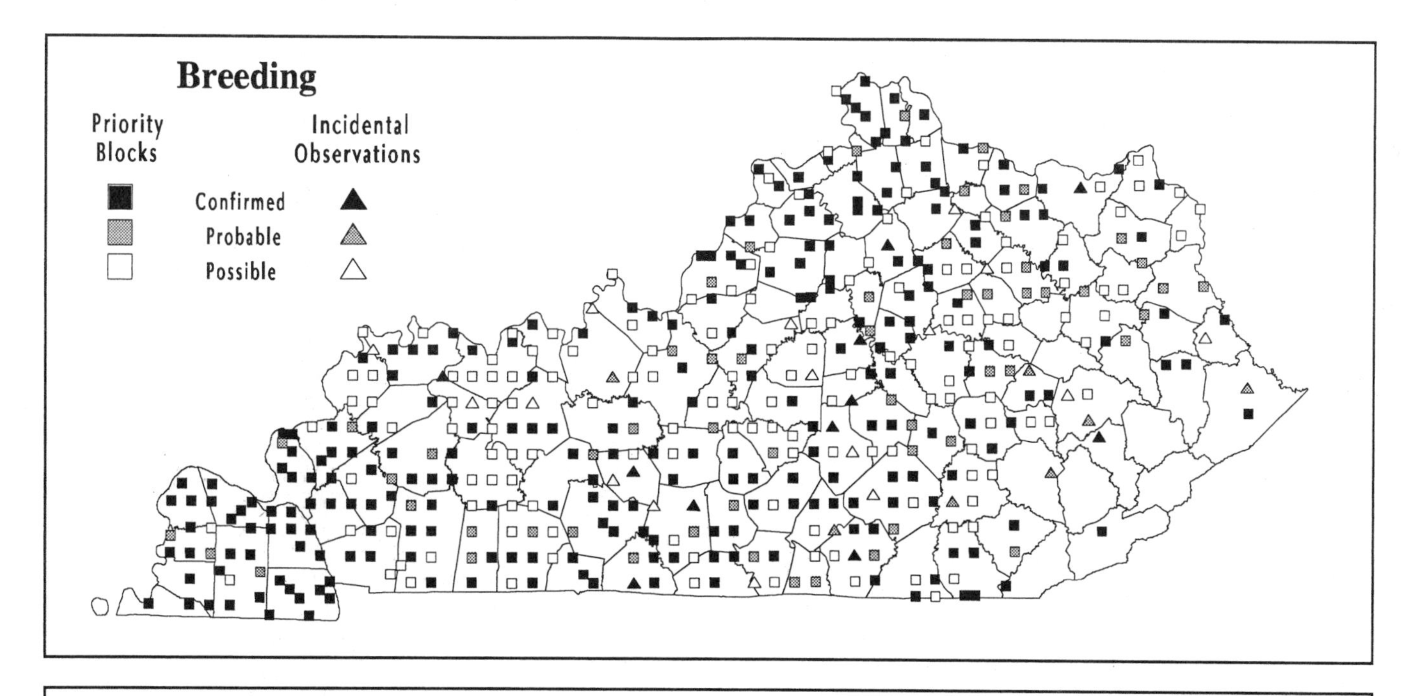

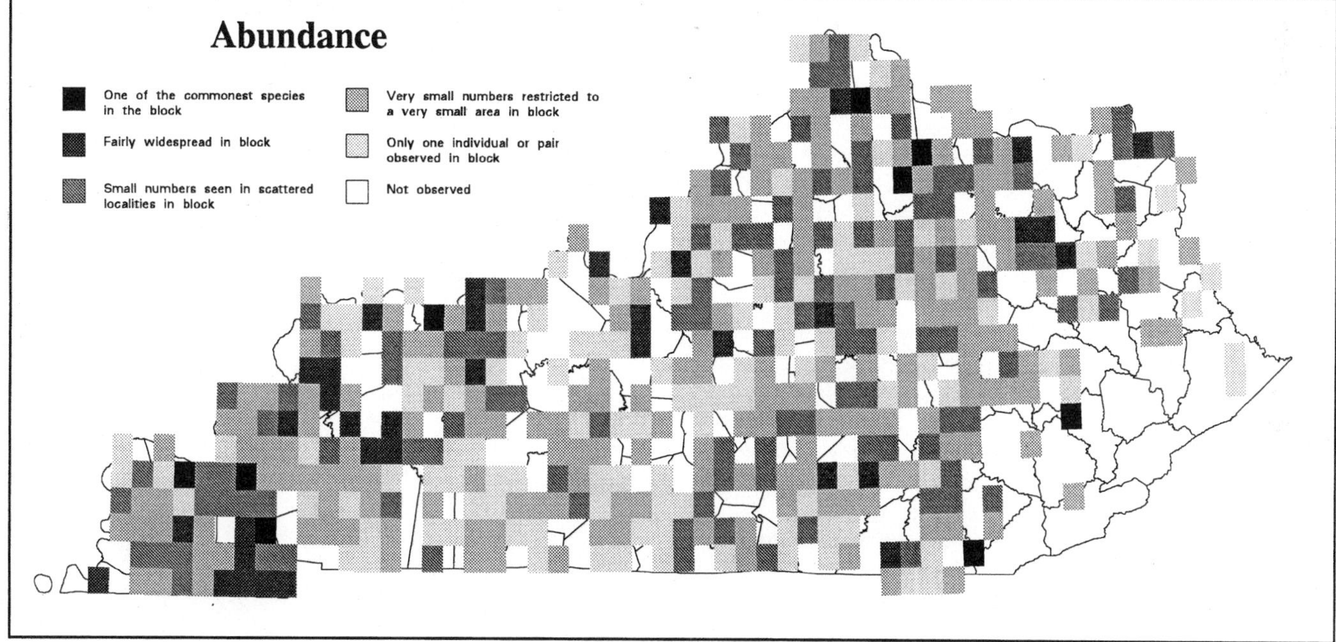

Analysis of Block Data by Physiographic Province Section

Physiographic Province Section	Total Blocks Surveyed	Blocks with Data	Avg Abund	% with Data	Section's % for State
Mississippi Alluvial Plain	14	7	2.3	50.0	1.6
East Gulf Coastal Plain	36	34	3.0	94.4	7.7
Highland Rim	139	98	2.0	70.5	22.2
Shawnee Hills	142	97	2.2	68.3	21.9
Blue Grass	204	145	2.3	71.1	32.8
Cumberland Plateau	173	59	2.2	34.1	13.3
Cumberland Mountains	19	2	3.5	10.5	0.5

Summary of Breeding Status

Number of Blocks in Which Species Was Recorded		
Total	**442**	**60.8%**
Confirmed	241	54.5%
Probable	57	12.9%
Possible	144	32.6%

Tree Swallow

Tachycineta bicolor

The Tree Swallow is a relatively recent addition to the breeding avifauna of Kentucky. As of the late 1950s, only a few vague summer records had been reported from the far western part of the state (e.g., Pindar 1925). In the mid-1960s and mid-1970s Tree Swallows were found nesting in Lyon County (Robinson 1965) and Jefferson County (Stamm 1976), respectively. These records apparently were part of a southward expansion in nesting range that was documented during the same period in adjacent states (e.g., Robinson 1990; Peterjohn and Rice 1991). Since these initial observations, the species has expanded dramatically throughout much of northern and western Kentucky, where it is now a regular and locally common breeder.

Tree Swallows use a variety of semi-open and open habitats during the nesting season. They are most frequently encountered near lakes, ponds, and marshes, especially where impoundment or other alterations of the water level have caused many trees to die, resulting in an abundance of nesting cavities. Although most common near water, the species is now being found sporadically in farmland or other open situations far from any substantial bodies of water. Most such records have originated in the Blue Grass, but at least one has come from Pulaski County in the Highland Rim (Stamm 1986a; J. Elmore, pers. comm.).

Exactly why nesting Tree Swallows have expanded southward is unknown, but the creation of an abundance of suitable nesting habitat has at least in part been responsible. Kentucky has 17 reservoirs of more than 1,000 acres, but before 1950 there were only three (Kleber 1992). While Tree Swallows sometimes use small bodies of water, the creation of habitat on the larger impoundments seems to have initiated colonization of many areas nearby. In addition, the recent return of beavers to large portions of formerly occupied range after decades of absence has contributed to the amount of optimal habitat. The availability of nest boxes erected for other species, especially bluebirds, likewise has assisted the expansion.

Tree Swallows are among the earliest of summer residents to return to their Kentucky breeding grounds. It is not uncommon for the first males to arrive by mid-March, and most individuals are probably on territory by April 1. Because the species was not known to nest in the state until the 1970s, Mengel (1965) includes no information on nesting. In a detailed study of an Ohio County population, B. Ferrell and D. Stephens (in prep.) have found that egg laying commences during the first week of April. Subsequently, two noticeable peaks in clutch completion occur, one during the middle of April and a second in late June. The average clutch size for early nestings in Ohio County was 5.24 (range of 4–7) compared with an average clutch size of 4.35 (range of 3–6) for late nestings (B. Ferrell and D. Stephens, in prep.). Interestingly, Ferrell discovered that birds involved in early nestings do not seem to nest again, instead yielding their nesting cavities to other individuals. Late nestings are usually completed by the middle of August, and family groups gather into large flocks before moving south in early fall.

The nest is built in a natural or artificial cavity, most commonly over permanent water. Although abandoned woodpecker holes are often used, the species also readily accepts natural tree cavities and nest boxes. Nest cavities have been reported as high as 20 feet (Stamm 1976), and B. Ferrell (pers. comm.) has located nests in natural cavities as low as about 3.5 feet and in nest boxes as low as 2.0 feet over normal water level. The female collects pieces of dead grass to form a shallow cup and lines the nest with an abundance of feathers (Harrison 1975). Although individual pairs sometimes nest alone, this species often nests in loose colonies when suitable cavities are abundant.

Maslowski Wildlife Productions

The atlas survey yielded records of Tree Swallows in nearly 9% of priority blocks statewide, and 22 incidental observations were reported. Occurrence was highest in the Mississippi Alluvial Plain, where suitable habitat is most abundant. Elsewhere occurrence was greater than 10% only in the Shawnee Hills, and the species was not found in the Cumberland Mountains.

More than 30% of priority block records were for confirmed breeding. Of the 19 confirmed records, most involved the observation of active nests. The observation of recently fledged young and nest building accounted for the remainder.

During the seven years of the atlas project, the expansion of nesting Tree Swallows continued. Because sites with optimal nesting habitat are scattered throughout most of the state, there is little reason to doubt that the nesting population will continue to grow in coming years, especially where nest boxes are available. In contrast, it appears that the availability of suitable nest cavities may eventually limit the species's numbers in some areas. Optimal habitat provided by stands of dead timber in reservoirs is temporary. Cavity snags may persist for only a decade or two following impoundment, after which little or no new nesting habitat becomes available. Since 1977 only three reservoirs larger than 1,000 acres have been constructed in Kentucky (Kleber 1992). As the number of dead trees in reservoirs declines, Tree Swallows are also likely to decline locally. For this reason, the presence of nest boxes may become more important.

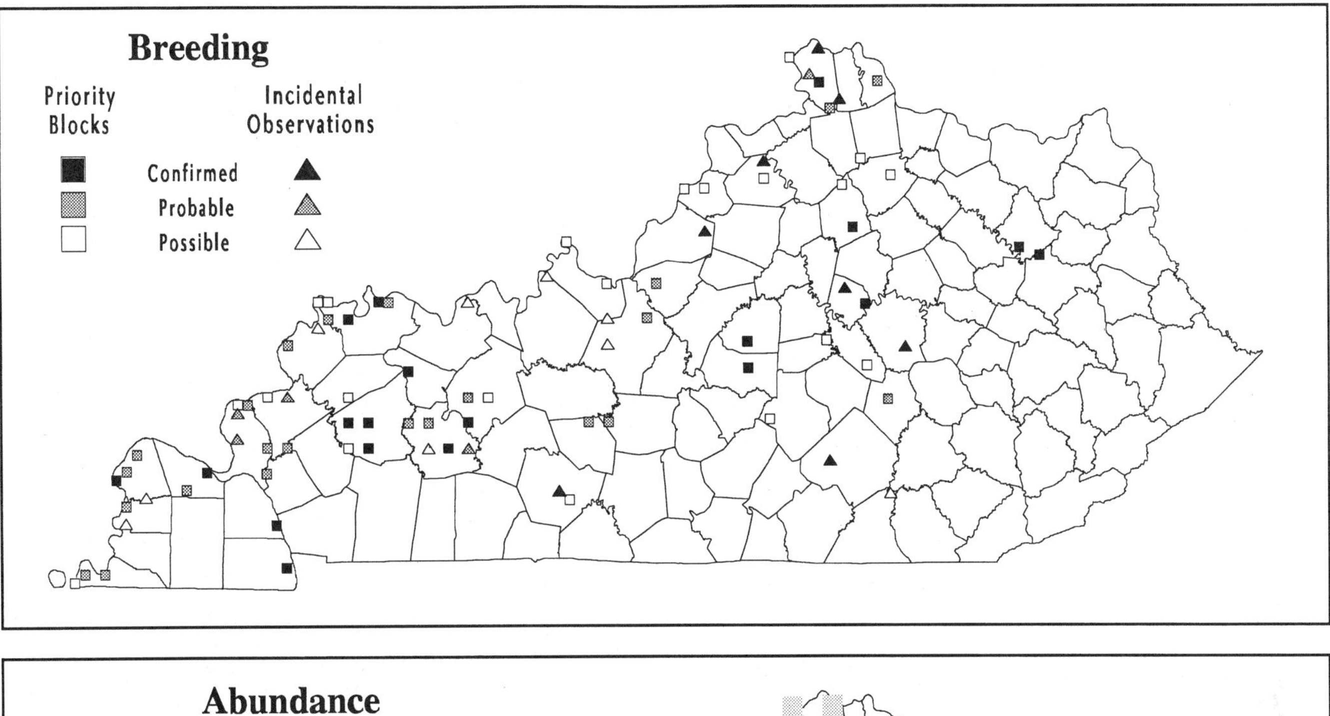

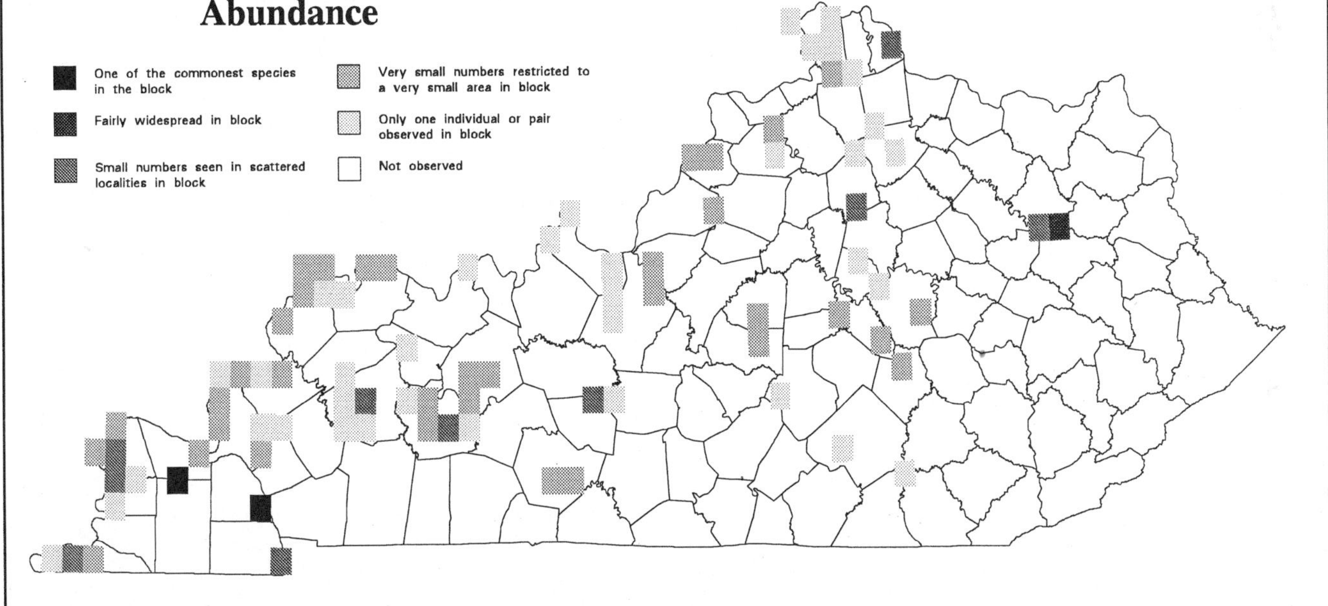

Analysis of Block Data by Physiographic Province Section

Physiographic Province Section	Total Blocks Surveyed	Blocks with Data	% with Data	Section's % for State
Mississippi Alluvial Plain	14	8	57.1	12.7
East Gulf Coastal Plain	36	3	8.3	4.8
Highland Rim	139	7	5.0	11.1
Shawnee Hills	142	25	17.6	39.7
Blue Grass	204	19	9.3	30.2
Cumberland Plateau	173	1	0.6	1.6
Cumberland Mountains	19	-	-	-

Summary of Breeding Status

Number of Blocks in Which Species Was Recorded		
Total	**63**	**8.7%**
Confirmed	19	30.2%
Probable	23	36.5%
Possible	21	33.3%

Northern Rough-winged Swallow

Stelgidopteryx serripennis

The Northern Rough-winged Swallow breeds throughout much of North America, including all of Kentucky. Although nowhere especially common, small numbers seem to occur wherever suitable nesting sites are present. Mengel (1965) regarded the species as a fairly common to common, but locally distributed, summer resident statewide.

Like other swallows, the Rough-winged is often found in fairly open situations, but this species is also regularly encountered in forested areas, especially along clifflines and streams. It is most conspicuous in rural countryside, but small numbers occur in a wide variety of habitats, including suburban stream corridors, quarries, reclaimed surface mines, and large rivers and reservoirs.

Northern Rough-winged Swallows likely have increased in Kentucky in response to the creation of artificial nest sites. Audubon drew the species at a cavity in a dirt bank at Henderson in the early 1800s (Wiley 1970), indicating that these swallows likely inhabited clifflines and stream banks across the state before European settlement. In many areas, however, the absence of suitable nest sites was likely limiting. Subsequently, the creation of an abundance of nest sites as a result of development has allowed for colonization of many areas where the species formerly must not have occurred.

Northern Rough-winged Swallows winter primarily in the extreme southern United States and Middle America (AOU 1983), and most birds return to their Kentucky breeding grounds by the third week of April. Noisy courtship activities commence immediately, and nest construction is often under way by the beginning of May. According to Mengel (1965), early clutches are completed during the first ten days of May, with a peak in clutch completion during the last ten days of the month. Later nestings are apparently undertaken by the first part of June (Mengel 1965), and most young fledge during July, when family groups are most conspicuous on rural utility lines. Most birds depart soon thereafter, although late summer and fall aggregations typically linger in far western Kentucky until mid-October. Kentucky data on clutch size are limited to one report of six eggs in a nest in Jefferson County (Mengel 1965). Harrison (1975) gives rangewide clutch size as 4–8, commonly 6 or 7.

Forest Cover

Value	% of Blocks	Avg Abund
All	49.4	1.8
1	33.7	2.1
2	51.6	1.7
3	53.0	1.6
4	51.3	1.8
5	50.0	1.8

Although Rough-wingeds often use natural sites for nesting, they also use artificially created habitats extensively. In fact, there seem to be few deep road cuts across the state that do not have at least a pair or two occupying them during the breeding season. Quarry walls, strip mine highwalls, and other artificial situations are also used frequently, and a pair or two can sometimes be found in nooks and crannies under bridges, on the sides of buildings, and in abandoned machinery. Frequently used natural sites include clifflines and rock or dirt banks along rivers and streams. Croft and Stamm (1967) included a report of a nest in a tree cavity over a stream in Franklin County.

Ron Austing

Nests are usually placed in the rear of a crevice in rocky substrates or at the end of a burrow in dirt or sandy banks. Although there have been reports of Rough-wingeds excavating their own burrows, abandoned burrows of other species seem to be most frequently used (Harrison 1975). The species is not as colonial as the Bank Swallow, and where several pairs breed together the nests are widely separated. The nest itself is a bulky structure of twigs, weed stems, and coarse grass, lined with finer grasses (Harrison 1975). Nest cavities are usually chosen that lie well up on larger banks and cliffs, although nests have been found as low as about four feet above the ground (Stamm 1963b).

The atlas survey yielded records of Northern Rough-winged Swallows in more than 49% of priority blocks statewide, and 40 incidental observations were reported. Occurrence in the four largest physiographic province sections was uniform (49–50%), and it was noticeably lower only in the East Gulf Coastal Plain. Average abundance also was uniform across the state. The wide variety of habitats used by these swallows for nesting probably explains the uniformity of occurrence and average abundance data.

Despite the relative abundance of this swallow, only 25% of priority block records were for confirmed breeding. Most were based on observation of recently fledged young, although some active nests were located, and nest building was also observed.

Northern Rough-winged Swallows are infrequently reported on most Kentucky BBS routes. The average number of individuals recorded per BBS route for the periods 1966–91 and 1982–91 was 2.24 and 1.94, respectively. Trend analysis of these data shows a nonsignificant increase of 3.1% per year for the period 1966–91 and a nonsignificant decrease of 1.0% per year for the period 1982–91.

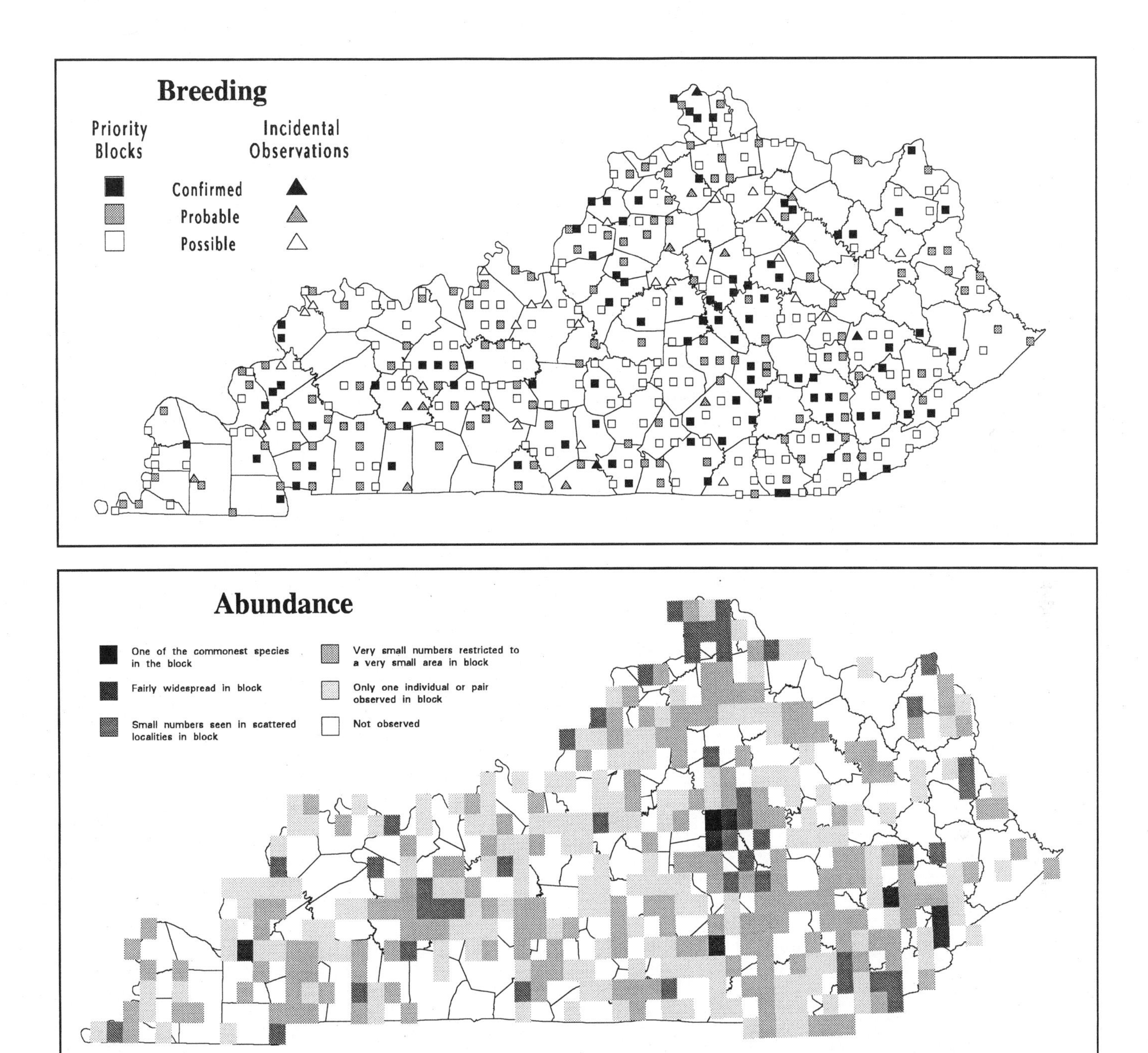

Analysis of Block Data by Physiographic Province Section

Physiographic Province Section	Total Blocks Surveyed	Blocks with Data	Avg Abund	% with Data	Section's % for State
Mississippi Alluvial Plain	14	9	1.8	64.3	2.5
East Gulf Coastal Plain	36	11	1.9	30.6	3.1
Highland Rim	139	70	1.6	50.4	19.5
Shawnee Hills	142	71	1.6	50.0	19.8
Blue Grass	204	100	1.9	49.0	27.9
Cumberland Plateau	173	85	1.8	49.1	23.7
Cumberland Mountains	19	13	1.9	68.4	3.6

Summary of Breeding Status

Number of Blocks in Which Species Was Recorded		
Total	**359**	**49.4%**
Confirmed	89	24.8%
Probable	120	33.4%
Possible	150	41.8%

Northern Rough-winged Swallow

Bank Swallow

Riparia riparia

The Bank Swallow is widely distributed across North America as a nesting bird, occurring throughout all but the extreme southeastern and northern portions of the continent (AOU 1983). In Kentucky the species occurs primarily along the state's northern border, although the possibility of nesting in other areas has been illustrated by the recent discovery of an active colony in Fayette County. Mengel (1965) regarded the Bank Swallow as a very rare to uncommon and locally distributed summer resident.

During migration Bank Swallows are reported in a variety of open habitats, but they are generally restricted to the immediate vicinity of major rivers during the nesting season. Before the 1991 discovery of a small nesting colony in southern Fayette County, all recent reports came from the Mississippi and Ohio rivers. The only published accounts from other parts of the state are given by Mengel (1965) as unconfirmed reports for Calloway, Nelson, and Warren counties during the late 1800s and the early 1900s. Mengel expressed some concern about the authenticity of these reports, in large part because of the treatment given to the more common and widespread Rough-winged Swallow in the same accounts.

Before the early 1980s Bank Swallows were known to nest only in natural banks in Kentucky. Since that time the species has been found nesting in vertical banks of sand and gravel quarries adjacent to the Ohio River in Boone County (Stamm 1981b; subsequent nesting season reports in *The Kentucky Warbler*), Carroll County (Stamm 1985, 1986a, 1987c, 1988b), and Meade County (Stamm 1988b). These artificial sites are all within a mile or so of the main river channel. In 1991 a small nesting colony was discovered in a storage pile of agricultural lime at a quarry site along the Kentucky River in southern Fayette County (S. Bonney, pers. comm.). In 1992 this colony contained approximately 25 active burrows (M. Burns, pers. comm.).

Numbers of nesting Bank Swallows have likely decreased in Kentucky during the last century. Audubon (1861) considered the species common along the Ohio River and some parts of the Mississippi River in the early 1800s. The impoundment of large rivers with navigation dams, especially on the lower Ohio River, has altered a substantial amount of suitable nesting habitat. Even though gravel pits have been colonized in recent years, the creation of suitable habitat at these sites has not made up for the loss of natural riverbank habitat.

Bank Swallows winter primarily in South America (AOU 1983). A few birds begin returning to their Kentucky breeding grounds by mid-April, although the entire nesting population may not return until mid-May, especially in northern Kentucky. Once pairs have formed and the initial stages of courtship are completed, the birds begin to excavate nesting burrows 15–47 inches into the chosen site (Harrison 1975). In natural banks burrows are often lined up along horizontal strata of sandier, more easily excavated soil. In artificial substrates, burrows are generally clustered loosely together. Some burrows from previous years may be reused or modified, but many new burrows must be excavated each year. Suitable sites sometimes persist for decades, but most do not last for more than a few years, and new sites are colonized as they become available. If spring rains are above average, nest building may be delayed along the major rivers until water levels drop to normal summer levels. Specific data on nesting in Kentucky are scarce, but young birds ready to fledge from the nest have been observed in Jefferson County in mid-July (Mengel 1965). Soon after most of the young fledge, large aggregations of birds stage in nearby areas, gathering on wires or bare tree limbs. By mid-August most have likely departed for their wintering grounds. Kentucky data on clutch size are lacking, but Harrison (1975) gives rangewide clutch size as 4–6, commonly 5.

Maslowski Wildlife Productions

The atlas survey yielded 21 records of Bank Swallows in priority blocks, and 12 incidental observations were reported. Most observations originated from previously known nesting areas, although several new locality records were obtained. Occurrence was primarily related to the major rivers, particularly the Ohio River, rather than physiographic province sections. Some birds begin their fall migration by early July, and some records denoted as possible, such as the Rowan County record, may pertain to wandering birds or migrants.

More than 52% of priority block records and nearly half of the incidental reports were for confirmed breeding. Almost all confirmed records were based on the observation of active nesting colonies, although one involved the observation of an adult carrying food.

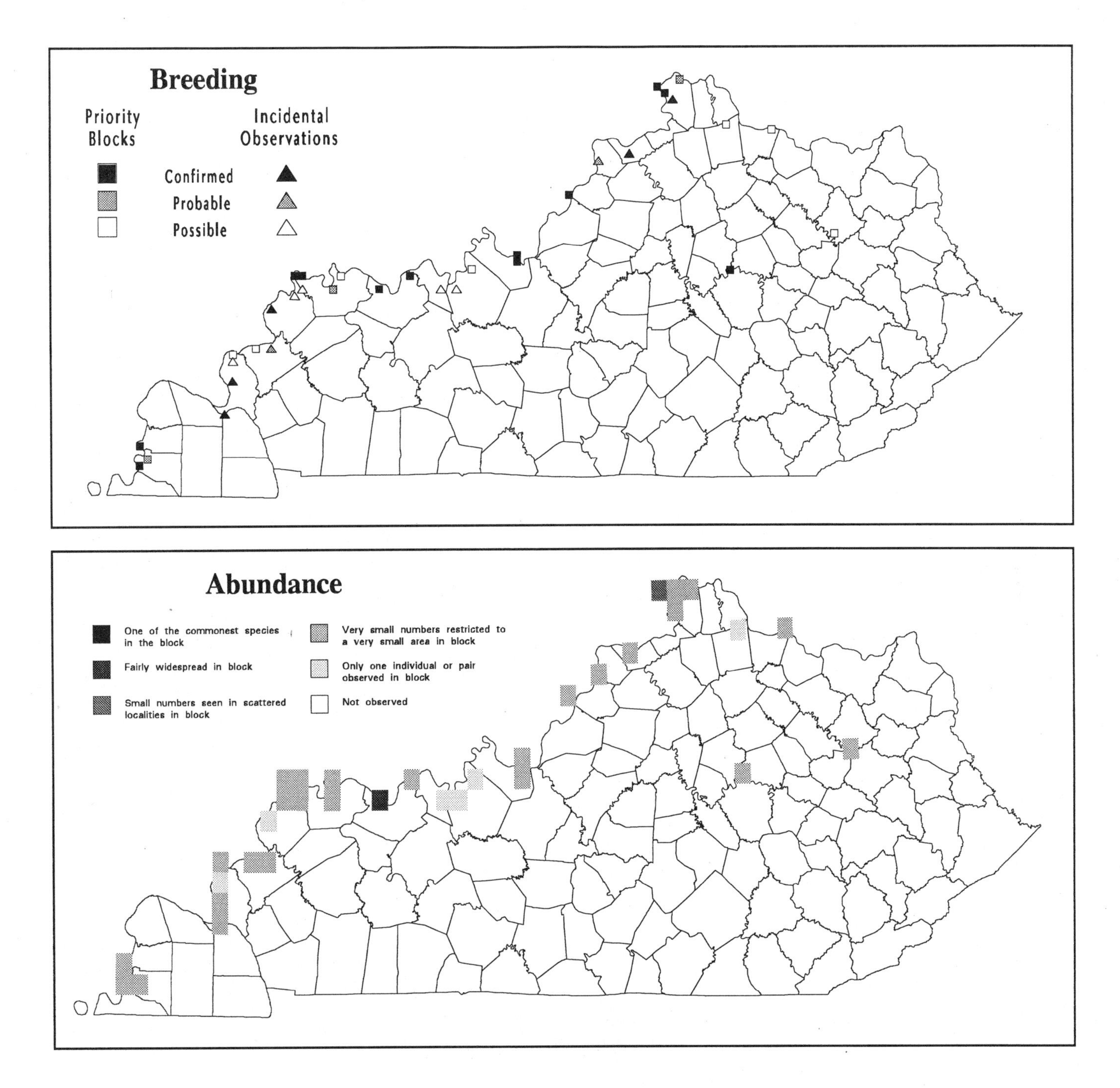

Analysis of Block Data by Physiographic Province Section

Physiographic Province Section	Total Blocks Surveyed	Blocks with Data	% with Data	Section's % for State
Mississippi Alluvial Plain	14	3	21.4	14.3
East Gulf Coastal Plain	36	-	-	-
Highland Rim	139	2	1.4	9.5
Shawnee Hills	142	9	6.3	42.9
Blue Grass	204	7	3.4	33.3
Cumberland Plateau	173	-	-	-
Cumberland Mountains	19	-	-	-

Summary of Breeding Status

Number of Blocks in Which Species Was Recorded		
Total	**21**	**2.9%**
Confirmed	11	52.4%
Probable	3	14.3%
Possible	7	33.3%

Bank Swallow

Cliff Swallow

Hirundo pyrrhonota

Although the Cliff Swallow nests sporadically throughout much of Kentucky today, the species's breeding status has fluctuated dramatically since the 19th century. During the 1800s this swallow likely nested regularly in towns and farmland, constructing nests on the sides of buildings and barns. According to Mengel (1965), the appearance of the House Sparrow in the late 1800s probably marked the beginning of the Cliff Swallow's disappearance from the region as a breeding bird, presumably because of competition for nests once built by the swallows. An increase in the practice of painting barns has also been implicated in contributing to the decline (Robbins et al. 1986). Regardless of the causes, between the early 1900s and about 1945 the Cliff Swallow nearly disappeared as a breeding bird in Kentucky (Mengel 1965).

Nesting Cliff Swallows reappeared in the mid-1940s, when they were found nesting on the structure of Kentucky Dam, in Livingston and Marshall counties, in western Kentucky (Mengel 1965). During the next two decades the nesting population expanded over Kentucky Lake and nearby Lake Barkley, and by 1969 the species was known from at least 25 sites in the area (Peterson 1970). By the mid-1970s the species had colonized bridges on Barren River Lake (Mason 1977), and since that time Cliff Swallows have gradually spread across much of the state, primarily on dams and under bridges associated with larger rivers and reservoirs. As the breeding population has increased, bridges over smaller streams and roadway overpasses also have been colonized, but only a few small colonies have become established far from large bodies of water (e.g., Stamm 1979a, 1980). Several new colonies were discovered each year of the atlas fieldwork, and known nesting localities now number close to 100.

Without the construction of large reservoirs, the Cliff Swallow would be far less numerous in Kentucky today. Before 1950 there were only three reservoirs of 1,000 or more acres, but today there are 17 reservoirs this size (Kleber 1992). The increase in the number of suitable nest sites resulting from reservoir construction seems to have greatly assisted the spread of Cliff Swallows. To date only a few reservoirs in the extreme southeastern quarter of the state have not been colonized.

Cliff Swallows begin returning to their nesting colonies during the first two weeks of April, although it may be early May before all of the breeding birds have returned statewide. The inaccessibility of most colonies has precluded detailed study of nesting behavior. Nest building has been observed as early as mid-April (Peterson 1970), and most clutches are probably completed by the end of May. Although unconfirmed, it appears that at least some pairs are double-brooded (Peterson 1970). Kentucky data on clutch size are limited to the observations of Audubon (1861), who examined a number of nests, each containing four eggs, at Newport in 1819. Harrison (1975) gives rangewide clutch size as 3–6, commonly 4 or 5. Soon after a majority of the young leave the nests, family groups gather together on wires and dead trees nearby. Transients appear in August, and it seems that most birds have departed for their wintering grounds in South America by the beginning of September.

Although Cliff Swallows originally may have used natural clifflines in Kentucky, they appear to use artificial sites exclusively today. Most nests are placed under bridges over permanent water, but dams and highway overpasses are also frequently used. The use of eaves on barns and other buildings was regularly observed until the 1920s (e.g., Dodge 1940), but today that behavior is rare in Kentucky. Use of protected eaves of a large hotel along the Ohio River at Paducah, in McCracken County, by a large nesting colony appears to be the only current example.

Alvin E. Staffan

Nests often persist for years and may be reused or repaired, but some new nests are constructed each year. Both members of the pair collect pellets of mud from the margins of puddles and plaster them on a vertical surface to form the nest. In contrast to the Barn Swallow, this species builds a nest that is usually composed entirely of mud. A tube-shaped entrance is typically placed on the side of the hollow globe of mud and usually directed downward. Inside the nest, the eggs are laid on a lining of grass and feathers (Harrison 1975). Individual nests are sandwiched together in tight rows or clusters, often sharing common walls. Although more than 900 nests have been counted at a single site, most colonies contain 12–100 nests (Peterson 1970, 1973).

The atlas survey yielded records of Cliff Swallows in nearly 5% of priority blocks, and 22 incidental observations were reported. The species is very locally distributed, in large part because of its association with larger bodies of water. All of the larger recreational reservoirs except those in the southeastern quarter of the state support substantial populations of nesting birds.

Almost 77% of priority block records and nearly all incidental observations were for confirmed breeding. This high percentage in large part results from the conspicuousness of active nests. Most all confirmed reports were based on the observation of active nests, although a few involved the observation of used nests, nest building, and recently fledged young.

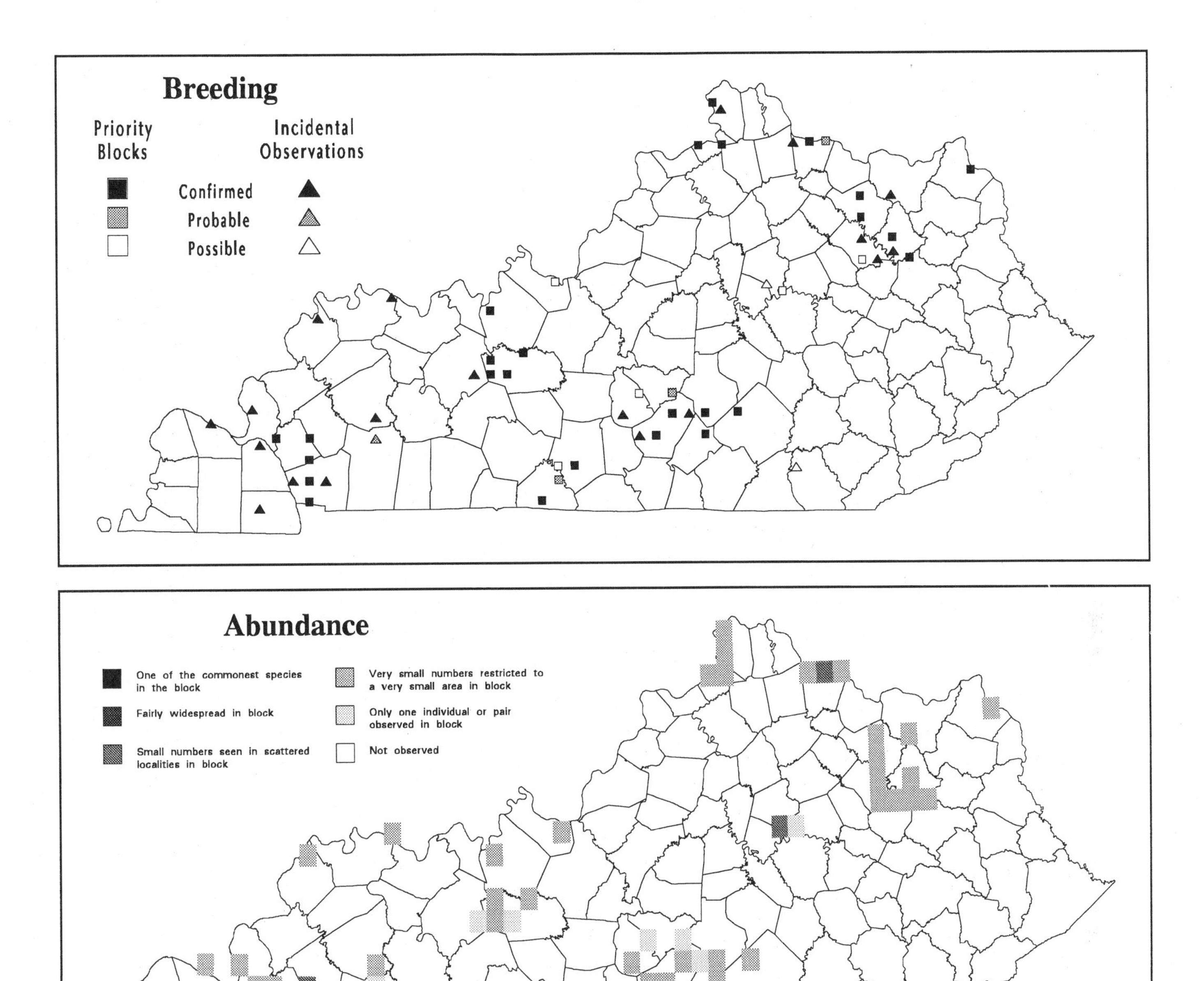

Analysis of Block Data by Physiographic Province Section

Physiographic Province Section	Total Blocks Surveyed	Blocks with Data	% with Data	Section's % for State
Mississippi Alluvial Plain	14	-	-	-
East Gulf Coastal Plain	36	-	-	-
Highland Rim	139	17	12.2	50.0
Shawnee Hills	142	5	3.5	14.7
Blue Grass	204	10	4.9	29.4
Cumberland Plateau	173	2	1.2	5.9
Cumberland Mountains	19	-	-	-

Summary of Breeding Status

Number of Blocks in Which Species Was Recorded		
Total	**34**	**4.7%**
Confirmed	26	76.5%
Probable	3	8.8%
Possible	5	14.7%

Cliff Swallow

Barn Swallow

Hirundo rustica

The darting images of foraging Barn Swallows are among the most conspicuous summer sights of Kentucky's rural countryside. In fact, according to continental BBS data for the period 1965–79, the species was as common in Kentucky as anywhere within its range (Robbins et al. 1986). Mengel (1965) regarded the Barn Swallow as a fairly common to common summer resident throughout the state, except in the heavily forested portions of southeastern Kentucky. Today the species's status is similar, although the progression of settlement in eastern Kentucky has resulted in the continued spread of Barn Swallows there.

The Barn Swallow occurs in a variety of semi-open to open habitats during the nesting season. The species has adapted so readily to nesting in artificial situations that it is found virtually everywhere that openings have been created and settlement has been established. Although these swallows are probably most frequent in rural farmland and settlement, they are also often found in roadway corridors, reclaimed surface mines, and other artificial openings where suitable nest sites are available.

It is likely that if Barn Swallows occurred in Kentucky before artificial nest sites were created, they must have been very locally distributed. Audubon (1861) noted their use of sinkholes for nesting in the barrens as of the early 1800s, but he also observed that they commonly used bridges and barns at that time. In general, it seems that the species must have colonized the state concurrently with Native American or European settlement (see Mengel 1965).

Barn Swallows winter primarily in South America (AOU 1983), and most birds return to Kentucky by the third week of April. Courtship behavior begins almost immediately, and nest building may be under way by the end of April. Early clutches are completed during the first week of May, and an early peak in clutch completion occurs during the middle of the month (Mengel 1965). Two broods are usually raised, and a second peak in clutch completion occurs sometime in early July. The average size of 45 clutches or broods reported by Croft and Stamm (1967) and Stamm and Croft (1968) was 4.7 (range of 3–7). Groups of young are conspicuous on roadside wires during mid-June and again in August. After the second broods fledge, nesting birds quickly disappear, and sightings of local birds are uncommon by the beginning of September.

Forest Cover

Value	% of Blocks	Avg Abund
All	91.5	3.2
1	97.0	3.4
2	97.9	3.5
3	97.0	3.3
4	85.1	2.7
5	51.9	1.9

Barn Swallows most frequently nest inside open barns and buildings, under bridges, and under covered eaves and porches on homes and other structures. Less frequently used sites include small road culverts, covered piers, and overhangs on artificial rock walls, including strip mine highwalls (Allaire 1979, 1981). Both members of the pair contribute to construction of the nest of mud and grass. It is typically somewhat layered and bulky and is lined with fine grass and feathers. The average height of 20 nests summarized by Croft and Stamm (1967) was 9.0 feet (range of 7–25 feet). Nests are usually used in successive years if they persist, and there are numerous reports of the same nest being used to raise consecutive broods (Guthrie 1961; Mengel 1965; Croft and Stamm 1967). Nests are seldom placed closer than several feet apart, although as many as 25–30 active nests have been located in a single barn (Powell 1961).

Alvin E. Staffan

The atlas fieldwork yielded records of Barn Swallows in nearly 92% of priority blocks statewide, and seven incidental observations were reported. The species ranked 15th according to the number of priority block records, 9th by total abundance, and 11th by average abundance. Occurrence was higher than 90% in all physiographic province sections except the Cumberland Mountains and the Cumberland Plateau. Average abundance varied somewhat similarly. Lower values in eastern Kentucky were probably most closely related to the higher amount of undeveloped, forested land.

More than 61% of priority block records were for confirmed breeding. This relatively high figure primarily resulted from the Barn Swallow's conspicuousness in most areas during the breeding season. Nest sites were relatively easy to find, and groups of recently fledged young were often observed perched on electrical wires near nest sites. Of the 407 confirmed records, most were based on the observation of recently fledged young, but many involved the observation of active or used nests as well as adults engaged in nest building.

Barn Swallows are very conspicuous in summer along rural Kentucky BBS routes, and the average number of individuals reported per BBS route for the periods 1966–91 and 1982–91 was 34.22 and 34.06, respectively. According to these data, the Barn Swallow ranked 11th in abundance on BBS routes during the period 1982–91. Trend analysis of these data indicates only small, nonsignificant decreases in the average number of individuals per BBS route over both periods.

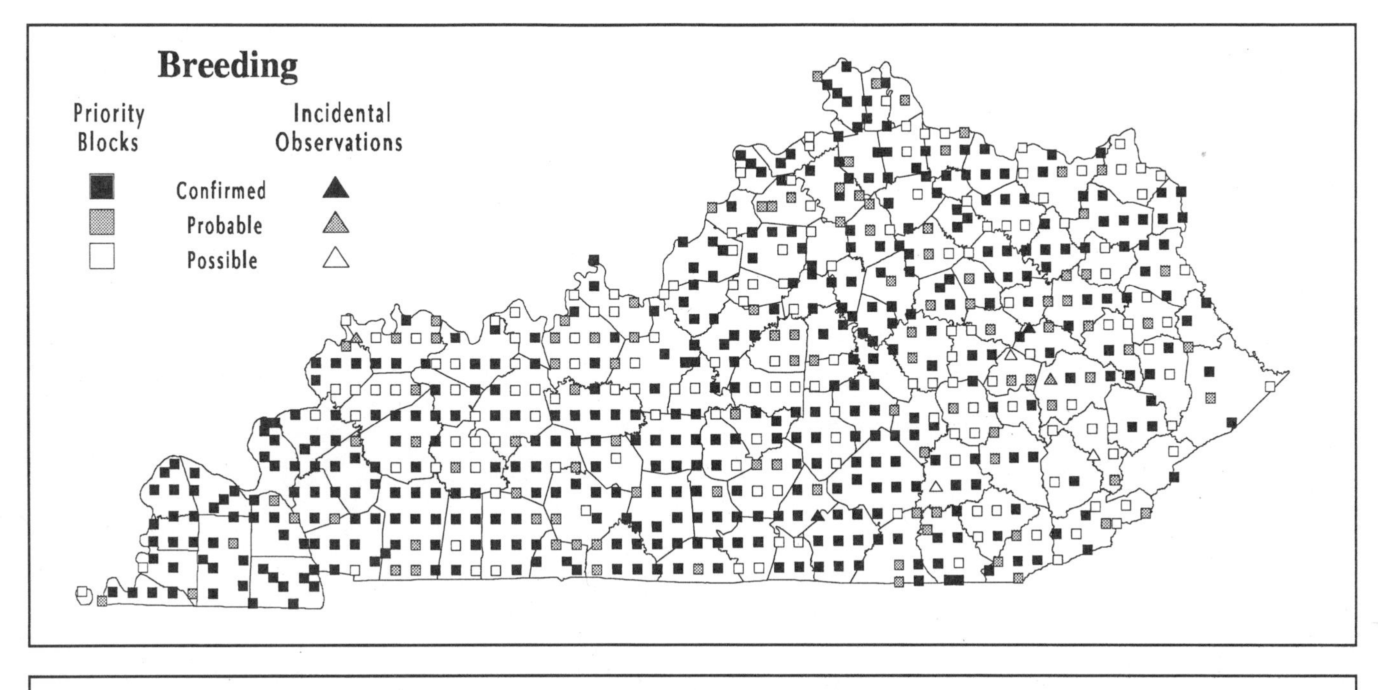

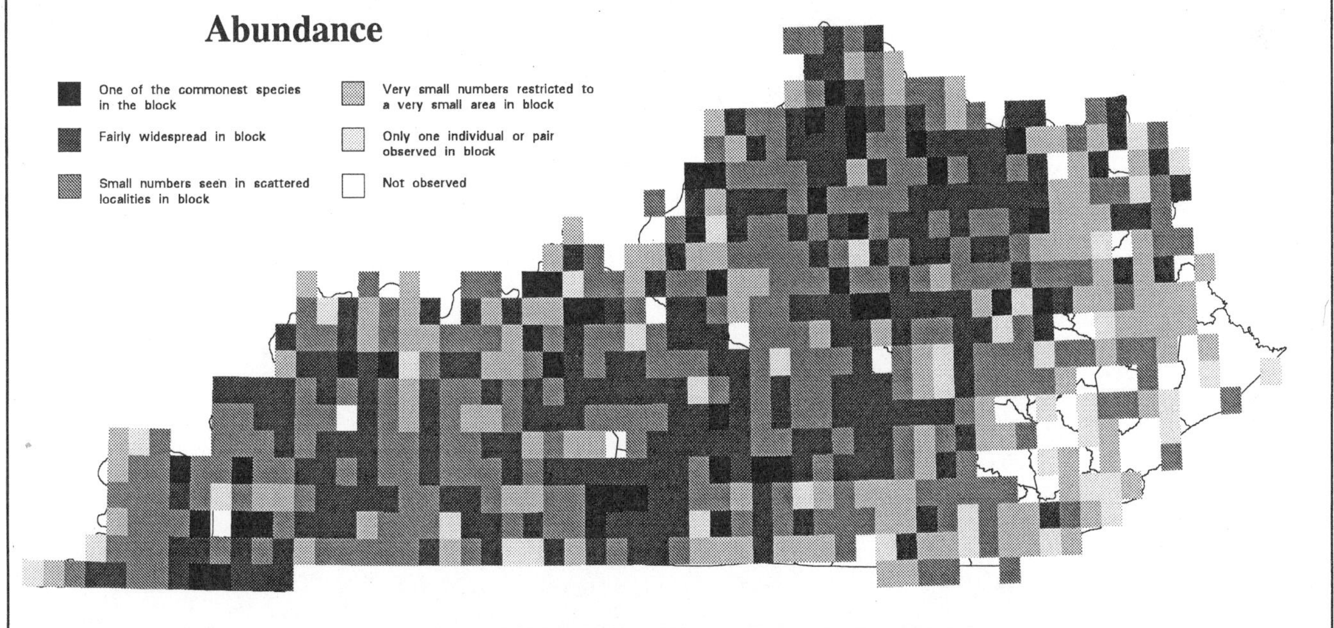

Analysis of Block Data by Physiographic Province Section

Physiographic Province Section	Total Blocks Surveyed	Blocks with Data	Avg Abund	% with Data	Section's % for State
Mississippi Alluvial Plain	14	13	2.5	92.9	2.0
East Gulf Coastal Plain	36	35	3.5	97.2	5.3
Highland Rim	139	137	3.4	98.6	20.6
Shawnee Hills	142	135	3.2	95.1	20.3
Blue Grass	204	199	3.4	97.5	29.9
Cumberland Plateau	173	131	2.6	75.7	19.7
Cumberland Mountains	19	15	2.1	78.9	2.3

Summary of Breeding Status

Number of Blocks in Which Species Was Recorded		
Total	**665**	**91.5%**
Confirmed	407	61.2%
Probable	105	15.8%
Possible	153	23.0%

Barn Swallow

Blue Jay

Cyanocitta cristata

The raucous calls of Blue Jays are a well-known sound that can be heard throughout the year in Kentucky. Although the species may become somewhat reclusive during the nesting season, it occurs widely across the state. Mengel (1965) regarded the Blue Jay as a common and uniformly distributed resident statewide.

Blue Jays are encountered in a great variety of forested and semi-open habitats. The species appears to be almost as numerous in regions covered by extensive mature forest as in open farmland with scattered woodlots and farmsteads. Having adapted well to the presence of humans, substantial numbers also nest in urban and suburban situations, as long as at least a few trees are present.

It is likely that the Blue Jay has always been widespread in Kentucky. Audubon (1861) considered the species to be plentiful across all of those parts of the United States that he had visited as of the early 1800s. A common inhabitant of wooded areas, the species must have been numerous before clearing and fragmentation of the once widespread forests. Moreover, the jay's ability to adapt to human alteration of the landscape has allowed it to remain among the state's most abundant nesting species.

Although the species's seasonal movements are not completely understood, birds from farther north pass through Kentucky during migration and perhaps replace or supplement the local breeding population in winter. About the time large numbers of transient jays are moving through on their way north in late April and early May, resident birds have already begun nesting. Mengel (1965) noted that clutches may be completed by the last week of March and as late as the last ten days of June. Although peaks in clutch completion are not evident, data suggest that some pairs raise two broods (Mengel 1965). The average size of twelve clutches reported by Mengel was 4.1 (range of 3–5).

Forest Cover

Value	% of Blocks	Avg Abund
All	97.5	2.9
1	97.0	2.7
2	99.5	3.0
3	99.1	2.9
4	96.8	2.8
5	86.5	2.6

Blue Jays use both deciduous and evergreen trees for nesting. The nest is usually constructed in a fairly substantial, upright fork of a fairly large tree, although smaller forks in the outer canopy of foliage are sometimes used, especially in conifers. Nest trees are just as likely to be located in deep forest as in open parklike situations. The average height of 18 nests reported by Mengel (1965) was 20.5 feet (range of 6–45 feet), although Stamm (1975) noted construction of a nest 90 feet above the ground in a tulip poplar tree, and nests have been observed as low as four feet (KOS Nest Cards). The nest is a bulky structure, composed primarily of small sticks and twigs, and it is lined with finer twigs and rootlets (Harrison 1975).

Blue Jays were found in nearly 98% of priority blocks statewide, and five incidental observations were reported. The species ranked 5th according to the number of priority block records, 11th by total abundance, and 19th by average abundance. As expected, little variation in occurrence and average abundance by physiographic province section was evident. Occurrence and average abundance were slightly lower in extensively forested areas.

Ron Austing

Despite the Blue Jay's general abundance, only about 22% of priority block records were for confirmed breeding. This was at least partially the result of the species's secretive behavior during the nesting season. While Blue Jays' characteristic calls often made atlasers aware of their presence, their behavior otherwise was not conspicuous, leading to a large number of possible records. Perhaps the most useful clue to nesting was the begging of fledgling birds, easily heard but not always recognized by atlasers. Although a few active nests with eggs or young were found, most confirmed records were based on the observation of recently fledged young, adults carrying food for young, and nest building.

Blue Jays are reported on most Kentucky BBS routes in good numbers. The average number of individuals per BBS route for the periods 1966–91 and 1982–91 was 15.83 and 14.31, respectively. According to these data, the Blue Jay ranked 17th in abundance on BBS routes during the period 1982–91. Trend analysis of these data shows a significant ($p<.05$) decrease of 1.5% per year for the period 1966–91, but a nonsignificant decrease of 1.6% per year for the period 1982–91.

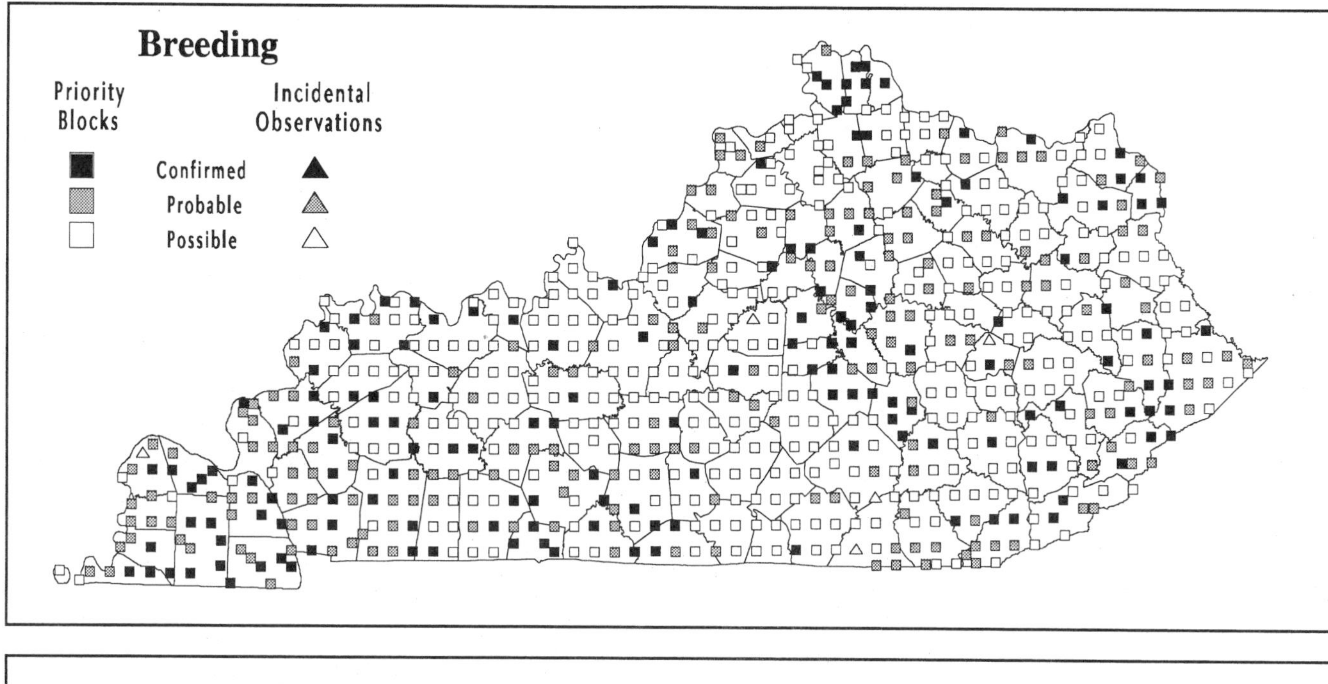

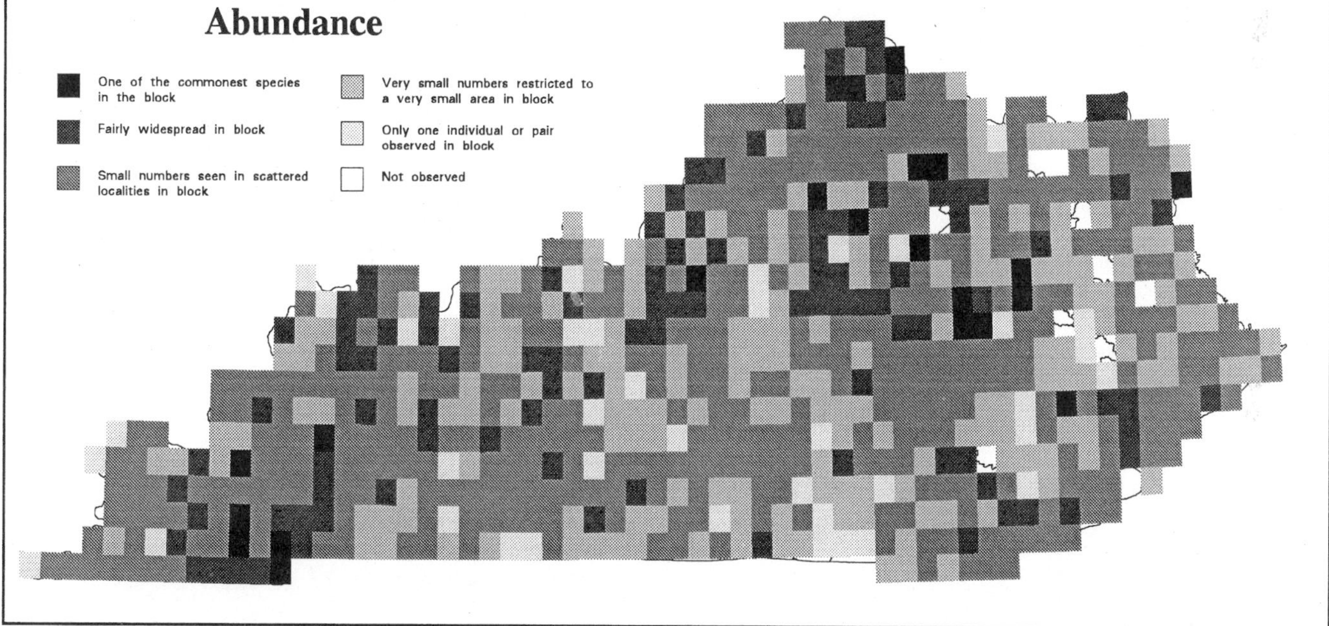

Analysis of Block Data by Physiographic Province Section

Physiographic Province Section	Total Blocks Surveyed	Blocks with Data	Avg Abund	% with Data	Section's % for State
Mississippi Alluvial Plain	14	13	2.7	92.9	1.8
East Gulf Coastal Plain	36	36	3.3	100.0	5.1
Highland Rim	139	139	2.7	100.0	19.6
Shawnee Hills	142	140	2.8	98.6	19.7
Blue Grass	204	200	3.1	98.0	28.2
Cumberland Plateau	173	163	2.7	94.2	23.0
Cumberland Mountains	19	18	3.1	94.7	2.5

Summary of Breeding Status

Number of Blocks in Which Species Was Recorded		
Total	**709**	**97.5%**
Confirmed	158	22.3%
Probable	201	28.3%
Possible	350	49.4%

American Crow

Corvus brachyrhynchos

Because of its distinctive calls and large size, the American Crow is probably one of the most widely recognized birds in Kentucky. Mengel (1965) regarded the species as a variably rare to common permanent resident, conspicuous throughout all of the state except the heavily forested portions of eastern Kentucky.

American Crows are encountered in a wide variety of habitats, from open farmland to predominantly forested regions with scattered clearings. During the breeding season the species is noticeably less numerous only in extremely open situations where cover for nesting is not sufficient, and in large tracts of unbroken forest where foraging habitat is limited. Being quite adaptable, crows regularly nest in urban and suburban parks, residential yards, and other settled areas, as long as some sort of cover for nesting is available.

It is possible that American Crows were fairly widespread in Kentucky before settlement, but the species's current distribution suggests that human alteration of the landscape has resulted in an increase in its abundance and distribution. While extensively forested regions appear to support a few crows, the presence of at least some small openings for foraging may be necessary. Areas of fragmented forest interspersed with farmland and settlement seem to harbor substantially greater numbers.

Outside the breeding season American Crows appear to move about seasonally, and large numbers may gather into flocks to forage and roost. Winter flocks may include birds that have migrated from more northerly breeding areas. At the onset of warm weather in early spring, local birds begin to break off into pairs, and territories are established soon thereafter. Nest building may be under way by the middle of March, and early clutches are apparently completed by the end of March (Mengel 1965). Following a peak in clutch completion during the first ten days of April, later clutches may be completed into the middle of May (Mengel 1965). The distinctive begging calls of juvenile birds become conspicuous by June, and family groups are often seen into July. Although two broods are sometimes raised in the southern part of the bird's range (Harrison 1975), evidence of such in Kentucky is lacking. The average size of 10 clutches given by Mengel (1965) was 3.9 (range of 3–5).

Forest Cover

Value	% of Blocks	Avg Abund
All	96.8	2.8
1	85.1	2.6
2	97.9	2.8
3	98.3	2.9
4	100.0	3.0
5	100.0	2.7

American Crows usually place their bulky nest in a substantial fork of a large tree (Mengel 1965). Both deciduous and evergreen trees are used, although the latter seems to be preferred for early nests. The nest tree may be located in heavy forest, forest edge, plantations, isolated woodlots, riparian corridors, or parks. The nest is constructed of small branches and twigs and lined with grass, fur, and other fine materials. The average height of nine nests given by Mengel (1965) was 28.6 feet (range of 15–60 feet).

The atlas survey yielded records of American Crows in nearly 97% of priority blocks statewide, and three incidental observations were reported. The species ranked 8th according to the number of priority block records, 13th by total abundance, and 21st by average abundance. Occurrence was below 90% only in the Mississippi Alluvial Plain, where the species is replaced to some degree by the Fish Crow. Average abundance was not highly variable. Occurrence was lowest in extensively cleared areas and increased as the percentage of forest cover increased.

Ron Austing

Despite the American Crow's abundance, only about 41% of priority block records were for confirmed breeding. Although a few active nests were located, most confirmed records resulted from the observation of family groups including recently fledged young. Family groups were easy to locate owing to the loud begging calls of young birds. Although it is likely that the FL code was sometimes applied to family groups that were relatively far from the actual nest site, all confirmed records based on the FL code were included in the data set.

Conspicuous by both sight and sound, and common in most rural situations, American Crows are frequently reported on most Kentucky BBS routes. The average number of individuals per BBS route for the periods 1966–91 and 1982–91 was 36.46 and 42.70, respectively. According to these data, the American Crow ranked 10th in abundance on BBS routes during the period 1982–91. Trend analysis of these data yields nonsignificant increases of 0.4% per year for the period 1966–91 and 0.5% per year for the period 1982–91.

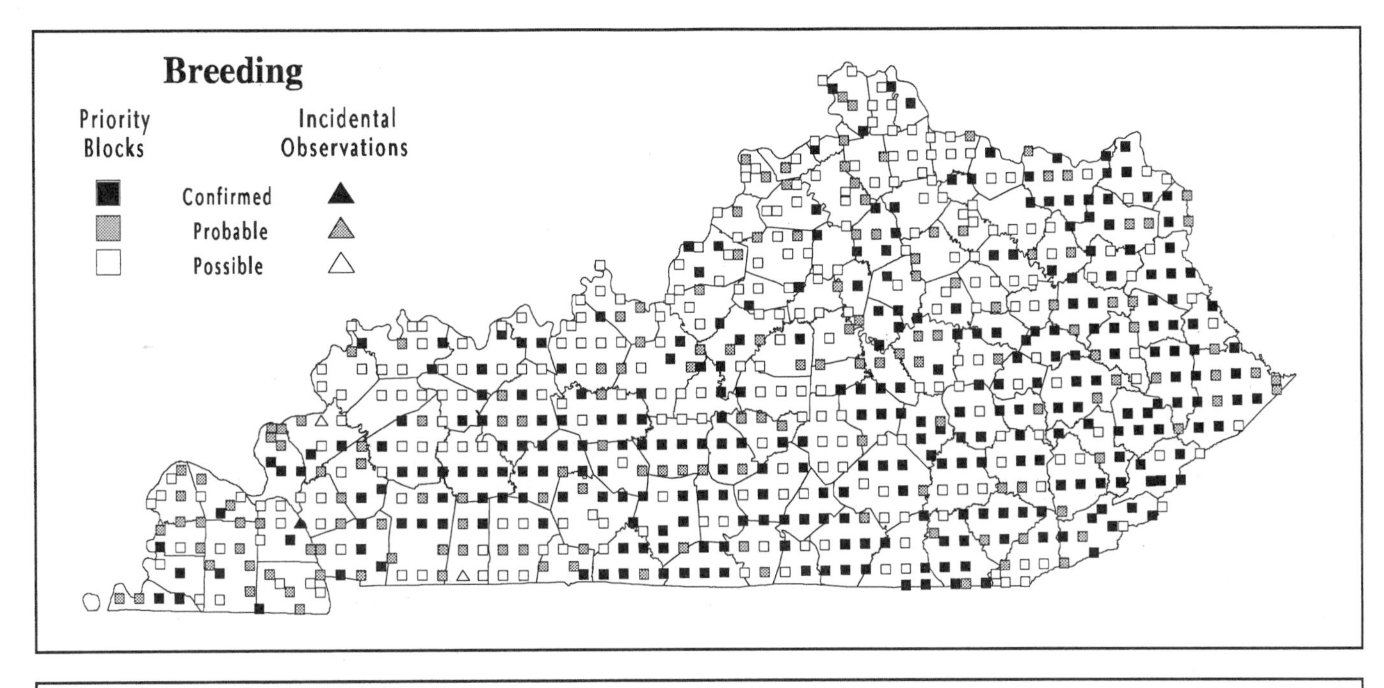

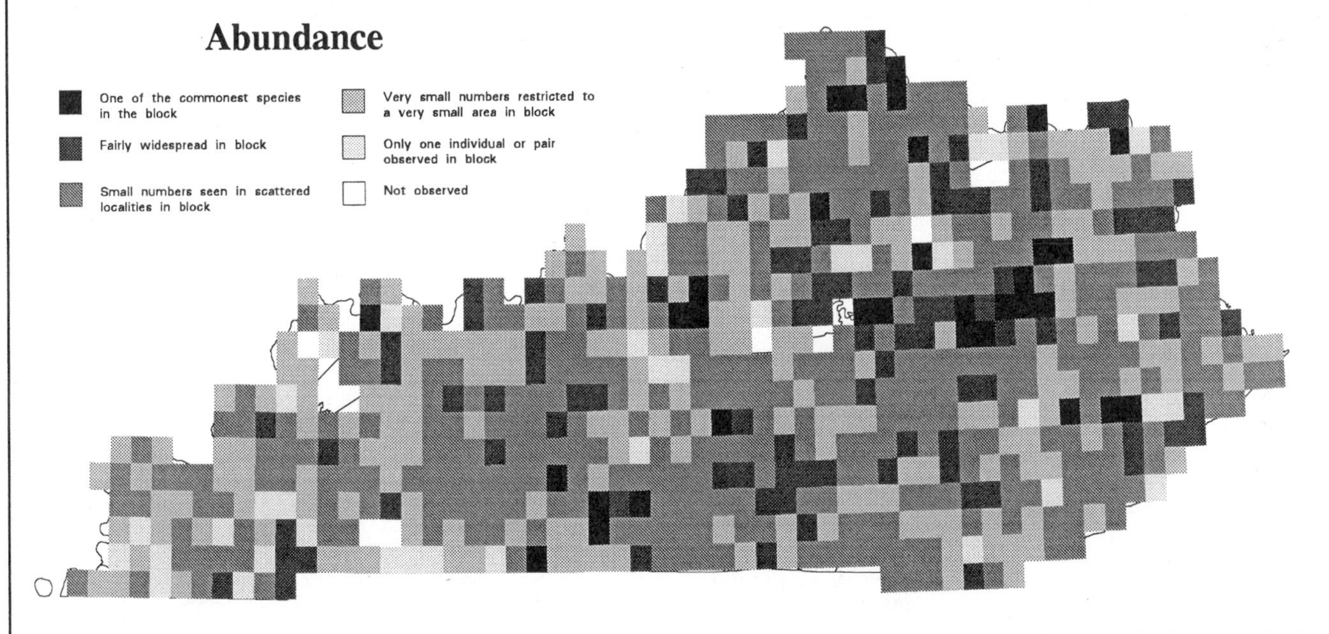

Analysis of Block Data by Physiographic Province Section

Physiographic Province Section	Total Blocks Surveyed	Blocks with Data	Avg Abund	% with Data	Section's % for State
Mississippi Alluvial Plain	14	11	2.4	78.6	1.6
East Gulf Coastal Plain	36	35	2.4	97.2	5.0
Highland Rim	139	136	2.8	97.8	19.3
Shawnee Hills	142	132	2.8	93.0	18.8
Blue Grass	204	198	3.0	97.1	28.1
Cumberland Plateau	173	173	2.9	100.0	24.6
Cumberland Mountains	19	19	2.5	100.0	2.7

Summary of Breeding Status

Number of Blocks in Which Species Was Recorded		
Total	**704**	**96.8%**
Confirmed	291	41.3%
Probable	156	22.2%
Possible	257	36.5%

Fish Crow

Corvus ossifragus

The Fish Crow is a relatively recent addition to the avifauna of Kentucky. The species was first reported on May 24, 1959, when two birds were observed at Hickman, in Fulton County (Coffey 1959). About the time of this observation, the species also appeared for the first time in Missouri (Easterla 1965). Interestingly, the Fish Crow's observation in both states occurred more than 25 years after it was first sighted in Tennessee, near Memphis, in August 1931 (Coffey 1942). Early ornithologists such as Dr. L. Otley Pindar, who conducted extensive fieldwork in extreme southwestern Kentucky during the late 19th century, did not report it. Likewise, Mengel (1965) and others who visited the area during the late 1940s and early 1950s did not detect its presence either. Thus it must be assumed that Fish Crows were not present in Kentucky before the mid- to late 1950s.

Although the 1959 report turned out to be the first of many in the far western counties, the Fish Crow apparently took several years to become established. The species was not reported again until April 1966, when several birds were seen at Kentucky Bend, in Fulton County (Croft and Rowe 1966). Monroe and Able (1968) were the first to report finding Fish Crows regularly in western Kentucky along the Mississippi River in Fulton, Hickman, Carlisle, and Ballard counties. Since that time the species has continued to be observed regularly along the Mississippi River and the lower portions of its major tributaries (Mayfield Creek, Obion Creek, Bayou du Chien) and the lower Ohio River upstream to the vicinity of Paducah, in McCracken County.

Through the mid-1980s the Fish Crow was known only from the limited range described above. During the summer of 1988 atlasers recorded the species at several new localities to the north and east, including the lower Tennessee and Cumberland rivers at Kentucky and Barkley dams; farther up the Ohio River in the vicinity of Smithland Dam, in Livingston County; above the Shawneetown, Illinois, bridge, in Union County; and near Henderson, in Henderson County. In addition, the species was observed along the lower Green River near Pleasant Valley, in Henderson and McLean counties. Since that time little summer fieldwork has been undertaken in these areas, but subsequent observations at Kentucky, Barkley, and Smithland dams, and in northern Union County, indicate that this range expansion has persisted. Moreover, the species has been found recently in Posey County in extreme southwestern Indiana, across the Ohio River from Union and Henderson counties (M. Homoya and R. Hedge, pers. comm.)

Fish Crows are seldom encountered far from the extensive floodplain corridors of the larger rivers and the lower portions of their tributaries. They are most frequently seen on exposed river bars and in agricultural fields immediately adjacent to river or stream channels. In addition, they also occur in forested floodplain swamps and tracts of mature bottomland forest, habitats that probably serve as nesting areas.

Although Fish Crows may overwinter in far western Kentucky during milder years, most birds likely retreat southward in winter. The species usually becomes conspicuous within its limited range by late March or early April, and it is likely that most of the nesting population has returned by early May. Despite observations of presumed breeding birds from many locations, an active nest has never been found in the state, and evidence of breeding is limited to the observation of presumed family groups. According to Harrison (1975), the nest is typically placed in the top of a tree, 20–80 feet above the ground, and it is constructed of sticks and twigs and lined with a variety of finer material. Rangewide clutch size is given as 4 or 5, rarely more (Harrison 1975).

James Parnell

The atlas fieldwork yielded 16 records of Fish Crows in priority blocks, and six incidental observations were reported. While most of the reports came from traditional summering areas, the reports from north and east of Paducah represent a significant range expansion.

Most atlas reports, including all records of birds in the new areas noted above, involved observation of a single bird (a pair on one occasion) and were treated as possible records. Probable breeding records were reported in six priority blocks, and there was one probable incidental observation. Probable records were based on the observation of pairs of birds in appropriate nesting habitat over an extended period of time, indicating the likelihood that the birds were occupying a territory.

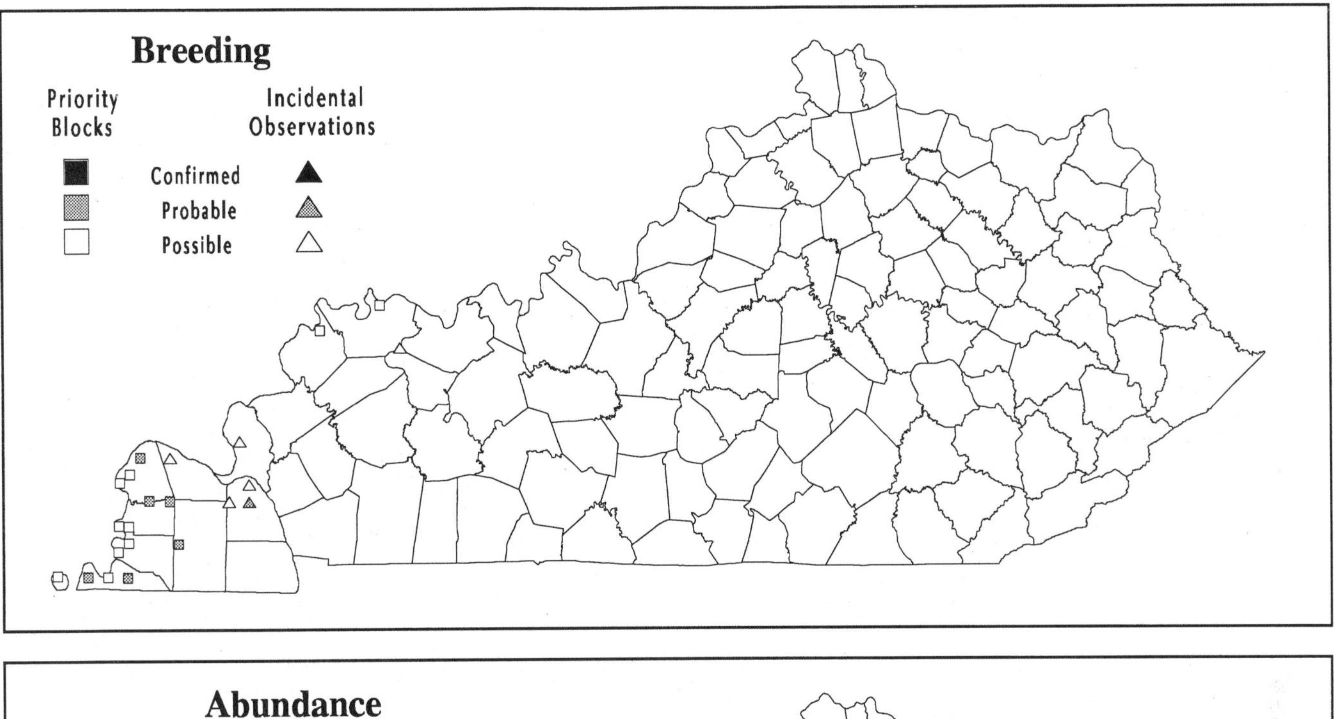

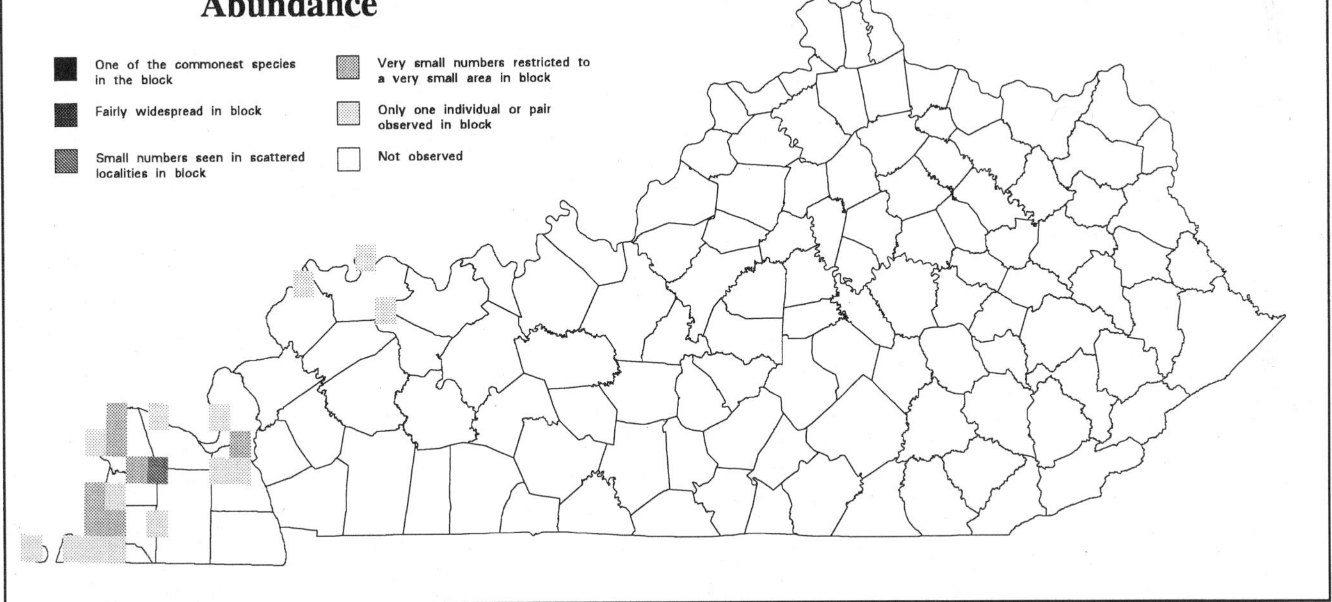

Analysis of Block Data by Physiographic Province Section

Physiographic Province Section	Total Blocks Surveyed	Blocks with Data	% with Data	Section's % for State
Mississippi Alluvial Plain	14	12	85.7	75.0
East Gulf Coastal Plain	36	2	5.6	12.5
Highland Rim	139	-	-	-
Shawnee Hills	142	2	1.4	12.5
Blue Grass	204	-	-	-
Cumberland Plateau	173	-	-	-
Cumberland Mountains	19	-	-	-

Summary of Breeding Status

Number of Blocks in Which Species Was Recorded		
Total	**16**	**2.2%**
Confirmed	-	-
Probable	6	37.5%
Possible	10	62.5%

Fish Crow

Common Raven

Corvus corax

Although the hoarse, croaking calls of Common Ravens are rarely heard in Kentucky today, at one time the species was apparently fairly widespread. Reports from the late 1700s and the 1800s indicate that ravens may have been at least occasional throughout much of the state. They appear to have occurred most frequently in the rugged forests of the Cumberland Mountains of extreme southeastern Kentucky, but small numbers were regularly found in the Cliff Section of the Cumberland Plateau as well (Mengel 1965).

Ravens gradually declined after settlement, and most, if not all, had disappeared by the 1930s. It appears that the species decreased as forested land was cleared and settled, and perhaps as the American Crow increased. Mengel (1965) regarded the Common Raven as extirpated from Kentucky as of the late 1950s. Croft (1970) rediscovered the species in the southeastern mountains in 1969, and ravens have been reported in small numbers there since that time (e.g., Hall 1976; Smith and Davis 1979; Stamm 1978b; Davis et al. 1980; Stamm 1981c). Reports have ranged from as far northeast as Breaks Interstate Park, in Pike County (Heilbrun et al. 1983), to Cumberland Gap National Park, in Bell County, along the Tennessee border. Before the atlas survey, all recent records were restricted to the immediate vicinity of the major mountain ridges, including Pine Mountain, Black Mountain, and Cumberland Mountain. Since at least 1982 an active nest has been present at Bad Branch State Nature Preserve on the southeast side of Pine Mountain in Letcher County (Fowler et al. 1985).

During the atlas survey ravens were found on six priority blocks, and two incidental observations were reported. Most records originated from traditionally known sites in the Cumberland Mountains; however, on July 12, 1989, two birds were seen flying over a ridge along Jones Fork, south of Mousie, and one bird was seen flying over a ridge north of Balls Fork, west of Vest, both in Knott County. These observations represent an extension of the known range of ravens some twenty miles northwestward into the Cumberland Plateau. Although nesting in the vicinity of these observations was possible, the records also could have represented birds ranging out following nesting somewhere on Pine Mountain. Curiously, the same year a raven was reported from Natural Bridge State Park, in Powell County, on May 5, 1989 (J. and P. Bell, pers. comm.; Stamm 1989b). This record was not included in the atlas results. The only confirmed record pertained to the well-known pair at Bad Branch, where young were heard calling in the nest on April 14, 1989.

Ravens may wander widely over large home ranges, but it appears that Kentucky's nesting birds are resident throughout the year. Nesting is initiated relatively early. Reports of young in the nest at Bad Branch on April 10, 1984 (Fowler et al. 1985), and April 14, 1989 (Stamm 1989b), indicate that clutch completion occurs sometime during early March. After young fledge sometime in late April or early May, family groups probably forage near the nest site for several weeks. Kentucky data on clutch size are limited to the observation of six young in the Bad Branch nest in 1982 (Fowler et al. 1985), but Harrison (1975) gives rangewide clutch size as 3–6, commonly 4 or 5.

Common Ravens nest in coniferous trees throughout much of their range in the northern United States and Canada (Harrison 1975), but the Bad Branch nest is typical of those constructed throughout the Appalachian Mountains. It is placed on a well-protected ledge, well up on a large cliff face. The nest is a bulky structure, composed chiefly of small branches and sticks and lined with finer plant material and fur (Harrison 1975).

Maslowski Wildlife Productions

Although ravens have adapted to human presence in some parts of their range, in Kentucky they are still birds of remote areas. Birds are rarely seen away from extensively forested portions of the mountains, where they usually can be found along or near the ridge crests. Although ravens nest and loaf along clifflines and exposed rock outcrops, they are most frequently seen flying along the ridges or soaring overhead. The abundance of suitable nest sites in eastern Kentucky indicates that other factors are responsible for the species's overall scarcity.

Common Ravens are increasing in West Virginia (Buckelew and Hall 1994), and if the current trend continues the population of ravens in southeastern Kentucky may continue to expand slowly. Furthermore, if birds can successfully reach the western Cliff Section of the Cumberland Plateau, the species might become reestablished somewhere within this region of extensive forests and precipitous cliffs.

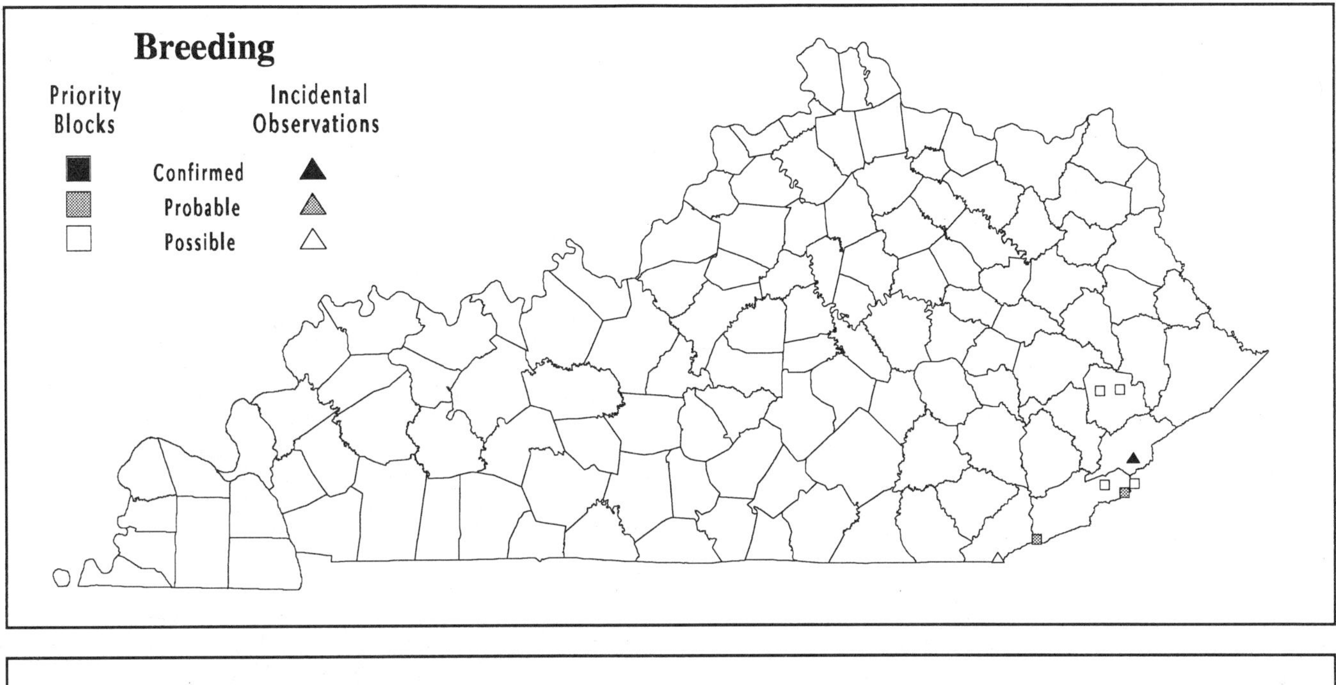

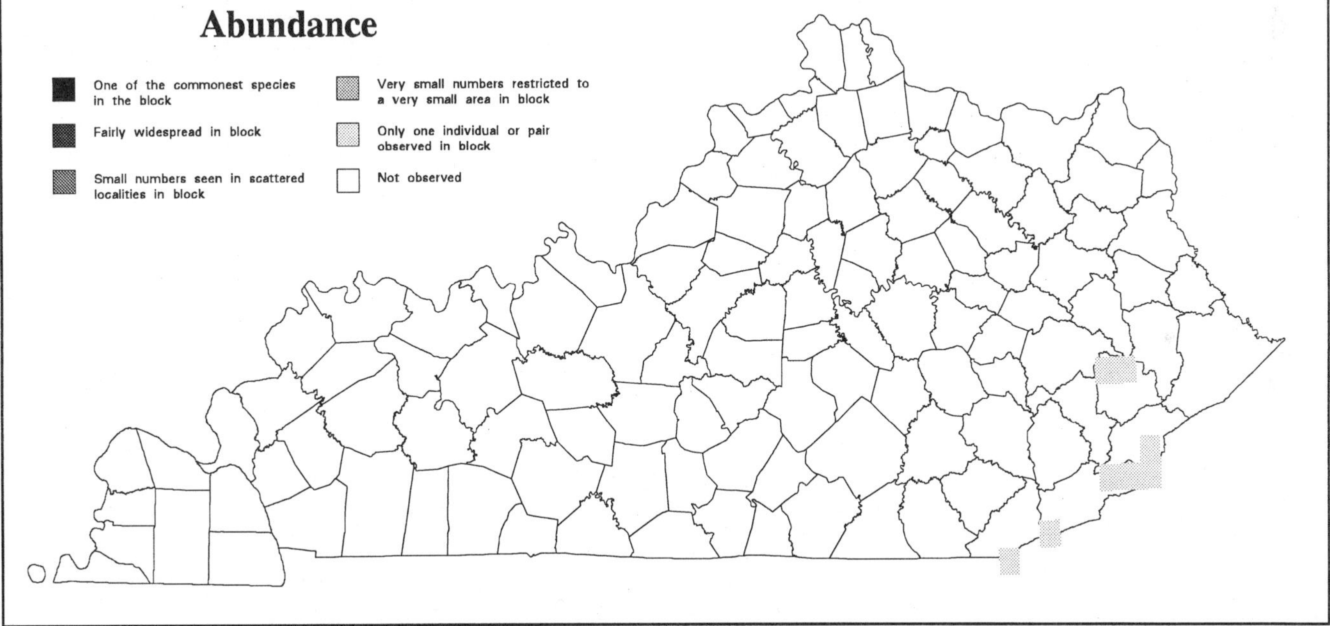

Analysis of Block Data by Physiographic Province Section

Physiographic Province Section	Total Blocks Surveyed	Blocks with Data	% with Data	Section's % for State
Mississippi Alluvial Plain	14	-	-	-
East Gulf Coastal Plain	36	-	-	-
Highland Rim	139	-	-	-
Shawnee Hills	142	-	-	-
Blue Grass	204	-	-	-
Cumberland Plateau	173	2	1.2	33.3
Cumberland Mountains	19	4	21.1	66.7

Summary of Breeding Status

Number of Blocks in Which Species Was Recorded		
Total	**6**	**0.8%**
Confirmed	-	-
Probable	2	33.3%
Possible	4	66.7%

Common Raven

Carolina Chickadee

Parus carolinensis

Throughout much of the year, roving bands of Carolina Chickadees in the company of other small passerines are a fairly conspicuous sight in Kentucky. Mengel (1965) regarded the species as a common resident of forested and edge habitats, being found in every woodland and forest type across the state.

Carolina Chickadees are encountered in a great variety of habitats, from small woodlots and wooded corridors in predominantly open areas to the largest tracts of unbroken forest. Having adapted well to settled conditions, they are quite numerous about suburban parks and yards as well as rural farmland.

Before European settlement Carolina Chickadees were probably common and widespread in Kentucky, inhabiting a variety of forest and edge habitats. The clearing and fragmentation of the once widespread forests of the state have likely resulted in an overall decrease in abundance in some areas. The species has continued to thrive amid most types of human alteration, however, and it probably has not declined significantly in most regions.

Carolina Chickadees are permanent residents in Kentucky. Although the first songs are often heard on warm late winter days, it is usually April before nesting activity commences. According to Mengel (1965), early clutches may be completed by the beginning of April, but a peak in clutch completion occurs during the last ten days of the month. Later clutches are laid into late May (Mengel 1965), but in the absence of specific evidence of double-brooding, it must be assumed that most late attempts represent renesting following earlier failure. The average size of 15 clutches or broods reported by Mengel (1965) was 4.8 (range of 4–6). Following the nestling period, family groups are commonly observed from the latter part of May through June, although some remain apparent throughout the summer.

Forest Cover

Value	% of Blocks	Avg Abund
All	97.1	3.2
1	90.1	2.6
2	97.4	3.0
3	97.4	3.2
4	100.0	3.6
5	100.0	3.6

Carolina Chickadees are cavity nesters, frequently using cavities constructed by other birds or otherwise formed naturally in limbs or posts. They also regularly excavate their own cavities out of the soft, rotting wood of a stump or snag, and the species readily accepts nest boxes. Nest sites are just as likely to be located within forest interior as along woodland edge, but chickadees occasionally choose a site relatively far from any tracts of large trees, especially in largely cultivated areas.

The bottom of the cavity is filled with moss and downy plant material. The eggs are laid in a neat cup, formed within the nest material and lined with animal hair. The average height of five nests reported by Mengel was 5.8 feet (range of 2–12 feet), although Stamm and Croft (1968) note a nest 20 feet high.

Chickadees were recorded in more than 97% of priority blocks statewide, and six incidental observations were reported. The species ranked 6th according to the number of priority block records, 6th by total abundance, and 10th by average abundance. Average abundance was relatively uniform across the state, but it was highest in eastern Kentucky. When occurrence is compared with percentage of forest cover, a perceptible trend appears. Chickadees were less frequently encountered and less abundant in extensively cleared areas, and they occurred more often and in greater numbers as forestation increased.

Ron Austing

Although the chickadee is a common summer resident, less than 26% of priority block records were for confirmed breeding. The species's relatively quiet, reclusive behavior during nesting is largely responsible for the lack of more confirmed reports. Although some active nests were located, most records of confirmed breeding were for the observation of family groups of recently fledged young. A few others were based on the observation of adults carrying food for young and nest building.

Carolina chickadees are reported in relatively small numbers on most Kentucky BBS routes, in large part because of their inconspicuousness during the breeding season. The average number of individuals per BBS route for the periods 1966–91 and 1982–91 was 5.61 and 6.80, respectively. Trend analysis of these data shows slight, nonsignificant increases of 0.8% per year for the period 1966–91 and 0.9% per year for the period 1982–91. Interestingly, the Carolina Chickadee did not rank among the top 30 species on Kentucky BBS routes for the period 1982–91.

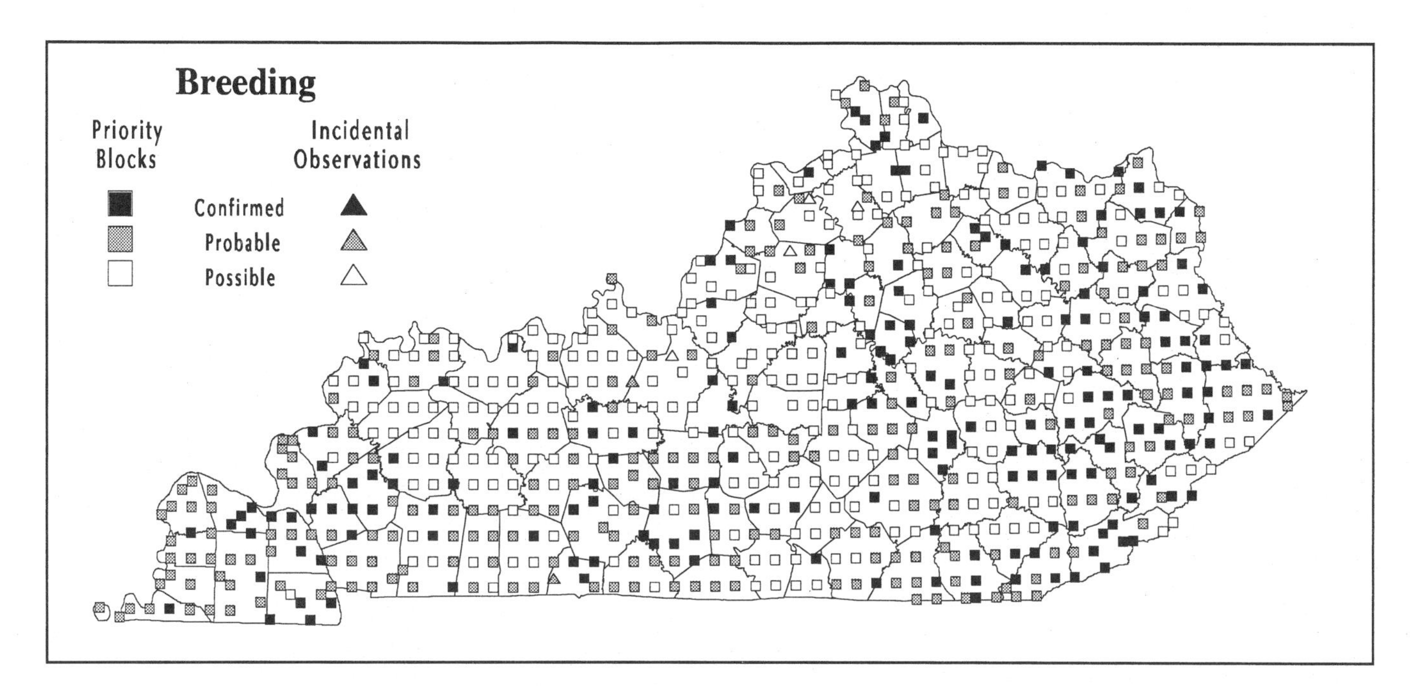

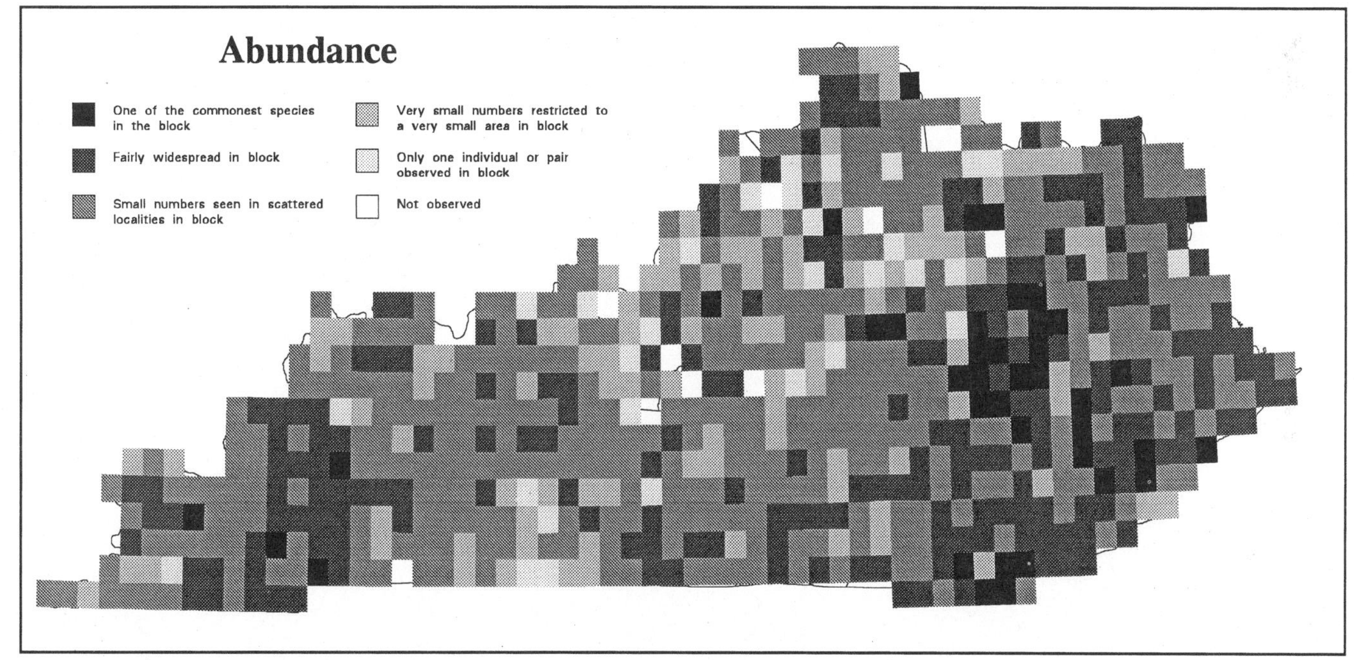

Analysis of Block Data by Physiographic Province Section

Physiographic Province Section	Total Blocks Surveyed	Blocks with Data	Avg Abund	% with Data	Section's % for State
Mississippi Alluvial Plain	14	14	3.0	100.0	2.0
East Gulf Coastal Plain	36	36	3.3	100.0	5.1
Highland Rim	139	133	3.0	95.7	18.8
Shawnee Hills	142	137	3.0	96.5	19.4
Blue Grass	204	194	2.9	95.1	27.5
Cumberland Plateau	173	173	3.6	100.0	24.5
Cumberland Mountains	19	19	3.6	100.0	2.7

Summary of Breeding Status

Number of Blocks in Which Species Was Recorded		
Total	**706**	**97.1%**
Confirmed	183	25.9%
Probable	257	36.4%
Possible	266	37.7%

Tufted Titmouse

Parus bicolor

Although not quite as abundant as the closely related Carolina Chickadee, the Tufted Titmouse is among the most widely distributed songbirds in Kentucky. Mengel (1965) considered the titmouse to be a common permanent resident. He found that it occurred in substantial numbers in all forest types across the state except those at higher elevations of Black Mountain, where the species was considerably less numerous in summer.

Tufted Titmice are encountered in a great variety of forested and semi-open habitats. Although most common in regions with a good supply of woodland, they seem to be slightly less numerous in large tracts of unbroken forest, favoring areas with numerous edges and openings. The species also occurs in considerable numbers in open habitats, as long as at least some edge habitat and scattered tracts of forest are present. The titmouse is also frequent in settled situations, and it often nests in suburban parks as well as rural farmland.

The status of the Tufted Titmouse in Kentucky may not have changed significantly in response to human alteration of the landscape. The species seems to be slightly less numerous in areas of extensive forest, and it may have benefited in some regions from fragmentation of once widespread forests. In contrast, in areas where forests have been extensively cleared, its overall abundance is probably much lower today.

The Tufted Titmouse is a permanent resident in Kentucky. Although song is occasionally heard during the winter, territorial behavior commences during the latter part of March, and according to Mengel (1965), nest building may be under way by the beginning of April. A peak in clutch completion occurs in late April, although a few later clutches are laid into early June (Mengel 1965). Through the latter half of May and June, family groups are conspicuous, as they forage noisily together. Although double-brooding has been reported in the southern part of the species's range (Ehrlich et al. 1988), evidence of such in Kentucky is lacking. The average size of eight clutches or broods reported by Mengel (1965) was 5.0 (range of 3–7).

Forest Cover

Value	% of Blocks	Avg Abund
All	97.0	3.0
1	92.1	2.5
2	96.8	3.0
3	97.8	3.1
4	99.3	3.3
5	96.2	3.2

Titmice are cavity nesters, using both natural cavities as well as those excavated by other birds. Unlike the Carolina Chickadee, titmice are not believed to excavate their own cavities (Harrison 1975), and they also seem to use artificial nest boxes less frequently. The bottom of the cavity is filled with moss, fine strips of bark, and other fine plant matter. The eggs are laid in a neat cup formed within the nest material and lined with animal hair. The average height of five nests reported by Mengel (1965) was 19 feet (range of 8–35 feet), although nests have been reported as low as 4.5 feet above the ground (KOS Nest Cards).

The atlas survey yielded records of Tufted Titmice in 97% of priority blocks statewide, and six incidental observations were reported. The species ranked 7th according to the number of priority block records, 8th by total abundance, and 15th by average abundance. Occurrence was very high in all physiographic province sections, and average abundance also was relatively consistent statewide. Occurrence and average abundance varied slightly by percentage of forest cover, being lower in extensively cleared areas.

Ron Austing

Despite the Tufted Titmouse's statewide abundance, only about 32% of priority block records were for confirmed breeding. A few active nests were located, but most confirmed reports resulted from the observation of foraging family groups. In addition, a few records were based on the observation of adults carrying food and nest building.

Probably because of the conspicuousness of their song and calls, Tufted Titmice are more frequently reported on BBS routes than the Carolina Chickadee. The average number of individuals per BBS route for the periods 1966–91 and 1982–91 was 11.39 and 10.31, respectively. According to these data, the species ranked 21st in abundance on BBS routes for the period 1982–91. Trend analysis of these data shows a nonsignificant decrease of 0.7% per year for the period 1966–91, but a significant ($p<.01$) increase of 5.0% per year for the period 1982–91.

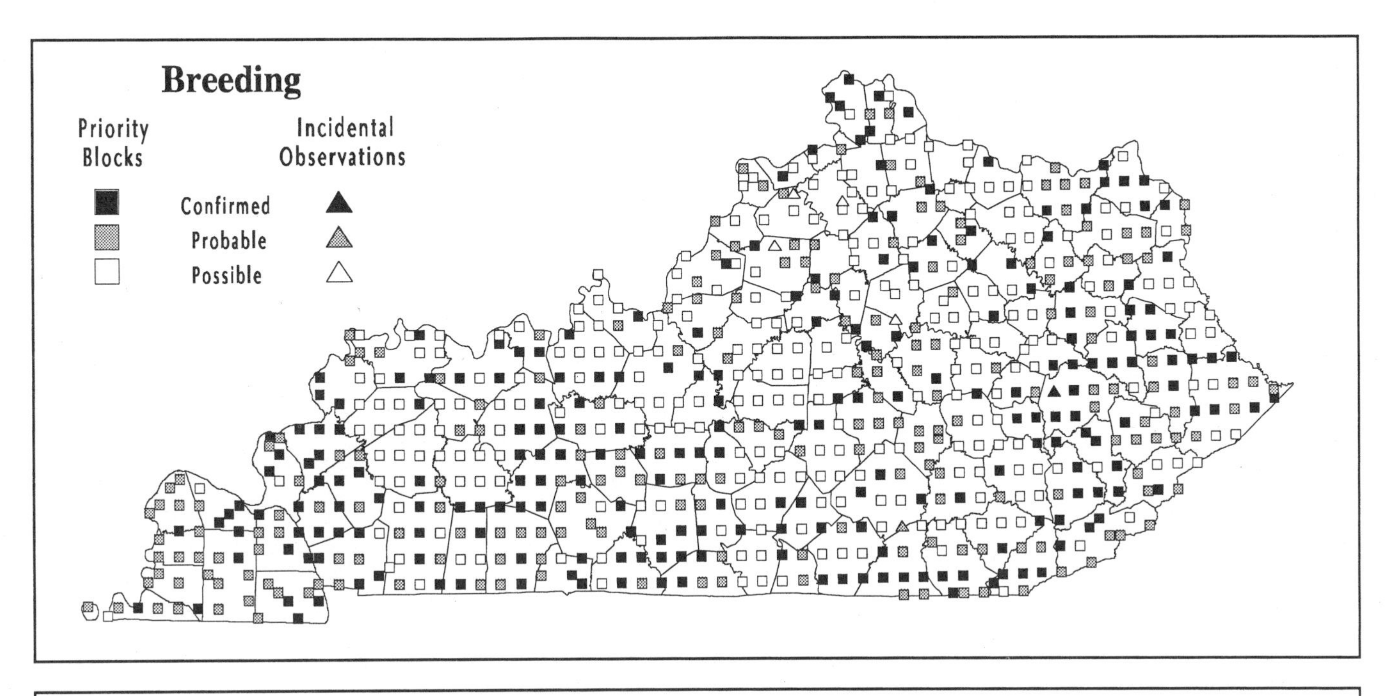

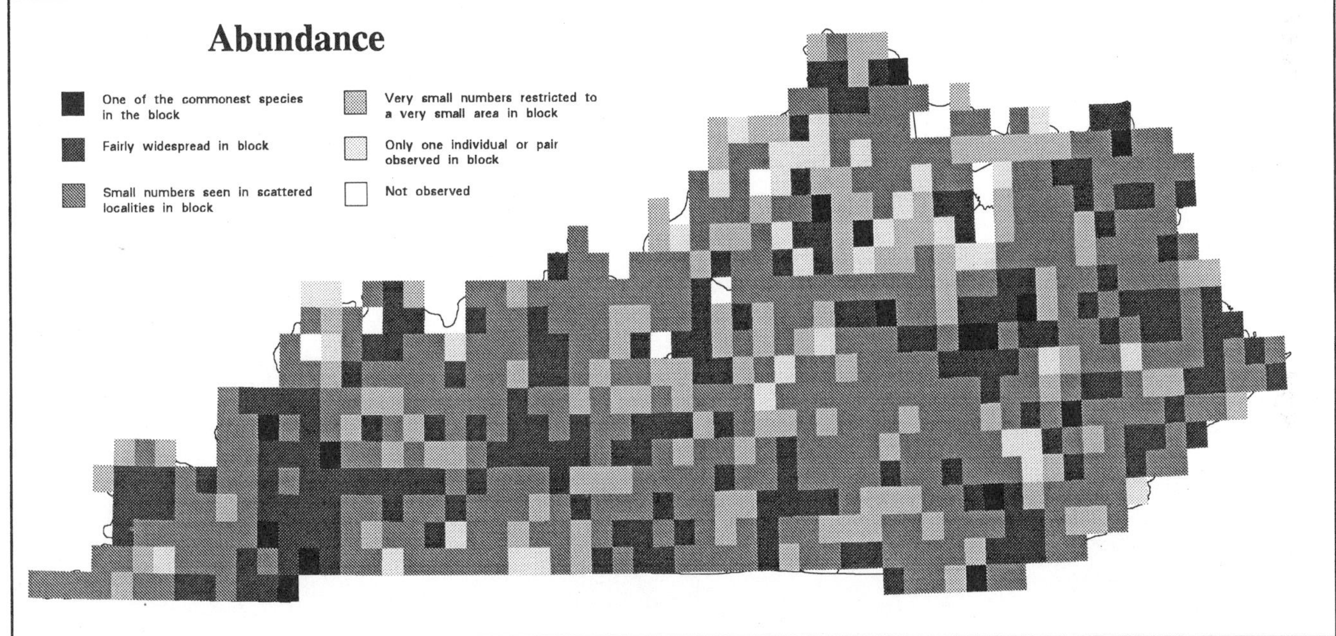

Analysis of Block Data by Physiographic Province Section

Physiographic Province Section	Total Blocks Surveyed	Blocks with Data	Avg Abund	% with Data	Section's % for State
Mississippi Alluvial Plain	14	14	3.1	100.0	2.0
East Gulf Coastal Plain	36	36	3.2	100.0	5.1
Highland Rim	139	138	3.0	99.3	19.6
Shawnee Hills	142	138	3.2	97.2	19.6
Blue Grass	204	190	2.8	93.1	27.0
Cumberland Plateau	173	171	3.2	98.8	24.3
Cumberland Mountains	19	18	3.0	94.7	2.5

Summary of Breeding Status

Number of Blocks in Which Species Was Recorded		
Total	**705**	**97.0%**
Confirmed	227	32.2%
Probable	212	30.1%
Possible	266	37.7%

White-breasted Nuthatch

Sitta carolinensis

The nasal call notes of the White-breasted Nuthatch are a fairly conspicuous year-round sound in forested habitats throughout Kentucky. Mengel (1965) regarded the species as an uncommon to fairly common permanent resident, varying somewhat in abundance from region to region. He found it to be most numerous in the lowland forests of central and western Kentucky and least common in the rich, mixed mesophytic forests of slopes and ravines in the Cumberland Plateau and the Cumberland Mountains.

White-breasted Nuthatches are seldom found far from forested areas. While the species is sometimes encountered in fairly open situations, in most such cases a tract of forest is located nearby. These nuthatches are also quite frequent in suburban parks and yards, as long as some large trees and woodlots are in the vicinity.

The White-breasted Nuthatch must have been more common and widespread in Kentucky before settlement. The clearing and fragmentation of the widespread forests that once covered much of the state have likely diminished numbers substantially, especially in regions where loss of forest has been extreme.

Little specific information on the nesting ecology of White-breasted Nuthatches in Kentucky has been published. No detailed records of nesting were available to Mengel (1965), and he included only a few dates for the observation of young out of the nest being fed by adults. The species apparently maintains foraging territories throughout the year (Ehrlich et al. 1988), and territorial calling is sometimes heard in late winter. Nest building has been observed by late March (Mengel 1965), but more frequently in April (KOS Nest Cards; Alsop 1971). Most clutches are likely completed by the middle of May, and family groups of recently fledged young have been observed most often in June. Kentucky data on clutch and brood size are limited to an apparently complete clutch of three eggs in Oldham County in 1972 (KOS Nest Cards). During atlas fieldwork, a family group including at least four young was observed in Simpson County on June 19, 1989. Harrison (1975) gives rangewide clutch size as 5–10, commonly 8.

Forest Cover

Value	% of Blocks	Avg Abund
All	72.5	2.0
1	56.4	1.6
2	68.4	2.0
3	76.5	2.1
4	82.5	2.2
5	71.2	1.8

White-breasted Nuthatches are cavity nesters. They usually choose natural tree cavities, but they also frequently use cavities excavated by woodpeckers, and occasionally they accept artificial nest boxes (Ehrlich et al. 1988). All reported Kentucky nests have been in deciduous trees. The bottom of the cavity is lined with bark shreds, twigs, grasses, rootlets, and animal hair (Harrison 1975). Nest trees are typically located within fairly mature to mature forest, but occasionally along forest edge. The average height of four reported nests is 21 feet (range of 9–35 feet) (Alsop 1971; KOS Nest Cards; L. McNeely, pers. comm.).

Alvin E. Staffan

The atlas survey yielded records of White-breasted Nuthatches in nearly 73% of priority blocks statewide, and 10 incidental observations were reported. Occurrence was highest in the Mississippi Alluvial Plain, where the forests are characteristic of those in which Mengel (1965) found the species to be most numerous. Occurrence in the Highland Rim, the Cumberland Plateau, the East Gulf Coastal Plain, and the Shawnee Hills was quite similar (75–81%). Occurrence was lowest in the Cumberland Mountains, where Mengel (1965) found that the species occurred in small numbers, and in the Blue Grass, probably because of the extent of deforestation within the Inner Blue Grass subsection. Average abundance was relatively consistent statewide. Both occurrence and average abundance were lowest in extensively cleared areas and highest in predominantly forested areas with some openings.

Despite this nuthatch's general abundance, only 8% of priority block records were for confirmed breeding. Although at least four active nests were reported, most of the 42 confirmed records were based on the observation of family groups including recently fledged young. The remainder pertained to adults carrying food.

White-breasted Nuthatches are infrequently reported on Kentucky BBS routes. The average number of individuals per BBS route for the periods 1966–91 and 1982–91 was 1.09 and 2.17, respectively. Trend analysis of these data reveals a significant ($p<.01$) increase of 5.1% per year for the period 1966–91, but a nonsignificant increase of 0.6% per year for the period 1982–91.

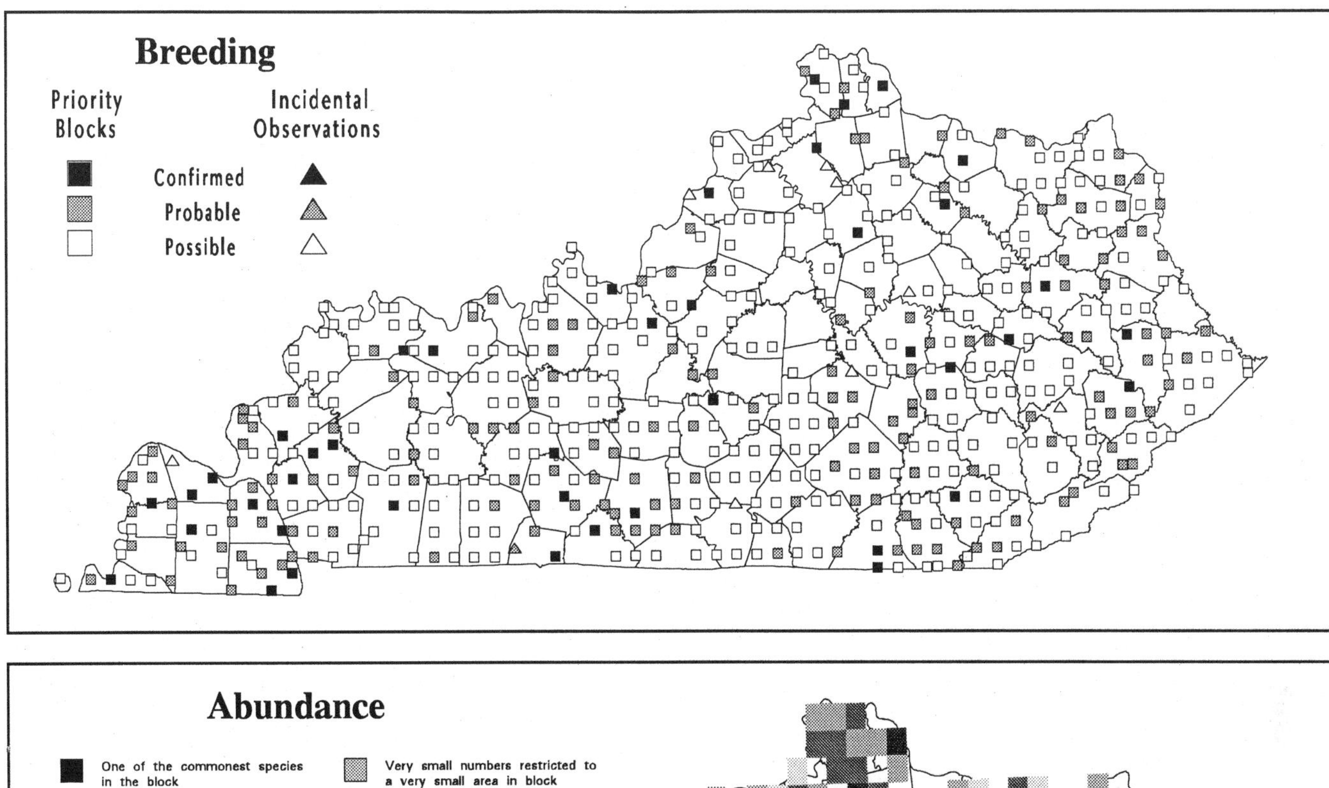

Analysis of Block Data by Physiographic Province Section

Physiographic Province Section	Total Blocks Surveyed	Blocks with Data	Avg Abund	% with Data	Section's % for State
Mississippi Alluvial Plain	14	13	2.1	92.9	2.5
East Gulf Coastal Plain	36	28	2.5	77.8	5.3
Highland Rim	139	113	2.0	81.3	21.4
Shawnee Hills	142	106	2.0	74.6	20.1
Blue Grass	204	116	2.0	56.9	22.0
Cumberland Plateau	173	139	2.0	80.3	26.4
Cumberland Mountains	19	12	2.1	63.2	2.3

Summary of Breeding Status

Number of Blocks in Which Species Was Recorded		
Total	**527**	**72.5%**
Confirmed	42	8.0%
Probable	161	30.5%
Possible	324	61.5%

White-breasted Nuthatch

Brown Creeper

Certhia americana

Brown Creepers nest across much of northeastern North America and southward through the Appalachian Mountains to western North Carolina (AOU 1983). Within this range the species typically occupies coniferous and mixed forest types. In addition, several disjunct populations occur in the Mississippi River Valley, including parts of central and southern Illinois (Bohlen 1989), southeastern Missouri (Robbins and Easterla 1992), and western Tennessee (Robinson 1990). Here the species usually occupies permanently inundated swamp forest dominated by bald cypress and water tupelo. It is within such habitat that the species was first confirmed nesting in Kentucky at Axe Lake Swamp, in Ballard County, in 1988 (Palmer-Ball and Haag 1989).

Gary Meszaros

Before the late 1980s the Brown Creeper was known only as a transient and winter resident, regularly occurring from late September to early May. Outside this period, creepers had been observed on only a few occasions: June 4, 1977, at Murphy's Pond, in Hickman County (B. Monroe Jr., pers. comm.); July 5–6, 1944, at Black Mountain, in Harlan County (Breiding 1947); August 3, 1981, at Audubon State Park, in Henderson County (Stamm 1982a); August 20, 1972, at Cumberland Falls State Park, in Whitley County (Kleen and Bush 1973); and August 23, 1980, at Louisville, in Jefferson County (Stamm 1981a). Mengel (1965) viewed the record from Black Mountain with skepticism, largely because of the inability of others to find the species during fieldwork conducted there both before and after 1944. It is likely that most of the August records pertain to early fall migrants, but the records from Murphy's Pond and Audubon State Park, as well as some May records, may represent locally nesting birds.

Specific Kentucky nesting records were all obtained during the period of the atlas project, and most were the direct result of atlas fieldwork. Following the initial discovery of nesting Brown Creepers in Ballard County, an active nest was found at Cypress Creek Swamp, in Marshall County, on May 5, 1991 (Hendricks et al. 1991). Soon thereafter, recently fledged young were observed on the Jenny Hole–Highland Creek Unit of Sloughs WMA, in Henderson County, and probable nesting birds were found at two other locations in the Union County portion of the WMA, both on May 22, 1991. In the latter county, a male was seen gathering food at one spot, and at least one bird was heard singing at another.

Although nesting Brown Creepers were originally thought to be restricted to permanently inundated swamp forests, the birds in Henderson and Union counties were found in seasonally inundated bottomland forest and the margins of open-water sloughs. These observations indicate that breeding creepers may be more widespread than formerly believed.

From the limited amount of information available, it appears that the breeding season may begin relatively early. Nesting activity has been documented as early as May 5 in Marshall County, with the observation of adults apparently carrying food to a nest site, indicating clutch completion sometime in mid-April. Likewise, the observation of young off the nest in Henderson County on May 22 indicates clutch completion before the first of May. Observations of family groups at Axe Lake Swamp in early June may indicate a slightly later date for clutch completion, as would the observation of a male carrying food to an undetermined location in Union County on May 22. The observation of a pair apparently involved in the final stages of nest building at Axe Lake Swamp on June 2 indicates that later clutches may be completed at least occasionally into June. Both Kentucky nests were typical of those the species builds in other parts of its range, being constructed behind slabs of outer bark separated from the core of large, dead trees. Both nest sites were in the midst of extensive, permanent water swamps dominated by water tupelo and bald cypress, and they were both located approximately six feet above water level.

The atlas results probably approximate the present nesting range of the Brown Creeper in Kentucky. Other scattered locations along the large river floodplains hold potential for nesting, and it would not be surprising to find the species elsewhere in western Kentucky. Based on early accounts of breeding in southeastern Missouri and southern Illinois, it seems likely that at least some nesting creepers went undetected in western Kentucky before the late 1980s. It is also possible that the species's appearance in at least a portion of western Kentucky coincided with a recent southward expansion in nesting range that has been documented in bordering states (Peterjohn and Rice 1991; Bohlen 1989; Ford 1987; Keller et al. 1986).

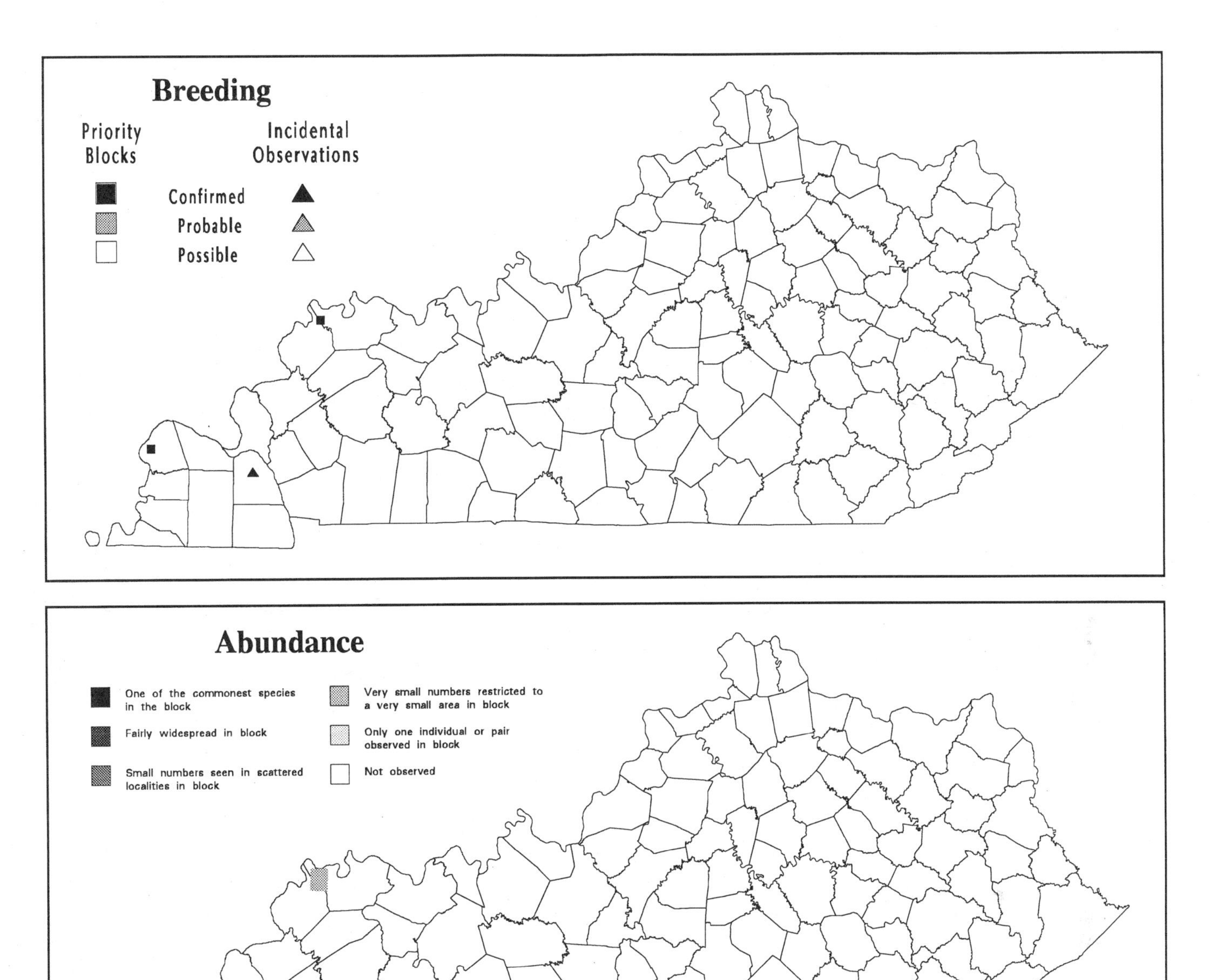

Analysis of Block Data by Physiographic Province Section

Physiographic Province Section	Total Blocks Surveyed	Blocks with Data	% with Data	Section's % for State
Mississippi Alluvial Plain	14	1	7.1	50.0
East Gulf Coastal Plain	36	-	-	-
Highland Rim	139	-	-	-
Shawnee Hills	142	1	0.7	50.0
Blue Grass	204	-	-	-
Cumberland Plateau	173	-	-	-
Cumberland Mountains	19	-	-	-

Summary of Breeding Status

Number of Blocks in Which Species Was Recorded		
Total	**2**	**0.3%**
Confirmed	2	100.0%
Probable	-	-
Possible	-	-

Carolina Wren

Thryothorus ludovicianus

The Carolina Wren's musical song is a conspicuous sound throughout most of Kentucky. Mengel (1965) regarded the species as a common resident statewide, except in extensively cleared areas such as the Inner Blue Grass, where he considered it rare, and at the summit of Black Mountain, where he found it to be virtually absent. This wren is very susceptible to severe winter weather, and its numbers often decline substantially following abnormally harsh winters (e.g., Wilson and Stamm 1960; Whitt 1977). In subsequent years the species builds back in numbers, and after recovery from a dramatic crash in the late 1970s, it appears to be as common and widespread as ever.

Carolina Wrens are found in a great variety of wooded and semi-open habitats, from extensive tracts of mature forest to fencerows and woodlots in fairly open areas. Although the species is abundant in natural situations, it has adapted readily to the presence of humans and commonly inhabits both suburban and rural settlement. As noted by Mengel (1965), high levels of deforestation may limit numbers, and some extensively cleared regions do not have enough cover to support a substantial population.

The status of the Carolina Wren in Kentucky may not have changed significantly since the time of settlement. Audubon (1861) considered the species to be common in the state as of the early 1800s, and its adaptation to settlement has likely helped to make up for decreases in many areas because of clearing and fragmentation of forests.

Carolina Wrens are permanent residents, and pairs typically remain together throughout the year. It is not rare to hear an occasional song during the winter months. In March, though, territorial behavior begins, and singing increases. Egg laying has been reported as early as late March, and an initial peak in clutch completion occurs during early April (Mengel 1965). Many pairs raise a second brood, and although a later peak in clutch completion is not evident, active nests are regularly observed in July. The raising of a third brood has not been documented in Kentucky, but active nests are not uncommon in August and have been reported occasionally into mid-September (KOS Nest Cards). Mengel (1965) reported the average size of 16 clutches as 4.7 (range of 3–6).

Forest Cover

Value	% of Blocks	Avg Abund
All	97.9	3.1
1	93.1	2.6
2	98.9	3.0
3	98.7	3.0
4	98.7	3.5
5	98.1	3.6

Carolina Wrens place their nests in a variety of natural and artificial situations. They often fill a cavity with nest material, but if a chosen site is too large to fill, then a bulky, domed nest is constructed. Natural nest sites include small recesses in rock outcrops and clifflines, protected sites at the base of trees or in fallen brush, and hollows in logs. Carolina Wrens also frequently nest in and about human structures, and they often use such situations as hanging baskets and flowerpots under protected eaves of houses, unused machinery, nest boxes, and various cubbyholes in garages, outbuildings, and abandoned homes. The nest is constructed primarily of dead leaves, other dead plant debris, and moss, and it is lined with finer, soft plant material. Most nests are placed within six to eight feet of the ground, and the species at least sometimes nests on the ground (Ehrlich et al. 1988).

Alvin E. Staffan

The atlas survey yielded reports of Carolina Wrens in nearly 98% of priority blocks statewide, and one incidental observation was reported. The species ranked 4th according to the number of priority block records, 7th by total abundance, and 13th by average abundance. Occurrence and average abundance were relatively uniform across the state. Both were lower only in extensively cleared areas.

Despite this wren's general abundance, less than 24% of priority block records were for confirmed breeding. A number of active nests were located, but most confirmed records were based on the observation of family groups including recently fledged young. Foraging groups of bobtailed young and their attentive parents were most conspicuous in May and June, but they were observed throughout the summer. Other confirmed records were based on the observation of nest building, used nests, and adults carrying food for young.

Carolina Wrens are reported in moderate numbers on most Kentucky BBS routes. The average number of individuals recorded per BBS route for the periods 1966–91 and 1982–91 was 8.54 and 8.96, respectively. According to these data, the species ranked 27th in abundance on BBS routes during the period 1982–91. Trend analysis of these data shows a nonsignificant increase of 2.0% per year for the period 1966–91, but a significant ($p<.01$) increase of 11.6% per year for the period 1982–91. The latter is a result of the Carolina Wren's gradual recovery following its virtual disappearance after the two severe winters of 1976–77 and 1977–78 (Monroe 1978; Robbins et al. 1986).

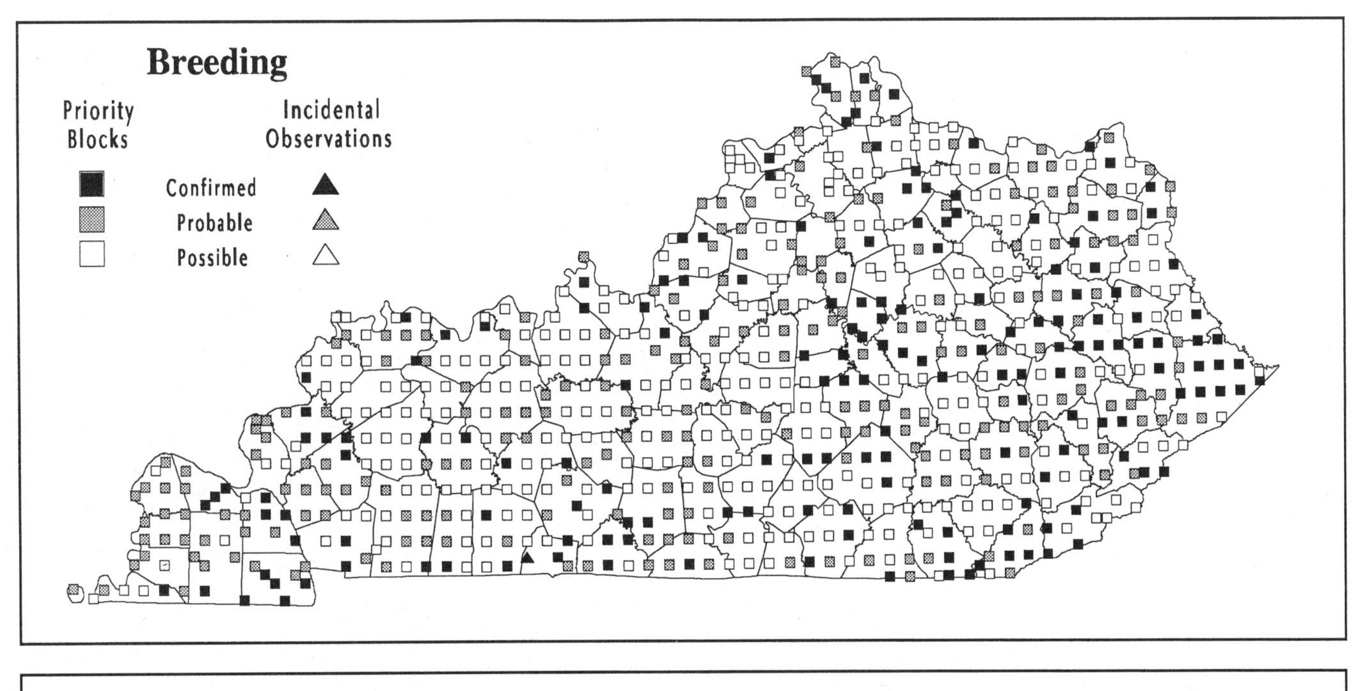

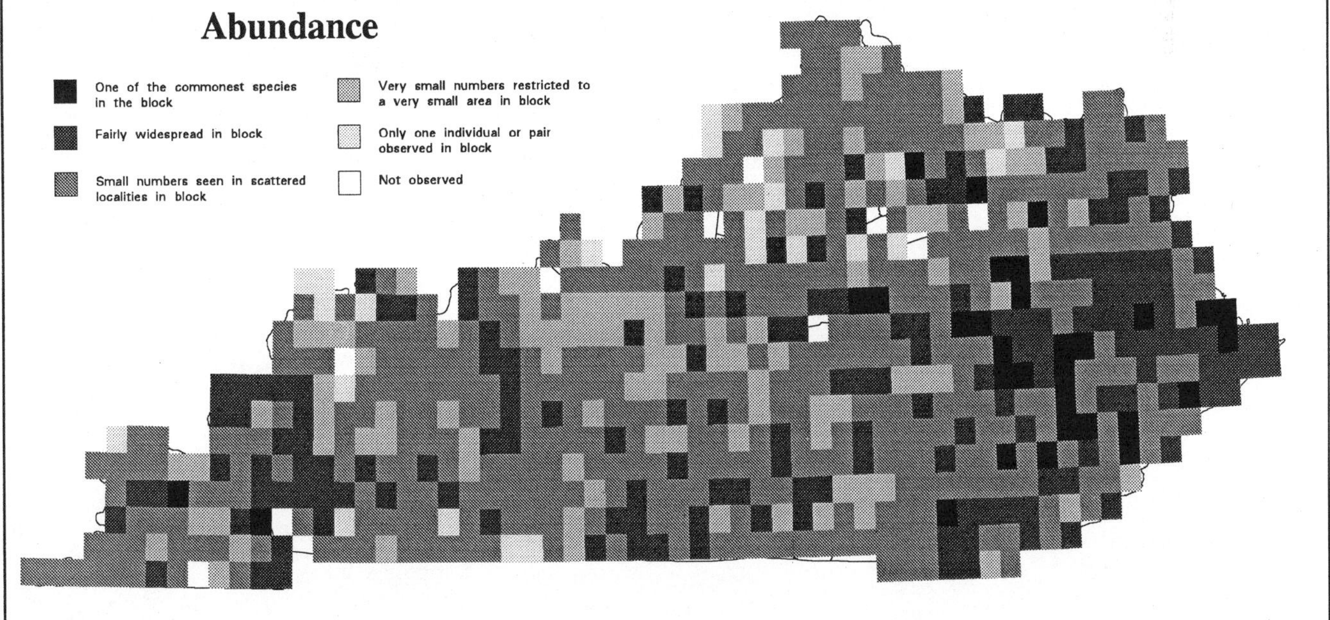

Analysis of Block Data by Physiographic Province Section

Physiographic Province Section	Total Blocks Surveyed	Blocks with Data	Avg Abund	% with Data	Section's % for State
Mississippi Alluvial Plain	14	14	2.9	100.0	2.0
East Gulf Coastal Plain	36	35	3.2	97.2	4.9
Highland Rim	139	136	3.0	97.8	19.1
Shawnee Hills	142	139	2.9	97.9	19.5
Blue Grass	204	197	2.9	96.6	27.7
Cumberland Plateau	173	173	3.6	100.0	24.3
Cumberland Mountains	19	18	3.2	94.7	2.5

Summary of Breeding Status

Number of Blocks in Which Species Was Recorded		
Total	**712**	**97.9%**
Confirmed	169	23.7%
Probable	237	33.3%
Possible	306	43.0%

Carolina Wren

Bewick's Wren

Thryomanes bewickii

The Bewick's Wren has exhibited a dramatic decline in Kentucky in the past several decades. As recently as the late 1950s the species was considered a common summer resident throughout the state (Mengel 1965), but today it must be regarded as locally distributed and rare to absent in most areas. Although reasons for this decline are not completely understood, some have suggested the influence of the southward spread of the House Wren and severe winter weather (Mengel 1965; Monroe 1978; Robbins et al. 1986).

Bewick's Wrens are encountered in a variety of semi-open habitats. Although evidence suggests that the species formerly inhabited natural forest openings, it is primarily a bird of altered habitats today. These wrens are most conspicuous in rural farmland and settlement, but small numbers also inhabit suburban yards of towns, brushy forest margins, and forest clear-cuts.

Although the recent decline of the Bewick's Wren has been well documented, the species's historical status is somewhat unclear. Audubon and other early ornithologists apparently did not encounter it in Kentucky (Mengel 1965). Local increases were documented between the late 1800s and early 1900s, and Mengel (1965) notes that it was not recorded at Cincinnati until 1879. The species may have invaded the midwestern United States from the south and west after settlement, responding to the fragmentation of extensive forests and the creation of an abundance of suitable nesting habitat.

Most detailed information on nesting Bewick's Wrens in Kentucky dates from before 1960. Before the species's marked decline, overwintering was reported regularly, but these wrens are rarely reported outside the breeding season today. Although a few birds may be on territory by mid-March, most of the nesting population appears to return from more southerly wintering grounds by mid-April. Mengel (1965) gives a range of dates of clutch completion from late March to early July, with a probable peak during the first ten days of May. Although a later peak is not evident, the raising of two broods has been documented (Mengel 1965). The average size of 10 clutches reported by Mengel was 5.1 (range of 3–8).

Bewick's Wrens are typically cavity nesters. Kentucky records are almost exclusively limited to artificial sites, including nooks and crannies in outbuildings, machinery, and woodpiles, as well as nest boxes. Although natural situations have been largely abandoned, a pair was found nesting in the hollow base of a fallen tree in Lyon County in 1992 (Palmer-Ball 1993). The nest is constructed of dead leaves, sticks, and other plant material and is lined with feathers, hair, moss, and dead leaves (Harrison 1975). Most nests are situated within six to eight feet of the ground.

The atlas survey yielded reports of Bewick's Wrens in slightly more than 4% of priority blocks, and 15 incidental observations were reported. Occurrence was highest in the Blue Grass and the East Gulf Coastal Plain, and the species was not located in the Mississippi Alluvial Plain, the Cumberland Plateau, and the Cumberland Mountains. Although Bewick's Wrens formerly occurred statewide, atlas results suggest that they are now limited in distribution to portions of central and western Kentucky. The species is encountered regularly in the hilly country of the Outer Blue Grass, occasionally across the central Highland Rim and Shawnee Hills, and more regularly again in the western Highland Rim and the East Gulf Coastal Plain. Outside these areas, Bewick's Wrens are quite rare, and there are no recent reports from eastern Kentucky, where Mengel (1965) considered the species to be common.

Ron Austing

The location of a substantial population of Bewick's Wrens in Trigg and Lyon counties during the atlas fieldwork was especially noteworthy (Palmer-Ball 1993). The species was found to be one of the characteristic breeding birds of large tracts of recently logged hillsides otherwise surrounded by extensive forest. Bewick's Wrens were found to be relatively common in similar habitat in western Tennessee in 1987 (Robinson 1989).

There were seven confirmed records for nesting in priority blocks and two confirmed incidental records. Most confirmed records involved the observation of family groups of recently fledged young in the company of adults, and only one active nest was located.

Kentucky BBS data illustrate the decline in Bewick's Wrens during the past twenty-five years. The average number of individuals reported per BBS route for the periods 1966–91 and 1982–91 was 0.53 and 0.17, respectively. From 1966 to 1991 the species declined by 7.0% per year ($p<.01$); however, it appears that Bewick's Wrens may have stabilized in the last decade or so. From 1982 to 1991 the species continued to show a decrease of 1.8% per year, but the trend was not statistically significant. The number of literature records for Bewick's Wrens during the same period has reflected this stabilization, although it is likely that increased summer birdwatching as a result of the atlas project has been at least partially responsible for the apparent increase.

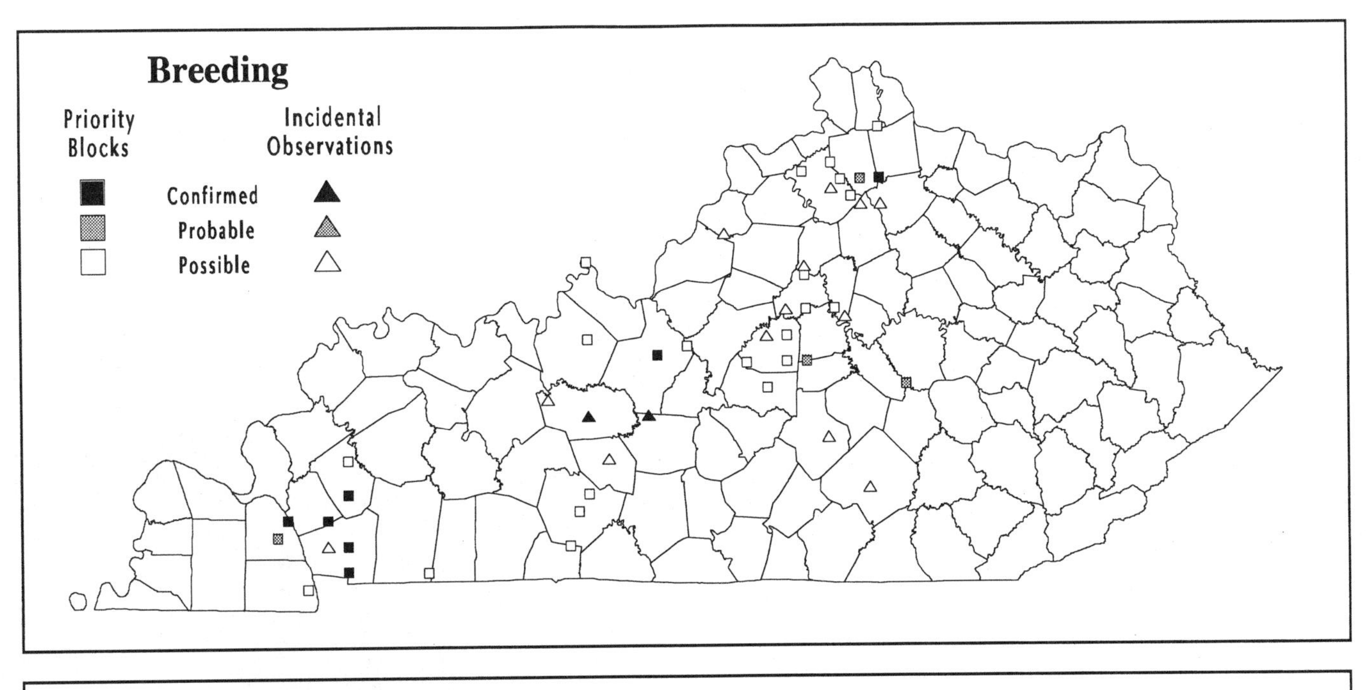

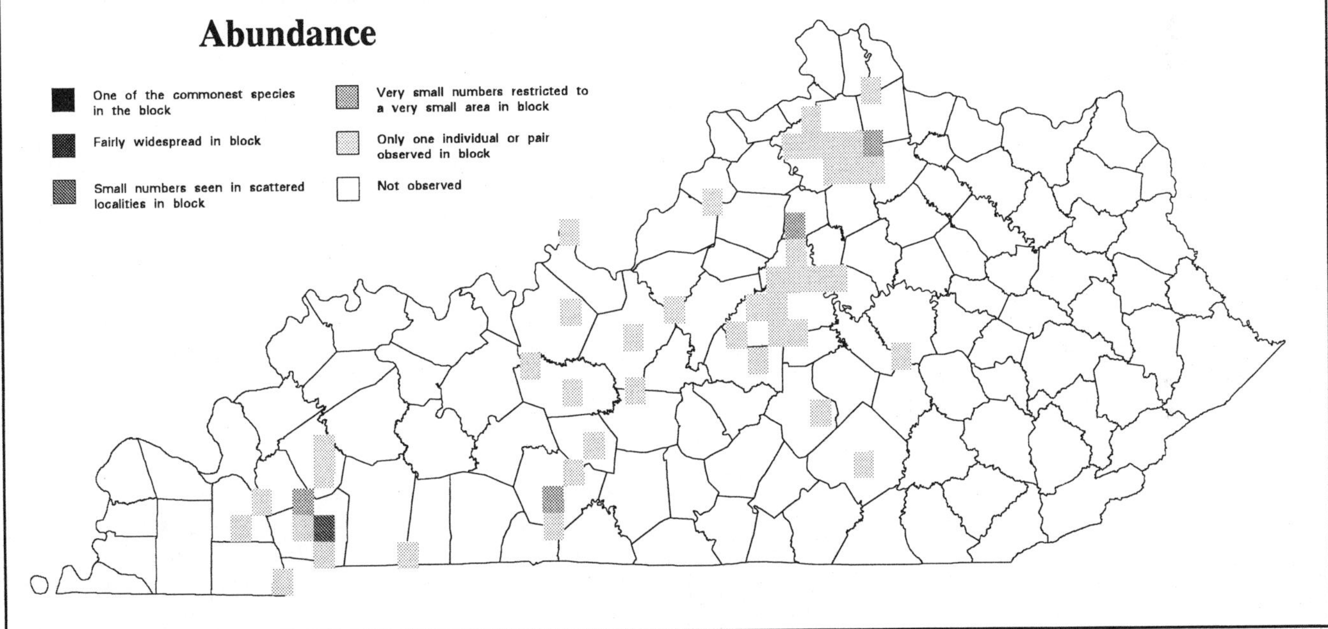

Analysis of Block Data by Physiographic Province Section

Physiographic Province Section	Total Blocks Surveyed	Blocks with Data	% with Data	Section's % for State
Mississippi Alluvial Plain	14	-	-	-
East Gulf Coastal Plain	36	3	8.3	9.4
Highland Rim	139	9	6.5	28.1
Shawnee Hills	142	3	2.1	9.4
Blue Grass	204	17	8.3	53.1
Cumberland Plateau	173	-	-	-
Cumberland Mountains	19	-	-	-

Summary of Breeding Status

Number of Blocks in Which Species Was Recorded		
Total	**32**	**4.4%**
Confirmed	7	21.9%
Probable	4	12.5%
Possible	21	65.6%

House Wren

Troglodytes aedon

The House Wren has invaded Kentucky as a nesting bird in relatively recent times. Audubon did not observe the species during the early 1800s, and even as late as the early 1900s it was not known to have definitely nested in the state (Mengel 1965). During the 1910s and 1920s the House Wren began nesting in extreme northern Kentucky, and by about 1950 it had been reported throughout much of central Kentucky (Stamm 1951a). Mengel (1965) described the species's southward expansion and increase in abundance through the 1950s. As of the early 1980s it was still increasing in western Kentucky (Stamm 1981b).

House Wrens occur in a variety of semi-open and open habitats. They are most often associated with settlement, being locally common in rural farmland and residential yards. These wrens also can be numerous in natural situations far from human habitations, especially in riparian zones along major rivers. On the Ohio River, for example, House Wrens are found as far south and west as Ballard County in summer. The species's preference for open habitats apparently limits its occurrence in forested areas and has likely slowed the progress of its expansion in some portions of the state (Mengel 1965).

House Wrens are occasionally observed in winter, but they normally retreat farther south. Most nesting birds return during the third week of April, although a few have been reported as early as mid-March. Territorial behavior begins immediately upon arrival, and nest building is initiated soon thereafter. Egg laying has been reported as early as April 3, although a peak in the completion of first clutches occurs during the middle of May (Mengel 1965). Following the fledging of young from first nestings, a second clutch is often, if not typically, laid. According to Mengel (1965), completion of second clutches peaks during the latter part of June, although egg laying has been noted as late as mid-July. A record of the raising of a third brood has been reported recently from Franklin County (Palmer-Ball 1992). The average size of 35 clutches or broods reported by Mengel (1965) was 5.8 (range of 4–8), with little variation in the size of early and late clutches.

Forest Cover

Value	% of Blocks	Avg Abund
All	31.2	2.2
1	58.4	2.5
2	42.1	2.3
3	31.7	2.0
4	9.1	1.6
5	1.9	1.0

House Wrens are cavity nesters, using a variety of natural and artificial sites. The species readily accepts nest boxes, although natural tree cavities and old woodpecker holes in trees and fence posts are also commonly used. The species frequently nests in settled situations, but it uses sites in garages and outbuildings less than does the Carolina Wren.

The nest is typically constructed of small sticks and twigs and lined with fine plant material, animal hair, and feathers. Typically, the cavity is filled with sticks, and a few protrude from the entrance. Although nests have been reported as high as about 30 feet, most nest sites are situated 3–10 feet above the ground (Mengel 1965; KOS Nest Cards).

Alvin E. Staffan

The atlas survey yielded records of House Wrens in more than 31% of priority blocks statewide, and 23 incidental observations were reported. Occurrence varied dramatically by physiographic province section, ranging from more than 70% in the Blue Grass to just over 5% in the Cumberland Mountains. Results of the atlas project show the extent to which the House Wren has continued to spread southward during the past 35 years, but the data also show that the species remains well distributed only in the north central part of the state. These wrens are probably most abundant in the Inner Blue Grass, where they are conspicuous in rural yards and along hedgerows and stone fences in the rolling, open farmland. Outside north central Kentucky the species remains very locally distributed, for the most part occurring regularly only in residential areas of cities and towns. Average abundance was relatively low across the state, but varied similarly. Although the House Wren's northern affinity probably influences its statewide range more than any other factor, the species's tendency to occur in open areas rather than in predominantly forested regions also influences its occurrence.

Nearly 19% of priority block records were for confirmed breeding. Of the 42 confirmed records, most were based on the observation of active nests, but adults carrying food for young, nest building, and recently fledged young were also observed.

House Wrens are reported in small numbers on most Kentucky BBS routes. The average number of individuals reported per BBS route for the periods 1966–91 and 1982–91 was 1.27 and 2.76, respectively. Interestingly, despite the low sample size, trend analysis of these data shows significant ($p<.01$) increases of 12.4% per year for the period 1966–91 and 5.7% per year for the period 1982–91.

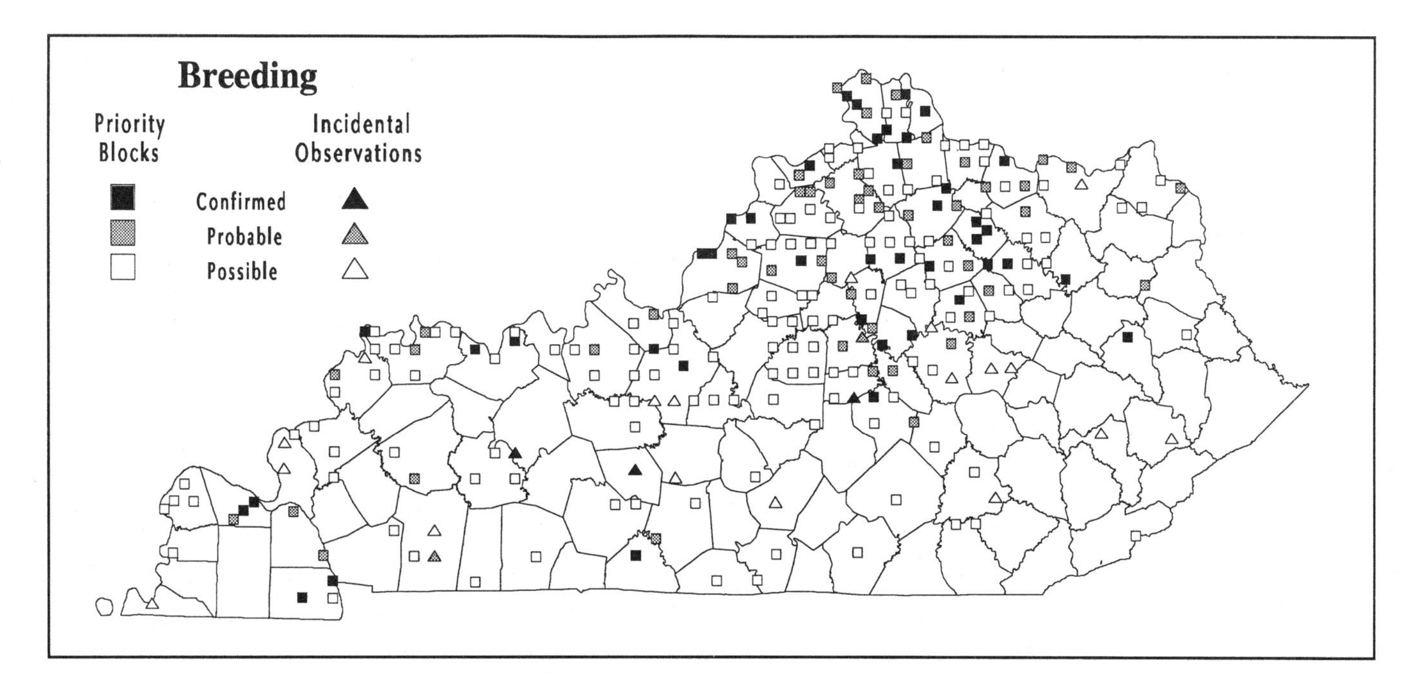

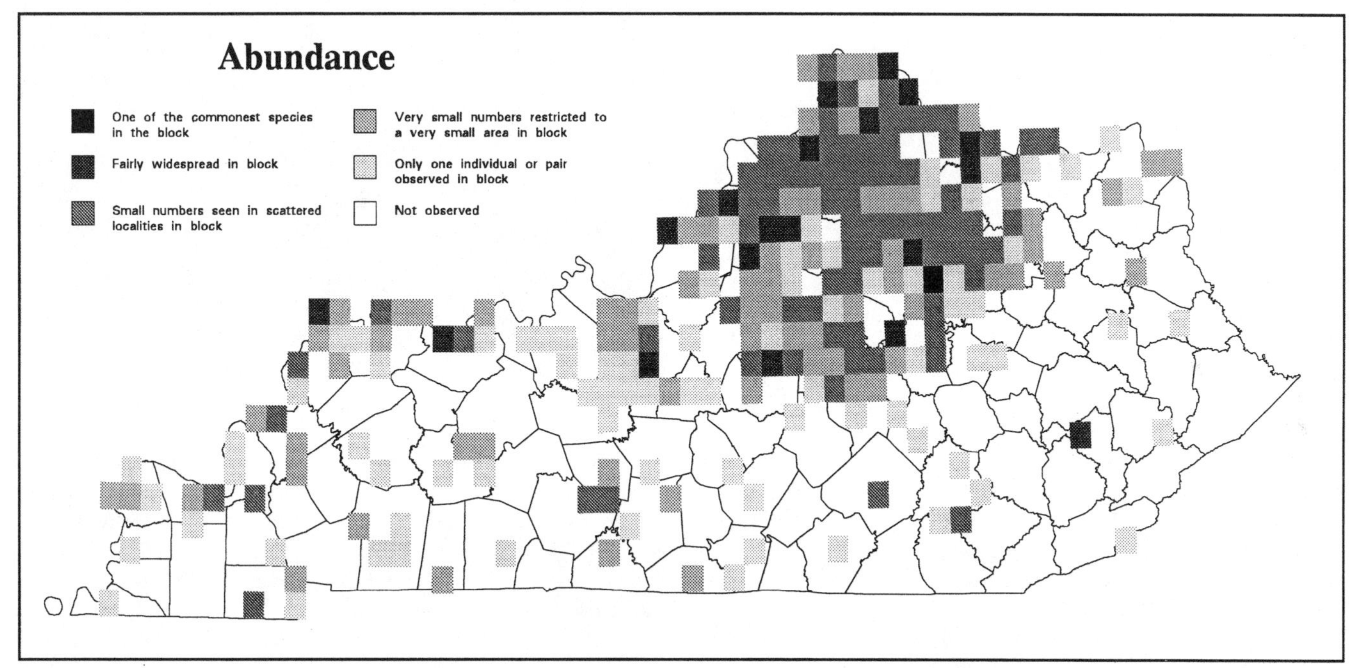

Analysis of Block Data by Physiographic Province Section

Physiographic Province Section	Total Blocks Surveyed	Blocks with Data	Avg Abund	% with Data	Section's % for State
Mississippi Alluvial Plain	14	5	1.8	35.7	2.2
East Gulf Coastal Plain	36	8	1.8	22.2	3.5
Highland Rim	139	27	1.8	19.4	11.9
Shawnee Hills	142	32	1.7	22.5	14.1
Blue Grass	204	143	2.5	70.1	63.0
Cumberland Plateau	173	11	1.5	6.4	4.8
Cumberland Mountains	19	1	1.0	5.3	0.4

Summary of Breeding Status

Number of Blocks in Which Species Was Recorded		
Total	**227**	**31.2%**
Confirmed	42	18.5%
Probable	51	22.5%
Possible	134	59.0%

House Wren

Sedge Wren

Cistothorus platensis

Kentucky lies at the southern limit of the breeding range of the Sedge Wren, and the species has been reported nesting only irregularly. Unlike most songbirds, which exhibit relatively consistent patterns of occurrence, these wrens are well known for their erratic seasonal movements and irregular occurrence from year to year (Ehrlich et al. 1988; Peterjohn and Rice 1991). Most years the species is reported at a few scattered localities, but birds may appear at any time during summer, and they seldom nest in any one area for more than a year or two.

Throughout parts of their nesting range, Sedge Wrens are encountered in naturally occurring habitats, including moist meadows of grasses and sedges and along the grassy margins of marshes and bogs (AOU 1983). In Kentucky, where such habitats are virtually absent, the species inhabits hayfields, overgrown pastures, and fallow fields. Sedge Wrens seem to prefer moist situations, but they typically avoid marshes. Wherever these wrens occur, they choose only areas with an abundance of thick, herbaceous cover.

It is unclear whether the Sedge Wren nested in Kentucky before recent times. The species has rarely been reported away from altered situations, and before 1900 it had been reported only as a transient. The Sedge Wren was not confirmed breeding in the state until 1934 (Hibbard 1935), but summer records from as early as 1908 probably involved nesting birds. During the past 30 years, confirmed and probable nesting reports have accumulated gradually, although no dramatic increase has been noted at any time. Mengel (1965) suggested that the species may have expanded southward in nesting range during more recent times, but numbers of breeding birds have apparently declined in Ohio since the early 1900s (Peterjohn and Rice 1991).

Little detailed information on the nesting of Sedge Wrens in Kentucky is available. Mengel (1965) included records of confirmed breeding from only four counties (Edmonson, Jefferson, Oldham, and Warren) and breeding season records from four others (Bath, Knox, Laurel, and Rowan). Since that time but before the atlas fieldwork, confirmed records were added for the counties of Meade (Rowe 1964), Nelson (Croft 1972), and Larue (Stamm 1983, 1984a). Breeding season records were published for the counties of Calloway, Green, Powell (Stamm 1984a), and Wayne (Stamm 1980).

Spring migrants pass through the state during late April and early May, and there are sporadic reports of birds lingering through the latter half of May and into early June. It is unclear how late in the season spring migrants normally occur in Kentucky, but birds observed after the beginning of June are likely summer residents. Although there are a few June records, the species is more often detected in July and August, when all records of confirmed nesting have been made. Because the movements of this species are not clearly understood, it is not known whether these late summer birds arrive from more southerly points or have returned south following earlier nestings to the north. Whichever the case, it is apparent that most or all nesting activity occurs during this period. The one active nest that has been observed in Kentucky contained a clutch of five eggs on August 18, 1950 (Mengel 1965). Recently fledged young have been observed most frequently during August, but as late as early October (Mengel 1965; author's notes).

Sedge Wrens typically nest in dense, grassy vegetation. In such habitat the birds are able to conceal their nest among the thick growth, usually a foot or so above the ground. The nest is a bulky, spherical structure with an entrance facing out to the side. It is constructed primarily of dead grasses and interwoven into the surrounding vegetation, and it is lined with feathers, fur, and soft plant down (Harrison 1975). Males often construct one or more unlined dummy nest within the territory in which eggs are not laid (Mengel 1965).

Ron Austing

The atlas survey yielded only four reports of Sedge Wrens in priority blocks, although six incidental observations were reported. Most occurrences were in the Shawnee Hills (two priority and two incidental), but two reports came from the Highland Rim (one priority and one incidental), two from the Mississippi Alluvial Plain (one priority and one incidental), and one incidental observation each from the Blue Grass and the Cumberland Plateau.

Only one incidental record was for confirmed breeding. On July 31, 1988, at least seven males were heard singing in appropriate nesting habitat on the Sauerheber Unit of Sloughs WMA, in Henderson County. Visits in August and September yielded continued observations of territorial birds, and a recently fledged bird was observed on October 9, 1988, confirming that nesting had occurred. Other reports were generated during atlas work on July 13, 1985 (Livingston County), August 11–September 1, 1985, and September 1, 1986 (Fulton County), June 6, 1988 (Breckinridge County), August 11, 1988 (Morgan County), July 22–August 22, 1989 (Warren County), June 25, 1991 (Pulaski County), June 29 and July 21, 1991 (Woodford County), and August 17, 1991 (Ballard County). All involved the observation of one or more singing males. Although no further evidence of nesting was noted, it is likely that all represented potential breeding individuals. The incidental record from Hopkins County pertained to a published report of a singing bird on August 25, 1987 (Stamm 1988a).

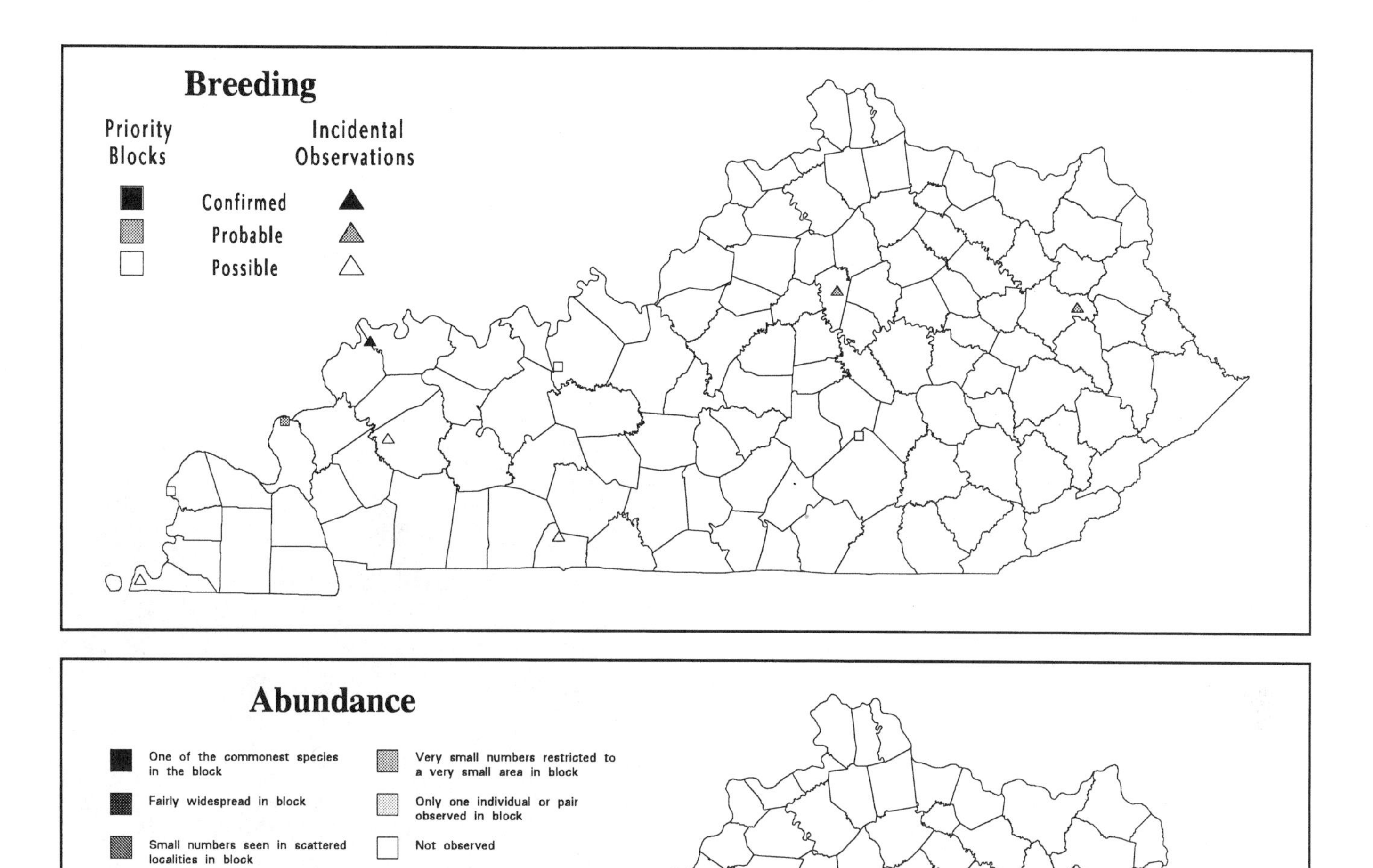

Analysis of Block Data by Physiographic Province Section

Physiographic Province Section	Total Blocks Surveyed	Blocks with Data	% with Data	Section's % for State
Mississippi Alluvial Plain	14	1	7.1	25.0
East Gulf Coastal Plain	36	-	-	-
Highland Rim	139	1	0.7	25.0
Shawnee Hills	142	2	1.4	50.0
Blue Grass	204	-	-	-
Cumberland Plateau	173	-	-	-
Cumberland Mountains	19	-	-	-

Summary of Breeding Status

Number of Blocks in Which Species Was Recorded		
Total	**4**	**0.6%**
Confirmed	-	-
Probable	1	25.0%
Possible	3	75.0%

Blue-gray Gnatcatcher

Polioptila caerulea

The Blue-gray Gnatcatcher's wheezy notes are a fairly conspicuous summer sound across most of Kentucky. Mengel (1965) regarded the species as a fairly common to common summer resident statewide. Gnatcatchers are found in a variety of forested and semi-open habitats. Although they occur in closed-canopy forest, these active mites are most frequently encountered in partially open situations, including woodland borders, selectively logged or naturally open forest, wooded parks, and riparian corridors. The species inhabits a variety of forest types, including upland deciduous and mixed associations, but it is most numerous in bottomland and riparian situations.

Human alteration of the landscape likely has resulted in a decrease in the numbers of Blue-gray Gnatcatchers in Kentucky. While the species appears to have adapted fairly well to the fragmentation of forested areas, extensive clearing has eliminated an abundance of suitable breeding habitat. Likewise, the replacement of old-growth forests by younger, secondary forests (with more closed-in midstory and fewer gaps in the canopy) may have caused a decrease in its abundance in present-day forests.

Gnatcatchers return from their wintering grounds in the southern United States and Middle America very early. A few birds often appear before the end of March, and by mid-April most of the nesting population has likely returned. Nest building commences soon thereafter, and according to Mengel (1965), early clutches are completed by mid-April. A peak in clutch completion occurs about May 1, although later nestings have been reported into mid-July (KOS Nest Cards). Although specific documentation is lacking, the range of dates suggests that at least some pairs raise a second brood (Mengel 1965). Gnatcatchers begin to move about soon after the young fledge, and most birds seem to have departed by early September. Kentucky data on clutch size are very limited; Mengel (1965) listed only a clutch of four eggs and a brood of five young, but a clutch of three eggs was observed in Oldham County in 1972 (KOS Nest Cards). Harrison (1975) gives rangewide clutch size as 4–5.

Forest Cover

Value	% of Blocks	Avg Abund
All	88.2	2.5
1	58.4	1.8
2	86.8	2.3
3	94.8	2.6
4	97.4	2.9
5	94.2	2.8

The nest is placed at the base of an upright or horizontal fork and attached firmly to supporting branches. It may be situated in a variety of aspects, from the inner branches of small trees (sometimes in crotches of the main stem) to well out near the outer canopy on more substantial branches of larger trees. In the latter circumstance, the nest is often situated over an opening such as a road, path, stream, or lakeshore. The average height of 25 nests reported by Mengel (1965) was 22.2 feet (range of 9–40 feet), although more recent nests have been observed as low as 2.5 feet (KOS Nest Cards) and as high as 50 feet (Stamm and Croft 1968) above the ground.

Blue-gray Gnatcatchers were recorded on more than 88% of priority blocks statewide, and ten incidental observations were reported. The species ranked 20th according to the number of priority block records, 28th by total abundance, and 29th by average abundance. Occurrence was highest in the Mississippi Alluvial Plain, where gnatcatchers are especially conspicuous in cypress swamps and bottomland forest surrounding floodplain sloughs, and lowest in the Blue Grass, where forest clearing has been extensive. Average abundance was relatively uniform across the state. Occurrence and average abundance were lowest in extensively cleared areas and highest where forest was predominant.

Ron Austing

Gnatcatchers nest quite early, and some birds may begin to wander by early June. Thus some atlas records may have pertained to birds that nested away from the point of observation. Nonetheless, all reports of gnatcatchers in suitable breeding habitat were included in the final data set.

Slightly more than 17% of priority block records were for confirmed breeding. A number of confirmed records were based on the observation of active nests, but most pertained to observation of recently fledged young. Other confirmed records were based on nest building, adults carrying food, and used nests.

Gnatcatchers are typically reported on Kentucky BBS routes in small numbers. The average number of individuals per BBS route for the periods 1966–91 and 1982–91 was 3.34 and 4.92, respectively. Trend analysis of these data indicates a nonsignificant decrease of 2.7% per year for the period 1966–91 and a nonsignificant increase of 2.4% per year for the period 1982–91.

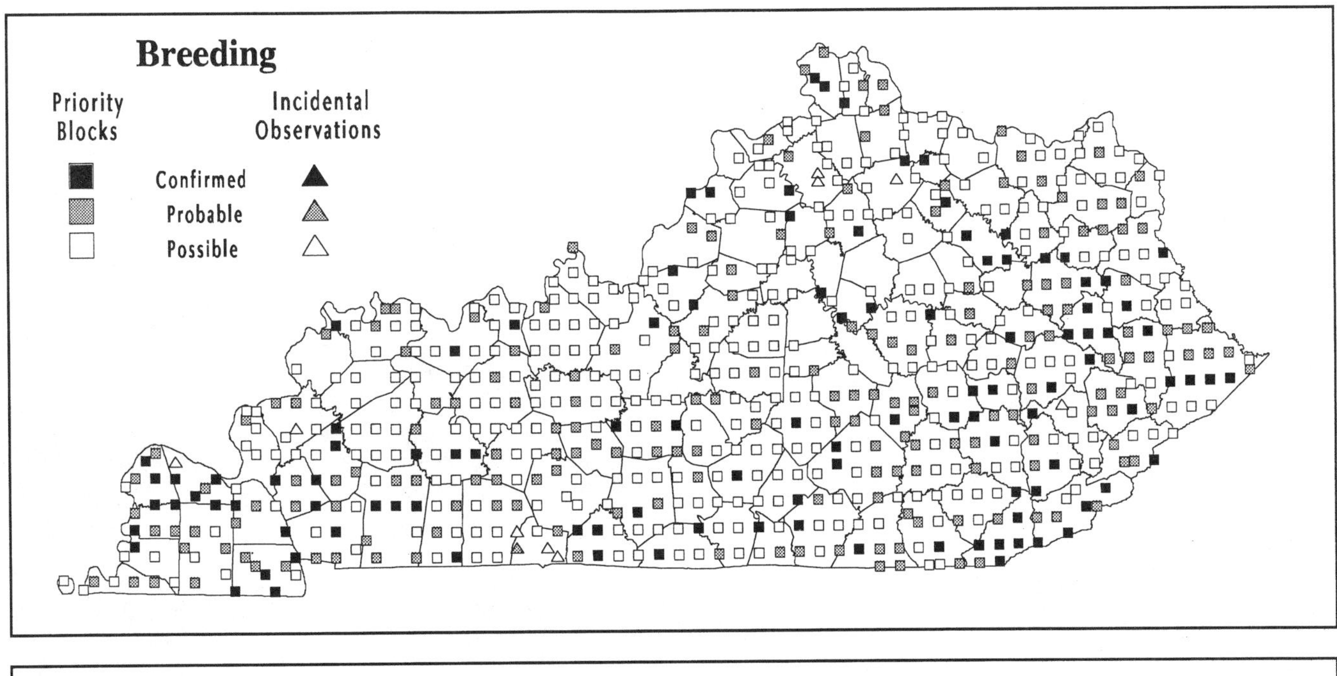

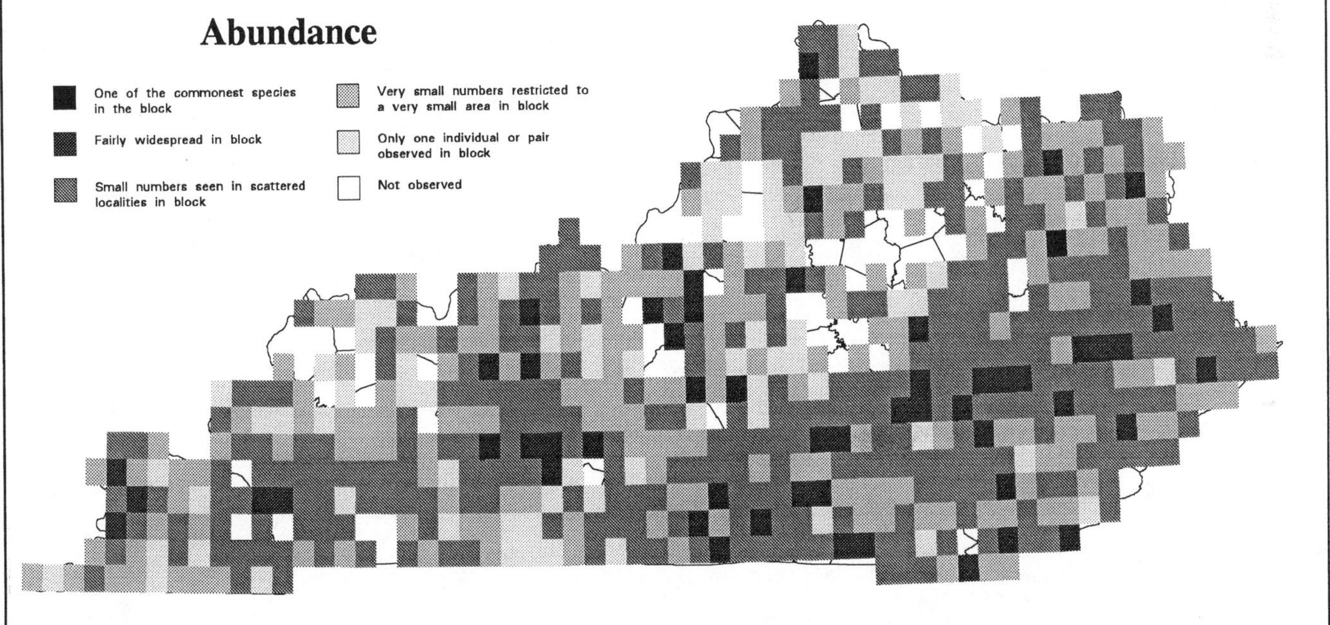

Analysis of Block Data by Physiographic Province Section

Physiographic Province Section	Total Blocks Surveyed	Blocks with Data	Avg Abund	% with Data	Section's % for State
Mississippi Alluvial Plain	14	14	2.6	100.0	2.2
East Gulf Coastal Plain	36	33	2.4	91.7	5.1
Highland Rim	139	125	2.6	89.9	19.5
Shawnee Hills	142	127	2.5	89.4	19.8
Blue Grass	204	157	2.3	77.0	24.5
Cumberland Plateau	173	169	2.8	97.7	26.4
Cumberland Mountains	19	16	2.8	84.2	2.5

Summary of Breeding Status

Number of Blocks in Which Species Was Recorded		
Total	**641**	**88.2%**
Confirmed	110	17.2%
Probable	190	29.6%
Possible	341	53.2%

Blue-gray Gnatcatcher

Eastern Bluebird

Sialia sialis

Family groups of Eastern Bluebirds on rural fences and utility wires are a characteristic sight throughout much of Kentucky. In fact, continental BBS data indicate that during the period 1965–79, bluebirds were as common in parts of central and western Kentucky as anywhere within their range (Robbins et al. 1986). Mengel (1965) regarded the species as a common resident, occurring in numbers throughout the state wherever openings were present. Unfortunately, bluebirds are susceptible to severe winter weather, and after unusually harsh winters their numbers typically crash. Such events were noted during the late 1950s (Wilson and Stamm 1960; Wilson 1962b) and the late 1970s (Monroe 1978; Stamm 1979a, 1979b). Numbers have always become reestablished within several years, however.

Eastern Bluebirds are typically encountered in semi-open and open habitats, but they are sometimes found in relatively small woodland openings and artificial clearings. Although the species is most frequent in rural farmland, many are observed in other settings, including parks, rural yards, forest clear-cuts, and reclaimed surface mines.

It is unclear to what extent Eastern Bluebirds occurred in Kentucky before settlement. Substantial numbers likely used the prairies of western Kentucky as well as open woodlands throughout the state. By the early 1800s people were already erecting nest boxes for them in settled areas (Audubon 1861). Furthermore, the fragmentation of forests that has occurred in the last two centuries has likely resulted in an expansion of bluebirds into many areas. More recently the species has suffered from competition with introduced cavity nesters such as the European Starling and the House Sparrow (Robbins et al. 1986).

While some of Kentucky's bluebirds probably retreat southward in winter, birds from farther north join with those that remain to form flocks that wander about the countryside. At the onset of warmer days in late winter, territorial behavior is initiated, and nest building may be under way by mid-March. Egg laying has been noted as early as March 26, with an early peak in clutch completion during early April (Mengel 1965). The species is typically double-brooded, and there are many reports of nesting into July (Stamm and Jones 1966; Croft and Stamm 1967; Stamm and Croft 1968). Even later nests have been noted into the fall, perhaps representing the raising of a third brood (Rippy 1974; Stamm 1977). The average size of 30 clutches given by Mengel (1965) was 4.4 (range of 3–6).

Forest Cover

Value	% of Blocks	Avg Abund
All	91.6	3.0
1	94.1	3.0
2	97.9	3.4
3	97.8	3.1
4	86.4	2.5
5	51.9	1.7

Bluebirds are typically cavity nesters. Natural tree cavities and old woodpecker holes are frequently used, but artificial sites, including nest boxes, holes in fence posts, and abandoned mailboxes, are also used regularly. Allaire (1976) documented nesting on open rock ledges on strip mine highwalls in eastern Kentucky, where suitable nest sites are limited.

The nest typically is constructed of dead grass, and the bottom of the cavity is filled with material to form a shallow cup. Nests in more open situations are formed in a corner with enough material mounded up to support the eggs. The average height of 36 nests summarized by Croft and Stamm (1967) and Stamm and Croft (1968) was 5.6 feet (range of 3–20 feet).

Ron Austing

Eastern Bluebirds were recorded on nearly 92% of atlas priority blocks statewide, and nine incidental observations were reported. The species ranked 14th according to the number of priority block records, 12th by total abundance, and 16th by average abundance. Occurrence was lower than 90% only in the Cumberland Plateau and the Cumberland Mountains. Generally, average abundance varied similarly. Occurrence and average abundance were lowest in forested areas and highest in areas with a mixture of openings and forest.

Fifty-five percent of priority block records were for confirmed breeding. Bluebirds are quite conspicuous during the breeding season, and both active nests and family groups of recently fledged young were observed often. Most other confirmed records were based on nest building and adults carrying food for young.

Eastern Bluebirds are reported on most Kentucky BBS routes in moderate numbers. The average number of individuals per BBS route for the periods 1966–91 and 1982–91 was 10.44 and 13.00, respectively. According to these data, the bluebird ranked 20th in abundance on BBS routes during the period 1982–91. Trend analysis of these data shows a nonsignificant increase of 1.7% per year for the period 1966–91, but a significant increase ($p<.01$) of 5.6% per year for the period 1982–91. The increase during the last decade can be attributed to the bluebird's recovery from a population crash that resulted from the harsh winters of the mid- to late 1970s (Monroe 1978).

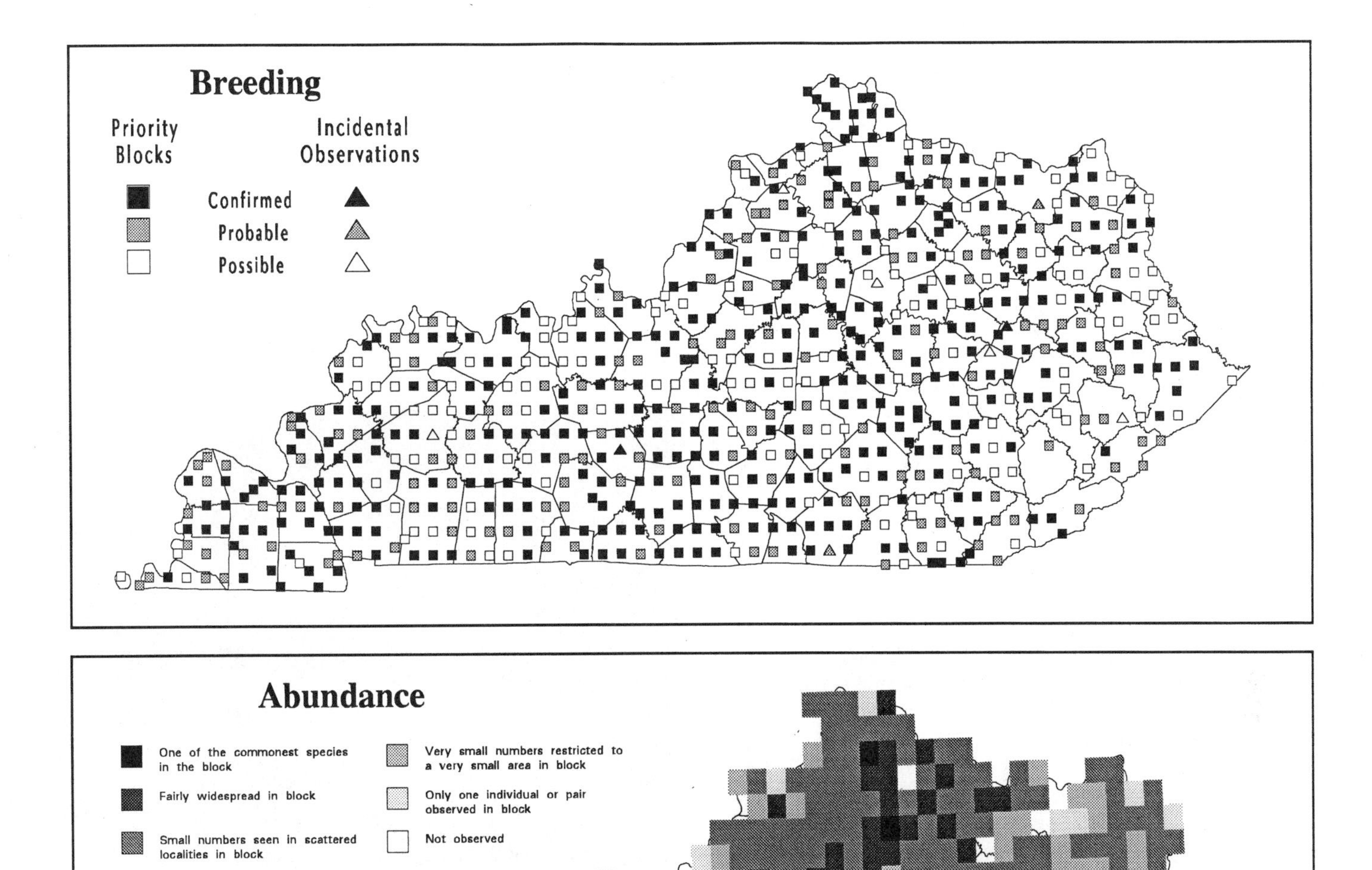

Analysis of Block Data by Physiographic Province Section

Physiographic Province Section	Total Blocks Surveyed	Blocks with Data	Avg Abund	% with Data	Section's % for State
Mississippi Alluvial Plain	14	13	2.2	92.9	2.0
East Gulf Coastal Plain	36	36	3.3	100.0	5.4
Highland Rim	139	139	3.4	100.0	20.9
Shawnee Hills	142	136	3.2	95.8	20.4
Blue Grass	204	193	3.1	94.6	29.0
Cumberland Plateau	173	139	2.4	80.3	20.9
Cumberland Mountains	19	10	1.7	52.6	1.5

Summary of Breeding Status

Number of Blocks in Which Species Was Recorded		
Total	**666**	**91.6%**
Confirmed	366	55.0%
Probable	165	24.8%
Possible	135	20.3%

Eastern Bluebird

Veery

Catharus fuscescens

Most of the Veery's nesting range lies well to the north of Kentucky, across southeastern Canada and the northeastern United States (AOU 1983). A small portion extends south through the Appalachian Mountains, barely reaching into the Cumberland Mountains in the southeastern corner of the state. The Veery was unknown in summer until the early 1900s, when Howell (1910) found it to be common near the summit of Black Mountain, in Harlan County. Subsequent fieldwork yielded additional summer records and the confirmation of breeding (Wetmore 1940; Barbour 1941; Breiding 1947; Lovell 1950a; Mengel 1965). According to Mengel (1965), Veeries are present on Black Mountain above about 3,200 feet, although they are rare up to about 3,600 feet. Above this elevation they become quite common, being one of the most numerous and conspicuous species near the summit. More recent fieldwork also has documented the species's presence on the Letcher County portion of the mountain (KSNPC, unpub. data).

For many years it was believed that nesting Veeries were restricted to the higher portions of Black Mountain, but in July 1969 a bird was found at about 3,200 feet on Cumberland Mountain in Cumberland Gap National Historical Park (Croft 1969). Subsequent fieldwork in the park has revealed that the species occurs regularly on the mountain in summer, primarily scattered along the ridge above about 2,900 feet (Davis et al. 1980). Veeries also have been reported in summer from Chunklick Spur of Little Black Mountain to the north of Cumberland Mountain at about 3,300 feet (Davis and Smith 1978).

The Veery is a bird of deciduous forest and associated openings and edges. In the Cumberland Mountains it is found primarily in the secondary forests previously disturbed by logging or fire. In such areas the understory layer is well developed.

Veeries winter primarily in South America (AOU 1983), and most birds return to their Kentucky breeding grounds during the first two weeks of May. Nest building probably commences soon thereafter, and it appears that clutches are completed from late May to mid-June (Mengel 1965). Kentucky data on clutch size are limited, but Mengel (1965) includes details of four reported nests. One nest contained three eggs, two contained four eggs, and one contained a brood of three young. Stamm (1981b) provides a report of a nest situated in a wild hydrangea bush that contained three eggs on June 4, 1980. Harrison (1975) gives rangewide clutch size as 3–5, commonly 4.

The nest is typically constructed on or near the ground and is well concealed amid the clutter of vines, brambles, ferns, and other herbaceous ground cover. Mengel (1965) described a typical nest as a bulky structure of dry leaves, tendrils, strips of inner bark, herbaceous stems, and grasses that is lined with rootlets and decayed leaves.

During the atlas survey Veeries were reported only in traditional breeding areas, including the summit of Black Mountain, where they remain one of the most conspicuous species. In addition, a few Veeries were found at several locations along the crest of Cumberland Mountain within Cumberland Gap National Historical Park at about 3,300 feet. Confirmed nesting was recorded only on Black Mountain, where an active nest was located and recently fledged young were observed on June 21–23, 1989.

Hal H. Harrison

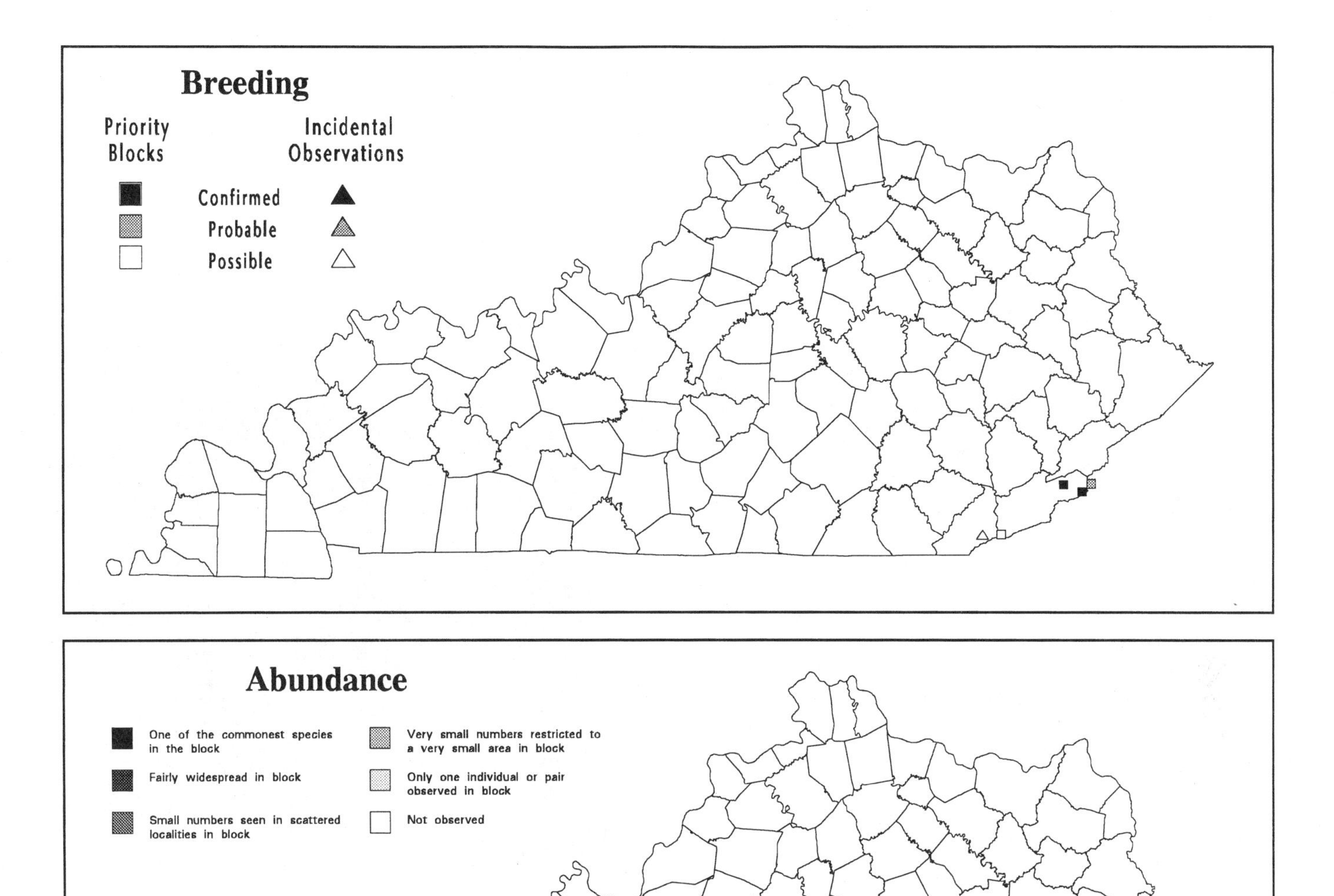

Abundance

One of the commonest species in the block
Fairly widespread in block
Small numbers seen in scattered localities in block
Very small numbers restricted to a very small area in block
Only one individual or pair observed in block
Not observed

Analysis of Block Data by Physiographic Province Section

Physiographic Province Section	Total Blocks Surveyed	Blocks with Data	% with Data	Section's % for State
Mississippi Alluvial Plain	14	-	-	-
East Gulf Coastal Plain	36	-	-	-
Highland Rim	139	-	-	-
Shawnee Hills	142	-	-	-
Blue Grass	204	-	-	-
Cumberland Plateau	173	-	-	-
Cumberland Mountains	19	4	21.1	100.0

Summary of Breeding Status

Number of Blocks in Which Species Was Recorded		
Total	**4**	**0.6%**
Confirmed	2	50.0%
Probable	1	25.0%
Possible	1	25.0%

Wood Thrush

Hylocichla mustelina

The flutelike notes of the Wood Thrush are one of the most conspicuous summer sounds in forested habitats throughout Kentucky. Mengel (1965) regarded the species as a common summer resident statewide, being less common only in the most mesic forest situations of the Cumberland Plateau and the Cumberland Mountains. Despite extensive clearing and fragmentation of native forests, this thrush remains one of the state's most widely distributed forest-nesting birds.

The Wood Thrush is common in most mesic and subxeric forest types with a well-developed shrub and midstory layer. Substantial numbers also occur in drier deciduous and mixed forests of ridges and slopes, as long as the understory is not too open. Occurrence is greatly reduced in very young forest as well as disturbed forest lacking understory cover. Although Wood Thrushes are most common in areas of extensive forest, they tolerate moderate disturbance and fragmentation. Owing to this adaptability, the species is often found in semi-open habitats, as long as forested tracts are not reduced to narrow strips or small, isolated woodlots.

Even though the Wood Thrush remains numerous today, it must have been much more common in Kentucky two centuries ago. The widespread clearing of land for conversion to agricultural use and settlement has resulted in the loss of much suitable nesting habitat. Moreover, the fragmentation of remaining forest apparently has resulted in declines caused by increased nest predation and cowbird parasitism (Hoover and Brittingham 1993).

Wood Thrushes winter primarily in Middle America (AOU 1983), and most birds return to Kentucky during the last two weeks of April. Egg laying has been reported by April 25 (Stamm and Jones 1966), although an early peak in clutch completion occurs about mid-May (Mengel 1965). Two broods are often raised, and records of active nests extend into mid-August (Mengel 1965; Pasikowski 1972). The average size of 18 clutches or broods reported by Mengel (1965) was 3.2 (range of 1–5), although some were incomplete because of cowbird parasitism.

Forest Cover

Value	% of Blocks	Avg Abund
All	78.3	2.4
1	48.5	1.9
2	74.2	2.1
3	82.6	2.3
4	91.6	2.8
5	92.3	2.9

Wood Thrushes typically place their nest in a crotch or fork of a limb of a shrub or small tree in the lower and midstory layers. The nest is constructed of dead plant material, including leaves and grass, and it sometimes contains paper or plastic (Pasikowski 1972; Harrison 1975). Mud may be incorporated into the structure, and it is typically lined with fine rootlets (Harrison 1975). The average height of 24 nests reported by Mengel (1965) was 10.5 feet (range of 3–30 feet), although Davis et al. (1980) reported a nest 45 feet above the ground on Cumberland Mountain.

The atlas survey yielded records of Wood Thrushes in more than 78% of priority blocks statewide, and 15 incidental observations were reported. The species is certainly more common than indicated by the atlas results, but its pronounced pattern of singing only during early morning and late afternoon contributed to its being missed on some blocks. Occurrence was higher than 90% in all sections except the Shawnee Hills, the Highland Rim, and the Blue Grass. Average abundance varied similarly, being slightly lower in the same three sections. As expected, both occurrence and average abundance were closely related to the degree of forest cover.

Ron Austing

Despite the abundance of Wood Thrushes, only 6% of priority block records were for confirmed breeding. Most confirmed records were based on the observation of active nests, but several pertained to recently fledged young, and a few were based on nest building, adults carrying food, and used nests. Cowbird parasitism of Wood Thrush nests is high in many areas, especially where forest fragmentation is pronounced. During atlas fieldwork both cowbird eggs and young were observed in Wood Thrush nests, as were recently fledged juvenile cowbirds being fed by adult thrushes.

Wood Thrushes are reported in small to moderate numbers on most Kentucky BBS routes. The average number of individuals per BBS route for the periods 1966–91 and 1982–91 was 11.09 and 9.00, respectively. Trend analysis of these data yields a nonsignificant increase of 0.4% per year for the period 1966–91 and a nonsignificant decrease of 1.9% per year for the period 1982–91. Although a decline has been detected in some parts of the species's range during the past 30 years (Hoover and Brittingham 1993; Robbins et al. 1989), Kentucky data do not reflect that trend.

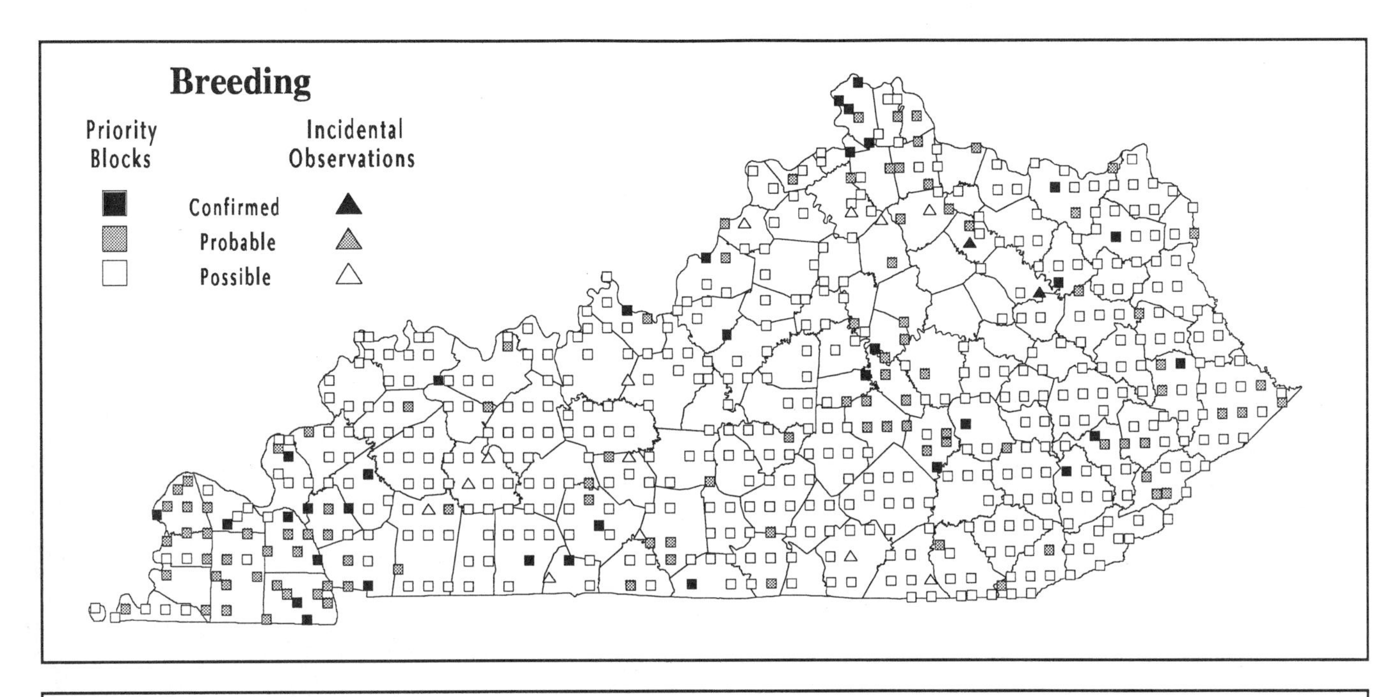

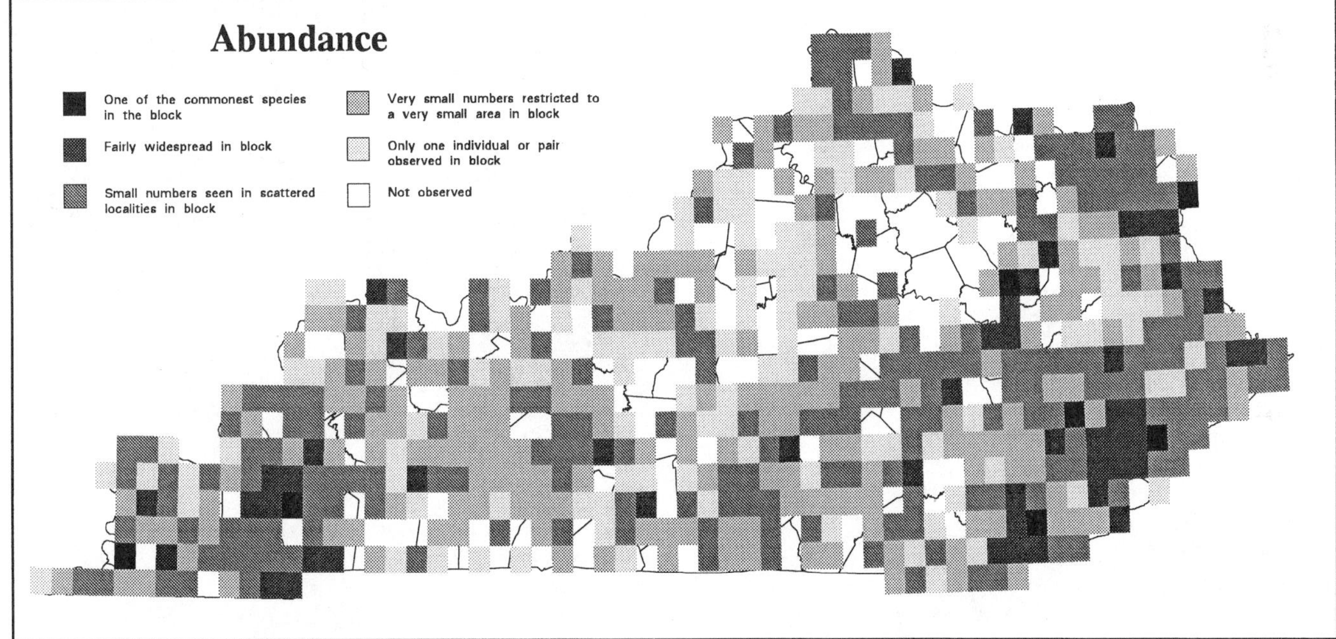

Analysis of Block Data by Physiographic Province Section

Physiographic Province Section	Total Blocks Surveyed	Blocks with Data	Avg Abund	% with Data	Section's % for State
Mississippi Alluvial Plain	14	13	2.7	92.9	2.3
East Gulf Coastal Plain	36	33	2.7	91.7	5.8
Highland Rim	139	107	2.3	77.0	18.8
Shawnee Hills	142	111	2.2	78.2	19.5
Blue Grass	204	130	2.2	63.7	22.8
Cumberland Plateau	173	157	2.7	90.8	27.6
Cumberland Mountains	19	18	2.7	94.7	3.2

Summary of Breeding Status

Number of Blocks in Which Species Was Recorded		
Total	**569**	**78.3%**
Confirmed	34	6.0%
Probable	103	18.1%
Possible	432	75.9%

Wood Thrush

American Robin
Turdus migratorius

The American Robin's melodic song is one of the most conspicuous summer sounds throughout much of Kentucky. The predawn chorus of robin song characterizes most settled areas, and the sight of robins hopping across residential lawns in search of food perhaps makes them the most widely recognized bird in the state. Mengel (1965) regarded the species as a locally uncommon to very common summer resident, being less numerous only where settlement had not encroached on the landscape.

The American Robin is found in a great variety of habitats, from mature forest and rural farmland to residential yards and urban parks. During the breeding season these large thrushes prefer to forage on bare soil or in short, well-manicured vegetation. Although the species is regularly encountered in undisturbed forest, the robin typically avoids habitats with thick ground cover during summer.

Robins are sometimes found in mature woodlands, and they may have been fairly common in Kentucky prior to settlement. Audubon (1861) considered the species to be "tolerably abundant" as of the early 1800s. Nonetheless, robins likely have increased in response to human alteration of the landscape. The clearing and fragmentation of vast forested areas for rural and suburban settlement have likely created even better habitat for robins, and very large populations have developed in settled areas throughout the state.

Although robins are seen throughout the year, many summer residents retreat southward for the winter. Once the weather starts to moderate, large numbers of robins begin returning to their Kentucky breeding grounds, often by late February. Territorial behavior commences immediately, and nest building may be under way by early March if the weather remains mild. Egg laying has been reported by mid-March, and an early peak in clutch completion occurs during early April (Mengel 1965). Robins usually raise two or three broods in Kentucky, and although later peaks in clutch completion are not apparent, active nests are commonly observed through July. Even later nests are occasionally observed well into August (Stamm 1964; Stamm and Jones 1966). The average size of 33 clutches or broods reported by Mengel (1965) was 3.5 (range of 2–5).

Forest Cover

Value	% of Blocks	Avg Abund
All	98.2	3.4
1	99.0	3.8
2	98.9	3.8
3	99.1	3.5
4	96.8	3.1
5	94.2	2.4

American Robins place their nests in a great variety of situations. While evergreen trees are commonly chosen for nests built early in the spring, deciduous trees are more frequently used for later ones. Nests are placed on horizontal forks of branches as well as upright crotches of larger branches and the main trunk. Robins also frequently nest on protected shelves of homes and buildings, using porches, eaves, gutters, and the like. The nest is composed of dead plant material, primarily grass, cemented together with mud, and lined with fine grass. The average height of 37 nests reported by Mengel (1965) was 13.3 feet (range of 1–50 feet), although Stamm (1964) reported a nest 80 feet above the ground.

The atlas survey yielded records of American Robins in more than 98% of priority blocks statewide, and five incidental observations were reported. The robin ranked 3rd by the number of records in priority blocks, by total abundance, and by average abundance. Occurrence was greater than 90% in all physiographic province sections, and it is likely that robins were simply missed in the 13 priority blocks in which they were not seen. Average abundance was 3.0 or greater statewide. Occurrence was slightly lower in extensively forested areas, but average abundance decreased more noticeably as forestation increased.

Alvin E. Staffan

Fifty percent of priority block reports were for confirmed breeding. This relatively high total can be attributed to the abundance of robins in settled areas as well as the conspicuousness of their nests and young. Many active nests were found during atlas work, but most confirmed records were based on the observation of recently fledged young. Other confirmed records were based on the observation of nest building, adults carrying food for young, and used nests.

American Robins are typically reported in large numbers on most Kentucky BBS routes. The average number of individuals recorded per BBS route for the periods 1966–91 and 1982–91 was 38.88 and 58.46, respectively. According to these data, the robin ranked 5th in abundance on BBS routes during 1982–91. Trend analysis of these data shows a significant ($p<.01$) increase of 3.7% per year for the period 1966–91 and a nonsignificant increase of 0.8% per year for the period 1982–91. The long-term increase can probably be attributed to an increase in settlement along BBS routes.

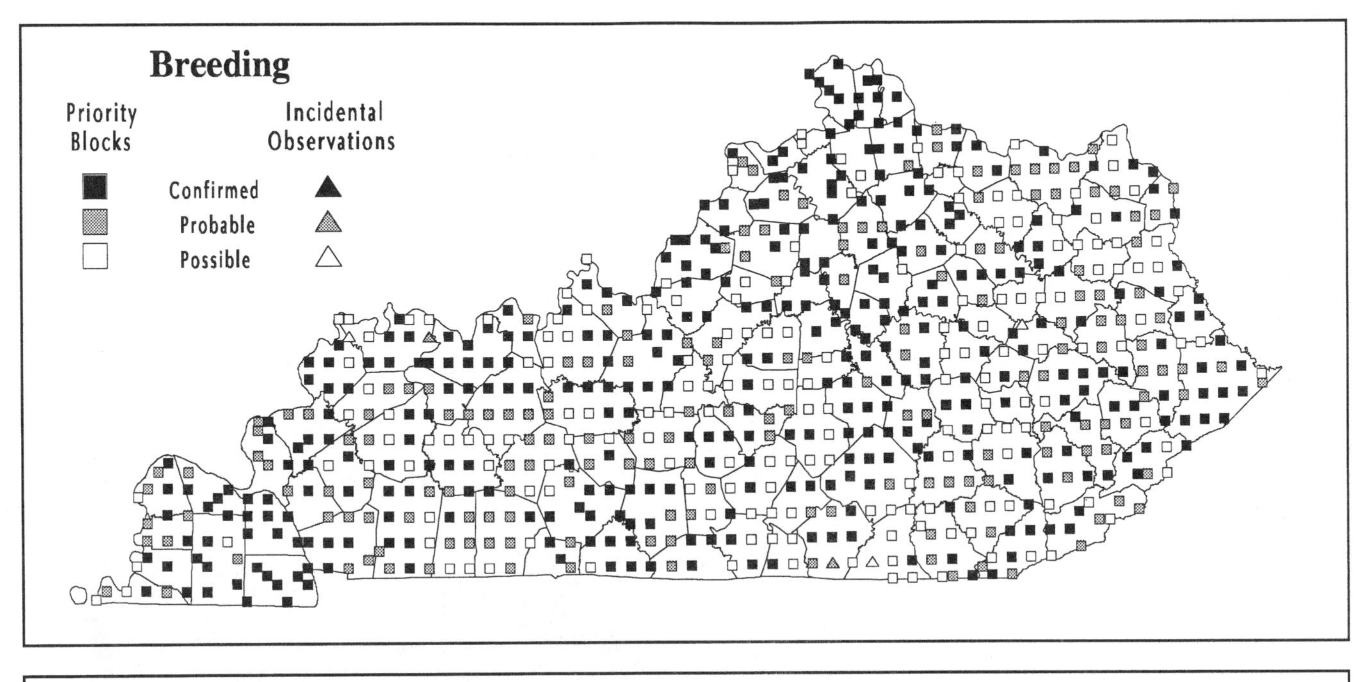

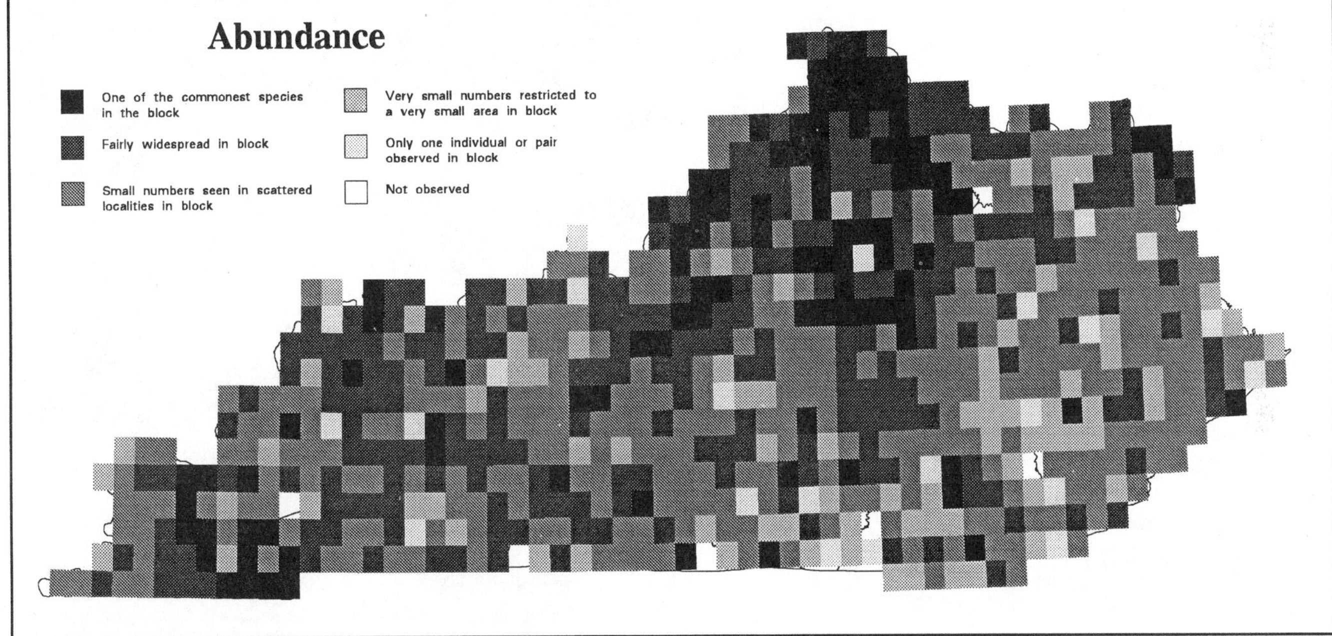

Analysis of Block Data by Physiographic Province Section

Physiographic Province Section	Total Blocks Surveyed	Blocks with Data	Avg Abund	% with Data	Section's % for State
Mississippi Alluvial Plain	14	13	3.1	92.9	1.8
East Gulf Coastal Plain	36	36	4.0	100.0	5.0
Highland Rim	139	136	3.4	97.8	19.0
Shawnee Hills	142	140	3.4	98.6	19.6
Blue Grass	204	203	3.9	99.5	28.4
Cumberland Plateau	173	168	3.0	97.1	23.5
Cumberland Mountains	19	18	3.0	94.7	2.5

Summary of Breeding Status

Number of Blocks in Which Species Was Recorded		
Total	**714**	**98.2%**
Confirmed	357	50.0%
Probable	179	25.1%
Possible	178	24.9%

American Robin

Gray Catbird

Dumetella carolinensis

Although the Gray Catbird is rather secretive in its habits and nowhere especially abundant during the nesting season, it is widely distributed in appropriate habitat throughout the state. Mengel (1965) regarded the species as a common summer resident, occurring wherever dense shrubbery of medium height was present in good supply.

The Gray Catbird is found primarily in semi-open habitats, but small numbers are regularly encountered in both very open and extensively forested areas. Wherever it occurs, the species is typically associated with dense, brushy cover. Most frequently inhabited situations include old fields, woodland edge, forest clearcuts, and rural settlement. Naturally occurring habitats are less frequently used today, but they include dense borders surrounding swampy areas, brushy riparian zones, and thick woodland openings created by fire or storm damage. Although less frequently reported in residential areas than the other two members of the family Mimidae that occur in Kentucky, catbirds are regularly encountered in parks and yards with hedgerows or other densely leaved plantings.

The Gray Catbird may have been less common and widespread in Kentucky two centuries ago. Some suitable habitat likely has been lost owing to the conversion of forested areas to agricultural use and settlement. Also, fire and storm damage may have contributed more significantly to maintenance of suitable habitat before settlement. In contrast, fragmentation of remaining forest and neglect of formerly cleared areas have resulted in the creation of an abundance of brushy habitat that likely has more than made up for the habitat lost.

Gray Catbirds return to Kentucky from more southerly wintering grounds during the last two weeks of April. According to Mengel (1965), nest building may be under way before the end of April, and early clutches are completed during the first ten days of May. Although a peak in clutch completion occurs during the middle of May, a second brood is sometimes raised, and there are numerous records of active nests extending into late July (Mengel 1965; Stamm and Jones 1966) and at least occasionally into mid-August (Croft and Stamm 1967). The average size of 31 clutches or broods reported by Mengel (1965) was 3.4 (range of 2–5).

Forest Cover

Value	% of Blocks	Avg Abund
All	80.3	2.2
1	70.3	2.1
2	76.3	2.0
3	83.5	2.2
4	86.4	2.5
5	82.7	2.5

Catbirds typically place their nests in dense shrubs, tangles of vines, or thickly leaved trees. Nests are sometimes situated in trees or shrubs standing alone or along a narrow corridor such as a fence- or hedgerow, but more often they occur along woodland edge. The nest usually is placed on a horizontal fork or supported by thick vines or small branches. It is constructed primarily of twigs, vines, leaves, and strips of bark and is lined with fine rootlets (Harrison 1975). The average height of 23 nests reported by Mengel (1965) was 7.0 feet (range of 2–15 feet).

The atlas fieldwork yielded records of Gray Catbirds in slightly more than 80% of priority blocks statewide, and 18 incidental observations were reported. Occurrence was highest in the Cumberland Plateau, the Blue Grass, and the Cumberland Mountains, and lowest in the East Gulf Coastal Plain, the Highland Rim, and the Mississippi Alluvial Plain. Average abundance varied somewhat similarly. This distribution probably results from a combination of the degree of forestation and the species's more northern breeding range. While catbirds were found in extensively cleared areas, they were more numerous in places where forest predominated. Also, the center of the catbird's breeding range lies to the north of Kentucky (AOU 1983), making it naturally more likely to occur in the slightly cooler northern and eastern parts of the state.

Alvin E. Staffan

Less than 13% of priority block records were for confirmed breeding, probably owing to the catbird's inconspicuousness during the breeding season. Although a few active nests were located, most confirmed records were based on the observation of recently fledged young and adults carrying food for young.

Gray Catbirds are not reported in substantial numbers on most Kentucky BBS routes. The average number of individuals reported per BBS route for the periods 1966–91 and 1982–91 was 7.06 and 5.34, respectively. Trend analysis of these data shows a nonsignificant decrease of 1.3% per year for the period 1966–91, but a significant ($p<.01$) decrease of 5.6% per year for the period 1982–91. The reason for the long-term decline is not known, but it may in part reflect mortality caused by the severe winters of the late 1970s (Robbins et al. 1986).

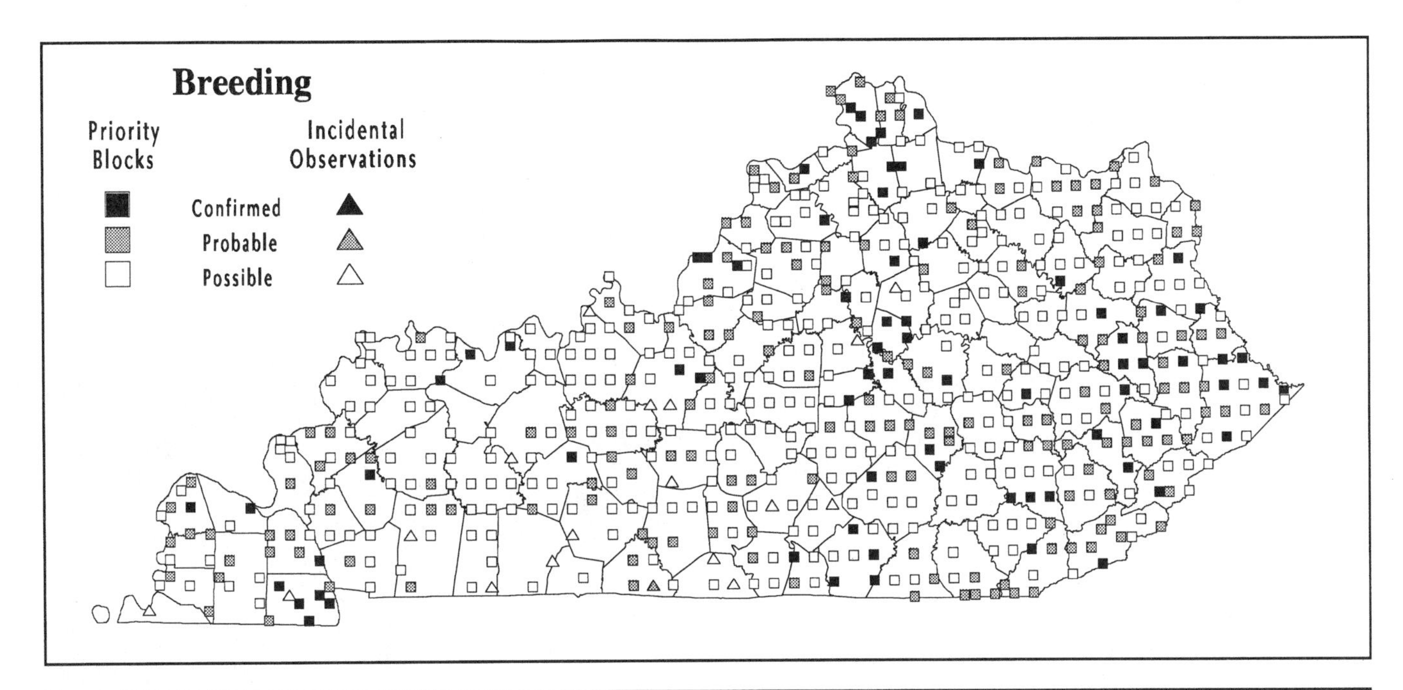

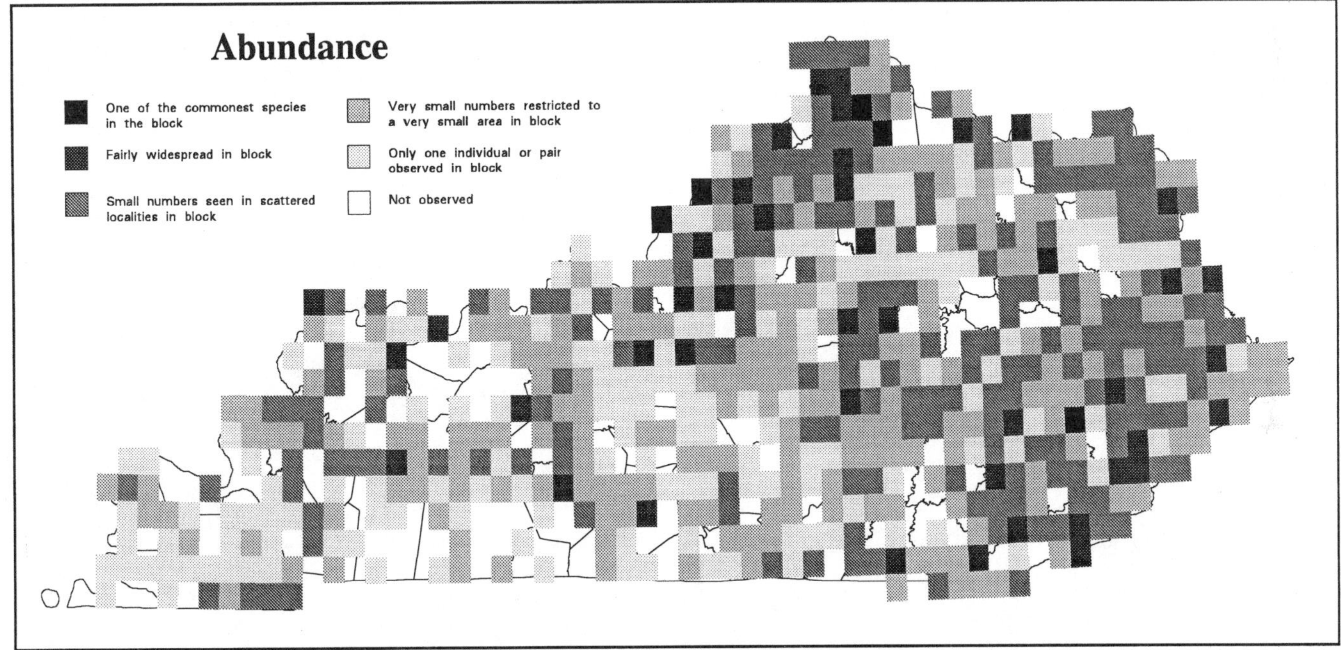

Analysis of Block Data by Physiographic Province Section

Physiographic Province Section	Total Blocks Surveyed	Blocks with Data	Avg Abund	% with Data	Section's % for State
Mississippi Alluvial Plain	14	9	1.7	64.3	1.5
East Gulf Coastal Plain	36	25	1.6	69.4	4.3
Highland Rim	139	95	1.9	68.3	16.3
Shawnee Hills	142	103	2.0	72.5	17.6
Blue Grass	204	180	2.4	88.2	30.8
Cumberland Plateau	173	156	2.5	90.2	26.7
Cumberland Mountains	19	16	2.7	84.2	2.7

Summary of Breeding Status

Number of Blocks in Which Species Was Recorded		
Total	**584**	**80.3%**
Confirmed	74	12.7%
Probable	176	30.1%
Possible	334	57.2%

Gray Catbird

Northern Mockingbird

Mimus polyglottos

The Northern Mockingbird is not the most widespread of the three species of mimids in Kentucky, but it is certainly the most conspicuous. Common in residential areas, the species is perhaps best known for its persistent, often nocturnal, summer song. Mengel (1965) regarded the mockingbird as common in central and western Kentucky, but uncommon to rare in the mountainous eastern part of the state.

Mockingbirds are typically found in semi-open and open habitats, being only rarely reported in areas dominated by extensive forest. Today the species appears to be restricted to areas altered by humans. It is most common in and near settlement, including rural farmsteads and towns, suburban yards, and urban parks, but it is also found in abandoned farmland, along rural roadway corridors, and on reclaimed surface mines. In heavily forested areas, mockingbirds seem to be restricted to openings surrounding rural homesteads.

The Northern Mockingbird likely has increased dramatically in Kentucky in response to human alteration of the landscape. While the species probably used open habitats in and around the native prairies, Audubon (1861) made no mention of it as of the early 1800s. Because of their absence in extensively forested areas, mockingbirds must have been relatively scarce in most areas. The subsequent conversion of forested habitats to residential settlement and farmland has likely resulted in a dramatic increase.

Northern Mockingbirds are present throughout the year, although some birds may retreat southward in winter (Mengel 1965). With the onset of warmer weather, numbers begin to increase, and song commences during March. Nest building has been observed in the latter part of March, and early clutches are completed during the first part of April (Mengel 1965). Mengel (1965) could not discern a peak in completion of early clutches among the nesting records he reviewed, but it likely occurs in the latter half of April. Many pairs raise a second brood, and later nests have been reported into late July and August (Stamm 1965; Stamm and Croft 1968). The average size of 16 clutches or broods reported by Mengel (1965) was 3.5 (range of 3–4), but Stamm and Croft (1968) listed an apparently complete clutch of only two eggs.

Forest Cover

Value	% of Blocks	Avg Abund
All	77.4	2.9
1	97.0	3.2
2	97.9	3.3
3	92.2	2.7
4	39.6	2.0
5	11.5	1.5

Mockingbirds typically place their nests among the dense foliage of a tree or shrub. Although evergreens are most commonly used for early nests, deciduous plants are used later in the year. Nests are typically built in relatively open areas such as yards and hedgerows, and only rarely along woodland edge. The nest is usually placed within a forking of several branches, but sometimes in a tangle of small branches or vines. It is constructed primarily of small twigs and lined with fine grass or rootlets (Harrison 1975). The average height of 16 nests reported by Mengel (1965) was 5.5 feet (range of 1.5–15.0 feet), although Stamm and Croft (1968) reported a nest 25 feet above the ground.

Alvin E. Staffan

The atlas survey yielded records of Northern Mockingbirds in more than 77% of priority blocks statewide, and nine incidental observations were reported. The species ranked 40th according to the number of priority block records, but 27th by total abundance and 17th by average abundance. Although the mockingbird is not as widespread as many other common species, where it does occur it is relatively numerous and conspicuous. Occurrence was distinctly higher in central and western Kentucky than in the east. Average abundance varied similarly. Both occurrence and average abundance were closely related to the percentage of forest cover, with substantially lower values in areas with a predominance of forest. This trend contrasts sharply with that of the Gray Catbird and is more pronounced than that of the Brown Thrasher.

Despite the conspicuousness of Northern Mockingbirds, less than 27% of priority block records were for confirmed breeding. Although a few active nests were reported, most confirmed records were based on the observation of recently fledged young. Others resulted from the observation of adults carrying food for young, nest building, and used nests.

Northern Mockingbirds are reported in fairly substantial numbers on most Kentucky BBS routes. The average number of individuals per BBS route for the periods 1966–91 and 1982–91 was 14.35 and 13.55, respectively. According to these data, the species ranked 19th in abundance on BBS routes during the period 1982–91. Trend analysis of these data yields a nonsignificant decrease of 0.8% per year for the period 1966–91, but a significant ($p<.01$) increase of 5.6% per year for the period 1982–91. The short-term increase reflects a return to more normal numbers following decreases experienced in the late 1970s because of severe winter weather (Robbins et al. 1986).

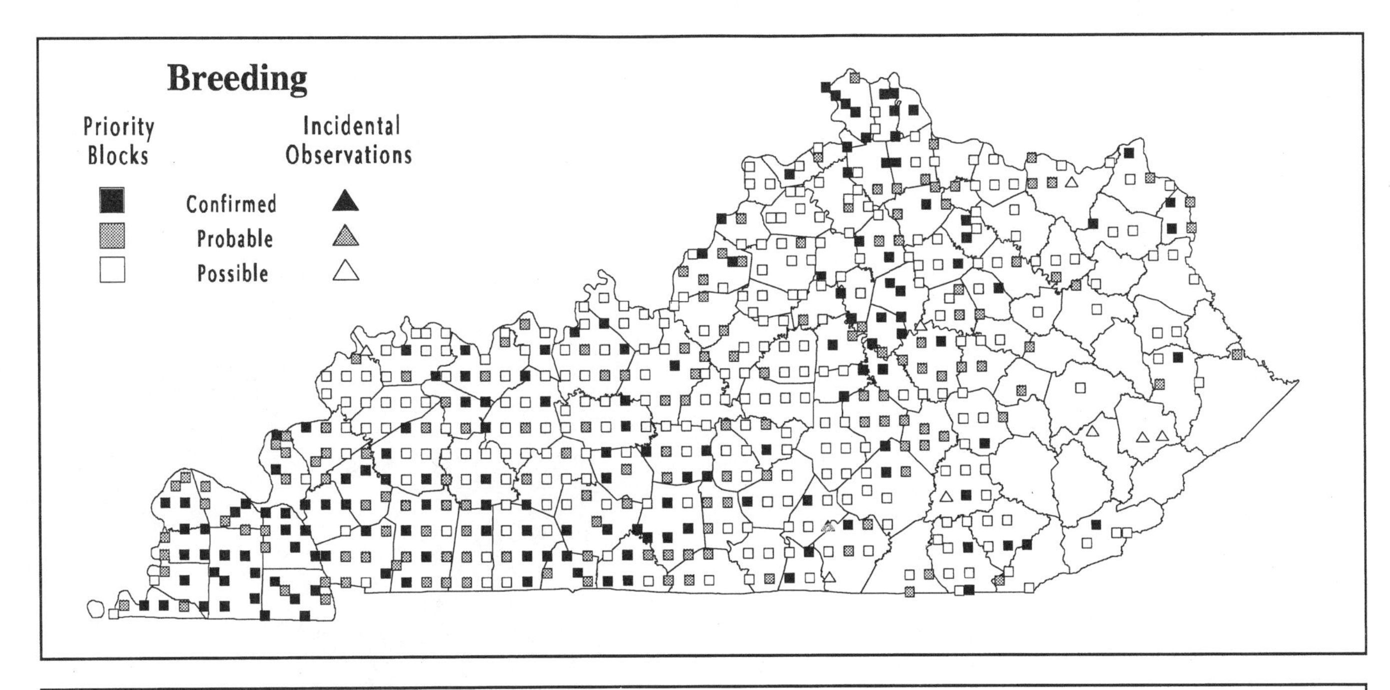

Analysis of Block Data by Physiographic Province Section

Physiographic Province Section	Total Blocks Surveyed	Blocks with Data	Avg Abund	% with Data	Section's % for State
Mississippi Alluvial Plain	14	12	3.2	85.7	2.1
East Gulf Coastal Plain	36	36	3.9	100.0	6.4
Highland Rim	139	132	3.0	95.0	23.4
Shawnee Hills	142	137	3.0	96.5	24.3
Blue Grass	204	180	2.8	88.2	32.0
Cumberland Plateau	173	59	2.0	34.1	10.5
Cumberland Mountains	19	7	1.6	36.8	1.2

Summary of Breeding Status

Number of Blocks in Which Species Was Recorded		
Total	**563**	**77.4%**
Confirmed	150	26.6%
Probable	153	27.2%
Possible	260	46.2%

Northern Mockingbird

Brown Thrasher

Toxostoma rufum

The Brown Thrasher's habitat preferences lie somewhere between those of the other two members of the family Mimidae that occur in Kentucky. For that reason it is probably the most widespread of the three. Mengel (1965) regarded the species as common and uniformly distributed across the state.

Brown Thrashers are found in a variety of habitats, from hedgerows and farmsteads in relatively open, rural areas to suburban parks and yards. The species is most common in semi-open rural situations with an abundance of brushy cover. These mimids also inhabit predominantly forested areas, although they are noticeably less numerous than catbirds. They tend to occur more frequently in drier habitats than catbirds, typically avoiding wetlands and riparian corridors. Brown Thrashers primarily inhabit artificially created situations, but they are sometimes found in naturally occurring habitats such as brushy forest margins and openings.

Brown Thrashers likely have increased in Kentucky since the 18th century. Audubon (1861) reported that thrashers were abundant in the barrens in the early 1800s, and they may have occurred in natural clearings and forest margins across the state before European settlement. In contrast, the species may have been relatively local in distribution overall, owing to the scarcity of open habitats in extensively forested regions. The conversion of native forests to agricultural use and settlement has resulted in the creation of an abundance of brushy nesting habitat.

Although a few Brown Thrashers seem to remain throughout the winter, most birds move farther south. The first males begin returning to their Kentucky breeding grounds in early March. Song commences immediately, and nest building may be under way by mid-March (Mengel 1965). Active nests have been observed as early as late March, but a peak in early clutch completion occurs during the middle of April (Mengel 1965). Most pairs appear to raise two broods, and a second peak in clutch completion occurs in early June (Mengel 1965). Even later nests have been recorded into mid-July (Stamm and Jones 1966; KOS Nest Cards). The average size of 27 clutches or broods reported by Mengel (1965) was 3.7 (range of 3–6), although a seemingly complete clutch of two eggs was observed in Franklin County on July 4, 1963 (Stamm and Jones 1966).

Forest Cover

Value	% of Blocks	Avg Abund
All	84.5	2.4
1	94.1	2.6
2	95.8	2.7
3	89.1	2.4
4	69.5	2.0
5	48.1	1.5

Thrashers typically place their nests in the dense cover of small trees, shrubs, and tangles of briars or vines, but a few nests have been found on the ground (Hancock 1951, 1954; Croft and Stamm 1967; Stamm and Croft 1968). Although nests are occasionally built in a tree or shrub standing alone, they are much more frequently constructed along a fencerow or woodland border. The nest is a loose, bulky structure built mostly of twigs and dead leaves and lined with fine grass or rootlets (Ehrlich et al. 1988). The average height of 22 above-ground nests reported by Mengel (1965) was 4.3 feet (range of 1.5–9.5 feet), but a nest observed in Daviess County in 1968 was constructed 15 feet above the ground (KOS Nest Cards).

Ron Austing

The atlas fieldwork yielded records of Brown Thrashers in nearly 85% of priority blocks statewide, and 10 incidental observations were reported. The species ranked 25th according to the number of records in priority blocks, 32nd by total abundance, and 32nd by average abundance. It was thus the most widely distributed mimid in the state, but it was not as abundant overall as the mockingbird. Occurrence was highest in the East Gulf Coastal Plain, the Shawnee Hills, and the Highland Rim, and lowest in the Cumberland Plateau and the Cumberland Mountains. This distribution was most similar to the Northern Mockingbird's, although the thrasher occurs more widely in eastern Kentucky. Average abundance varied somewhat similarly. Both occurrence and average abundance were highest in cleared areas with some forest cover and decreased as forestation increased.

Nearly 23% of priority block records were for confirmed breeding. Active nests accounted for a number of confirmed records, but most were based on the observation of recently fledged young. Adults carrying food for young, nest building, and used nests accounted for most others.

Brown Thrashers are reported in small to moderate numbers on most Kentucky BBS routes. The average number of individuals per BBS route for the periods 1966–91 and 1982–91 was 5.79 and 7.11, respectively. According to these figures, the thrasher ranked 33rd in abundance on BBS routes during 1982–91. Trend analysis of these data indicates a nonsignificant increase of 0.1% per year for the period 1966–91, but a significant ($p<.05$) decrease of 2.9% per year for the period 1982–91. Reasons for the short-term decrease are not known.

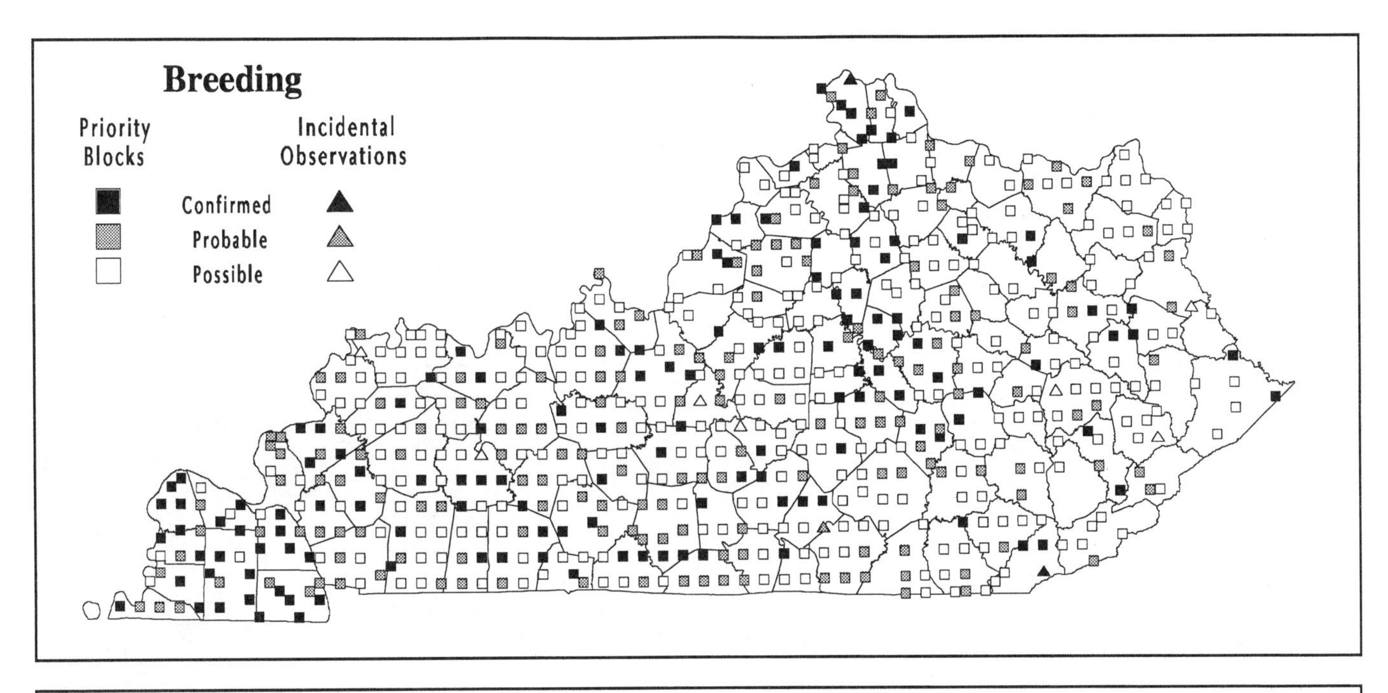

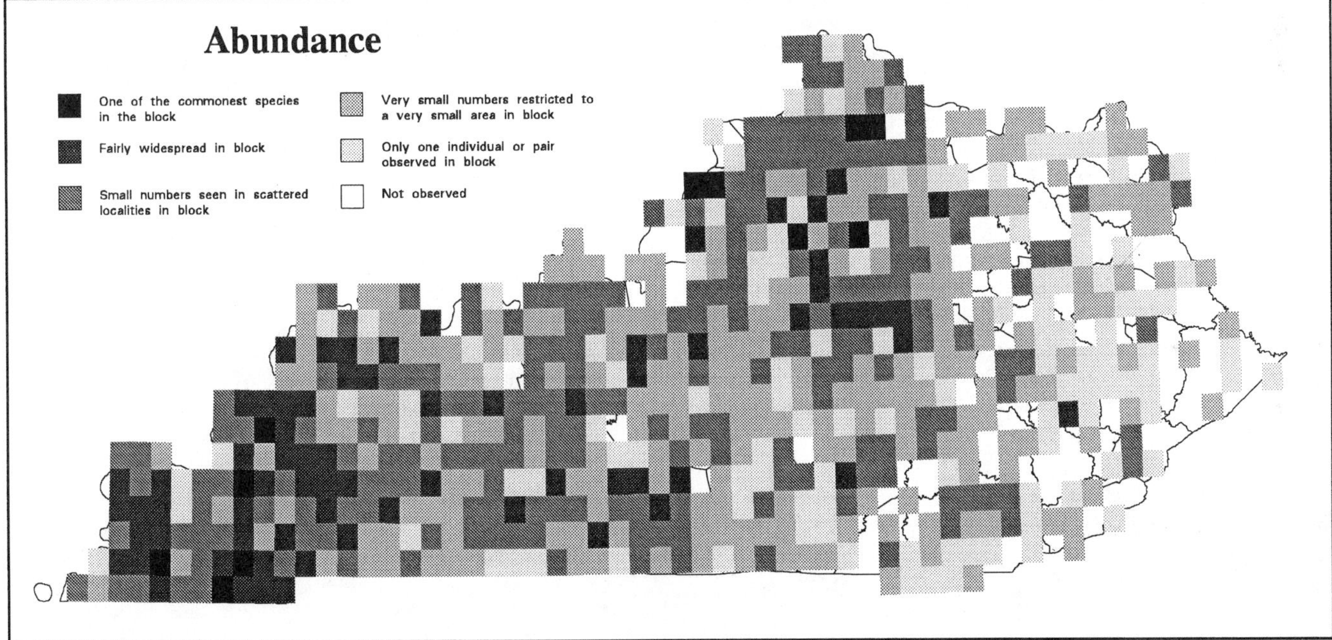

Analysis of Block Data by Physiographic Province Section

Physiographic Province Section	Total Blocks Surveyed	Blocks with Data	Avg Abund	% with Data	Section's % for State
Mississippi Alluvial Plain	14	11	3.1	78.6	1.8
East Gulf Coastal Plain	36	36	3.4	100.0	5.9
Highland Rim	139	133	2.4	95.7	21.7
Shawnee Hills	142	136	2.6	95.8	22.1
Blue Grass	204	176	2.5	86.3	28.7
Cumberland Plateau	173	113	1.9	65.3	18.4
Cumberland Mountains	19	9	1.4	47.4	1.5

Summary of Breeding Status

Number of Blocks in Which Species Was Recorded		
Total	**614**	**84.5%**
Confirmed	139	22.6%
Probable	174	28.3%
Possible	301	49.0%

Brown Thrasher

Cedar Waxwing

Bombycilla cedrorum

The Cedar Waxwing is a relatively recent addition to the breeding avifauna of Kentucky. The first nest was discovered in 1934 (Monroe 1946), and as of the late 1950s fewer than twenty documented nesting records existed (Stamm 1951b; Mengel 1965). Based primarily on breeding season sightings, Mengel (1965) regarded the species as a locally rare to common summer resident, breeding regularly only in the northern Blue Grass, the Cumberland Plateau, and the Cumberland Mountains. As recently as the mid-1970s waxwings were still regarded as rare in parts of central Kentucky in summer (Monroe 1976), but during the 1980s they expanded throughout the central and western parts of the state, appearing in many areas where they had never before been reported (e.g., C. Peterson 1980; Mason and Ferrell 1983).

Ron Austing

Cedar Waxwings are found in a great variety of habitats in summer. Although occasionally encountered in extensively forested areas, they are much more frequently found in semi-open and open areas with scattered trees. The species is most conspicuous in association with settlement, whether it be parks and residential neighborhoods of cities and towns or rural cemeteries and homesteads.

Waxwings are irregular in their occurrence and abundance from year to year, but spring migrants typically pass through Kentucky from mid-April to mid-May. Flocks of late migrants or nonbreeding birds can be observed well into June, but nesting may commence much earlier. During the atlas fieldwork, nest building was noted in late May, and early clutches may be completed by June 1–10 (Mengel 1965). Although a peak in early clutch completion occurs in mid-June, some pairs may raise two broods, and a later peak occurs in late July (Mengel 1965). The average size of eight clutches or broods reported by Mengel (1965) was 3.5 (range of 2–5).

Although waxwings sometimes place their nest in a small tree or shrub, they more typically choose medium-sized to large trees. Deciduous trees are commonly used, but evergreens, especially pines, seem to be preferred. Trees situated in semi-open, parklike habitats are chosen most frequently, although the species likely nests occasionally along natural woodland borders. The nest is invariably situated on a horizontal or slightly upright fork, usually some distance out on a larger branch near the outer crown of leaves.

Forest Cover

Value	% of Blocks	Avg Abund
All	49.9	1.7
1	56.4	1.8
2	56.3	1.7
3	53.0	1.7
4	43.5	1.8
5	17.3	1.9

The typical nest is somewhat bulky and is constructed primarily of coarse grasses. It often contains debris such as string, paper, or cellophane and is lined with finer material, including grass, rootlets, and plant down (Harrison 1975). The average height of 12 nests reported by Mengel (1965) was 25.2 feet (range of 8–60 feet).

The atlas survey yielded records of Cedar Waxwings in nearly 50% of priority blocks statewide, and 34 incidental observations were reported. Occurrence was lowest in the Mississippi Alluvial Plain and the East Gulf Coastal Plain, but surprisingly, it was below average in the Cumberland Mountains and the Cumberland Plateau, where waxwings were expected to be most common. In contrast, occurrence was highest in the Blue Grass, the Highland Rim, and the Shawnee Hills. Average abundance was relatively low and unvaried across the state. Occurrence was highest in relatively open areas and lowest in completely forested areas. At the onset of the atlas project, many early summer observations were cautiously regarded as nonbreeding birds; however, after a number of late May and June reports of nest building and active nests accumulated, all records of single birds or pairs in suitable nesting habitat were recorded as representative of possible breeding.

Nearly 14% of priority block records were for confirmed breeding. Although at least 15 active nests were observed, most confirmed records were based on the observation of recently fledged young. Additional confirmed records were based on nest building and adults carrying food for young.

Kentucky BBS data have documented the increase of nesting waxwings since the mid-1960s. The average number of individuals per BBS route for the periods 1966–91 and 1982–91 was 0.56 and 1.78, respectively. Trend analysis of these data shows nonsignificant increases of 2.4% per year for the period 1966–91 and 7.4% per year for the period 1982–91. The reasons for this increase are not fully understood, but it mirrors a trend that has occurred in surrounding states (Robbins et al. 1986; Robinson 1990; Peterjohn and Rice 1991). It seems that a shift in the breeding population has occurred in Kentucky, where the species now nests more commonly in the central part of the state than in the east.

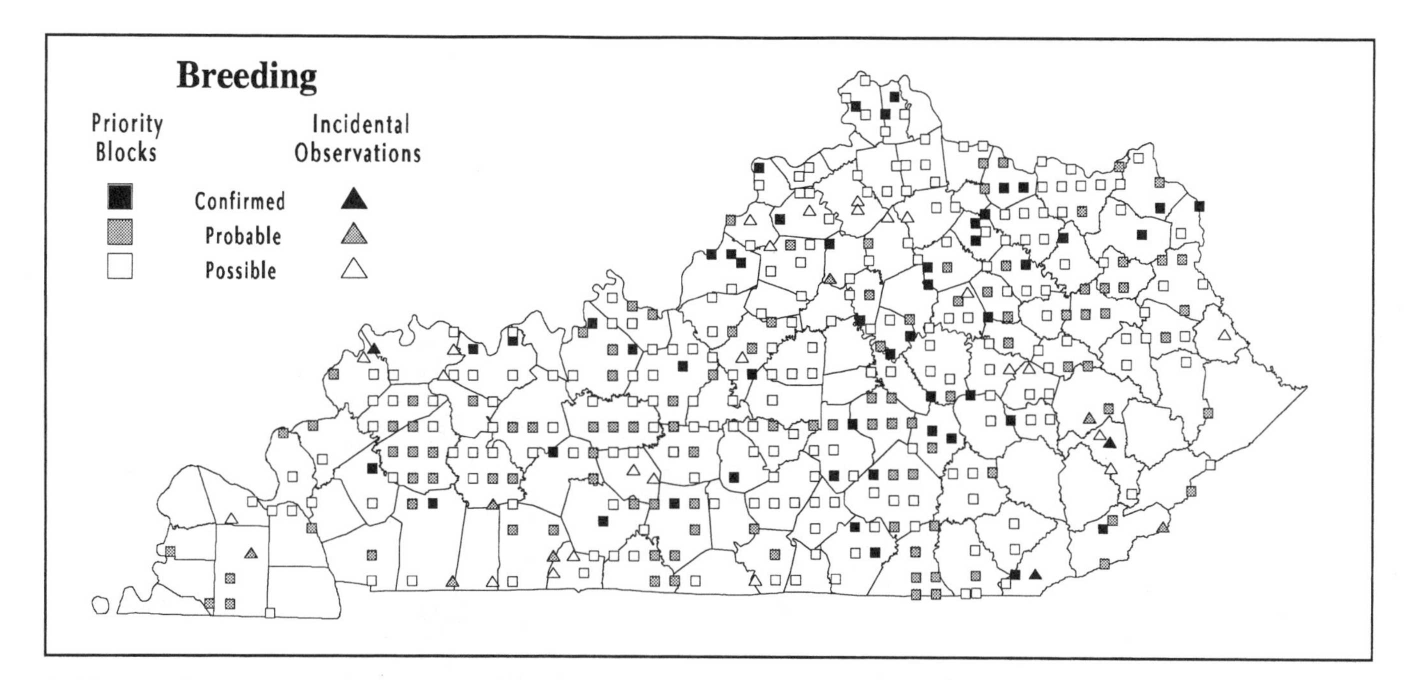

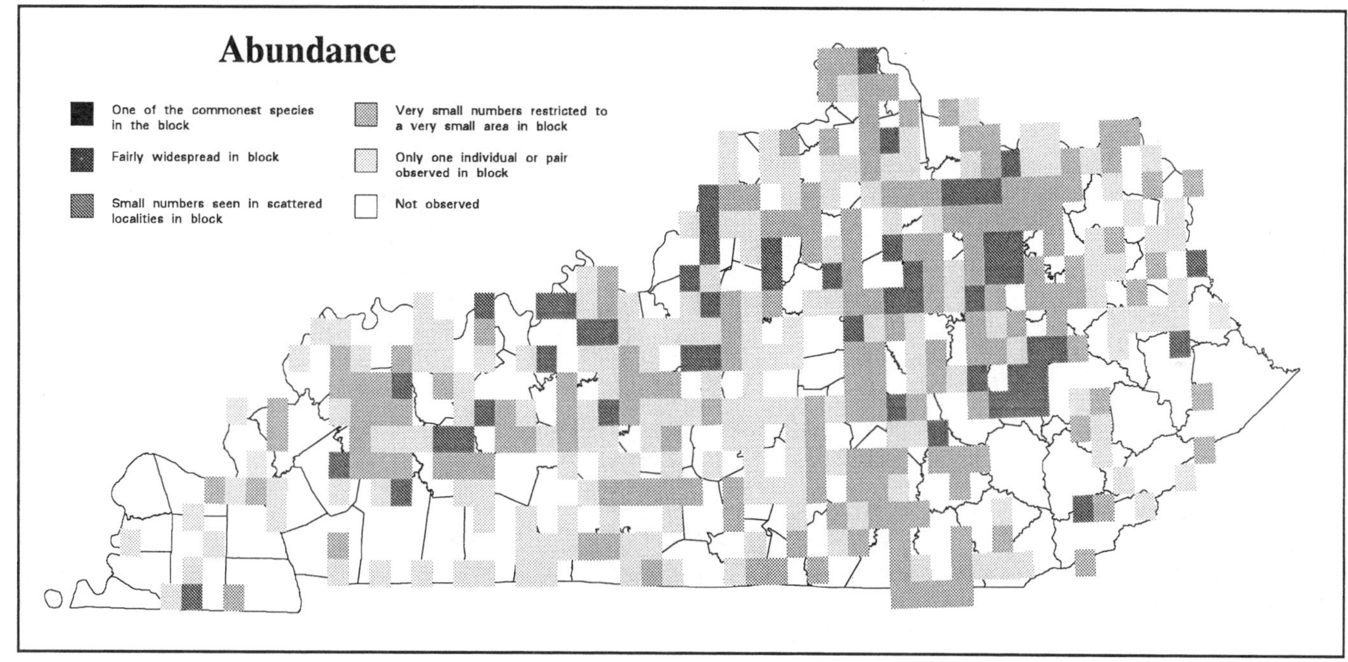

Analysis of Block Data by Physiographic Province Section

Physiographic Province Section	Total Blocks Surveyed	Blocks with Data	Avg Abund	% with Data	Section's % for State
Mississippi Alluvial Plain	14	2	1.5	14.3	0.6
East Gulf Coastal Plain	36	7	1.6	19.4	1.9
Highland Rim	139	76	1.5	54.7	20.9
Shawnee Hills	142	71	1.7	50.0	19.6
Blue Grass	204	135	1.9	66.2	37.2
Cumberland Plateau	173	67	1.8	38.7	18.5
Cumberland Mountains	19	5	1.6	26.3	1.4

Summary of Breeding Status

Number of Blocks in Which Species Was Recorded		
Total	**363**	**49.9%**
Confirmed	49	13.5%
Probable	109	30.0%
Possible	205	56.5%

Cedar Waxwing

Loggerhead Shrike

Lanius ludovicianus

As of the late 1950s the Loggerhead Shrike was regarded as a rare to fairly common and locally distributed resident in Kentucky, occurring primarily west of the Cumberland Plateau (Mengel 1965). Since that time the species has declined substantially throughout much of its range (Anderson and Duzan 1978; Robbins et al. 1986). In 1982 it was designated as a candidate for federal listing (USFWS 1982). A similar decline has been noted in Kentucky (Monroe 1978), but while the shrike is nowhere especially numerous, it remains a characteristic summer bird throughout a large portion of south central and western Kentucky.

The Loggerhead Shrike is a bird of open and semi-open habitats, being only rarely reported in areas of extensive forest. The species seems to favor areas with short or sparse ground cover, usually avoiding habitats dominated by tall, thick vegetation. In Kentucky shrikes are most frequently encountered in rural farmland, where they forage primarily in bare fields, pastures, mowed hayfields, yards, and roadsides. In addition, the species can be found in developed habitats, including airports, industrial parks, and rural roadway corridors and residential areas.

It is unclear to what extent Loggerhead Shrikes inhabited Kentucky before settlement. Audubon (1861) did not observe the species while residing in the state in the early 1800s. While shrikes may have occurred in the prairies of southern and western Kentucky, as well as the savannas of the Inner Blue Grass, it is possible that grasses of these habitats were tall and thick enough to exclude them. Outside these regions, the species must have been largely or entirely absent. The conversion of vast forested areas to agriculture and settlement surely resulted in the expansion of birds into many areas.

Loggerhead Shrikes are permanent residents in Kentucky, although some seasonal movement seems to occur within the state (Mengel 1965). Territorial behavior begins during March, and nest building may be under way by late in the month. According to Mengel (1965), early clutches are completed during the first ten days of April, with a peak during the latter half of the month. Evidence suggests that at least some pairs raise two broods, and young from later nests may not fledge until late July (Stamm 1982b). The average size of 11 clutches or broods reported by Mengel (1965) was 4.9 (range of 4–6), but a brood of 7 young was observed in Trigg County in 1978 (Stamm 1978a).

Forest Cover

Value	% of Blocks	Avg Abund
All	31.5	1.6
1	49.5	1.7
2	54.2	1.6
3	30.9	1.4
4	3.2	1.0
5	—	—

Shrikes typically nest in a dense tree or shrub, often using thorny species if present. Although both deciduous and evergreen trees are used, red cedars appear to be favored in Kentucky, especially for early nests. Nest trees are often situated alone in a rural yard or pasture, but they also occur among other scattered trees in yards or along fencerows. The nest is typically well concealed within the thick cover of twigs and leaves.

The nest is a bulky structure, composed primarily of sticks and lined with fine grass, rootlets, feathers, or cottony material (Harrison 1975). The average height of 10 nests reported by Mengel (1965) was 9.4 feet (range of 5–12 feet), although at least one nest observed during atlas work was about 25 feet above the ground.

Ron Austing

The atlas survey yielded records of Loggerhead Shrikes in almost 32% of priority blocks, and 27 incidental observations were reported. As expected, occurrence was highest in the Mississippi Alluvial Plain and the East Gulf Coastal Plain, and decreased to the north and east. Only one report came from the Cumberland Plateau, and none from the Cumberland Mountains. Average abundance was relatively low across the state. Occurrence and average abundance were highest in open areas and decreased as the percentage of forest cover increased.

Thirty-one percent of priority block records were for confirmed breeding. Although at least 11 active nests were located, most confirmed records were based on the observation of recently fledged young. Additional confirmed records were based on adults carrying food for young, nest building, and distraction displays.

The decline in shrikes has been documented by BBS data in central and western Kentucky. The average number of individuals per BBS route for the periods 1966–91 and 1982–91 was 1.24 and 1.01, respectively. Although sample size is quite low, trend analysis of these data shows a highly significant ($p<.01$) decrease of 3.8% per year for the period 1966–91 and a significant ($p<0.1$) decrease of 6.6% per year for the period 1982–91. Pesticide use and intensive farming practices are most frequently implicated as the primary causes of this decline (Anderson and Duzan 1978; Robbins et al. 1986; Graber et al. 1973).

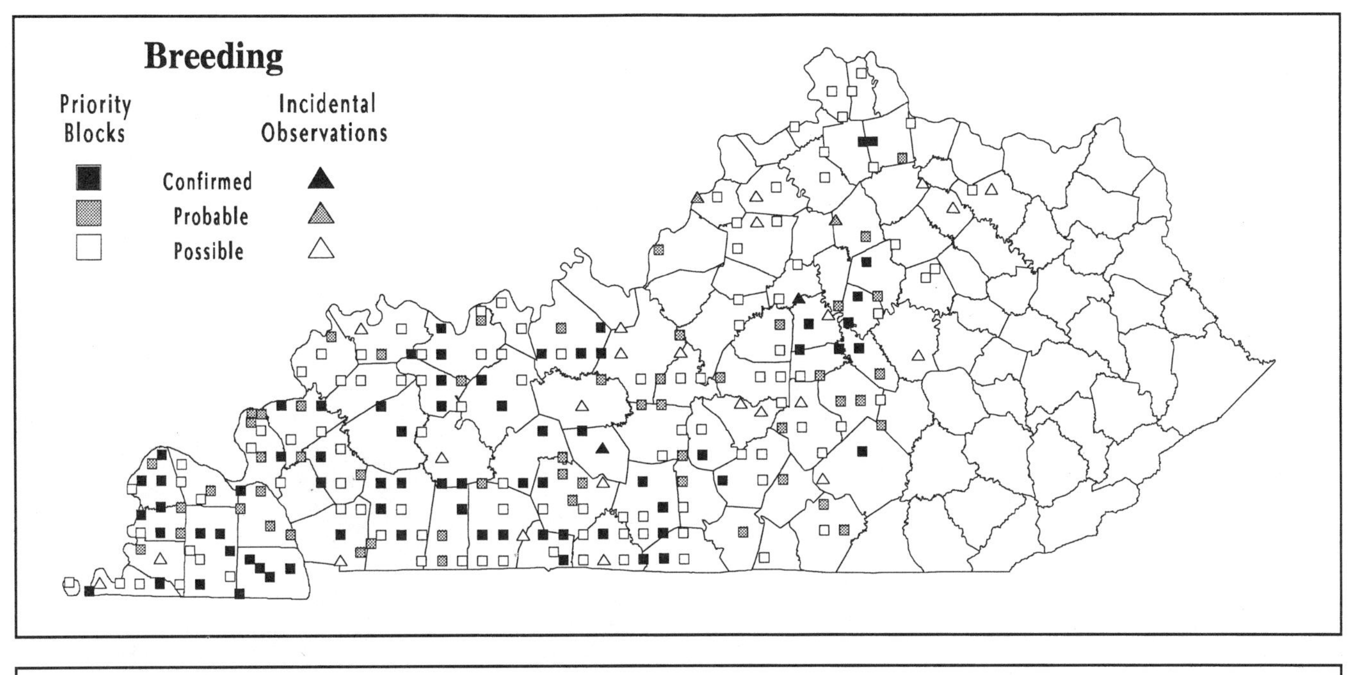

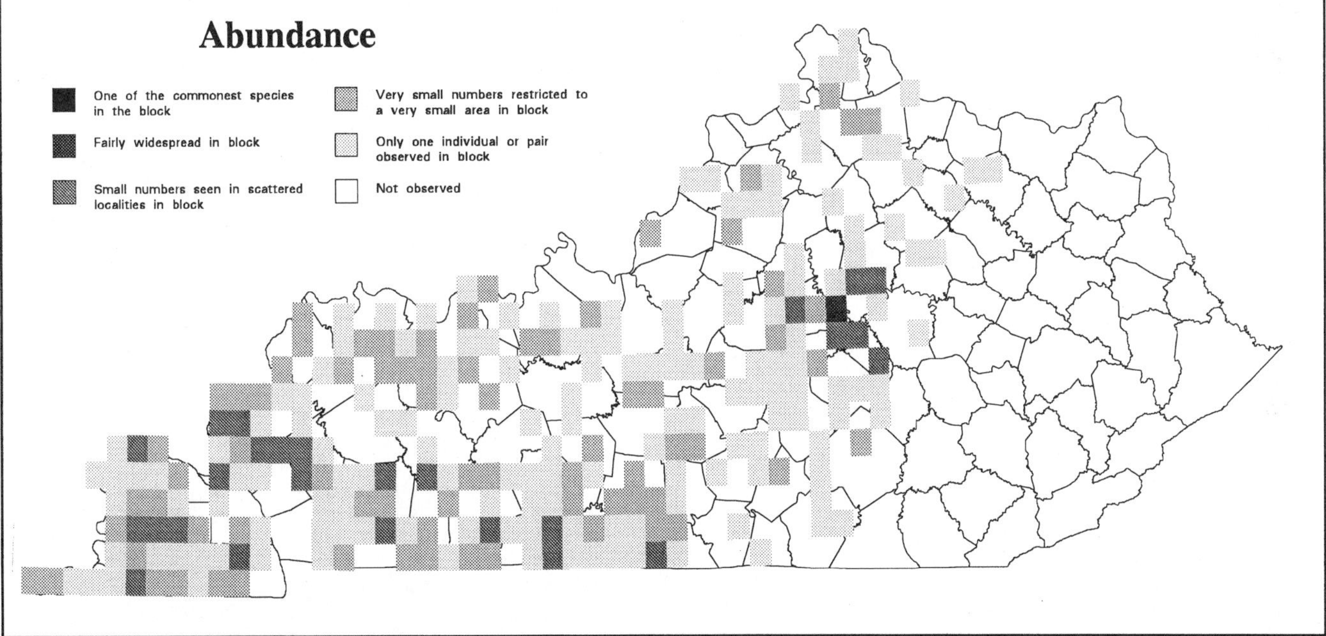

Analysis of Block Data by Physiographic Province Section

Physiographic Province Section	Total Blocks Surveyed	Blocks with Data	Avg Abund	% with Data	Section's % for State
Mississippi Alluvial Plain	14	11	1.4	78.6	4.8
East Gulf Coastal Plain	36	28	1.8	77.8	12.2
Highland Rim	139	73	1.5	52.5	31.9
Shawnee Hills	142	67	1.6	47.2	29.3
Blue Grass	204	49	1.5	24.0	21.4
Cumberland Plateau	173	1	1.0	0.6	0.4
Cumberland Mountains	19	-	-	-	-

Summary of Breeding Status

Number of Blocks in Which Species Was Recorded		
Total	**229**	**31.5%**
Confirmed	71	31.0%
Probable	54	23.6%
Possible	104	45.4%

Loggerhead Shrike

European Starling

Sturnus vulgaris

Introduced into North America in the late 1800s, the European Starling quickly colonized the eastern United States. It was apparently first reported in Kentucky in 1919 (Dodge 1951), and it became common in most areas by the mid-1930s (Lovell 1942). As of the late 1950s, Mengel (1965) regarded the species as a common to abundant resident, and it remains so today.

European Starlings are found primarily about settled areas, although they are reported almost anywhere that suitable foraging habitat and nest sites are available. The species typically avoids extensive areas of forest, but it does invade forest edge to use nest sites and frequently inhabits settled areas in predominantly forested areas. When foraging, the starling avoids thick vegetation, preferring bare soil or short, manicured vegetation, further limiting its occurrence to altered habitats.

Starlings are present throughout the year, but there appears to be much seasonal movement. Many birds from farther north join with local birds to form large roosting and foraging flocks in winter (Monroe and Cronholm 1977). As soon as winter weather begins to subside, the large flocks begin to break up, and territorial behavior commences. Nest building has been observed by late March, and the earliest egg date is March 30 (Hancock 1954). Active nests are not uncommon by the second week of April, and a peak in early clutch completion occurs during the middle of April (Twedt and Oddo 1984). Although Mengel (1965) noted no evidence of the species's being double-brooded, many pairs evidently raise two broods. A second peak in clutch completion occurs in late May (Twedt and Oddo 1984), and most late clutches are apparently completed by early June (Twedt and Oddo 1984). Noisy flocks of young remain conspicuous throughout the summer, foraging in farmland and other open habitats. The average size of eight clutches reported by Mengel (1965) was 4.4 (range of 3–6), while the average size of 97 clutches studied by Twedt and Oddo (1984) in 1981 and 1983 was 4.75 (range of 1–8).

Forest Cover

Value	% of Blocks	Avg Abund
All	83.2	3.3
1	95.0	3.9
2	96.8	3.6
3	89.1	3.2
4	65.6	2.5
5	36.5	1.8

Starlings are cavity nesters, using a great variety of natural and artificial sites, from old woodpecker holes and natural tree cavities to nest boxes, crevices in road cuts, and nooks and crannies in various human structures. The birds construct a bulky, untidy nest, filling the cavity with dead grass, weed stalks, trash, and other materials, and line it with fine grass and feathers. Nest height is variable, but most nests are situated 5–50 feet above the ground, with an average height of approximately 30 feet (Mengel 1965).

The atlas survey yielded records of starlings in more than 83% of priority blocks statewide, and 13 incidental observations were reported. The species ranked 28th according to the number of records in priority blocks, but 15th by total abundance and 6th by average abundance. While the starling is not as widespread as some other common species, it is quite abundant where it does occur. Occurrence was highest in central and western Kentucky, although the species was present in more than 60% of priority blocks even in the east. Average abundance was highest in the Blue Grass and lowest in the Cumberland Mountains, but higher than 2.0 statewide. Occurrence and average abundance were highest in areas with an abundance of cleared habitat and decreased substantially as forestation increased.

Maslowski Wildlife Productions

Two-thirds of priority block records were for confirmed breeding. This relatively high rate is attributable to a combination of the species's abundance in settled areas and its conspicuousness during nesting. Many active nests were reported, but most confirmed records were based on the observation of recently fledged young. Most other confirmed records were based on nest building and adults carrying food for young.

European Starlings are recorded in substantial numbers on most Kentucky BBS routes. The average number of individuals per BBS route for the periods 1966–91 and 1982–91 was 69.87 and 78.07, respectively. According to these data, the starling ranked 2nd in abundance on BBS routes during the period 1982–91. Although trend analysis of these data shows a significant ($p<0.1$) decrease of 1.3% per year for the period 1966–91, a slight increase of 0.4% per year for the period 1982–91 is not statistically significant. Robbins et al. (1986) suggest that the long-term decrease may have been caused by the severe winter weather of the mid- to late 1970s.

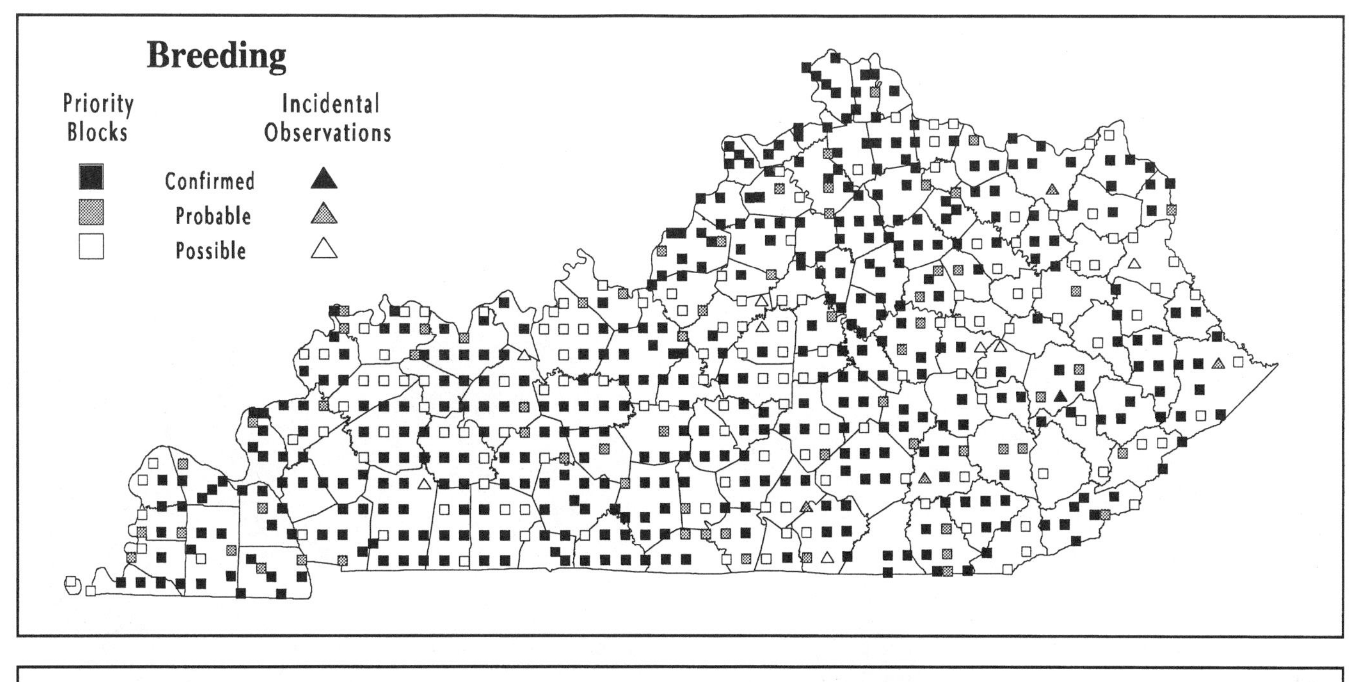

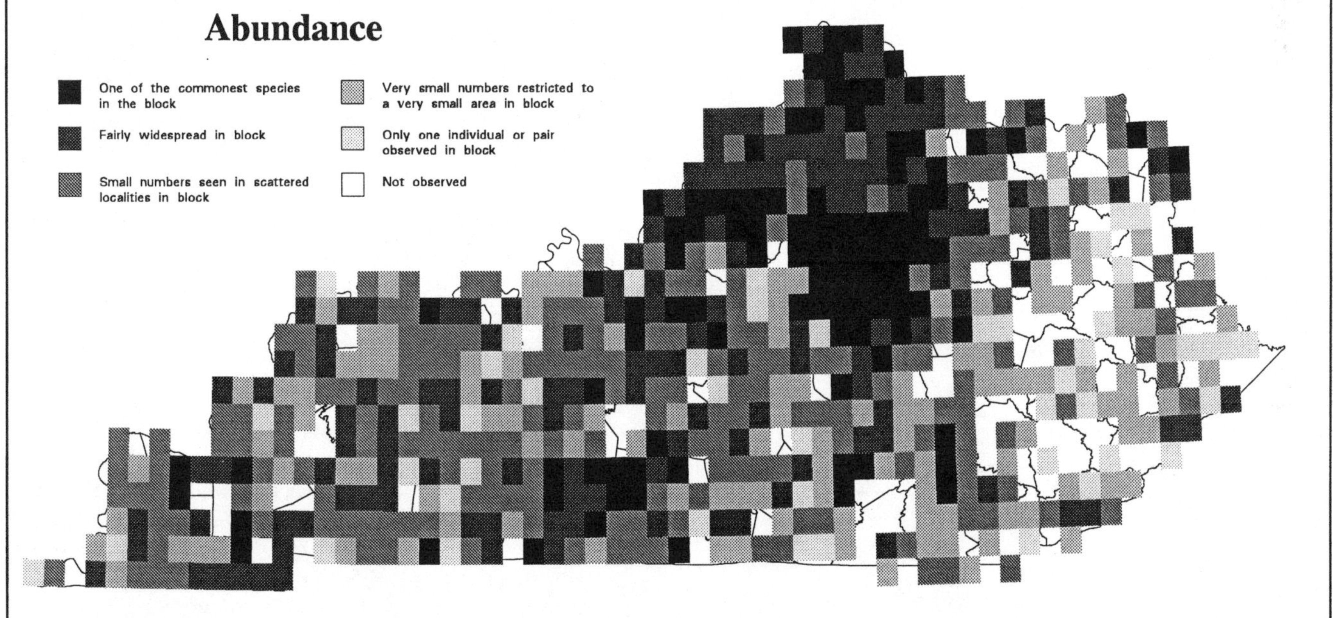

Analysis of Block Data by Physiographic Province Section

Physiographic Province Section	Total Blocks Surveyed	Blocks with Data	Avg Abund	% with Data	Section's % for State
Mississippi Alluvial Plain	14	12	2.6	85.7	2.0
East Gulf Coastal Plain	36	31	3.5	86.1	5.1
Highland Rim	139	128	3.3	92.1	21.2
Shawnee Hills	142	126	2.9	88.7	20.8
Blue Grass	204	188	3.9	92.2	31.1
Cumberland Plateau	173	107	2.5	61.8	17.7
Cumberland Mountains	19	13	2.4	68.4	2.1

Summary of Breeding Status

Number of Blocks in Which Species Was Recorded		
Total	**605**	**83.2%**
Confirmed	403	66.6%
Probable	64	10.6%
Possible	138	22.8%

White-eyed Vireo

Vireo griseus

The distinctive song of the White-eyed Vireo is a fairly conspicuous summer sound across most of Kentucky. Mengel (1965) regarded the species as a common summer resident in suitable habitat statewide except above 2,600 feet on Black Mountain, where he found it to be absent. Likewise, Davis et al. (1980) did not observe White-eyed Vireos high on Cumberland Mountain during the late 1970s, but Croft (1969) located several birds above 3,700 feet on Black Mountain in early June 1969.

The White-eyed Vireo is a bird of early successional habitats with an abundance of dense, shrubby cover. Today the species is encountered most often in altered habitats, including abandoned fields and overgrown drainages, regenerating clear-cuts or heavily logged areas, the brushy margins of woodlots, and reclaimed surface mines. This vireo is also found in several types of natural edge habitat, such as the margins of swampy areas, the edges of glades, and forest openings created by fire or storm damage.

It is likely that the White-eyed Vireo has been affected in several ways by human alteration of Kentucky's landscape. Large areas formerly containing suitable habitat have been cleared, and others previously maintained in an early successional condition by fire have become forested because of the suppression of wildfire. In contrast, the neglect of many cleared sites, as well as the fragmentation of intact forest tracts by logging, has served to create an abundance of prime habitat.

White-eyed Vireos winter from the southern United States south through Middle America (AOU 1983). A few birds return to Kentucky by the middle of April, but it may be late in the month before the full complement of breeding birds has returned. Nest building is usually under way by the middle of May, and early clutches are completed during the last two weeks of the month (Mengel 1965). No peak in clutch completion is evident, but a few clutches may be completed as late as early July (Mengel 1965). Although double-brooding has been reported in some parts of the species's range (Ehrlich et al. 1988; Peterjohn and Rice 1991), evidence of such in Kentucky is absent. Some late nestings likely represent efforts to renest following earlier failure because of predation or storms. The average size of seven clutches reported by Mengel (1965) was 3.3 (range of 2–4). According to Mengel, this species is among those most heavily parasitized by Brown-headed Cowbirds in Kentucky.

Forest Cover

Value	% of Blocks	Avg Abund
All	77.4	2.4
1	39.6	1.9
2	72.6	2.1
3	84.3	2.4
4	94.2	2.7
5	88.5	2.5

The nest is usually placed low in shrubby vegetation, most often near a woodland edge but sometimes in more open situations. It is typically suspended within a fork formed by two or more small, horizontal branches just inside the outer canopy of leaves of a deciduous shrub or small tree. The somewhat elongated nest is constructed of a variety of plant materials, including small pieces of soft wood and bark shreds held together with cobwebs; it is lined with fine plant stems and grass (Harrison 1975). The average height of six nests reported by Mengel (1965) was 2.4 feet (range of 1.5–3.5 feet), although Stamm and Croft (1968) included details of a nest in Daviess County that was constructed four feet above the ground.

Alvin E. Staffan

White-eyed Vireos were found on more than 77% of priority blocks statewide, and 14 incidental observations were reported. Occurrence was highest in the Cumberland Plateau, where the mixture of forested areas and brushy edges may be optimal. In contrast, occurrence was lowest in the Cumberland Mountains, where the transition to a more northern avifauna at high elevations involves a corresponding decrease in this southern species, and the Blue Grass, where suitable habitat is scarce owing to the extent of clearing. Average abundance was relatively low and unvaried across the state. Occurrence and average abundance were closely related to the percentage of forest cover, being lowest in extensively cleared areas and highest in predominantly wooded areas with some openings. This vireo's habit of singing regularly into late summer probably resulted in a few more records than were obtained for many other species that cease singing earlier in the season.

Despite the species's overall abundance, less than 9% of priority block records were for confirmed breeding. Although a few active nests were located, most confirmed records were based on the observation of recently fledged young. In addition, young cowbirds were observed being fed by White-eyed Vireos on at least one occasion. Other confirmed records were based on adults carrying food for young, distraction displays, and nest building.

White-eyed Vireos are typically reported on Kentucky BBS routes in small numbers. The average number of individuals recorded per BBS route for the periods 1966–91 and 1982–91 was 5.61 and 3.53, respectively. Trend analysis of these data reveals a significant ($p<0.1$) decrease of 1.7% per year for the period 1966–91, but a nonsignificant decrease of 1.4% per year for the period 1982–91.

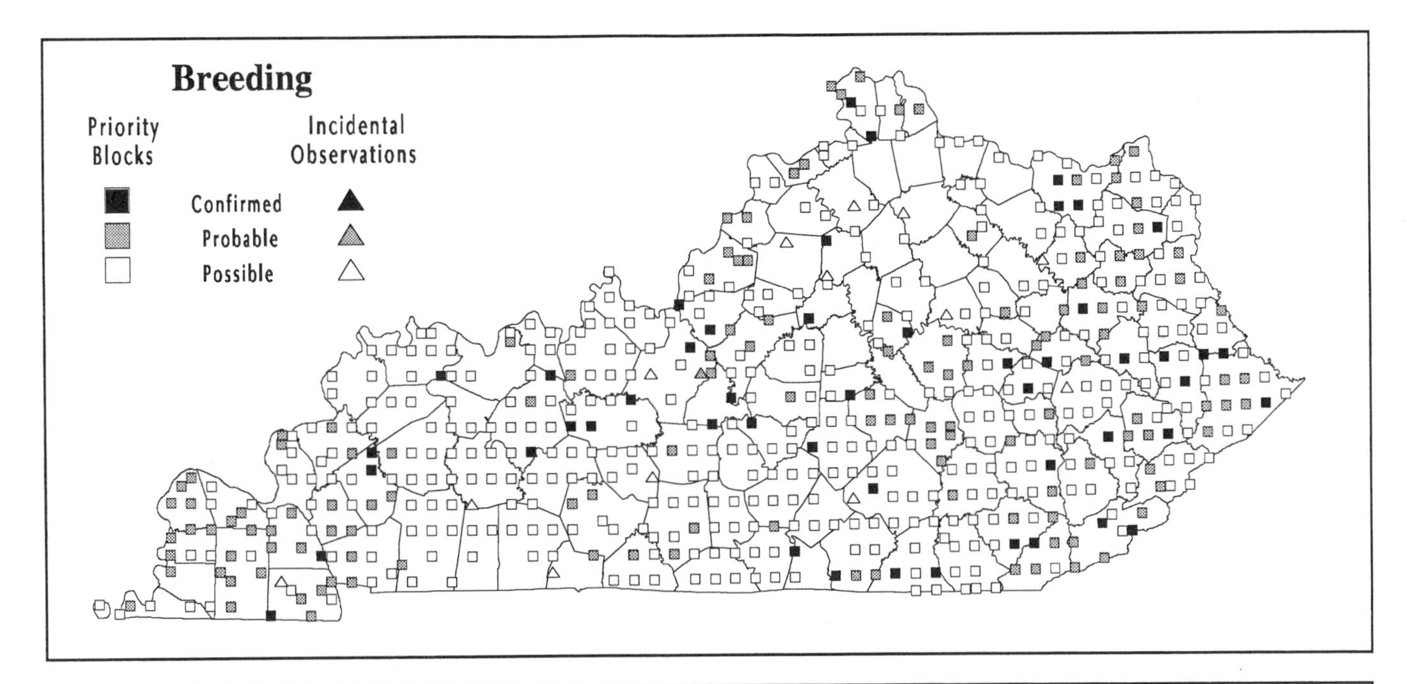

Analysis of Block Data by Physiographic Province Section

Physiographic Province Section	Total Blocks Surveyed	Blocks with Data	Avg Abund	% with Data	Section's % for State
Mississippi Alluvial Plain	14	11	2.6	78.6	2.0
East Gulf Coastal Plain	36	32	2.5	88.9	5.7
Highland Rim	139	109	2.3	78.4	19.4
Shawnee Hills	142	115	2.4	81.0	20.4
Blue Grass	204	119	2.2	58.3	21.1
Cumberland Plateau	173	164	2.6	94.8	29.1
Cumberland Mountains	19	13	2.2	68.4	2.3

Summary of Breeding Status

Number of Blocks in Which Species Was Recorded		
Total	**563**	**77.4%**
Confirmed	49	8.7%
Probable	134	23.8%
Possible	380	67.5%

Bell's Vireo

Vireo bellii

The Bell's Vireo is a recent addition to the list of Kentucky's breeding birds. Mengel (1965) considered its occurrence as hypothetical based on several sight records and one unverifiable specimen record. During the 1970s the species apparently began a fairly substantial expansion in nesting range into the Midwest that still appears to be under way today. Bell's Vireos now breed regularly in small numbers across southern Illinois, much of Indiana, and western Ohio (Bohlen 1978; Keller et al. 1986; Peterjohn and Rice 1991).

Subsequent to Mengel's work, little additional information was published concerning the species's occurrence in Kentucky before 1980, when several pairs were found nesting at the Tennessee Valley Authority's Shawnee Steam Plant (Nicholson 1981) and on the adjacent West Kentucky WMA (Palmer-Ball and Barron 1982), both in McCracken County. Following these initial observations, small numbers have continued to nest on the wildlife management area (Stamm 1984c), and the species was confirmed as breeding there during the atlas fieldwork.

Atlas fieldwork also yielded probable and confirmed records from three additional areas, as far east as Ohio County. In 1988 a population of at least a dozen territorial males was discovered on the U.S. Job Corps Center outside Morganfield, in Union County, although confirmed evidence of breeding was not obtained (Stamm 1988b). In 1989–90 the species was found to be fairly well distributed on recently reclaimed surface mines in Muhlenberg and Ohio counties. Small clusters of one to five singing males were found at four locations, and nesting was confirmed at two sites. Finally, in late June 1990 an active nest was located along a shrubby power line right-of-way near the eastern end of Cottonpatch Ridge, in rural Crittenden County. Possible breeding birds were also found at two places farther east. A singing male was heard for more than a week from late May to early June 1988, about one and a half miles west-northwest of Ekron, in Meade County, while another singing bird was observed on July 30, 1991, about one mile west of Crenshaw, in western Spencer County.

Like Bell's Vireos in other parts of the species's breeding range, locally nesting birds have been found in large tracts of early successional habitat dominated by deciduous shrubs and small trees. All such sites represent altered habitats that have been cleared and are in early stages of reforestation. In contrast to the White-eyed Vireo, Bell's Vireos seem to prefer more open situations, and none of the nesting pairs have been found along mature woodland edge.

Although a few records of Bell's Vireos date from mid-April, it is typically early May before most birds have returned to their local breeding grounds. Nest building has been observed as early as May 8 (no nest was observed, but a male was carrying nest material) and as late as June 27 (partially completed nest), but nearly or seemingly complete nests have also been seen on May 18, May 26, June 2, June 20–21, and June 26 (Nicholson 1981; Palmer-Ball and Barron 1982, 1990; KBBA data). Active nests have been reported from May 20 and May 26 (complete clutches being incubated), June 14 (large young), June 23 (small young), and July 15 (young only a few days old) (Palmer-Ball and Barron 1990; KBBA data). The latter nest also contained a nestling Brown-headed Cowbird. Only two complete clutches have been recorded in Kentucky, and both have contained four eggs (Palmer-Ball and Barron 1990). The species may be double-brooded in Kentucky, but documentation of this is lacking, and some late nests likely represent attempts to renest following earlier failure because of predation or storms.

Hal H. Harrison

The nest is typically placed just inside the outer crown of leaves of a small tree or shrub, and fairly low to the ground. Most nests have been located in young black locusts, but sumacs and indigo bush also have been used. The nest is constructed of a variety of plant materials, including strips of soft bark, and is lined with fine grass. It is invariably suspended from a small, horizontal fork of a low branch. The average height of seven nests reported in the literature and atlas notes is 2.5 feet (range of 2.0–3.5 feet).

The Bell's Vireo has experienced a dramatic decline throughout a portion of its range (Robbins et al. 1986), but there is no reason to expect that the species will not continue to colonize early successional habitats of western Kentucky. Recently reclaimed mines of the Shawnee Hills provide especially good habitat, although succession probably causes abandonment of nesting areas after several years.

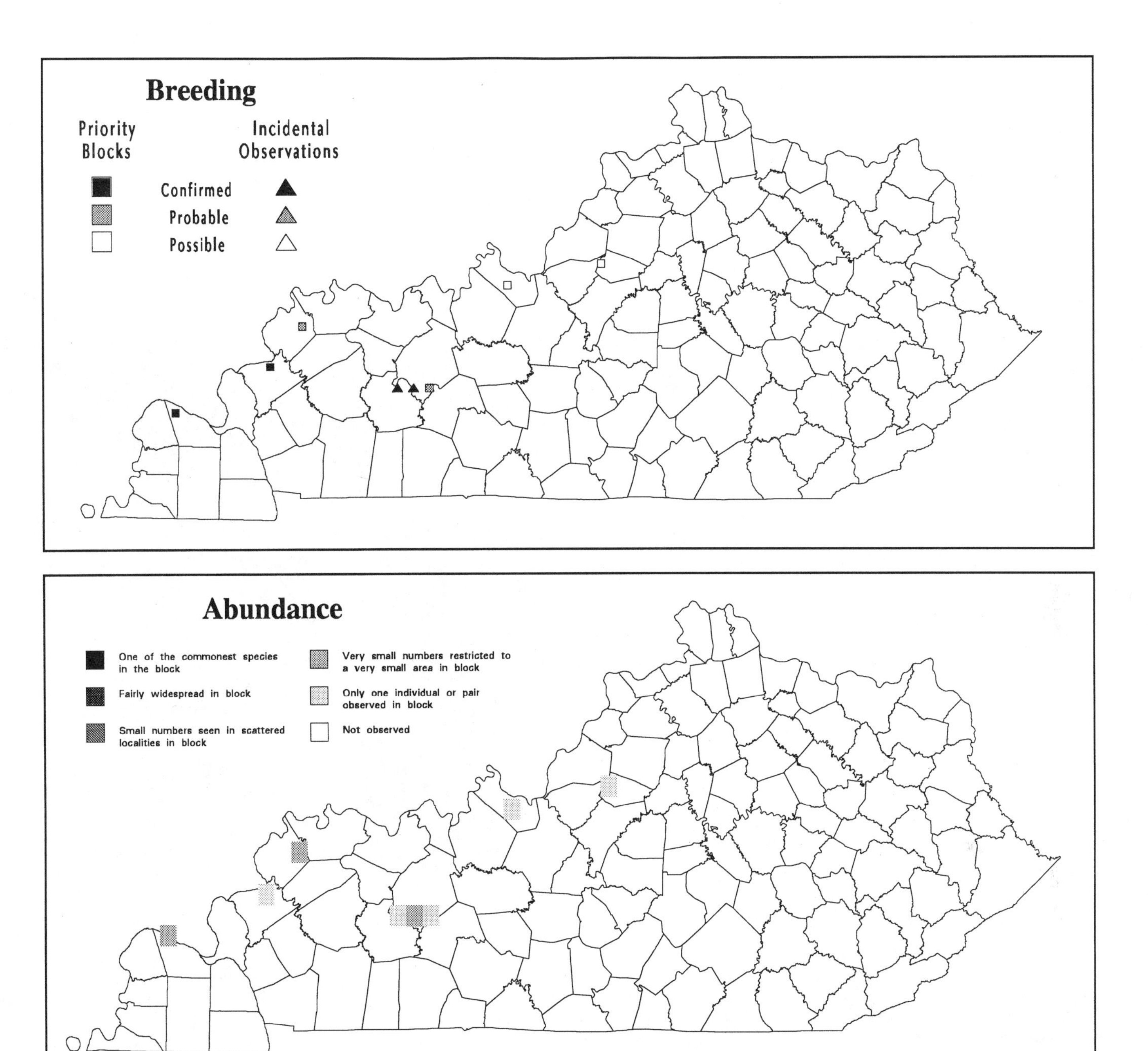

Analysis of Block Data by Physiographic Province Section

Physiographic Province Section	Total Blocks Surveyed	Blocks with Data	% with Data	Section's % for State
Mississippi Alluvial Plain	14	-	-	-
East Gulf Coastal Plain	36	1	2.8	16.7
Highland Rim	139	1	0.7	16.7
Shawnee Hills	142	3	2.1	50.0
Blue Grass	204	1	0.5	16.7
Cumberland Plateau	173	-	-	-
Cumberland Mountains	19	-	-	-

Summary of Breeding Status

Number of Blocks in Which Species Was Recorded		
Total	**6**	**0.8%**
Confirmed	2	33.3%
Probable	2	33.3%
Possible	2	33.3%

Solitary Vireo

Vireo solitarius

The Solitary Vireo's clear, whistled song is frequent in Kentucky in summer only at higher elevations in the Cumberland Mountains. Mengel (1965) reported the species as common on Black Mountain above 3,200 feet, but also scattered about on Pine Mountain at elevations of 1,800 to 3,000 feet. Croft (1969, 1971) added nearby Log and Cumberland mountains to the list of sites where the species regularly summered, and Davis et al. (1980) also reported it as fairly common at higher elevations on Cumberland Mountain. The atlas survey reaffirmed these observations, but it also resulted in the discovery of Solitary Vireos at more than a dozen sites on the Cumberland Plateau, extending the species's known summer range well beyond the Cumberland Mountains.

Solitary Vireos inhabit a variety of fairly mature to mature deciduous and mixed forest types, seeming to prefer situations with a relatively open midstory as well as forest edge. They are most frequently encountered in drier forests of ridges and slopes, but these vireos also occur in more mesic situations, especially at lower elevations in the Cumberland Mountains and on the Cumberland Plateau. Here the species has been found with greatest regularity in association with stands of hemlock or white pine in mixed forest types.

Solitary Vireos winter primarily from the Gulf Coast states southward through most of Central America (AOU 1983). A few birds return to their Kentucky breeding grounds during the last week of March, although it may be mid-April before most are on territory. Nest building may be under way by the end of April at lower elevations, but it is likely that retarded leaf development delays nesting until mid-May at higher elevations. Little has been published concerning the timing of nesting activities, especially at lower elevations, but the observation of a nest at Pine Mountain State Park on May 6, 1972, represents the earliest nesting date (KOS Nest Cards). Mengel (1965) observed a newly completed nest on Black Mountain on May 15, 1952, and Croft (1971) located a newly completed nest on Cumberland Mountain on May 22, 1970. A later nest, which contained three heavily incubated eggs on July 2–3, 1950, was observed by Mengel (1965) on Black Mountain. The only published record concerning clutch or brood size in Kentucky is for the above-noted clutch of three eggs, but Harrison (1975) gives rangewide clutch size as 3–5, commonly 4.

The nest is generally placed in a semi-open situation within the midstory level of the forest, and it is usually suspended from a small, horizontal fork among the outer leaves. The nest collected by Mengel (1965) was a shallow cup, loosely constructed, and beautifully decorated with lichens, spiderwebs, and strips of bark lining. The average height of four reported nests is 12.8 feet (range of 10-18 feet) (Mengel 1965; Croft 1971; KOS Nest Cards).

As expected, the atlas survey documented the continued presence of Solitary Vireos at higher elevations throughout the Cumberland Mountains. Somewhat surprisingly, the species was also found to occur regularly at scattered localities on the Cumberland Plateau, especially in the southern half of the western Cliff Section, and within the southeastern portion adjacent to the Cumberland Mountains. A probable report also came from the Red River Gorge area, in Wolfe County, in 1988. Because the atlas project constitutes the most comprehensive survey of nesting birds ever undertaken in the state, it is possible that small numbers of breeding birds have been previously overlooked on the Cumberland Plateau. In West Virginia the species has been known for some time to occur on the Cumberland Plateau, just to the east of Kentucky (Hall 1983). In contrast, an expansion of nesting range has occurred elsewhere in the region in recent years (Robbins et al. 1986; Peterjohn and Rice 1991), and it is possible that the discovery of birds in new areas could be associated with that trend.

Ron Austing

In addition to records from the Cumberland Plateau and the Cumberland Mountains, probable breeding birds were reported from two locations in the Highland Rim. Just west of the Cumberland Plateau, summering birds were observed in mature deciduous forest of eastern Lincoln County, and a pair of birds was observed twice and nearly a month apart (June 19 and July 15, 1990) in Pennyrile State Forest, in Christian County, well west of the species's normal summer range. Further evidence of nesting could not be obtained at the latter site, but the singing male and his mate responded to pishing as if they were territorial, and it can only be assumed that nesting was attempted at this locality. The birds' territory seemed to include a large stand of white pines in otherwise mixed, subxeric forest.

Of the 22 records in priority blocks, five were for confirmed breeding. All five were based on the observation of recently fledged young. Probable records were based on the observation of pairs of birds or the presence of a single bird over an extended period of time.

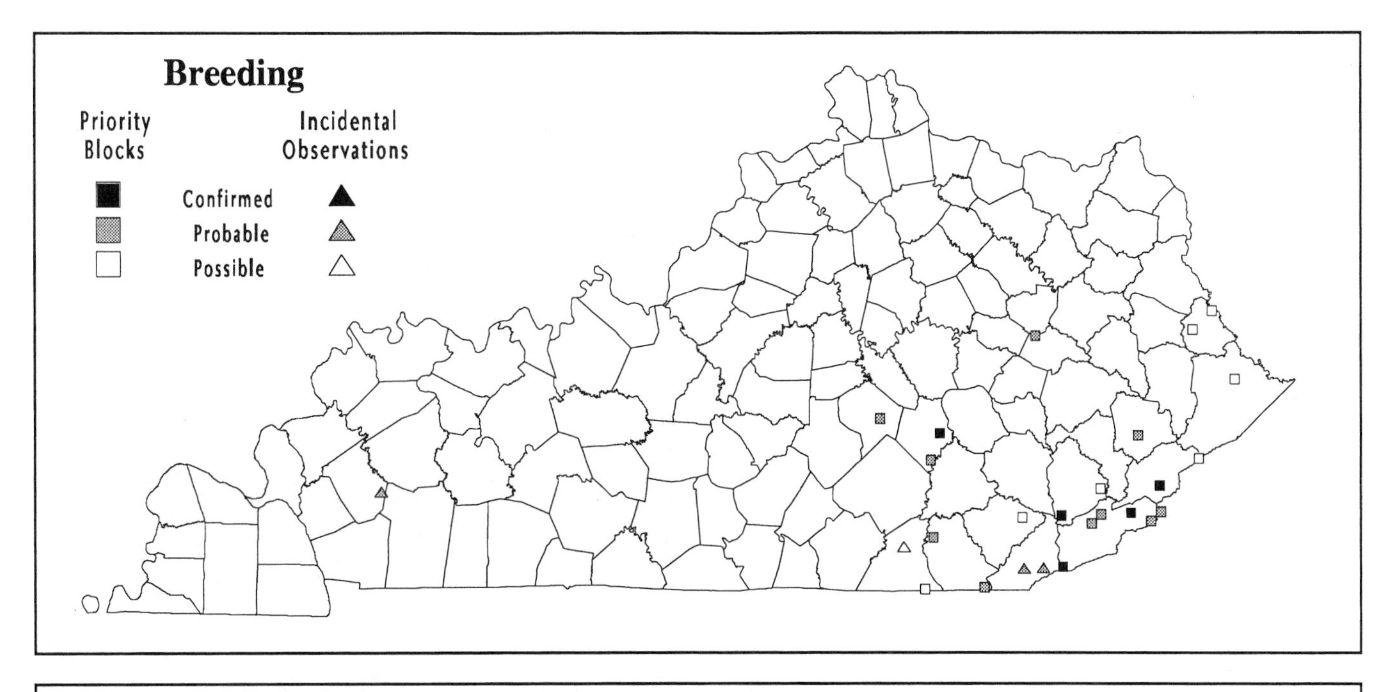

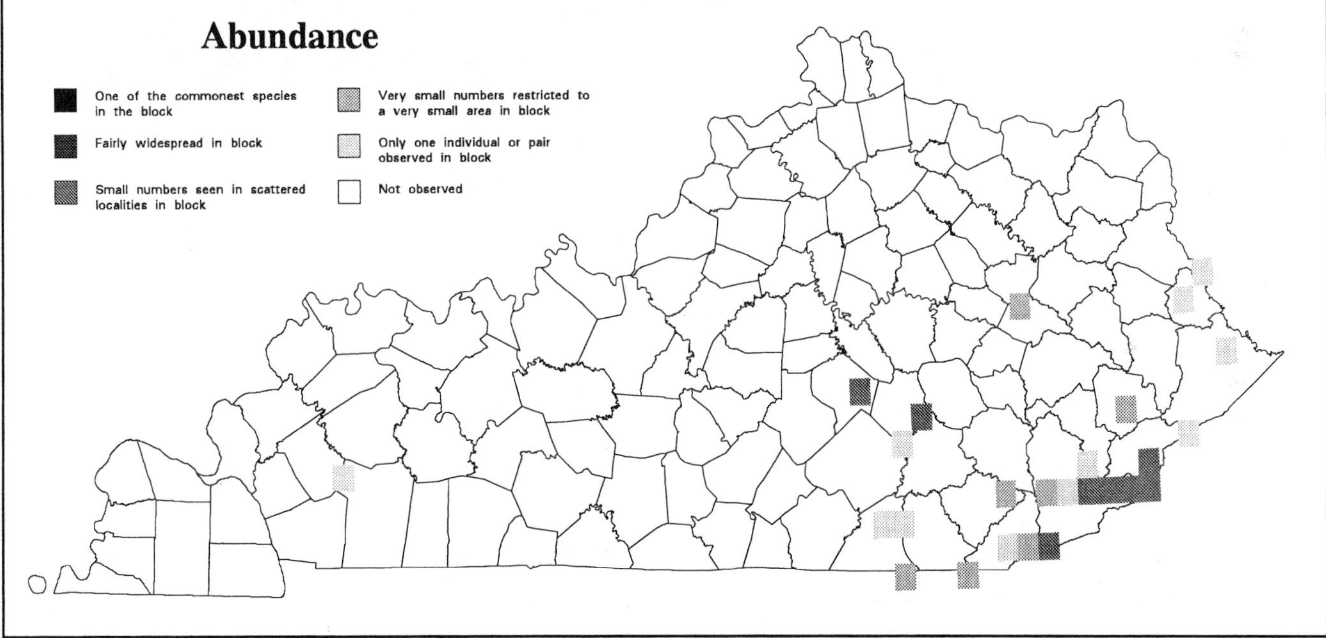

Analysis of Block Data by Physiographic Province Section

Physiographic Province Section	Total Blocks Surveyed	Blocks with Data	% with Data	Section's % for State
Mississippi Alluvial Plain	14	-	-	-
East Gulf Coastal Plain	.36	-	-	-
Highland Rim	139	1	0.7	4.5
Shawnee Hills	142	-	-	-
Blue Grass	204	-	-	-
Cumberland Plateau	173	14	8.1	63.6
Cumberland Mountains	19	7	36.8	31.8

Summary of Breeding Status

Number of Blocks in Which Species Was Recorded		
Total	**22**	**3.0%**
Confirmed	5	22.7%
Probable	10	45.5%
Possible	7	31.8%

Solitary Vireo

Yellow-throated Vireo

Vireo flavifrons

The Yellow-throated Vireo is irregularly distributed throughout most of Kentucky in summer. Mengel (1965) regarded the species as a variably rare to common summer resident across the state, being most common on Pine Mountain, through most of the Cumberland Plateau, in parts of the Shawnee Hills, and in the Mississippi Alluvial Plain. In contrast, he considered it to be slightly less numerous in most of the Knobs subsection of the Blue Grass and the western Shawnee Hills, and decidedly less numerous in portions of the Highland Rim and the remainder of the Blue Grass. In general, its distribution is closely related to the amount of mature forest; however, this vireo is less numerous at the predominantly forested higher elevations in the Cumberland Mountains, perhaps because of competition with the Solitary Vireo (Mengel 1965; Davis et al. 1980).

The Yellow-throated Vireo is encountered in a variety of forested habitats, especially those with a fairly open midstory. While the species is typically an inhabitant of fairly mature woodlands, it seems that a certain amount of natural or induced disturbance (e.g., from fire or selective logging) does not affect numbers dramatically. On the other hand, this vireo appears to avoid small, isolated tracts of mature forest as well as highly dissected and younger forest tracts. Although Yellow-throated Vireos seem to be most frequent in various upland forests dominated by oaks and hickories (including those shared with pines), they are also fairly widespread in more mesic slope forests, as well as bottomland forests in the western part of the state.

The Yellow-throated Vireo is certainly less numerous in Kentucky today than before settlement. The widespread clearing and dissection of the once expansive forests that occurred across much of the state surely have resulted in a dramatic loss of suitable habitat. While fragmentation and selective logging of some forests may have replaced natural forest openings used by this vireo, the high frequency of logging in many forests has resulted in the exclusion of the species because of alteration of the forest structure to types with more closed-in midstory levels.

Forest Cover

Value	% of Blocks	Avg Abund
All	54.7	1.9
1	12.9	1.2
2	43.2	1.6
3	63.0	1.9
4	76.0	2.0
5	78.8	2.1

Yellow-throated Vireos typically return to Kentucky from their wintering grounds in the tropics by the third week of April, although it may be the end of the month before the entire nesting population has arrived. Few specific details on nesting activity in Kentucky have been published, but it is apparently not uncommon for nest building to be under way by the first week of May. Mengel (1965) notes that clutches are completed from early May to early June, with no clear peak. Kentucky data on clutch size are absent, but Harrison (1975) gives rangewide clutch size as 3–5, commonly 4.

Yellow-throated Vireos usually place their nest within the midstory level of fairly mature to mature forest. The nest is typically constructed in a fairly open setting, often along a roadway corridor or other natural or artificial opening. The nest is a compact cup, normally suspended from a slender fork and constructed of grasses, strips of inner bark, and other soft plant material, decorated on the outside with moss and lichens, and lined with fine grass (Harrison 1975). The average height of three nests reported by Mengel (1965) was 21.7 feet (range of 10–40 feet).

Hal H. Harrison

Yellow-throated Vireos were recorded on nearly 55% of priority blocks statewide, and 29 incidental observations were reported. Occurrence was highest in the Cumberland Plateau and Cumberland Mountains, and lowest in East Gulf Coastal Plain and the Blue Grass. Average abundance was relatively uniform across the state. Although this vireo may be less numerous in the Cumberland Mountains because of the transition to a more northern avifauna, variation in occurrence across the remainder of the state is probably most closely related to the amount of fairly mature to mature forest present.

Despite the Yellow-throated Vireo's general abundance, less than 2% of priority block records were for confirmed breeding. Most of the seven confirmed records were based on the observation of recently fledged young, but an active nest containing young was found in Carroll County. Another confirmed record was based on the observation of an adult carrying food.

Yellow-throated Vireos are typically reported on Kentucky BBS routes in very small numbers. The average number of individuals reported per BBS route for the periods 1966–91 and 1982–91 was 1.59 and 1.09, respectively. Trend analysis of these data reveals a nonsignificant increase of 0.3% per year for the period 1966–91 and a nonsignificant decrease of 3.0% per year for the period 1982–91.

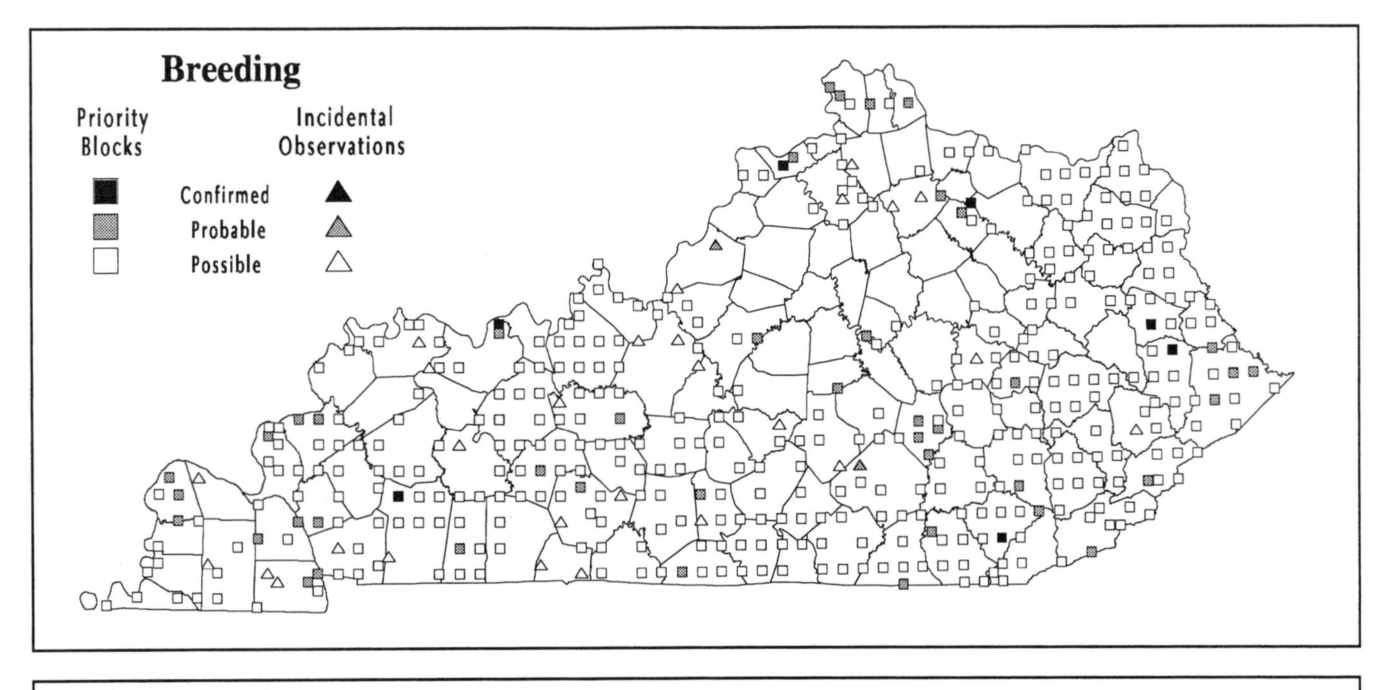

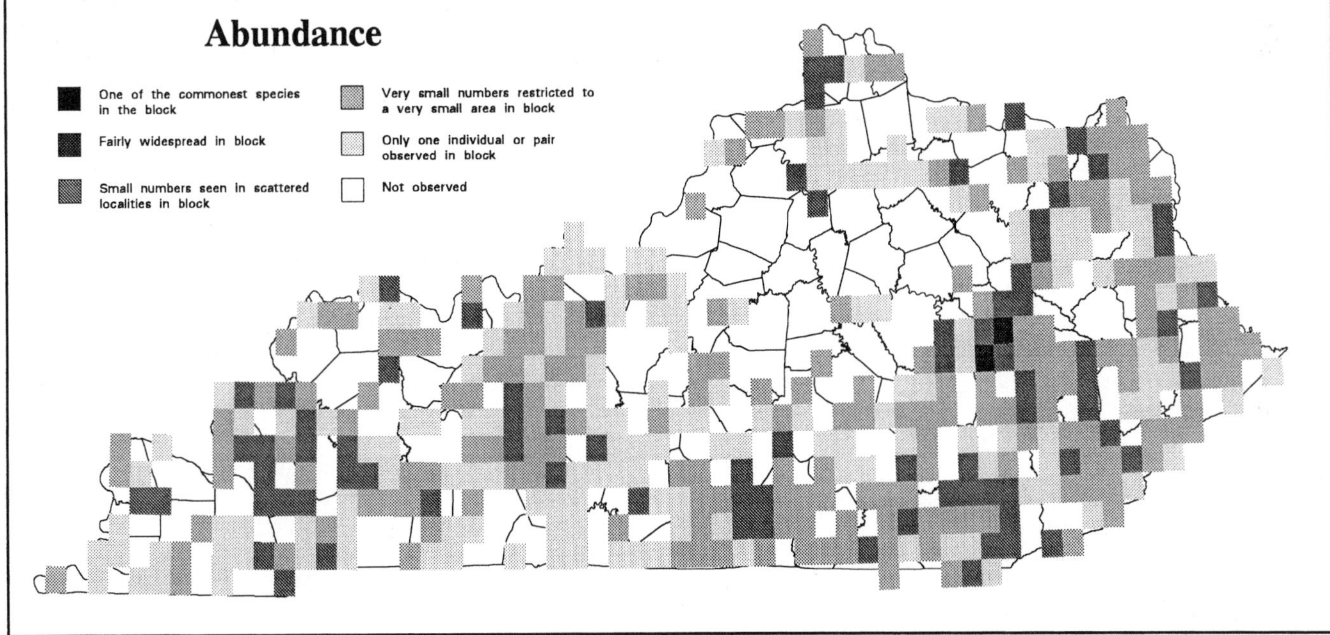

Analysis of Block Data by Physiographic Province Section

Physiographic Province Section	Total Blocks Surveyed	Blocks with Data	Avg Abund	% with Data	Section's % for State
Mississippi Alluvial Plain	14	8	1.6	57.1	2.0
East Gulf Coastal Plain	36	15	1.9	41.7	3.8
Highland Rim	139	78	1.7	56.1	19.6
Shawnee Hills	142	91	1.8	64.1	22.9
Blue Grass	204	61	1.8	29.9	15.3
Cumberland Plateau	173	131	2.0	75.7	32.9
Cumberland Mountains	19	14	1.9	73.7	3.5

Summary of Breeding Status

Number of Blocks in Which Species Was Recorded		
Total	**398**	**54.7%**
Confirmed	7	1.8%
Probable	43	10.8%
Possible	348	87.4%

Yellow-throated Vireo

Warbling Vireo

Vireo gilvus

The Warbling Vireo's distinctive song is a locally uncommon to fairly common summer sound throughout much of the western two-thirds of Kentucky. In contrast, the species is absent throughout most of the Cumberland Plateau and Cumberland Mountains, where suitable nesting habitat is scarce. Mengel (1965) regarded this vireo as a fairly common summer resident west of the Cumberland Plateau but considered it rare in eastern Kentucky.

The Warbling Vireo is a bird of semi-open and open habitats with scattered large trees, and it is rarely encountered in extensively forested areas. Consequently, in eastern Kentucky the species is generally found only in cleared bottomland situations. Warbling Vireos are sometimes found in naturally occurring habitats, such as riparian zones along larger rivers, but they primarily occur in altered situations, including rural farmland, parks, cemeteries, lakeshores, and other openings amid settlement.

It is possible that the Warbling Vireo is more common in Kentucky today than before European settlement. Audubon (1861) did not observe the species in the early 1800s, but an abundance of suitable nesting habitat must have occurred in the barrens and savannas of the western and central parts of the state. Today most of these areas have been cleared or otherwise altered, but the rural farmland replacing them still offers an abundance of suitable habitat. Furthermore, the dissection of large expanses of native forest has created an abundance of habitat where formerly the species was likely absent.

Warbling Vireos winter primarily in Middle America (AOU 1983), and most birds return to their Kentucky breeding grounds by the last week of April. Territorial behavior seems to begin immediately, and nests are often under construction by the beginning of May, especially in western Kentucky. Specific details of nesting activity are scarce, but Mengel (1965) reports that most clutches are probably completed from early May to early June. Singing becomes much less frequent after the end of June, and nesting birds seem to disappear by mid-August. Kentucky data on clutch size are absent except for two records by A. Stamm, of three and four young in nests in Jefferson County (Mengel 1965). Harrison (1975) gives rangewide clutch size as 3–5, commonly 4.

Forest Cover

Value	% of Blocks	Avg Abund
All	29.0	1.8
1	44.6	1.8
2	45.8	1.8
3	29.6	1.7
4	6.5	1.8
5	1.9	1.0

The nest is generally placed in a solitary tree, or along the outer edge of a grove of trees in an otherwise open to semi-open situation. It is typically suspended in a slender fork among the outer leaves of a branch, and far out from the trunk. Deciduous trees seem to be used exclusively in Kentucky. The nest is a compact cup constructed of strips of bark, leaves, grasses, feathers, and plant down, held together with spider silk, and lined with fine plant stems and hair (Harrison 1975). The average height of seven nests summarized by Mengel (1965) was 23.6 feet (range of 12–35 feet), although two more recent nests have been estimated to be 50 feet above the ground (Croft and Stamm 1967; KOS Nest Cards).

Ron Austing

Warbling Vireos were recorded on 29% of priority blocks statewide, and 29 incidental observations were reported. Occurrence by physiographic province section varied dramatically. The species was found in more than half of priority blocks in the Mississippi Alluvial Plain and the East Gulf Coastal Plain, and it was reported in at least 33% of priority blocks throughout the remainder of central and western Kentucky. In contrast, it was recorded in less than 3% of priority blocks on the Cumberland Plateau and was missed entirely in the Cumberland Mountains. Average abundance varied similarly. Differences in regional occurrence were probably related most closely to the amount of suitable habitat, which is inversely related to the percentage of forest cover.

Less than 7% of priority block records were for confirmed breeding. Although a few of the 14 confirmed records were based on the observation of active nests, most involved adults carrying food for young, nest building, and observations of recently fledged young being accompanied or fed by parents.

Warbling Vireos are reported in small numbers on central and western Kentucky BBS routes. The average number of individuals reported per BBS route for the periods 1966–91 and 1982–91 was 1.29 and 2.07, respectively. Trend analysis of these data shows a nonsignificant increase of 1.9% per year for the period 1966–91 and a nonsignificant decrease of 2.8% per year for the period 1982–91.

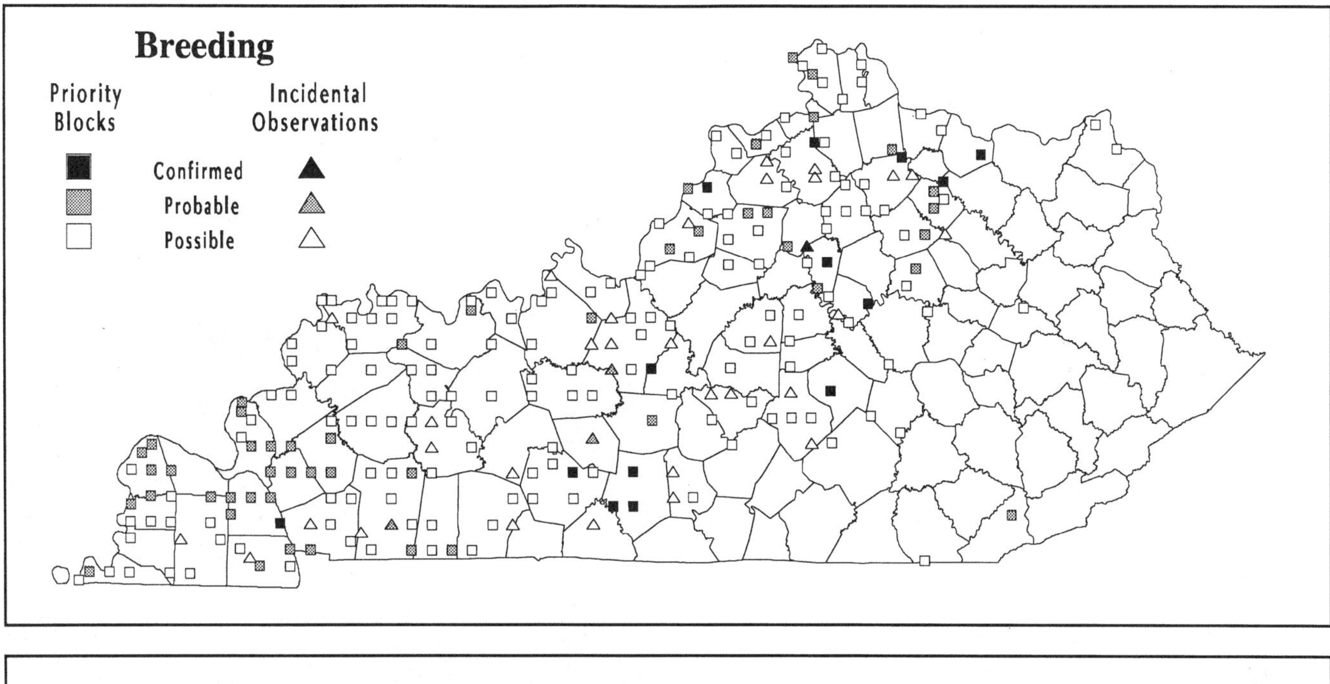

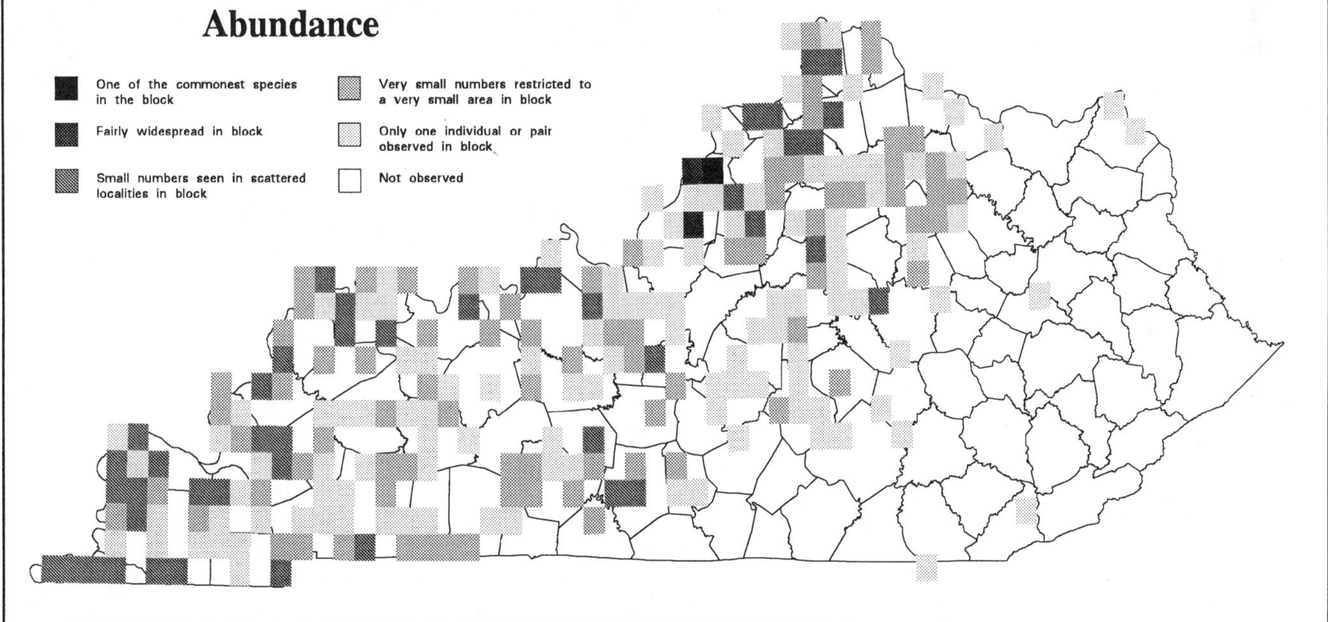

Analysis of Block Data by Physiographic Province Section

Physiographic Province Section	Total Blocks Surveyed	Blocks with Data	Avg Abund	% with Data	Section's % for State
Mississippi Alluvial Plain	14	10	2.5	71.4	4.7
East Gulf Coastal Plain	36	20	2.0	55.6	9.5
Highland Rim	139	46	1.7	33.1	21.8
Shawnee Hills	142	58	1.8	40.8	27.5
Blue Grass	204	72	1.7	35.3	34.1
Cumberland Plateau	173	5	1.0	2.9	2.4
Cumberland Mountains	19	-	-	-	-

Summary of Breeding Status

Number of Blocks in Which Species Was Recorded		
Total	**211**	**29.0%**
Confirmed	14	6.6%
Probable	49	23.2%
Possible	148	70.1%

Red-eyed Vireo

Vireo olivaceus

The Red-eyed Vireo's monotonous song is a common summer sound in forests and woodlots across Kentucky. While the species is most common in the extensive woodlands of eastern Kentucky, it is also quite frequent in forested habitats in other regions. Mengel (1965) regarded this vireo as a common summer resident statewide and (along with the Wood Thrush) perhaps one of the two most typical birds of the eastern deciduous forest.

The Red-eyed Vireo is probably most abundant statewide in mesic slope forests, but it also occurs frequently in subxeric deciduous and mixed forest types in the Cumberland Plateau, as well as in bottomland forests of western Kentucky. Unlike less adaptable forest-nesting species, this vireo seems to use smaller tracts of forest in more highly dissected areas, as well as tracts of younger forest regenerating from clear-cutting or selective logging activity.

Even though it remains numerous, the Red-eyed Vireo must be much less common in Kentucky now than it was two centuries ago. The clearing of vast areas of forest for agricultural use and settlement has resulted in a significant reduction in the amount of available nesting habitat. Furthermore, the dissection of remaining forest has likely resulted in an increase in cowbird parasitism, further reducing nesting success and overall abundance (Brittingham and Temple 1983).

Red-eyed Vireos winter in South America (AOU 1983), and many birds return to Kentucky by the end of April. Territorial behavior seems to commence soon thereafter, and nest building may be under way by the first week of May. According to Mengel (1965), early clutches are completed during the first ten days of May, although a marked peak in clutch completion occurs during the last ten days of the month. The raising of a second brood has not been documented in Kentucky, but data from Ohio suggest that at least some late clutches laid into the last week or so of June represent double-brooding (Peterjohn and Rice 1991). Kentucky data on clutch size are scarce, but Mengel (1965) gives the average of eight clutches as 2.6 (range of 2–3). Harrison (1975) gives rangewide clutch size as 2–4, commonly 4. According to Mengel (1965), this species is among those most heavily parasitized by Brown-headed Cowbirds in Kentucky.

Forest Cover

Value	% of Blocks	Avg Abund
All	86.4	2.9
1	59.4	1.8
2	78.9	2.3
3	92.2	2.7
4	100.0	3.7
5	100.0	3.9

Red-eyed Vireos usually place their nest within the midstory of fairly mature to mature forest. They use a variety of situations, however, and nests are often found over logging roads and stream corridors, and sometimes along forest edge. Although the species often breeds in mixed pine-hardwood forests, nests appear to be placed exclusively in deciduous trees in Kentucky. The nest is usually suspended in a small fork well out from the trunk, but usually not at the very tip of a branch. It is neatly constructed of soft strips of bark, plant stems, grass, and other plant materials, adorned with lichens or other soft, downy plant material, and lined with fine grass (Harrison 1975). The average height of 18 nests reported by Mengel (1965) was 13.3 feet (range of 5–40 feet), although a few more recent nests have been reported to be as low as 2–4 feet above the ground (KOS Nest Cards).

Alvin E. Staffan

The Red-eyed Vireo was recorded on more than 86% of priority blocks statewide, and 14 incidental observations were reported. The species ranked 22nd according to the number of priority block records, 22nd by total abundance, and 18th by average abundance. Occurrence was highest in the Cumberland Mountains and the Cumberland Plateau, but greater than 70% throughout the state. Average abundance varied similarly. Both occurrence and average abundance were closely related to the percentage of forest cover, being lowest in predominantly cleared areas and increasing substantially as forestation increased.

Despite this vireo's general abundance, less than 10% of priority block records were for confirmed breeding. Although about a dozen active nests were located, most of the 61 confirmed records were based on the observation of recently fledged young. In addition, young cowbirds were observed being fed by Red-eyed Vireos on several occasions. Other confirmed records were based on the observation of adults carrying food, nest building, and distraction displays.

Red-eyed Vireos are typically reported in small to moderate numbers on most Kentucky BBS routes. The average number of individuals per BBS route for the periods 1966–91 and 1982–91 was 8.53 and 8.50, respectively. According to these data, the species ranked 29th in abundance on BBS routes during the period 1982–91. Trend analysis of these data yields slight, though nonsignificant, increases of 0.9% per year for the period 1966–91 and 0.1% per year for the period 1982–91.

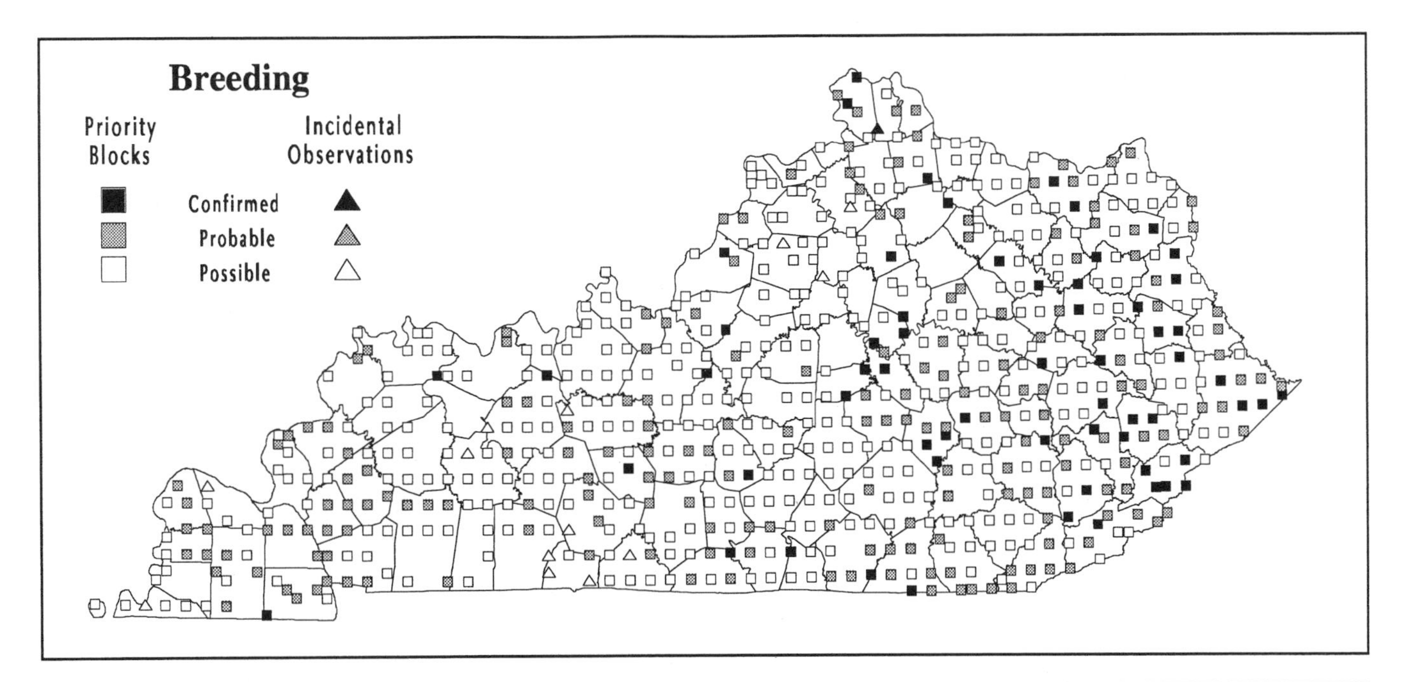

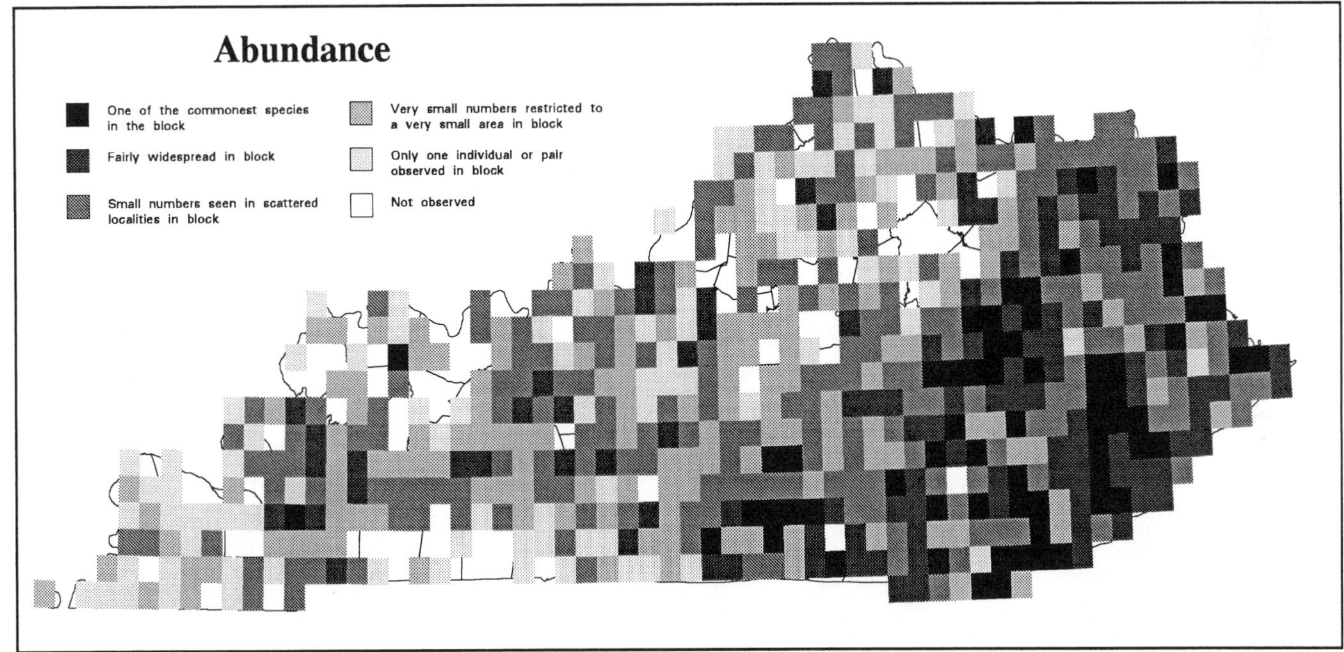

Analysis of Block Data by Physiographic Province Section

Physiographic Province Section	Total Blocks Surveyed	Blocks with Data	Avg Abund	% with Data	Section's % for State
Mississippi Alluvial Plain	14	10	1.6	71.4	1.6
East Gulf Coastal Plain	36	27	1.8	75.0	4.3
Highland Rim	139	120	2.6	86.3	19.1
Shawnee Hills	142	113	2.5	79.6	18.0
Blue Grass	204	167	2.6	81.9	26.6
Cumberland Plateau	173	172	3.7	99.4	27.4
Cumberland Mountains	19	19	3.9	100.0	3.0

Summary of Breeding Status

Number of Blocks in Which Species Was Recorded		
Total	**628**	**86.4%**
Confirmed	61	9.7%
Probable	192	30.6%
Possible	375	59.7%

Red-eyed Vireo

Blue-winged Warbler

Vermivora pinus

The Blue-winged Warbler is a relatively inconspicuous and locally distributed breeding bird across much of Kentucky. In central and western parts of the state it occurs sporadically, and it is nowhere especially numerous, but in the predominantly forested Cumberland Plateau it is fairly common and much more widely distributed. Mengel (1965) regarded the species as variably rare to fairly common throughout the state. Interestingly, none of several authors list any records for the species during fieldwork conducted in the Cumberland Mountains (see Mengel 1965; Croft 1969; Davis et al. 1980).

Blue-winged Warblers are typically found in early successional habitats. They are most frequent in low, moist areas, but they also can be found on drier slopes. The species formerly used natural forest openings and woodland borders, which may have been rather common because of fire and storm damage. Today, altered situations, including overgrown fields, reclaimed strip mines, and regenerating forest clear-cuts, provide most nesting habitat. This warbler does not occupy these areas unless a good scattering of small trees, shrubs, and dense herbaceous growth is present. In contrast, the species is typically absent if the ground cover is reduced by grazing or shading from a closed tree canopy.

While Blue-winged Warblers have adapted to some of the changes humans have brought to Kentucky's landscape, the species may have been more widespread before European settlement. In the early 1800s Audubon (1861) noted that it was frequent in the barrens, and suitable habitat was likely scattered across much of the state. As vast areas were converted to intensive agricultural use and settlement, numbers likely decreased in some regions. Conversely, the neglect of some cleared areas and maintenance of artificial edges have certainly resulted in the presence of much suitable habitat, especially in the eastern part of the state.

Blue-winged Warblers winter throughout much of Middle America (AOU 1983), and most birds typically return to their Kentucky breeding grounds during the last two weeks of April. Mengel (1965) listed only one record of an active nest for May 1938 in Rowan County, and more recent records suggest that most clutches are completed during the latter half of May (Mason 1978; KOS Nest Cards; K. Caminiti, pers. comm.). Family groups are fairly conspicuous during late June, and birds begin to turn up in nonbreeding areas in mid-July. Kentucky data on clutch size are scarce, but the average size of four reported clutches or broods is 4.5 (range of 4-5) (Mason 1978; KOS Nest Cards; K. Caminiti, pers. comm.). Harrison (1975) gives rangewide clutch size as 4–5, rarely 6.

Forest Cover

Value	% of Blocks	Avg Abund
All	24.3	1.9
1	2.0	1.5
2	5.8	1.3
3	22.6	1.7
4	54.5	2.0
5	53.8	2.0

Blue-winged Warblers nest on or near the ground. The nest is concealed among weeds and grasses, usually near the base of a small tree or shrub that is situated along a wooded edge. It is constructed of coarse grasses, dead leaves, and shreds of bark and is lined with finer grasses, shreds of bark, and hair (Harrison 1975).

Kathy Caminiti

The atlas survey yielded records of Blue-winged Warblers in about 24% of priority blocks statewide, and 11 incidental observations were reported. There were three Cumberland Plateau reports of hybrid Blue-winged × Golden-winged Warblers, and details of these can be found in the Nonbreeding Species section (pp. 327–28). Occurrence was substantially higher in the Cumberland Plateau than in any other physiographic province section; it was lower than 20% throughout central and western Kentucky. The species was not found in the Cumberland Mountains or the Mississippi Alluvial Plain. Average abundance was low across the state but varied similarly. Blue-winged Warblers are more common in predominantly forested areas than in extensively cleared land. This is especially noticeable in central and western Kentucky, where good numbers occupy suitable habitat in largely forested areas such as Mammoth Cave National Park and Land Between the Lakes. In contrast, Blue-wingeds are virtually absent in surrounding areas, where forests have been mostly cleared. The species's absence in the Cumberland Mountains likely is the result of the transition to a more northern avifauna, with the resultant loss of southern species like the Blue-winged Warbler.

More than 28% of priority block records were for confirmed breeding. Although one active nest was located, most confirmed records were based on the observation of recently fledged young being attended to by parents. Begging young were conspicuous because of their buzzy calls, and many family groups were detected by following these characteristic notes. The remaining confirmed records were based on adults carrying food.

Blue-winged Warblers do not turn up often on Kentucky BBS routes. The average number of individuals per BBS route for the periods 1966–91 and 1982–91 was 0.46 and 0.42, respectively. In part due to small sample sizes, trend analysis of these data does not reveal statistically significant results.

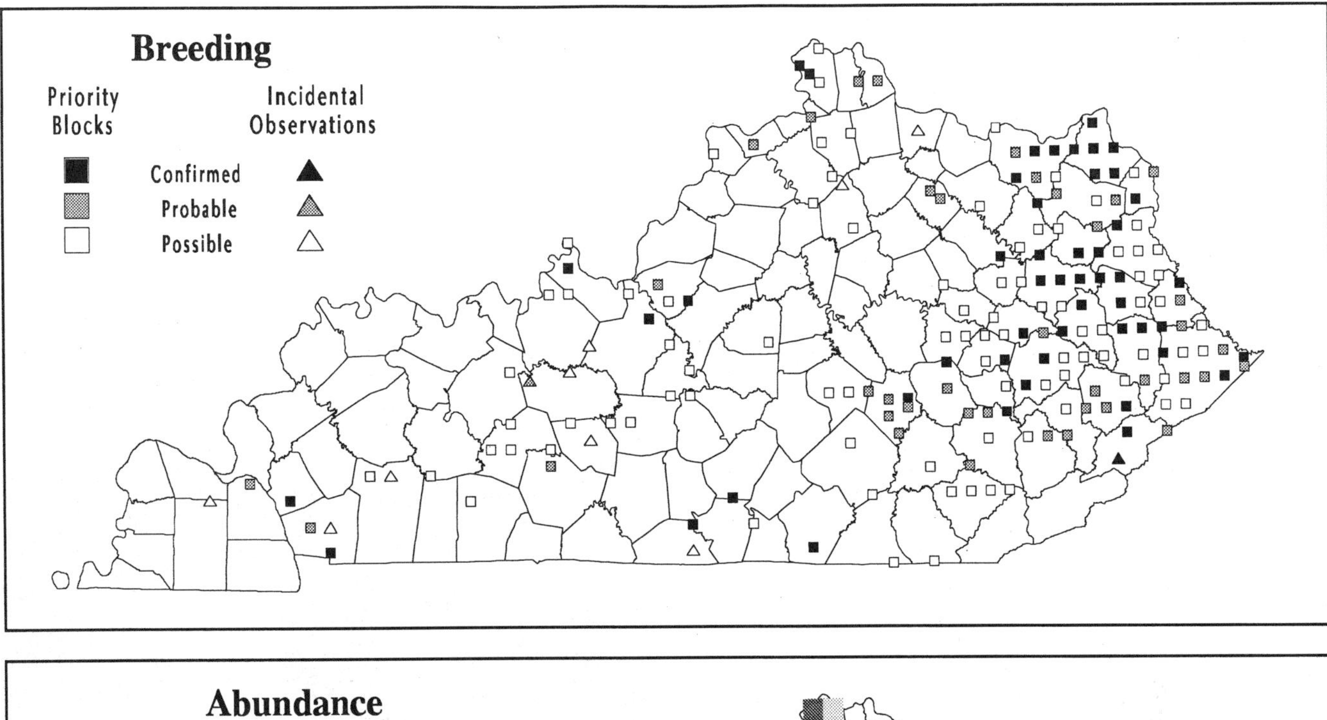

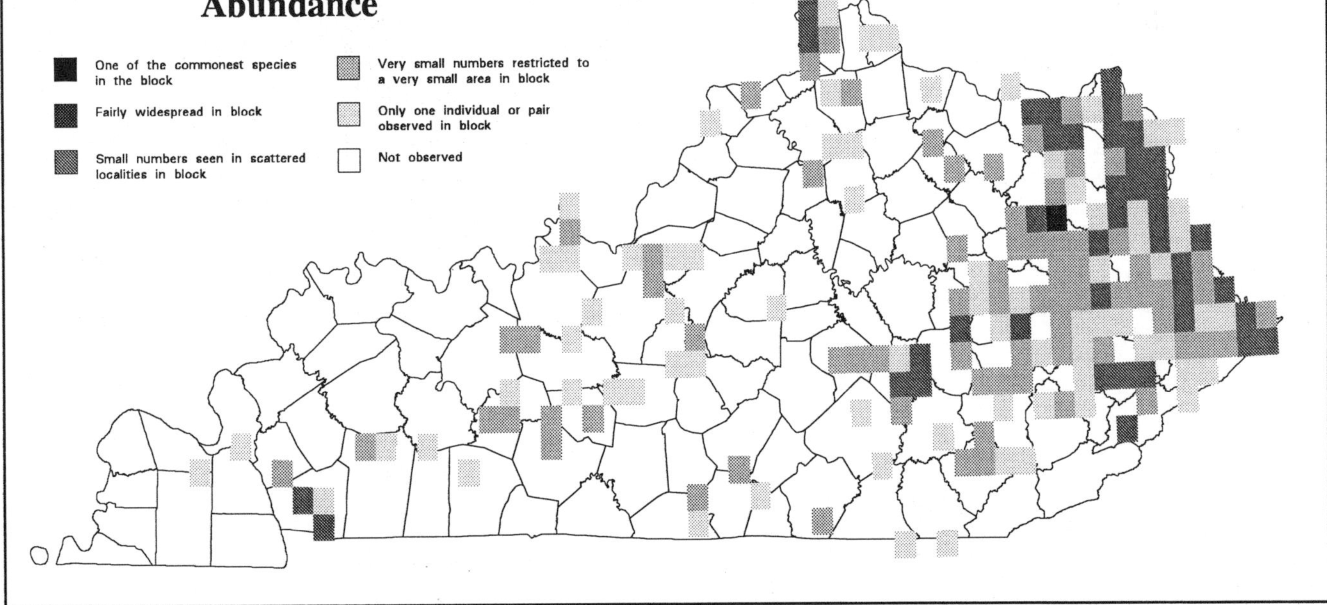

Analysis of Block Data by Physiographic Province Section

Physiographic Province Section	Total Blocks Surveyed	Blocks with Data	Avg Abund	% with Data	Section's % for State
Mississippi Alluvial Plain	14	-	-	-	-
East Gulf Coastal Plain	36	1	1.0	2.8	0.6
Highland Rim	139	14	1.7	10.1	7.9
Shawnee Hills	142	16	1.4	11.3	9.0
Blue Grass	204	40	1.8	19.6	22.6
Cumberland Plateau	173	106	2.0	61.3	59.9
Cumberland Mountains	19	-	-	-	-

Summary of Breeding Status

Number of Blocks in Which Species Was Recorded		
Total	**177**	**24.3%**
Confirmed	50	28.2%
Probable	39	22.0%
Possible	88	49.7%

Blue-winged Warbler

Golden-winged Warbler

Vermivora chrysoptera

Despite the Golden-winged Warbler's striking plumage, little is known about its breeding status in Kentucky. The species's secretive nature and inconspicuous, insectlike song have contributed to the virtual absence of published nesting data. Although there is a summer record for Hopkins County, in the western Shawnee Hills (Hancock 1947), this warbler is generally restricted to the southeastern portion of the state, in and near the Cumberland Mountains. Mengel (1965) listed no definite breeding records but reported the species's presence in small numbers at higher elevations on Black Mountain, where it was first reported in 1944 (Breiding 1947). Since the late 1950s Golden-wingeds have been found at several additional localities in and near the Cumberland Mountains, including Pine and Cumberland mountains and adjacent valleys (Croft 1969, 1971); Pike County (BBS data); along Brownies Creek, in Bell County (Stamm 1978a); and near Cumberland, in Harlan County (author's notes). The latter observation, involving at least one recently fledged young bird being fed by an adult male, constitutes the most substantial evidence of nesting that has been reported in the state.

Golden-winged Warblers are typically encountered in early successional habitats with a predominance of shrubs or small trees. In addition, the presence of a dense layer of herbaceous vegetation appears to be critical, and the species is not found in areas where the ground cover is grazed or the tree canopy is closed to the point that weeds and grasses are substantially reduced. In Kentucky the species is generally a bird of the drier slopes that have been cleared in the recent past, including reverting clear-cuts and old fields, reclaimed strip mines, and utility corridors. Natural fire or storm damage may result in the creation of suitable habitat, and the bird also has been reported in such areas (Croft 1969).

Golden-winged Warblers winter from southern Mexico south into northern South America (AOU 1983). Most spring migrants pass through Kentucky during late April and early May, and locally nesting birds are likely on territory by mid-May. Specific details of nesting activity in Kentucky are virtually absent, but a peak in clutch completion likely occurs in late May or early June. A nest has never been located, but recently fledged young have been observed on June 17, 1981, in Harlan County (author's notes), and during the atlas fieldwork on July 9, 1991, in Pike County.

This secretive warbler builds its nest on or near the ground, usually in the dense cover of grasses and weeds at the base of a shrub or small tree (Harrison 1975). The bulky nest is built on a base of dead leaves and constructed of grass, tendrils, and shreds of bark. It is lined with finer grasses and hair (Harrison 1975). Kentucky data on clutch size are lacking, but Harrison (1975) gives the rangewide clutch size as 3–6, commonly 4 or 5.

Four records of Golden-winged Warblers were reported during the atlas fieldwork. Two came from higher elevations on Black Mountain in Harlan County, where the species has been observed traditionally. A pair was observed on June 23 and 25, 1989, on a spur of the mountain west of Cloverlick Creek. The birds were seen in shrubby growth along a power line corridor at approximately 2,800 feet. The second record was of a bird observed during the summer of 1985 near the summit of the mountain, at about 3,800 feet.

Interestingly, the other two reports came from lower elevations on the Cumberland Plateau adjacent to Pine Mountain. A female or freshly molted immature was observed in early July 1991 at approximately 1,600 feet along Jones Branch near Burdine, in Letcher County. The second record was of a freshly molted immature bird in the company of an adult female in worn plumage, observed on July 9, 1991, in the shrubby growth of a reclaimed strip mine above the head of Solomon Fork near Sulphur Knob, in Pike County, at about 1,700 feet. The latter record was interpreted as probable rather than confirmed because of the age of the bird and the July date.

Hal H. Harrison

In addition to the four records of Golden-wingeds, there were three Cumberland Plateau reports of hybrid Blue-winged × Golden-winged Warblers. Details of these observations and comments concerning hybridization and its effect on the number of Golden-winged Warblers nesting in Kentucky can be found in the Nonbreeding Species section (pp. 327–28).

The Golden-winged Warbler has decreased dramatically across much of its nesting range during the 20th century, largely as a result of competition with the expanding Blue-winged Warbler (Gill 1980). In some parts of states near Kentucky, expanding populations of Blue-winged Warblers have actually replaced Golden-wingeds (Brauning 1992; Peterjohn and Rice 1991; Buckelew and Hall 1994). Kentucky's small nesting population has been documented so poorly that an assessment of trends within the state is impossible. Owing to the scarcity of Blue-winged Warblers in the Cumberland Mountains, it appears that competition between the two species must be limited, and we can only hope that a small nesting population of Golden-wingeds will persist in southeastern Kentucky.

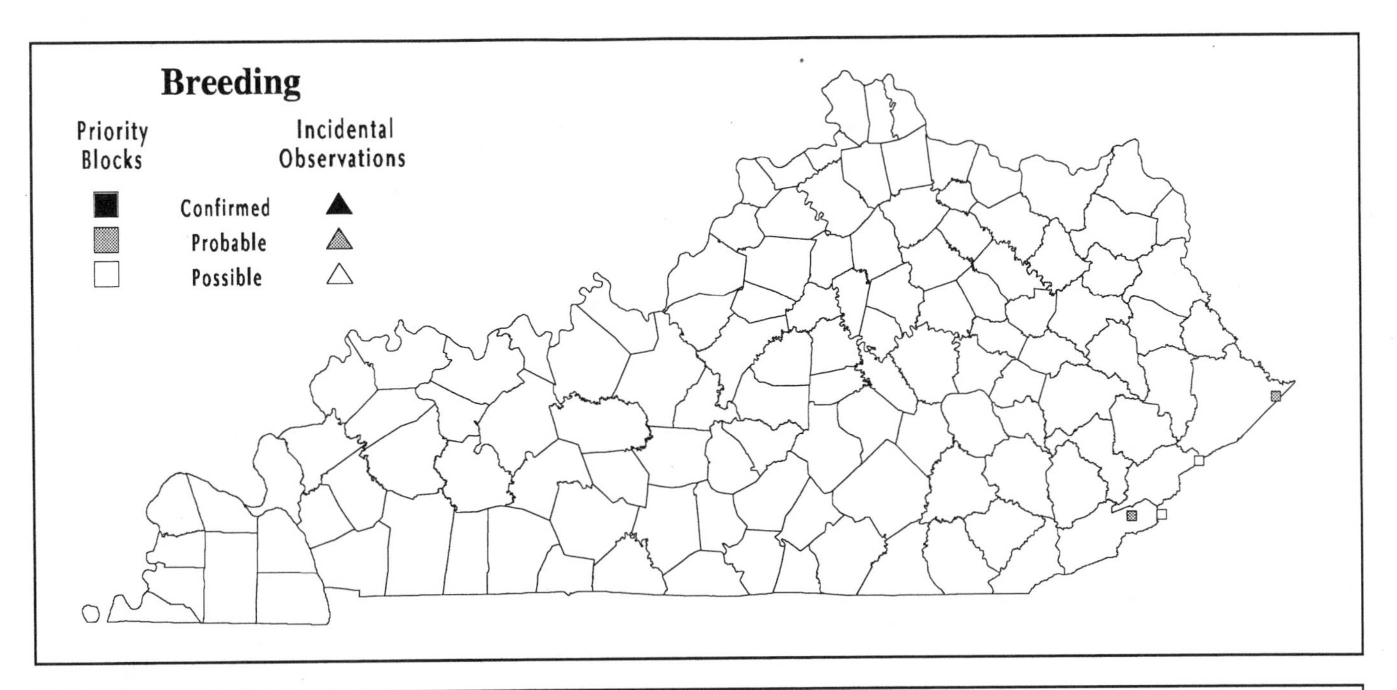

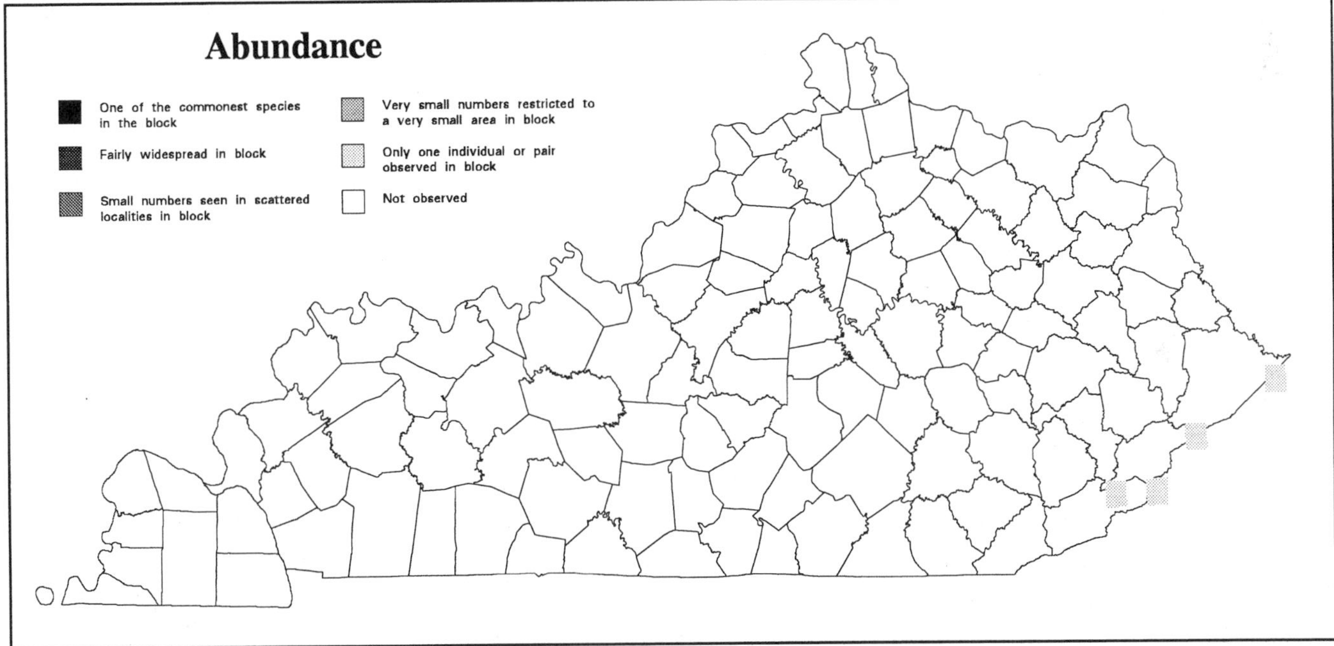

Analysis of Block Data by Physiographic Province Section

Physiographic Province Section	Total Blocks Surveyed	Blocks with Data	% with Data	Section's % for State
Mississippi Alluvial Plain	14	-	-	-
East Gulf Coastal Plain	36	-	-	-
Highland Rim	139	-	-	-
Shawnee Hills	142	-	-	-
Blue Grass	204	-	-	-
Cumberland Plateau	173	2	1.2	50.0
Cumberland Mountains	19	2	10.5	50.0

Summary of Breeding Status

Number of Blocks in Which Species Was Recorded		
Total	**4**	**0.6%**
Confirmed	-	-
Probable	2	50.0%
Possible	2	50.0%

Northern Parula

Parula americana

Although the easily recognized song of the Northern Parula can be heard throughout most of Kentucky in summer, the species is irregularly distributed across the state. Mengel (1965) described this warbler as common in the Cumberland Mountains, the Cumberland Plateau, and the lowlands of western Kentucky, less numerous and rather local in the Shawnee Hills, and very rare or absent elsewhere. He also noted that the species was apparently absent from the higher elevations of Black Mountain.

While the Northern Parula is generally regarded as a bird of fairly mature forest, it uses a variety of forest types (Wilson 1947; Mengel 1965; Rowe 1967). In western Kentucky the species is typically found in bottomland forests and swamps, especially where bald cypress is present, but it also occurs along riparian corridors. Across central Kentucky this warbler is usually encountered along riparian corridors, especially where sycamores abound, but it also can be found at least occasionally in oaks. In eastern Kentucky the Northern Parula is observed most often in ravines and on mesic slopes of the mixed mesophytic forest, usually in association with hemlocks, although it also can be found along river and stream corridors where sycamores are prevalent.

The Northern Parula is probably less common and widespread in Kentucky today than before settlement. The species appears to tolerate moderate forest disturbance, and small numbers can be found in fragmented woodland tracts and corridors of forest. In contrast, the substantial loss of mature forest, especially in bottomland situations, has undoubtedly caused a dramatic decrease in overall abundance, especially in central and western Kentucky.

Northern Parulas winter primarily in Middle America (AOU 1983). The first males typically return to western Kentucky by early April, and it is likely that most nesting birds are present by the end of the month. The only active nest ever reported in Kentucky was found high in a sycamore tree in Franklin County on May 28, 1966 (Jones 1966). Additional information on nesting activity is limited to the observation of presumed nest building in mid-June (Mengel 1965) and of recently fledged young being fed from early June to early July (Stamm 1961b; Mengel 1965; KBBA data). It appears that most clutches are completed by mid-May, and most fledglings leave the nest by early July. Kentucky data on clutch size are lacking, although Stamm (1961b) reported that five young were observed being fed in Knott County in early July 1960. Harrison (1975) gives rangewide clutch size as 3–7, commonly 4 or 5.

Forest Cover

Value	% of Blocks	Avg Abund
All	26.1	1.6
1	4.9	1.4
2	23.2	1.5
3	26.1	1.6
4	34.4	1.6
5	53.8	1.9

Northern Parulas generally nest in large trees. Sycamores appear to be favored across much of the state, but bald cypress and hemlock are likely used most often in western and eastern Kentucky, respectively. The only reported Kentucky nest was estimated to be about 50 feet above the ground (Jones 1966), and Harrison (1975) gives a range of nest height of 10–40 feet. In some parts of the species's range, the nest is hidden in festoons of lichens, Spanish moss, or bromeliads (see photo; Harrison 1975). In the absence of such situations in Kentucky, nests are probably hung in a pendant fashion from the leaf clumps of the outer portions of branches and may be woven into natural or artificial material (Harrison 1975). For example, flood debris left hanging in the lower limbs of trees may be used along streams (Ehrlich et al. 1988).

Ron Austing

The Northern Parula was recorded in about 26% of the priority blocks statewide, and 22 incidental observations were reported. Occurrence was highest in the Mississippi Alluvial Plain and the East Gulf Coastal Plain, where remaining blocks of forest often are large and swampy. In contrast, occurrence was lowest in the Highland Rim and the Blue Grass, where optimal bottomland habitat has been nearly eliminated. Relatively low occurrence in the Cumberland Plateau may be the result of biased atlas coverage. Average abundance was relatively low statewide but highest in the Mississippi Alluvial Plain. Both occurrence and average abundance were lowest in extensively cleared areas and increased as the amount of forest increased.

Less than 5% of priority block records were for confirmed breeding. This low figure resulted from a combination of the species's habit of foraging high in forest cover and the typically inconspicuous placement of the nest. Most of the nine confirmed records were based on the observation of recently fledged young. The remainder involved the observation of adults carrying food.

Northern Parulas are scarcely reported on Kentucky BBS routes. The average number of individuals per BBS route for the periods 1966–91 and 1982–91 was 0.34 and 0.31, respectively. In part due to small sample sizes, trend analysis of these data does not reveal statistically significant results.

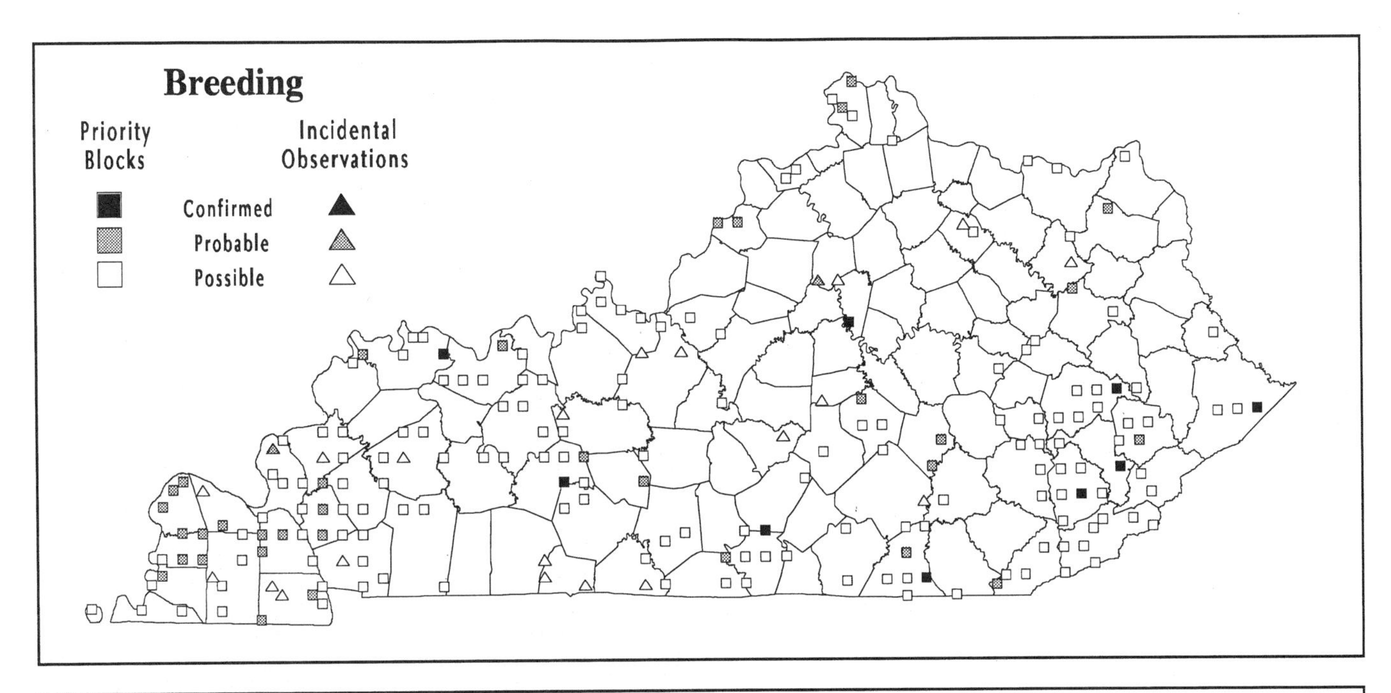

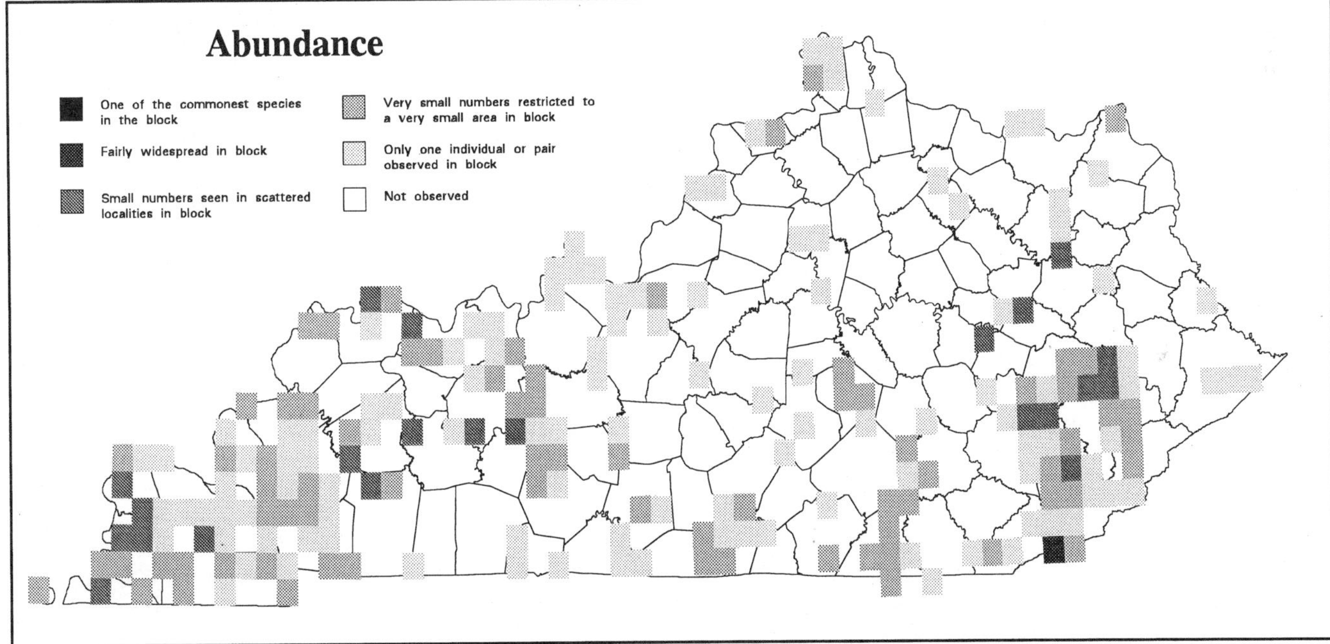

Analysis of Block Data by Physiographic Province Section

Physiographic Province Section	Total Blocks Surveyed	Blocks with Data	Avg Abund	% with Data	Section's % for State
Mississippi Alluvial Plain	14	8	2.5	57.1	4.2
East Gulf Coastal Plain	36	20	1.5	55.6	10.5
Highland Rim	139	34	1.4	24.5	17.9
Shawnee Hills	142	48	1.7	33.8	25.3
Blue Grass	204	20	1.3	9.8	10.5
Cumberland Plateau	173	50	1.7	28.9	26.3
Cumberland Mountains	19	10	1.6	52.6	5.3

Summary of Breeding Status

Number of Blocks in Which Species Was Recorded		
Total	**190**	**26.1%**
Confirmed	9	4.7%
Probable	34	17.9%
Possible	147	77.4%

Northern Parula

Yellow Warbler

Dendroica petechia

The spirited song of the Yellow Warbler is a fairly common summer sound throughout the northern and eastern sections of Kentucky. In contrast, the species decreases dramatically to the south and west, where many areas that appear to offer suitable habitat are unoccupied. Mengel (1965) regarded this warbler as a variably uncommon to common summer resident throughout the state except in the higher elevations of the Cumberland Mountains, where it appeared to be absent.

The Yellow Warbler occurs in a variety of semi-open and open habitats, typically avoiding mature forest. These bright warblers are usually found near water or low, swampy ground, where they inhabit thickets of young trees, but they also occur in upland situations. In Kentucky most suitable habitat is artificially created and includes the margins of reservoirs, reverting fields, rural yards, and reclaimed strip mines. Stands of young willows and other deciduous trees along the major rivers and surrounding natural lakes and ponds apparently provide the only naturally occurring nesting habitat.

It is likely that the Yellow Warbler is more numerous and widespread in Kentucky as a result of human alteration of the landscape. Audubon drew the species at the Falls of the Ohio in the early 1800s (Wiley 1970), and suitable nesting habitat was likely scattered across the state in wetlands and along riparian corridors. The clearing and fragmentation of extensive floodplain forests, however, has resulted in the creation of an abundance of suitable habitat, especially in eastern Kentucky, where the extensive cover of mature forest likely precluded nesting in most areas.

Yellow Warblers return from their wintering grounds in the tropics during the latter half of April. The first males usually arrive in Kentucky during the third week of the month, and most of the breeding population is likely in place by the first week of May. The earliest nesting date known is April 27 (Stamm and Monroe 1990a), but most clutches are probably completed during the latter half of May (Mengel 1965). The average size of four clutches or broods given by Mengel (1965) was 3.8 (range of 3–4), but two more recent nests have contained broods of five young (KOS Nest Cards). Cowbird parasitism has been noted on several occasions (e.g., Bowne 1972; Smith 1972). Most young apparently leave the nest sometime in June, and most of the breeding population seems to depart by early August.

Forest Cover

Value	% of Blocks	Avg Abund
All	31.1	1.8
1	22.8	1.8
2	16.8	1.6
3	25.6	1.8
4	51.3	2.1
5	63.5	1.7

Nests are usually placed in small to medium-sized trees and shrubs. A variety of species are used, but willows, boxelder, buttonbush, and other lowland plant species are favored. The nest is most often placed in an upright fork, usually just inside the outer crown of leaves. It is largely constructed of cottony plant material and lined with fine grasses and hair. The average height of four nests given by Mengel (1965) was 10.1 feet (range of 4–20 feet), although most recent nests have been below this average height (KOS Nest Cards).

Alvin E. Staffan

The atlas survey yielded records of Yellow Warblers in more than 31% of priority blocks statewide, and 36 incidental observations were reported. Occurrence was highest in the Cumberland Plateau and lowest in the Shawnee Hills and the East Gulf Coastal Plain. Nearly half of the state's nesting population was found in the Cumberland Plateau, where the species ranks among the most abundant members of the genus *Dendroica*. In this region these warblers chiefly inhabit riparian corridors of young trees in farmed or settled floodplains. Most observations within the Cumberland Mountains occurred along stream drainages, and none came from high elevations. Average abundance was relatively low statewide. Although occurrence was highest in areas with an abundance of forest and relatively little open ground, this is largely a result of the species's general abundance in eastern Kentucky, where forestation is greater overall. In other parts of the state Yellow Warblers are usually associated with fairly open, early successional stands of trees along rivers and lakeshores.

Nearly 21% of priority block records were for confirmed breeding. Although at least four active nests were located, most confirmed records were based on the observation of recently fledged young in the company of adults, and a few were based on adults carrying food.

Kentucky's BBS surveys do not adequately sample Yellow Warbler occurrence because most of the Cumberland Plateau routes are not run. The average number of individuals per BBS route for the periods 1966–91 and 1982–91 was 2.31 and 1.46, respectively. In part due to small sample sizes, trend analysis of these data does not reveal statistically significant results.

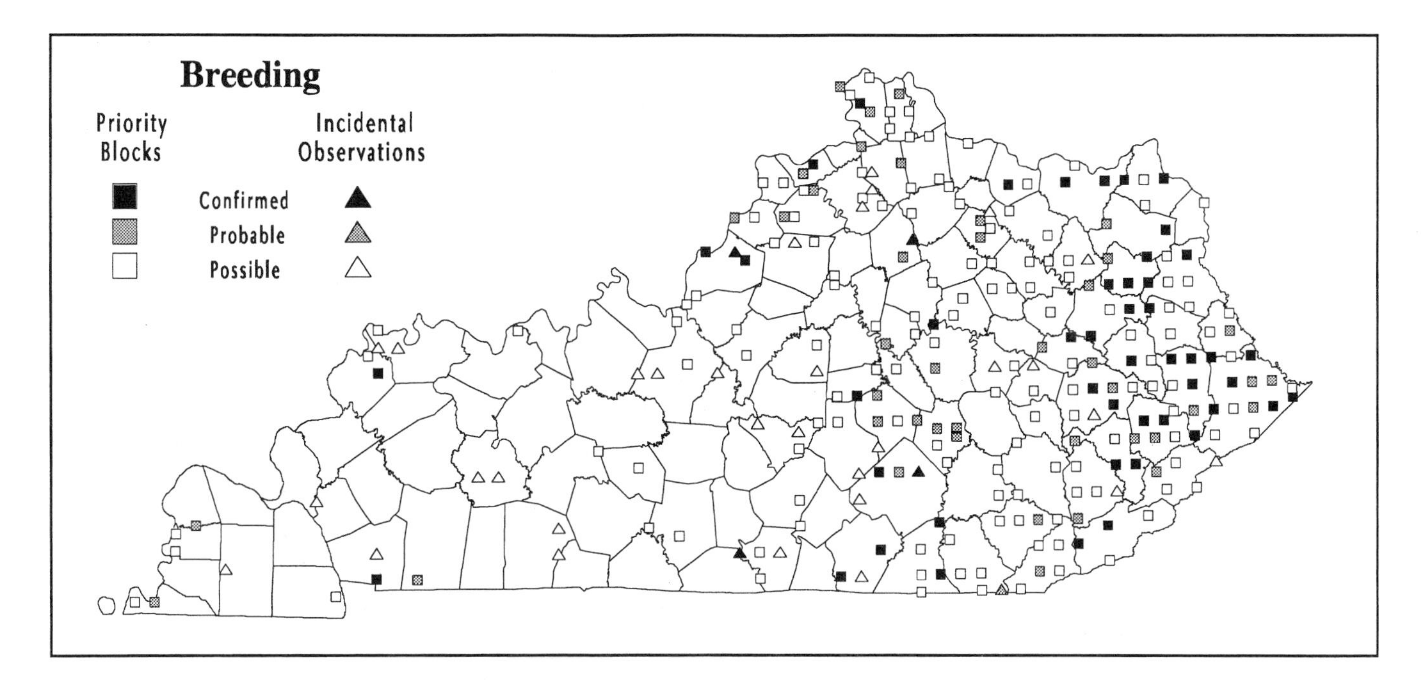

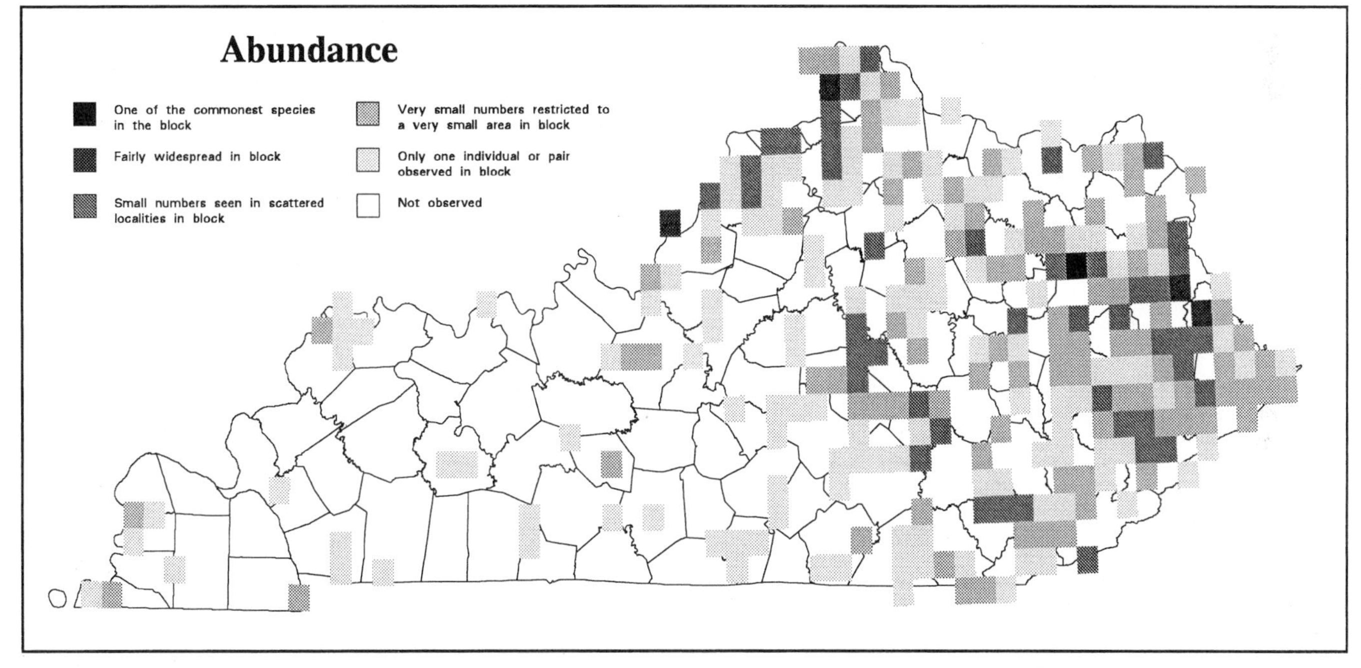

Analysis of Block Data by Physiographic Province Section

Physiographic Province Section	Total Blocks Surveyed	Blocks with Data	Avg Abund	% with Data	Section's % for State
Mississippi Alluvial Plain	14	5	1.4	35.7	2.2
East Gulf Coastal Plain	36	1	2.0	2.8	0.4
Highland Rim	139	15	1.2	10.8	6.6
Shawnee Hills	142	6	1.3	4.2	2.7
Blue Grass	204	84	1.8	41.2	37.2
Cumberland Plateau	173	107	2.0	61.8	47.4
Cumberland Mountains	19	8	1.5	42.1	3.5

Summary of Breeding Status

Number of Blocks in Which Species Was Recorded		
Total	**226**	**31.1%**
Confirmed	47	20.8%
Probable	42	18.6%
Possible	137	60.6%

Chestnut-sided Warbler

Dendroica pensylvanica

Most of the Chestnut-sided Warbler's nesting range lies well to the north of Kentucky, in the northeastern United States and southern Canada (AOU 1983), but it also dips south through the Appalachian Mountains, barely reaching the mountainous southeastern corner of the state. The species has been known to occur in summer near the summit of Black Mountain, in Harlan County, since the early 1900s (Howell 1910; Wetmore 1940; Barbour 1941), and several observers have reported breeding there (e.g., Lovell 1950a, 1950b; Mengel 1965). More recently the bird also has been found on the mountain's Chunklick Spur in Harlan County (Davis and Smith 1978), as well as along its crest in Letcher County (KSNPC, unpub. data). Fieldwork undertaken since the late 1960s on adjacent ridges in the Cumberland Mountains has revealed that the species also summers and likely breeds along the crest of Log Mountain (Croft 1969) and Cumberland Mountain (Croft 1969, 1971; Davis et al. 1980) in Bell and Harlan counties.

Ron Austing

The Chestnut-sided Warbler is a bird of early successional openings and forest edge where a dense shrub layer of weeds, briars, and young trees predominates. At the summit of Black Mountain, where maximum breeding density is reached, the species inhabits early successional woodland, the margins of roadway and utility corridors, and other abandoned clearings. In addition, periodic disturbance by selective logging or fire has created numerous pockets of early successional habitat, especially on or near the crest of the mountain. Similar disturbance is present along the crest of Cumberland Mountain.

Chestnut-sided Warblers winter primarily in Middle America (AOU 1983), and birds breeding in southeastern Kentucky seem to return by the second week of May (Stamm 1978b). Retarded leaf development at the higher elevations may inhibit commencement of nesting activity until the latter half of May. Mengel (1965) indicated that on Black Mountain, clutches were completed chiefly from May 21 to June 10, and family groups are prevalent there during the second half of June and early July. The average size of five clutches reported by Lovell (1950a) and Mengel (1965) from Black Mountain was 3.6 (range of 3–4).

Chestnut-sided Warblers nest in thicketlike openings and along forest margins. The nest is placed relatively low, usually in a thick tangle of brambles and weeds. Mengel (1965) observed three nests on Black Mountain that ranged 2.5-5.0 feet above the ground, but Lovell (1950a) found a nest only 21 inches above the ground. The nest is constructed of grasses and cottony plant material and is lined with fine grass and rootlets (Lovell 1950a).

The atlas survey yielded a significant amount of new information concerning the breeding range of the Chestnut-sided Warbler. Five records were obtained from traditionally known high elevation areas in the Cumberland Mountains, but breeding was confirmed on Cumberland Mountain for the first time. On June 17, 1989, a pair of adults was observed carrying food and engaging in distraction displays near the eastern end of Bailes Meadows in Cumberland Gap National Historical Park. In addition, a few seemingly territorial birds were observed in a burned-over area along the crest of Pine Mountain west of Shell Gap, in Harlan County, on June 15, 1989. This represents the first time Chestnut-sideds have been reported from Pine Mountain in summer.

Of greater significance were the first summer records for the Chestnut-sided Warbler outside the Cumberland Mountains. On July 7, 1989, a recently fledged young bird was observed being fed by an adult in suitable breeding habitat east of Sizerock, in Leslie County, on the Cumberland Plateau. The greatest surprise was the observation of Chestnut-sideds at four locations in three Lewis County blocks in the northeastern Blue Grass on successive days in mid-July 1991. One of these records noted a young bird, capable of flight but being attended by a parent. The other three observations involved single birds in suitable breeding habitat. These reports were unexpected, but they apparently are related to a southward expansion in nesting range that has been documented in Ohio (Peterjohn and Rice 1991). Further documentation of the status of these new summer populations is needed.

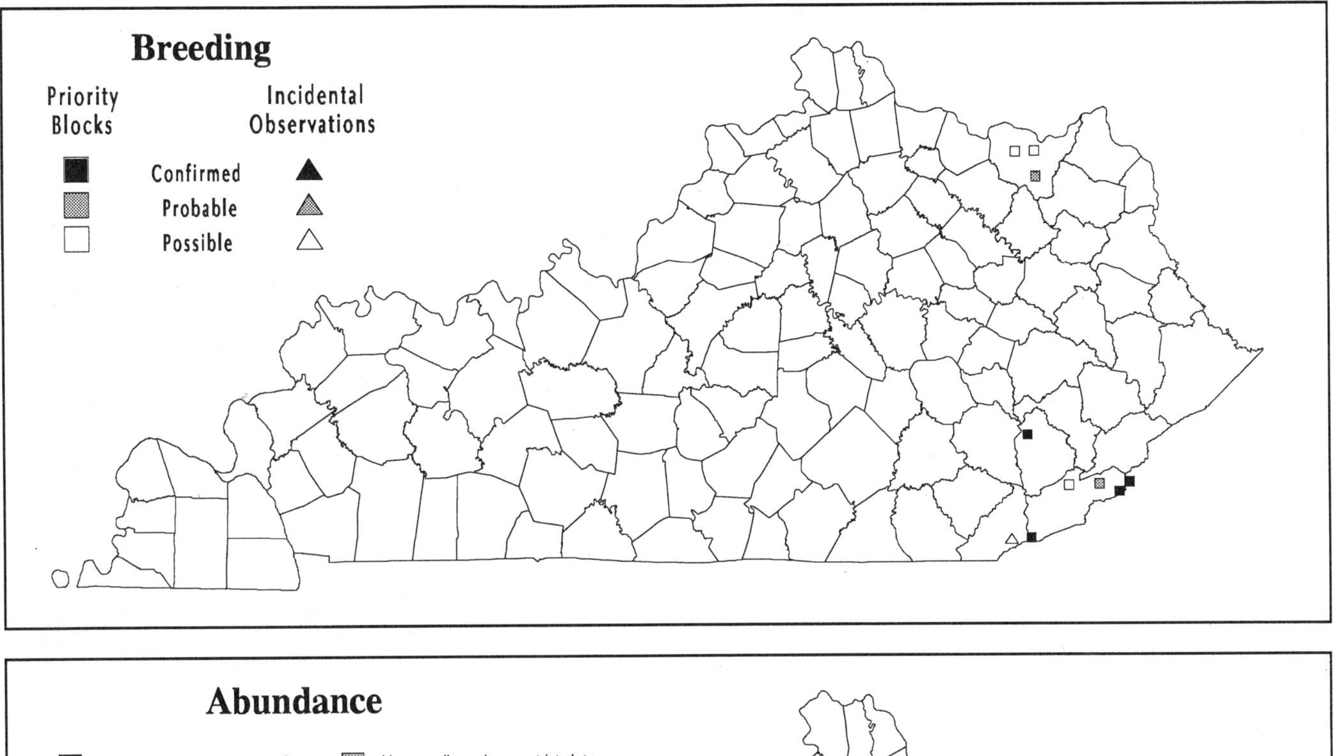

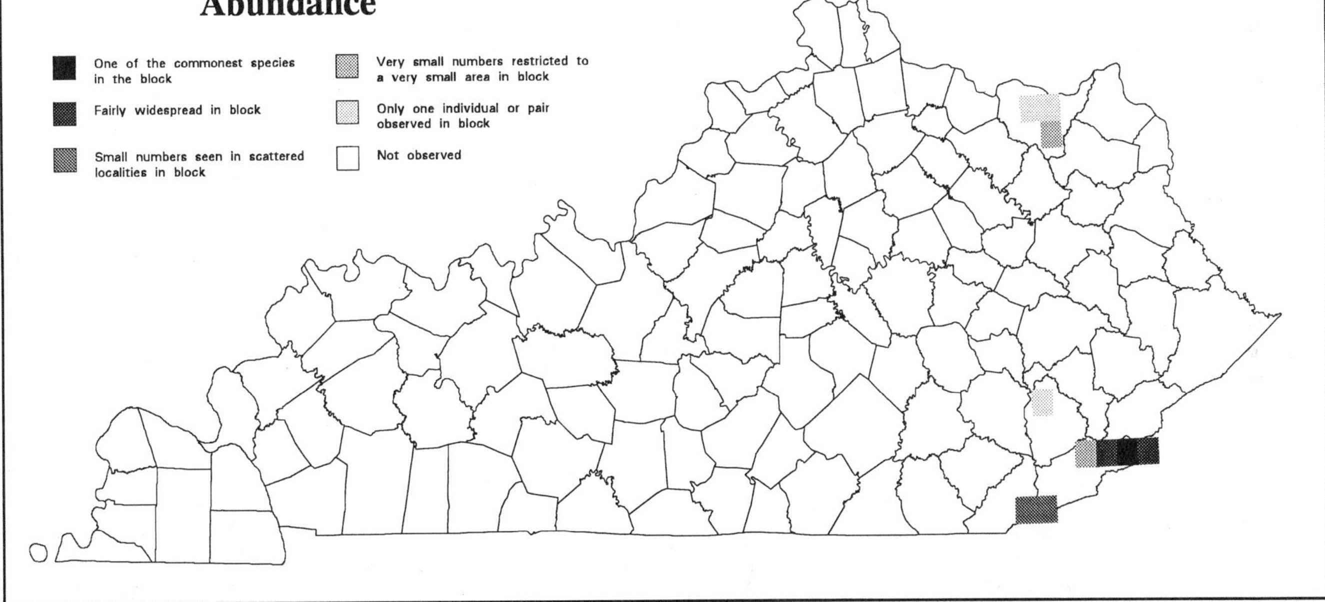

Analysis of Block Data by Physiographic Province Section

Physiographic Province Section	Total Blocks Surveyed	Blocks with Data	% with Data	Section's % for State
Mississippi Alluvial Plain	14	-	-	-
East Gulf Coastal Plain	36	-	-	-
Highland Rim	139	-	-	-
Shawnee Hills	142	-	-	-
Blue Grass	204	3	1.5	33.3
Cumberland Plateau	173	1	0.5	11.1
Cumberland Mountains	19	5	26.3	55.6

Summary of Breeding Status

Number of Blocks in Which Species Was Recorded		
Total	**9**	**1.2%**
Confirmed	4	44.5%
Probable	2	22.2%
Possible	3	33.3%

Chestnut-sided Warbler

Black-throated Blue Warbler

Dendroica caerulescens

While most of the Black-throated Blue Warbler's breeding range lies well to the north of Kentucky, in the northeastern United States and southern Canada, the race *D. c. cairnsi* is also found southward through the Appalachian Mountains into northern Georgia (Mengel 1965; AOU 1983). This southern extension of the range barely reaches into extreme southeastern Kentucky. The wheezy song of the species is one of the more conspicuous sounds of the higher elevations of Black Mountain, although it also can be heard along other ridges in the Cumberland Mountains.

Before the late 1960s the Black-throated Blue Warbler had been documented only from higher elevations of Black Mountain in Harlan and Letcher counties. Howell (1910) was apparently the first to report the species there, and it has been reported regularly since (Wetmore 1940; Breiding 1947; Lovell 1950a; Mengel 1965; KSNPC, unpub. data). Croft (1969, 1971) first reported the species from higher points on nearby Cumberland Mountain, where subsequent observations have occurred (Davis et al. 1980). The only other summer locality records have involved single birds on Chunklick Spur of Black Mountain, in Harlan County, on June 20, 1978 (Davis and Smith 1978), and on the north slope of Pine Mountain, above Pine Mountain Settlement School, in Harlan County, at about 1,700 feet on June 24, 1984 (Stamm 1984b; author's notes).

The Black-throated Blue Warbler is a bird of forested habitats, although it is also found along forest edge. On Black Mountain, where it is one of the most abundant breeding birds above about 3,400 feet, it is found in mixed mesophytic forest and forest edge with an open to moderately dense understory (Mengel 1965). On Cumberland and Pine mountains, it seems to occur primarily along stream ravines with an abundance of rhododendron and hemlock. Regardless of the forest type, these warblers are almost always found foraging in the shrub layer or midstory trees, well beneath the forest canopy.

Black-throated Blue Warblers apparently return to Kentucky from their wintering grounds in Middle America and the Caribbean in early May (Stamm 1978b), but retarded leaf development at higher elevations probably results in a delay in the commencement of nesting until sometime later in the month. Egg laying apparently occurs primarily during the latter part of May and early June. Family groups are conspicuous on Black Mountain during late June and early July. The average size of three clutches reported by Breiding (1947), Lovell (1950a), and Mengel (1965) from Black Mountain was 3.0 (range of 2–4).

The nest is usually placed low in the understory layer, most often in an upright fork of a small tree or shrub. While both forest and forest edge situations are used, most nests are placed where the understory is relatively thick. Two nests found on Black Mountain were located in a small chestnut sprout (Mengel 1965) and a tiny buckeye tree (Lovell 1950a). The nest is constructed of dead leaves and other dead plant material and is lined with finer materials, often blackish rootlets (Lovell 1950b). Mengel (1965) gives the heights of two nests as 6 and 18 inches above the ground.

Ron Austing

The atlas survey yielded four records of Black-throated Blue Warblers in priority blocks, and two incidental observations were reported. As expected, five of the six reports came from higher elevations on Black and Cumberland mountains. One of these involved an adult bird that was found on June 17, 1989, along the Ridge Trail in Cumberland Gap National Historical Park. The bird was carrying food and exhibited agitated behavior, apparently providing a first confirmed nesting record for Cumberland Mountain. Two of the three records from Black Mountain were for confirmed breeding. One was based on the observation of adults carrying food, and the other on the observation of recently fledged young. In addition, the species was found in the upper portion of Bad Branch on Pine Mountain, in Letcher County. Five territorial males and an agitated pair of birds, including one bird carrying food, were observed there on June 7, 1991. Although these birds were likely nesting, the observation was conservatively recorded as probable. An apparently vagrant male was found in the northern Blue Grass near McKinneysburg, in southern Pendleton County, on June 9, 1990 (Stamm and Monroe 1990a). Although this individual was singing in appropriate nesting habitat, it was not seen again and was not included in the final atlas data set.

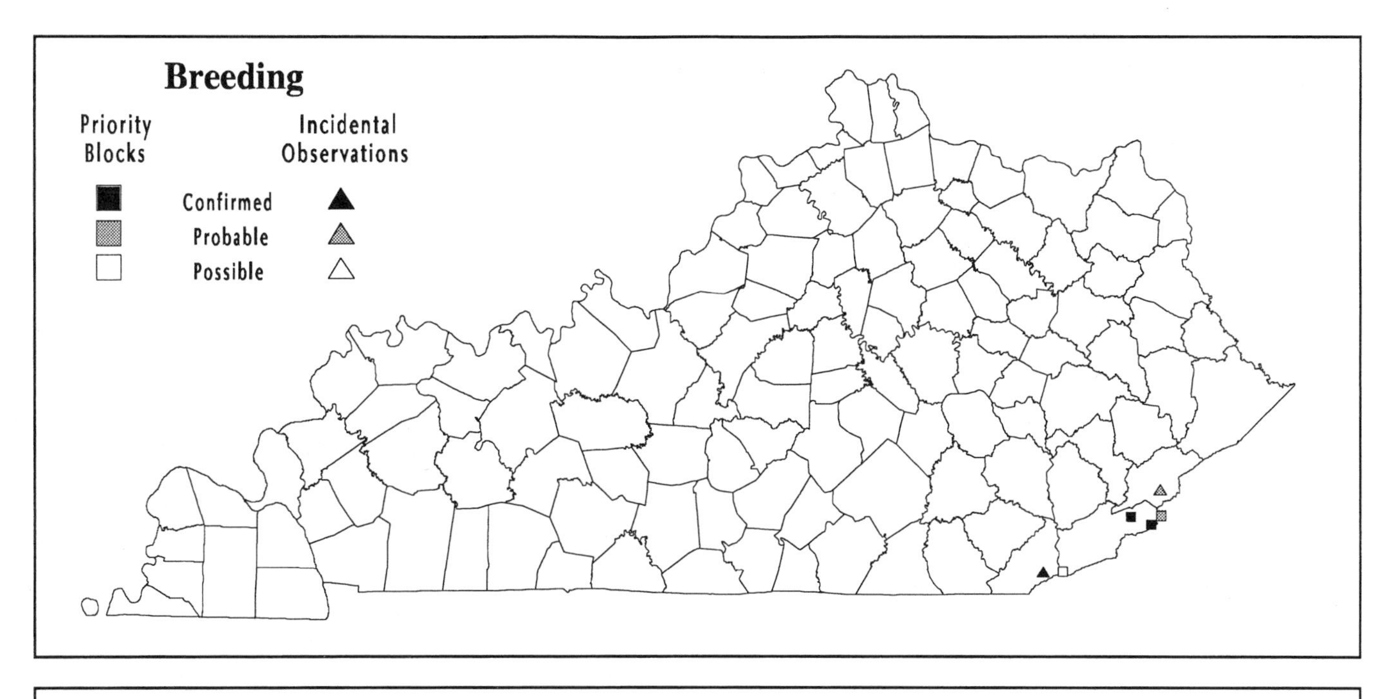

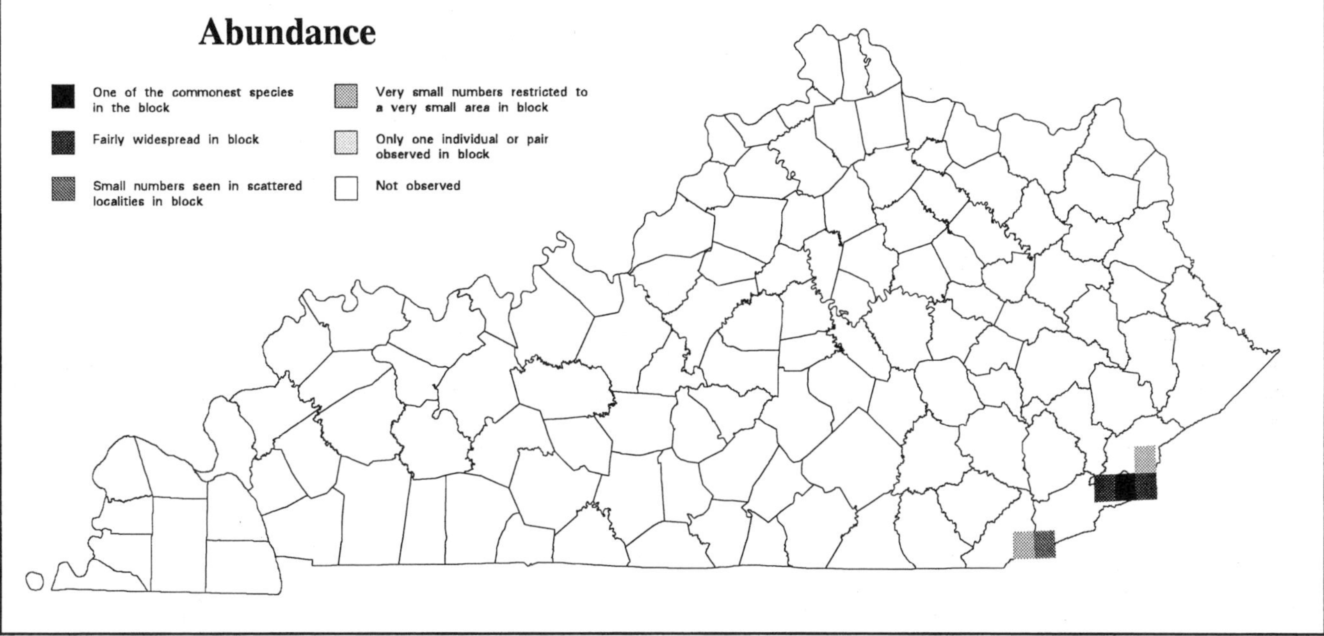

Analysis of Block Data by Physiographic Province Section

Physiographic Province Section	Total Blocks Surveyed	Blocks with Data	% with Data	Section's % for State
Mississippi Alluvial Plain	14	-	-	-
East Gulf Coastal Plain	36	-	-	-
Highland Rim	139	-	-	-
Shawnee Hills	142	-	-	-
Blue Grass	204	-	-	-
Cumberland Plateau	173	-	-	-
Cumberland Mountains	19	4	21.1	100.0

Summary of Breeding Status

Number of Blocks in Which Species Was Recorded		
Total	**4**	**0.6%**
Confirmed	2	50.0%
Probable	1	25.0%
Possible	1	25.0%

Black-throated Blue Warbler

Black-throated Green Warbler

Dendroica virens

The predominantly northern nesting range of the Black-throated Green Warbler extends south through the Appalachian Mountains and includes much of the eastern third of Kentucky (Mengel 1965; AOU 1983). Mengel (1965) regarded the species as a rare to common, but somewhat locally distributed, summer resident throughout the Cumberland Plateau and the Cumberland Mountains. In addition, birds have been reported a few times in central Kentucky in summer (Mengel 1965; Croft and Lawrence 1970; Alsop 1971). Although some suitable breeding habitat is present west of the Cumberland Plateau, further evidence of breeding has not been obtained there.

Black-throated Green Warblers usually occur in coniferous and mixed forest types (AOU 1983). In eastern Kentucky the species is typically found in association with stands of hemlock, although it has been found occasionally in deciduous or mixed pine-hardwood forest types (Mengel 1965; Hudson 1971). These warblers are most numerous in fairly mature forest, but they also use regenerating second-growth forests and forest edge. When foraging, they are usually found in the outer canopy of the forest trees, actively pursuing insects at or near the tips of branches.

It is likely that the Black-throated Green Warbler has decreased only slightly in distribution and abundance in eastern Kentucky during the last two centuries. Although native forests have been cleared and fragmented, the steep-sided ravines that harbor good stands of hemlock have not been affected as much as bottomlands and uplands. While some habitat has been destroyed by logging, the distribution of hemlocks in eastern Kentucky has likely remained relatively unchanged.

Black-throated Greens are among the first warblers to return to Kentucky from their more southerly wintering grounds in early spring. It is not rare to hear their characteristic song before the first of April, and most birds seem to arrive by mid-April (Stamm 1978b). Specific data on nesting activity in Kentucky are virtually absent. An active nest has never been found, but Mengel (1965) includes a few records of young observed out of the nest during July. It is likely that nest building is under way by late April and that a peak in clutch completion occurs by mid-May. Harrison (1975) gives rangewide clutch size as 4–5.

Although no active nest has been located in Kentucky, it is likely that hemlocks (where present) typically serve as nest trees. The nest is generally placed near the ends of horizontal branches of the tree and within dense foliage. Nests are constructed of a variety of plant materials and are lined with a thick layer of hair and other fine materials, including feathers (Harrison 1975). Harrison gives a range in nest height of 3–80 feet, and it is likely that most Kentucky nests are placed high in the larger hemlocks, which are often well above 60 feet tall.

Somewhat surprisingly, the atlas survey produced records of Black-throated Green Warblers in less than 6% of priority blocks, and only one incidental observation was reported. It is likely that if more time had been devoted to off-road coverage of blocks, more Black-throated Greens would have been reported from their hemlock ravine habitats in eastern Kentucky. As expected, the species was locally distributed throughout the Cumberland Plateau and the Cumberland Mountains. The distribution of records obtained during the atlas fieldwork correlates well with that illustrated by Mengel (1965): the species occurred most often within the Cliff Section of the Cumberland Plateau and the Cumberland Mountains and was less frequently observed throughout the remainder of the Cumberland Plateau.

Ron Austing

Although forested ravines with hemlocks are locally distributed in the eastern Shawnee Hills, the Black-throated Green Warbler does not seem to have colonized suitable habitat in this part of the state. Several atlas reports originated in the eastern Blue Grass, but all were single-date observations made late in the spring or after mid-June. All were presumed to represent vagrant individuals and were not included in the final atlas data set.

Given the low number of reports and the difficulty of finding nests of the species, it was not surprising that breeding was confirmed in only eight priority blocks. Two confirmed records were based on the observation of adults carrying food, and the other six involved observations of recently fledged young.

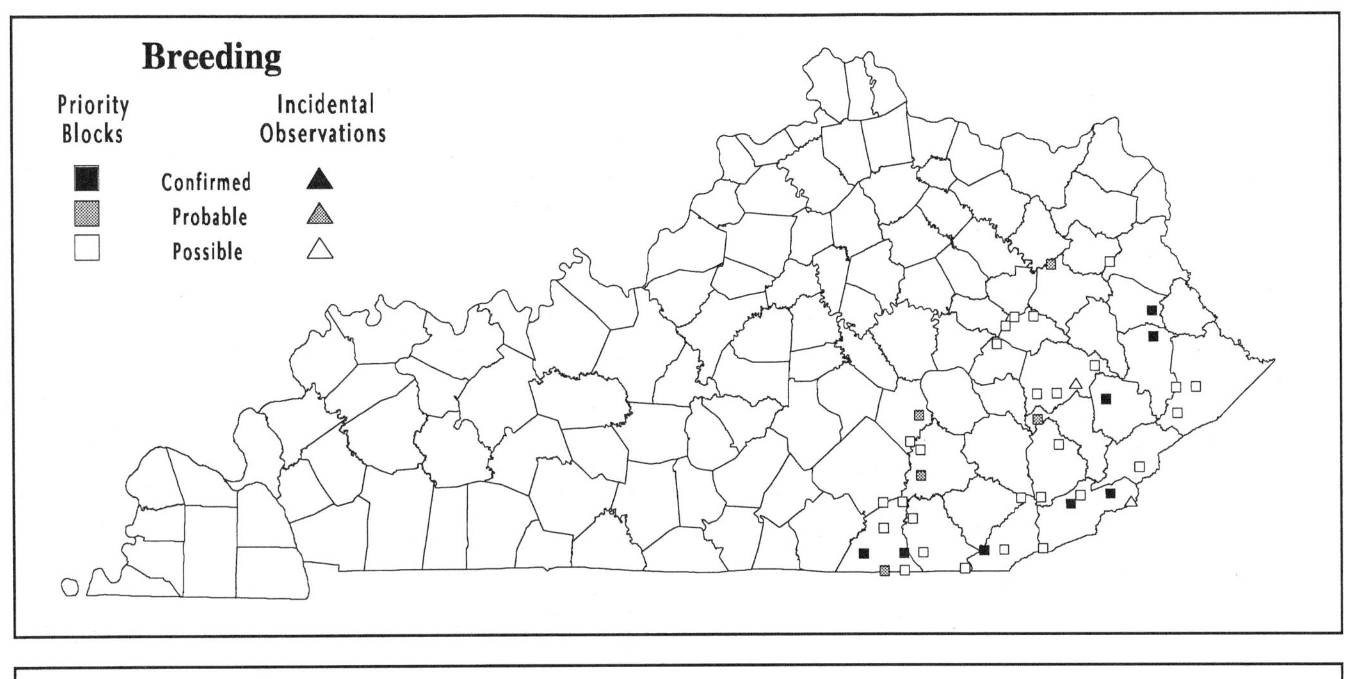

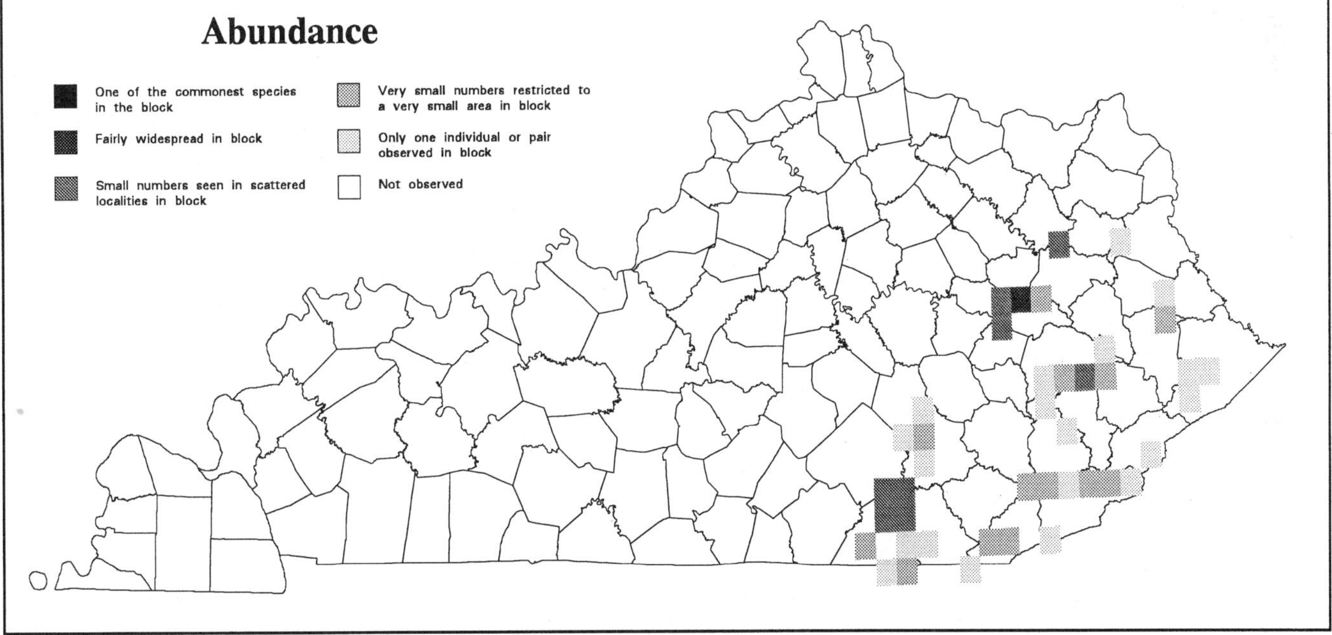

Analysis of Block Data by Physiographic Province Section

Physiographic Province Section	Total Blocks Surveyed	Blocks with Data	% with Data	Section's % for State
Mississippi Alluvial Plain	14	-	-	-
East Gulf Coastal Plain	36	-	-	-
Highland Rim	139	-	-	-
Shawnee Hills	142	-	-	-
Blue Grass	204	-	-	-
Cumberland Plateau	173	33	19.1	82.5
Cumberland Mountains	19	7	36.8	17.5

Summary of Breeding Status

Number of Blocks in Which Species Was Recorded		
Total	**40**	**5.5%**
Confirmed	8	20.0%
Probable	5	12.5%
Possible	27	67.5%

Black-throated Green Warbler

Blackburnian Warbler

Dendroica fusca

Most of the breeding range of the Blackburnian Warbler lies well to the north of Kentucky, in forests of the northern United States and southern Canada (AOU 1983). A narrow band, however, extends south through the Appalachian Mountains to northern Georgia, barely reaching into extreme southeastern Kentucky (AOU 1983). As far as is known, the species is restricted to the higher elevations of Black Mountain in Harlan and Letcher counties, where it has been reported regularly.

Howell (1910) first reported Blackburnian Warblers from Black Mountain in July 1908. At that time he reported the bird as common in heavy timber on the summit. Most subsequent researchers either have missed the species or have encountered greater difficulty in locating it there (Wetmore 1940; Barbour 1941; Lovell 1950a; Croft 1969). Mengel (1965), however, found Blackburnian Warblers to be relatively numerous on upper slopes at elevations of 2,800–3,800 feet on the mountain (but not near the summit) during the late 1940s and early 1950s. Breeding bird surveys conducted along the crest of the mountain in early June 1980 and 1981 revealed the species's presence in small numbers (KSNPC, unpub. data). It is likely that these colorful warblers are overlooked to some extent, perhaps in part because of a shift in song type in late May (Mengel 1965), as well as their apparent presence in greatest abundance below the summit, where human access is extremely limited.

Throughout most of its range, the Blackburnian Warbler is typically a bird of coniferous and mixed coniferous-deciduous forests (AOU 1983). On Black Mountain, however, the breeding population occurs in entirely deciduous forest. This situation has been noted elsewhere in the southern part of the species's nesting range and is discussed briefly by Mengel (1965). Although the species occurs in disturbed and regenerating forests, it seems most numerous in fairly mature forest of maple-beech-basswood associations on the mountain (Mengel 1965). This factor may be responsible for the species's scarcity near the summit, where fires and habitat manipulation have greatly altered the forests.

Blackburnian Warblers winter primarily in South America (AOU 1983), and birds begin returning to Black Mountain by early May. Most may not arrive until the middle of the month (Mengel 1965), and retarded leaf development at high elevations may delay the commencement of nesting until the end of May. Specific Kentucky data on nesting activity are virtually absent, but most clutches are likely completed by early June. Mengel (1965) observed two young birds being fed on June 29, 1951, and recently fledged young were observed on June 21–23, 1989, during the atlas fieldwork. Kentucky data on clutch size are lacking, but Harrison (1975) gives rangewide clutch size as 4–5.

According to Harrison (1975), nests are usually placed very high in the crowns of large trees. Generally the nest is saddled to smaller branches at the outer portion of a horizontal limb and hidden among the foliage. Elsewhere in the species's range, many nests are placed in conifers, especially spruce (see photo), but on Black Mountain nests must be located in deciduous trees. The nest is constructed of a variety of plant materials, including small twigs, and is lined with hair, black rootlets, and fine grass (Harrison 1975).

Hal H. Harrison

Blackburnian Warblers were found in all three priority blocks situated within the highest elevations of Black Mountain. Observations of recently fledged young were the basis for the confirmed report from the Louellen quadrangle. This portion of the mountain reaches high points of 3,600–3,700 feet and is in Harlan County, south of Sand Hill and Hiram. The other two reports were recorded as possible because they represented singing males observed on single dates.

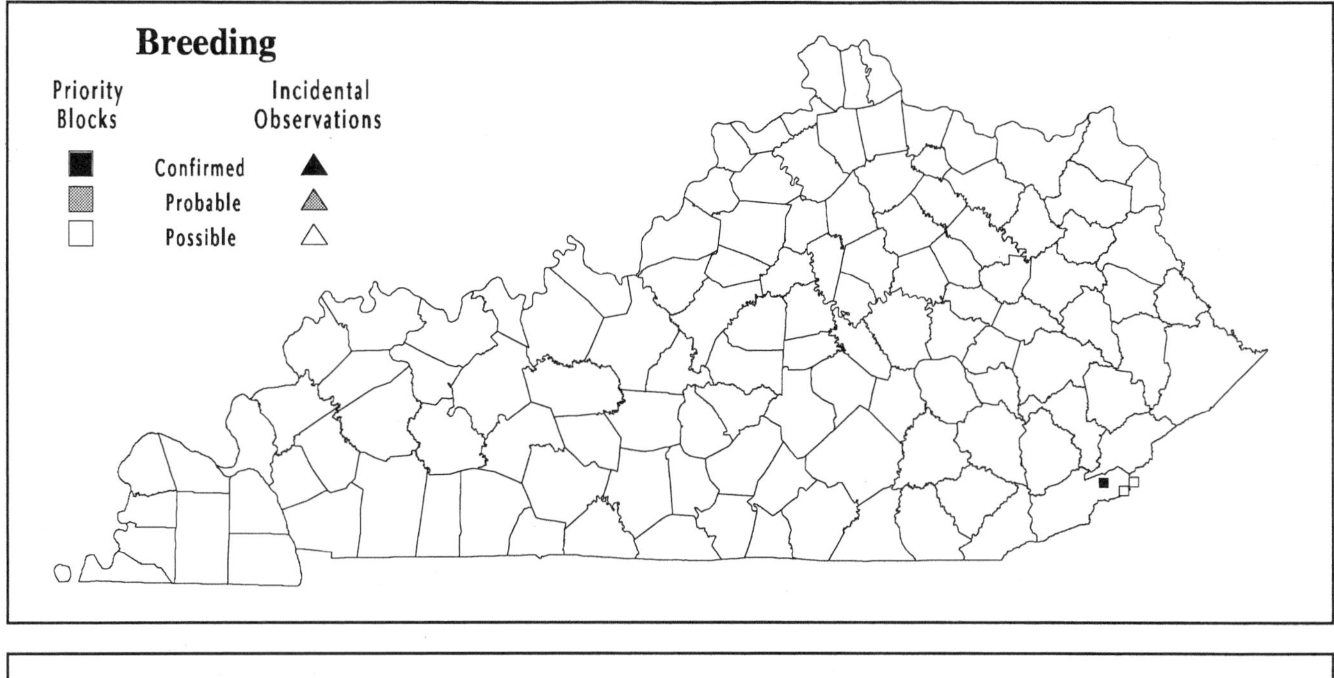

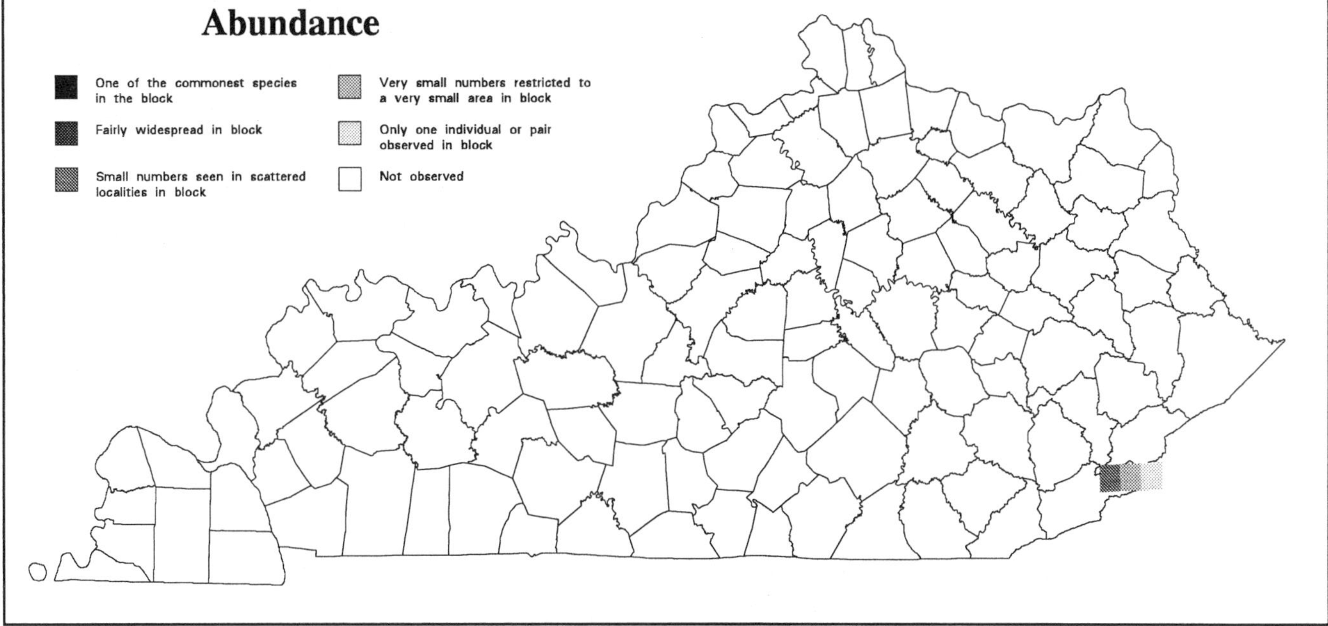

Analysis of Block Data by Physiographic Province Section

Physiographic Province Section	Total Blocks Surveyed	Blocks with Data	% with Data	Section's % for State
Mississippi Alluvial Plain	14	-	-	-
East Gulf Coastal Plain	36	-	-	-
Highland Rim	139	-	-	-
Shawnee Hills	142	-	-	-
Blue Grass	204	-	-	-
Cumberland Plateau	173	-	-	-
Cumberland Mountains	19	3	15.8	100.0

Summary of Breeding Status

Number of Blocks in Which Species Was Recorded		
Total	**3**	**0.4%**
Confirmed	1	33.3%
Probable	-	-
Possible	2	66.7%

Blackburnian Warbler

Yellow-throated Warbler

Dendroica dominica

The Yellow-throated Warbler is one of the more widespread of Kentucky's nesting warblers. The species is nowhere especially abundant, but its clear, melodic song can be heard across most of the state in summer. Among members of the genus *Dendroica,* in fact, it probably ranks only behind the Prairie Warbler in statewide abundance. Mengel (1965) regarded the Yellow-throated Warbler as uncommon in the east and common in the central and western parts of the state in summer.

The Yellow-throated Warbler's widespread distribution at least in part results from its use of a wide variety of habitat types (Mengel 1965). In the western part of the state, the species is associated with bald cypress and generally is found wherever that tree occurs. In some places in the east and west, but especially in central Kentucky, the species is associated with sycamores, chiefly along riparian corridors. In eastern Kentucky these warblers are found chiefly in upland forests in association with pines of various species, but they also occur in deciduous forest dominated by oaks (Croft 1969; Hudson 1971). Interestingly, as plantations of introduced pines have become more prevalent in western Kentucky, Yellow-throated Warblers have become established in these artificial habitats. In fact, it is not uncommon to hear neighboring territorial males singing from any two of the three habitats described above, especially where pine plantations are situated on upland sites immediately adjacent to cypress swamps or riparian corridors with sycamores.

Yellow-throated Warblers appear to have adapted fairly well to human alteration of the landscape. While deforestation certainly must have resulted in an overall decline, fragmentation of remaining forest apparently has had less of an effect on it than on many other woodland species. While Yellow-throated Warblers are most frequent where extensive forest cover is present, they will also use fairly dissected corridors and tracts of forest.

Yellow-throated Warblers are among the first spring migrants to return to Kentucky from their more southerly wintering grounds. Males are usually heard by the last week of March in western Kentucky, and it is likely that most nesting birds arrive throughout the state by mid-April (Mengel 1965). Specific Kentucky data on nesting activity are scarce, but nest building has been observed as early as April 8 (Mengel 1965), and a peak in clutch completion likely occurs by early May. Nesting activity extends into July, and some pairs may raise two broods, although this is undocumented. Late nests also may represent renesting following earlier failure owing to predation or severe weather. Young being fed out of the nest have been observed most frequently during June, but as late as early August (Mengel 1965). Kentucky data on clutch size are lacking, but Harrison (1975) gives rangewide clutch size as 4, sometimes 5.

Forest Cover

Value	% of Blocks	Avg Abund
All	42.1	1.7
1	11.9	1.6
2	32.6	1.5
3	47.0	1.5
4	60.4	2.0
5	59.6	2.1

Nests are usually placed high in large trees. They are typically constructed in a fork or saddled upon a branch, often among the foliage at or near the tip of a horizontal branch (Chapman 1968). Wilson (1922) recorded three nests in the same sycamore tree in Warren County during successive years at heights varying from 20 to 50 feet. Nests are constructed of plant material, often including strips of bark, grass, weed stems, and cottony plant down; they are lined with hair, plant down, and feathers (Chapman 1968; Harrison 1975).

Hal H. Harrison

The atlas survey yielded records of Yellow-throated Warblers in more than 42% of priority blocks statewide, and 29 incidental observations were reported. Occurrence was highest in the Cumberland Mountains, the Mississippi Alluvial Plain, and the Cumberland Plateau, where suitable habitat may be most abundant, and lowest in the East Gulf Coastal Plain and the Blue Grass, where deforestation is greatest. Average abundance was relatively uniform across the state. Occurrence was likely influenced by the amount of forest cover, being lowest in extensively cleared areas and generally increasing as forestation increased.

Despite this warbler's widespread distribution, only about 6% of priority block records were for confirmed breeding. One active nest was located, but most of the 18 confirmed records involved the observation of recently fledged young. Other confirmed records were based on adults carrying food, nest building, and distraction displays.

Yellow-throated Warblers are reported in small numbers on Kentucky BBS routes. The average number of individuals per BBS route for the periods 1966–91 and 1982–91 was 0.80 and 1.18, respectively. Trend analysis of these data reveals a nonsignificant increase of 2.2% per year for the period 1966–91, but a significant ($p<.01$) increase of 7.2% per year for the period 1982–91. The short-term increase reflects a trend that has been occurring in the region since the mid-1960s (Robbins et al. 1986).

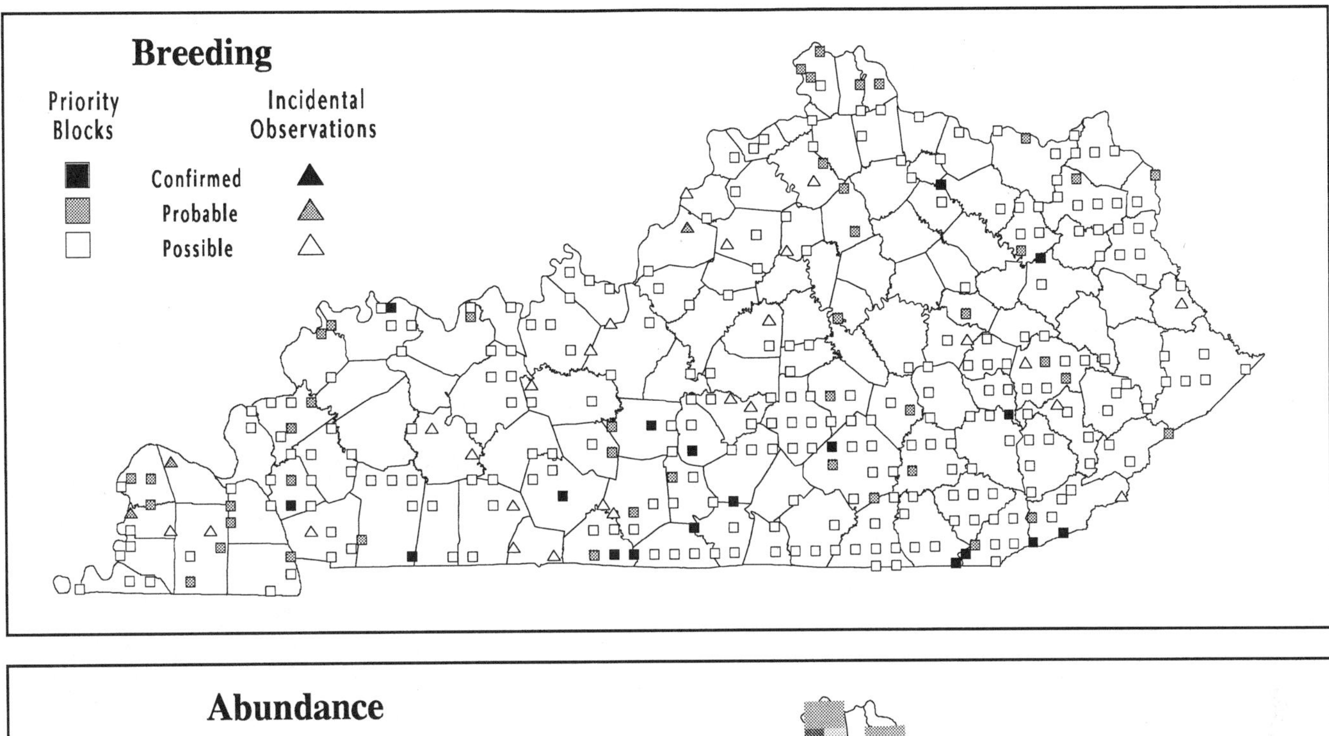

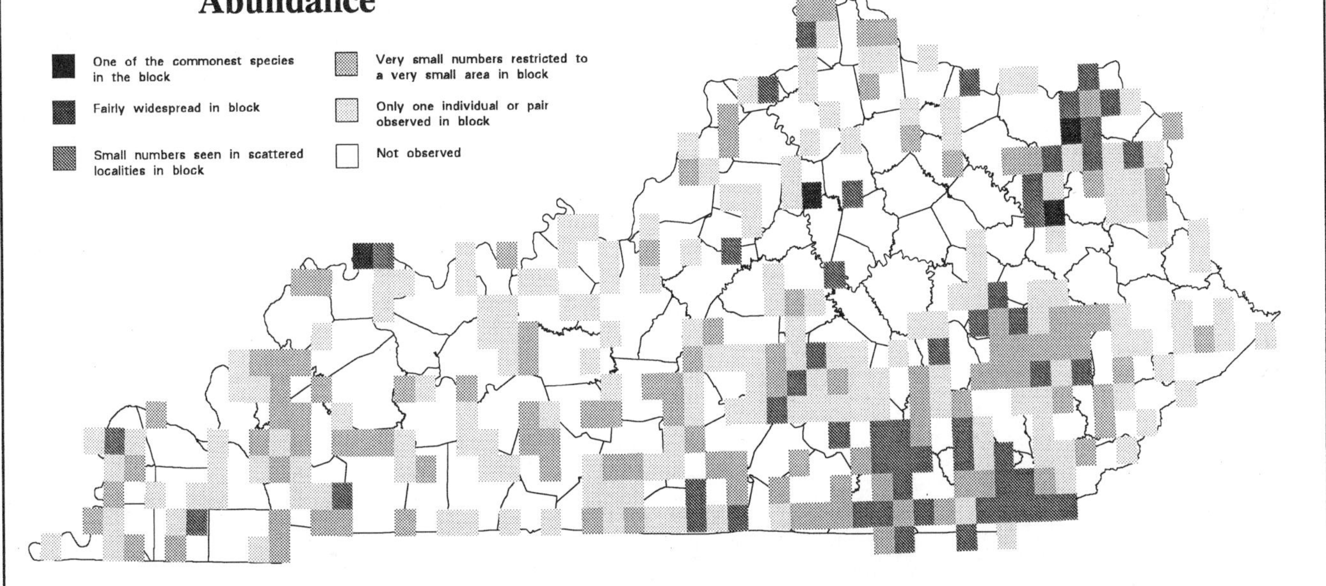

Analysis of Block Data by Physiographic Province Section

Physiographic Province Section	Total Blocks Surveyed	Blocks with Data	Avg Abund	% with Data	Section's % for State
Mississippi Alluvial Plain	14	8	1.6	57.1	2.6
East Gulf Coastal Plain	36	12	1.5	33.3	3.9
Highland Rim	139	66	1.6	47.5	21.6
Shawnee Hills	142	52	1.5	36.6	17.0
Blue Grass	204	60	1.8	29.4	19.6
Cumberland Plateau	173	96	1.9	55.5	31.4
Cumberland Mountains	19	12	2.1	63.2	3.9

Summary of Breeding Status

Number of Blocks in Which Species Was Recorded		
Total	**306**	**42.1%**
Confirmed	18	5.9%
Probable	44	14.4%
Possible	244	79.7%

Yellow-throated Warbler

Pine Warbler

Dendroica pinus

The musical song of the Pine Warbler is heard locally throughout much of Kentucky where stands of native or introduced pines occur. Mengel (1965) regarded the species as common in parts of the Cumberland Plateau and the Cumberland Mountains, but rare and local in the Knobs of the Blue Grass and the Shawnee Hills. More recently the introduction of pines in western Kentucky has resulted in the expansion of nesting birds into many new areas.

As its name suggests, the Pine Warbler is a bird of pine forests and mixed pine-hardwood forests. In Kentucky, where true pine forest types are virtually absent, the species inhabits the mixed pine-hardwood forests of the Cumberland Plateau and the Cumberland Mountains, the Knobs of the Blue Grass, and the Shawnee Hills. By far the most widespread native species of tree used is the Virginia pine, although pitch pine, shortleaf pine, and white pine are used wherever they occur. In contrast, it appears that neither hemlock nor red cedar is used extensively, if at all.

The Pine Warbler is likely more common and widespread in Kentucky today than before settlement. While the suppression of fire has contributed to the reduction of native pines in some areas, clearing of forests has resulted in the spread of pines into many places. In regions where Virginia pine is fairly common, old fields and other artificial openings are often colonized by young pines during early succession. When these stands begin to mature, they are commonly used by Pine Warblers for nesting. In relatively recent times the planting of pines, especially loblolly pine, for pulpwood production has resulted in the colonization of many areas by nesting birds, especially in western Kentucky. Pines also have been planted widely as a result of surface mine reclamation in the Shawnee Hills.

Although it appears that, at least during milder winters, a few Pine Warblers winter in Kentucky, most birds retreat southward to the southern United States (AOU 1983). The first males typically return by early March, and the full complement of breeding birds is probably present by early April. Presumed incubation has been observed as early as April 7 in Christian County (Hancock 1966) and April 10 in Bullitt County (KOS Nest Cards). Mengel (1965) included only one report of a nest, which contained small young on June 13, 1948 (Lovell 1948). A nest observed during atlas fieldwork contained young on June 14, 1991. Most records of young off the nest are for June. Although presently unconfirmed, it is probable that late nests involve the raising of a second brood. Kentucky data on clutch size are limited to the observation of three young in a nest in Bullitt County on May 16, 1971 (KOS Nest Cards). Harrison (1975) gives rangewide clutch size as 3–5, commonly 4.

Forest Cover

Value	% of Blocks	Avg Abund
All	17.6	1.8
1	1.0	3.0
2	10.5	1.6
3	19.1	1.6
4	33.8	1.8
5	23.1	2.1

Nests are placed in pines, usually at or near the tips of branches and among the denser clumps of needles. Fairly mature trees are usually used, either standing singly in mixed forest or as part of native or introduced stands. It appears that thickets of young pines are avoided until they have become fairly mature and have attained a more open structure. The average height of four reported nests is 38.7 feet (range of 15–70 feet) (Lovell 1948; Hancock 1966; KOS Nest Cards; KBBA data).

Hal H. Harrison

The atlas survey yielded records of Pine Warblers in nearly 18% of priority blocks, and 14 incidental observations were reported. Somewhat unexpectedly, occurrence was highest in the East Gulf Coastal Plain, being slightly greater than in the Cumberland Plateau and the Cumberland Mountains. In addition, the species was recorded in more than 15% of the blocks in the Highland Rim and the Shawnee Hills. As already noted, the Pine Warbler's expansion into the western half of the state is largely the result of human activity.

Despite the bird's conspicuous song and fairly widespread occurrence, only about 17% of priority block records were for confirmed breeding. Although one active nest was located, most of the 22 confirmed records were based on the observation of recently fledged young being attended by parents. In addition, on June 14, 1989, an adult Pine Warbler was observed feeding a young Brown-headed Cowbird in Butler County. Other confirmed records were based on the observation of adults carrying food for young.

Pine Warblers are seldom recorded on Kentucky BBS routes. The average number of individuals per BBS route for the periods 1966–91 and 1982–91 was 0.35 and 0.37, respectively. In part due to small sample sizes, trend analysis of these data does not reveal statistically significant results.

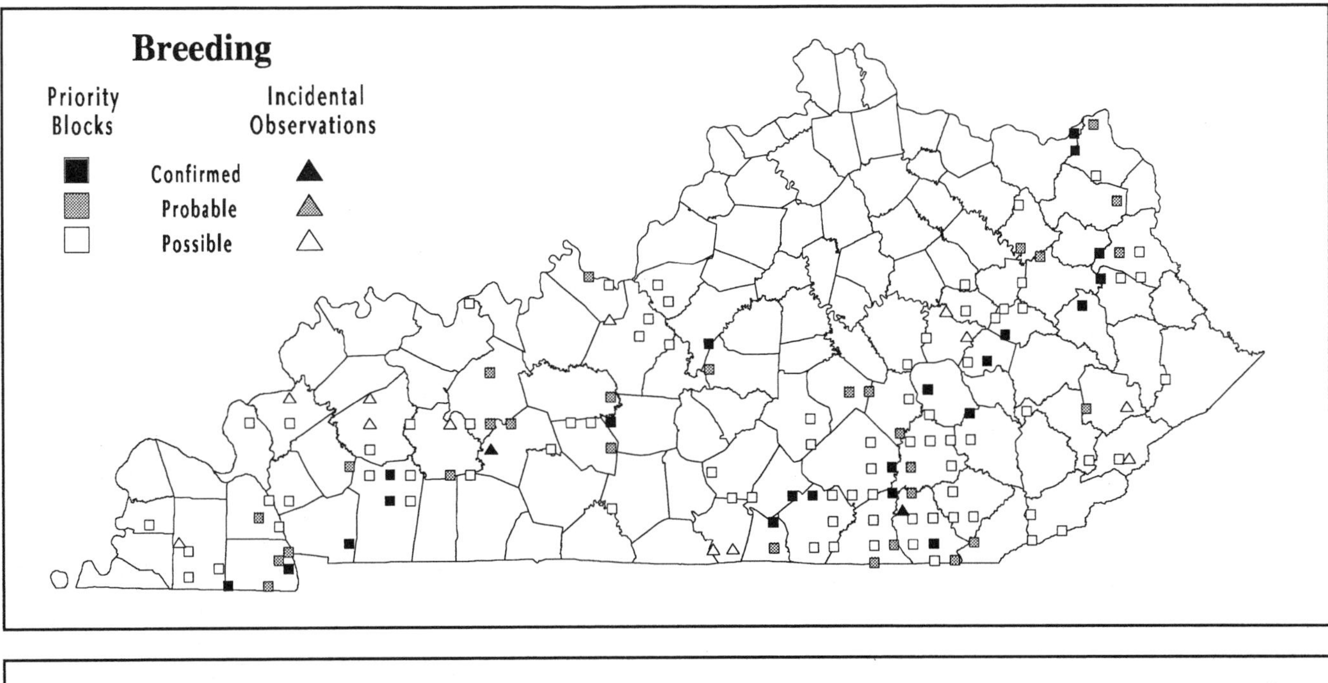

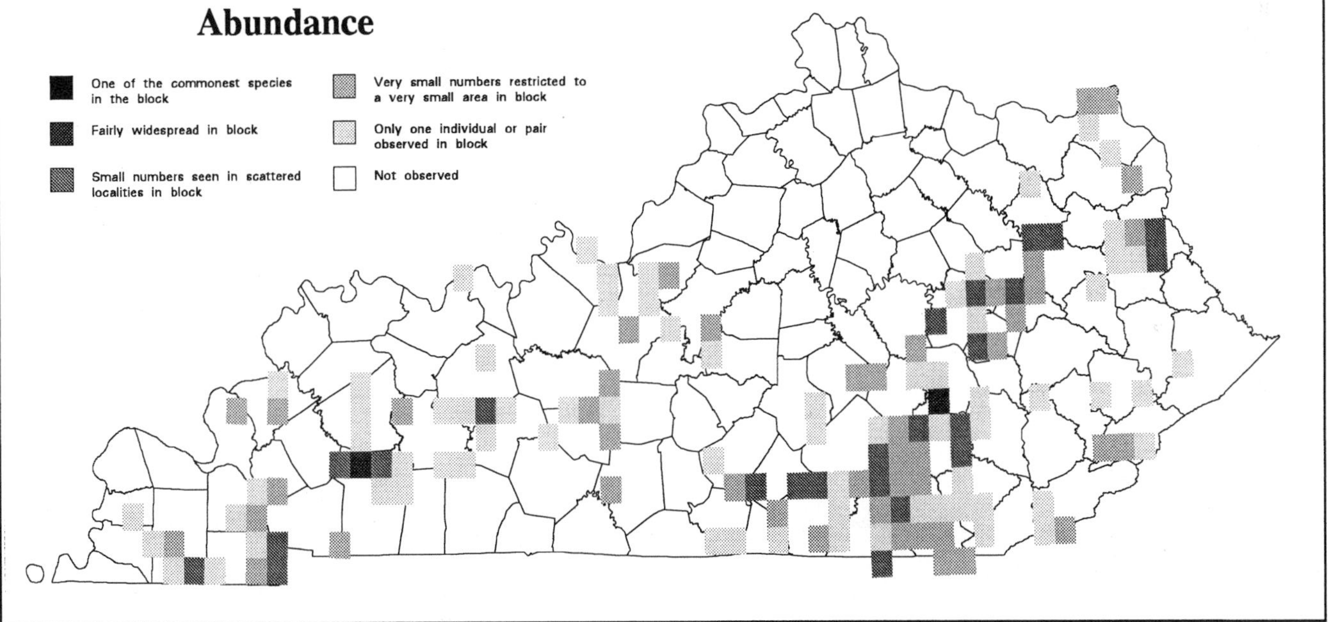

Analysis of Block Data by Physiographic Province Section

Physiographic Province Section	Total Blocks Surveyed	Blocks with Data	Avg Abund	% with Data	Section's % for State
Mississippi Alluvial Plain	14	-	-	-	-
East Gulf Coastal Plain	36	12	1.8	33.3	9.4
Highland Rim	139	21	1.7	15.1	16.4
Shawnee Hills	142	23	1.6	16.2	18.0
Blue Grass	204	14	1.9	6.9	10.9
Cumberland Plateau	173	53	1.8	30.6	41.4
Cumberland Mountains	19	5	1.4	26.3	3.9

Summary of Breeding Status

Number of Blocks in Which Species Was Recorded		
Total	**128**	**17.6%**
Confirmed	22	17.2%
Probable	29	22.7%
Possible	77	60.2%

Pine Warbler

Prairie Warbler

Dendroica discolor

The Prairie Warbler's conspicuous song is a fairly frequent summer sound throughout much of Kentucky. In fact, the species is undoubtedly the most abundant and widespread member of the genus *Dendroica* nesting in Kentucky. Mengel (1965) regarded the Prairie Warbler as common and generally distributed across all of eastern Kentucky except on Black Mountain, where he did not observe it above 2,800 feet. In contrast, he considered the species to be rare and more local in the western part of the state. Results of the atlas fieldwork indicate that today it is generally distributed, occurring wherever suitable habitat is present.

The Prairie Warbler inhabits a variety of semi-open, often successional habitats, including brushy forest edge, but it typically avoids mature forest. Although the species can be found in deciduous vegetation, it occurs most frequently in mixed community types where pines or red cedars are present or dominant. Small numbers are sometimes found in natural situations, such the margins of cedar glades, but these warblers occur primarily in artificially created habitats, including reverting agricultural fields and pastures, regenerating forest clear-cuts, reclaimed strip mines, and young pine plantations.

Before altered habitats became available, the Prairie Warbler was probably restricted to natural openings in cedar glades and barrens, although fires and storm damage may have created suitable nesting habitat in forested areas. The species likely increased in response to the clearing and fragmentation of forested areas. These activities have resulted in the creation of an abundance of suitable early successional habitat across much of the state where formerly the species must have been largely or wholly absent.

Prairie Warblers winter primarily in the Caribbean (AOU 1983), and most birds return to the state during the latter half of April. Nesting activities commence soon thereafter, and egg laying begins by early May (Mengel 1965). According to Mengel (1965), an early peak in clutch completion occurs during mid-May, followed by a second peak in late June. Later nests may represent the raising of a second brood, but documentation of double-brooding in Kentucky is lacking. Mengel (1965) gives the average size of 14 clutches or broods known or thought to be complete as 3.5 (range of 3–4). The species is sometimes parasitized by Brown-headed Cowbirds (Mengel 1965).

Forest Cover

Value	% of Blocks	Avg Abund
All	55.4	2.0
1	20.8	1.7
2	52.1	1.7
3	66.1	2.2
4	68.8	2.2
5	48.1	1.9

The nest is often placed in a small tree or shrub, although tangles of briars or vines are also used. Most nests have been located in deciduous trees, but cedars and small pines are also chosen. The nest is either constructed in an upright fork or suspended between branches of a horizontal fork, and it is typically hidden by surrounding foliage. The nest is constructed of a variety of soft plant materials, including cottony plant down, and is lined with finer materials (Harrison 1975). The average height of 13 nests given by Mengel (1965) was 4.7 feet (range of 2.5–8.0 feet).

Ron Austing

Prairie Warblers were recorded in more than 55% of priority blocks statewide, and 19 incidental observations were reported. Occurrence was lowest in the Mississippi Alluvial Plain and highest in the Highland Rim, the Cumberland Plateau, and the Shawnee Hills. Average abundance was relatively uniform across the state. Occurrence was probably most closely related to the amount of available habitat, which is scarce in highly agricultural areas like the Inner Blue Grass, the southwestern Highland Rim, the northwestern Shawnee Hills, and the Mississippi Alluvial Plain. Low occurrence in the Cumberland Mountains likely can be explained by the transition to a more northern avifauna there and a corresponding loss of more southern species like the Prairie Warbler. Occurrence and average abundance were highest in forested areas with some openings.

About 12% of priority block records were for confirmed breeding. Although at least two active nests were located, most confirmed records were based on the observation of recently fledged young. Other confirmed records were based on adults carrying food and nest building. Prairie Warblers were also seen feeding young cowbirds on several occasions.

Prairie Warblers are reported in small numbers on most Kentucky BBS routes. The average number of individuals recorded per BBS route for the periods 1966–91 and 1982–91 was 3.34 and 1.67, respectively. Trend analysis of these data reveals a significant ($p<.01$) decrease of 3.1% per year for the period 1966–91, but a nonsignificant decrease of 6.1% per year for the period 1982–91. Although habitat loss may be responsible for the long-term decline (Robbins et al. 1986), some loss of early successional habitat is attributable simply to the reversion of forests since the mid-1900s (U.S. Department of Agriculture 1978).

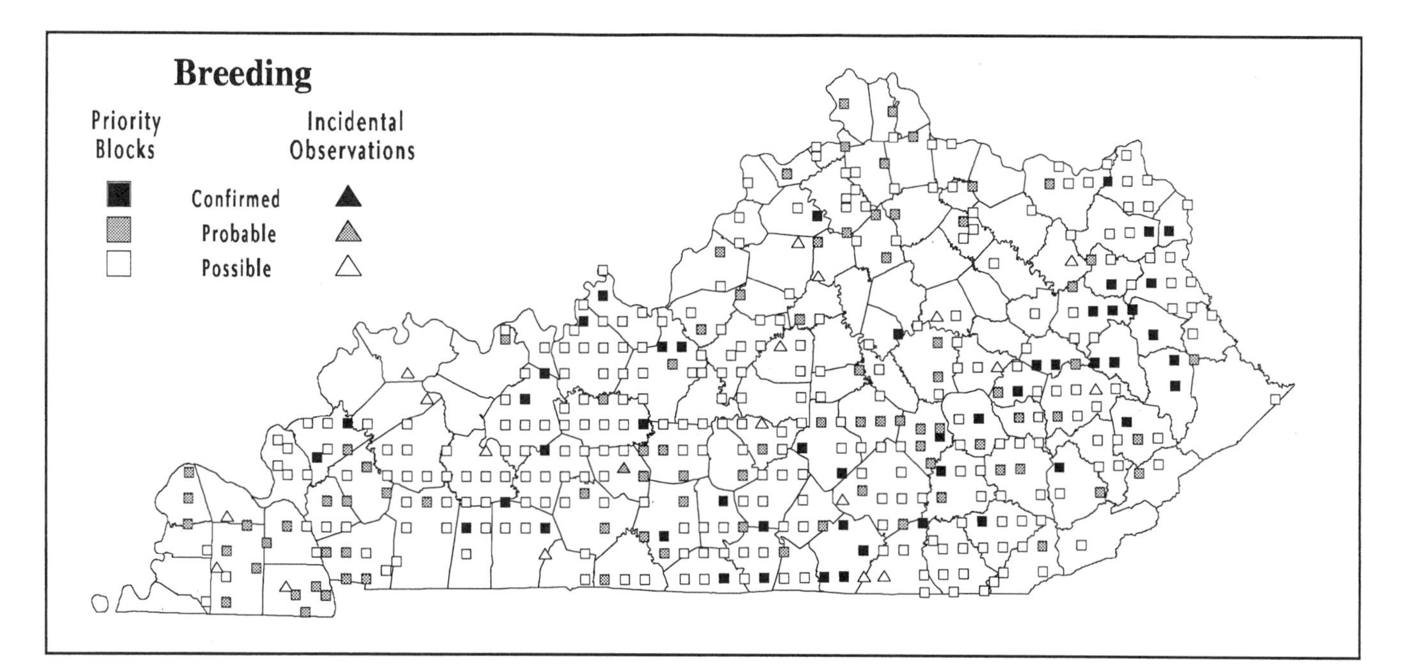

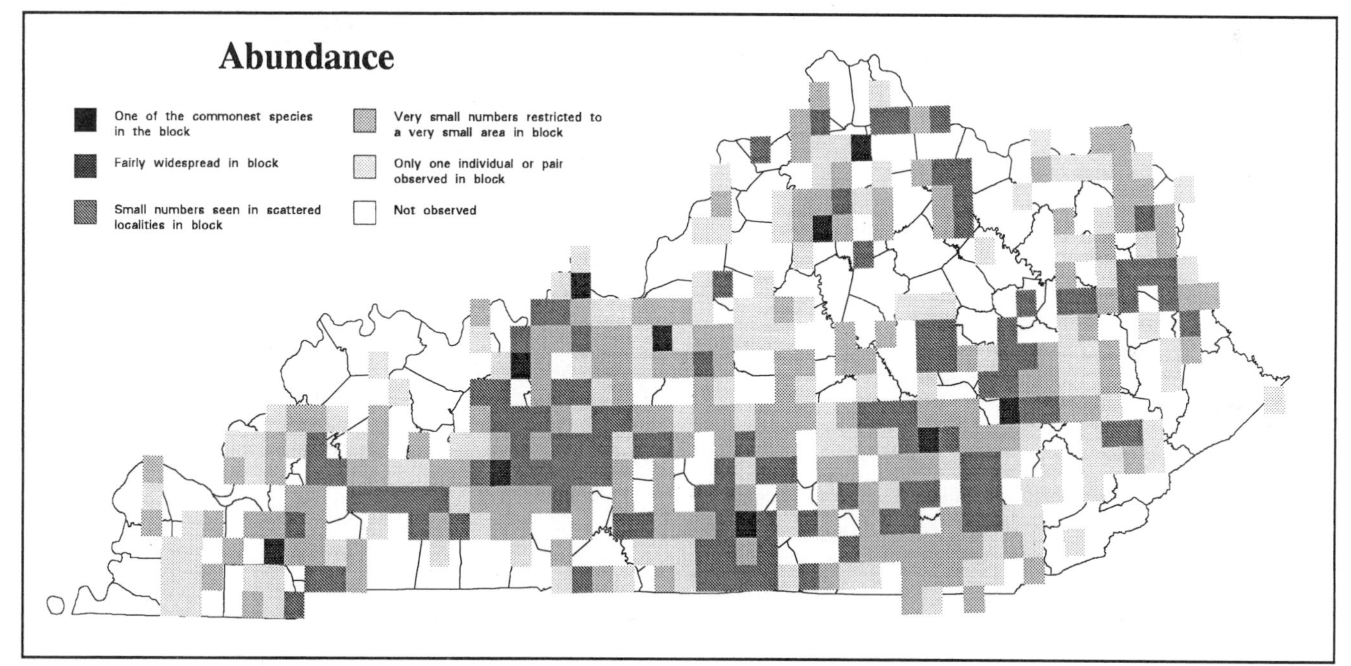

Analysis of Block Data by Physiographic Province Section

Physiographic Province Section	Total Blocks Surveyed	Blocks with Data	Avg Abund	% with Data	Section's % for State
Mississippi Alluvial Plain	14	1	2.0	7.1	0.2
East Gulf Coastal Plain	36	17	1.7	47.2	4.2
Highland Rim	139	90	2.1	64.7	22.3
Shawnee Hills	142	90	2.2	63.4	22.3
Blue Grass	204	89	1.9	43.6	22.1
Cumberland Plateau	173	111	2.0	64.2	27.5
Cumberland Mountains	19	5	1.4	26.3	1.2

Summary of Breeding Status

Number of Blocks in Which Species Was Recorded		
Total	**403**	**55.4%**
Confirmed	49	12.2%
Probable	87	21.6%
Possible	267	66.2%

Prairie Warbler

Cerulean Warbler

Dendroica cerulea

The Cerulean Warbler has apparently declined substantially in Kentucky as a breeding bird during the past few decades. Mengel (1965) regarded the species as "fairly common to common in western and central Kentucky, somewhat less numerous and more local eastward." This warbler's current status is somewhat the reverse: it is now very locally distributed in summer over much of Kentucky west of the Cumberland Plateau, and it seems to be fairly widespread only in the Cumberland Plateau and the Cumberland Mountains. The Cerulean Warbler has declined not only in Kentucky but also throughout much of its range (Robbins et al. 1989; Robbins et al. 1992). Recently it was added as a candidate for federal listing (USFWS 1991).

The Cerulean Warbler is typically found in mature deciduous forest, where it forages high in the canopy. The species is encountered most frequently in mesic situations, especially bottomland forests, although mesophytic and subxeric forests of slopes are also used, especially in eastern Kentucky. Cerulean Warblers seem to tolerate light forest disturbance, but forested areas moderately or heavily disturbed by logging or other activity are often abandoned. While this warbler is occasionally found in relatively isolated tracts of suitable habitat, it occurs with much greater regularity in extensively forested areas.

The Cerulean Warbler was certainly more common and widespread in Kentucky two centuries ago. Because the species typically inhabits mature forest, human alteration of the landscape has had a devastating impact on its numbers across much of the state. This warbler has declined most directly as a result of the nearly complete deforestation of the bottomlands of central and western Kentucky. In addition, the fragmentation and frequent logging of remaining forest have likely contributed to the reduction in its abundance.

Cerulean Warblers generally return to Kentucky from their wintering grounds in South America during the last two weeks of April, although it may be early May before all of the nesting birds have returned. Specific Kentucky data on nesting activity are scarce, but four nests observed near Mammoth Cave were under construction during the first week of May (Mengel 1965), and it is likely that most clutches are completed by the middle of the month. Family groups are most conspicuous during June, and a few birds begin to wander by early July. Kentucky data on clutch size are lacking, but Harrison (1975) gives rangewide clutch size as 3–5, commonly 4.

Forest Cover

Value	% of Blocks	Avg Abund
All	16.4	1.7
1	2.0	1.0
2	6.3	1.1
3	13.0	1.5
4	33.8	1.9
5	44.2	1.8

Nests are typically placed high in deciduous trees, often over roads, streams, or other artificial or natural clearings (Mengel 1965). The nest is usually situated inside the outer crown of leaves at the base of a fork on a thin, horizontal branch. The average height of eight nests reported by Mengel (1965) was 38 feet (range of 18–60 feet). The nest is constructed of a variety of plant materials, bound with spiderwebs, and lined with fine grasses and hair (Harrison 1975).

Results of the atlas survey yielded a considerably different picture of the status of the Cerulean Warbler than that summarized by Mengel (1965). The species was recorded in only about 16% of the priority blocks statewide, and 12 incidental observations were reported. Occurrence was greatest in the Cumberland Plateau and the Cumberland Mountains, but even there the species was found in only about one-third of the blocks. Moreover, Cerulean Warblers were found in less than 13% of the blocks throughout central and western Kentucky except in the Mississippi Alluvial Plain, where some large tracts of bottomland forest remain. Average abundance was relatively low statewide. As expected, occurrence and average abundance were lowest in extensively cleared areas and highest in predominantly forested areas.

Hal H. Harrison

Although it is likely that survey limitations resulted in Cerulean Warblers' being missed on numerous blocks, the paucity of atlas data is believed to indicate the decrease in the nesting population previously noted. Because the species begins dispersing from the breeding grounds so early, it is possible that some late season records pertained to early transients or postbreeding dispersers. Nonetheless, all records of birds in suitable breeding habitat were included in the atlas data set.

Not surprisingly, less than 10% of priority block records were for confirmed breeding. Most of the 11 confirmed records were based on the observation of recently fledged young, but one involved adults carrying food.

Cerulean Warblers are not adequately surveyed by Kentucky BBS surveys. The average number of individuals per BBS route for the periods 1966–91 and 1982–91 was 0.95 and 1.23, respectively. Trend analysis of these data reveals a nonsignificant decrease of 2.6% per year for the period 1966–91, but a significant ($p<.01$) increase of 12.1% per year for the period 1982–91. The explanation for the apparent short-term increase is not clear; the data may simply reflect sampling bias.

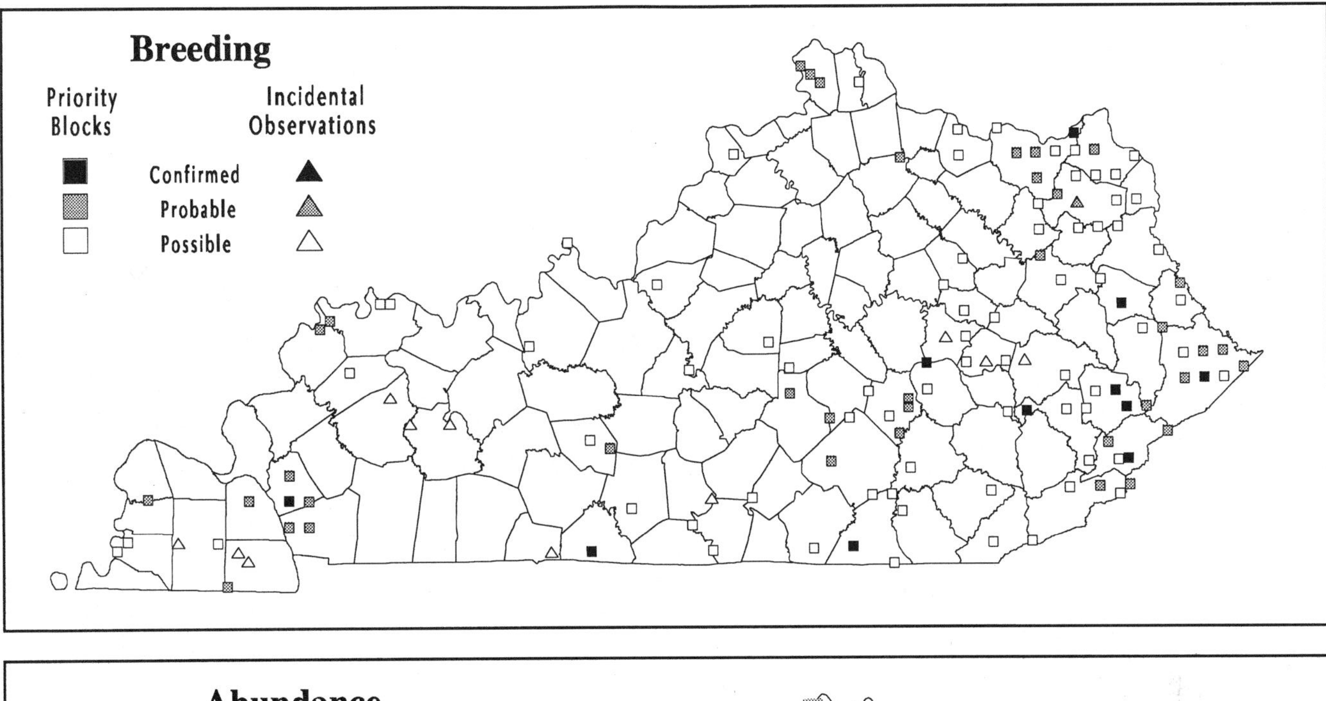

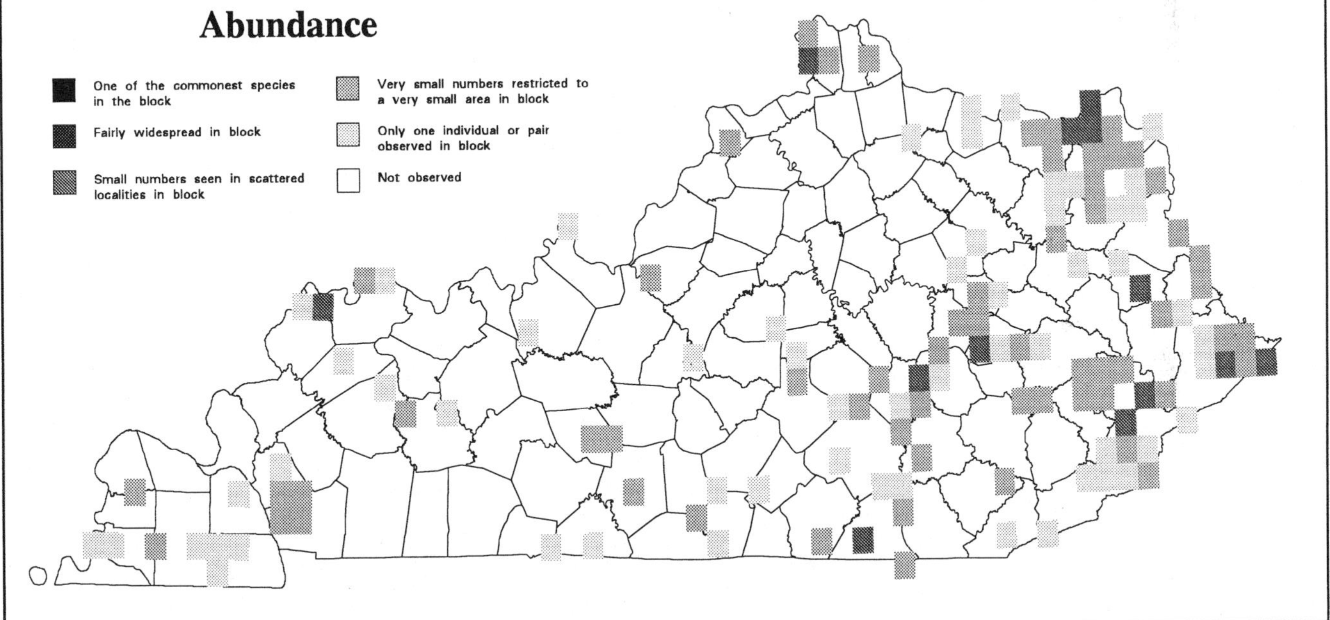

Analysis of Block Data by Physiographic Province Section

Physiographic Province Section	Total Blocks Surveyed	Blocks with Data	Avg Abund	% with Data	Section's % for State
Mississippi Alluvial Plain	14	3	1.3	21.4	2.5
East Gulf Coastal Plain	36	3	1.0	8.3	2.5
Highland Rim	139	13	1.5	9.4	10.9
Shawnee Hills	142	9	1.6	6.3	7.6
Blue Grass	204	26	1.8	12.7	21.8
Cumberland Plateau	173	59	1.8	34.1	49.6
Cumberland Mountains	19	6	1.2	31.6	5.0

Summary of Breeding Status

Number of Blocks in Which Species Was Recorded		
Total	**119**	**16.4%**
Confirmed	11	9.2%
Probable	37	31.1%
Possible	71	59.7%

Cerulean Warbler

Black-and-white Warbler

Mniotilta varia

The Black-and-white Warbler's high-pitched song is a fairly common summer sound in the forests of eastern Kentucky, but it is seldom encountered west of the Cumberland Plateau. Mengel (1965) regarded the species as a common summer resident in the Cumberland Mountains and the Cumberland Plateau, but less numerous westward, being rare to very rare in most areas and locally common only in the Knobs of the Blue Grass and the Shawnee Hills.

Black-and-white Warblers are found in a variety of deciduous and mixed forest types. They are observed most frequently in midstory and understory forest levels, where they creep along on trunks, branches, and vines below the main forest canopy. The species occurs most frequently in subxeric to mesic woodlands, and it seems to be more common on slopes than in forests with little or no relief. These warblers are most frequent in mature or fairly mature woodlands, but they also inhabit younger forest as well as forest recovering from selective logging.

Black-and-white Warblers certainly have decreased across much of Kentucky as human alteration of the landscape has progressed. While the species has adapted to forest disturbance to some degree, the extensive clearing of vast woodland areas for agricultural use and settlement has resulted in a tremendous loss of suitable nesting habitat, especially in the central and western parts of the state.

Black-and-white Warblers winter from the extreme southern United States, southward into northern South America (AOU 1983), and they are among the first of Kentucky's nesting warblers to return in early spring. The first males typically arrive by early April, although the full complement of breeding birds may not return until late in the month. Specific Kentucky data on nesting activity are scarce, but young capable of flight have been observed as early as mid-May in the Land Between the Lakes (author's notes). Most clutches are likely completed by late May, and most young fledge by mid-June. The average size of seven reported clutches is 4.3 (range of 3–5) (Stamm and Slack 1957; Mengel 1965; Davis et al. 1980; KOS Nest Cards). Family groups are conspicuous during June, in large part because of the noisy calling of the young as they move about the forest in the company of their parents. A few birds usually begin to wander by mid-June, and many appear in nonbreeding areas during July.

Forest Cover

Value	% of Blocks	Avg Abund
All	24.9	2.0
1	1.0	1.0
2	5.3	1.5
3	13.9	1.6
4	59.1	1.9
5	90.4	2.6

The nest is placed among the leaf litter on the forest floor, usually on a moderate to steep slope (Mengel 1965). It is typically situated next to and sheltered by a fallen tree limb or the base of a rock outcrop, tree stump, or shrub. The nest is constructed primarily of dead leaves and is lined with fine grasses, weed fibers, strips of bark, and rootlets (Mengel 1965; Harrison 1975).

Alvin E. Staffan

The atlas survey yielded records of Black-and-white Warblers in nearly 25% of priority blocks, and nine incidental observations were reported. As expected, occurrence was highest in the Cumberland Mountains and the Cumberland Plateau, and low throughout the rest of the state. Across central Kentucky the species is primarily restricted to areas affording the greatest amount of forest, including large tracts of public land like Mammoth Cave National Park. Farther west it is scarce even in extensively forested areas like the Land Between the Lakes and the Pennyrile State Forest, where only a few possible reports were obtained. Most Blue Grass records came from the Knobs subsection, while most in the Highland Rim came from its eastern transition to the Cumberland Plateau. The species was not recorded in the Mississippi Alluvial Plain, where periodic flooding and the lack of relief are likely responsible for its absence. For the most part, average abundance varied similarly. Occurrence and average abundance were closely related to the percentage of forest cover.

Black-and-white Warblers seem to begin dispersing from breeding areas very early, and it is unclear to what extent this behavior affected the atlas results. The species was occasionally recorded in areas unsuitable for nesting as early as mid-June, and although most of these reports were properly noted by atlasers in the observed category, a few may have been listed as indicating nesting. An effort was made to check as many of these reports as possible, especially those from areas that do not support documented nesting populations.

Only about 11% of priority block records were for confirmed breeding. Almost all of the 19 confirmed records were based on the observation of recently fledged young, but a few involved adults carrying food.

Black-and-white Warblers are not recorded frequently on most Kentucky BBS routes. The average number of individuals per BBS route for the periods 1966–91 and 1982–91 was 0.58 and 0.41, respectively. In part due to small sample sizes, trend analysis of these data does not reveal statistically significant results.

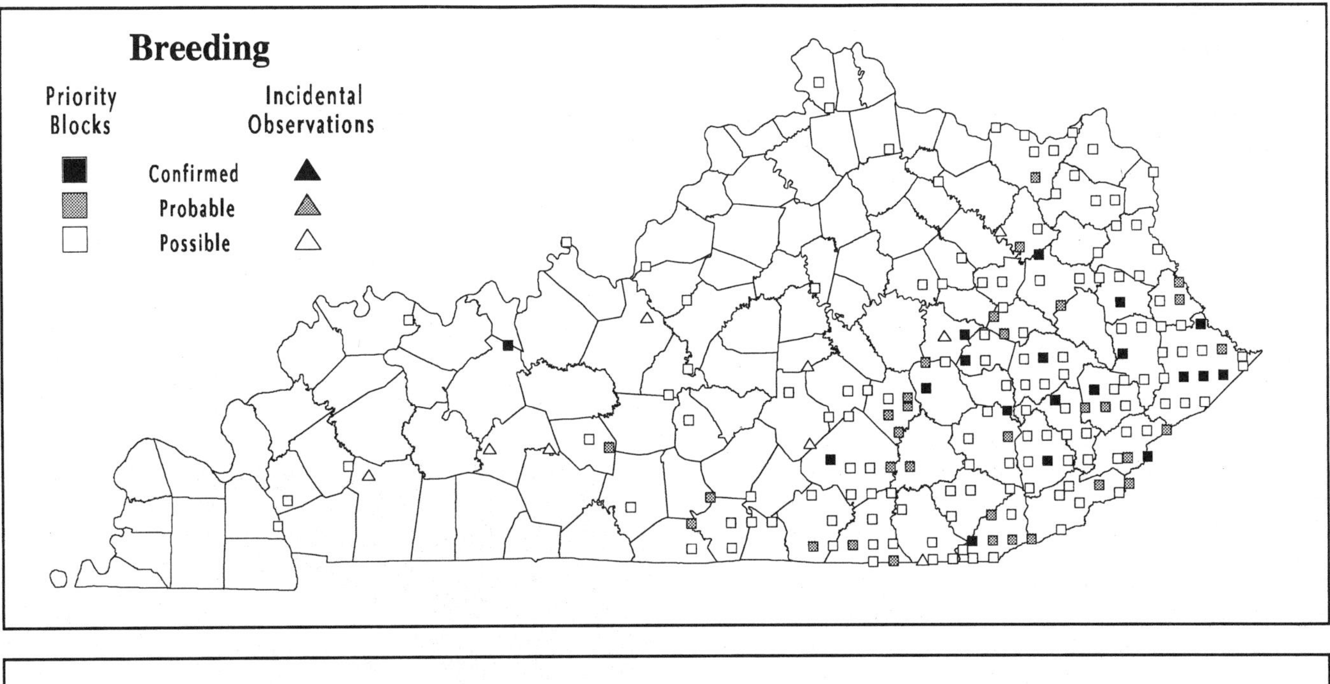

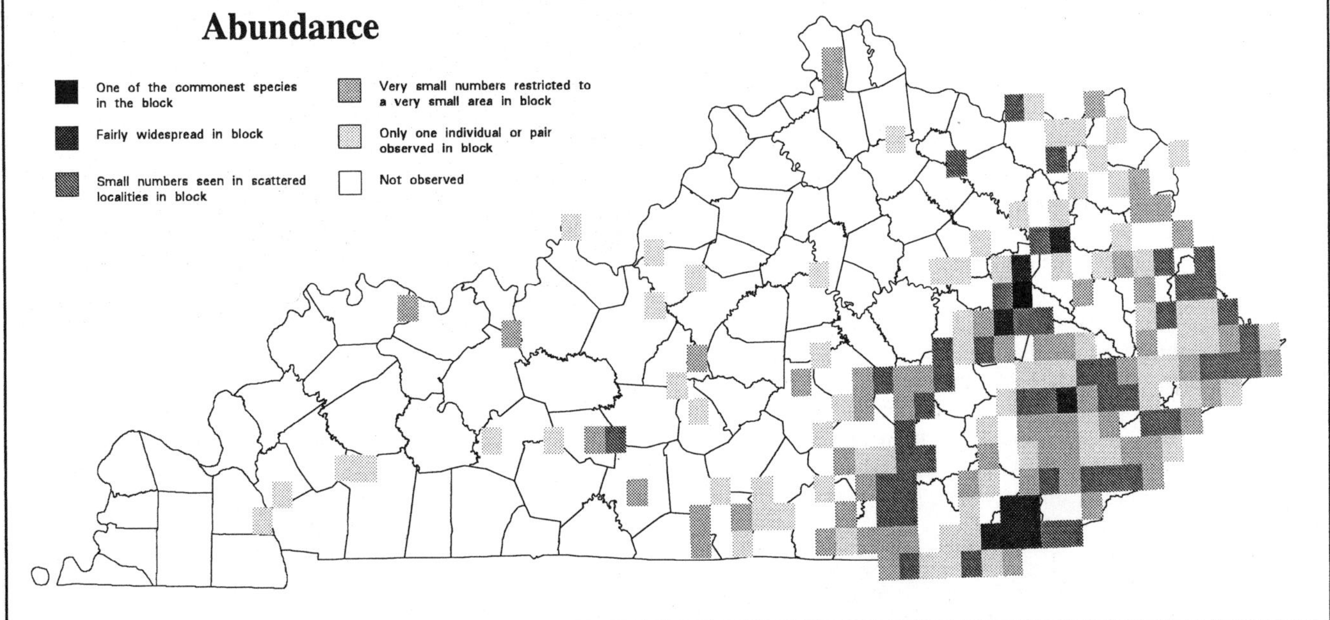

Analysis of Block Data by Physiographic Province Section

Physiographic Province Section	Total Blocks Surveyed	Blocks with Data	Avg Abund	% with Data	Section's % for State
Mississippi Alluvial Plain	14	-	-	-	-
East Gulf Coastal Plain	36	1	1.0	2.8	0.6
Highland Rim	139	19	1.4	13.7	10.5
Shawnee Hills	142	6	1.8	4.2	3.3
Blue Grass	204	22	1.8	10.8	12.2
Cumberland Plateau	173	117	2.1	67.6	64.6
Cumberland Mountains	19	16	2.5	84.2	8.8

Summary of Breeding Status

Number of Blocks in Which Species Was Recorded		
Total	**181**	**24.9%**
Confirmed	19	10.5%
Probable	32	17.7%
Possible	130	71.8%

Black-and-white Warbler

American Redstart

Setophaga ruticilla

Although the American Redstart's breeding range is considered to include all of Kentucky, the species is very locally distributed throughout most of the state in summer. Mengel (1965) summarized its statewide occurrence as common to fairly common in the mixed mesophytic forests of the mountainous east, rare to completely absent in the drier forests of central Kentucky, and again locally common in the mesic forests of the western lowlands. Primarily because of continued deforestation in central and western Kentucky, this colorful warbler is even more locally distributed today.

The American Redstart is typically found in mesic forests of floodplains and riparian zones, although it also occurs on mesic slopes, especially in the Cumberland Mountains. In contrast, the species is typically absent in drier forests of slopes and uplands. Redstarts are sometimes found in fairly mature forest situations, but they occur primarily in younger forests or along forest margins. Although this warbler can be found in natural forest and edge habitats, it occurs primarily in altered situations including regenerating fields and clearcuts, roadway and utility corridors, and the margins of rural yards.

While the redstart possibly was never common across much of the state, deforestation (especially in the bottomlands of central and western Kentucky) probably has resulted in a decrease in its abundance during the last two centuries. The species has remained fairly widespread in much of eastern Kentucky, where selective logging in floodplain habitats may have resulted in local increases, as forests have proceeded through succession.

Redstarts winter throughout the tropics of Middle America and northern South America (AOU 1983). A few birds appear in Kentucky by late April, although it may be the middle of May before most of the nesting population has returned. According to Mengel (1965), clutches are completed from mid-May to mid-June, without an evident peak. A nest observed in Leslie County during the atlas fieldwork still contained young on July 7, 1989. Kentucky data on clutch size are lacking, but Harrison (1975) gives rangewide clutch size as 4, sometimes 2 or 3, rarely 5.

Forest Cover

Value	% of Blocks	Avg Abund
All	19.4	1.9
1	1.0	1.0
2	4.2	1.7
3	8.3	1.3
4	48.1	1.8
5	75.0	2.3

Nests are built in a variety of deciduous tree species. They may be situated at midstory levels beneath the canopy of larger trees in mature forest or along forest edge, but they usually are found in stands of younger trees in successional or disturbed forest. The nest is usually placed in an upright fork in the upper portion of a small tree, but sometimes in the lower branches of a large tree. The neat, cup-shaped structure is composed of grasses, rootlets, plant fibers, and other plant materials and is lined with finer grasses (Mengel 1965; Harrison 1975). The average height of four nests reported by Mengel (1965) was 20.5 feet (range of 12–30 feet).

The atlas survey yielded records of redstarts in more than 19% of priority blocks statewide, and seven incidental observations were reported. As expected, occurrence was highest in the Cumberland Mountains and the Cumberland Plateau, and redstarts were found in more than one-fourth of the blocks in the Mississippi Alluvial Plain. Throughout the rest of the state the species was distributed very locally, occurring in less than 6% of priority blocks. In these areas it was usually found in tracts of floodplain or riparian forest along the larger rivers. Average abundance was relatively low statewide. Occurrence and average abundance were lowest in extensively cleared areas and highest in forested areas.

Ron Austing

Almost 26% of priority block records were for confirmed breeding. One active nest was located in northern Leslie County, but most of the 36 confirmed records were based on the observation of recently fledged young. Family groups were relatively conspicuous as they foraged about woodland openings and edges, especially in southeastern Kentucky, where rural roads often run through suitable nesting habitat in stream valleys.

American Redstarts are not reported on most Kentucky BBS routes. The average number of individuals per BBS route for the periods 1966–91 and 1982–91 was 0.46 and 0.41, respectively. Trend analysis of these data reveals a nonsignificant decrease of 2.3% per year for the period 1966–91, but a significant ($p<.01$) decrease of 33.8% per year for the period 1982–91. Reasons for the short-term decline are not known, but a small sample size is likely responsible for its magnitude.

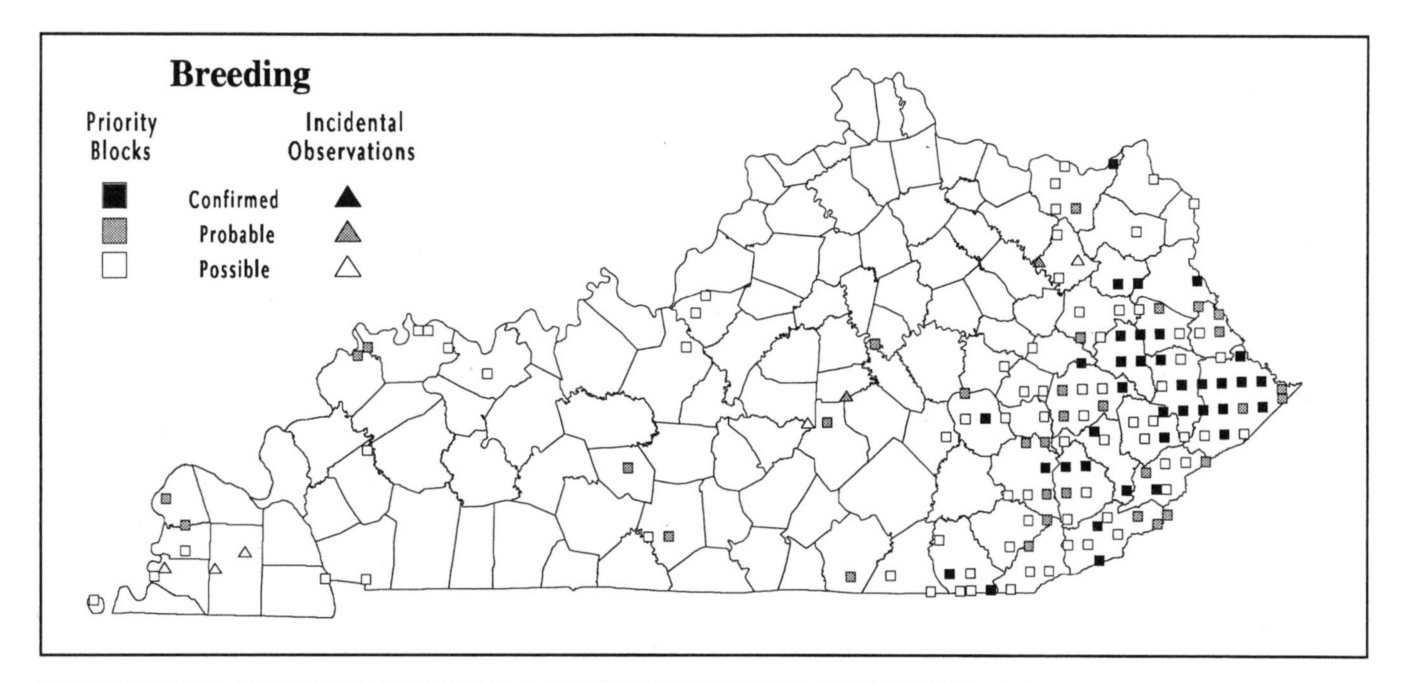

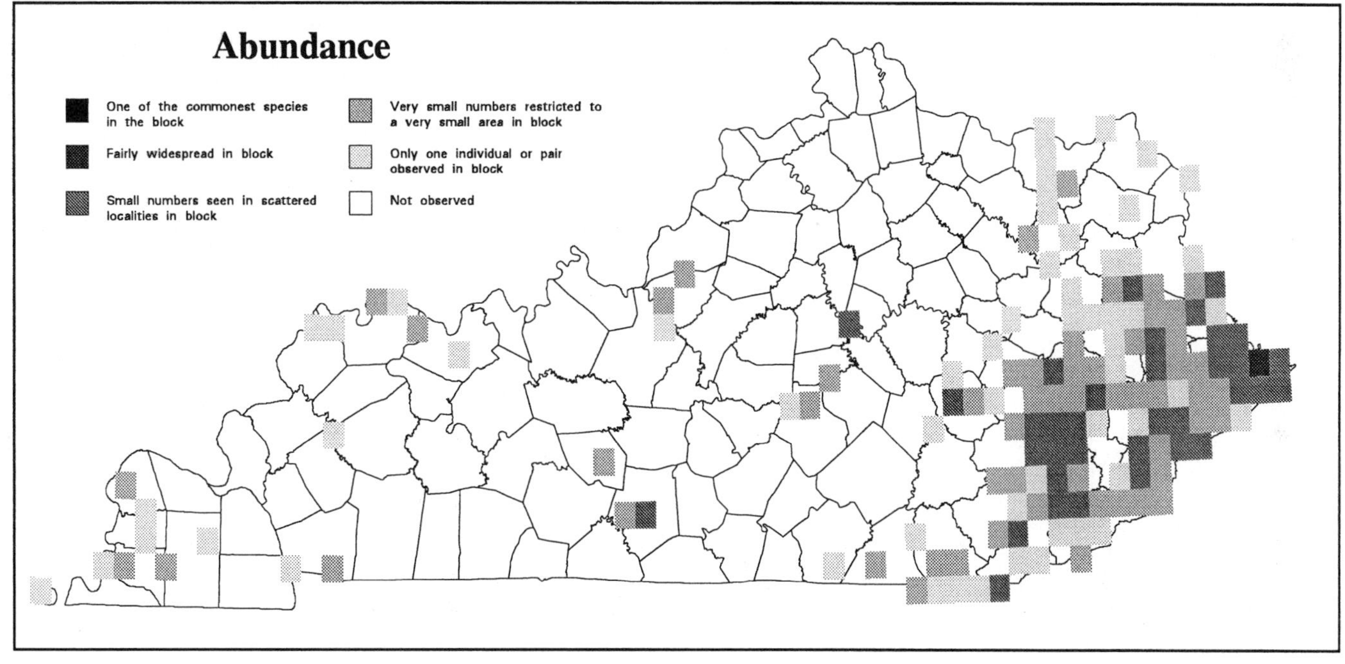

Analysis of Block Data by Physiographic Province Section

Physiographic Province Section	Total Blocks Surveyed	Blocks with Data	Avg Abund	% with Data	Section's % for State
Mississippi Alluvial Plain	14	4	1.2	28.6	2.8
East Gulf Coastal Plain	36	2	1.0	5.6	1.4
Highland Rim	139	4	2.0	2.9	2.8
Shawnee Hills	142	8	1.4	5.6	5.7
Blue Grass	204	12	1.5	5.9	8.5
Cumberland Plateau	173	98	2.0	56.6	69.5
Cumberland Mountains	19	13	1.7	68.4	9.2

Summary of Breeding Status

Number of Blocks in Which Species Was Recorded		
Total	**141**	**19.4%**
Confirmed	36	25.5%
Probable	33	23.4%
Possible	72	51.1%

American Redstart

Prothonotary Warbler

Protonotaria citrea

Easily recognized by both voice and plumage, the Prothonotary Warbler is a variably rare to common summer resident throughout Kentucky west of the Cumberland Plateau. In its favored floodplain swamp habitat in far western Kentucky, the species ranks among the most numerous nesting birds, but it decreases eastward. Mengel (1965) considered this warbler to be common along the streams and in the lowlands of southern and western Kentucky and as far up the Ohio River Valley as Louisville, but uncommon to rare farther up the Ohio and its larger tributaries to various points west of the Cumberland Plateau.

Prothonotary Warblers are seldom encountered far from water. They frequent a great variety of natural and artificial habitats, including riparian corridors along rivers and streams, floodplain sloughs, swamps, and the margins of reservoirs. In addition, the species uses seasonally flooded bottomland forest that may be dry throughout the summer, as well as residential areas near bodies of water.

It is possible that the Prothonotary Warbler was more widespread and numerous in Kentucky before settlement. Audubon (1861) found the species along the Ohio River regularly to Henderson and rarely above Louisville in the early 1800s, and prime habitat was likely scattered in river floodplains across central and western Kentucky. Much wetland habitat has subsequently been lost owing to draining and clearing of bottomland forests and swamps. In contrast, human changes to the landscape also have resulted in the creation of much suitable habitat. Some activities have resulted in raised water levels, killing timber and creating optimal nesting habitat, and the construction of reservoirs probably has contributed to the species's spread in more recent times.

Although Prothonotary Warblers often arrive in Kentucky from their wintering grounds in the tropics by early April, it is probably late in the month before the entire nesting population has returned. Nest site selection and egg laying begin soon thereafter, and a peak in clutch completion occurs in mid-May. A second brood is raised at least occasionally, and egg laying may continue into early July (Mengel 1965). The average size of eight clutches given by Mengel was 4.5 (range of 4–5), although two more recent nests have apparently held complete clutches of only two eggs (Stamm and Jones 1966; KOS Nest Cards). Even though the species nests in cavities, it is sometimes parasitized by Brown-headed Cowbirds (Mengel 1965).

Forest Cover

Value	% of Blocks	Avg Abund
All	18.3	1.9
1	20.8	1.7
2	27.4	1.9
3	19.1	2.0
4	9.1	2.1
5	3.8	2.0

Prothonotary Warblers are cavity nesters, choosing a variety of natural and artificial sites. They typically nest over or near standing water, but occasionally choose a site relatively far from water. In addition to natural cavities in trees and old woodpecker holes, nest boxes and other artificial sites may be used. Off and on for many years a pair has nested in a cavity on the Mammoth Cave ferry on the Green River (Binnewies 1943; J. Elmore, pers. comm.). The nest typically consists of a shallow cup of mosses, rootlets, twigs, and leaves. It is lined with fine grass, leaf stems, and feathers (Harrison 1975). The average height of 12 nests given by Mengel (1965) was 6.8 feet (range of 3–18 feet).

Ron Austing

The atlas survey yielded records of Prothonotary Warblers in more than 18% of priority blocks, and 18 incidental observations were reported. As expected, occurrence was highest in western Kentucky and decreased eastward. The species occurred locally across central Kentucky up to the edge of the Cumberland Plateau. Most of the easternmost records were obtained along the larger rivers, including the Cumberland, the Kentucky, the Licking, and the Ohio. The species was found in only one block on the Cumberland Plateau and was missed in the Cumberland Mountains. Average abundance was relatively low and unvaried. Occurrence was probably more closely related to physiography and the proximity of water than to the percentage of forest cover.

The Prothonotary Warbler was slightly less difficult to confirm as nesting than most other warblers. Nearly 23% of priority block records were for confirmed breeding. A number of active nests were located, some documented only by the observation of adults carrying food into nest cavities. Other confirmed records were based on the observation of recently fledged young, adults carrying food, nest building, and distraction displays.

Prothonotary Warblers are regularly recorded on only about one-third of Kentucky's BBS routes. The average number of individuals per BBS route for the periods 1966–91 and 1982–91 was 0.45 and 0.71, respectively. In part due to small sample sizes, trend analysis of these data does not reveal statistically significant results.

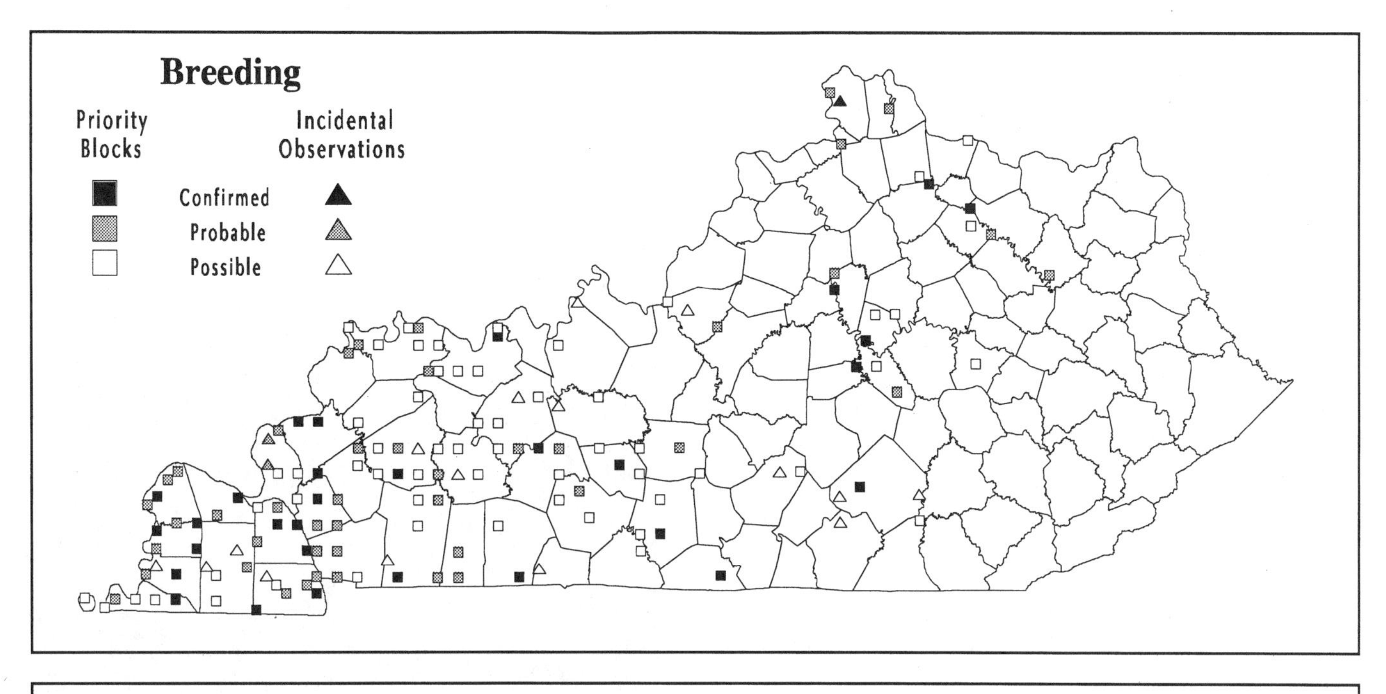

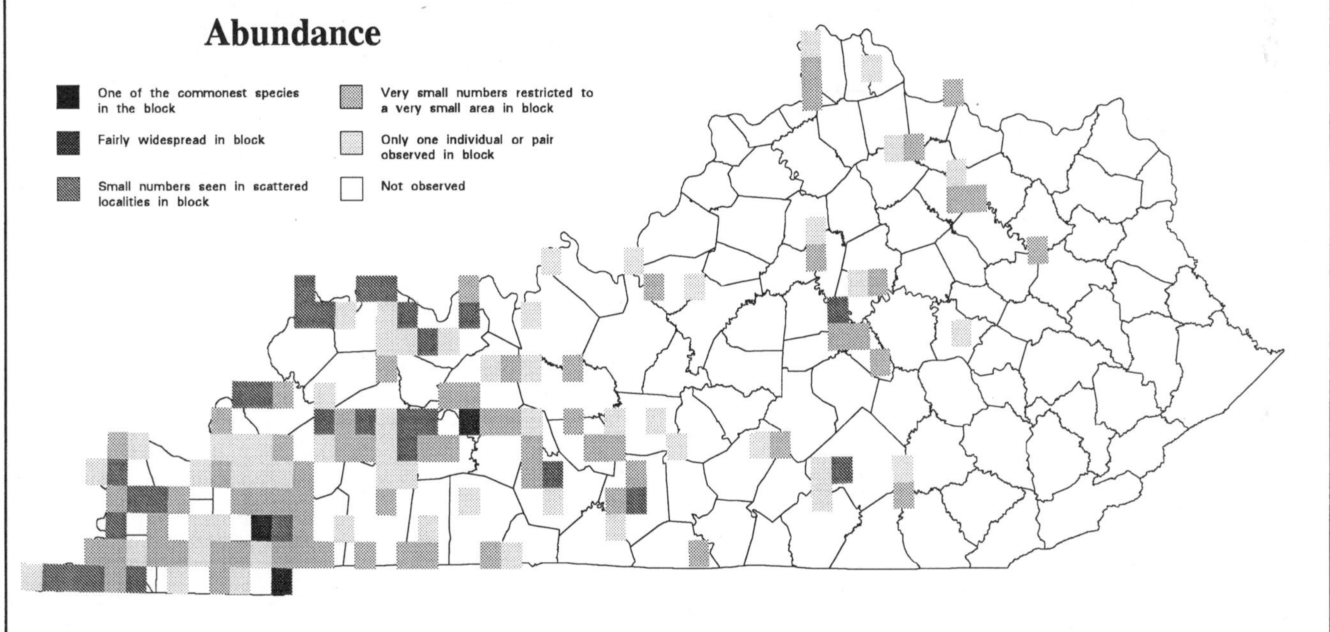

Analysis of Block Data by Physiographic Province Section

Physiographic Province Section	Total Blocks Surveyed	Blocks with Data	Avg Abund	% with Data	Section's % for State
Mississippi Alluvial Plain	14	13	2.2	92.9	9.8
East Gulf Coastal Plain	36	21	1.9	58.3	15.8
Highland Rim	139	26	1.8	18.7	19.5
Shawnee Hills	142	51	2.1	35.9	38.3
Blue Grass	204	21	1.6	10.3	15.8
Cumberland Plateau	173	1	2.0	0.6	0.8
Cumberland Mountains	19	-	-	-	-

Summary of Breeding Status

Number of Blocks in Which Species Was Recorded		
Total	**133**	**18.3%**
Confirmed	30	22.5%
Probable	44	33.1%
Possible	59	44.4%

Prothonotary Warbler

Worm-eating Warbler

Helmitheros vermivorus

The Worm-eating Warbler's insectlike song is a fairly common summer sound throughout the forests of eastern Kentucky, but the species is encountered only locally throughout the rest of the state. Mengel (1965) considered this warbler to be common at lower elevations in the Cumberland Mountains and throughout the Cumberland Plateau. West of the Cumberland Plateau, however, he found that it seemed to be restricted to the Knobs subsection of the Blue Grass and the Shawnee Hills, becoming increasingly more local and less numerous westward. More recent fieldwork seems to indicate that the species is scarce at the highest elevations in the Cumberland Mountains (Mengel 1965; Croft 1971; Davis et al. 1980).

The Worm-eating Warbler is a bird of forests, especially favoring moderate to steep slopes. The species uses a wide variety of forest types, including subxeric oak-hickory and mixed pine-hardwood communities, but it is most common in more mesic deciduous and mixed types of lower slopes and ravines. Like many other ground nesters, this warbler typically avoids floodplain forests. While the Worm-eating Warbler usually inhabits mature or fairly mature forest, it also uses younger forest and forest edge created by natural or artificial disturbance. The species is regularly encountered in areas of dissected woodland, but it generally avoids small, isolated tracts.

It is likely that Worm-eating Warblers have decreased across Kentucky in the last two centuries. While the native prairies, barrens, and savannas did not provide suitable habitat, vast areas in the central and western parts of the state were covered in forests that have been cleared for agricultural use and settlement. These forests probably once supported a substantial nesting population that has largely disappeared. In contrast, a history of periodic disturbance to remaining woodlands probably has not affected the species significantly, especially in eastern Kentucky, where second-growth forests suitable for nesting dominate the landscape today.

Forest Cover

Value	% of Blocks	Avg Abund
All	23.4	1.8
1	4.9	1.2
2	4.2	1.4
3	22.6	1.5
4	46.1	1.8
5	65.4	2.3

Worm-eating Warblers winter primarily in Middle America (AOU 1983), and most birds return to their Kentucky breeding grounds during the second half of April. Nest building has been observed as early as May 3 (KOS Nest Cards), and a peak in clutch completion likely occurs before the beginning of June. Family groups of fledged young are most conspicuous during June, and later nests have not been documented. Kentucky data on clutch size are scarce, but the average size of four clutches or broods noted by Mengel (1965) was 3.8 (range of 3–5). Harrison (1975) gives rangewide clutch size as 3–6, commonly 4 or 5.

The nest is placed on the ground, usually on a slope amid the leaf litter. It is often situated next to the base of a tree, a stump, or a fallen log and is typically hidden beneath the cover of overhanging sapling branches or herbaceous plants (Harrison 1975). The nest is constructed of dead leaves and lined with mosses, fine grass, pine needles, or hair (Mengel 1965; Harrison 1975).

Hal H. Harrison

The atlas fieldwork yielded records of Worm-eating Warblers in more than 23% of priority blocks statewide, and six incidental observations were reported. Occurrence was highest in the Cumberland Mountains and the Cumberland Plateau and lowest in the East Gulf Coastal Plain, but the species was missed entirely in the Mississippi Alluvial Plain. Across central Kentucky, Worm-eating Warblers were found in only about 17% of priority blocks. Most Highland Rim records came from the region's eastern transition to the Cumberland Plateau, while most records from the Blue Grass originated in the Knobs subsection. Average abundance was relatively low, but generally it varied similarly. As expected, occurrence and average abundance were lowest in extensively cleared areas and increased as the degree of forestation increased.

Slightly more than 25% of priority block records were for confirmed breeding. This relatively high figure in large part resulted from the conspicuousness of family groups including recently fledged young. Adults and begging young keep in contact with each other using conspicuous notes that are easily heard. Worm-eating Warblers are sometimes parasitized by Brown-headed Cowbirds, and adult warblers were observed feeding young cowbirds on at least two occasions. The few remaining confirmed records were based on the observation of adults carrying food and distraction displays.

Worm-eating Warblers are reported in small numbers on most Kentucky BBS routes. The average number of individuals per BBS route for the periods 1966–91 and 1982–91 was 0.40 and 0.38, respectively. Trend analysis of these data reveals a significant ($p<0.1$) increase of 3.1% per year for the period 1966–91 and a nonsignificant decrease of 1.0% per year for the period 1982–91.

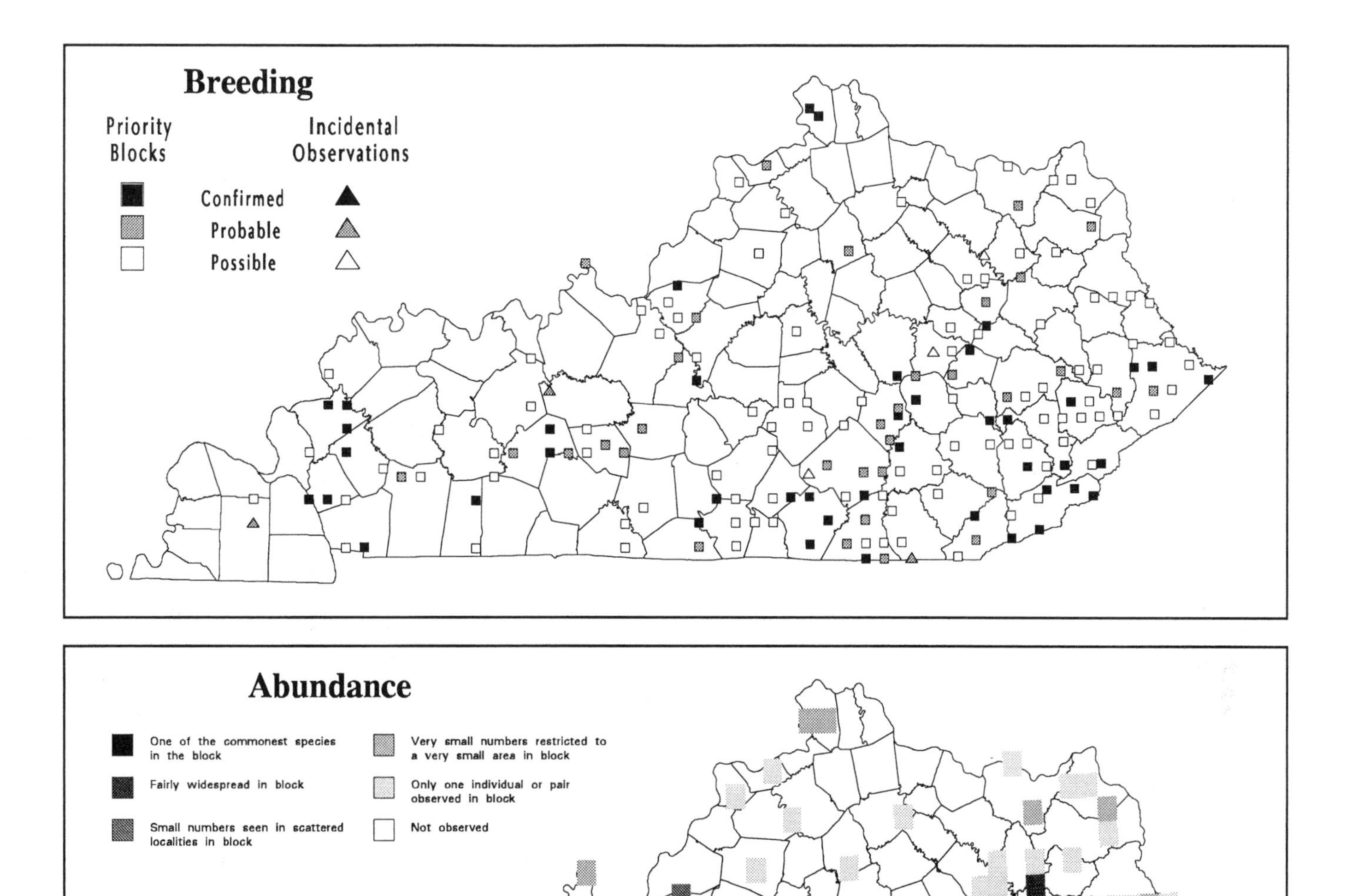

Analysis of Block Data by Physiographic Province Section

Physiographic Province Section	Total Blocks Surveyed	Blocks with Data	Avg Abund	% with Data	Section's % for State
Mississippi Alluvial Plain	14	-	-	-	-
East Gulf Coastal Plain	36	2	1.5	5.6	1.2
Highland Rim	139	31	1.6	22.3	18.2
Shawnee Hills	142	24	1.6	16.9	14.1
Blue Grass	204	26	1.5	12.7	15.3
Cumberland Plateau	173	78	2.0	45.1	45.9
Cumberland Mountains	19	9	1.9	47.4	5.3

Summary of Breeding Status

Number of Blocks in Which Species Was Recorded		
Total	**170**	**23.4%**
Confirmed	43	25.3%
Probable	33	19.4%
Possible	94	55.3%

Worm-eating Warbler

Swainson's Warbler

Limnothlypis swainsonii

The Swainson's Warbler is seldom encountered in Kentucky, but it can be found locally in forested habitats in the southeastern and southwestern portions of the state, and at least occasionally in south central Kentucky. Mengel (1965) regarded the species as fairly common locally in lowland forests of extreme western Kentucky, rare in swamp forests of the Highland Rim and the Shawnee Hills, and locally present in mixed mesophytic forest in the Cumberland Mountains, and possibly elsewhere in eastern Kentucky.

Since the late 1950s Swainson's Warblers seem to have nearly disappeared from southwestern Kentucky, apparently because of continued clearing and dissection of bottomland forests. In the Highland Rim and the Shawnee Hills, a few additional records have been published for the counties of Christian (Stamm 1983), Edmonson (Wilson 1962a), and Hopkins (Stamm 1983; Stamm 1984b), but the species continues to occur sporadically. In contrast, other recent fieldwork has added to our knowledge concerning the occurrence of the Swainson's Warbler in eastern Kentucky. Coskren (1979) summarized a number of reports at several localities on the Cumberland Plateau in Floyd, Lee, Menifee, and Powell counties, and other reports have come from the counties of Elliott (Greene 1978), Letcher (Clark 1962; Hudson 1971), and Pike (Clark 1962; BBS data).

The Swainson's Warbler is a bird of mesic forests with a dense understory. In western and south central Kentucky, the species is typically found in lowland situations, especially floodplain forests with an abundance of giant cane. These warblers also have been found in thickets of young trees in wet bottomlands, regenerating after heavy logging or agricultural use. In southeastern Kentucky, the species frequents forested ravines and lower slopes, and it is most often encountered where a dense understory of rhododendron is present. As in southwestern Kentucky, however, Swainson's Warblers are also found in regenerating forest where the understory is thick and dense (e.g., Hudson 1971).

Little is known about the nesting habits of this reclusive species in Kentucky. Swainson's Warblers winter in Middle America and the Caribbean (AOU 1983), and locally nesting birds seem to return by early May. Nest building was observed a few hundred yards south of the Kentucky state line in Henry County, Tennessee, on May 7, 1986 (author's notes). Mengel (1965) noted nest building in Fulton County on May 17, 1949, and he located a completed but empty nest he believed to belong to this species on May 25 of the same year. During the atlas fieldwork a family group including recently fledged young was observed on June 27, 1988, in Wolfe County. The species has been heard singing through June, although little is known about its habits after the beginning of July. Harrison (1975) gives rangewide clutch size as 3, sometimes 4, rarely 5.

According to Harrison (1975), the nest of this species may lie outside of its singing territory. Nests are usually placed near the ground in dense growth. In the Deep South, Swainson's Warblers often build in palmetto (see photo) or giant cane. The Henry County, Tennessee, nest was situated about two feet above the ground and supported by a tangle of greenbriar vines. Nests are typically constructed of dead leaves, grasses, and plant stems and are lined with finer materials (Harrison 1975). Harrison gives a range in nest height of 2–10 feet.

Hal H. Harrison

The atlas survey yielded only 10 records of Swainson's Warblers in priority blocks, and six incidental observations were reported. The species is fairly well distributed in western West Virginia (Buckelew and Hall 1994), and it certainly must be more widespread in Kentucky than atlas data indicate, especially in eastern Kentucky, where its favored habitat was not well sampled from roadsides. Eastern Kentucky reports mostly involved single records of singing birds, but the species was more widespread in the Cumberland Mountains. In this region, two to five singing males, as well as territorial pairs, were found at three locations along Pine Mountain. In central and western Kentucky the species likely is also distributed locally, even though only one incidental record was obtained during the atlas survey. In Calloway County, single territorial birds were heard at four separate locations, all in the same general area.

Of the 16 total records, only one was confirmed. On June 27, 1988, a family group including recently fledged young was observed in fairly mature forest near the mouth of Chester Branch in Wolfe County. Probable records were based on the observation of birds at a specific location on more than one occasion during the breeding season.

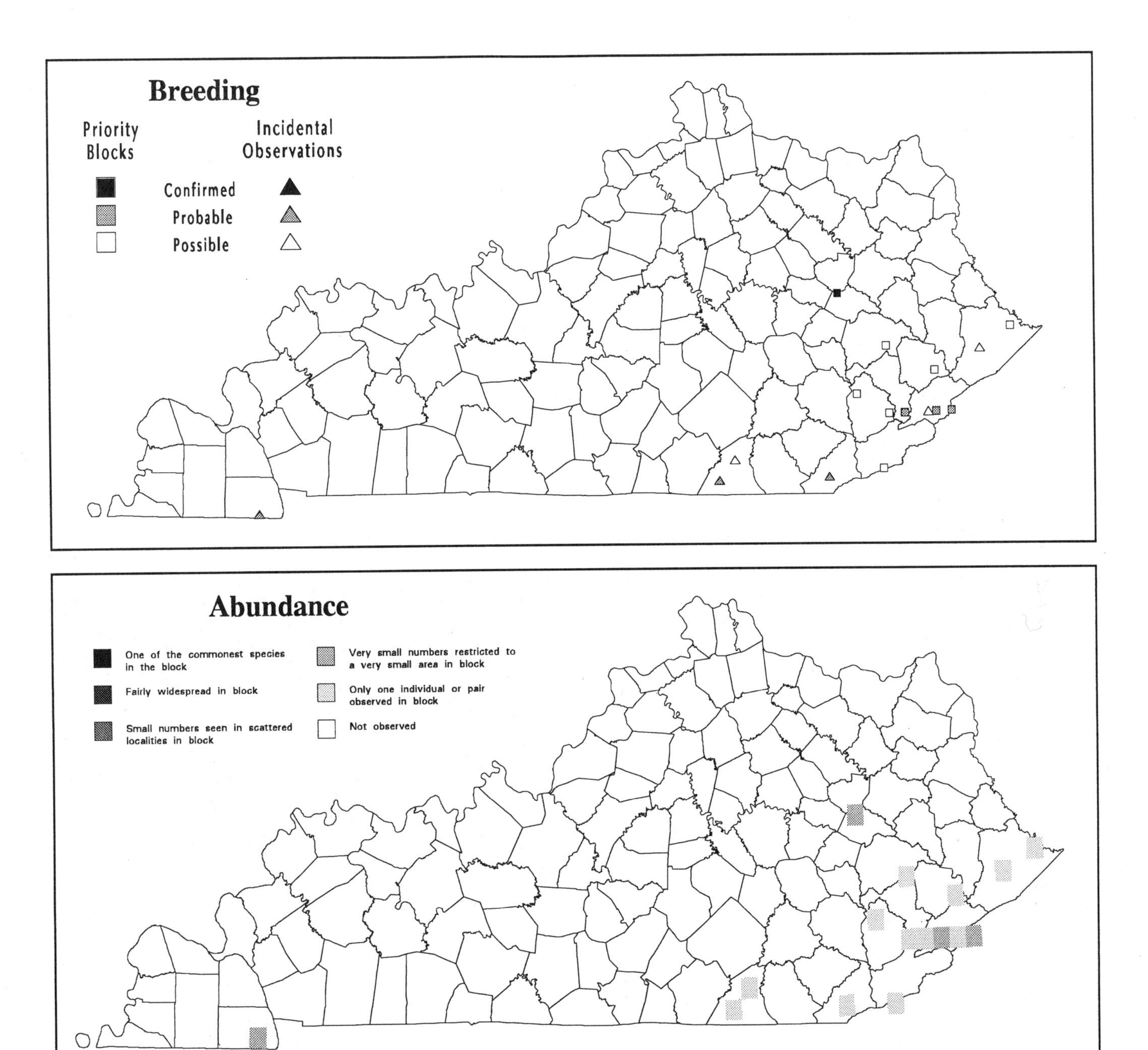

Analysis of Block Data by Physiographic Province Section

Physiographic Province Section	Total Blocks Surveyed	Blocks with Data	% with Data	Section's % for State
Mississippi Alluvial Plain	14	-	-	-
East Gulf Coastal Plain	36	-	-	-
Highland Rim	139	-	-	-
Shawnee Hills	142	-	-	-
Blue Grass	204	-	-	-
Cumberland Plateau	173	7	4.0	70.0
Cumberland Mountains	19	3	15.8	30.0

Summary of Breeding Status

Number of Blocks in Which Species Was Recorded		
Total	**10**	**1.4%**
Confirmed	1	10.0%
Probable	3	30.0%
Possible	6	60.0%

Ovenbird

Seiurus aurocapillus

The emphatic song of the Ovenbird is a characteristic summer sound of the forests of eastern Kentucky, and within its favored habitats this ground-foraging warbler ranks among the commonest of nesting birds. In contrast, the species is very locally distributed in the central and western portions of the state. Mengel (1965) summarized its breeding status as common in the Cumberland Plateau and Mountains, locally distributed and uncommon to common in parts of the eastern Knobs subsection of the Blue Grass and the Shawnee Hills, and elsewhere rare and local to absent.

The Ovenbird occupies a wide range of forest types. It is most abundant in subxeric forest without a well-developed understory, including upland oak-hickory and mixed pine-hardwood associations, as well as moderately mesic types in transition to moister conditions. In contrast, it is seldom encountered in the most mesic forest types and is typically absent in floodplain situations. The species is decidedly more numerous in eastern Kentucky forests than in similar habitat in the central and western parts of the state, probably owing to the degree of forest fragmentation west of the Cumberland Plateau. Although Ovenbirds are numerous in mature or fairly mature forests, they also use younger second-growth forest. For this reason, the species remains relatively numerous in eastern Kentucky, despite the regular disturbance of forests by periodic logging.

Although Ovenbirds are common in the forests of eastern Kentucky, the species likely has declined statewide since the time of settlement. These warblers probably were common in the upland forests that once covered much of central Kentucky. The vestiges of this widespread nesting population are still apparent in the few areas that remain well forested. Interestingly, the paucity of summer records from suitable habitat in the western Highland Rim and the East Gulf Coastal Plain suggests that these warblers may never have been common in these areas.

Ovenbirds winter primarily from the Gulf Coast south through Middle America (AOU 1983). Returning males typically appear during the third week of April, and according to Mengel (1965), clutches are completed from mid-May to mid-June, without an evident peak. Kentucky data on clutch size are limited, but the average size of five clutches and broods given by Mengel (1965) was 3.6 (range of 3–4). The average clutch size of six more recent nests is 4.5 (range of 3–6) (Croft and Stamm 1967; Croft 1971; KOS Nest Cards). Harrison (1975) gives rangewide clutch size as 3–6, commonly 4 or 5.

Forest Cover

Value	% of Blocks	Avg Abund
All	28.6	2.3
1	5.0	2.4
2	7.9	1.3
3	21.3	2.0
4	61.7	2.4
5	84.6	2.8

Ovenbird nests are placed on the ground and are typically sheltered by herbaceous vegetation of the forest floor. Dead leaves, coarse grasses, and plant stems are used to construct the bulk of the domed nest, and it is lined with fine rootlets, fibers, and hair (Harrison 1975).

The atlas survey yielded records of Ovenbirds in nearly 29% of priority blocks, and 12 incidental observations were reported. Occurrence was highest in the Cumberland Mountains and the Cumberland Plateau, and lowest in the East Gulf Coastal Plain. The species was not found in the Mississippi Alluvial Plain, probably because of the prevalence of seasonally flooded bottomland forest. Average abundance varied similarly. Occurrence in central Kentucky was highest in areas with the greatest forest cover and relief, such as the Knobs subsection of the Blue Grass, the Highland Rim's eastern transition to the Cumberland Plateau, and the southern Shawnee Hills. Interestingly, while Ovenbirds are fairly widespread in most parts of central and western Kentucky where forested areas predominate, such as Mammoth Cave National Park and Pennyrile State Forest, they appear to be locally distributed in Land Between the Lakes, where one would expect them to be fairly well distributed. As expected, occurrence and average abundance were lowest in extensively cleared areas and increased as the percentage of forest cover increased.

Alvin E. Staffan

Less than 12% of priority block records were for confirmed breeding. There were two incidental reports of active nests during the atlas fieldwork, but most of the 24 confirmed records in priority blocks were based on the observation of recently fledged young. Family groups of Ovenbirds were not as conspicuous as those of some other warblers, but adults are very defensive of fledglings, and the presence of young was usually given away by the actions of parents. Additional confirmed records involved adults carrying food, nest building, and distraction displays. The species is sometimes parasitized by cowbirds (Mengel 1965), and adult Ovenbirds were observed feeding fledgling cowbirds on at least two occasions.

Ovenbirds are reported in small numbers on most Kentucky BBS routes. The average number of individuals per BBS route for the periods 1966–91 and 1982–91 was 1.96 and 3.08, respectively. Trend analysis of these data shows a significant ($p<0.1$) decrease of 6.9% per year for the period 1966–91, but only a nonsignificant decrease of 0.7% per year for the period 1982–91. These trends probably are related to continued alteration of forested habitats along BBS routes.

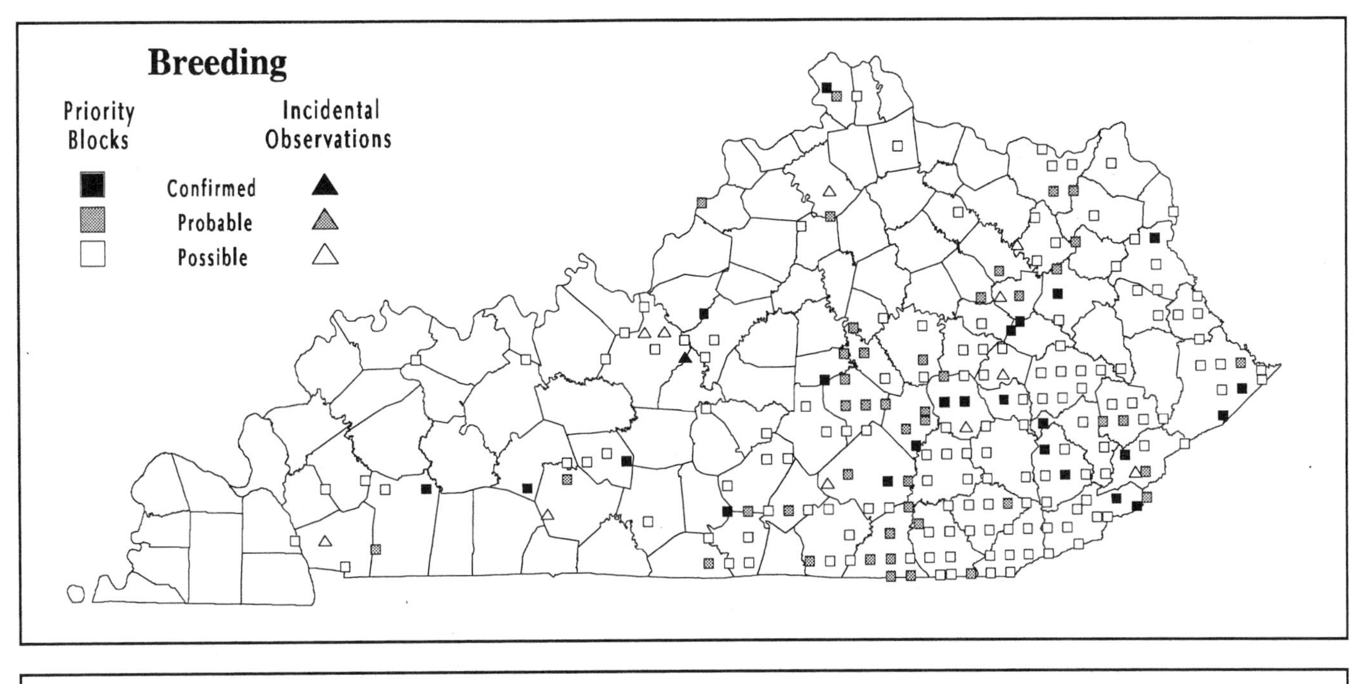

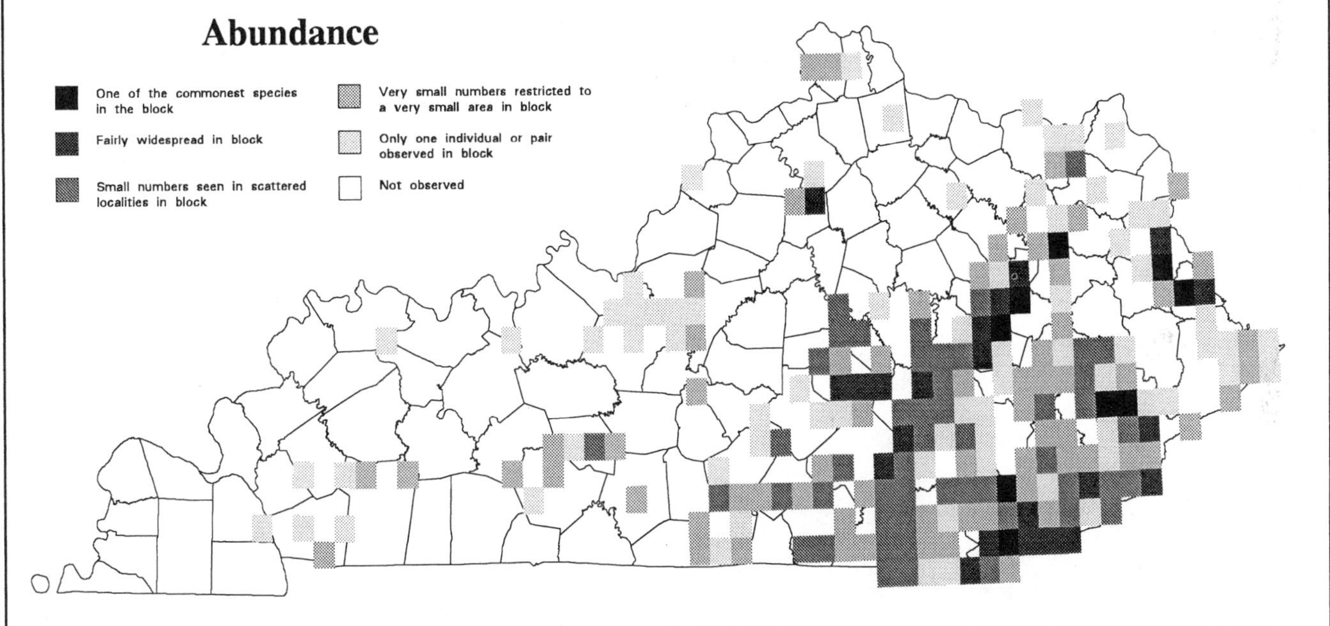

Analysis of Block Data by Physiographic Province Section

Physiographic Province Section	Total Blocks Surveyed	Blocks with Data	Avg Abund	% with Data	Section's % for State
Mississippi Alluvial Plain	14	-	-	-	-
East Gulf Coastal Plain	36	1	1.0	2.8	0.5
Highland Rim	139	30	1.9	21.6	14.4
Shawnee Hills	142	12	1.7	8.5	5.8
Blue Grass	204	36	2.0	17.6	17.3
Cumberland Plateau	173	112	2.4	64.7	53.8
Cumberland Mountains	19	17	3.2	89.5	8.2

Summary of Breeding Status

Number of Blocks in Which Species Was Recorded		
Total	**208**	**28.6%**
Confirmed	24	11.5%
Probable	44	21.2%
Possible	140	67.3%

Louisiana Waterthrush

Seiurus motacilla

The clear, whistled song of the Louisiana Waterthrush is a fairly widespread sound along Kentucky's streams during spring and early summer. This warbler occupies a wide variety of habitat types across the state, but it is usually encountered near water. Mengel (1965) regarded the species as a fairly common to common summer resident statewide.

In eastern and central Kentucky the Louisiana Waterthrush is usually encountered along rills and streams with steep to moderate gradients, but it is fairly widespread along slow-moving creeks and swampy areas with standing water in the western part of the state. These warblers also may occur in woodlands rather far from permanent water, especially along stream drainages that are dry for most of the year. Louisiana Waterthrushes seem to avoid larger streams, perhaps because of the magnitude of flooding that occurs regularly enough to preclude successful nesting. While the birds are sometimes encountered foraging along narrow forested riparian corridors through otherwise cleared land, it appears that they do not use streams for nesting unless there is a tract of forest along at least one side.

It is likely that the Louisiana Waterthrush is somewhat less common and widespread in Kentucky today than two centuries ago. Audubon (1861) observed the species in swampy areas of western Kentucky east to Henderson in the early 1800s, and it surely must have been common along the many miles of pristine streams. Today, segments of many streams have been cleared and channelized, making habitat less attractive, and pollution of streams has certainly had some effect. In addition, the impounding of rivers and streams to create reservoirs has reduced the amount of suitable habitat.

The Louisiana Waterthrush is among the first of Kentucky's nesting warblers to return in early spring. The first songs are often heard before April 1, and it is likely that nearly all nesting birds return by the end of the month. Nesting begins soon thereafter, and early clutches may be completed during the last week of April (Mengel 1965; KOS Nest Cards). A peak in clutch completion occurs during the first ten days of May, but later clutches are completed into early June (Mengel 1965). Song becomes infrequent by mid-June, and most birds seem to depart for the wintering grounds by early August. Kentucky data on clutch size are scarce, but Mengel (1965) included details of three nests, two containing four eggs and the other six eggs. Four nests observed during atlas fieldwork contained the following: five eggs on May 5, 1985, and then five young on May 11, 1985; six eggs on May 8–11, 1985; three young waterthrushes and a young cowbird on June 10, 1989; and four young on June 25, 1990. Harrison (1975) gives rangewide clutch size as 4–6.

Forest Cover

Value	% of Blocks	Avg Abund
All	32.2	1.5
1	11.9	1.4
2	29.5	1.3
3	38.3	1.5
4	40.9	1.7
5	28.8	1.5

The nest is placed on the ground, typically on a slope or bank immediately above a creek or rill but occasionally on artificial roadbanks, especially along gravel lanes through woods (KBBA data). The nest is usually sheltered by overhanging rocks, roots, or plants. It is constructed of dead leaves, grasses, and other coarse plant material and is lined with fine rootlets and hair (Harrison 1975).

Ron Austing

The atlas fieldwork yielded records of Louisiana Waterthrushes in more than 32% of priority blocks statewide, and 23 incidental observations were reported. As expected, the species was distributed fairly evenly across the state, occurring in 21–37% of priority blocks in all physiographic province sections except the East Gulf Coastal Plain, where it was recorded in two-thirds of priority blocks. A high level of participation by volunteer atlasers in this area may explain the higher occurrence. Average abundance was relatively low statewide. Although this warbler is fairly widespread and its song is quite distinctive, it probably should have been recorded more often. Nesting occurs very early, and when most atlas fieldwork was undertaken in June and July, many birds sang only infrequently and some may have dispersed from breeding areas. Also, the roadside surveys employed for most of the atlas work were not especially effective at detecting the species, especially in blocks with only a few streams.

More than 14% of priority block records were for confirmed breeding. Although a few active nests were located, most confirmed records were based on the observation of recently fledged young. Other confirmed records involved the observation of adults carrying food and distraction displays.

Louisiana Waterthrushes are typically reported in small numbers on Kentucky BBS routes. The average number of individuals per BBS route for the periods 1966–91 and 1982–91 was 0.37 and 0.38, respectively. In part due to small sample sizes, trend analysis of these data does not reveal statistically significant results.

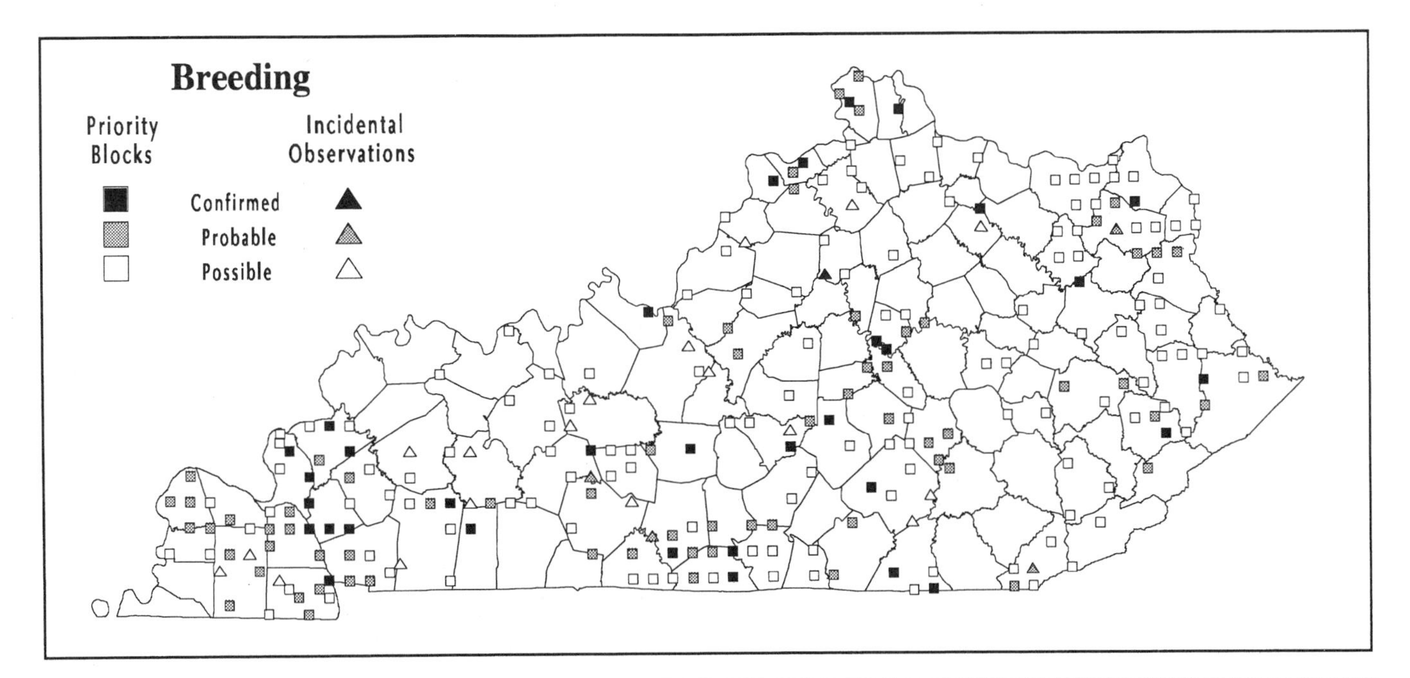

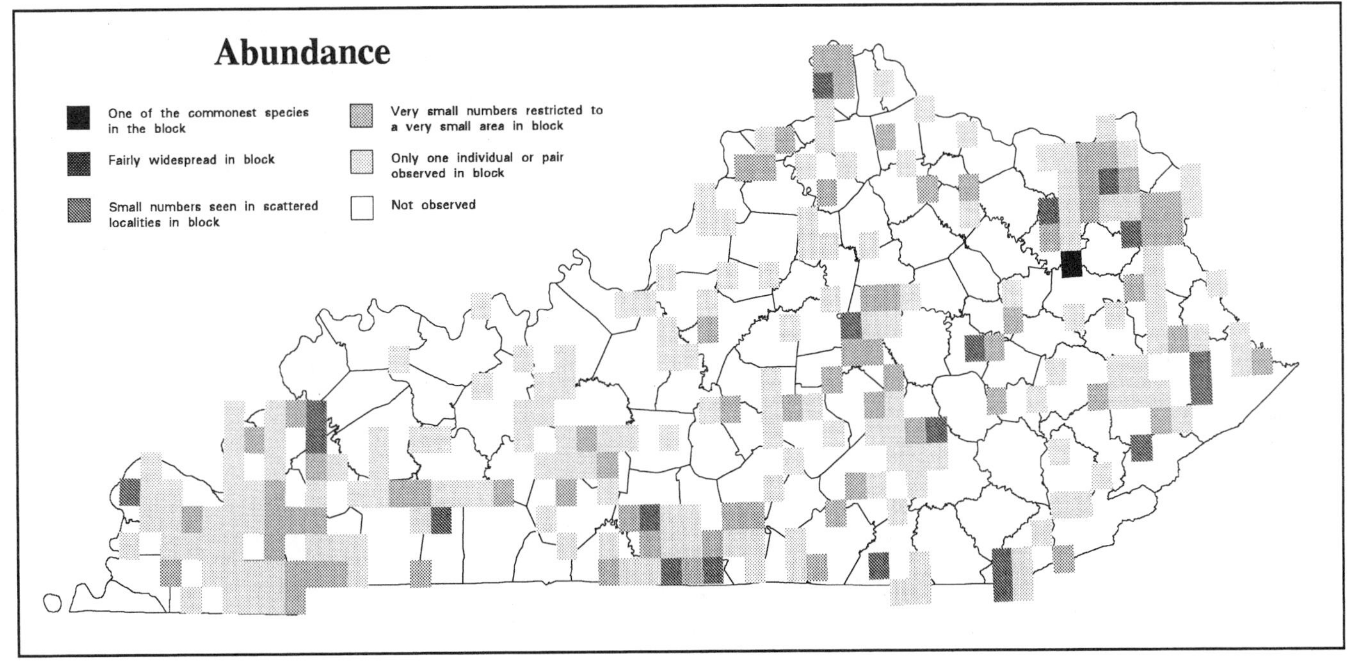

Analysis of Block Data by Physiographic Province Section

Physiographic Province Section	Total Blocks Surveyed	Blocks with Data	Avg Abund	% with Data	Section's % for State
Mississippi Alluvial Plain	14	3	1.7	21.4	1.3
East Gulf Coastal Plain	36	24	1.2	66.7	10.3
Highland Rim	139	51	1.4	36.7	21.8
Shawnee Hills	142	38	1.4	26.8	16.2
Blue Grass	204	57	1.5	27.9	24.4
Cumberland Plateau	173	56	1.6	32.4	23.9
Cumberland Mountains	19	5	2.0	26.3	2.1

Summary of Breeding Status

Number of Blocks in Which Species Was Recorded		
Total	**234**	**32.2%**
Confirmed	33	14.1%
Probable	69	29.5%
Possible	132	56.4%

Louisiana Waterthrush

Kentucky Warbler

Oporornis formosus

The subdued song of the Kentucky Warbler is one of the more widespread summer sounds in forested habitats across the state. Mengel (1965) regarded the species as a common summer resident and one of the most widespread and generally numerous forest-breeding warblers. Only in extensively settled areas and at higher elevations in the Cumberland Mountains is it especially difficult to find (Mengel 1965; Davis et al. 1980).

The Kentucky Warbler is generally a bird of forests with a moderate to dense shrub layer. Although the species occurs predominantly in deciduous forest, mixed forest types with pines or hemlocks are also used. A great variety of mesic to subxeric forests is inhabited, although more xeric forests are often avoided, apparently because of the lack of a well-developed shrub layer. Unlike several other ground-nesting warblers, the Kentucky also occurs regularly in bottomland forests along major river floodplains, apparently nesting successfully despite periodic flooding.

The Kentucky Warbler was probably more numerous in the state before settlement, but it appears that the species has adapted better to human alteration of the landscape than most other woodland warblers. Because it prefers forest with a well-developed shrub layer, many forest tracts that have been disturbed by selective logging are suitable for nesting even though the canopy has been disrupted. Only a year or two after selective logging, the species may inhabit these areas if the disturbance has not been too great. Moreover, Kentucky Warblers seem to be much more likely to inhabit smaller tracts of dissected forest than other woodland warblers.

Kentucky Warblers winter from southern Mexico south to northern South America (AOU 1983). Birds begin arriving in the state during the third week of April, although it is probably early May before the full complement of nesting birds has returned. According to Mengel (1965), early clutches are completed during the first week of May, and a peak in clutch completion occurs in mid-May. Evidence of double-brooding in Kentucky is lacking, and most later nests observed into mid-June likely represent renesting following earlier failure because of predation or storms. The average size of seven clutches or broods given by Mengel (1965) was 4.4 (range of 4–5).

Forest Cover

Value	% of Blocks	Avg Abund
All	65.5	2.0
1	24.8	1.7
2	54.7	1.8
3	76.1	2.0
4	82.5	2.2
5	86.5	2.4

The nest is placed on or just above the ground, often on a gentle slope. It is typically concealed by vegetation on the forest floor and may be situated at the base of a small tree or shrub, or beneath a fallen tree branch (Chapman 1968). The nest is constructed of dead leaves, grasses, and other coarse plant material and is lined with fine grass, rootlets, and weed stalks (Harrison 1975).

The atlas survey yielded records of Kentucky Warblers in nearly 66% of priority blocks statewide, and 18 incidental observations were reported. The species ranked behind only the Common Yellowthroat and the Yellow-breasted Chat in total abundance among nesting warblers, but it ranked highest among warblers nesting in woodlands. As expected, the species was well distributed across the state, occurring in more than 50% of blocks in all sections except the Blue Grass. Average abundance was relatively uniform across the state. Occurrence and average abundance probably were related most closely to the amount of forested habitat. This explains the species's general scarcity across the Inner Blue Grass, the northwestern Shawnee Hills, and the southern Highland Rim.

Alvin E. Staffan

Although the Kentucky Warbler is relatively widespread and numerous, only about 11% of priority block records were for confirmed breeding. Only two active nests were located, and most of the 53 confirmed records involved the observation of recently fledged young. Adults were quite conspicuous in their defense of young and responded to the presence of potential harm by calling loudly. This behavior helped in the confirmation of many observations. Other confirmed records were based on the observation of adults carrying food, distraction displays, and nest building. The species is frequently parasitized by Brown-headed Cowbirds (Mengel 1965), and adult warblers were observed feeding young cowbirds on at least two occasions.

Despite their relative abundance, Kentucky Warblers are not reported in large numbers on most Kentucky BBS routes. The average number of individuals per BBS route for the periods 1966–91 and 1982–91 was 2.21 and 1.47, respectively. Trend analysis of these data shows a nonsignificant increase of 1.5% per year for the period 1966–91 and a nonsignificant decrease of 1.3% per year for the period 1982–91.

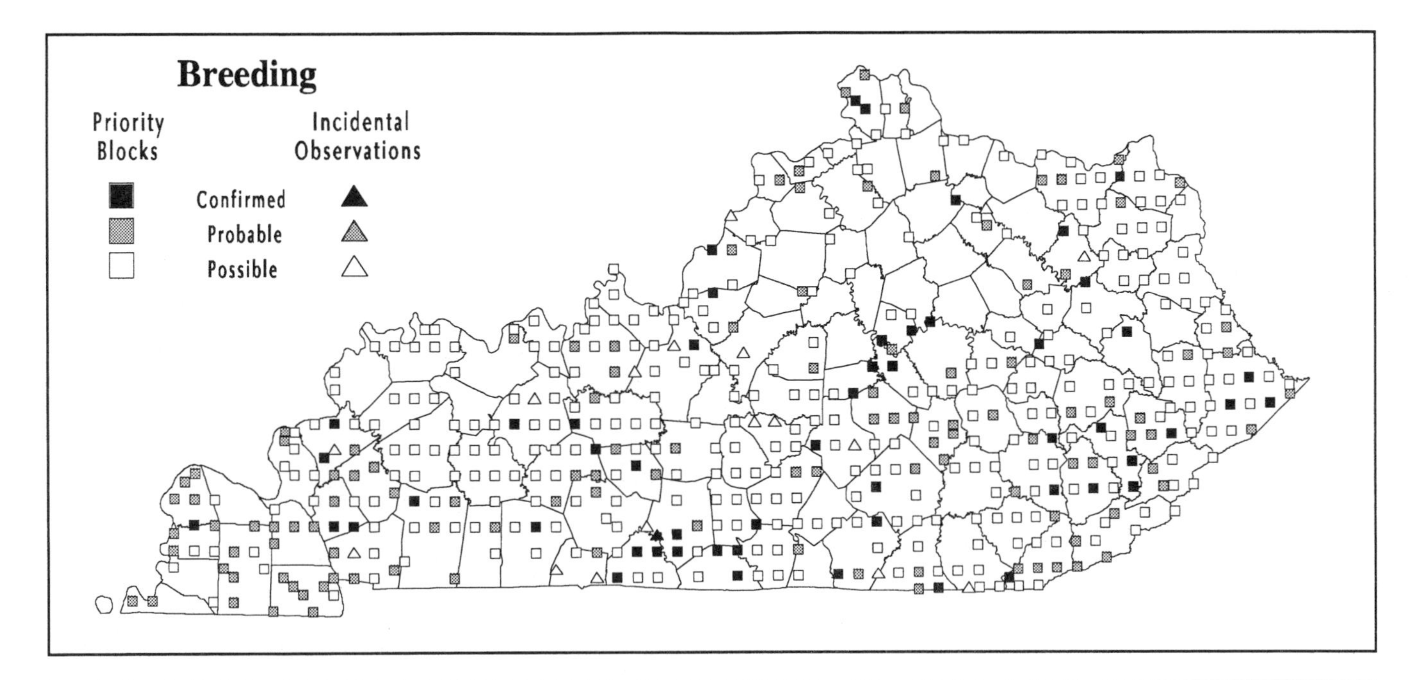

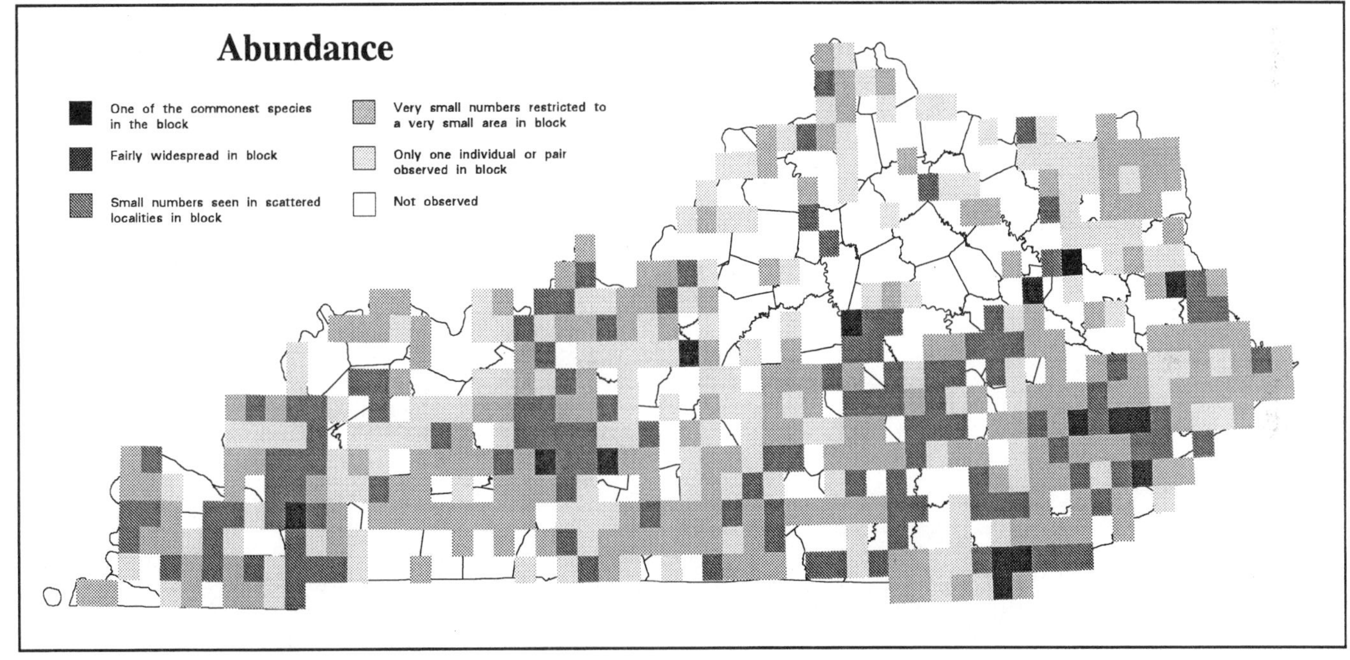

Analysis of Block Data by Physiographic Province Section

Physiographic Province Section	Total Blocks Surveyed	Blocks with Data	Avg Abund	% with Data	Section's % for State
Mississippi Alluvial Plain	14	8	2.2	57.1	1.7
East Gulf Coastal Plain	36	27	2.1	75.0	5.7
Highland Rim	139	93	2.0	66.9	19.5
Shawnee Hills	142	108	2.0	76.1	22.7
Blue Grass	204	90	1.8	44.1	18.9
Cumberland Plateau	173	133	2.2	76.9	27.9
Cumberland Mountains	19	17	2.5	89.5	3.6

Summary of Breeding Status

Number of Blocks in Which Species Was Recorded		
Total	**476**	**65.5%**
Confirmed	53	11.1%
Probable	117	24.6%
Possible	306	64.3%

Kentucky Warbler

Common Yellowthroat

Geothlypis trichas

The Common Yellowthroat is the most widespread and common of Kentucky's nesting warblers. Mengel (1965) regarded the species as a common summer resident statewide, and its occurrence remains similar today. Only in areas of intensively cleared or completely forested land does this warbler become difficult to find.

The Common Yellowthroat is a bird of weedy, grassy, or brushy habitats. Although the species is sometimes found along brushy woodland borders, it typically avoids the interior of canopied forest. Yellowthroats are most common in rural situations where land has been set aside or neglected. Abandoned fields, fencerows, woodland borders, unmowed road or utility corridors, weedy drainages, reclaimed strip mines, and other similar altered habitats are most frequently inhabited. The species is less often encountered in natural habitats, but grassy or shrubby borders to marshes, low damp meadows with a profusion of rank growth, and remnants of tallgrass prairie also are used.

During the 20th century it is likely that yellowthroats have become more widespread and numerous in Kentucky than ever before. Audubon (1861) thought that the species bred "in the barrens, swamps and briar patches" in the early 1800s, and it is likely that suitable breeding habitat occurred throughout the state. Although vast areas of naturally open and edge habitats have been lost through conversion to agricultural land and settlement, overall human alteration of the landscape likely has resulted in the creation of much more suitable habitat than has been lost. The species does not inhabit extensively forested areas that lack openings with weedy cover, and deforestation has resulted in the creation of much land that is left in borders or, through neglect, reverts back to early successional vegetation suitable for nesting.

Common Yellowthroats return from more southerly wintering grounds during the latter half of April, although it may be early May before the entire nesting population has returned. According to Mengel (1965), early clutches are completed by about May 21, although later nests have been reported into mid-June. A peak in clutch completion is not evident, and it is likely that at least some pairs raise a second brood. Kentucky data on clutch size are scarce, but Mengel (1965) gives the average size of four clutches and broods as 4.8 (range of 4–5). This warbler is regularly parasitized by the Brown-headed Cowbird (Mengel 1965; Stamm and Croft 1968).

Forest Cover

Value	% of Blocks	Avg Abund
All	96.3	3.2
1	86.1	2.9
2	98.4	3.1
3	97.4	3.2
4	98.7	3.5
5	96.2	2.9

The nest is placed on or near the ground and is typically hidden amid thick herbaceous growth. In open areas nests are often placed near a hedgerow or shrub. The nest is woven out of grasses and other coarse plant material and anchored in the surrounding vegetation. It is lined with fine grass, bark fibers, and hair (Harrison 1975).

Ron Austing

Common Yellowthroats were found in more than 96% of priority blocks statewide, and five incidental observations were reported. As expected, the species ranked highest in abundance among nesting warblers and was found in more than 84% of blocks in every physiographic province section. The species ranked 10th among Kentucky's nesting birds according to the number of priority block records, 5th by total abundance, and 9th by average abundance. Occurrence was lowest in the Mississippi Alluvial Plain, where extensive clearing of bottomland habitats for cropland has excluded the species from many areas, and in the Cumberland Mountains, where forested habitats predominate. The species was also difficult to locate in the highly manicured farmland of the Inner Blue Grass. Average abundance was relatively uniform across the state. Occurrence and average abundance were slightly lower in extensively cleared and predominantly forested areas.

Despite the species's conspicuousness, less than 14% of priority block records were for confirmed breeding. Although a few active nests were located, most confirmed reports were based on the observation of recently fledged young and adults carrying food. In addition, adult yellowthroats were observed feeding young cowbirds on several occasions. Other confirmed records involved the observation of nest building, distraction displays, and adults carrying fecal sacs.

Common Yellowthroats are reported in relatively high numbers on most Kentucky BBS routes. The average number of individuals per BBS route for the periods 1966–91 and 1982–91 was 19.48 and 19.63, respectively. According to these data, the species ranked 14th in abundance on BBS routes during the period 1982–91. Trend analysis reveals a nonsignificant increase of 0.7% per year for the period 1966–91, but a significant (p<.01) decrease of 7.6% per year for the period 1982–91. The short-term increase shown by BBS data for the period 1967–77 (Monroe 1979) apparently has been reversed, but for unknown reasons.

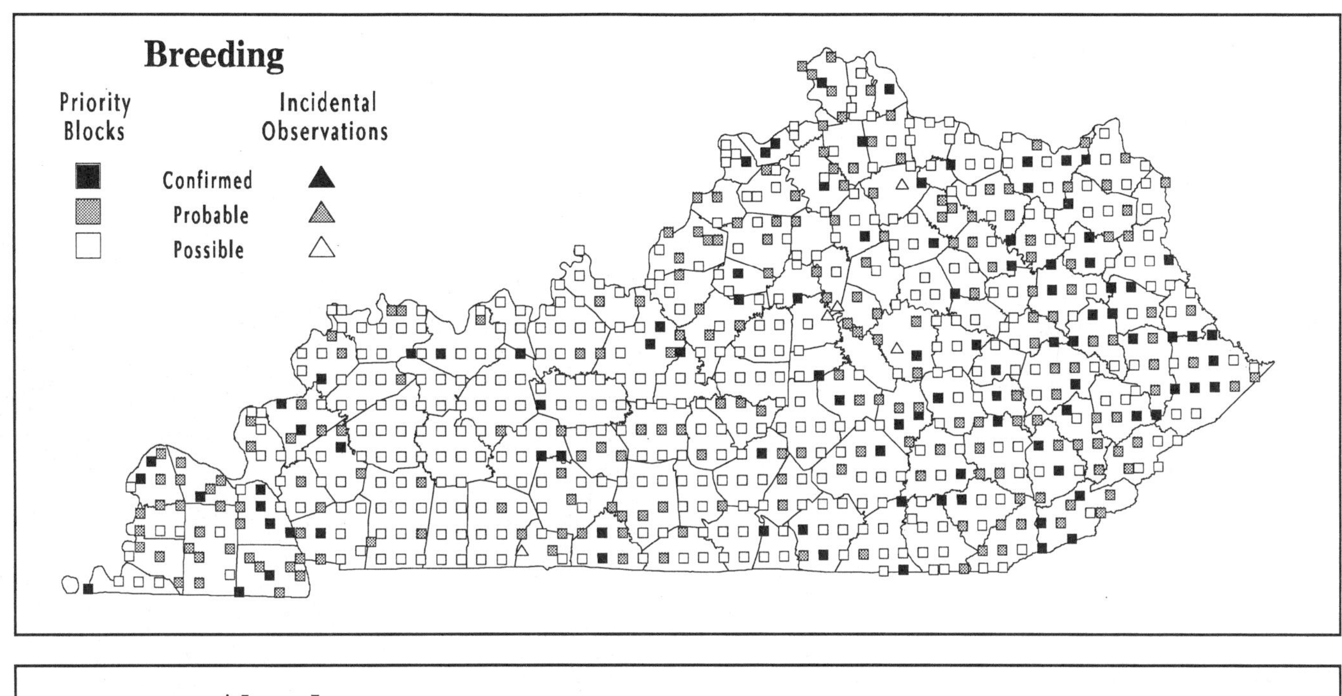

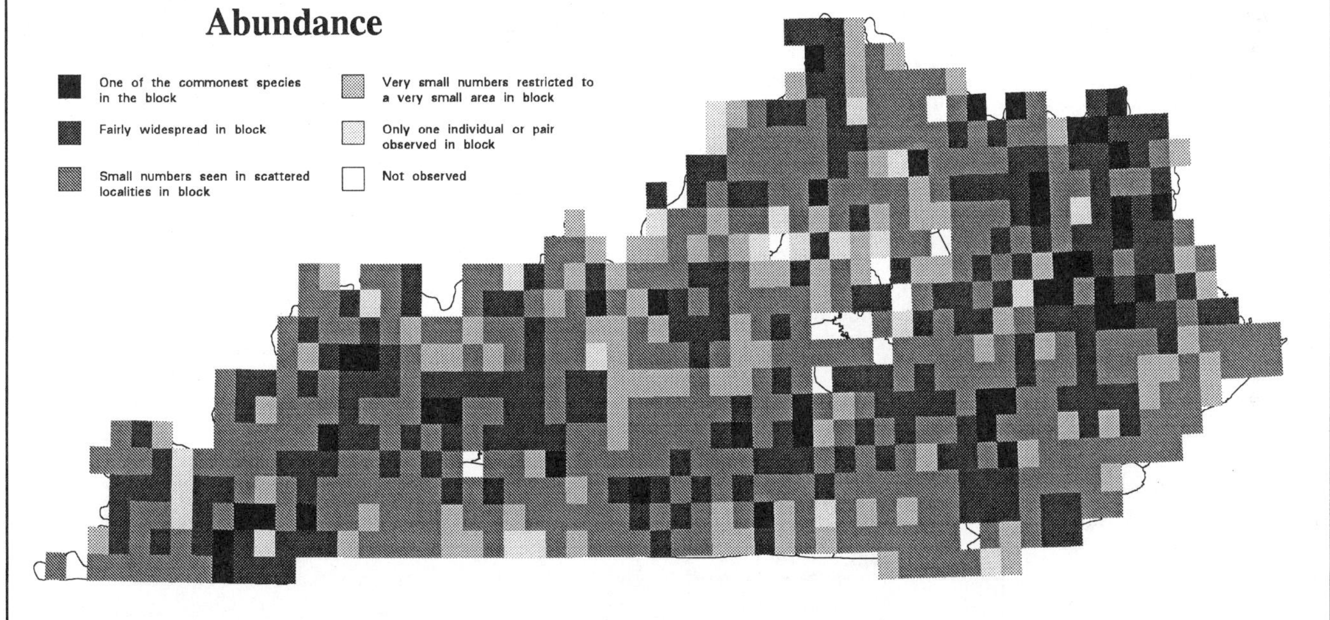

Analysis of Block Data by Physiographic Province Section

Physiographic Province Section	Total Blocks Surveyed	Blocks with Data	Avg Abund	% with Data	Section's % for State
Mississippi Alluvial Plain	14	12	3.2	85.7	1.7
East Gulf Coastal Plain	36	36	3.3	100.0	5.1
Highland Rim	139	138	3.1	99.3	19.7
Shawnee Hills	142	139	3.2	97.9	19.9
Blue Grass	204	188	3.1	92.2	26.9
Cumberland Plateau	173	171	3.3	98.8	24.4
Cumberland Mountains	19	16	2.9	84.2	2.3

Summary of Breeding Status

Number of Blocks in Which Species Was Recorded		
Total	**700**	**96.3%**
Confirmed	97	13.9%
Probable	212	30.3%
Possible	391	55.9%

Common Yellowthroat

Hooded Warbler

Wilsonia citrina

Despite the Hooded Warbler's reclusive habits, its emphatic song is one of the more widespread summer sounds in the forests of eastern Kentucky. In fact, according to BBS data compiled for the period 1965–79, the species was as numerous in the Cumberland Plateau of Kentucky as anywhere within its range (Robbins et al. 1986). In contrast, this warbler is quite locally distributed throughout the rest of the state, where it is restricted to areas with extensive forest cover (Mengel 1965).

The Hooded Warbler occurs in a variety of deciduous and mixed forest types, as long as a well-developed shrub layer is present. In eastern Kentucky, forested ravines and slopes with a predominance of rhododendron or mountain laurel are most frequently used. Across the rest of the state the species is typically found in rich deciduous forest with an understory layer of small saplings and shrubs. Because of its preference for a dense understory layer, this warbler often inhabits naturally or artificially disturbed forest that is regenerating. Thus it can be found along the margins of old clear-cuts and selectively logged areas, as well as natural openings created by fire or windstorms.

While Hooded Warblers may be as abundant in eastern Kentucky today as they were two centuries ago, they must have been more widespread and common throughout the rest of the state before European settlement. The species was likely absent in the native prairies, barrens, and savannas of the region, but Audubon (1861) encountered it regularly in forests along the Ohio River in the early 1800s. Considerable numbers likely occurred in forested areas statewide. The clearing and fragmentation of these forests for agricultural use and settlement have resulted in a dramatic reduction in the amount of suitable habitat, especially west of the Cumberland Plateau.

Hooded Warblers winter primarily in Central America (AOU 1983), and many birds reappear in Kentucky during the third week of April. Nest building has been reported as early as May 7 (Croft and Stamm 1967), and Mengel (1965) gives a peak in clutch completion during the last ten days of May. There is no evidence of the raising of a second brood, and most late nests likely represent renesting following earlier failure because of predation or storms. The average size of seven clutches and one brood listed by Mengel (1965) was 3.3 (range of 2–4). A nest observed during the atlas survey in Boone County contained one young warbler and a cowbird egg on June 7, 1991. Harrison (1975) gives rangewide clutch size as 3–4, rarely 5.

Forest Cover

Value	% of Blocks	Avg Abund
All	33.7	2.2
1	2.0	2.0
2	3.7	1.4
3	23.0	1.6
4	85.7	2.3
5	98.1	2.8

The nest is placed in the understory layer, usually in the upright fork of a small tree or shrub, but sometimes in a sturdy herbaceous plant closer to the ground. It is beautifully constructed of dead leaves, strips of bark, and other coarse plant material and is lined with fine grass and hair (Mengel 1965). The average height of four nests given by Mengel (1965) was 1.5 feet (range of 0.5–3.0 feet).

Alvin E. Staffan

The atlas survey yielded records of Hooded Warblers in nearly 34% of priority blocks, and five incidental observations were reported. As expected, occurrence was highest in the Cumberland Mountains and the Cumberland Plateau and was significantly lower across central and western Kentucky. The demarcation between regions of general occurrence and sporadic distribution was remarkable. Most of the records from the Blue Grass and the Highland Rim came from their eastern transitions to the Cumberland Plateau. Today Hooded Warblers occur in substantial numbers in central and western Kentucky only in extensive tracts of public land reverting to forest (e.g., Mammoth Cave National Park and Pennyrile State Forest) and in areas of considerable relief (e.g., the western Knobs subsection of the Blue Grass and ravine forests along the Ohio and Kentucky rivers). It appears that a general decrease occurs westward, as fairly comprehensive coverage of the Land Between the Lakes yielded only a few observations of Hooded Warblers there. As noted by Mengel (1965), the species regularly occurs in bottomland forests of western Kentucky, but it can no longer be regarded as common. Average abundance varied similarly to occurrence, and both were closely related to the amount of forest cover.

Less than 10% of priority block records were for confirmed breeding. Four active nests were located, but most of the 24 confirmed records were based on the observation of recently fledged young. The remaining confirmed records involved the observation of adults carrying food.

The Hooded Warbler is seldom reported on most Kentucky BBS routes. The average number of individuals per BBS route for the periods 1966–91 and 1982–91 was 1.01 and 0.29, respectively. In part due to small sample sizes, trend analysis of these data does not reveal statistically significant results.

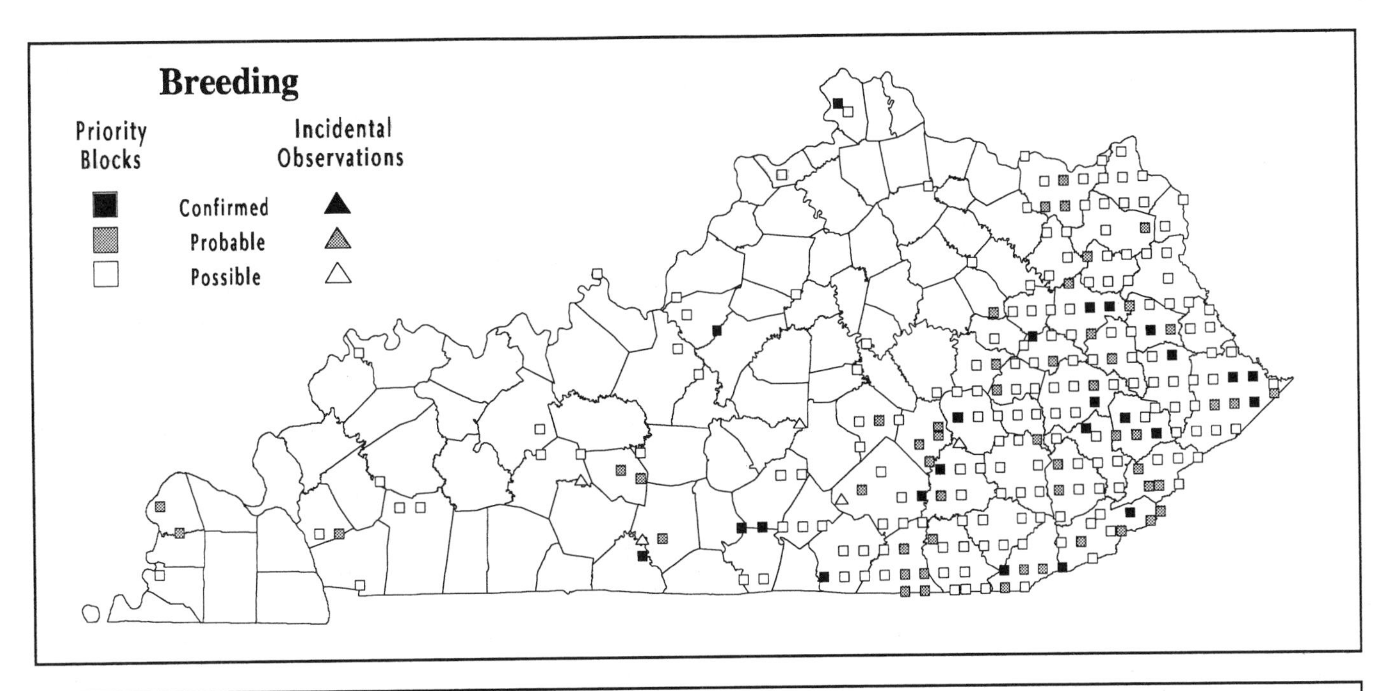

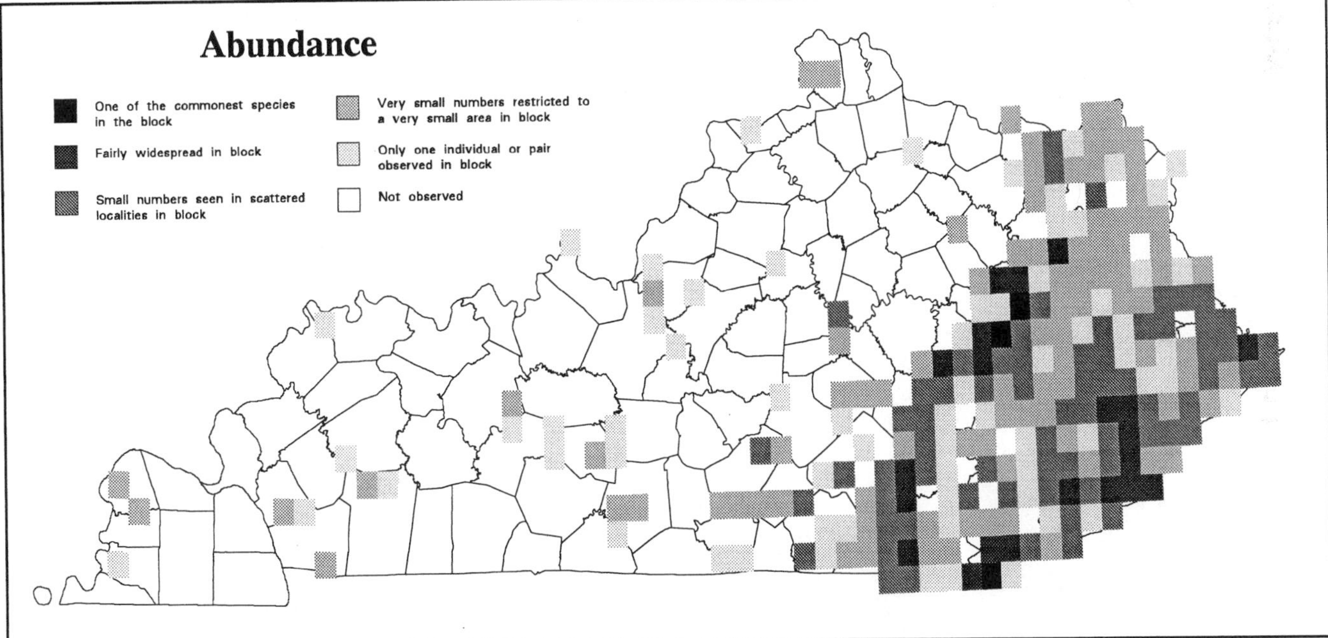

Analysis of Block Data by Physiographic Province Section

Physiographic Province Section	Total Blocks Surveyed	Blocks with Data	Avg Abund	% with Data	Section's % for State
Mississippi Alluvial Plain	14	3	1.7	21.4	1.2
East Gulf Coastal Plain	36	-	-	-	-
Highland Rim	139	23	1.7	16.5	9.4
Shawnee Hills	142	11	1.3	7.7	4.5
Blue Grass	204	32	1.8	15.7	13.1
Cumberland Plateau	173	158	2.4	91.3	64.5
Cumberland Mountains	19	18	3.0	94.7	7.3

Summary of Breeding Status

Number of Blocks in Which Species Was Recorded		
Total	**245**	**33.7%**
Confirmed	24	9.8%
Probable	52	21.2%
Possible	169	69.0%

Hooded Warbler

Canada Warbler

Wilsonia canadensis

Most of the Canada Warbler's nesting range lies to the north of Kentucky, in the northeastern United States and southern Canada, but it also extends southward through the Appalachian Mountains, barely reaching into the southeastern part of the state (AOU 1983). The species is apparently limited in nesting range to the higher elevations of the Cumberland Mountains and is common only near the top of the highest peak, Black Mountain.

Canada Warblers were first reported in summer by Howell (1910), who observed numerous individuals near the summit of Black Mountain. Subsequent fieldwork has yielded additional records of summering and confirmed breeding (Wetmore 1940; Breiding 1947; Lovell 1950a; Mengel 1965), including observations from the Letcher County portion of the mountain (KSNPC, unpub. data). Mengel (1965) noted that the species occurred on Black Mountain primarily above 3,800 feet, although occasionally it was seen as low as 3,500 feet. He also determined an indicated density of breeding birds of about 25 singing males per 100 acres on the higher portions of the mountain. This density makes the Canada second only to the Black-throated Blue among nesting warblers at higher elevations there.

Before the late 1960s it was thought that the Canada Warbler was restricted to the higher elevations of Black Mountain. Croft (1969, 1971), however, discovered that the species was present in small numbers on adjacent Cumberland Mountain. Subsequent reports have confirmed that the species inhabits the higher elevations of the mountain in Cumberland Gap National Historical Park, in Bell and Harlan counties, although nesting has not been confirmed (Davis et al. 1980).

Within its limited nesting range in Kentucky, the Canada Warbler is a bird of mesic forest with a fairly well-developed understory layer. On Cumberland Mountain it is most often found in association with rhododendron (Davis et al. 1980). The summit of Black Mountain has entirely deciduous vegetation, however, and birds are found in the understory of mature deciduous forest as well as younger, cut-over forest and forest edge.

Canada Warblers winter primarily in South America (AOU 1983), and most birds return to southeastern Kentucky during the first two weeks of May. At higher elevations nesting may be delayed by retarded leaf development, but Mengel (1965) reports that clutches are completed from mid-May to early July, with a peak in late May. Of two nests observed on Black Mountain by Mengel, one contained five eggs, and the other contained five young. Harrison (1975) gives rangewide clutch size as 3-5, commonly 4.

Nests described by Mengel (1965) were placed on the ground among herbaceous ground cover or moss and were well concealed by overhanging vegetation. One was on a steep slope, while the other was on relatively flat ground. Nests are constructed of dead leaves, grasses, and other coarse plant material and are lined with rootlets, plant down, and hair (Harrison 1975).

Alvin E. Staffan

The atlas survey yielded four records of Canada Warblers in priority blocks, and one incidental observation was reported. All five reports came from previously known breeding or summering areas on Black and Cumberland mountains in Harlan and Bell counties. Although no confirmed records were obtained, the presence of birds throughout the summer on both mountains represented probable evidence of breeding.

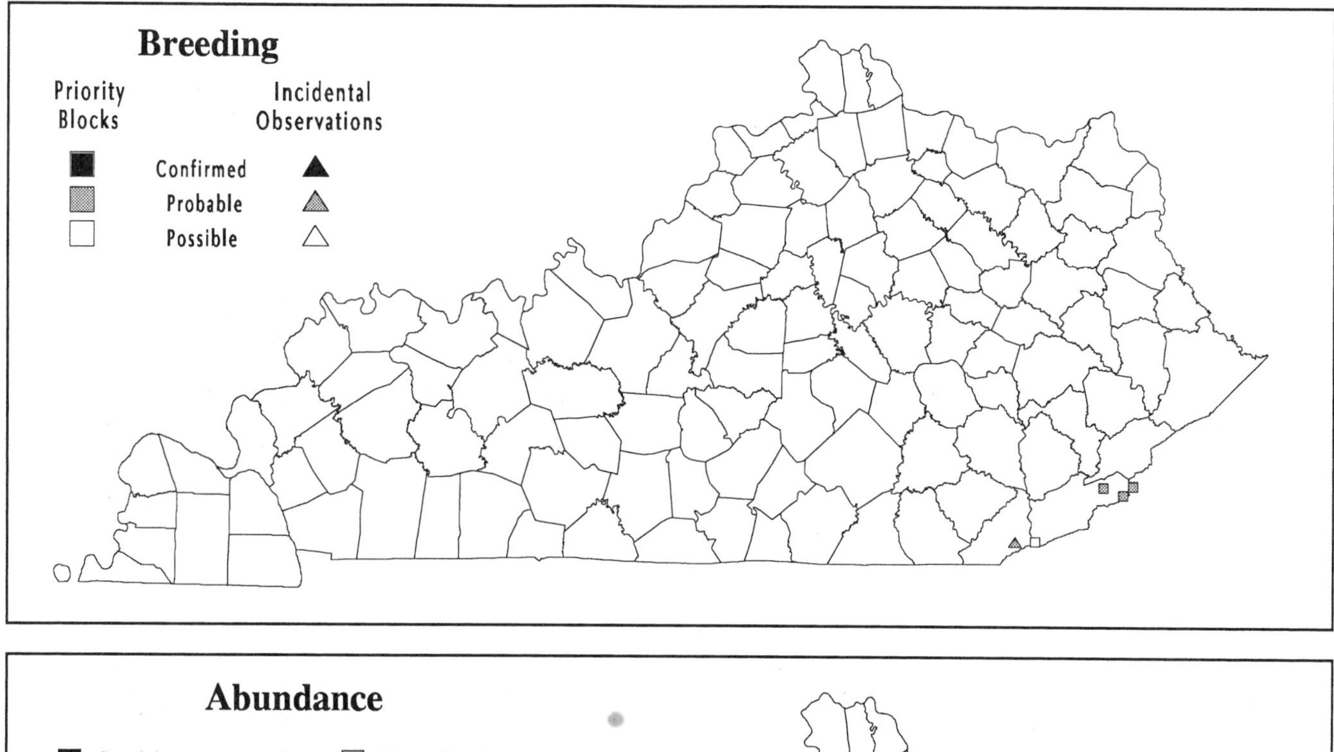

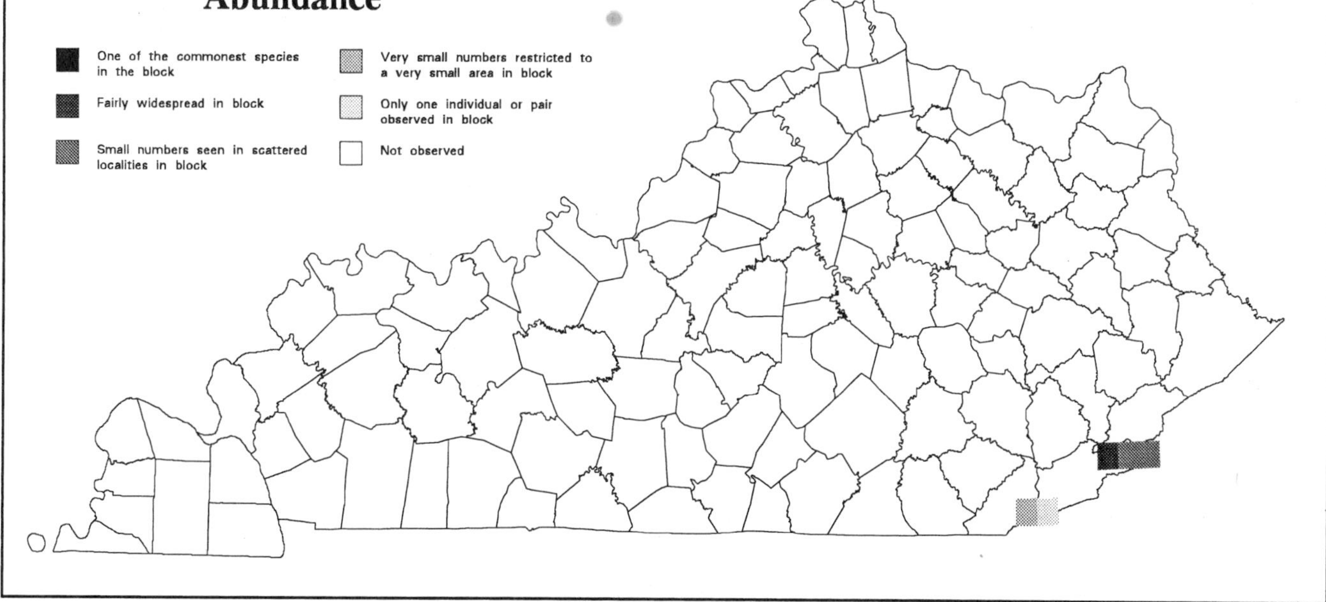

Analysis of Block Data by Physiographic Province Section

Physiographic Province Section	Total Blocks Surveyed	Blocks with Data	% with Data	Section's % for State
Mississippi Alluvial Plain	14	-	-	-
East Gulf Coastal Plain	36	-	-	-
Highland Rim	139	-	-	-
Shawnee Hills	142	-	-	-
Blue Grass	204	-	-	-
Cumberland Plateau	173	-	-	-
Cumberland Mountains	19	4	21.1	100.0

Summary of Breeding Status

Number of Blocks in Which Species Was Recorded		
Total	**4**	**0.6%**
Confirmed	-	-
Probable	3	75.0%
Possible	1	25.0%

Canada Warbler

Yellow-breasted Chat

Icteria virens

The unwarblerlike song of the Yellow-breasted Chat is a characteristic summer sound across Kentucky wherever brushy habitat occurs. Mengel (1965) regarded the species as a common summer resident that was evenly distributed across the state. BBS data for the period 1965–79 indicated that the chat was as common in Kentucky as anywhere within its range (Robbins et al. 1986). In contrast, trend analysis of the same data revealed a steady decline in the overall breeding population that also has been evident in Kentucky (Monroe 1978; BBS data).

The Yellow-breasted Chat is a bird of early successional habitats with an abundance of weedy cover and scattered trees. Although these large warblers often inhabit forest margins and openings, they avoid the interior of mature forest. Artificially created habitats are used most frequently and include abandoned fields, regenerating forest clear-cuts, young pine plantations, roadway and utility corridors, and reclaimed strip mines. The species is seldom found in natural habitats, but forest recovering from damage because of windstorms or fire is sometimes open enough to be suitable for nesting.

It is likely that the Yellow-breasted Chat is more widespread and numerous in Kentucky today as a result of human alteration of the landscape. Audubon (1861) considered the species to be abundant in the barrens in the early 1800s, and it is likely that other naturally open situations supported small numbers of birds across much of the state. While human alteration of the landscape has resulted in the loss of native prairies, the widespread clearing and dissection of forested habitats for agricultural use and settlement have created a large amount of suitable nesting habitat, especially where cleared areas have reverted to early successional vegetation.

Most chats winter in Middle America (AOU 1983), and birds begin to return to their Kentucky breeding grounds during the third week of April. According to Mengel (1965), early clutches are completed during the last ten days of April, with a peak in early clutch completion during the last ten days of May. Later egg dates ranging into mid-July suggest the possibility of the raising of a second brood, although documentation of this in Kentucky is lacking. Most later nests probably represent renesting following earlier failure because of predation or storms. The average size of 27 clutches and broods thought to be complete and listed by Mengel (1965) was 3.6 (range of 2–5). The chat is frequently parasitized by Brown-headed Cowbirds (Mengel 1965).

Forest Cover

Value	% of Blocks	Avg Abund
All	85.1	2.5
1	65.3	2.1
2	89.5	2.3
3	87.8	2.7
4	87.7	2.6
5	88.5	2.2

Nests are placed low in dense vegetation, most often in briars or other thick, weedy growth, but sometimes in small trees or shrubs. The nest is constructed of grass, weed stems, and other coarse plant material and is lined with finer grass and plant stems (Harrison 1975). The average height of 10 nests listed by Mengel (1965) was 3.1 feet (range of 2–4 feet), although more recent nests have been observed as low as 1 foot and as high as 7 feet above the ground (KOS Nest Cards; Stamm and Croft 1968).

Kathy Caminiti

The atlas survey yielded records of chats in more than 85% of priority blocks statewide, and 23 incidental observations were reported. The species was found in more than 85% of blocks in all physiographic province sections except the Cumberland Mountains and the Blue Grass. Within the Blue Grass, chats are absent from much of the extensively cleared Inner Blue Grass subsection, where suitable habitat is not abundant. In contrast, forests in the Cumberland Mountains are so extensive that early successional habitat is not as abundant as in most regions. Average abundance was relatively uniform across the state. Occurrence was lower only in extensively cleared areas.

Although the Yellow-breasted Chat is quite common throughout the state, less than 11% of priority block records were for confirmed breeding. The dense habitats in which chats are typically found make location of nests and observation of recently fledged young quite difficult. Of the 66 confirmed records, only a few involved active nests. Most were based on the observation of recently fledged young. The remaining confirmed records involved the observation of adults carrying food, distraction displays, and nest building. Interestingly, evidence of cowbird parasitism was not reported.

Chats are recorded in moderate numbers on most Kentucky BBS routes. The average number of individuals per BBS route for the periods 1966–91 and 1982–91 was 15.91 and 11.33, respectively. According to these data, the species ranked 22nd in abundance on BBS routes during the period 1982–91. Trend analysis of these data reveals a significant ($p<.01$) decline of 3.4% per year for the period 1966–91 and a significant ($p<.05$) decline of 3.7% per year for the period 1982–91. Reasons for these declines are unclear, but they may include pesticide use (Monroe 1978), reversion of early successional habitat to forest, and cowbird parasitism.

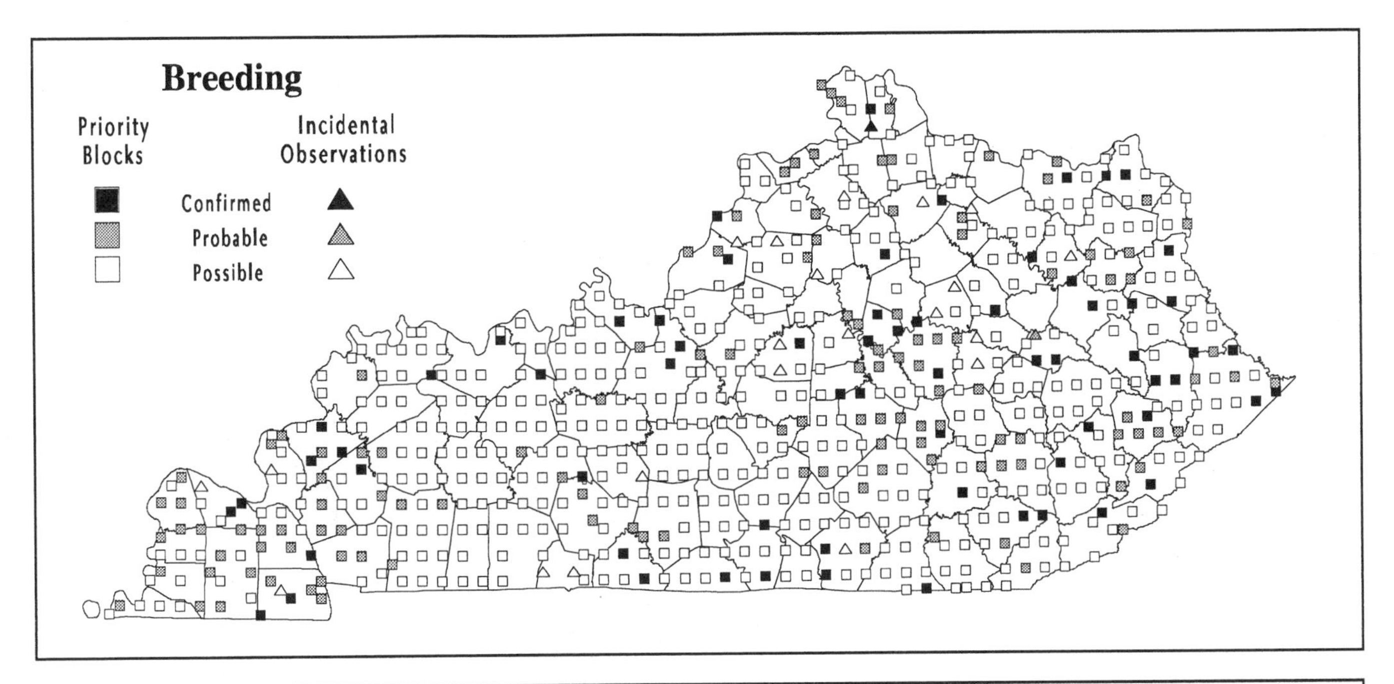

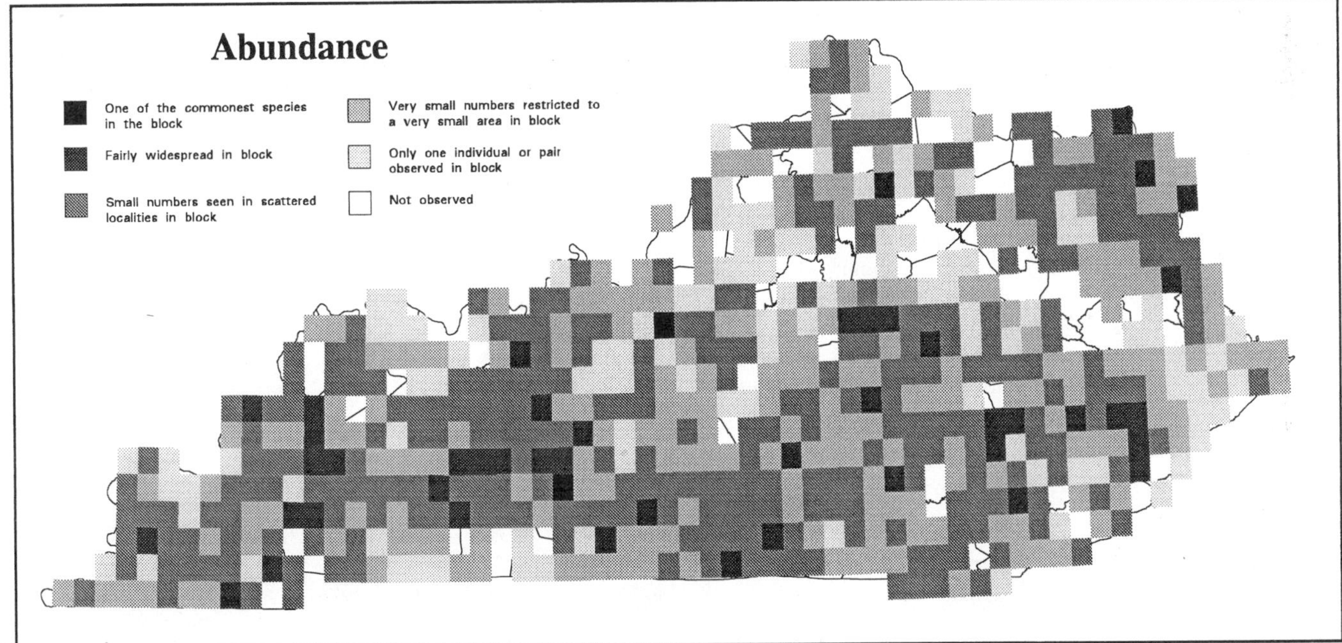

Analysis of Block Data by Physiographic Province Section

Physiographic Province Section	Total Blocks Surveyed	Blocks with Data	Avg Abund	% with Data	Section's % for State
Mississippi Alluvial Plain	14	12	2.3	85.7	1.9
East Gulf Coastal Plain	36	33	2.6	91.7	5.3
Highland Rim	139	130	2.5	93.5	21.0
Shawnee Hills	142	127	2.6	89.4	20.5
Blue Grass	204	149	2.3	73.0	24.1
Cumberland Plateau	173	154	2.4	89.0	24.9
Cumberland Mountains	19	14	2.0	73.7	2.3

Summary of Breeding Status

Number of Blocks in Which Species Was Recorded		
Total	**619**	**85.1%**
Confirmed	66	10.7%
Probable	125	20.2%
Possible	428	69.1%

Yellow-breasted Chat

Summer Tanager

Piranga rubra

The Summer Tanager is widely distributed across much of Kentucky as a breeding bird. The species is less numerous only in extensively cleared areas and in the more mesic forest types of eastern Kentucky, where typically it is replaced by the Scarlet Tanager (Mengel 1965). In addition, it appears to be nearly absent from higher elevations in the Cumberland Mountains (Mengel 1965; Croft 1971; Davis et al. 1980).

Summer Tanagers are birds of forested habitats, although their abundance varies substantially depending upon forest type. The species occurs in a variety of deciduous and mixed pine-hardwood associations, but it is most common in subxeric, upland forests of oak-hickory. In contrast, this tanager is least numerous in mesic forests of bottomlands, slopes, and ravines. The Summer Tanager apparently favors forests with semi-open canopy or mid-story, so it is abundant in drier situations. This preference also explains the species's use of altered situations that simulate naturally open habitats, including isolated woodlots, wooded parks, and forest disturbed by selective logging.

Although Summer Tanagers seem to have adapted to some of the changes humans have brought to the landscape, overall the species has probably decreased in abundance in Kentucky over the past two centuries. Vast upland areas formerly blanketed in mature forest have been cleared for agricultural use and settlement, especially in the central and western parts of the state. In addition, the savannas and barrens that once covered much of the Blue Grass, the Highland Rim, and the East Gulf Coastal Plain are now largely cleared for pastures and farm fields. These open woodland habitats probably once supported an abundant nesting population that has been greatly reduced.

Summer Tanagers return from their wintering grounds in the tropics during the last two weeks of April, and nesting activities appear to commence at once. Early clutches are completed during the first week of May, with a peak in clutch completion occurring during the last ten days of the month (Mengel 1965). Later clutches are regularly laid into early July and at least occasionally into late July, indicating that at least some pairs may raise a second brood (Mengel 1965). The average size of 14 clutches or broods given by Mengel (1965) was 3.2 (range of 2–4). According to data presented by Mengel, the Summer Tanager is more heavily parasitized by Brown-headed Cowbirds than any other of Kentucky's nesting birds.

Forest Cover

Value	% of Blocks	Avg Abund
All	79.2	2.2
1	53.5	1.8
2	83.2	2.1
3	85.2	2.4
4	85.1	2.1
5	71.2	1.8

Summer Tanagers typically nest in open woodland or along a forest margin. The nest is usually placed in the outer portion of a long branch in a fairly large tree. It is typically constructed in a small fork among or just inside the outer crown of leaves and often over a natural or artificial opening (frequently over road corridors). Although deciduous trees (most often oaks) are usually chosen, pines are used occasionally (Mengel 1965). The nest itself is a loosely constructed, shallow cup that may be so thin that the eggs are visible from below. It is constructed primarily of weed stems, bark, leaves, and grasses and is lined with fine grass (Harrison 1975). The average height of 25 nests given by Mengel (1965) was 17.5 feet (range of 5–45 feet).

Ron Austing

Summer Tanagers were recorded in more than 79% of priority blocks statewide, and twelve incidental observations were reported. The species ranked among the top three forest-nesting birds according to the number of priority block records. Summer Tanagers were found most frequently in the Highland Rim and the Shawnee Hills, where upland, subxeric forests are most common. The species was least numerous in the Mississippi Alluvial Plain, where the predominant bottomland forests are not favored for nesting; the Blue Grass, where deforestation is especially pronounced; and the Cumberland Mountains, where the abundance of mesic forest does not favor the species's presence. Average abundance was relatively uniform across most of the state, but lowest in the Cumberland Mountains. Both occurrence and average abundance were highest in areas with a good mix of forest and openings.

Despite the widespread occurrence of the Summer Tanager, less than 8% of priority block records were for confirmed breeding. Although several active nests were located, most of the 44 confirmed records involved the observation of recently fledged young. Cowbird parasitism was quite evident, and several instances of adult tanagers feeding young cowbirds were reported. Other confirmed records were based on the observation of adults carrying food for young and nest building.

Summer Tanagers are typically recorded in small numbers on Kentucky BBS routes. The average number of individuals per BBS route for the periods 1966–91 and 1982–91 was 4.63 and 3.38, respectively. Trend analysis of these data reveals a slight, nonsignificant decrease of 1.1% per year for the period 1966–91, but a significant ($p<.05$) decrease of 4.5% per year for the period 1982–91. Reasons for the short-term decrease are unknown.

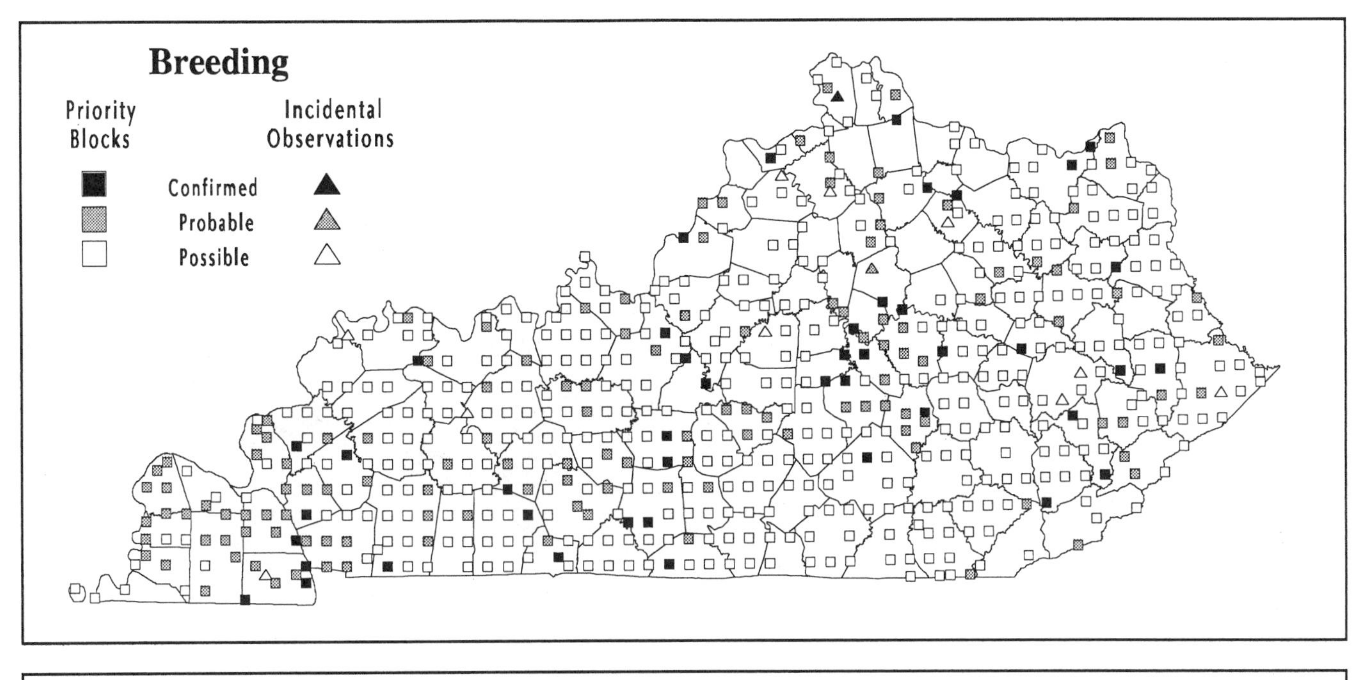

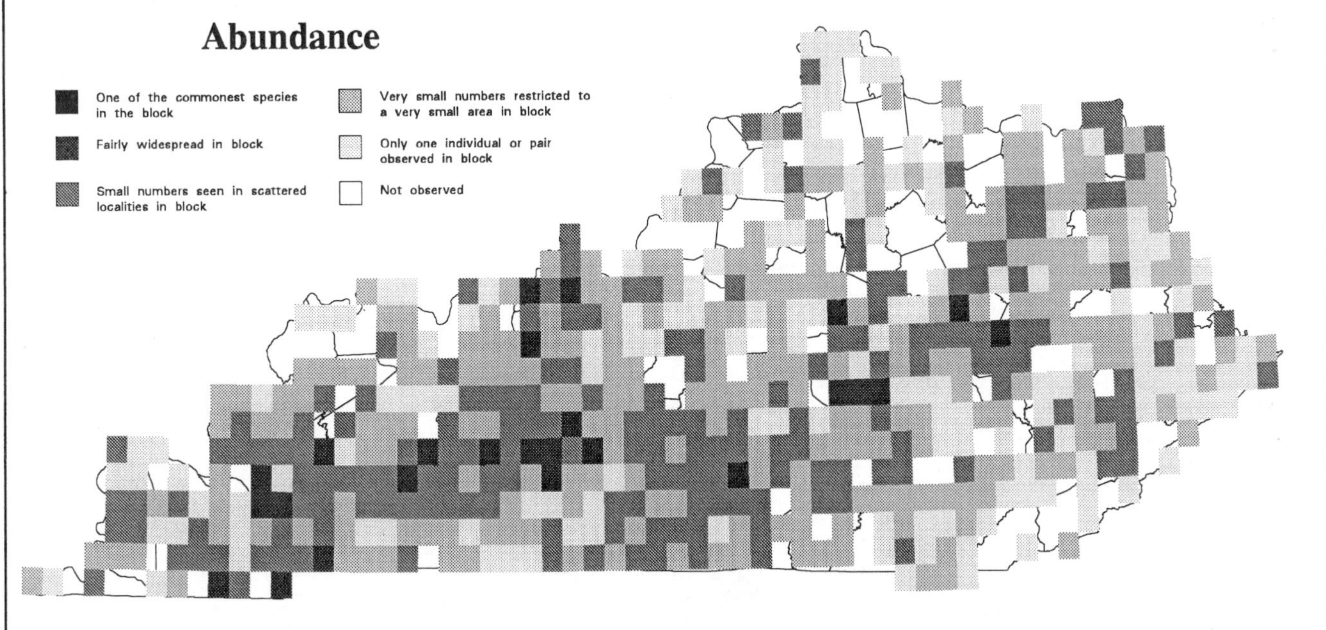

Analysis of Block Data by Physiographic Province Section

Physiographic Province Section	Total Blocks Surveyed	Blocks with Data	Avg Abund	% with Data	Section's % for State
Mississippi Alluvial Plain	14	10	2.3	71.4	1.7
East Gulf Coastal Plain	36	29	2.2	80.6	5.0
Highland Rim	139	131	2.3	94.2	22.7
Shawnee Hills	142	123	2.5	86.6	21.4
Blue Grass	204	133	2.0	65.2	23.1
Cumberland Plateau	173	143	1.9	82.7	24.8
Cumberland Mountains	19	7	1.1	36.8	1.2

Summary of Breeding Status

Number of Blocks in Which Species Was Recorded		
Total	**576**	**79.2%**
Confirmed	44	7.6%
Probable	129	22.4%
Possible	403	70.0%

Summer Tanager

Scarlet Tanager

Piranga olivacea

The burry song of the Scarlet Tanager is an uncommon to common summer sound in forests across Kentucky. The center of the species's breeding range lies to the north of the state, and like other more northern forest-nesting songbirds, it occurs in greatest abundance in eastern Kentucky. Mengel (1965) considered the Scarlet Tanager to be fairly common to common in eastern Kentucky and increasingly rare and local westward. Results of the atlas survey indicate that the species has apparently increased in central and western parts of the state.

The Scarlet Tanager inhabits a wide variety of deciduous and mixed forest types. The species occurs in greatest abundance in moist, mixed mesophytic forests of slopes and ravines in eastern Kentucky, but it is regularly encountered in mesic slope forests of central and western parts of the state. There, too, the species is locally common in extensive tracts of subxeric oak-hickory forests such as are found in and near Land Between the Lakes. This tanager prefers fairly mature to mature, relatively undisturbed forest, although it sometimes uses younger second-growth or slightly disturbed situations.

The Scarlet Tanager likely was more widespread and common in Kentucky before settlement. Audubon (1861) observed the species "throughout Kentucky" in the early 1800s, and the mature forests that once covered much of the state likely harbored a substantial population. Unlike the Summer Tanager, this species does not frequently inhabit dissected forest tracts and forest moderately to heavily disturbed by logging. For this reason it is restricted to larger tracts of extensive forest, and its numbers have been greatly reduced not only by forest clearing but also by fragmentation and periodic logging of remaining woodlands.

Scarlet Tanagers winter primarily in South America (AOU 1983), and birds usually begin to arrive on their Kentucky breeding grounds during the third week of April. According to Mengel (1965), early clutches are completed during the first week of May, and a peak in clutch completion probably occurs sometime during the middle of the month. Later clutch dates into mid-July suggest that some pairs may raise a second brood (Mengel 1965). Kentucky data on clutch size are scarce, but the average size of three clutches given by Mengel (1965) was 3.7 (range of 3–5). Harrison (1975) gives rangewide clutch size as 3–5, commonly 4.

Forest Cover

Value	% of Blocks	Avg Abund
All	63.3	2.1
1	14.9	1.4
2	45.8	1.6
3	73.9	2.0
4	90.9	2.3
5	92.3	2.6

Scarlet Tanagers often nest within mature forest, but they also choose sites along natural and artificial forest margins. The nest is usually placed far out from the trunk in the midlevel or lower branches of large trees. It is most often constructed at the base of a small horizontal fork, and among or just inside the leaves of the outer crown. Although deciduous trees are usually chosen, conifers may be used on occasion (Davis et al. 1980; KOS Nest Cards). The nest consists of a shallow cup, somewhat loosely constructed of twigs and rootlets, and is lined with fine weed stems and grasses (Harrison 1975). Kentucky data on nest height are scarce, but the average height of five reported nests is 23.2 feet (range of 8–45 feet) (Mengel 1965; Davis et al. 1980; KOS Nest Cards; L. McNeely, pers. comm.).

Ron Austing

The atlas survey yielded records of Scarlet Tanagers in more than 63% of priority blocks statewide, and 21 incidental observations were reported. This tanager was most widespread in the Cumberland Mountains and the Cumberland Plateau, and it was distributed rather uniformly across central and western Kentucky. West of the Cumberland Plateau, the species typically occurs in areas of high relief, such as the Knobs of the Blue Grass, dissected ravine systems, and bluffs along the margins of river floodplains. For the most part, average abundance varied similarly. Both occurrence and average abundance were closely related to percentage of forest cover, being lowest in predominantly cleared areas and increasing with the degree of forestation.

Although Scarlet Tanagers are fairly widespread, less than 6% of priority block records were for confirmed breeding. A few active nests were located, but most of the 26 confirmed records were based on the observation of recently fledged young and adults carrying food. Brown-headed Cowbird parasitism was evidenced by the observation of adult tanagers feeding young cowbirds on a number of occasions, especially in central and western Kentucky.

Scarlet Tanagers are not sampled well along Kentucky BBS routes, in large part because most routes in the eastern part of the state are not run. The average number of individuals recorded per BBS route for the periods 1966–91 and 1982–91 was 2.00 and 0.81, respectively. Trend analysis of these data yields a significant increase ($p<.01$) of 3.8% per year for the period 1966–91, but only a nonsignificant increase of 0.5% per year for the period 1982–91.

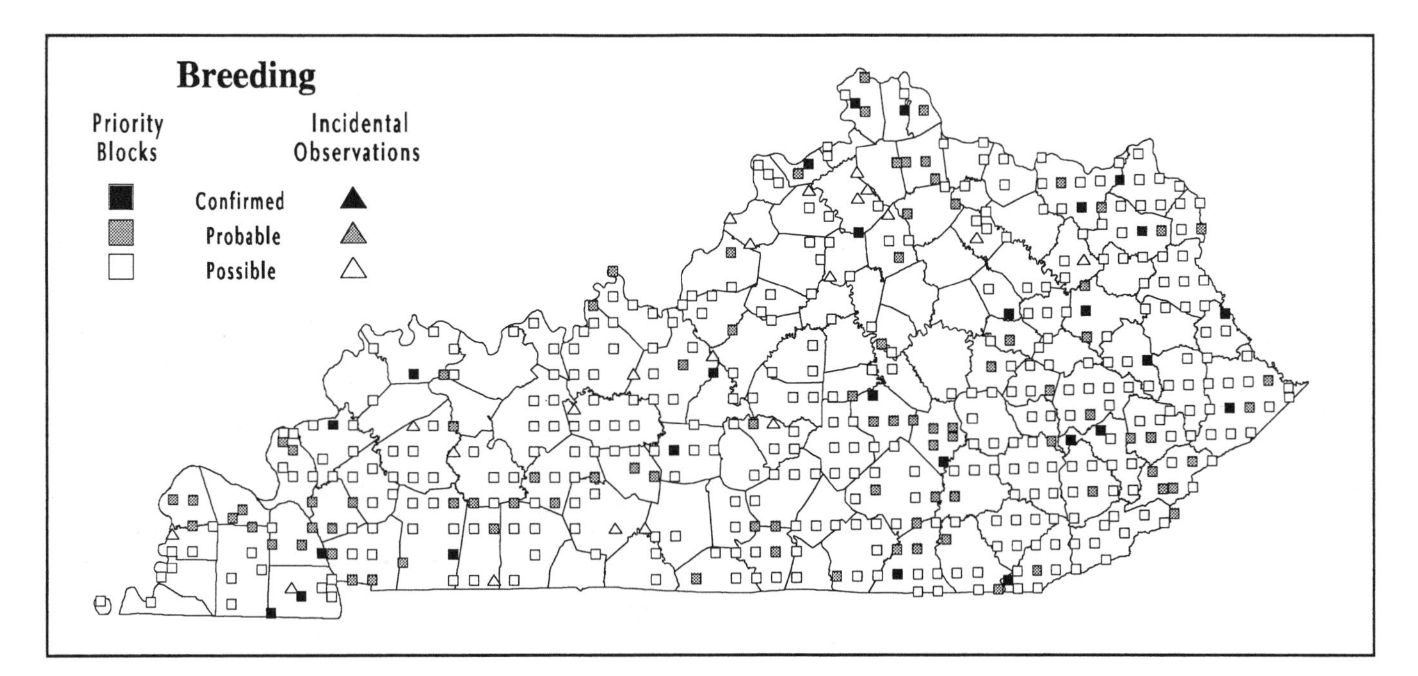

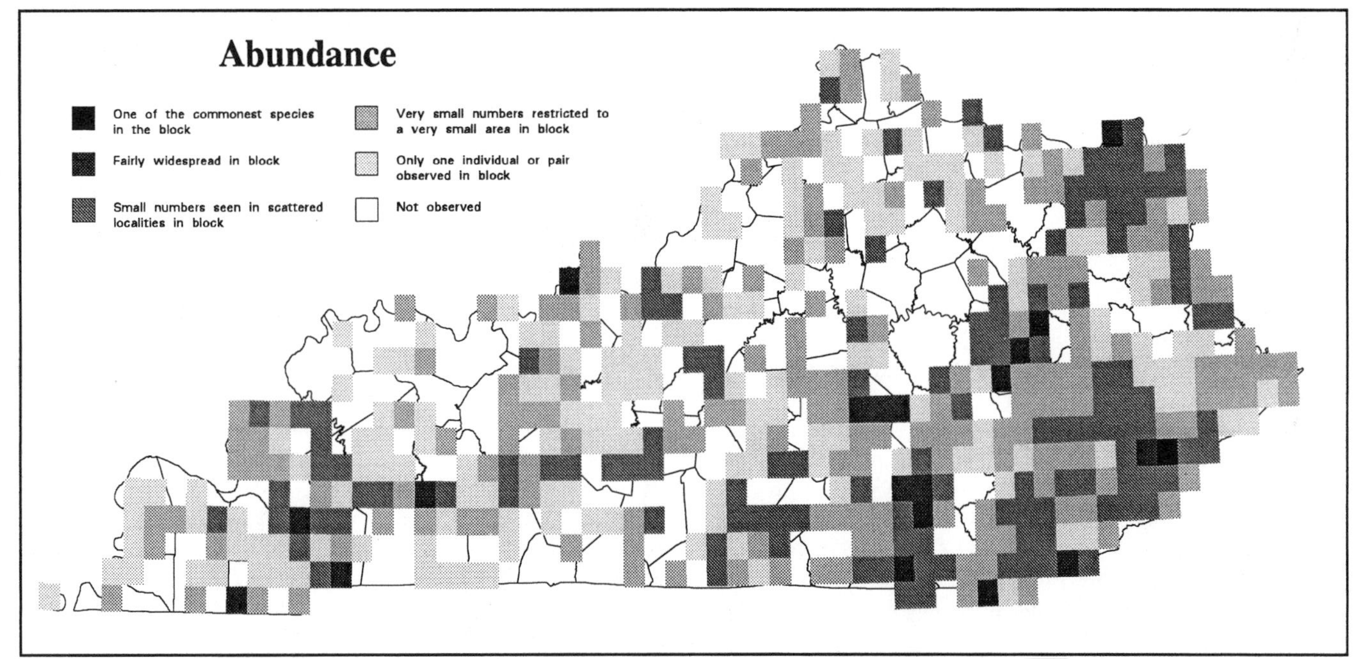

Analysis of Block Data by Physiographic Province Section

Physiographic Province Section	Total Blocks Surveyed	Blocks with Data	Avg Abund	% with Data	Section's % for State
Mississippi Alluvial Plain	14	7	1.3	50.0	1.5
East Gulf Coastal Plain	36	20	1.6	55.6	4.3
Highland Rim	139	79	2.0	56.8	17.2
Shawnee Hills	142	86	1.9	60.6	18.7
Blue Grass	204	98	2.0	48.0	21.3
Cumberland Plateau	173	152	2.3	87.9	33.0
Cumberland Mountains	19	18	2.4	94.7	3.9

Summary of Breeding Status

Number of Blocks in Which Species Was Recorded		
Total	**460**	**63.3%**
Confirmed	26	5.7%
Probable	90	19.6%
Possible	344	74.8%

Northern Cardinal

Cardinalis cardinalis

The Northern Cardinal is among the most abundant and widely recognized of Kentucky's nesting birds. The species is found in a great variety of natural and altered habitats with some sort of dense cover, being considerably less numerous only in extensively cleared and predominantly forested areas. For the most part cardinals have adapted well to the dramatic changes humans have made to the landscape, and they are frequent in a variety of artificial habitats, including urban and suburban parks and yards, brushy forest edge, old fields, overgrown fencerows, rural homesteads, and reclaimed strip mines. Although the species is not as abundant in extensively forested areas, small numbers are typically encountered wherever there is some dense cover or where natural disturbance has created brushy openings.

Cardinals may have been common in Kentucky two centuries ago, but it is likely that human alteration of the landscape has resulted in an overall increase in their abundance. Dissection of extensive forests that once covered much of the state has resulted in the creation of an abundance of early successional habitat that is now occupied by substantial numbers of birds. The exceptions to this trend are probably regions where clearing of land has been extraordinarily intense. Such areas are especially conspicuous in parts of the Inner Blue Grass, the south central Highland Rim, and the Mississippi Alluvial Plain, where low relief and clean farming practices often result in a scarcity of suitable nesting habitat.

Northern Cardinals are resident in Kentucky throughout the year, and males begin to sing as soon as day length increases in midwinter. Territorial behavior often commences during March, and early clutches are completed by early April (Mengel 1965). An initial peak in clutch completion probably occurs sometime in late April or early May. Many pairs are at least double-brooded, and later nesting activity is common through July, with clutches known as late as August 21–31 (Mengel 1965). The regularity of later dates suggests the possibility of the raising of a third brood by some pairs. The average size of 63 clutches or broods given by Mengel (1965) was 2.8 (range of 1–4). Occasional nests have been shared with other cardinals (Mengel 1965) and American Robins (W. Schoettler, pers. comm. and photographs). Brown-headed Cowbird parasitism is frequent, especially in suburban areas and rural farmland and settlement.

Forest Cover

Value	% of Blocks	Avg Abund
All	99.6	3.9
1	99.0	3.6
2	100.0	3.9
3	100.0	3.9
4	100.0	4.1
5	96.2	4.0

The nest is typically built amid some sort of cover, usually in tangles of vines, brambles, or other dense vegetation. Ornamental trees and shrubs are often used in urban and suburban areas. The nest is a relatively shallow, loosely woven cup constructed of a few dead leaves and many small twigs or vines. It is lined with fine grass or hair. The average height of 58 nests given by Mengel (1965) was 5.5 feet (range of 2–20 feet).

The Northern Cardinal was recorded in all but three priority blocks statewide, making it second only to the Indigo Bunting, which was recorded in all but one block. The species also ranked 2nd according to both total abundance and average abundance. Average abundance was relatively uniform across the state, but it was slightly lower in extensively cleared areas.

Alvin E. Staffan

More than 38% of priority block records were for confirmed breeding. Although many active nests were located, most confirmed records were based on the observation of recently fledged young. Other confirmed records involved the observation of adults carrying food, distraction displays, and nest building. Cowbird parasitism was noted frequently; many active nests containing cowbird eggs or young were observed, as well as fledgling cowbirds being fed by adult cardinals.

Northern Cardinals are recorded in substantial numbers on most Kentucky BBS routes. The average number of individuals recorded per BBS route for the periods 1966–91 and 1982–91 was 33.98 and 43.61, respectively. According to these data, the Northern Cardinal ranked 9th in abundance on BBS routes during the period 1982–91. Trend analysis of these data reveals a slight, nonsignificant decrease of 0.30% per year for the period 1966–91 and a small, nonsignificant increase of 1.2% per year for the period 1982–91.

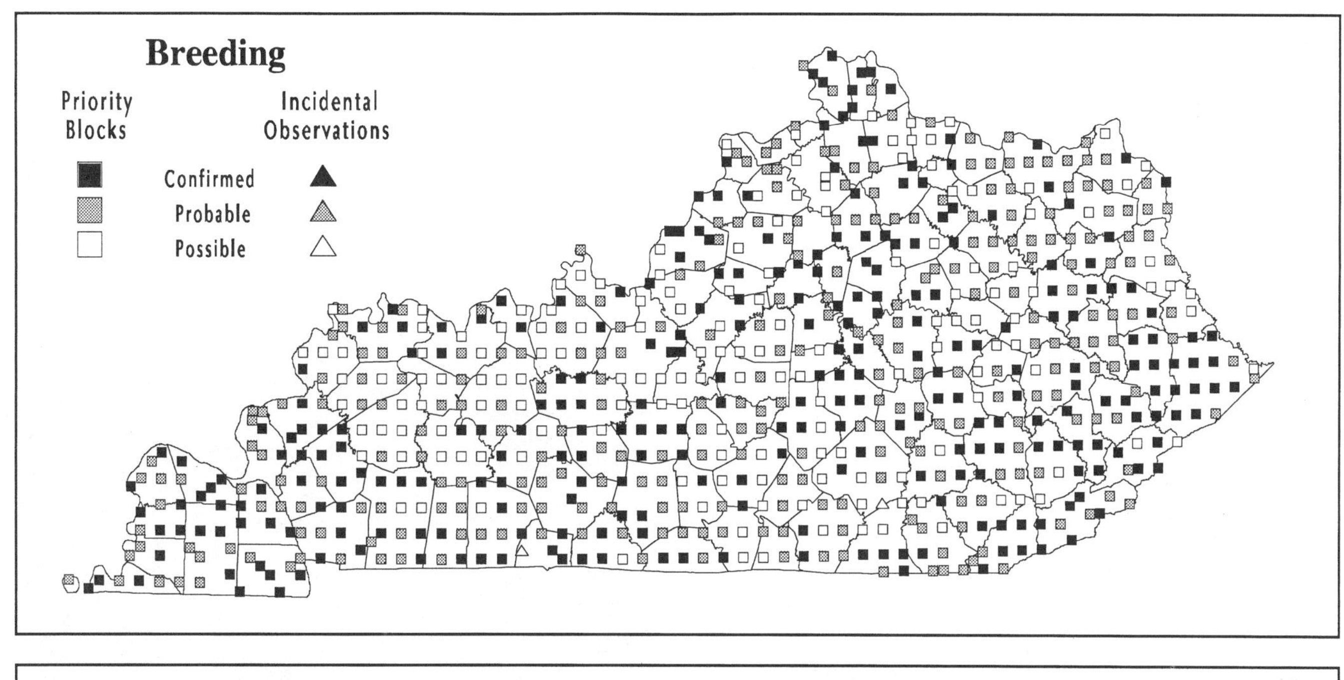

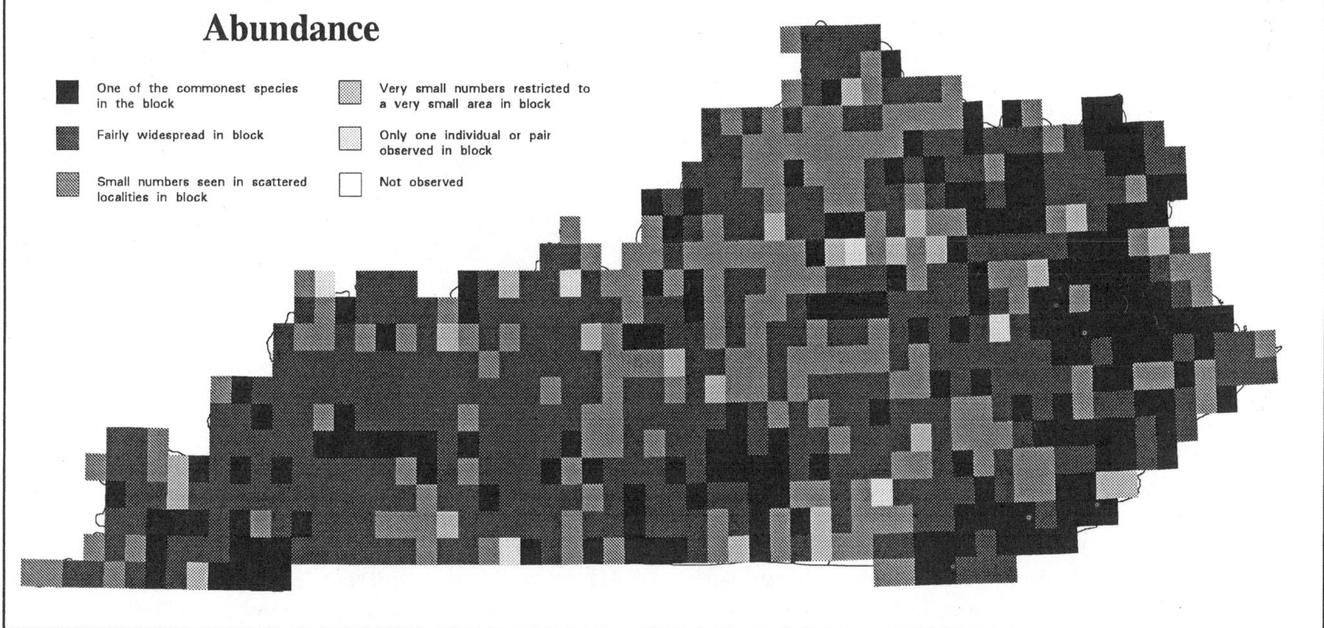

Analysis of Block Data by Physiographic Province Section

Physiographic Province Section	Total Blocks Surveyed	Blocks with Data	Avg Abund	% with Data	Section's % for State
Mississippi Alluvial Plain	14	14	3.8	100.0	1.9
East Gulf Coastal Plain	36	36	4.1	100.0	5.0
Highland Rim	139	138	3.9	99.3	19.1
Shawnee Hills	142	142	3.9	100.0	19.6
Blue Grass	204	204	3.8	100.0	28.2
Cumberland Plateau	173	172	4.1	99.4	23.8
Cumberland Mountains	19	18	4.3	94.7	2.5

Summary of Breeding Status

Number of Blocks in Which Species Was Recorded		
Total	**724**	**99.6%**
Confirmed	278	38.4%
Probable	289	39.9%
Possible	157	21.7%

Northern Cardinal

Rose-breasted Grosbeak

Pheucticus ludovicianus

Although the Rose-breasted Grosbeak breeds primarily to the north of Kentucky in the northeastern United States and southeastern Canada, its nesting range extends south through the Appalachian Mountains, barely reaching into the extreme southeastern corner of the state (Mengel 1965; AOU 1983). For many years the species has been known to occur above 3,000 feet on Black Mountain, in Harlan and Letcher counties, where its mellow, robinlike song is a fairly common summer sound (Mengel 1965).

Rose-breasted Grosbeaks were first reported on Black Mountain in 1908 (Howell 1910). Subsequent fieldwork on the mountain has produced continued documentation of summering and a few records of confirmed nesting, primarily above 3,000 feet (Wetmore 1940; Breiding 1947; Lovell 1950a; Mengel 1965; Croft 1969; KSNPC, unpub. data). Before the atlas fieldwork, the species was known elsewhere in summer only on the basis of a few records of presumed vagrants.

Rose-breasted Grosbeaks are typically found in deciduous forest and forest edge, where they forage mostly at upper and midstory levels. At higher elevations of Black Mountain, the species is numerous in fairly mature, relatively undisturbed forest as well as in forest edge and regenerating, younger forest disturbed by selective logging, resource extraction, or fire. Numbers there probably have fluctuated somewhat, as land use changes have occurred on the mountain. At present, the species is probably about as numerous as ever, given the diversity of forest habitats present near the summit.

Rose-breasted Grosbeaks return from their wintering grounds in the tropics during the last two weeks of April, although it may be mid-May before the entire nesting population has returned. Of four nests observed by Mengel (1965) on Black Mountain, all were probably completed during the last ten days of May. Family groups have been observed most frequently in June and July. Kentucky data on clutch size are lacking, but Harrison (1975) gives rangewide clutch size as 3–6, commonly 4.

Nests described by Mengel (1965) were placed in relatively small trees, including witch hazel, yellow birch, and red maple. Two trees were in second-growth forest, one was situated in a brushy clearing, and one was standing in the shaded understory of disturbed climax forest. The typical nest is a bulky structure, loosely constructed of a variety of plant materials, including fine twigs, rootlets, plant fibers, and grape tendrils with no special lining (Mengel 1965). All four nests described by Mengel (1965) were placed in forks of the outer portions of branches, and their average height was about 20 feet (range of 9–30 feet).

The atlas survey yielded a surprisingly large amount of new information about the breeding status of the Rose-breasted Grosbeak. While only eight priority block records and one incidental observation were reported, only three of the nine records came from traditionally known nesting areas on Black Mountain. The other six came from four widely separated places, and they included confirmed records from two locations.

Before the atlas survey, Rose-breasted Grosbeaks had not been reported in the Cumberland Mountains away from Black Mountain (Croft 1969, 1971; Davis et al. 1980). During atlas fieldwork the species was recorded for the first time along the crest of Cumberland Mountain in Cumberland Gap National Historical Park, in Bell and Harlan counties. On June 17, 1989, singing males were observed at two points along the Ridge Trail between Hensley Settlement and White Rocks.

Ron Austing

Potential breeding populations were also found in two areas of northern Kentucky in the summer of 1991. In Boone County, a pair of grosbeaks was observed carrying nesting material in suitable breeding habitat along Middle Creek Road on May 11 (Stamm and Monroe 1991b; McNeely et al. 1991). This pair remained in the area through mid-June, and a singing male was reported nearby at Dinsmore Woods on June 1 (McNeely et al. 1991). In rural Lewis County, a single female and a single male were noted at widely separated locations near Charters on July 13, but further evidence of breeding was not obtained. The appearance of these birds is certainly related to a southward expansion in nesting range that has been documented in Ohio (Peterjohn and Rice 1991).

Outside northern and southeastern Kentucky, nesting was confirmed in Daviess County near Maceo in the Shawnee Hills, where fledglings were observed being fed by an adult male in the summer of 1985. Although this record would appear to be extralimital, Rose-breasted Grosbeaks have been recorded nesting in Indiana near Daviess County, Kentucky (Mumford and Keller 1984), and at several localities in southern Illinois (Bohlen 1989).

Although this new information is intriguing, additional work is needed to determine the regularity and extent of nesting away from Black Mountain. Heretofore unknown nesting populations probably exist in other parts of the state, but a diligent effort will be required to locate them.

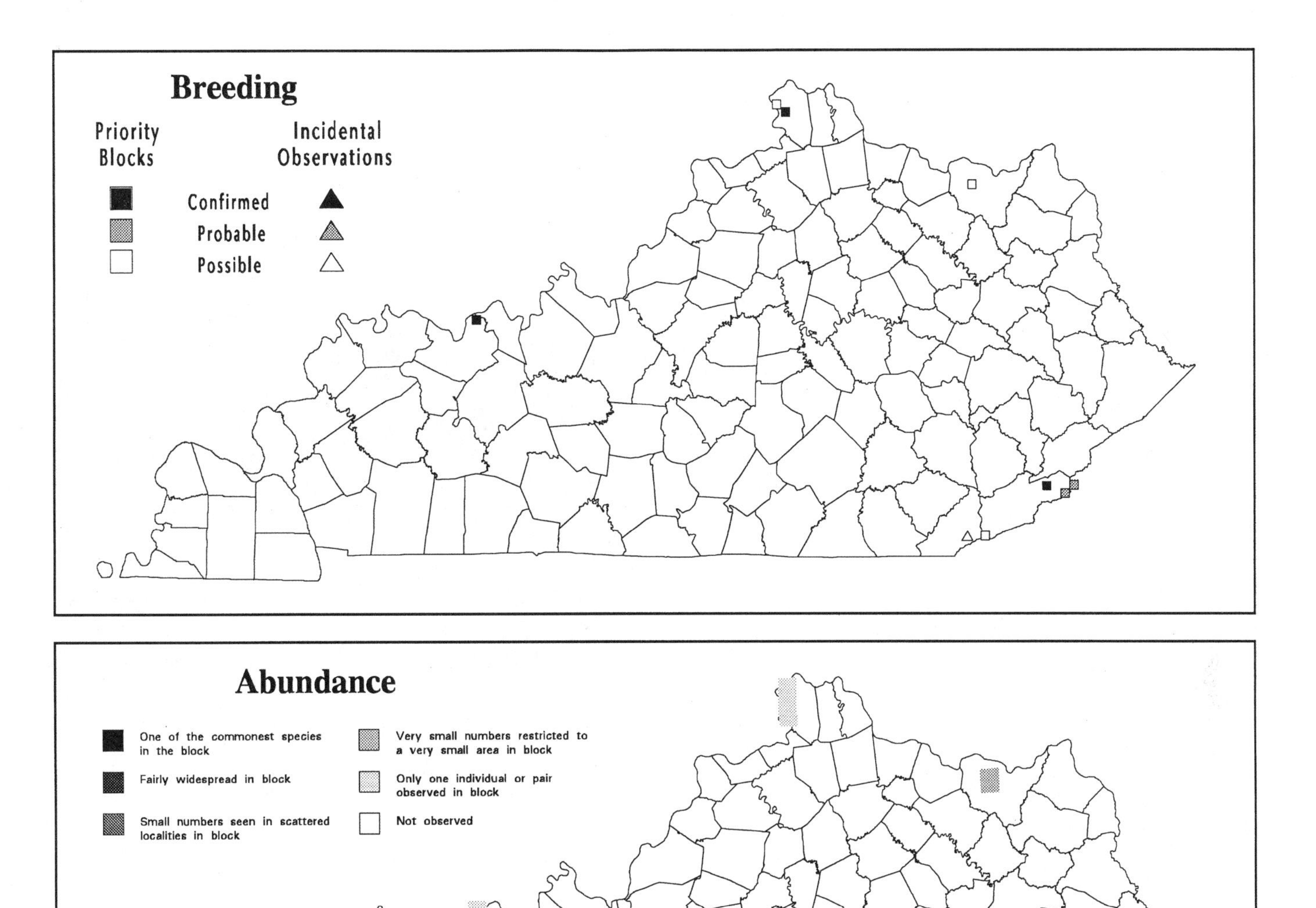

Analysis of Block Data by Physiographic Province Section

Physiographic Province Section	Total Blocks Surveyed	Blocks with Data	% with Data	Section's % for State
Mississippi Alluvial Plain	14	-	-	-
East Gulf Coastal Plain	36	-	-	-
Highland Rim	139	-	-	-
Shawnee Hills	142	1	0.7	12.5
Blue Grass	204	3	1.5	37.5
Cumberland Plateau	173	-	-	-
Cumberland Mountains	19	4	21.1	50.0

Summary of Breeding Status

Number of Blocks in Which Species Was Recorded		
Total	**8**	**1.1%**
Confirmed	3	37.5%
Probable	2	25.0%
Possible	3	37.5%

Rose-breasted Grosbeak

Blue Grosbeak

Guiraca caerulea

As recently as the late 1950s the Blue Grosbeak was considered a casual summer resident in western Kentucky (Mengel 1965). During the past 35 years, however, the species has expanded into all of the state except the southeastern section. These grosbeaks are now a regular constituent of the summer avifauna across much of central and western Kentucky, and small numbers are scattered throughout the Cumberland Plateau.

Mengel (1965) listed less than a dozen records for the Blue Grosbeak, and he considered the summering population as of the late 1950s to be negligible. Lancaster and Wilson (1964) first reported nesting in June 1964 in Warren County on the basis of fledglings being fed by adults, and Dubke (1966a) apparently located the first active nest in June 1966 in Taylor County. Wilson (1967) gave a brief summary of early records for south central Kentucky, and several other published accounts were part of the flurry of reports at the onset of the invasion during the mid- to late 1960s (Croft 1967; Russell et al. 1967; Gray 1968a; Monroe and Able 1968). More recently, Ritchison (1984) provided a summary of the species's status as of the early 1980s. Its increase in Kentucky is part of a regional expansion that also has been noted in adjacent states (Robinson 1990; Peterjohn and Rice 1991; Buckelew and Hall 1994).

The Blue Grosbeak is a bird of semi-open and open habitats with some brushy or weedy cover. Although the species occurs in natural grassland and shrubland habitats in parts of its range, it is primarily restricted to artificial habitats in Kentucky. These grosbeaks occur most frequently in rural farmland and settlement, where they inhabit abandoned fields, brushy fencerows, and other margins with brushy or weedy cover and scattered young trees. They are also regularly encountered on reclaimed surface mines, young pine plantations, and other open habitats in early stages of revegetation.

Why the Blue Grosbeak has only recently expanded into the eastern United States is unclear. Suitable habitat was present in Kentucky before settlement, especially in the prairies and transitional habitats that dominated much of the East Gulf Coastal Plain and the Highland Rim. Moreover, as human alteration of the landscape accelerated during the 19th century, early successional habitat became widespread.

Forest Cover

Value	% of Blocks	Avg Abund
All	53.8	1.9
1	59.4	1.7
2	75.8	2.1
3	60.9	2.0
4	29.2	1.6
5	1.9	2.0

Blue Grosbeaks often arrive in western Kentucky by mid-April, but it is probably mid-May before the full complement of breeding birds has returned statewide. Specific Kentucky nesting data are scarce, but early clutches likely are not completed until late May or early June. Many nests are apparently active into July (Ritchison 1984; Stamm and Monroe 1991a; KBBA data), and there are two reports of young fledging in late August (Ritchison 1984; KOS Nest Cards). Based on the number of late dates, it must be supposed that at least some pairs raise two broods. Kentucky data on clutch size are scarce, but the average size of six clutches or broods thought to be complete is 3.3 (range of 3–5) (Dubke 1966a; Gray 1968a; Ritchison 1984; KOS Nest Cards; KBBA data).

Ron Austing

The nest is typically placed in a fairly open situation, often along a fencerow or the margin of a field. It is usually constructed low to the ground in the dense cover of a small tree or shrub or in a tangle of vines, brambles, or weeds. The nest is a bulky cup constructed of grasses and weed stems, often with some snakeskin or clear plastic material incorporated into the structure, and lined with fine grass. The average height of seven reported nests is 2.9 feet (range of 1.5–6.0 feet) (Dubke 1966a; Gray 1968a; Ritchison 1984; KOS Nest Cards; KBBA data).

The atlas survey yielded records of Blue Grosbeaks in nearly 54% of priority blocks, and 24 incidental observations were reported. Occurrence was highest in the Highland Rim, the Shawnee Hills, and the East Gulf Coastal Plain. In contrast, the species was found in less than one-fourth of the priority blocks in the Cumberland Plateau, and it was not observed in the Cumberland Mountains. Blue Grosbeaks are still increasing in northern Kentucky, and it is expected that they will soon occupy suitable habitat throughout the Blue Grass. Average abundance was relatively uniform across the state. Occurrence was highest in open areas with some forest, and lowest in forested areas.

Less than 10% of priority block records were for confirmed breeding. Three active nests were located, but most of the 37 confirmed records were based on the observation of adults carrying food. Other confirmed records involved recently fledged young, nest building, and distraction displays.

BBS data illustrate the Blue Grosbeak's increase in Kentucky. The average number of individuals recorded per BBS route for the periods 1966–91 and 1982–91 was 1.78 and 3.90, respectively. Trend analysis of these data reveals a significant ($p<.01$) increase of 10.4% per year for the period 1966–91. In contrast, an increase of 2.7% per year for the period 1982–91 is not statistically significant, perhaps indicating that the Blue Grosbeak's population is leveling off, at least in areas where BBS routes are being run.

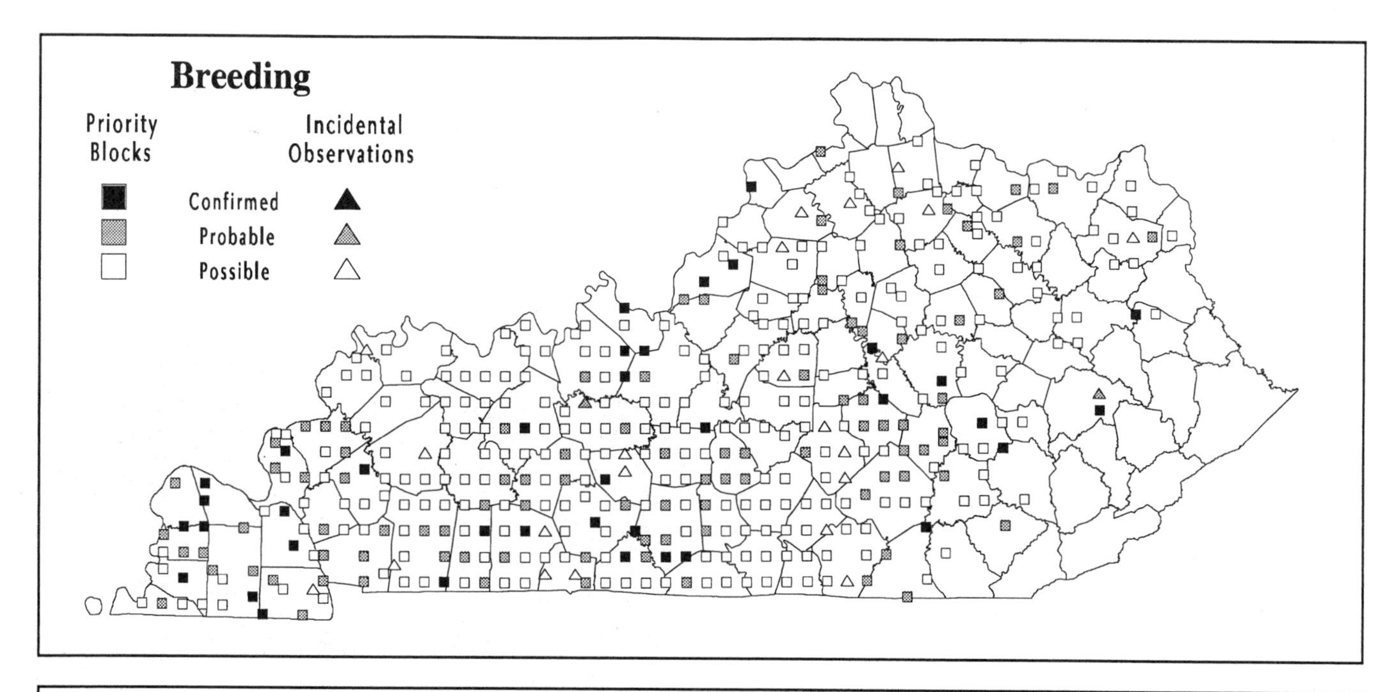

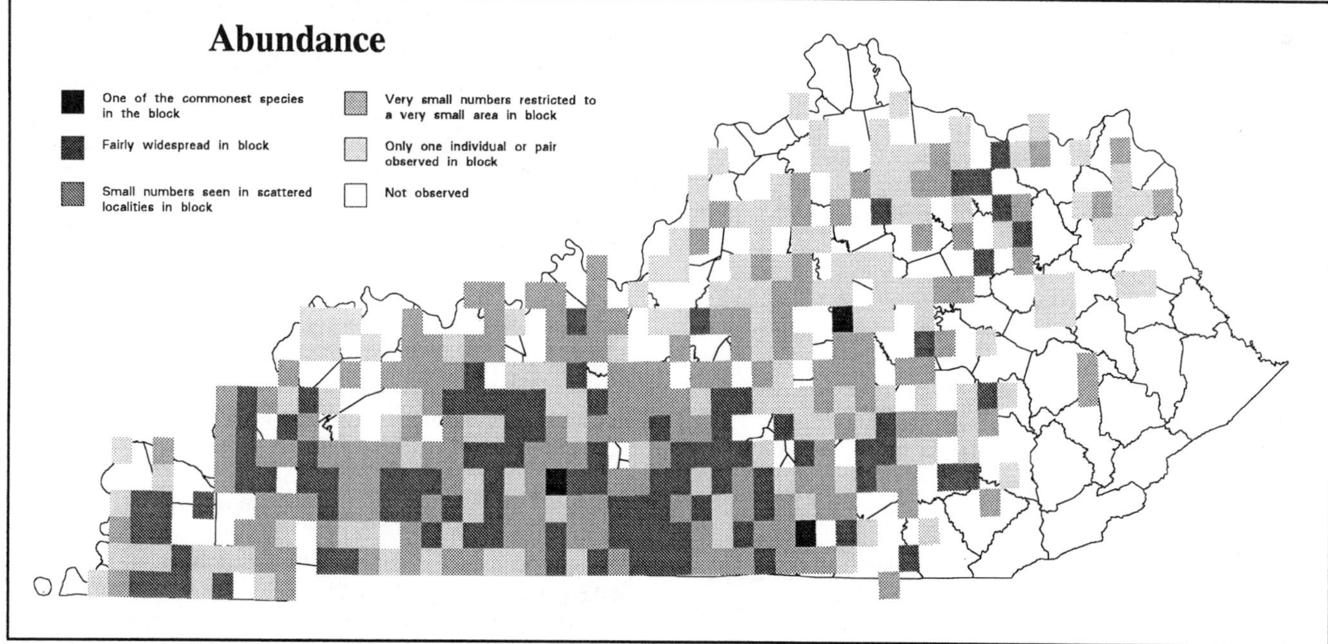

Analysis of Block Data by Physiographic Province Section

Physiographic Province Section	Total Blocks Surveyed	Blocks with Data	Avg Abund	% with Data	Section's % for State
Mississippi Alluvial Plain	14	7	1.6	50.0	1.8
East Gulf Coastal Plain	36	25	2.0	69.4	6.4
Highland Rim	139	118	2.2	84.9	30.2
Shawnee Hills	142	102	2.1	71.8	26.1
Blue Grass	204	98	1.6	48.0	25.1
Cumberland Plateau	173	41	1.6	23.7	10.5
Cumberland Mountains	19	-	-	-	-

Summary of Breeding Status

Number of Blocks in Which Species Was Recorded		
Total	**391**	**53.8%**
Confirmed	37	9.5%
Probable	96	24.6%
Possible	258	66.0%

Indigo Bunting

Passerina cyanea

The Indigo Bunting is one of the most widespread and abundant of Kentucky's nesting birds. In fact, according to BBS data for the period 1965–79, the Indigo was as common in parts of central and western Kentucky as anywhere within its range (Robbins et al. 1986). While it is possible that several nesting species may be more abundant, it is likely that none are as widely distributed. Whether the scene is a forest opening in the Cumberland Mountains or the brushy border of a floodplain slough in the far western part of the state, the Indigo Bunting is probably the most likely nesting bird to be encountered.

Indigo Buntings can be found in a great variety of habitats with at least some dense herbaceous cover. Frequently used habitats include brushy forest edge, the weedy margins of agricultural fields, roadway and utility corridors, overgrown drainages and fencerows, old fields, reclaimed strip mines, recent clear-cuts, young pine plantations, and open forest disturbed by selective logging. Perhaps the only areas in which the species is fairly rare are those virtually devoid of some weedy or brushy cover, such as many suburban parks and yards, excessively clean farms, and other intensively managed sites. Although the species is found primarily in artificial habitats, it also occurs in natural situations, such as woodland openings resulting from fire or windstorms, openings along stream corridors, the shrubby borders of floodplain sloughs, and the brushy margins of prairie remnants.

Although it is likely that the Indigo Bunting always has been fairly common in Kentucky, the species has certainly benefited from human alteration of the landscape. The extensive clearing and dissection of forests that once covered much of the state have resulted in the creation of an abundance of suitable nesting habitat. Only in areas where manipulation has been extraordinarily intense is suitable habitat scarce and the species less common.

Indigo Buntings return to Kentucky from their wintering grounds in the tropics during the last two weeks of April, although it may be the middle of May before nesting activity is under way. According to Mengel (1965), early clutches are completed in mid-May, with an early peak in clutch completion from late May through early June. A later peak in clutch completion occurs during the first ten days of July, indicating that many pairs apparently raise a second brood (Mengel 1965). Nesting activity continues at least occasionally into early August (Stamm and Jones 1966). The average size of 30 clutches or broods reported by Mengel (1965) was 2.9 (range of 2–4). The Indigo Bunting is among those species most often parasitized by Brown-headed Cowbirds (Mengel 1965).

Forest Cover

Value	% of Blocks	Avg Abund
All	99.9	4.2
1	99.0	3.9
2	100.0	4.2
3	100.0	4.2
4	100.0	4.5
5	100.0	4.1

Indigo Buntings most frequently choose a nest site along a forest margin, fencerow, or similar feature. The nest is typically placed fairly low to the ground in weeds, vines, or low shrubbery, and occasionally in the lower branches of small trees. Excluding an unusually high nest situated 15 feet above the ground, the average height of 32 nests given by Mengel (1965) was 2.8 feet (range of 2 inches to 6 feet). The nest is a sturdy cup constructed of grass, weed stems, and a few leaves and lined with fine grass. It is woven into place among supporting stems of the surrounding vegetation.

Alvin E. Staffan

The Indigo Bunting was the most frequently recorded species during the atlas fieldwork, located in all but one priority block statewide. The species also ranked 1st according to total abundance and average abundance. Average abundance by percentage of forest cover indicated that the species was slightly more common in areas with a relatively high, but not complete, forest cover and was least common in extensively cleared areas.

About 29% of priority block records were for confirmed breeding. This relatively high figure probably can be attributed to the species's overall abundance and the consequently greater likelihood of observing nesting activity. Although a number of active nests were located, most of the 212 confirmed records were based on the observation of recently fledged young. Other confirmed records involved adults carrying food, nest building, distraction displays, and used nests. Cowbird parasitism was evidenced by a number of records of cowbird eggs and young in nests, as well as adult buntings feeding young cowbirds.

Indigo Buntings are typically reported in substantial numbers on most Kentucky BBS routes. The average number of individuals recorded per BBS route for the periods 1966–91 and 1982–91 was 52.00 and 46.66, respectively. According to these data, the species ranked 6th in abundance on BBS routes during the period 1982–91. Trend analysis of these data shows significant ($p<.01$) decreases of 2.1% per year for the period 1966–91 and 3.3% per year for the period 1982–91. Reasons for these declines are not known.

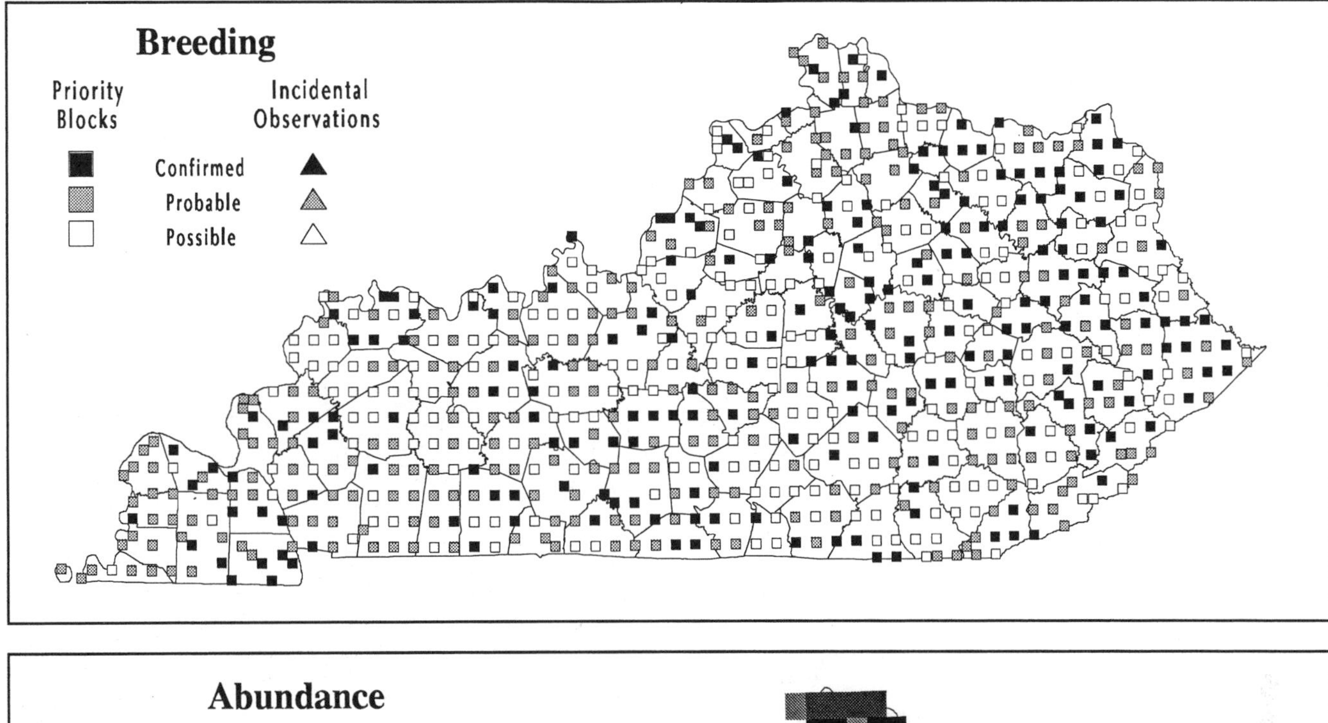

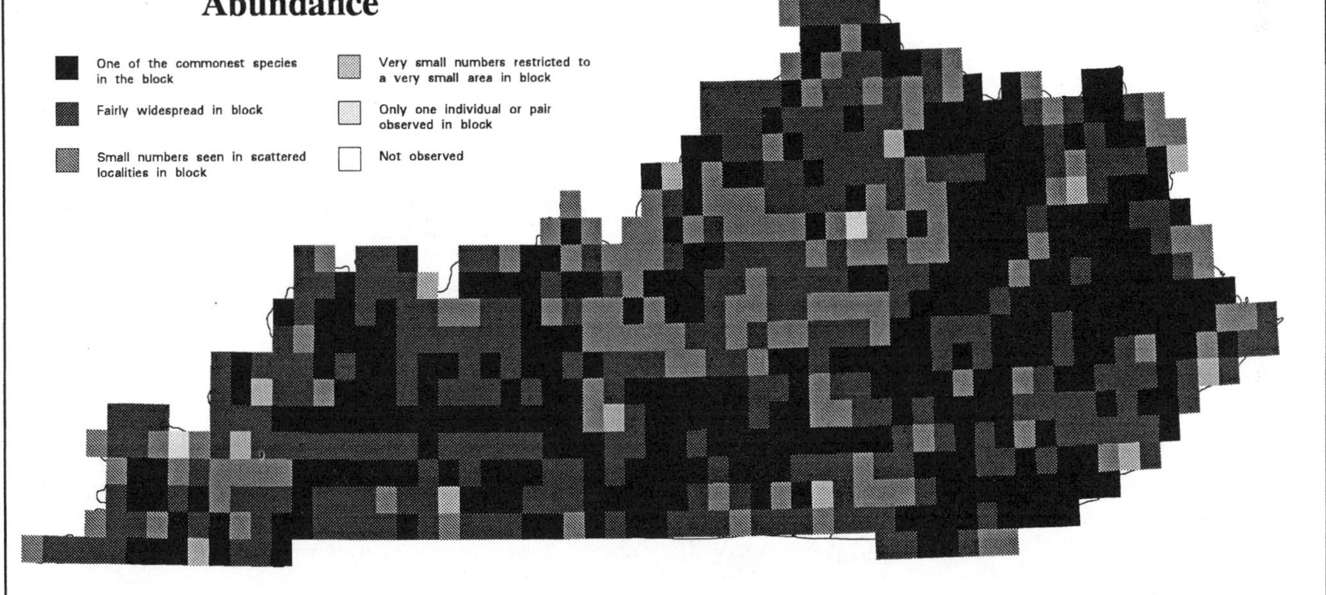

Analysis of Block Data by Physiographic Province Section

Physiographic Province Section	Total Blocks Surveyed	Blocks with Data	Avg Abund	% with Data	Section's % for State
Mississippi Alluvial Plain	14	14	3.7	100.0	1.9
East Gulf Coastal Plain	36	36	3.9	100.0	5.0
Highland Rim	139	139	4.3	100.0	19.1
Shawnee Hills	142	141	4.3	99.3	19.4
Blue Grass	204	204	4.1	100.0	28.1
Cumberland Plateau	173	173	4.4	100.0	23.8
Cumberland Mountains	19	19	4.3	100.0	2.6

Summary of Breeding Status

Number of Blocks in Which Species Was Recorded		
Total	**726**	**99.9%**
Confirmed	212	29.2%
Probable	285	39.3%
Possible	229	31.5%

Dickcissel

Spiza americana

The Dickcissel is an irregularly distributed breeding bird across the western two-thirds of Kentucky. Mengel (1965) noted that the species chiefly occurred west of the Cumberland Plateau and that it varied from being uncommon and locally distributed in central Kentucky to common and general in distribution in western Kentucky. The species is well known for the irregular fluctuations that occur in its distribution and abundance (Mengel 1965; Robbins et al. 1986), but it remains a well-established breeding bird throughout the western third of the state today.

The Dickcissel is a bird of open habitats with an abundance of low herbaceous vegetation. Today, natural habitats affording this type of cover have been virtually eliminated, and Dickcissels occur only in artificially created habitats. The species is most frequently found in rural farmland, but it also occurs in other open situations, such as reclaimed strip mines, the unmowed margins of airports, and similarly idle land. Even in areas where they are fairly common, Dickcissels typically are distributed irregularly. Loose colonies are often established in tracts of optimal habitat, while suitable habitat in surrounding areas goes unused, resulting in a very patchy distribution (Whitt 1969). Dickcissels sometimes inhabit grassy fields, but they are most common in habitats with an abundance of forbs, such as fields of clover and alfalfa, as well as fields of small grains (especially wheat). Within such areas, territorial males sing from scattered trees, power lines, and tall weed stems.

The Dickcissel is a bird of the Great Plains that, if it nested in Kentucky before settlement, likely occurred only in the native prairies of the East Gulf Coastal Plain and the Highland Rim (Mengel 1965). Audubon (1861) considered the species to be scarce in the early 1800s, although he drew a pair at Henderson on May 14, 1811, suggesting the possibility of local breeding (Wiley 1970). While natural grasslands have been virtually eliminated in the last two centuries, open habitats created by human alteration of the landscape have served to replace them. Furthermore, the widespread clearing of native forests for agricultural use and settlement has resulted in the species's expansion into other parts of the state. Thus, it is likely that in more recent times the Dickcissel has become more widespread than ever before.

Forest Cover

Value	% of Blocks	Avg Abund
All	23.2	2.2
1	49.5	2.5
2	39.5	2.2
3	19.1	2.0
4	—	—
5	—	—

Dickcissels winter primarily in the tropics (AOU 1983), and they begin returning to Kentucky during the last week of April. Most birds arrive by early May, but nesting activity probably does not commence until the middle of the month (Stamm 1962). According to Mengel (1965), early clutches are completed in mid-May, and a peak in clutch completion occurs during the last ten days of the month. While some pairs may raise two broods, harvest operations must result in the destruction of many nests in agricultural fields. The average size of eight clutches and two broods given by Mengel (1965) was 4.4 (range of 3–5).

The nest is placed low to or on the ground, amid dense grasses or weeds. It is usually situated well away from large trees or other tall vegetation but sometimes along or in low hedgerows. The nest is constructed of dead leaves, coarse grasses, and weed stems and is lined with fine grasses (Stamm 1962). The average height of nine nests given by Mengel (1965) was 8.8 inches above the ground (range of 2–20 inches), although a nest in Daviess County was found two feet above the ground (Croft and Stamm 1967).

Gary Meszaros

The atlas survey yielded records of Dickcissels in more than 23% of priority blocks, and 30 incidental observations were reported. As expected, the species was most widespread in the Mississippi Alluvial Plain and the East Gulf Coastal Plain, and it was not reported from the Cumberland Mountains. Reclaimed surface mines now provide suitable nesting areas on the Cumberland Plateau, and two incidental observations were reported in such habitat during the atlas period (Claus et al. 1988; Stamm and Monroe 1991b). Occurrence was highest in open areas, and the species was absent in predominantly forested priority blocks.

Less than 10% of priority block records were for confirmed breeding. An active nest was not located, and most of the 16 confirmed records were based on the observation of adults carrying food. Other confirmed records involved the observation of distraction displays, recently fledged young, and nest building.

Dickcissels are reported regularly on only about half of Kentucky's BBS routes. The average number of individuals recorded per BBS route for the periods 1966–91 and 1982–91 was 4.21 and 4.12, respectively. Trend analysis of these data yields a significant ($p<.01$) decrease of 5.4% per year for the period 1966–91, but a slight, nonsignificant increase of 2.3% per year for the period 1982–91. It is likely that natural variability in the nesting population is responsible for these trends (Robbins et al. 1986).

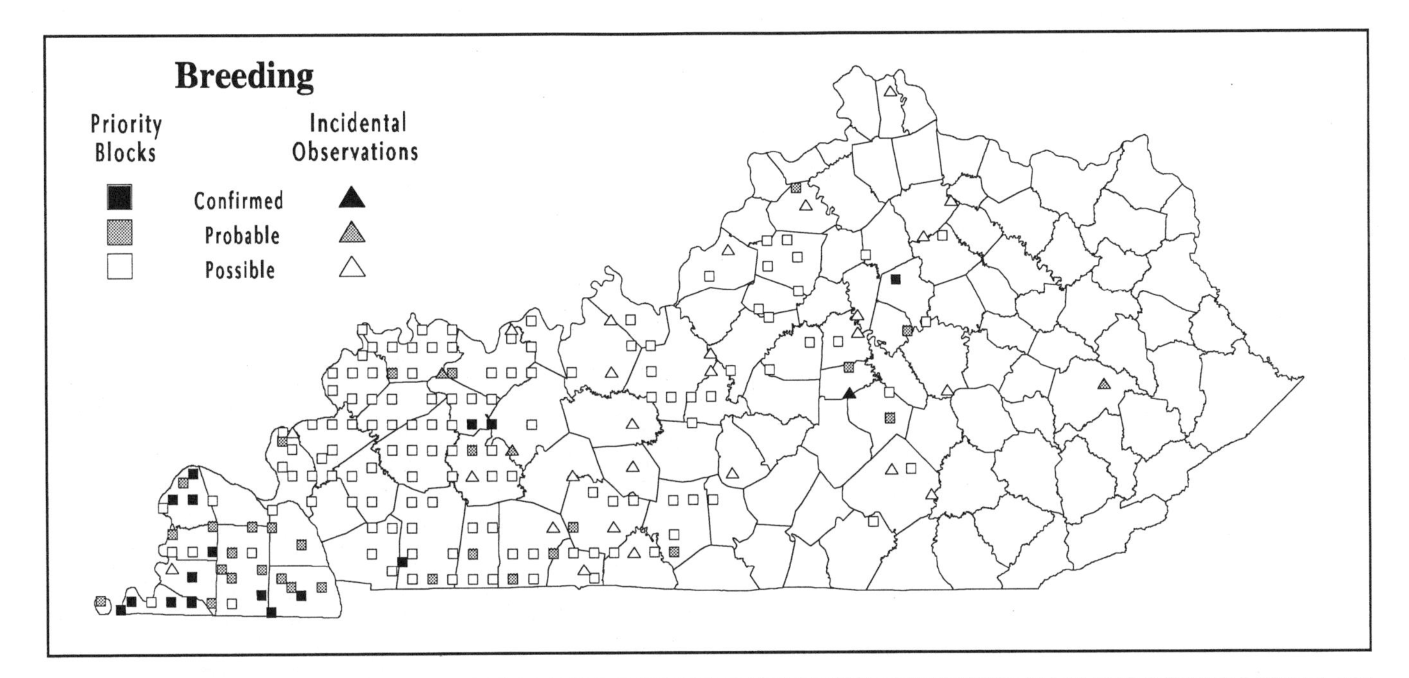

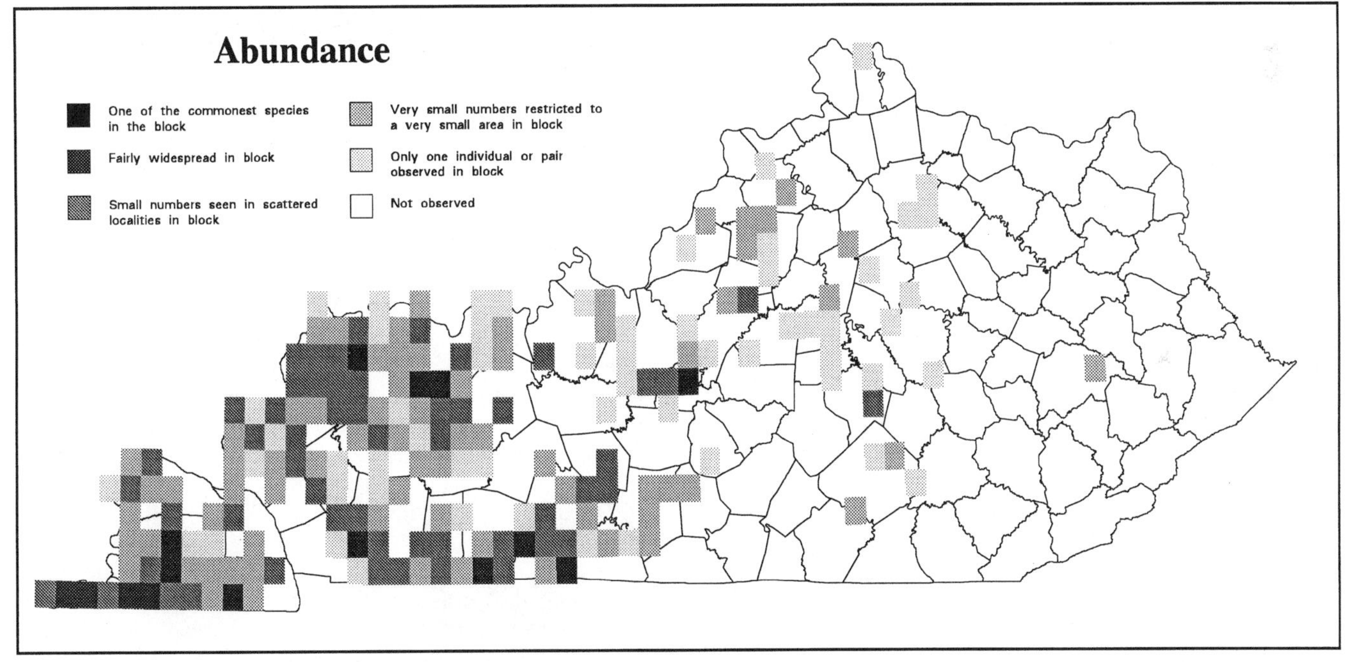

Analysis of Block Data by Physiographic Province Section

Physiographic Province Section	Total Blocks Surveyed	Blocks with Data	Avg Abund	% with Data	Section's % for State
Mississippi Alluvial Plain	14	10	2.8	71.4	5.9
East Gulf Coastal Plain	36	25	2.5	69.4	14.8
Highland Rim	139	47	2.5	33.8	27.8
Shawnee Hills	142	67	2.2	47.2	39.6
Blue Grass	204	20	1.4	9.8	11.8
Cumberland Plateau	173	-	-	-	-
Cumberland Mountains	19	-	-	-	-

Summary of Breeding Status

Number of Blocks in Which Species Was Recorded		
Total	**169**	**23.2%**
Confirmed	16	9.5%
Probable	29	17.1%
Possible	124	73.4%

Rufous-sided Towhee

Pipilo erythrophthalmus

The distinctive song of the Rufous-sided Towhee is a widespread summer sound throughout Kentucky. Mengel (1965) regarded the species as an uncommon to common summer resident statewide. Nowhere especially abundant, it is nonetheless about as widely distributed as any of the state's resident birds.

Rufous-sided Towhees are encountered in a variety of semi-open and forested habitats with some dense cover of weeds, brambles, or shrubbery. They occur most often in artificial situations, including brushy forest edge, regenerating clear-cuts, forest disturbed by selective logging, reclaimed strip mines, abandoned fields, overgrown fencerows, and other neglected areas. Towhees also can be found in considerable numbers in some naturally occurring habitats. They are most conspicuous in subxeric deciduous and mixed pine-hardwood forests with a dense shrub layer of blueberry or mountain laurel, but small numbers also occur in more mesic forests with a thick growth of cane, forest openings created by fire or windstorms, and the shrubby borders of floodplain sloughs and prairie remnants.

Audubon (1861) found towhees "in the greatest abundance" in the barrens of Kentucky in the early 1800s, and the species probably occurred locally across the state before settlement. It is likely, however, that its numbers have increased as a result of human alteration of the landscape. The clearing and dissection of forested areas have resulted in the creation of an abundance of suitable early successional habitat that has been maintained through subsequent periods of disturbance and neglect.

Towhees are present in Kentucky throughout the year, but it appears that some birds move about seasonally (Mengel 1965). The first songs of territorial males are heard during the first warm spells of March. According to Mengel (1965), early clutches are completed during the first 10 days of April, with an early peak of clutch completion occurring sometime during the latter half of the month. The species appears to be double-brooded, as suggested by a second poorly defined peak in clutch completion during the last week or so of May, and later clutches have been noted into early August (Mengel 1965). According to Mengel, the average size of 20 clutches or broods thought to be complete was 3.5 (range of 3–4), but a nest containing six eggs was reported in Jefferson County in 1979 (KOS Nest Cards).

Forest Cover

Value	% of Blocks	Avg Abund
All	94.6	2.8
1	81.2	2.6
2	95.3	2.9
3	98.3	2.9
4	96.8	2.8
5	96.2	2.6

The nest is well concealed among the dense cover of vines, shrubs, or small trees (including cedars) or on the ground beneath the cover of weeds, vines, or brambles (Mengel 1965). Early nests tend to be placed on the ground, while most later nests are elevated above it (Mengel 1965). The nest is a bulky cup constructed of dead leaves, grasses, weed stems, and other plant material and lined with fine grasses (Mengel 1965). The average height of the nine above-ground nests noted by Mengel was 3.8 feet (range of 2–6 feet), but more recent nests have been observed up to 11 feet above the ground (Stamm and Croft 1968; KOS Nest Cards).

Ron Austing

Rufous-sided Towhees were recorded in nearly 95% of priority blocks statewide, and nine incidental observations were reported. The species ranked 12th according to the number of priority block records, 17th by total abundance, and 23rd by average abundance. Little variation in occurrence by physiographic province section was noted, although towhees were recorded least often in the Cumberland Mountains. Average abundance was relatively uniform across the state, but it was also lowest in the Cumberland Mountains. Occurrence was highest in areas with a good mixture of forest and open areas, and lowest in extensively cleared areas.

Despite the towhee's widespread occurrence, only about 11% of priority block records were for confirmed breeding. Although a few active nests were located, most of the 78 confirmed records were based on the observation of recently fledged young. The species is sometimes parasitized by Brown-headed Cowbirds, and adult towhees were observed attending fledgling cowbirds on a few occasions. Other confirmed records involved the observation of adults carrying food, nest building, and distraction displays.

Rufous-sided Towhees are typically reported in moderate numbers on most Kentucky BBS routes. The average number of individuals recorded per BBS route for the periods 1966–91 and 1982–91 was 11.50 and 10.09, respectively. According to these data, the towhee ranked 24th in abundance on BBS routes during the period 1982–91. Trend analysis of these data reveals a significant decrease ($p<.01$) of 2.5% per year for the period 1966–91, but a small, nonsignificant increase of 0.3% per year for the period 1982–91.

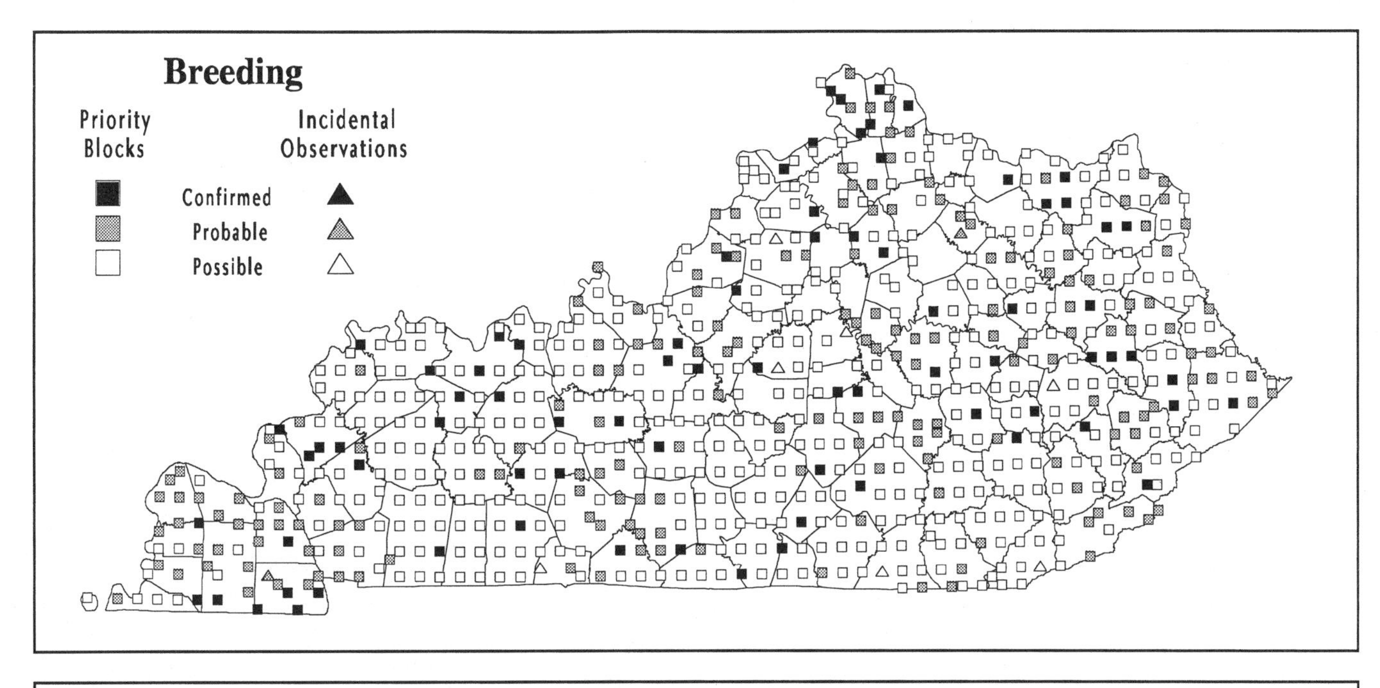

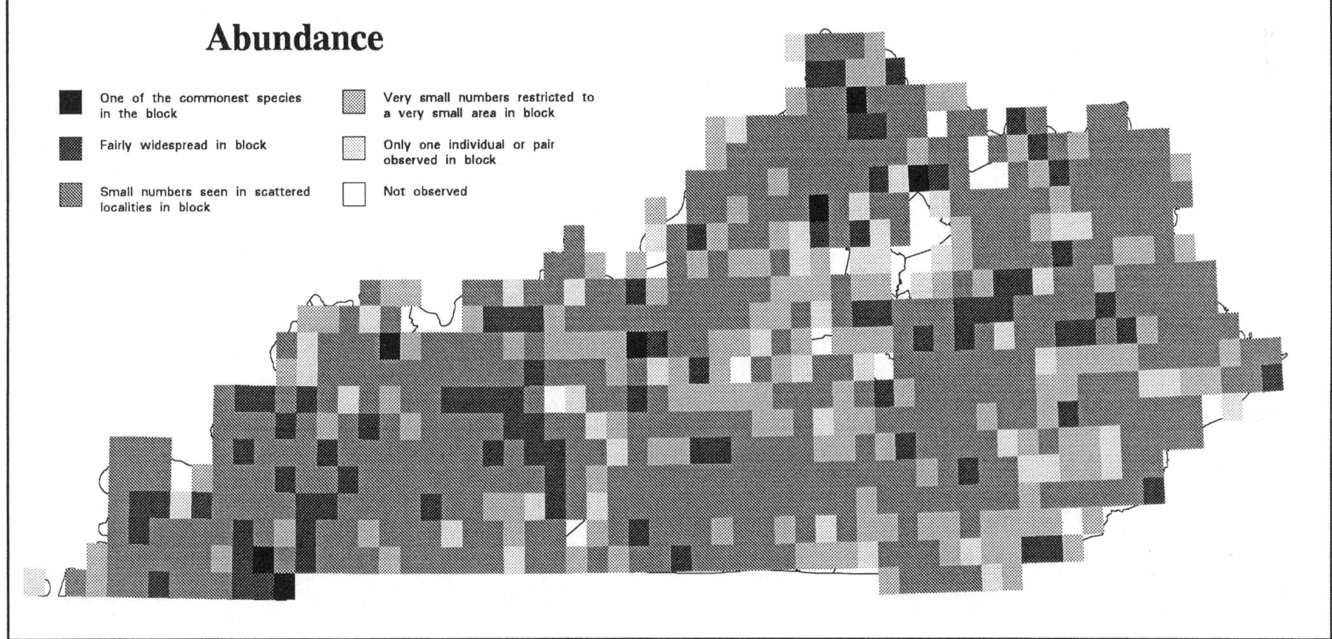

Analysis of Block Data by Physiographic Province Section

Physiographic Province Section	Total Blocks Surveyed	Blocks with Data	Avg Abund	% with Data	Section's % for State
Mississippi Alluvial Plain	14	12	2.7	85.7	1.7
East Gulf Coastal Plain	36	34	3.3	94.4	4.9
Highland Rim	139	137	2.8	98.6	19.9
Shawnee Hills	142	136	2.9	95.8	19.8
Blue Grass	204	184	2.8	90.2	26.7
Cumberland Plateau	173	170	2.7	98.3	24.7
Cumberland Mountains	19	15	2.5	78.9	2.2

Summary of Breeding Status

Number of Blocks in Which Species Was Recorded		
Total	**688**	**94.6%**
Confirmed	78	11.3%
Probable	182	26.5%
Possible	428	62.2%

Rufous-sided Towhee

Bachman's Sparrow

Aimophila aestivalis

Regarded as a locally distributed but uncommon to fairly common summer resident in Kentucky as recently as the mid- to late 1950s (Mengel 1965), the Bachman's Sparrow has declined substantially in the past 35 years. During that time the species has virtually disappeared from the state, and it may now be extirpated. Similar trends have occurred in Tennessee (Robinson 1990), West Virginia (Hall 1983), and much of the rest of the bird's range, prompting its designation as a candidate for federal listing (USFWS 1982).

After the late 1950s, reports of Bachman's Sparrows rapidly declined. Wilson (1962a) noted that the decrease was under way in south central Kentucky by the early 1960s. By about 1980 the species had become very difficult to find, and summer reports were reduced to a few areas in the south central and western portions of the state. In the years immediately before the atlas survey, Bachman's Sparrows were reported from only a few localities in the counties west of Kentucky Lake, where young pine plantations provided suitable nesting habitat. During the atlas period, two of the three reports originated from sites in eastern Calloway County. The third involved an individual heard on a BBS route in rural farmland along State Highway 307 south of Beulah in northeastern Hickman County on June 4, 1989. Since the summer of 1989 the Bachman's Sparrow has gone unreported, and it may have disappeared altogether. It is possible that a few birds remain, but numbers must be quite low.

Bachman's Sparrows formerly inhabited a variety of early successional habitats in Kentucky, including old fields and pastures, young pine plantations, and regenerating clear-cuts. Mengel (1965) described the species as having rather rigid habitat requirements, inhabiting an area only if it possessed the right combination of conditions: preferably (but not always) a hillside, some bare ground, some native grasses and forbs, patches of blackberry briars, and scattered small trees. Although the presence of pines was not essential, red cedars and other evergreens were frequently associated with such habitats. In the young pine plantations where the species was last found, the trees ranged from 1 to 10 feet in height, and ground cover included a mixture of grasses, forbs, and patches of bare ground.

Bachman's Sparrows typically retreated southward to overwinter in the southern United States (AOU 1983), and most birds returned to their Kentucky breeding grounds during the first half of April. According to Mengel (1965), clutches were completed from late April to late July. Although evidence suggests that two broods were sometimes raised, a peak in clutch completion occurred around the first of June. The nest was typically constructed of coarse grasses and placed on the ground, usually but not always in dense cover (Mengel 1965).

It is likely that if Bachman's Sparrows occurred in Kentucky before settlement, they were very locally distributed. The extensive, closed-canopy forests that once covered much of the state would not have been suitable for nesting, and potential habitat was likely restricted to transitional zones of open woodland, especially in areas frequently affected by fire.

During the last century it appears that Bachman's Sparrows gradually invaded habitats created by human activity. Early reports from the late 1800s and early 1900s summarized by Mengel (1965) suggest that the species was not numerous at that time. By the mid-1900s this sparrow was being reported more frequently, and Mengel (1965) listed numerous reports from all physiographic province sections, including the Cumberland Mountains. Clearing of forests during settlement and subsequent neglect of the land apparently created an abundance of suitable nesting habitat for Bachman's Sparrows. The introduction of pines and the spread of native pines in response to disturbance may have enhanced the attractiveness of many areas.

Kathy Caminiti

While the expansion of the Bachman's Sparrow into Kentucky may have occurred for relatively obvious reasons, theories to explain its recent decline are largely lacking. Today, one encounters suitable nesting situations often enough to consider that habitat is not limiting, although it is impossible to determine what amount of suitable habitat is sufficient to support a viable population. Because of loss and fragmentation of appropriate habitat, the species may have been unable to sustain population levels that perpetuated its presence at the northern limit of its range.

The future of the Bachman's Sparrow in Kentucky is unknown. Small numbers may exist in undiscovered areas, but it is also possible that the species has disappeared. Scattered populations persist just to the south, in western Tennessee (Robinson 1990), and it is possible that birds could reinvade suitable nesting habitat in Kentucky if the decline does not continue in areas to the south.

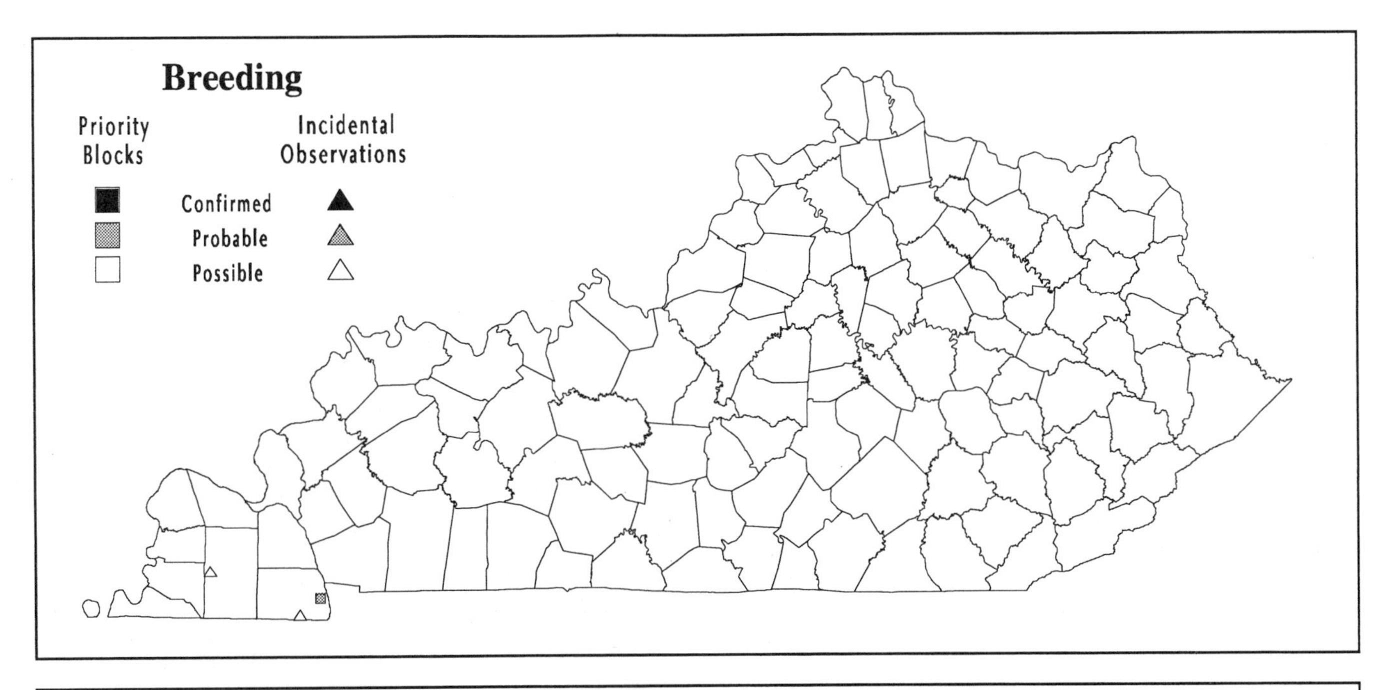

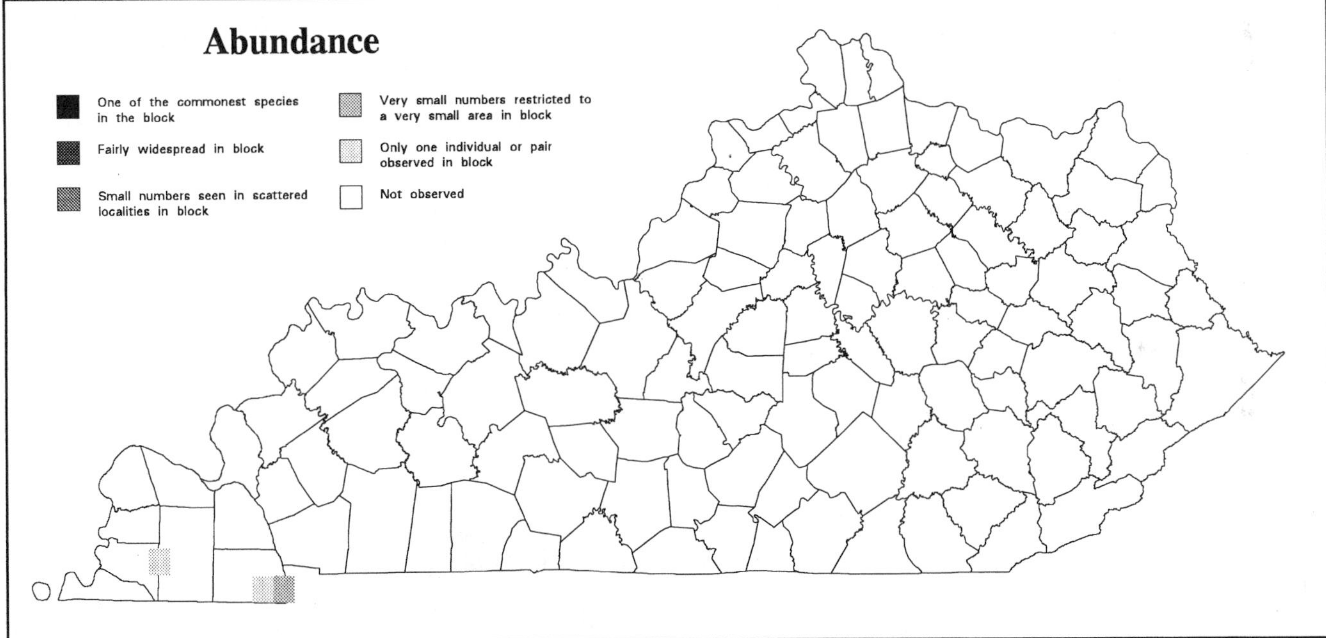

Analysis of Block Data by Physiographic Province Section

Physiographic Province Section	Total Blocks Surveyed	Blocks with Data	% with Data	Section's % for State
Mississippi Alluvial Plain	14	-	-	-
East Gulf Coastal Plain	36	1	2.8	100.0
Highland Rim	139	-	-	-
Shawnee Hills	142	-	-	-
Blue Grass	204	-	-	-
Cumberland Plateau	173	-	-	-
Cumberland Mountains	19	-	-	-

Summary of Breeding Status

Number of Blocks in Which Species Was Recorded		
Total	**1**	**0.1%**
Confirmed	-	-
Probable	1	100.0%
Possible	-	-

Bachman's Sparrow

Chipping Sparrow

Spizella passerina

The Chipping Sparrow's unmusical song is a common spring and summer sound throughout much of Kentucky. Mengel (1965) considered the species to be a common summer resident in suitable habitat statewide. Largely because of its adaptation to human alteration of the landscape, it ranks among the state's most widespread nesting sparrows.

The Chipping Sparrow is typically a bird of semi-open habitats, occurring most frequently in forested areas dissected by numerous small to moderate-sized openings with short or sparse ground cover. Although the species has been reported in natural situations, such as open pine-oak woodland (Mengel 1965), it is chiefly encountered in artificial habitats. Chipping Sparrows seem to be most common in association with rural settlement, being quite numerous about rural farmsteads, pastures, yards, and parks. The species also occurs frequently in other altered situations, including selectively logged forests, reclaimed strip mines, residential areas of cities and towns, and roadway corridors.

Although some naturally occurring habitats probably harbored Chipping Sparrows in Kentucky two centuries ago, it is likely that the species is much more common and widespread today. At one time the species must have been restricted to dry, open forests and transitional zones surrounding prairie habitats, situations that likely were once much more widespread because of the influence of fire. In contrast, the conversion of vast areas of closed-canopy forest to agricultural use and settlement has resulted in the creation of an abundance of suitable habitat where formerly the species was absent.

Chipping Sparrows winter primarily in the southern United States (AOU 1983), and they are among the earliest of Kentucky's summer residents to return in early spring. The first males often appear before the middle of March, and all or most of the nesting population has probably returned by the middle of April. Nest building has been observed by April 8, and early clutches are completed by mid-April (Mengel 1965). A peak in clutch completion occurs around May 1, but nesting activity continues into late July, indicating that at least some pairs raise a second brood (Mengel 1965). Family groups are fairly conspicuous throughout the summer, and they often remain together well into fall. The average size of 23 clutches or broods known or thought to be complete and summarized by Mengel (1965) was 3.5 (range of 2–5). The species is at least occasionally parasitized by Brown-headed Cowbirds (Mengel 1965).

Forest Cover

Value	% of Blocks	Avg Abund
All	86.9	2.7
1	60.4	2.2
2	83.7	2.7
3	93.9	2.9
4	96.1	2.8
5	92.3	2.3

Chipping Sparrows typically nest in trees or shrubs, securing the nest to a horizontal fork among the outer branches. Evergreens are used most frequently, especially for the early nestings, but deciduous trees and shrubs are also used in later nestings. The nest is a beautifully woven cup of fine dead grass, lined with animal hair. The average height of 18 nests reported by Mengel was 7.0 feet (range of 2–15 feet), although Stamm and Croft (1968) reported a nest 30 feet above the ground.

Bill Schoettler

The atlas survey yielded records of Chipping Sparrows in nearly 87% of priority blocks statewide, and 12 incidental observations were reported. The species ranked 21st according to the number of priority block records, 25th by total abundance, and 26th by average abundance. Occurrence was highest in the Cumberland Plateau, and it was greater than 75% in all physiographic province sections except the Mississippi Alluvial Plain. Average abundance was relatively uniform across most of central and western Kentucky, but it was slightly lower in the East Gulf Coastal Plain, the Cumberland Mountains, and the Mississippi Alluvial Plain. Both occurrence and average abundance were lowest in extensively cleared areas and highest in forested areas with a good mix of openings.

More than 28% of priority block records were for confirmed breeding. Although a number of active nests were located, most confirmed records were based on the observation of recently fledged young in the company of adults. The conspicuous call of fledgling Chipping Sparrows revealed the birds' presence on many occasions. In addition, young cowbirds were observed being fed by adult sparrows on a few occasions. Other confirmed records involved the observation of adults carrying food, nest building, and distraction displays.

Chipping Sparrows are reported on most Kentucky BBS routes in small to moderate numbers. The average number of individuals per BBS route for the periods 1966–91 and 1982–91 was 9.09 and 10.82, respectively. According to these data, the Chipping Sparrow ranked 23rd in abundance on BBS routes during the period 1982–91. Trend analysis of these data shows a nonsignificant decrease of 1.3% per year for the period 1966–91, but a significant ($p<.01$) increase of 3.9% per year for the period 1982–91.

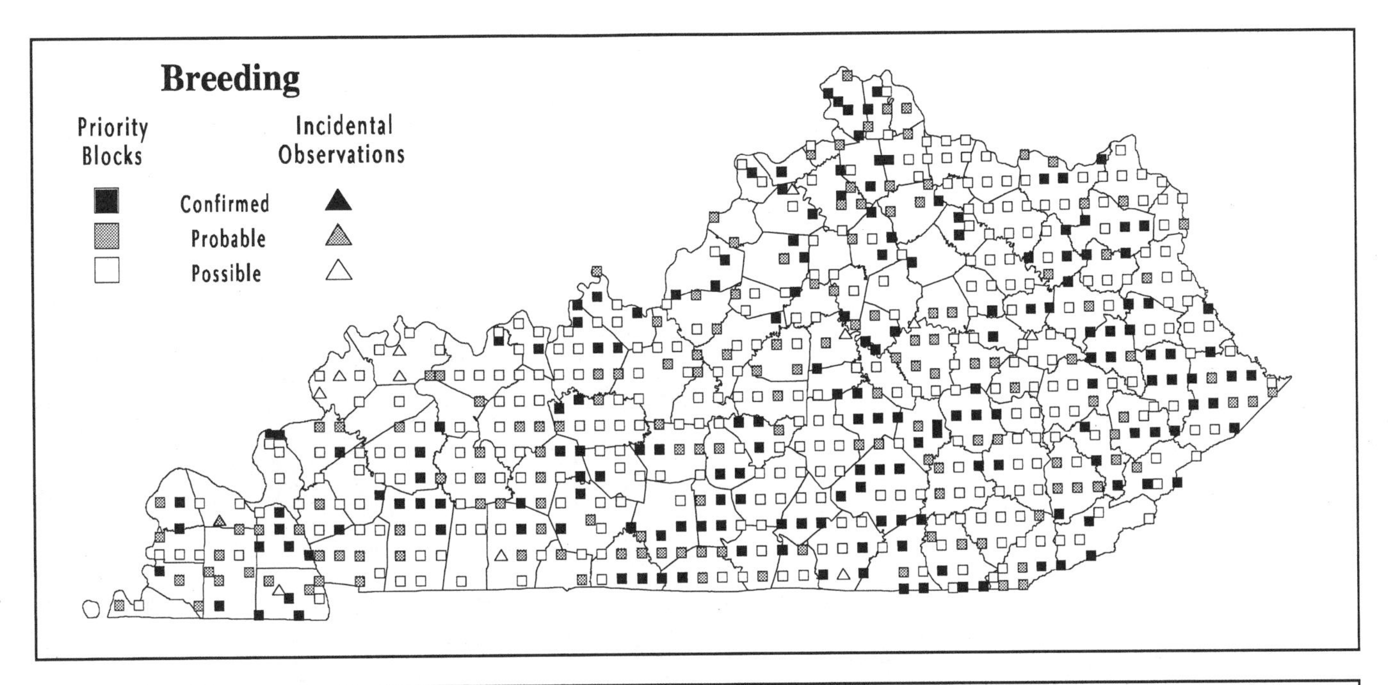

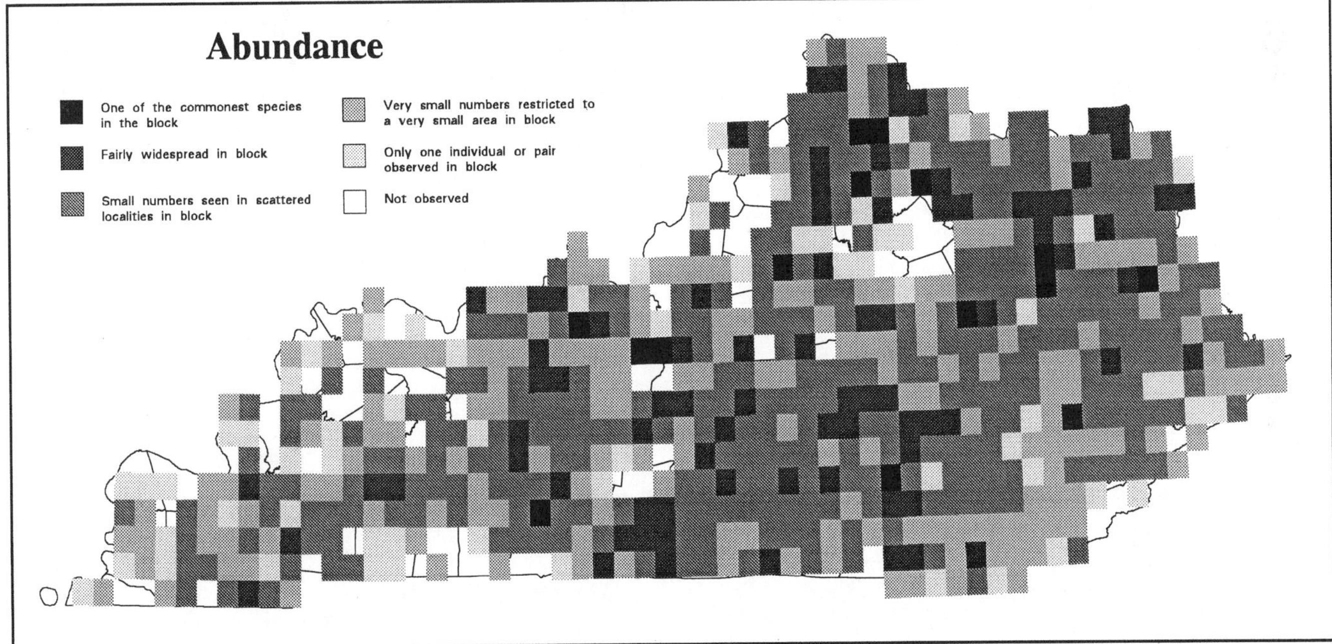

Analysis of Block Data by Physiographic Province Section

Physiographic Province Section	Total Blocks Surveyed	Blocks with Data	Avg Abund	% with Data	Section's % for State
Mississippi Alluvial Plain	14	8	1.8	57.1	1.3
East Gulf Coastal Plain	36	28	2.2	77.8	4.4
Highland Rim	139	124	2.9	89.2	19.6
Shawnee Hills	142	114	2.6	80.3	18.0
Blue Grass	204	174	2.9	85.3	27.5
Cumberland Plateau	173	169	2.7	97.7	26.7
Cumberland Mountains	19	15	2.1	78.9	2.4

Summary of Breeding Status

Number of Blocks in Which Species Was Recorded		
Total	**632**	**86.9%**
Confirmed	179	28.3%
Probable	150	23.7%
Possible	303	47.9%

Chipping Sparrow

Field Sparrow

Spizella pusilla

Although the Field Sparrow is a widespread summer resident across much of the eastern and central United States, BBS data for the period 1965–79 indicated that it was nowhere more common than in the Blue Grass and the Highland Rim of Kentucky (Robbins et al. 1986). Mengel (1965) regarded the species as a common permanent resident in suitable habitat statewide, and it is probably the most widespread nesting sparrow in the state.

Field Sparrows are birds of semi-open and open habitats with some tall grass and scattered shrubs or other thick growth. The species is not numerous in predominantly forested areas, but a few birds usually can be found in isolated openings and grassy corridors. These sparrows are also less numerous in areas where brushy vegetation has been reduced by intensive land use. Today they primarily occupy altered habitats, such as abandoned or idle fields and pastures, brushy forest margins, utility corridors, reclaimed surface mines, and other early successional openings with scattered weeds and shrubs. In contrast, a few birds can be found in natural situations, such as the margins of glade openings and remnant prairie patches.

Field Sparrows must have been less common and widespread in Kentucky before altered habitats became available. Audubon (1861) considered the species to be resident in the early 1800s, and substantial numbers must have inhabited the native savannas and prairies of central and western Kentucky. Subsequently, the widespread conversion of forested areas to agricultural use and settlement has resulted in the creation of an abundance of early successional habitat, and the species now occurs in many areas where formerly it must have been absent.

Although Field Sparrows are present in the state throughout the year, some seasonal movement appears to occur. The species becomes somewhat locally distributed in winter, when it typically occurs in flocks. These flocks break up during March, and males begin to stake out territories, singing frequently. According to Mengel (1965), a few clutches may be completed by mid-April, with a peak in early clutch completion during May. Although a later peak in clutch completion is not evident, the species is typically double-brooded, and active nests have been reported into mid-August. Mengel (1965) gives the average of 51 clutches or broods known or thought to be complete as 3.6 (range of 2–5). According to Mengel, Field Sparrows are frequently parasitized by Brown-headed Cowbirds.

Forest Cover

Value	% of Blocks	Avg Abund
All	90.8	2.8
1	91.1	2.8
2	97.9	3.1
3	96.1	3.1
4	87.7	2.4
5	50.0	1.9

Field Sparrows typically nest in open, brushy situations, although they sometimes use woodland edge. The nest is often built on the ground, especially early in the season. It is typically placed at the base of a shrub, a small tree, or a clump of grasses (Mengel 1965). Later nests are usually constructed above the ground in briars, weeds, small trees, and other low vegetation. The nest is a neat, shallow cup of fine grass, weed stems, and a few leaves, lined with fine grasses, rootlets, and hair (Harrison 1975). The average height of 32 above-ground nests summarized by Mengel (1965) was 2.0 feet (range of 4 inches to 5.5 feet).

Alvin E. Staffan

The atlas survey yielded reports of Field Sparrows in nearly 91% of priority blocks statewide, and seven incidental observations were reported. The species ranked 17th according to the number of priority block records, 19th by total abundance, and 20th by average abundance. These sparrows were found in more than 90% of priority blocks in physiographic province sections across most of central and western Kentucky, and occurrence was less than 75% only in the Cumberland Mountains. Average abundance varied similarly. Both occurrence and average abundance were highest in areas with a good mix of forest and openings, and lowest in heavily forested areas.

Despite the Field Sparrow's overall abundance, only about 16% of priority block records were for confirmed breeding. At least seven active nests were found, but most of the 107 confirmed records were based on the observation of recently fledged young. In addition, fledgling cowbirds were observed being fed by adult sparrows on a few occasions. Other confirmed records involved the observation of adults carrying food, nest building, and distraction displays.

Field Sparrows are reported in fairly substantial numbers on most Kentucky BBS routes. The average number of individuals per BBS route for the periods 1966–91 and 1982–91 was 21.61 and 16.40, respectively. According to these data, the species ranked 16th in abundance on BBS routes during the period 1982–91. Trend analysis of these data reveals significant ($p<.01$) declines of 4.1% per year for the period 1966–91 and 7.7% per year for the period 1982–91. The reasons for these declines are not completely understood, but severe winters in the late 1970s and intensive land use with a resulting decrease in acreage of reverting fields may be partially responsible (Monroe 1978; Robbins et al 1986).

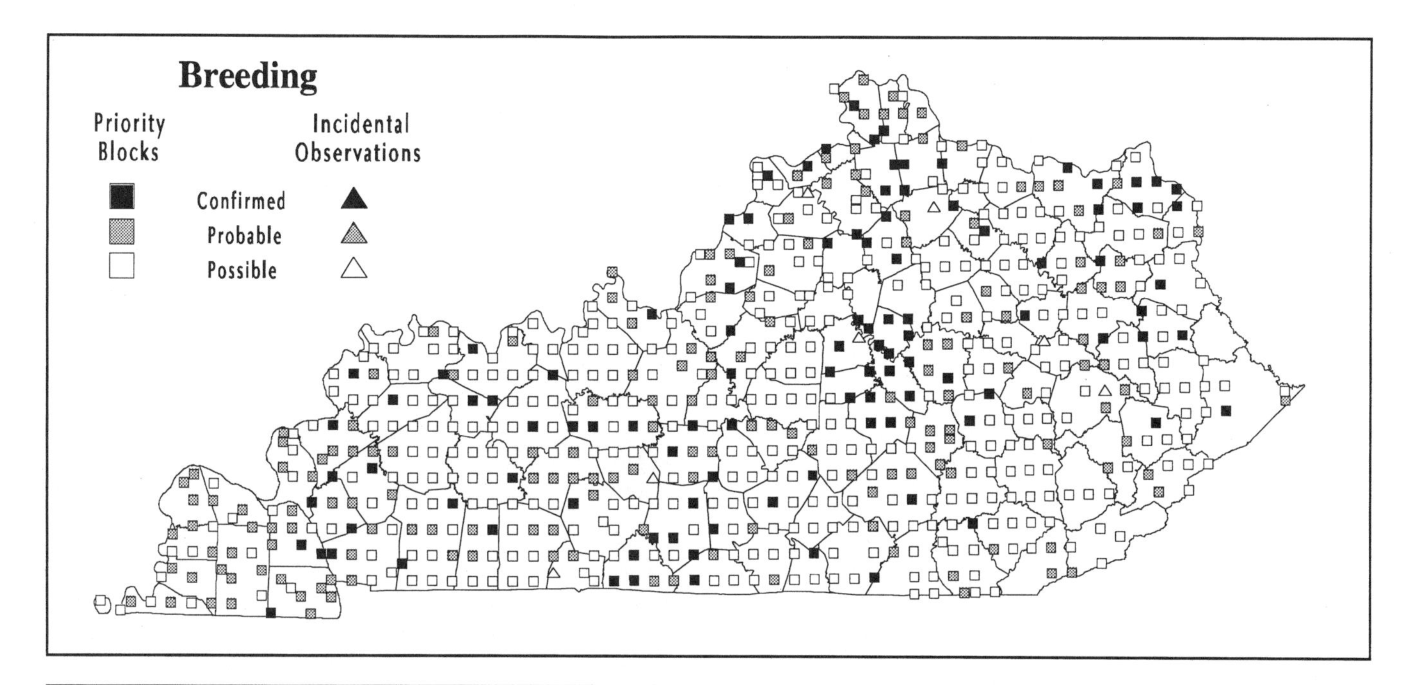

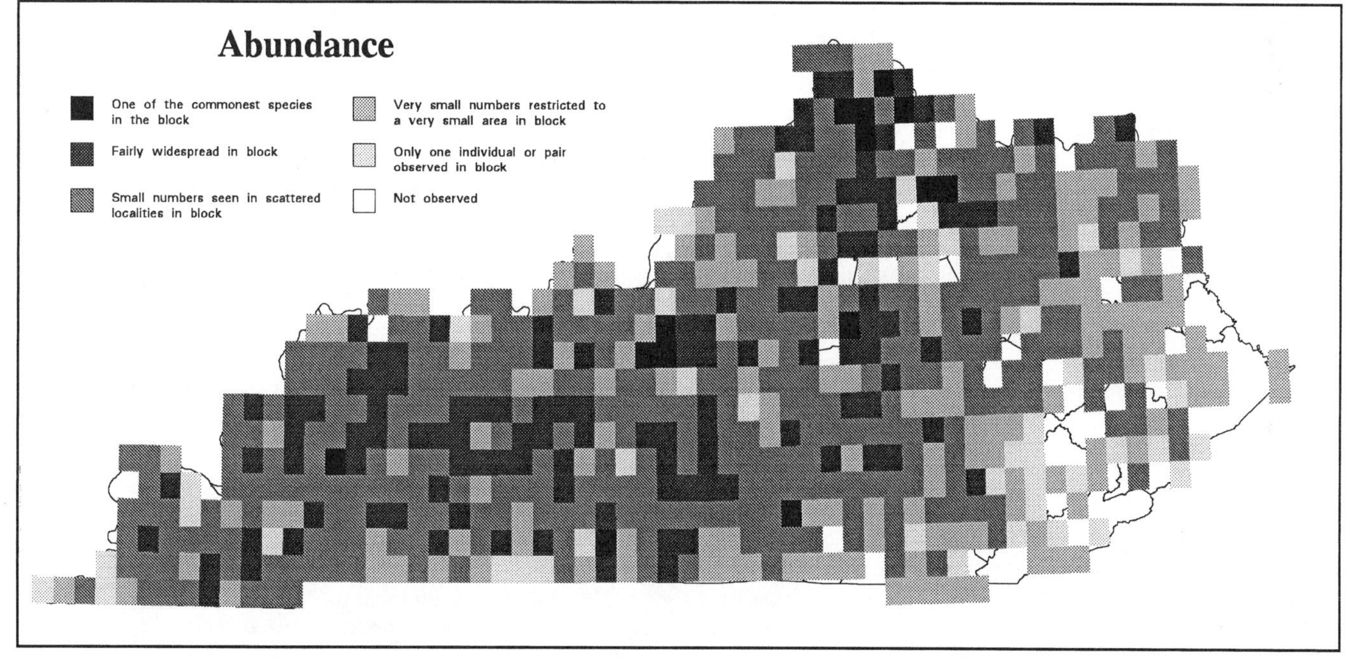

Analysis of Block Data by Physiographic Province Section

Physiographic Province Section	Total Blocks Surveyed	Blocks with Data	Avg Abund	% with Data	Section's % for State
Mississippi Alluvial Plain	14	11	2.3	78.6	1.7
East Gulf Coastal Plain	36	36	2.9	100.0	5.5
Highland Rim	139	137	3.0	98.6	20.8
Shawnee Hills	142	137	3.1	96.5	20.8
Blue Grass	204	191	3.0	93.6	28.9
Cumberland Plateau	173	140	2.3	80.9	21.2
Cumberland Mountains	19	8	1.6	42.1	1.2

Summary of Breeding Status

Number of Blocks in Which Species Was Recorded		
Total	**660**	**90.8%**
Confirmed	107	16.2%
Probable	172	26.1%
Possible	381	57.7%

Vesper Sparrow
Pooecetes gramineus

Although the breeding range of the Vesper Sparrow lies primarily to the north and west of Kentucky, in the northern United States and southern Canada (AOU 1983), its southern limit was documented by Mengel (1965) to extend southward into the north central part of the state. As of 1950 this sparrow was considered to be locally fairly common and distributed throughout a large part of the northern Blue Grass (Mengel 1965). Unfortunately, it appears that since that time the species has virtually disappeared as a breeding bird.

The earliest information available on the summer status of the Vesper Sparrow in Kentucky came from Beckham (1885), who regarded the species as a common summer resident in Nelson County. During the early 1900s the species was reported at various times in summer in central Kentucky, including records in the counties of Mercer (Van Arsdall 1949), Pulaski (Cooke 1911), and Woodford (Howell 1910). Mengel (1965) provided a wealth of information on the species's former breeding range when he surveyed the "hill country" of the Outer Blue Grass of northern Kentucky in the summer of 1950. There he found small numbers scattered across portions of Boone, Campbell, Gallatin, Grant, Harrison, Kenton, and Pendleton counties.

Subsequent to Mengel's effort, little has been reported concerning Vesper Sparrows in summer. The species was recorded occasionally on BBS routes from 1966 to 1970, and an active nest, the only one ever reported in the state, was found in Franklin County in 1968 (Moore 1969). Despite excellent coverage of most priority blocks in northern Kentucky, the atlas fieldwork yielded no reports of Vesper Sparrows in the counties where Mengel found them to be widely distributed in 1950. The only record generated by the atlas survey came from open farmland approximately three miles southeast of Paris, in Bourbon County, where a singing male was observed on June 22, 1991. This site was not revisited, so further evidence of breeding was not obtained.

Although Vesper Sparrows may have been overlooked to some extent since the early 1950s, it is more likely that they have largely or entirely disappeared since then. BBS data for the period 1965–79 indicated that the species appeared to be decreasing throughout much of its range (Robbins et al. 1986), and a gradual decline has been documented in Ohio since the 1930s, especially in the southern part of the state (Peterjohn and Rice 1991). Reasons for these declines are not fully understood, although many observers point to the increase in mowing of hayfields and the reversion of active farmland to second-growth forest (Robbins et al. 1986; Peterjohn 1989). Because the species nests primarily in agricultural situations, nesting success depends upon the timing of planting and harvesting operations.

Vesper Sparrows winter to the south of Kentucky, and spring migrants are conspicuous from mid-March through mid-April. Other than a few late reports of possibly lingering spring migrants and a couple of June reports from BBS routes, recent records of summering are absent. Historical information on nesting activity in Kentucky includes only general references to the species's presence in suitable nesting habitat throughout the summer and lacks specific details of breeding. Interestingly, the only active nest ever reported contained four eggs on the late date of July 25, 1968 (Moore 1969). In Ohio, nesting activity peaks in May and early June (Peterjohn and Rice 1991), but the species is double-brooded, and later nests have been reported into mid-July. Although most nesting activity in Kentucky must have occurred in May and June, Mengel's observations and the 1968 nest record indicate that nesting activity extended well into summer as well. Harrison (1975) gives rangewide clutch size as commonly 4, often 5, sometimes 3 or 6.

Maslowski Wildlife Productions

According to Harrison (1975), the nest is constructed in a slight depression in the ground, typically beneath the cover of a clump of weeds or grass. It is composed of dry grasses, weed stalks, and rootlets and is lined with finer grasses, rootlets, and occasionally hair. This description matches the nest found in Franklin County in 1968 (Moore 1969).

Mengel (1965) found that the Vesper Sparrow typically occupied well-grazed pastures on moderate to steep slopes with scattered rock outcrops and patches of bare ground, a few scattered trees, and usually a nearby pond. Although similar habitat is used in other parts of its range, Vesper Sparrows also inhabit flatter situations in farmland where row-crop fields predominate. Such situations occur frequently in the glaciated regions to the north of Kentucky, where the species is common. The site in Bourbon County where a Vesper Sparrow was found during the atlas fieldwork was more similar to these latter situations. There, a well-grazed pasture, a hedgerow, and cornfields were situated in relatively flat, very open terrain.

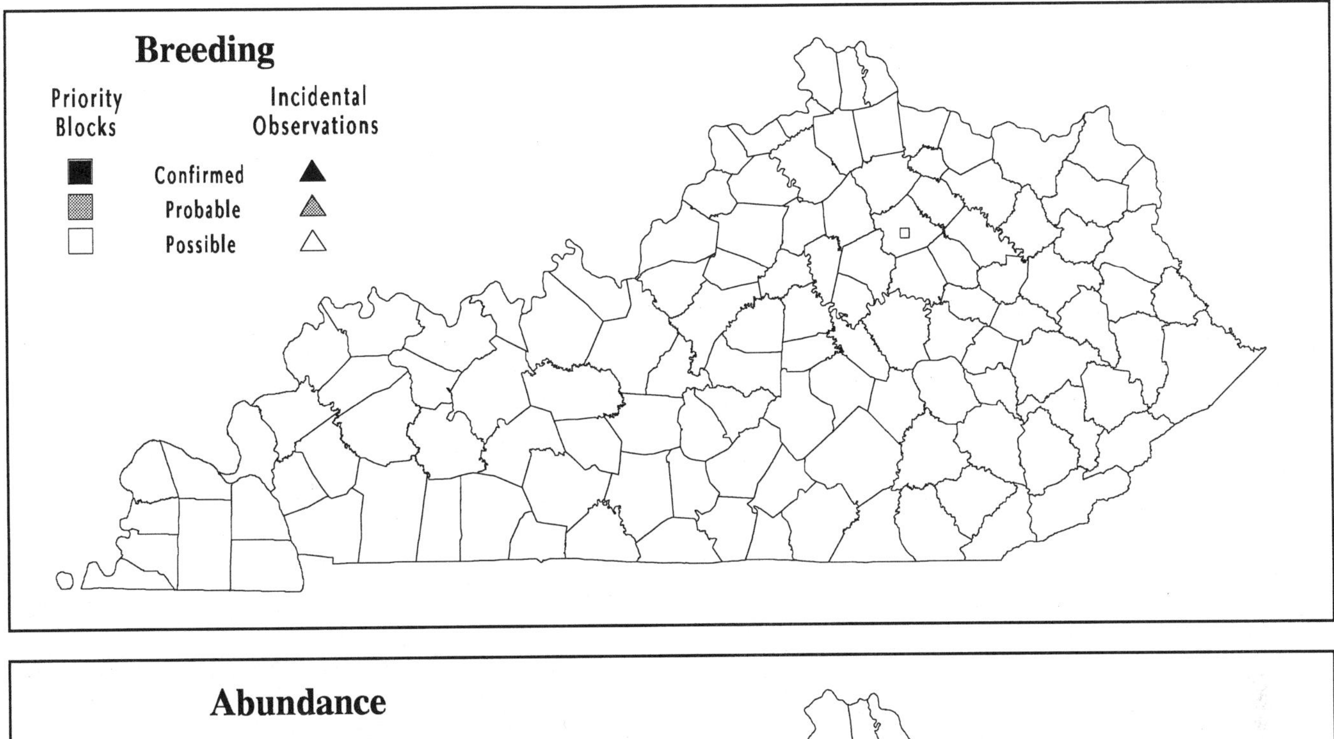

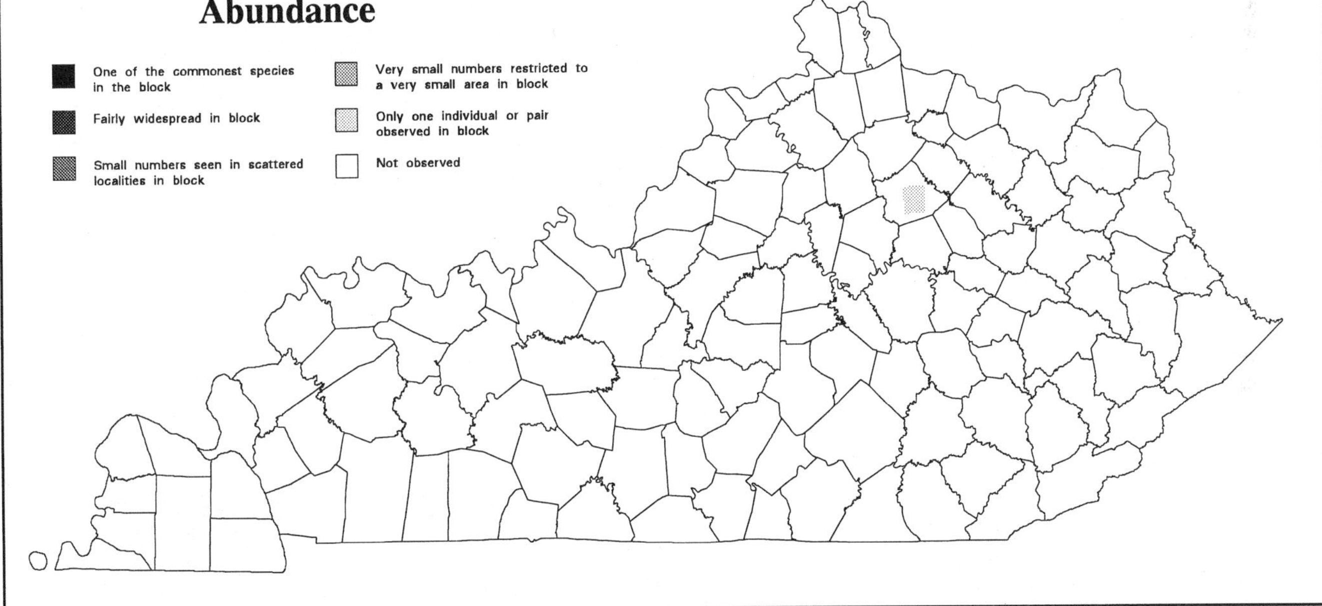

Analysis of Block Data by Physiographic Province Section

Physiographic Province Section	Total Blocks Surveyed	Blocks with Data	% with Data	Section's % for State
Mississippi Alluvial Plain	14	-	-	-
East Gulf Coastal Plain	36	-	-	-
Highland Rim	139	-	-	-
Shawnee Hills	142	-	-	-
Blue Grass	204	1	0.5	100.0
Cumberland Plateau	173	-	-	-
Cumberland Mountains	19	-	-	-

Summary of Breeding Status

Number of Blocks in Which Species Was Recorded		
Total	**1**	**0.1%**
Confirmed	-	-
Probable	-	-
Possible	1	100.0%

Vesper Sparrow

Lark Sparrow

Chondestes grammacus

The Lark Sparrow predominantly nests in the southern Great Plains and the western United States, but it also occurs locally east of the Mississippi River (AOU 1983). Although their numbers always have been relatively low and their presence fairly unpredictable, these beautiful sparrows have been documented as occurring across much of Kentucky, primarily west of the Cumberland Plateau. Mengel (1965) listed breeding season records from approximately 20 counties in central and western Kentucky, the easternmost of which was Whitley County.

Lark Sparrows are encountered in semi-open and open habitats with sparse ground cover. Although they have been found in natural cedar glades and prairie openings, these sparrows are most often observed in altered habitats today. The species occurs most frequently in rural farmland, where it typically inhabits well-grazed pastures with patches of bare ground or rocks, as well as scattered trees.

While Lark Sparrows may have inhabited the native prairies, glades, and barrens of central and western Kentucky, the vegetation was probably too thick to suit them in most areas. For this reason, if the species was present at all, it was likely distributed sporadically in pockets of drier habitats with sparse ground cover. Historical information is so sketchy that a clear picture of how the Lark Sparrow's occurrence might have changed after settlement is impossible to obtain. It is apparent from the species's current distribution, however, that altered habitats have replaced the naturally occurring ones that may have been lost.

Lark Sparrows winter to the south of Kentucky, but a few transients, and presumably the few nesting birds, return north during the last two weeks of April. Little precise data on nesting are available, but early clutches are likely completed by mid-May. According to Mengel (1965), a peak in clutch completion probably occurs around the first of June, although later nests have been reported into July. Evidence suggests that some pairs raise two broods (Mengel 1965). Kentucky data on clutch size are limited to one clutch of four eggs (Mengel 1965), but Harrison (1975) gives rangewide clutch size as 4–5, sometimes 3, rarely 6.

Nests are usually situated in a shallow depression on the ground beneath the cover of a clump of grass or other vegetation, but they are sometimes placed low to the ground in a tree or shrub. The nest is constructed primarily of dead grass and is lined with rootlets and finer grass (Harrison 1975).

During the atlas survey Lark Sparrows were found on five priority blocks, and four incidental observations were reported. Of these nine records, one came from the East Gulf Coastal Plain, three from the Highland Rim, one from the Shawnee Hills, and four from the Blue Grass. Eight of the sites were in rural farmland, most in well-grazed pastures; the other was in a natural prairie remnant opening in Lewis County.

Four of the nine atlas reports were for possible breeding, based simply on the observation of a single bird in appropriate breeding habitat during the breeding season, but the other five records involved more substantial evidence of nesting. The observation of at least one recently fledged young bird in the company of both parents in a heavily grazed pasture in southwestern Monroe County on July 11, 1990, provided the only confirmed breeding record. One bird of a pair in eastern Calloway County was observed gathering nest material on May 4, 1990, but further confirmation of nesting could not be obtained on subsequent visits, so the observation was conservatively regarded as probable. Other probable breeding records involved a singing male that held a territory in the same location along the Harrison/Pendleton county line in 1985 and 1986; a pair of birds observed in southwestern Washington County on July 5, 1990; and a pair of birds observed in southeastern Christian County on June 13, 1991.

Gary Meszaros

Atlas coverage was insufficient to detect most nesting Lark Sparrows. The records that did turn up, however, seem to indicate the presence of a small and scattered, but persistent, nesting population that roughly approximates previous assessments of occurrence and abundance. It is interesting that in the northern Blue Grass, where Mengel found small numbers scattered about in the summer of 1950, only one report was generated during atlas fieldwork. In contrast, the Lewis County observation represents the easternmost record ever documented in the state. Suitable nesting habitat appears to exist in abundance in much of central and western Kentucky, indicating that the population is probably not limited by habitat availability. More likely, Kentucky's placement along the eastern edge of the species's overall nesting range probably explains its low density.

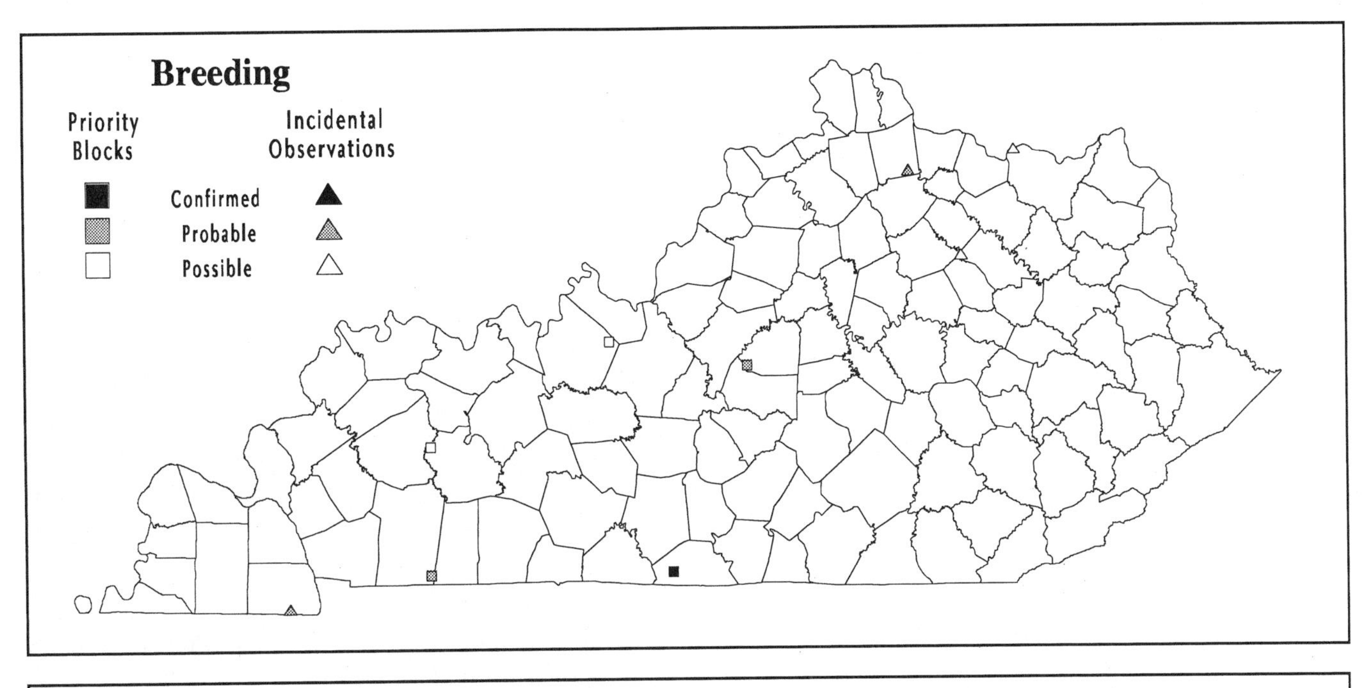

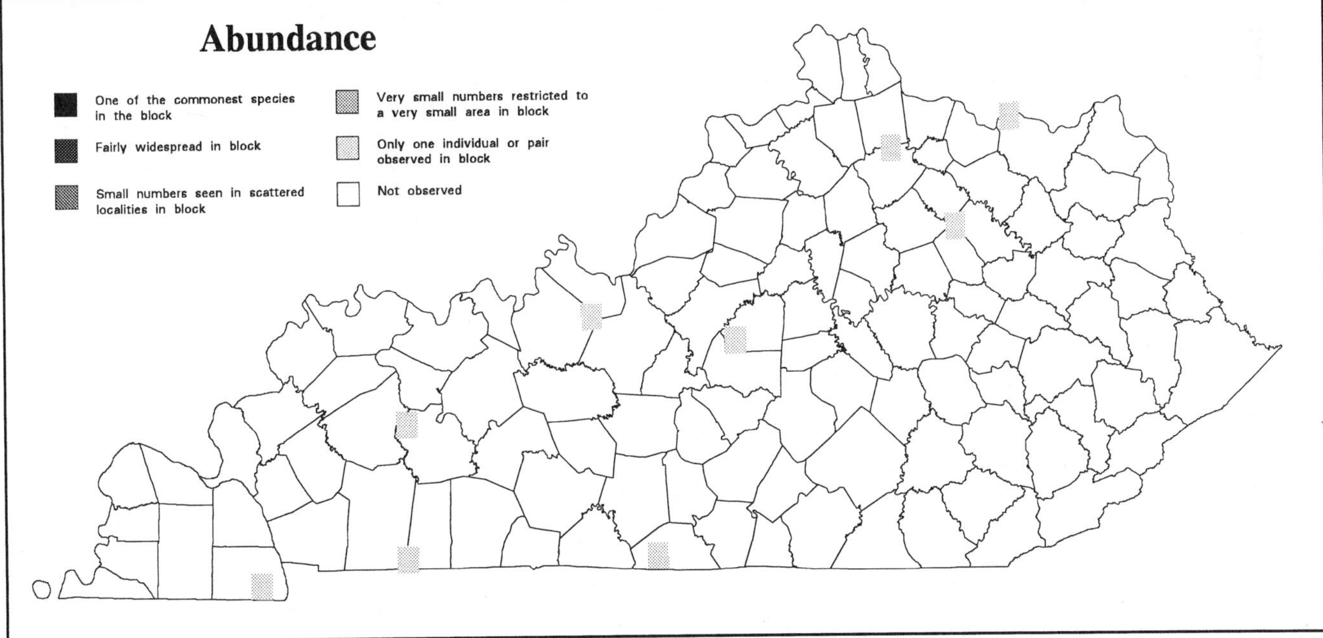

Analysis of Block Data by Physiographic Province Section

Physiographic Province Section	Total Blocks Surveyed	Blocks with Data	% with Data	Section's % for State
Mississippi Alluvial Plain	14	-	-	-
East Gulf Coastal Plain	36	-	-	-
Highland Rim	139	3	2.2	60.0
Shawnee Hills	142	1	0.7	20.0
Blue Grass	204	1	0.5	20.0
Cumberland Plateau	173	-	-	-
Cumberland Mountains	19	-	-	-

Summary of Breeding Status

Number of Blocks in Which Species Was Recorded		
Total	**5**	**0.7%**
Confirmed	1	20.0%
Probable	2	40.0%
Possible	2	40.0%

Lark Sparrow

Savannah Sparrow

Passerculus sandwichensis

The Savannah Sparrow is a relatively recent addition to Kentucky's breeding avifauna. Documentation of its presence in summer before the mid–20th century is lacking, but Mengel (1965) regarded the species as a casual summer resident. This status was based only on reports from southern Ohio and a record of a territorial male in Jefferson County in 1960 (Monroe and Monroe 1961).

Subsequent to the 1960 report from Jefferson County, summer records have accumulated gradually. During the late 1960s and 1970s a few birds were found scattered about the northern Blue Grass. In 1969 Stamm and McConnell (1971) observed a fledgling being fed on a farm near Goshen, in Oldham County, substantiating successful breeding for the first time. In 1972 Savannah Sparrows were confirmed breeding at Masterson Station Park, in Fayette County (Kleen 1973), and by the mid-1970s the species was considered regular in the Louisville area (Croft 1972; Monroe 1976). The atlas survey yielded additional records from scattered localities in the northern and central Blue Grass as well as single records from the Highland Rim and the Shawnee Hills.

Although the Savannah Sparrow occupies natural grassland habitats in some parts of its range, the species is restricted to altered habitats in Kentucky. These small sparrows are found most frequently in hayfields, pastures, and other grassy habitats where the vegetation is not especially tall or thick. In this regard the Savannah is most similar to the Grasshopper Sparrow. Although the species has been reported on grassy, reclaimed surface mines in Ohio (Peterjohn and Rice 1991), it has not been found nesting in similar habitats in Kentucky. These sparrows are sometimes locally common in favored situations: Croft (1972) counted 12 singing males in fields on one farm in Oldham County in 1971.

Although small numbers of Savannah Sparrows winter in Kentucky, the species is most conspicuous during late April and early May, when most spring transients pass through. Although arrival dates for locally nesting birds have not been reported, it appears that most return by mid-May. Specific Kentucky data on nesting activity are limited to observations of a fledgling being fed on June 28, 1969 (Stamm and McConnell 1971) and a bird carrying food near Tollesboro, in Lewis County, on July 13, 1988 (KBBA data). Most clutches are probably completed by early June, but song continues well into July. It is unclear whether late nesting activity represents the raising of a second brood or renesting following earlier failure. Because the species nests almost exclusively in hayfields and pastures, mowing and grazing must result in the destruction of many nests. Kentucky data on clutch size are lacking, but Harrison (1975) gives rangewide clutch size as 3–6, commonly 4–5.

An active nest has not been found in Kentucky, but according to Harrison (1975), the nest is typically situated within a shallow depression on the ground and well concealed by overhanging vegetation. It is constructed chiefly of coarse grasses and lined with finer grass, and sometimes rootlets and hair (Harrison 1975).

During the atlas fieldwork Savannah Sparrows were found on 10 priority blocks, and four incidental observations were reported. The species was observed at the two locations where nesting had been documented previously, but the remaining twelve records were all from new localities. Most records originated in the Blue Grass, especially in the intensively cleared farmland of the Inner Blue Grass subsection. Only two records originated outside the Blue Grass, one in southern Lincoln County and the other in Livingston County. Both involved the observation of singing males on a single date and no further evidence of nesting.

Ron Austing

Although most of the twelve Blue Grass records were also for possible breeding, eight involved the observation of more than one bird. Loose colonies of two to four singing males or pairs were found at seven sites, in the counties of Bourbon (two sites), Boyle, Fayette (Stamm and Monroe 1991b), Jefferson, Oldham, and Woodford. All sites were in closely mowed pastures in very open situations, indicating that these intensively managed situations provide Savannah Sparrows with suitable nesting habitat. Confirmed records were obtained from two localities, and both were based on the observation of birds carrying food. At least six pairs were observed at Masterson Station Park, in Fayette County, on several occasions during the atlas period (no specific dates), and one bird was observed in Lewis County on July 13, 1988.

While it is possible that Savannah Sparrows nested in Kentucky's native grasslands, it is much more likely that the species has invaded suitable altered habitats in recent times. Human influence on the landscape has resulted in the conversion of large areas of forest to grassland, and the amount of suitable nesting habitat has increased dramatically in relatively recent times. Apparently in response to the availability of nesting habitat, Savannah Sparrows have spread southward from their traditional breeding range to the north. An increase in overall abundance and southward spread have been documented in Ohio (Peterjohn and Rice 1991), and the appearance of nesting birds in northern Kentucky must represent an extension of those trends. Although data are certainly incomplete, the records generated by the atlas survey suggest that a regularly occurring, and probably increasing, nesting population is established in the northern Blue Grass.

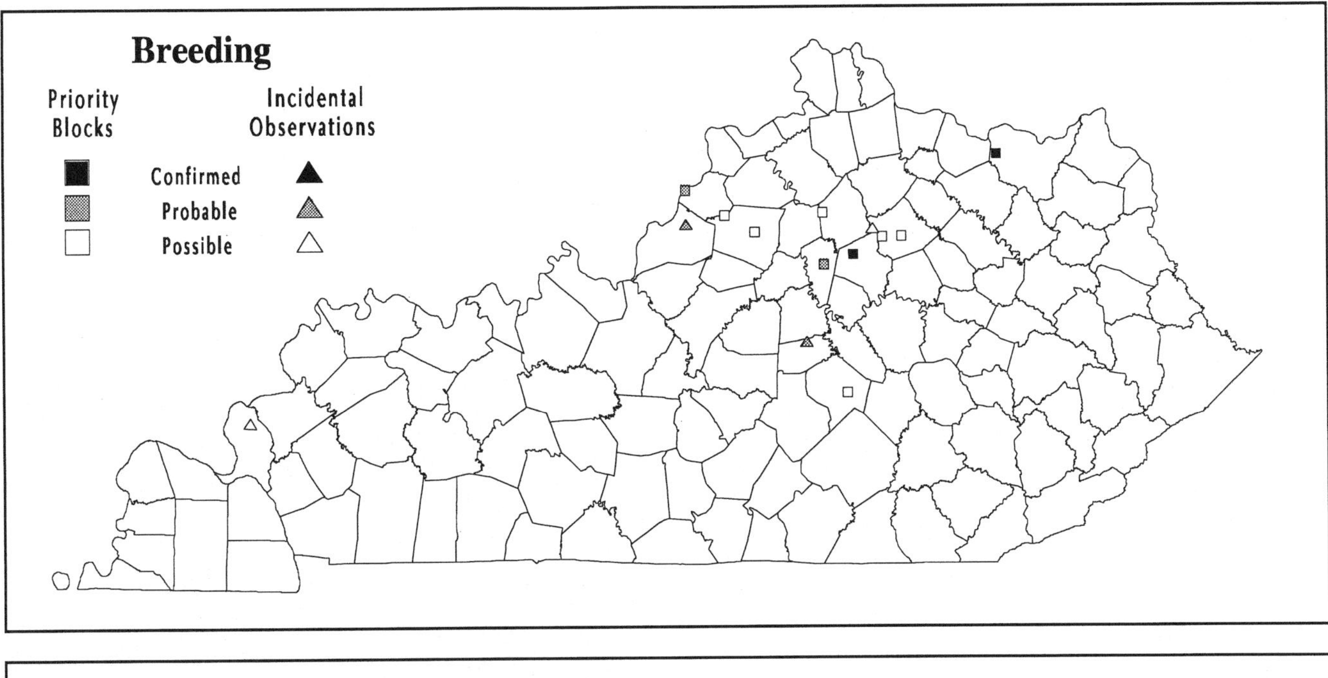

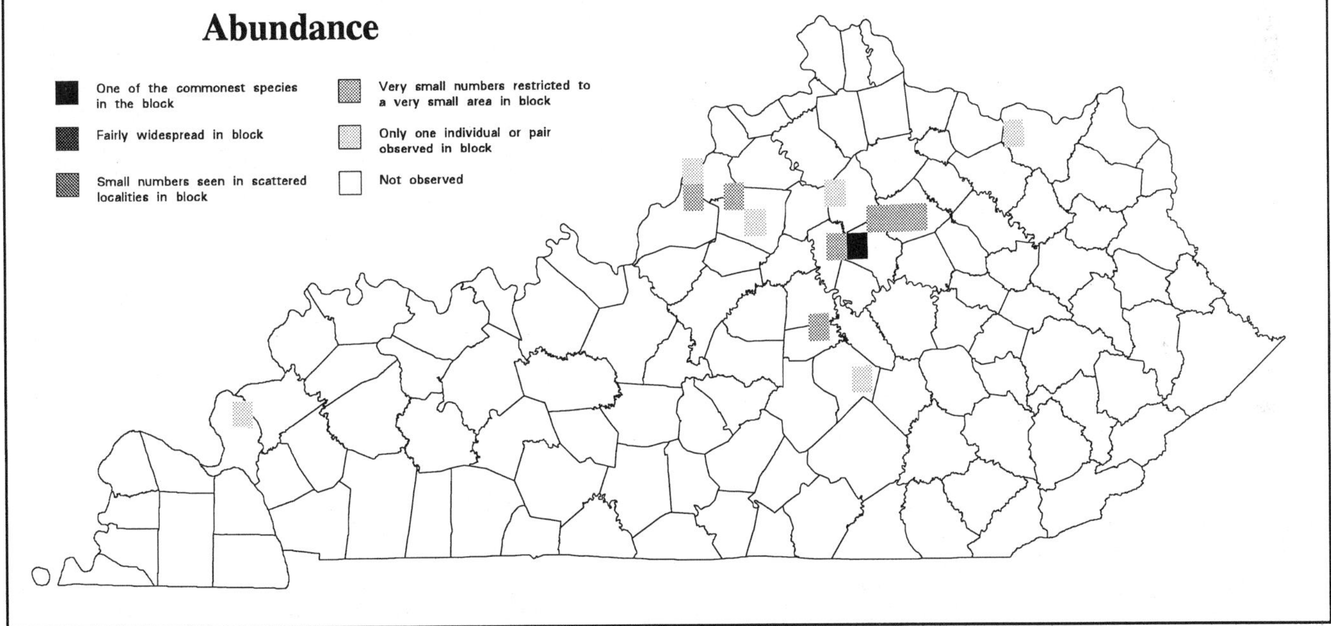

Analysis of Block Data by Physiographic Province Section

Physiographic Province Section	Total Blocks Surveyed	Blocks with Data	% with Data	Section's % for State
Mississippi Alluvial Plain	14	-	-	-
East Gulf Coastal Plain	36	-	-	-
Highland Rim	139	1	0.7	10.0
Shawnee Hills	142	-	-	-
Blue Grass	204	9	4.4	90.0
Cumberland Plateau	173	-	-	-
Cumberland Mountains	19	-	-	-

Summary of Breeding Status

Number of Blocks in Which Species Was Recorded		
Total	**10**	**1.4%**
Confirmed	2	20.0%
Probable	2	20.0%
Possible	6	60.0%

Grasshopper Sparrow

Ammodramus savannarum

The Grasshopper Sparrow's insectlike song is a rather conspicuous summer sound of open, grassy habitats throughout much of Kentucky. Mengel (1965) regarded the species as a fairly common to common summer resident, although less numerous in eastern Kentucky because of the scarcity of suitable habitat. As of the late 1970s BBS data indicated that the species was decreasing at an alarming rate (Monroe 1978), but more recent data suggest that the decline has leveled off (BBS data).

The Grasshopper Sparrow occurs in natural grasslands throughout parts of its range, but it is restricted to altered situations in Kentucky. These sparrows inhabit grasslands that are dominated by relatively sparse or short vegetation, and they are typically absent if the vegetation is tall and thick. Consequently, they are most numerous in lightly grazed pastures, hayfields on dry slopes and poor soil, reclaimed surface mines, and other situations in which vegetation is not profuse. Small numbers also occur in fallow row-crop fields where grassy vegetation is beginning to recolonize bare soil.

Although Grasshopper Sparrows may have inhabited Kentucky's native prairies, large portions of these grasslands may have been dominated by vegetation that was too thick and tall to have been favored. Audubon (1861), who spent much time in the native prairies in the early 1800s, apparently did not observe the species in the state. Human alteration of the landscape has resulted in a dramatic increase in suitable habitat during the past two centuries. As a result, this sparrow has become established throughout much of the state where formerly it must have been absent. For example, substantial numbers now occur on grassy, reclaimed surface mines in the Cumberland Plateau where native forests once covered essentially all of the landscape (Allaire 1981).

Grasshopper Sparrows winter from the southern United States southward into Central America (AOU 1983), and they typically reappear in Kentucky by the third week of April. Singing begins immediately upon arrival, but it may be the beginning of May before nest building is under way. According to Mengel (1965), clutches are completed from mid-May to early August, with a peak in clutch completion during the last ten days of May. Many pairs may be double-brooded, although documentation of such in Kentucky is absent. Because these sparrows nest predominantly in hayfields and pastures, it is likely that much nesting failure occurs because of mowing and grazing. Mengel (1965) gave the average size of five clutches as 4.8 (range of 4–5), although a brood of six young was observed in a nest in Franklin County in 1965 (Stamm and Jones 1966).

Forest Cover

Value	% of Blocks	Avg Abund
All	39.9	1.9
1	62.4	2.2
2	59.5	1.9
3	42.6	1.7
4	9.7	1.6
5	1.9	1.0

Grasshopper Sparrows place their nest on the ground, often in a shallow depression (Harrison 1975). It is usually situated on relatively flat terrain, but sometimes on slopes or embankments (Stamm and Slack 1955). The nest is constructed primarily of coarse grass and weed stems, lined with finer grass, and often slightly domed at the rear (Harrison 1975). Nests are typically well concealed among clumps of grass and are usually very difficult to locate (Stamm and Slack 1955).

Ron Austing

The atlas survey yielded reports of Grasshopper Sparrows in nearly 40% of priority blocks statewide, and 28 incidental observations were reported. Occurrence was highest in the Highland Rim, the Shawnee Hills, and the Blue Grass, and lowest in the Cumberland Mountains and the Cumberland Plateau. Average abundance was relatively low statewide. Both occurrence and average abundance were highest in extensively cleared areas and decreased substantially as the percentage of forest cover increased.

Only about 6% of priority block records were for confirmed breeding. Although Grasshopper Sparrows were relatively easy to detect, obtaining evidence of nesting was much more difficult. An incidental report in Ohio County was based on the observation of a nest containing five eggs on May 20, 1990, but all of the 17 confirmed records in priority blocks were based on the observation of adults carrying food and recently fledged young.

Grasshopper Sparrows are reported regularly on most Kentucky BBS routes. The average number of individuals per BBS route for the periods 1966–91 and 1982–91 was 3.00 and 2.45, respectively. Trend analysis of these data reveals a significant ($p<.01$) decrease of 8.0% per year for the period 1966–91, but a nonsignificant decrease of 2.0% per year for the period 1982–91. The reasons for the long-term decline are not fully understood but may be related to pesticide use and mowing of nesting areas (Monroe 1978).

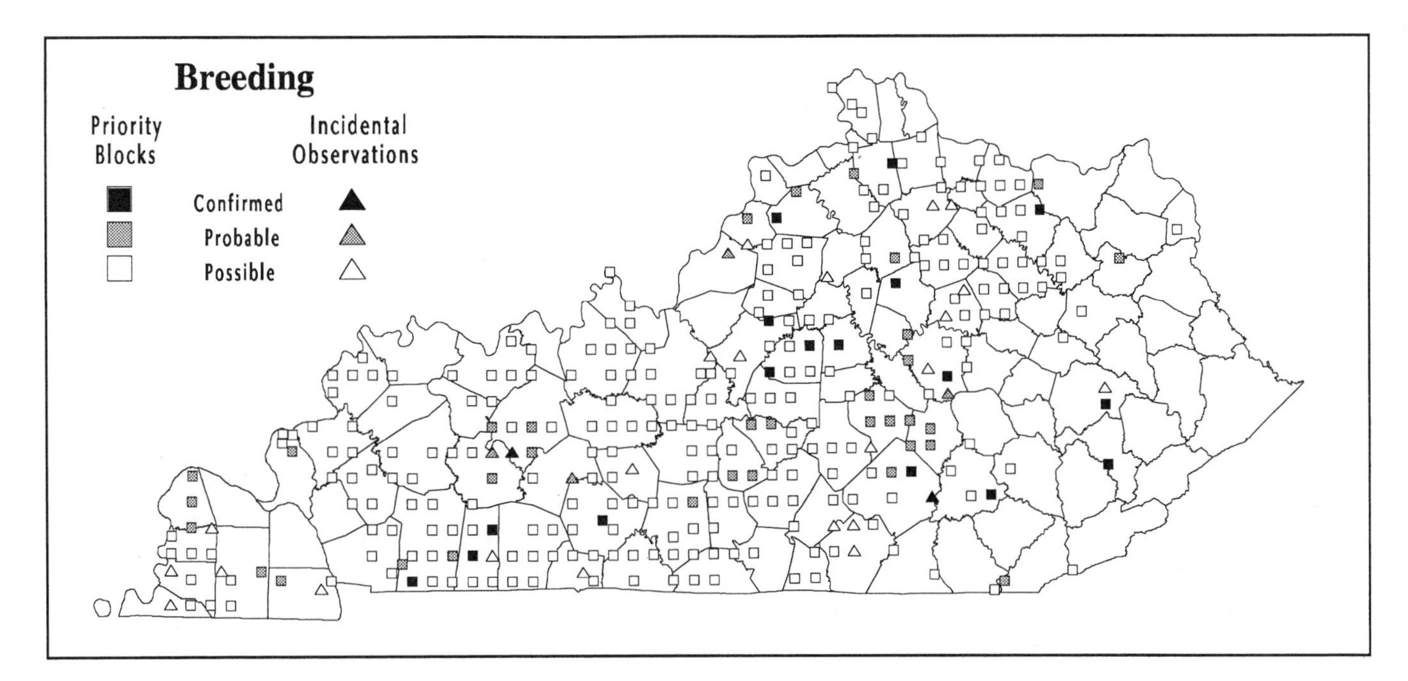

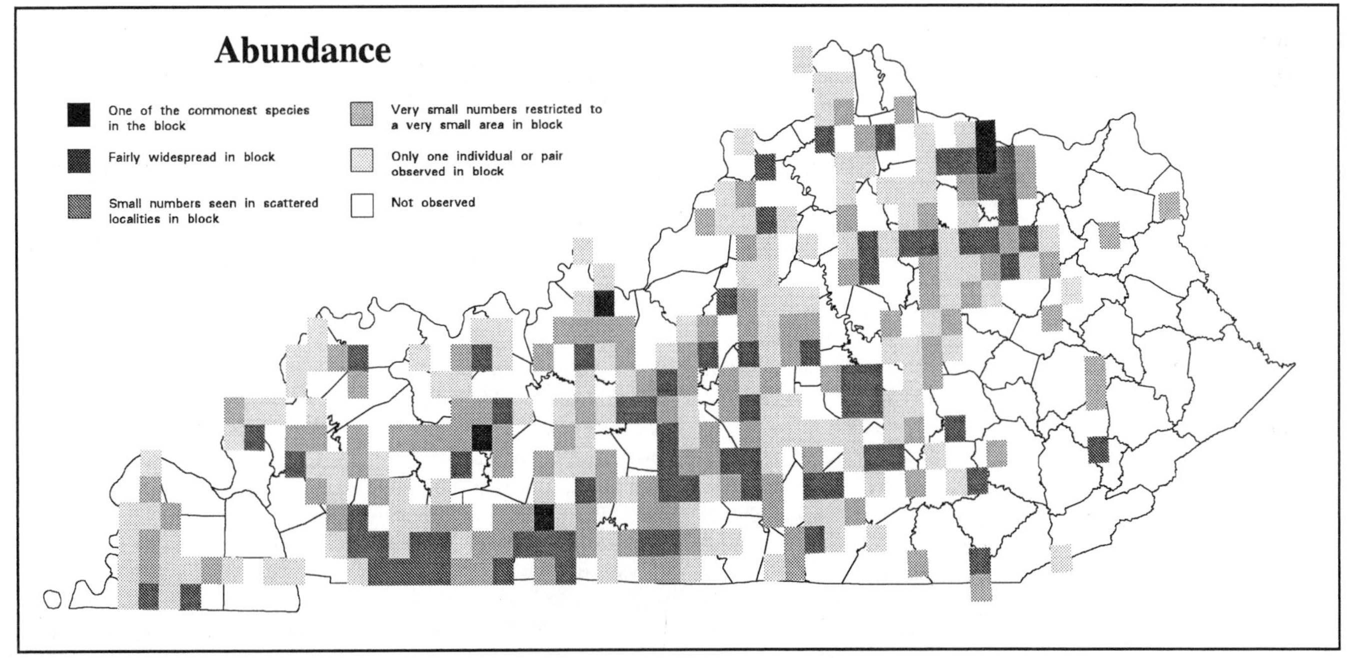

Analysis of Block Data by Physiographic Province Section

Physiographic Province Section	Total Blocks Surveyed	Blocks with Data	Avg Abund	% with Data	Section's % for State
Mississippi Alluvial Plain	14	3	1.0	21.4	1.0
East Gulf Coastal Plain	36	12	1.7	33.3	4.1
Highland Rim	139	98	2.1	70.5	33.8
Shawnee Hills	142	66	1.6	46.5	22.8
Blue Grass	204	94	1.9	46.1	32.4
Cumberland Plateau	173	14	1.9	8.1	4.8
Cumberland Mountains	19	3	2.0	15.8	1.0

Summary of Breeding Status

Number of Blocks in Which Species Was Recorded		
Total	**290**	**39.9%**
Confirmed	17	5.9%
Probable	34	11.7%
Possible	239	82.4%

Henslow's Sparrow

Ammodramus henslowii

The Henslow's Sparrow is a very locally distributed summer resident across Kentucky. The species has been reported from the western Highland Rim east through central Kentucky to the northern Cumberland Plateau, but typically it is reported at not more than a few locations in any year. Moreover, numbers of birds fluctuate from year to year, perhaps in part in response to habitat availability.

Henslow's Sparrows are typically found in open habitats dominated by thick, grassy vegetation. The species favors areas that have been neglected for a year or two and have accumulated a layer of dead plant material at the base of the current year's growth. The lack of disturbance also allows for the presence of dead weed stalks, young saplings, and briars, which are used as singing perches (Wiley and Croft 1964). Henslow's Sparrows may have occurred at least locally in the native prairies of the East Gulf Coastal Plain and the Highland Rim, but documentation of their presence in such habitat is absent. Today native grasslands have been virtually eliminated, and the species occurs entirely in altered situations. Although fallow fields and pastures provide most of the habitat used by Henslow's Sparrows in Kentucky, the species is also found on reclaimed surface mines, the margins of airfields, and other unmowed grassy habitats. Hayfields of tall, thick grasses like orchard grass and timothy are also used, although mowing results in abandonment (Wiley and Croft 1964; Mengel 1965).

The historical status of the Henslow's Sparrow in Kentucky is obscure. The type specimen was collected by Audubon in northern Kentucky in 1820, but otherwise the species was almost unknown until the mid-1940s (Mengel 1965). In 1946 these sparrows appeared in numbers in the Louisville area, being found at 16 sites in Jefferson and Oldham counties. In subsequent years the species was reported more regularly, and between 1946 and the early 1950s summer records were obtained for a number of localities in the central part of the state, from Clinton County north to Boone County (Mengel 1965). The appearance of Henslow's Sparrows in Kentucky roughly coincided with a shift in Ohio's breeding population from the largely agricultural northern and central parts of the state to the southeast (Peterjohn and Rice 1991). This change occurred during a period when nesting habitat in the north was being converted to more intensive agricultural use, and suitable habitat was being created in the southeast as a result of strip mining.

Since the early 1950s Henslow's Sparrows have continued to turn up sporadically during summer fieldwork. The species has been reported occasionally on BBS routes, and summer records have been published for 10 counties scattered across central Kentucky and extending westward as far as Muhlenberg County (Wiley and Croft 1964; Stamm 1979b, 1982b, 1984a, 1984b, 1985, 1987b). Only one of these reports has documented confirmed nesting (Wiley and Croft 1964).

Largely because of this sparrow's secretive habits, detailed information on nesting activity in Kentucky is scarce. The species winters in the southern United States (AOU 1983), and summer residents often return to the state by mid-April. Records of birds presumed to be migrants are few, and most spring records likely represent territorial individuals. By the end of April birds are usually singing persistently in traditional nesting areas, and a peak in nesting activity seems to occur in May and June. Juveniles have been collected in late June and early July, and a nest containing three small young was located in Oldham County on June 17, 1950 (Mengel 1965). Wiley and Croft (1964) netted a female that possessed brood patches on July 14, 1963, indicating that some pairs may nest later. The species is known to be double-brooded in Ohio (Peterjohn and Rice 1991), suggesting that at least some Kentucky pairs may raise a second brood. Kentucky data on clutch size are limited to the Oldham County record of a nest with three young, but Harrison (1975) gives rangewide clutch size as 3–5.

Ron Austing

Henslow's Sparrows typically nest on or near the ground in thick, grassy vegetation. The nest is well concealed from above by overhanging vegetation, and it is very difficult to locate. It is constructed of dead grasses and lined with finer grass and some hair (Harrison 1975). Nesting pairs sometimes occur singly, but more often the species is loosely colonial, especially in large fields of optimal habitat (Wiley and Croft 1964).

The atlas survey yielded 24 records of Henslow's Sparrows in priority blocks, and 10 incidental observations were reported. All 34 reports originated in the Blue Grass, the Highland Rim, the Shawnee Hills, and the Cumberland Plateau, with highest occurrence in the Blue Grass. Atlas records extended the species's known summer range eastward to Carter, Lewis, and Morgan counties, and westward to Crittenden and Caldwell counties. Some reports involved single singing males, but, more typically, small groups of two to five singing birds were encountered. The largest colony involved at least six singing males near Knoxville, in Pendleton County (Stamm 1989a).

Few sites were revisited at later dates or in subsequent years, so most reports resulted only in possible breeding records. Only three records (two in priority blocks and one incidental) were for confirmed breeding, and all involved the observation of adults carrying food. Records of probable breeding involved birds that were considered territorial based on multiple observations within one breeding season.

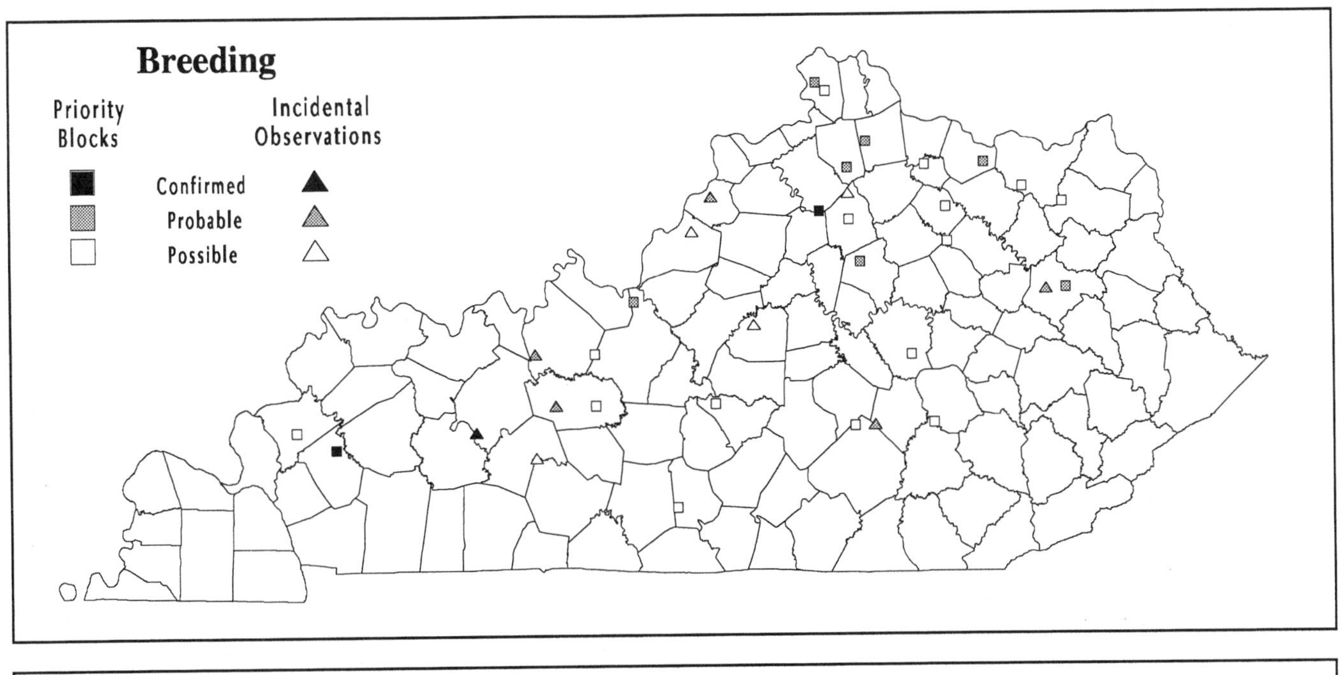

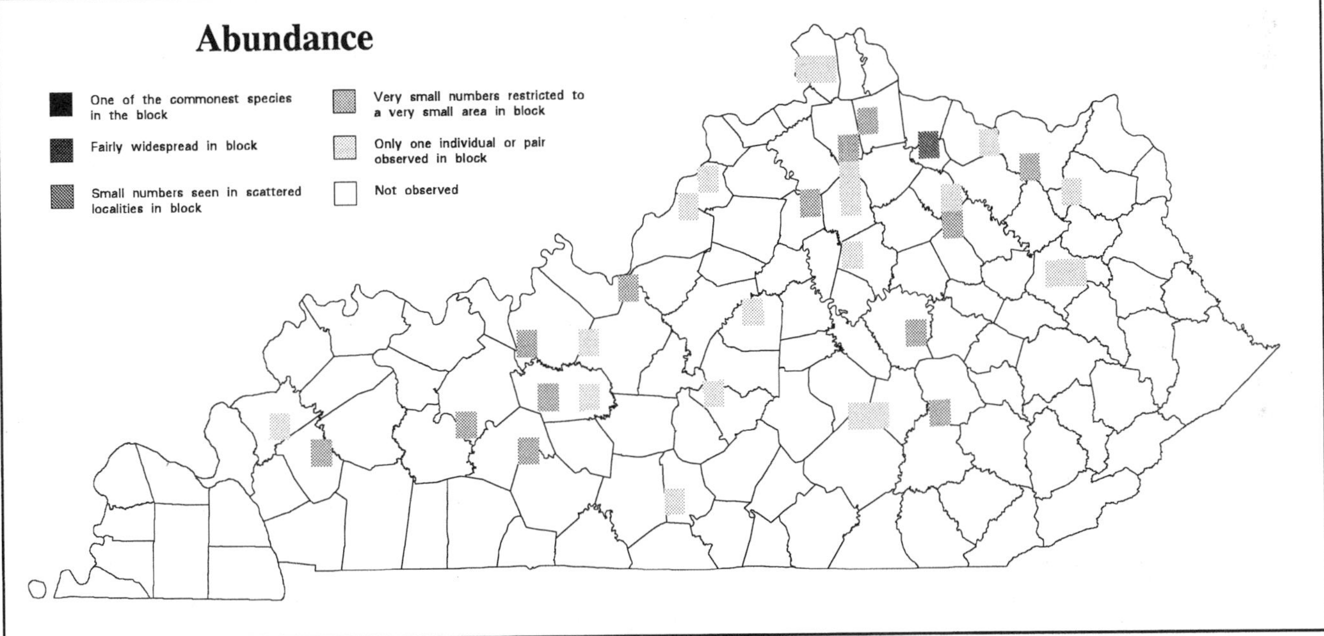

Analysis of Block Data by Physiographic Province Section

Physiographic Province Section	Total Blocks Surveyed	Blocks with Data	% with Data	Section's % for State
Mississippi Alluvial Plain	14	-	-	-
East Gulf Coastal Plain	36	-	-	-
Highland Rim	139	4	2.9	16.7
Shawnee Hills	142	4	2.8	16.7
Blue Grass	204	13	6.4	54.2
Cumberland Plateau	173	3	1.7	12.5
Cumberland Mountains	19	-	-	-

Summary of Breeding Status

Number of Blocks in Which Species Was Recorded		
Total	**24**	**3.3%**
Confirmed	2	8.3%
Probable	7	29.2%
Possible	15	62.5%

Henslow's Sparrow

Song Sparrow

Melospiza melodia

The musical song of the Song Sparrow is a characteristic spring and summer sound across much of Kentucky. The species has spread southward as a breeding bird during the past century (Mengel 1965). Audubon (1861) did not observe it in those parts of the state he frequented during the early 1800s, and it was apparently regular only in north central and northeastern Kentucky during the late 19th century (Mengel 1965). By the 1950s these sparrows were absent from only the Mississippi Alluvial Plain, the East Gulf Coastal Plain, the southern Shawnee Hills, and the western Highland Rim, and since that time small numbers have continued to move into new areas. For example, Hancock (1973) considered the species to be rare and irregular in Hopkins County about 1950, but by the early 1970s he regarded it as locally uncommon. Although Song Sparrows remain locally distributed in much of the Mississippi Alluvial Plain, the East Gulf Coastal Plain, and the western Highland Rim, they must be considered to occur statewide today.

Song Sparrows occupy a great variety of habitats, from small woodland openings in forested regions to brushy corridors in extensively cleared areas. The species is most frequently found in low, moist areas, especially along streams, where it typically occurs in great abundance. Although these sparrows can be found in natural habitats, such as the brushy margins of wetlands and remnant prairie patches, they are typically found in artificial habitats, such as suburban and rural yards, abandoned fields, utility and roadway corridors, reclaimed surface mines, and unmowed margins and fencerows in farmland.

Song Sparrows are found throughout the state in winter, but they typically disappear from nonbreeding areas in the southern part of the state by late April (Mengel 1965). Song is heard occasionally throughout the winter, but as soon as the first warm days of March arrive, territorial males begin to sing in earnest. Early clutches have been reported by April 1–10, and a peak in early clutch completion typically occurs about April 20. The species is regularly double-brooded, and although a later peak in clutch completion is not apparent, nesting continues well into August (Mengel 1965). The average size of 27 clutches reported by Mengel was 4.2 (range of 3–5).

Forest Cover

Value	% of Blocks	Avg Abund
All	77.6	3.1
1	81.2	3.0
2	73.2	2.7
3	71.7	2.9
4	85.7	3.6
5	88.5	3.7

Early nests are typically placed either on the ground or within the cover of an evergreen tree or shrub (Nice 1964). Ground nests are hidden amid the cover of grasses, weeds, and other thick vegetation. Later nests are usually built above the ground and can be found in a great variety of situations, including small trees, tangles of vines, and shrubs. Of 20 nests reported by Mengel (1965), six were built on the ground. Above-ground nests are usually placed on a horizontal fork and concealed from above by overhanging branches. The nest is typically constructed of dead grasses and weed stalks and is lined with fine grass, rootlets, and sometimes hair. The average height of 16 above-ground nests summarized by Mengel (1965) was 2.6 feet (range of 0.5–5.0 feet).

Alvin E. Staffan

The atlas survey yielded records of Song Sparrows in nearly 78% of priority blocks statewide, and 17 incidental observations were reported. The species ranked 39th according to the number of records in priority blocks, but 24th by total abundance and 14th by average abundance. These figures indicate that although the Song Sparrow is not as widespread as many other common species, it is relatively numerous and conspicuous where it does occur. Occurrence was highest in the Blue Grass, the Cumberland Plateau, and the Cumberland Mountains and decreased to the south and west. Average abundance varied similarly. In regions where Song Sparrows were uncommon, they were typically found only along the floodplains of rivers and streams. This pattern is typical of the species, especially in areas that it has recently colonized (Mengel 1965).

About 20% of priority block records were for confirmed breeding. Although more than a dozen active nests were located, most confirmed records were based on the observation of recently fledged young. Other confirmed records involved adults carrying food for young, nest building, and distraction displays. Brown-headed Cowbird parasitism of the Song Sparrow is relatively high, especially in settled areas. Cowbird eggs and young were found in a few sparrow nests, and adult sparrows were observed feeding young cowbirds on several occasions.

Song Sparrows are reported in low to moderate numbers on Kentucky BBS routes. The average number of individuals per BBS route for the periods 1966–91 and 1982–91 was 15.23 and 18.16, respectively. According to these data, the Song Sparrow ranked 15th in abundance on BBS routes for the period 1982–91. Trend analysis of these data shows a significant ($p<.05$) increase of 2.0% per year for the period 1966–91, but a nonsignificant increase of 0.9% per year for the period 1982–91.

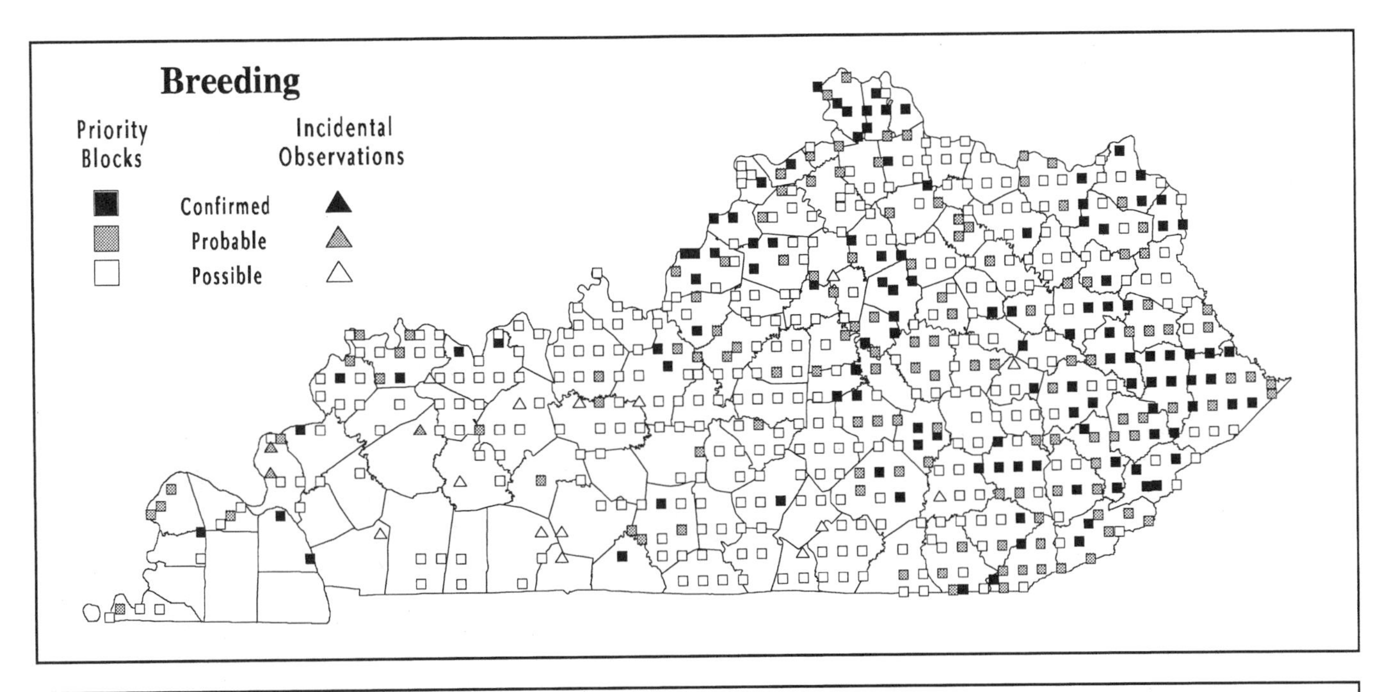

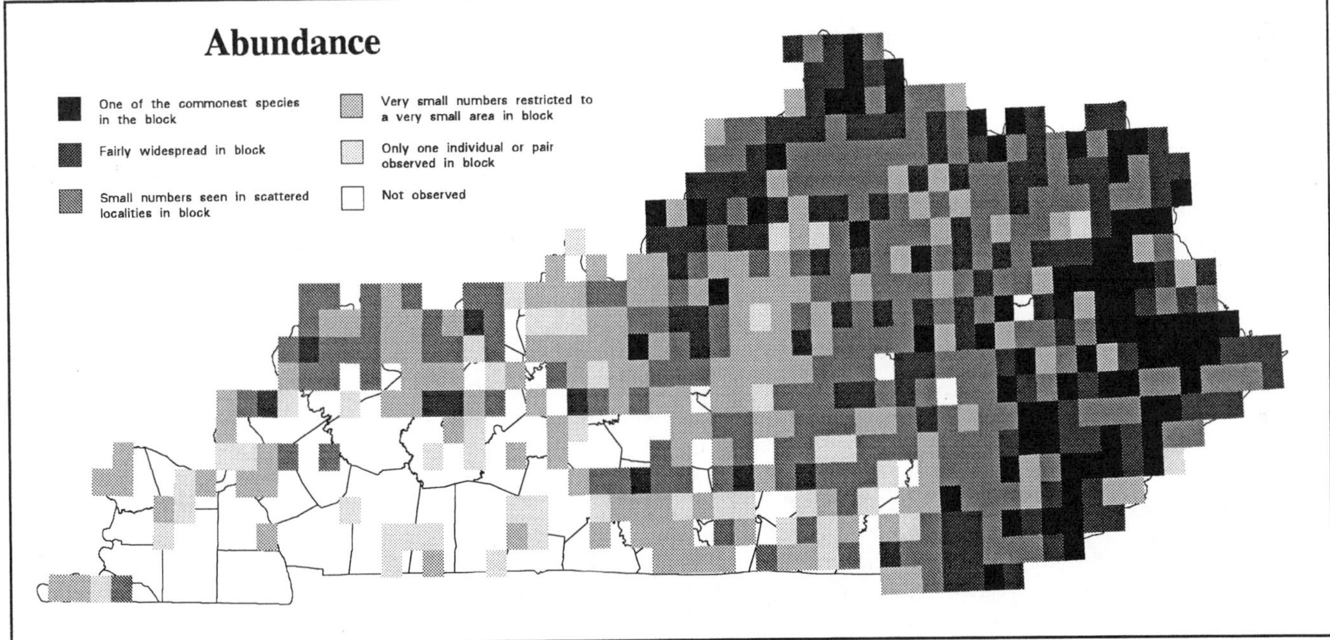

Analysis of Block Data by Physiographic Province Section

Physiographic Province Section	Total Blocks Surveyed	Blocks with Data	Avg Abund	% with Data	Section's % for State
Mississippi Alluvial Plain	14	8	2.0	57.1	1.4
East Gulf Coastal Plain	36	6	1.5	16.7	1.1
Highland Rim	139	92	2.3	66.2	16.3
Shawnee Hills	142	74	2.3	52.1	13.1
Blue Grass	204	202	3.3	99.0	35.8
Cumberland Plateau	173	164	3.7	94.8	29.1
Cumberland Mountains	19	18	3.8	94.7	3.2

Summary of Breeding Status

Number of Blocks in Which Species Was Recorded		
Total	**564**	**77.6%**
Confirmed	115	20.4%
Probable	129	22.9%
Possible	320	56.7%

Dark-eyed Junco

Junco hyemalis

In eastern North America the Dark-eyed Junco nests predominantly to the north of Kentucky, across the northeastern United States and Canada, but the race *J. h. carolinensis* breeds south through the Appalachian Mountains as far as northern Georgia (Mengel 1965; AOU 1983). This portion of the range barely reaches into southeastern Kentucky, where juncos nest at higher elevations of Black Mountain.

Juncos were first reported near the summit of Black Mountain, in Harlan County, in the summer of 1908 (Howell 1910). Breeding was not confirmed there, however, until Lovell (1950b) found a nest containing eggs on June 17, 1947. Mengel (1965) regarded the species as a common summer resident above 3,600 feet on the mountain, and subsequent fieldwork there has yielded additional observations, including a few from the Letcher County portion of the mountain (KSNPC, unpub. data).

On Black Mountain the Dark-eyed Junco occurs in a variety of semi-open habitats, including both natural and artificial forest margins and openings (Lovell 1950a). Mengel (1965) found that territories were typically centered around tangled masses of fallen timber or steep, root-lined banks along narrow woodland trails and logging roads.

Land use changes near the mountain's summit have likely resulted in fluctuations in the abundance of juncos there during the past two centuries. Although fires and windstorms may have maintained a certain amount of suitable nesting habitat on the mountain, the construction of roads, logging and mining activities, and other induced disturbance have resulted in the creation of a relative abundance of habitat. For this reason, the species is probably as numerous on the mountain today as at any time.

Mengel (1965) reported that, unlike most other northern species breeding on Black Mountain, which are regularly found as low as about 3,000 feet, juncos are more restricted to the higher elevations, rarely being found below 3,800 feet and never below 3,500 feet. There are at least occasional exceptions, however, as a nest was located during the atlas fieldwork at approximately 3,200 feet. Nevertheless, this tendency to prefer high elevations may explain the species's absence on nearby Cumberland Mountain (small parts of which reach 3,500 feet), where suitable habitat exists and other northern species are found in small numbers.

Dark-eyed Juncos from more northerly breeding populations commonly overwinter throughout much of Kentucky. Most of these birds depart during the second half of March, and they are rare by the middle of April. The species remains to nest only at the higher elevations of Black Mountain, where Mengel (1965) found breeding birds in song on May 17, 1952. According to Mengel, egg laying occurs chiefly in late May and June, with a peak in clutch completion during early June. Kentucky data on clutch size are limited to two nests described by Lovell (1950a, 1950b), both of which contained three eggs. A nest observed during the atlas fieldwork contained two eggs on June 21, 1989, but the clutch may have been incomplete. In contrast, Harrison (1975) gives rangewide clutch size as 4–5, occasionally 3, rarely 6. Family groups including young are commonly seen from about the middle of June through July (Mengel 1965).

Juncos usually nest on the ground, typically concealing their nest in a natural or artificial recess beneath the cover of a fallen branch, exposed roots, or vegetation. Most of the few Kentucky nests have been found on embankments along old roads, but they also occur on natural slopes. Nests on Black Mountain have been constructed of fine grasses and bits of moss and lined with the sporophytes of mosses (Lovell 1950b).

Ron Austing

The atlas survey yielded four records of Dark-eyed Juncos in priority blocks. Three of these came from traditionally known breeding areas on Black Mountain, but the fourth involved a male that was found about three miles southwest of Willard, in Carter County, on July 12, 1988. This individual was interpreted as a nonbreeding vagrant, and the record was omitted from the final data set. The three records from Black Mountain involved two reports of confirmed breeding. An active nest was located at approximately 3,200 feet just southwest of Cave Spur on June 21, 1989. Also, recently fledged young were observed higher on the mountain, near the summit, on the same day.

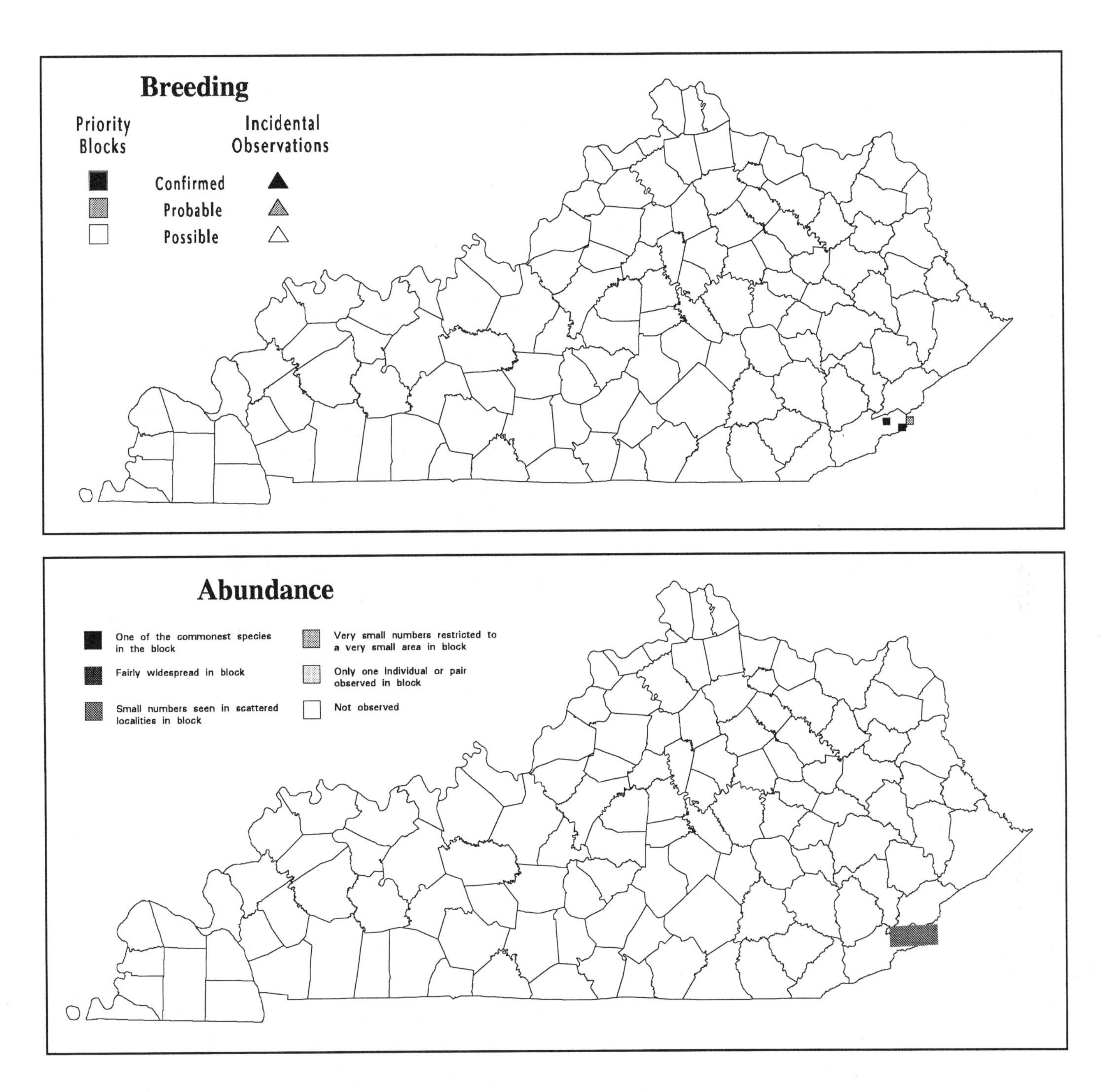

Analysis of Block Data by Physiographic Province Section

Physiographic Province Section	Total Blocks Surveyed	Blocks with Data	% with Data	Section's % for State
Mississippi Alluvial Plain	14	-	-	-
East Gulf Coastal Plain	36	-	-	-
Highland Rim	139	-	-	-
Shawnee Hills	142	-	-	-
Blue Grass	204	-	-	-
Cumberland Plateau	173	-	-	-
Cumberland Mountains	19	3	15.8	100.0

Summary of Breeding Status

Number of Blocks in Which Species Was Recorded		
Total	**3**	**0.4%**
Confirmed	2	66.7%
Probable	1	33.3%
Possible	-	-

Dark-eyed Junco

Bobolink

Dolichonyx oryzivorus

Kentucky lies at the southern limit of the breeding range of the Bobolink, and until relatively recent times it was not known to nest in the state. Early historical documentation of the Bobolink's presence is absent, but small numbers possibly occurred in the native grasslands of the central and western parts of the state. The species was first reported as a transient in the early 1900s, and by the late 1950s it was occurring regularly (Mengel 1965). In 1949 the Bobolink was discovered nesting in southwestern Ohio (Kemsies and Randle 1953), and in 1969 nesting was first documented in Kentucky in Oldham County (Croft 1972). These records occurred during a period when a southward expansion in breeding range was documented across the region (Croft 1972; Peterjohn and Rice 1991). Since 1969, breeding season reports have originated from nearly a dozen scattered localities in central Kentucky, although confirmed nesting has been documented at only one additional location.

Although Bobolinks have been reported summering at several localities in recent years, they have been confirmed breeding at only two locations in the state. In June 1969 small numbers were discovered nesting in "grainfields" (perhaps hayfields?) on a slightly rolling, upland farm in Oldham County (Croft 1972). This population persisted until the mid-1980s, but subsequently it seems to have disappeared, perhaps because of habitat alteration (Stamm 1987b). Although a nest was never located at this site, adults carrying food were observed on several occasions, and juveniles were seen there on June 21–24, 1972 (Croft 1972). In 1975 Bobolinks were found nesting in rolling, unmowed grasslands at Masterson Station Park in western Fayette County (Kleen 1975). Active nests have been located there on a few occasions, and one contained four young (R. Morris, pers. comm.). As many as "15–18 pairs" have been reported during early June (Stamm and Monroe 1990a), and adults carrying food have been observed on several occasions (e.g., Stamm 1985, 1986a).

The atlas survey further documented the presence of nesting Bobolinks in central Kentucky, yielding five records in priority blocks and three incidental observations. Although the species was confirmed nesting only at Masterson Station Park, apparently territorial birds were reported at four new locations: at least five or six males and some females just northeast and southwest of Centerville, in Bourbon County, on May 31, 1988; at least six males and a few females about two miles southwest of Hutchison, in Bourbon County, on June 1, 1986, and again on May 31, 1988; small groups of two to four males at four scattered localities a few miles southeast of Paris, in Bourbon County, on June 22, 1991; and two males in northern Shelby County, about 3.5 miles southwest of Eminence, on June 16, 1988. The latter observation was included as possible because the site was not revisited. In addition, there was an incidental report of a few pairs of apparently territorial birds west of Danville, in Boyle County, in June and early July 1989 (W. Kemper, pers. comm.; Stamm 1989a). All reports were from extensive hayfields or pastures in open farmland.

Priority block records and incidental reports of single birds observed without further evidence of nesting were made in Woodford County on June 29, 1991, and on a BBS route in Oldham County on June 8, 1986. Other occurrences reported during the atlas fieldwork included late May observations from Carroll, Breathitt, and Morgan counties. These latter records were not included in the final atlas results because of the possibility that the birds were migrants.

Alvin E. Staffan

Transient Bobolinks are conspicuous in hayfields and other open, grassy habitats from late April through mid-May. From the accumulated records, it appears that a few migrants regularly linger through late May, and occasionally into early June. For this reason, early summer records generally have been noted with caution. Detailed information on nesting activity in Kentucky is scarce, but nest building is likely under way by early June, and most clutches are probably completed during the middle of the month. Most young seem to fledge by mid-July, and early fall migrants sometimes appear by mid-August. Because the species nests primarily in hayfields and pastures, it is likely that mowing activities disturb or destroy many nests. The effect of mowing on nesting success is presently unknown, but disappearance of nesting birds following mowing has been noted several times (e.g., Stamm 1989a). Kentucky data on clutch size are lacking, but Harrison (1975) gives rangewide clutch size as 4–7, commonly 5 or 6.

During the breeding season, Bobolinks occur in a variety of grassy habitats. In Kentucky, the species is found entirely in artificial situations, including hayfields, pastures, and other unmowed or infrequently mowed fields of grasses and forbs. Very open situations seem to be favored, as is vegetation that is neither especially thick nor closely mowed or grazed. The nest is typically concealed on or near the ground among the thick growth of grasses and forbs. It is constructed of coarse grasses and weed stalks and is lined with finer grass (Harrison 1975).

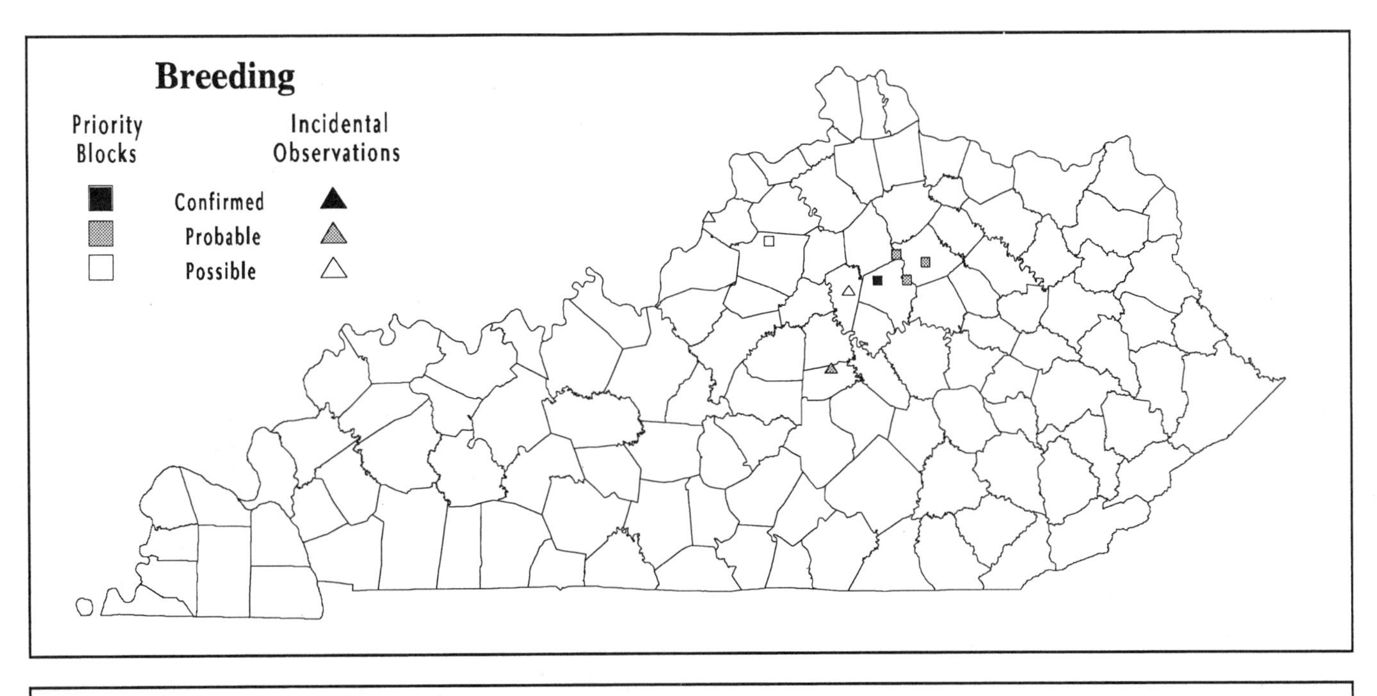

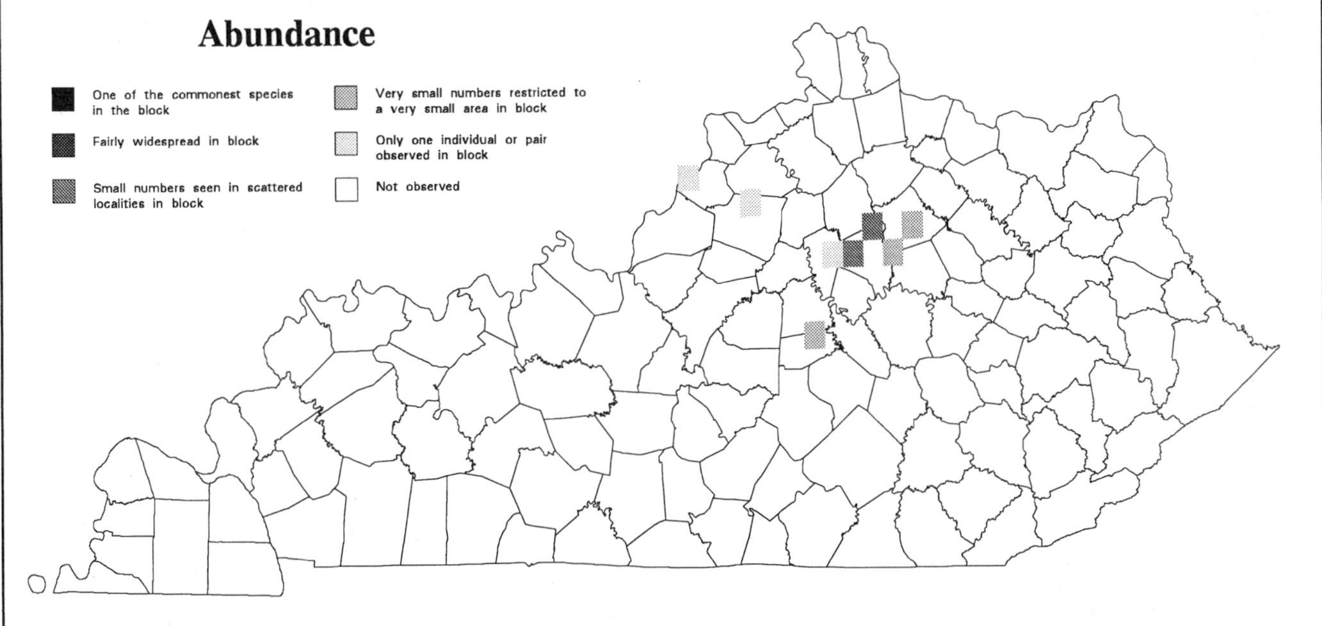

Analysis of Block Data by Physiographic Province Section

Physiographic Province Section	Total Blocks Surveyed	Blocks with Data	% with Data	Section's % for State
Mississippi Alluvial Plain	14	-	-	-
East Gulf Coastal Plain	36	-	-	-
Highland Rim	139	-	-	-
Shawnee Hills	142	-	-	-
Blue Grass	204	5	2.5	100.0
Cumberland Plateau	173	-	-	-
Cumberland Mountains	19	-	-	-

Summary of Breeding Status

Number of Blocks in Which Species Was Recorded		
Total	**5**	**0.7%**
Confirmed	1	20.0%
Probable	3	60.0%
Possible	1	20.0%

Red-winged Blackbird
Agelaius phoeniceus

The Red-winged Blackbird is one of Kentucky's most conspicuous and widely recognized nesting birds. Common in open situations, the species is most numerous in western Kentucky, where wetlands and agricultural land provide an abundance of nesting habitat. Mengel (1965) regarded the Red-winged Blackbird as a common summer resident throughout the state except in the heavily forested portions of eastern Kentucky.

Red-winged Blackbirds are found in a great variety of semi-open and open habitats with an abundance of thick vegetation in which to nest. The species is most common in marshes with rank herbaceous vegetation and shrub swamps dominated by buttonbush, but it is rare to find a cattail-ringed farm pond that does not harbor a pair or two. These blackbirds also can be found far from water in various open situations, including hayfields, fields of small grains such as wheat and rye, utility and roadway corridors, and reclaimed surface mines.

Although human alteration of the landscape has resulted in the loss of some suitable wetland nesting habitat for Red-winged Blackbirds, the species has certainly increased dramatically in Kentucky in the past two centuries. The clearing of vast areas of forest for human use has created an abundance of suitable nesting habitat. This trend is perhaps nowhere more apparent than on the Cumberland Plateau, where conversion of habitat because of coal mining has resulted in the establishment of a substantial nesting population (Allaire 1979).

Although Red-winged Blackbirds occur year-round in Kentucky, their occurrence is somewhat irregular. While some birds move south in winter, others join with birds from farther north to forage in large flocks (Monroe and Cronholm 1977). During the first warm days of late winter, a few males separate from winter flocks and start staking out territories. In areas where Red-wingeds have been absent all winter, males return in late February and early March (Mengel 1965). Females return soon thereafter, and as soon as there is sufficient cover for nesting, nest building commences. Clutch completion has been noted as early as mid-April, with an early peak in nesting activity during the latter part of May. Most birds appear to raise a second brood, and there are numerous records of nesting into July (Mengel 1965). Mengel gave the average size of 29 clutches as 3.2 (range of 2–4).

Forest Cover

Value	% of Blocks	Avg Abund
All	84.5	3.2
1	96.0	3.8
2	98.4	3.7
3	97.8	3.1
4	60.4	2.3
5	23.1	1.7

Red-winged Blackbirds often nest in loose colonies of a few to more than a dozen territorial males, but in optimal habitat as many as 30–35 nests have been found in a relatively small area (Mengel 1965). Males may be polygynous, mating with two or more females within one territory. Red-wingeds typically place their nest quite low and conceal it among the cover of dense vegetation. In marshy areas the species seems to use whatever plants are abundant, including both herbaceous and woody growth. In agricultural situations the bird typically chooses sturdier weeds and grasses. The nest is constructed primarily of grasses, woven together and attached to the supporting vegetation to form a sturdy cup, and is lined with finer grass. The average height of 16 nests reported by Mengel (1965) was 2.4 feet (range of 0.5–3.5 feet), although more recent nests have been reported as high as 15 feet (Stamm and Croft 1968).

Ron Austing

The atlas survey yielded records of Red-winged Blackbirds in nearly 85% of priority blocks, and 14 incidental observations were reported. The species ranked 25th according to the number of records in priority blocks, but 14th by total abundance and 7th by average abundance. Although Red-wingeds are not as widespread as some other common species, they are relatively common and conspicuous where they are present. Occurrence was greater than 94% in all physiographic province sections except the Cumberland Plateau and the Cumberland Mountains, where habitat is limited. Average abundance varied similarly. Occurrence and average abundance were highest in areas with an abundance of cleared habitat and lowest in forested areas.

Nearly 33% of priority block records were for confirmed breeding. Active nests were observed on a number of blocks, but most confirmed records were based on the observation of recently fledged young and adults carrying food. Additional confirmed records involved nest building, used nests, and distraction displays.

Red-winged Blackbirds are reported in substantial numbers on most Kentucky BBS routes. The average number of individuals reported per BBS route for the periods 1966–91 and 1982–91 was 63.99 and 68.89, respectively. According to these data, the species ranked 3rd in abundance on BBS routes during the period 1982–91. Trend analysis of these data reveals a slight, nonsignificant decrease of 0.6% per year for the period 1966–91, but a significant ($p<.01$) decrease of 4.8% per year for the period 1982–91. Reasons for the short-term decrease are not known.

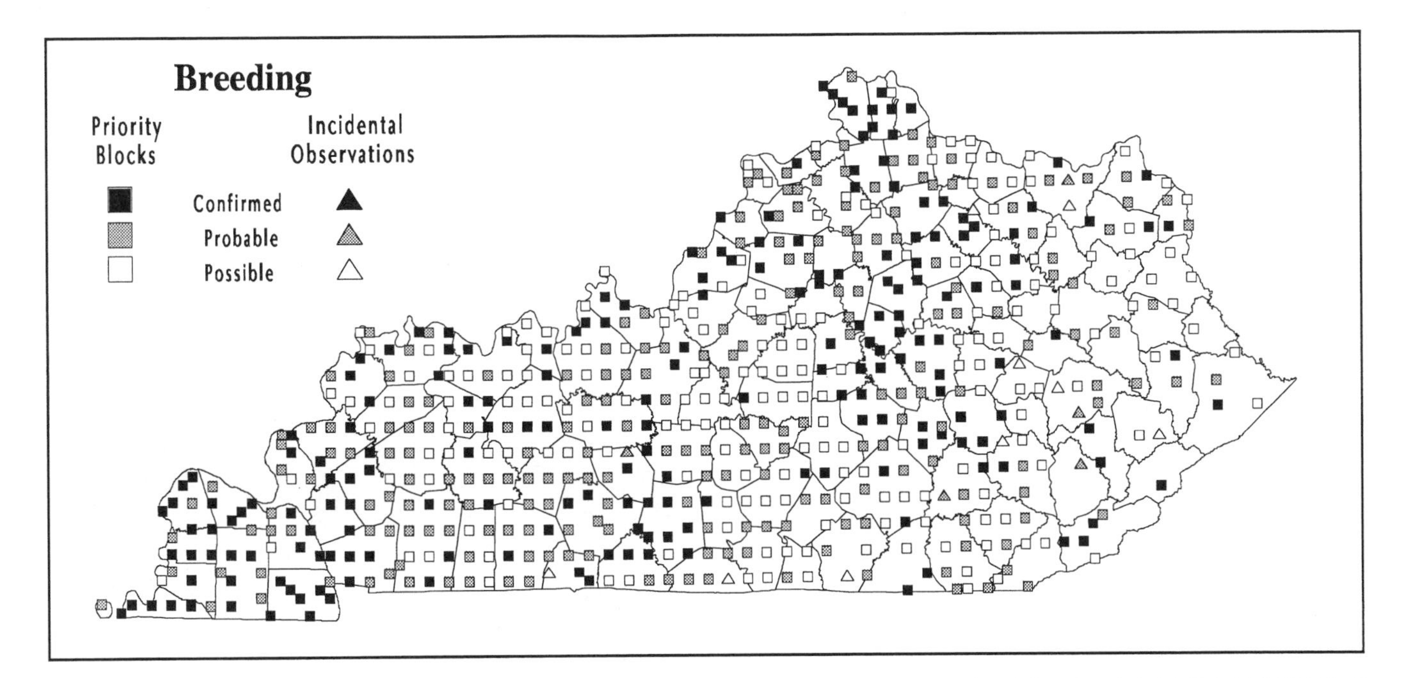

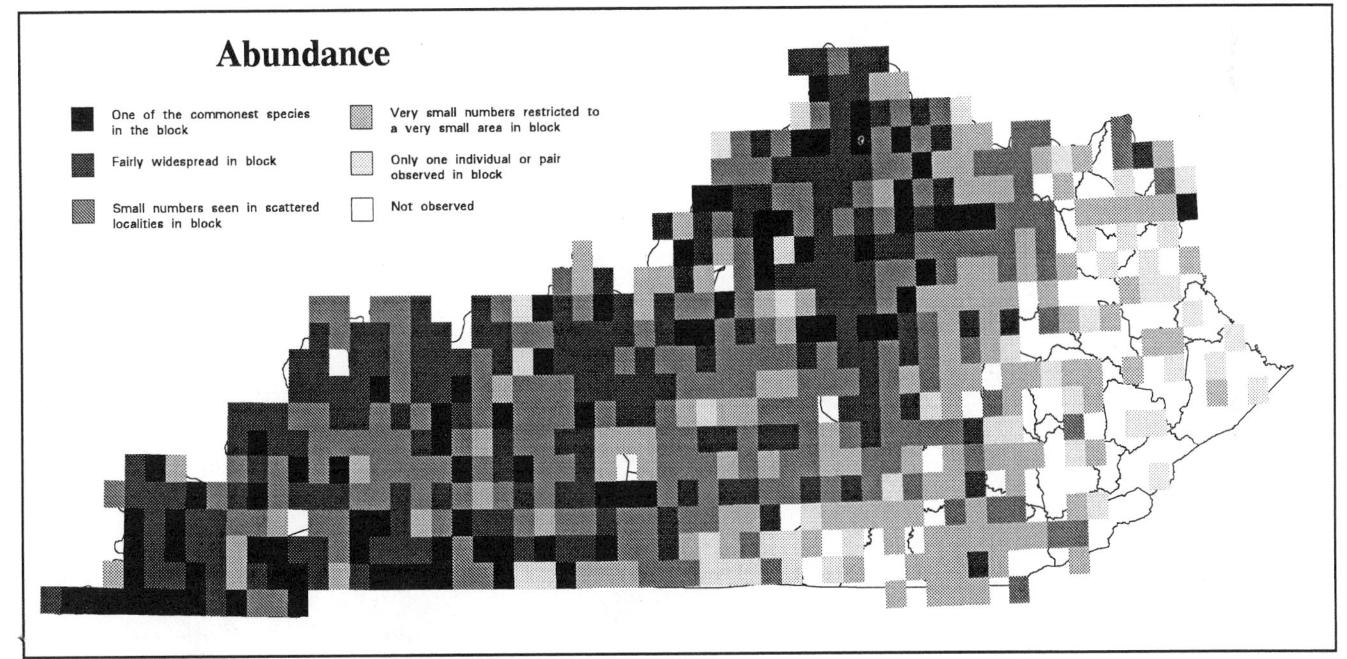

Analysis of Block Data by Physiographic Province Section

Physiographic Province Section	Total Blocks Surveyed	Blocks with Data	Avg Abund	% with Data	Section's % for State
Mississippi Alluvial Plain	14	14	4.2	100.0	2.3
East Gulf Coastal Plain	36	36	4.1	100.0	5.9
Highland Rim	139	134	3.3	96.4	21.8
Shawnee Hills	142	138	3.5	97.2	22.5
Blue Grass	204	193	3.4	94.6	31.4
Cumberland Plateau	173	90	2.0	52.0	14.7
Cumberland Mountains	19	9	2.4	47.4	1.5

Summary of Breeding Status

Number of Blocks in Which Species Was Recorded		
Total	**614**	**84.5%**
Confirmed	200	32.6%
Probable	225	36.6%
Possible	189	30.8%

Red-winged Blackbird

Eastern Meadowlark

Sturnella magna

The Eastern Meadowlark's musical song is a widespread summer sound throughout much of Kentucky. Although limited in occurrence in eastern Kentucky by the scarcity of grassland nesting habitat, the species still ranks among the state's most abundant nesting birds. Mengel (1965) regarded the meadowlark as a fairly common to common resident year-round, nesting in appropriate habitat statewide.

The Eastern Meadowlark is a bird of open, grassy habitats. Although the species sometimes nests in tall, thick grasses, shorter and sparser vegetation is used primarily for foraging. In the absence of natural grasslands today, meadowlarks are now primarily restricted to altered habitats. They commonly nest in hayfields and lightly grazed pastures, on airports and reclaimed surface mines, along roadway corridors, and in a variety of other grassy situations in which disturbance is infrequent enough to allow successful nesting.

Along with several other birds that nest in open, grassy habitats, the Eastern Meadowlark appears to have increased in Kentucky as a result of human alteration of the landscape. Audubon (1861) noted meadowlarks in the barrens during the early 1800s, and they may have occurred throughout the native grasslands of central and western Kentucky. While some of the original prairies and savannas have been replaced by settlement and row-crop fields, many areas have been converted to hayfields and pastures that simulate naturally occurring grasslands. Moreover, the conversion of vast areas of forest to agricultural use and settlement has resulted in the creation of an abundance of suitable nesting habitat in areas where formerly the species was excluded.

Although the Eastern Meadowlark is a permanent resident in Kentucky, some birds apparently move south in winter, and those that remain usually gather into flocks (Mengel 1965). Song is not infrequent on bright winter days, and the first warm days of early spring elicit territorial behavior. Nests containing eggs have been observed as early as April 22 and as late as July 31, with apparent peaks in clutch completion occurring in early May and late June (Mengel 1965). Later nestings have been reported, and young have been observed in the nest as late as the second week of August (Stamm 1961a). Some late nests likely represent the raising of a second brood, but because meadowlarks commonly nest in hayfields and pastures, many nests are probably destroyed by mowing. Mengel (1965) gives the average size of 16 clutches as 4.2 (range of 3–5).

Forest Cover

Value	% of Blocks	Avg Abund
All	80.5	3.3
1	96.0	3.9
2	99.5	3.7
3	93.0	3.2
4	52.6	2.0
5	7.7	1.3

Meadowlarks usually nest well out in the open, but territories are occasionally chosen within fairly small patches or corridors of grasses in otherwise closed-in situations. The nest is usually placed in a slight depression in the ground, often beneath a hummock of grass, well concealed by overhanging vegetation. It is constructed of dead grass, woven into the surrounding vegetation, and domed over to create a side entrance.

Alvin E. Staffan

The atlas survey yielded records of Eastern Meadowlarks in almost 81% of priority blocks statewide, and 11 incidental observations were reported. The species ranked 32nd according to the number of priority block records, but 18th by total abundance and 4th by average abundance. Although meadowlarks are not as widespread as some other common species, they are relatively common and conspicuous where they are present. Occurrence and average abundance were much lower in the Cumberland Plateau and the Cumberland Mountains than in central and western Kentucky. These trends are related to the amount of available habitat, which generally is inversely related to the degree of forestation.

Although Eastern Meadowlarks were widely distributed, only about 21% of priority block records were for confirmed breeding. Observers located several active nests, but most confirmed records were based on the observation of recently fledged young and adults carrying food. Other confirmed records involved the observation of nest building and distraction displays.

Eastern Meadowlarks are reported in substantial numbers on most Kentucky BBS routes. The average number of individuals per BBS route for the periods 1966–91 and 1982–91 was 52.04 and 45.83, respectively. According to these data, the species ranked 7th in abundance on BBS routes during the period 1982–91. Trend analysis of these data shows a significant ($p<.01$) decrease of 2.1% per year for the period 1966–91 and a significant ($p<0.1$) decrease of 2.4% per year for the period 1982–91. Reasons for these declines are not completely understood, but the long-term decrease is likely related to losses sustained during the severe winters of the late 1970s (Monroe 1978; Robbins et al. 1986).

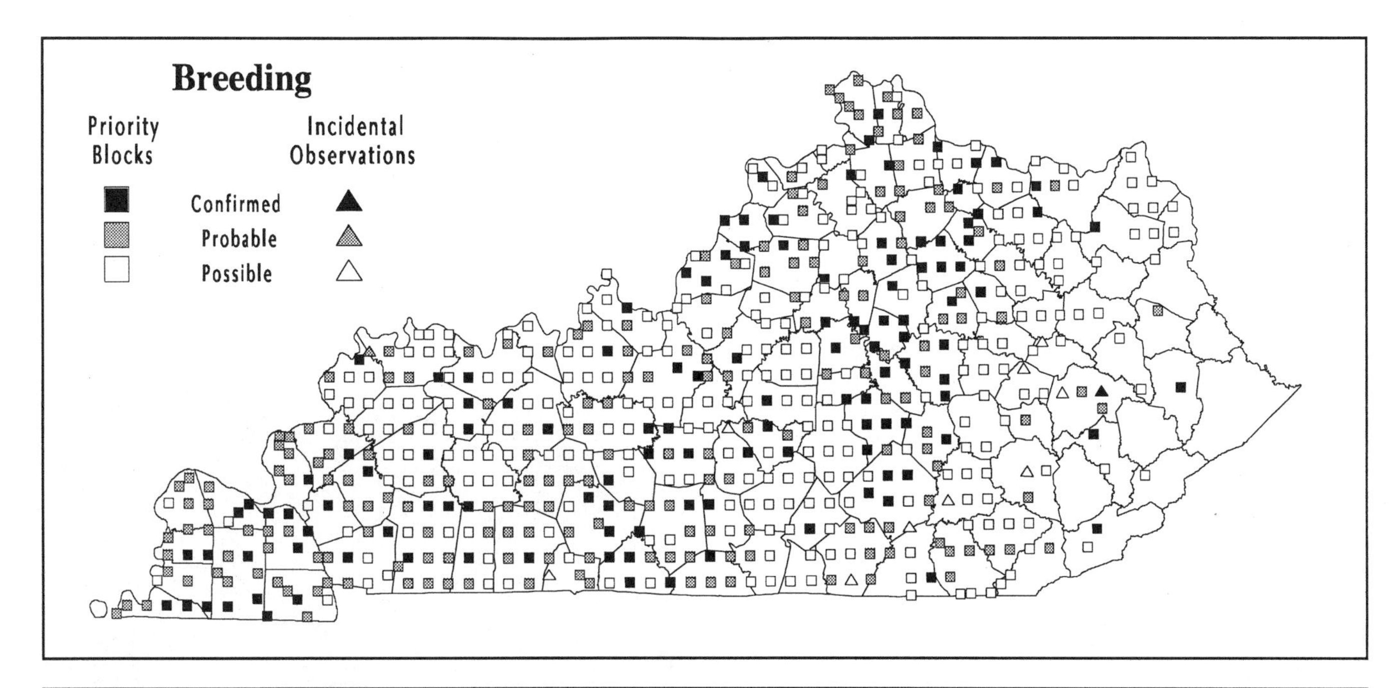

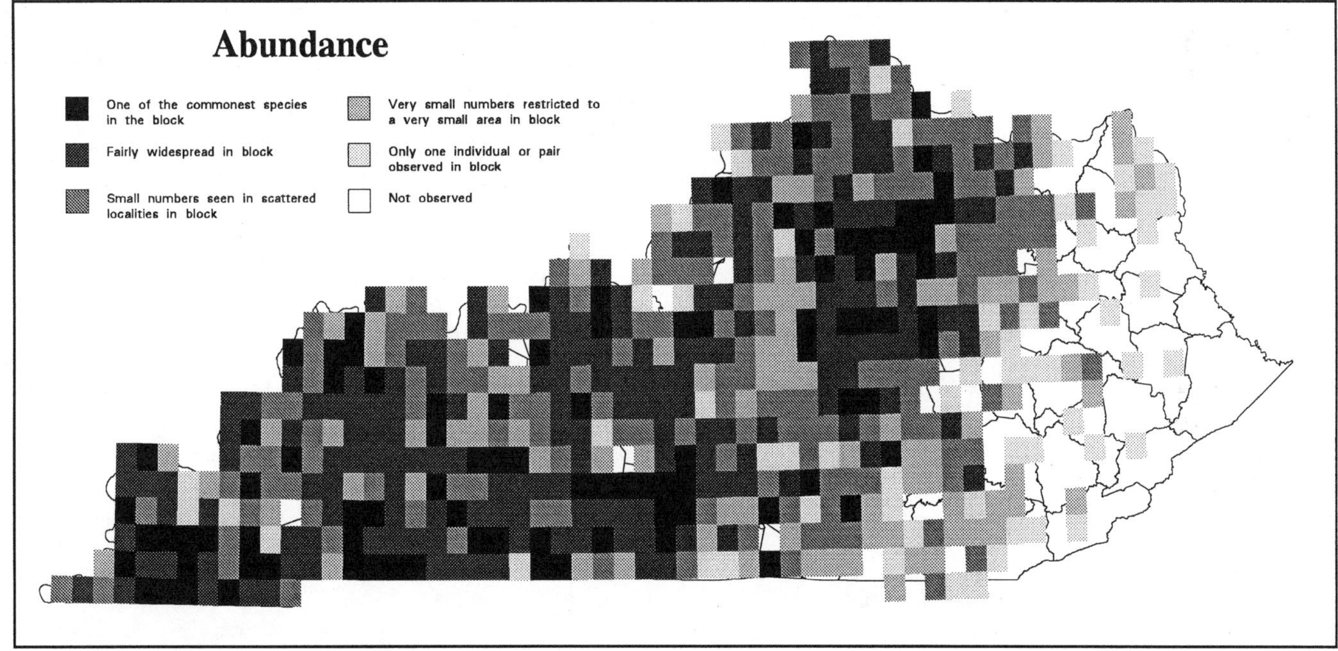

Analysis of Block Data by Physiographic Province Section

Physiographic Province Section	Total Blocks Surveyed	Blocks with Data	Avg Abund	% with Data	Section's % for State
Mississippi Alluvial Plain	14	12	3.6	85.7	2.0
East Gulf Coastal Plain	36	36	3.8	100.0	6.1
Highland Rim	139	135	3.6	97.1	23.1
Shawnee Hills	142	135	3.5	95.1	23.1
Blue Grass	204	193	3.3	94.6	33.0
Cumberland Plateau	173	69	1.9	39.9	11.8
Cumberland Mountains	19	5	1.6	26.3	0.9

Summary of Breeding Status

Number of Blocks in Which Species Was Recorded		
Total	**585**	**80.5%**
Confirmed	124	21.2%
Probable	200	34.2%
Possible	261	44.6%

Eastern Meadowlark

Common Grackle

Quiscalus quiscula

Although largely absent from heavily forested areas, the Common Grackle is a conspicuous summer bird throughout much of Kentucky where humans have settled. Mengel (1965) regarded the species as a common summer resident throughout most of the state, being local and rare to uncommon only in southeastern Kentucky.

Grackles are found in a great variety of semi-open and open habitats. The species probably is most common in rural farmland, but it is also generally distributed in other altered situations, including rural towns, roadway corridors, the margins of reservoirs, and suburban parks and yards. Although these large blackbirds primarily occupy artificial situations, they are sometimes encountered in naturally occurring habitats, including the margins of open wetlands and riparian zones along the larger rivers.

The Common Grackle has certainly increased in abundance in Kentucky in response to human alteration of the landscape. Although the species likely occurred at least locally in prairies, savannas, large river corridors, and wetlands, it probably was not widely distributed two centuries ago. The conversion of vast areas of forest to agricultural use and settlement has resulted in the creation of an abundance of suitable nesting habitat. Furthermore, most of the naturally occurring habitats that supported grackles have been converted to situations that afford as much potential for nesting as was present earlier.

Although Common Grackles are present year-round, many nesting birds move south in winter, and those that remain usually gather in large flocks with other blackbirds to feed and roost (Monroe and Cronholm 1977). These flocks begin to break up as soon as early spring arrives, and by the middle of March many locally nesting grackles have returned to their breeding grounds. Territorial behavior commences immediately upon arrival, and nest building may be under way by the end of March (Jones 1969). Egg laying has been noted by early April, with a peak in clutch completion by mid-April (Jones 1969). It is not known whether or not the species is double-brooded in Kentucky, but most late nestings are initiated by late May (Mengel 1965). In an extensive study of nesting biology in Franklin County in 1968, Jones (1969) found the average size of 38 clutches to be 4.7 (range of 3–6).

Forest Cover

Value	% of Blocks	Avg Abund
All	81.2	3.1
1	99.0	3.7
2	96.3	3.4
3	88.3	3.1
4	56.5	2.3
5	30.8	1.9

Grackles typically construct open nests in trees and shrubs. Evergreens, especially red cedars, are most frequently chosen, but a variety of deciduous trees are also used (Jones 1969). Less frequently the species uses cavities, including nest boxes, natural tree hollows, and nooks and crannies in buildings and bridges. Although individual pairs sometimes nest alone, grackles often nest in loose colonies, which sometimes consist of up to 25 or more pairs (Mengel 1965; Jones 1969).

The nest is a bulky structure composed of sticks, dead grass, weed stalks, and other debris. An inner layer of mud is often added to the nest before it is lined with finer grass (Jones 1969). The nest is usually placed in an upright fork, well inside the outer crown of foliage, or constructed to fill a cavity or recess. Mengel (1965) gave the average height of six nests as 20 feet (range of 15-30 feet), while Jones (1969) found the average height of 90 nests studied in Franklin County to be 12.5 feet (range of 2.5–35.0 feet).

Alvin E. Staffan

The atlas survey yielded records of Common Grackles in more than 81% of priority blocks statewide, and 16 incidental observations were reported. The species ranked 31st according to the number of priority block records, 20th by total abundance, and 12th by average abundance. Occurrence was greater than 89% in all physiographic province sections except the Cumberland Mountains and the Cumberland Plateau. Average abundance varied similarly, being considerably lower in eastern Kentucky. Both occurrence and average abundance were inversely related to the amount of forest cover, being highest in open areas and decreasing as forestation increased.

Nearly 45% of priority block records were for confirmed breeding. Although several active nests were located, most of the 263 confirmed records were based on the observation of recently fledged young. Juvenile grackles beg loudly for the first few weeks following fledging, and they were easily detected during atlas fieldwork. Other confirmed records involved the observation of adults carrying food and fecal sacs, nest building, and used nests.

Common Grackles are recorded in substantial numbers on many Kentucky BBS routes. The average number of individuals reported per BBS route for the periods 1966–91 and 1982–91 was 125.50 and 97.33, respectively. Mostly as a result of its abundance in rural farmland, where most BBS routes are run, the species ranked 1st in abundance on BBS routes during the period 1982–91. Trend analysis of these data reveals significant ($p<.01$) decreases of 2.6% per year for the period 1966–91 and 5.0% per year for the period 1982–91. Reasons for these declines are not completely understood, but they may involve severe winter weather and blackbird control operations (Robbins et al 1986).

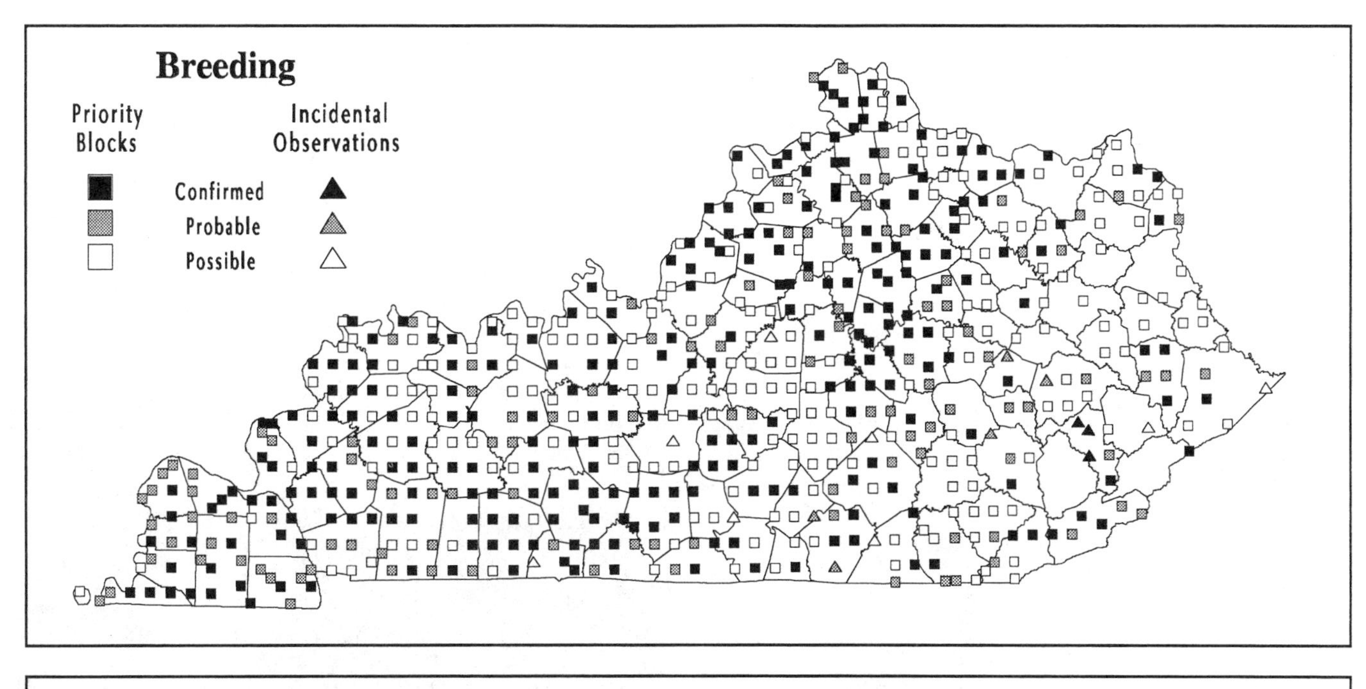

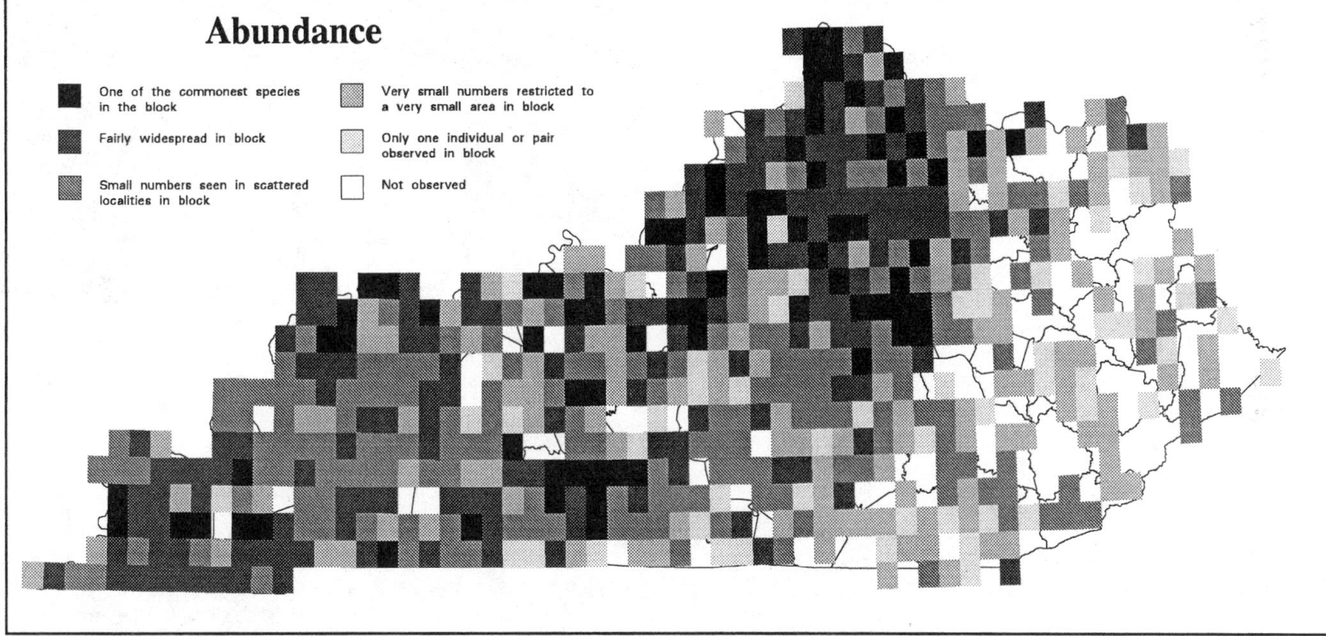

Analysis of Block Data by Physiographic Province Section

Physiographic Province Section	Total Blocks Surveyed	Blocks with Data	Avg Abund	% with Data	Section's % for State
Mississippi Alluvial Plain	14	14	3.5	100.0	2.4
East Gulf Coastal Plain	36	35	3.7	97.2	5.9
Highland Rim	139	124	3.1	89.2	21.0
Shawnee Hills	142	129	3.2	90.8	21.9
Blue Grass	204	189	3.5	92.6	32.0
Cumberland Plateau	173	88	2.1	50.9	14.9
Cumberland Mountains	19	11	2.4	57.9	1.9

Summary of Breeding Status

Number of Blocks in Which Species Was Recorded		
Total	**590**	**81.2%**
Confirmed	263	44.6%
Probable	121	20.5%
Possible	206	34.9%

Common Grackle

Brown-headed Cowbird

Molothrus ater

The Brown-headed Cowbird is a fairly common to common summer resident across much of Kentucky, especially where the land has been settled or cleared for farming. Mengel (1965) regarded the species as common west of the Cumberland Plateau, uncommon on the Plateau, and apparently absent from the higher elevations of the Cumberland Mountains. The cowbird is unique among Kentucky's nesting birds in that it is a brood parasite, laying its eggs in the nests of other songbirds so that they will raise its young.

Cowbirds are encountered in a great variety of semi-open and open habitats. The species is most abundant in rural farmland, but substantial numbers can be found in suburban parks and yards, rural towns, roadway and utility corridors, and other habitats with short or sparse vegetation. These small blackbirds typically avoid deep forest except to parasitize nests, but small numbers range widely during the nesting season. In areas where corridors and openings dissect large tracts of forest, they may be quite numerous. For example, Claus et al. (1988) found cowbirds to be relatively common on reclaimed mines on the Cumberland Plateau.

The Brown-headed Cowbird may have occurred locally in Kentucky before settlement, especially in association with the great herds of bison that roamed across the native prairies and savannas. The species was probably excluded from the remainder of the state by the extent of forestation. Subsequent conversion of the land to agricultural use and settlement has benefited the cowbird greatly. In fact, the clearing of vast areas of forest and the introduction of livestock have more than made up for the loss of native grasslands and grazing animals, and the cowbird is certainly more common and widespread today than ever before.

Although Brown-headed Cowbirds are present in Kentucky throughout the year, most nesting birds move south in winter (Dolbeer 1982). Those that remain join with other blackbirds to forage and roost in large flocks; for this reason the species is rare or absent throughout much of the state in winter. When the large blackbird flocks break up during March, cowbirds return to breeding areas and initiate courtship behavior. Egg laying commences at least occasionally by early April and extends into early July, with a peak during May (Mengel 1965; Stamm and Croft 1968). Females apparently observe host songbirds during nest building and typically replace the victims' eggs with their own during the hosts' laying period. Female cowbirds may lay up to 40 eggs in a single season, parasitizing a number of nests (Ehrlich et al. 1988).

Forest Cover

Value	% of Blocks	Avg Abund
All	83.9	2.6
1	93.1	2.8
2	95.3	2.8
3	87.8	2.7
4	66.9	2.3
5	57.7	1.7

Cowbirds parasitize a variety of resident and migratory songbirds in Kentucky. Mengel (1965) listed 25 species that were known to have served as hosts. To these, more recent publications have added Gray Catbird, Red-winged Blackbird, and Orchard Oriole (Stamm and Croft 1968), Eastern Wood-Pewee (Stamm 1970), and Louisiana Waterthrush (Stamm 1980). During the atlas fieldwork at least 21 species were documented as hosts, including most of those noted above as well as Bell's Vireo, Yellow-throated Warbler, and Worm-eating Warbler. In suburban settings, cowbirds probably parasitize Northern Cardinals and Song Sparrows most frequently. In rural farmland and settlement, species such as Warbling and White-eyed vireos, Prairie Warbler, Indigo Bunting, and Chipping and Field sparrows are frequent hosts. In woodland areas, Acadian Flycatcher, Wood Thrush, Red-eyed and Yellow-throated vireos, several species of warblers, and Scarlet and Summer tanagers are most often victimized.

Hal H. Harrison

The atlas survey yielded records of Brown-headed Cowbirds in nearly 84% of priority blocks statewide, and nine incidental observations were reported. The species ranked 27th by the number of records in priority blocks, 29th by total abundance, and 28th by average abundance. Occurrence was greater than 86% in all physiographic province sections except the Cumberland Mountains and the Cumberland Plateau, but even across eastern Kentucky cowbirds were found in more than 60% of priority blocks. Average abundance varied similarly. Both occurrence and average abundance were highest in predominantly open areas and decreased as the percentage of forest cover increased.

Cowbirds were confirmed breeding in more than 30% of priority blocks statewide. Most of the 184 confirmed records were based on the observation of host species feeding recently fledged cowbirds. The remaining confirmed records involved nests of host species containing cowbird eggs or young.

Brown-headed Cowbirds are typically reported in moderate numbers on Kentucky BBS routes. The average number of individuals reported per BBS route for the periods 1966–91 and 1982–91 was 13.49 and 14.17, respectively. According to these data, the species ranked 18th in abundance on BBS routes during the period 1982–91. Trend analysis of these data reveals a nonsignificant increase of 1.0% per year for the period 1966–91, but a significant ($p<.01$) increase of 7.5% per year for the period 1982–91.

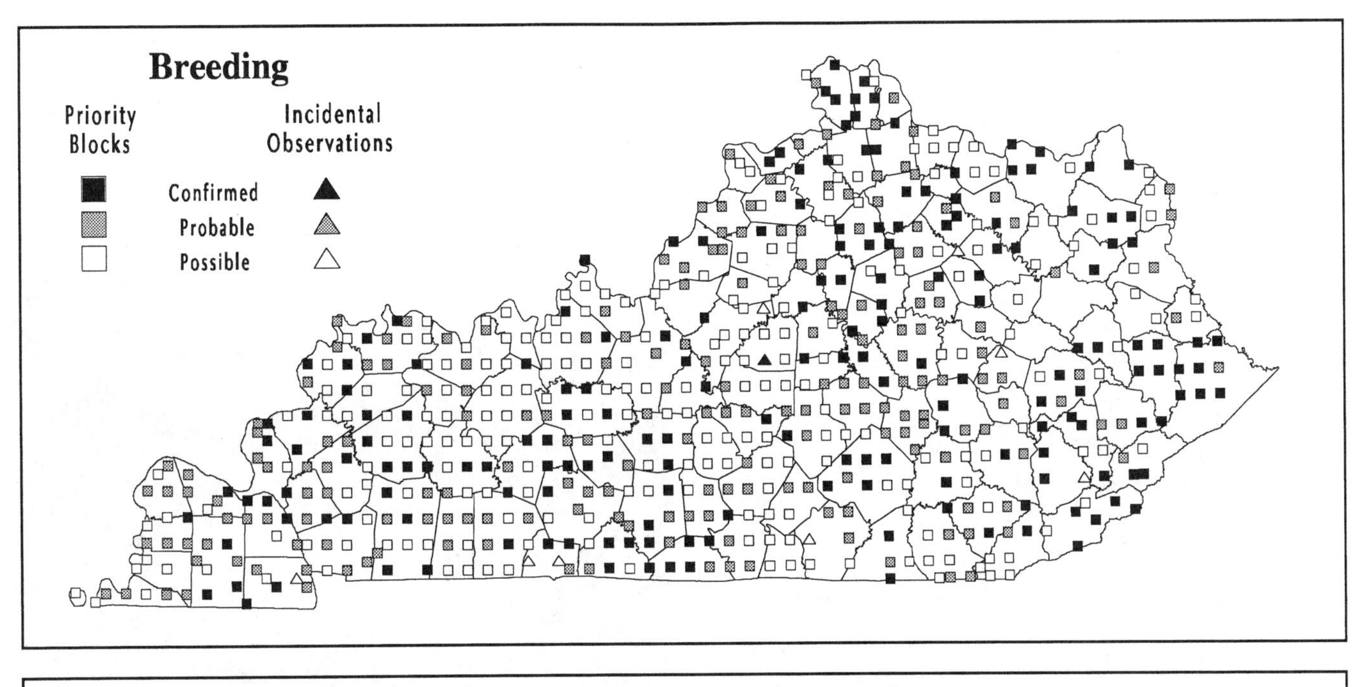

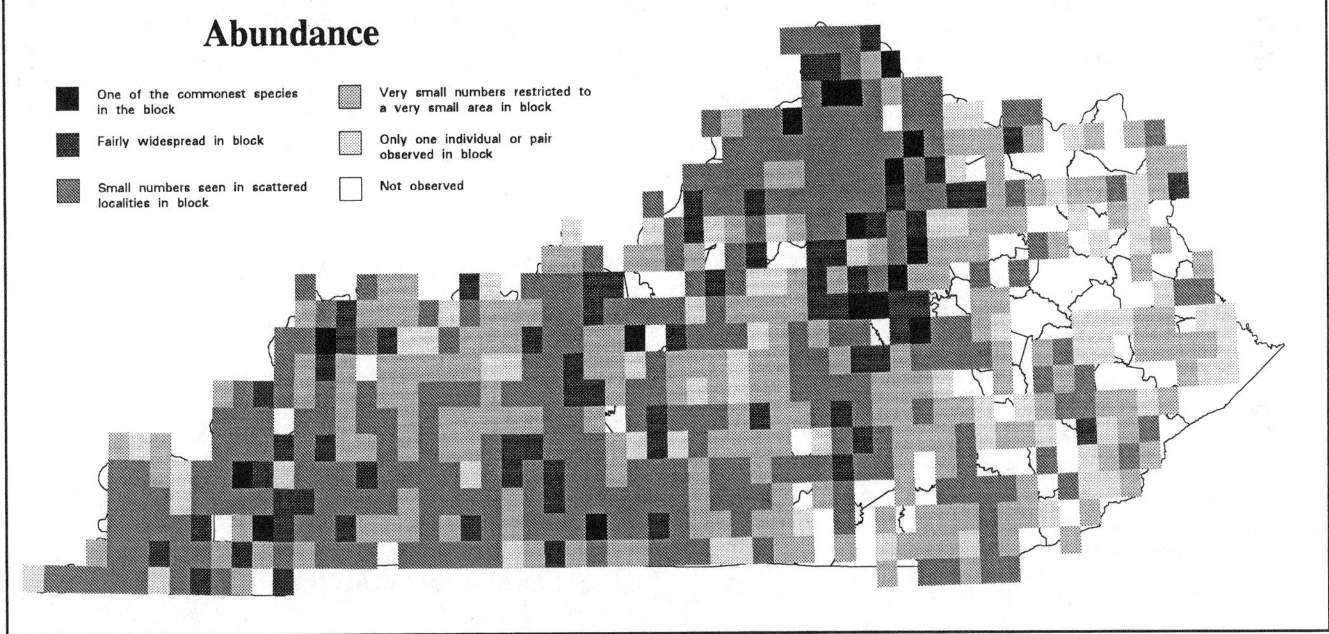

Analysis of Block Data by Physiographic Province Section

Physiographic Province Section	Total Blocks Surveyed	Blocks with Data	Avg Abund	% with Data	Section's % for State
Mississippi Alluvial Plain	14	13	2.7	92.9	2.1
East Gulf Coastal Plain	36	33	2.9	91.7	5.4
Highland Rim	139	129	2.8	92.8	21.1
Shawnee Hills	142	137	2.6	96.5	22.5
Blue Grass	204	176	2.9	86.3	28.9
Cumberland Plateau	173	109	2.1	63.0	17.9
Cumberland Mountains	19	13	2.1	68.4	2.1

Summary of Breeding Status

Number of Blocks in Which Species Was Recorded		
Total	**610**	**83.9%**
Confirmed	184	30.2%
Probable	201	32.9%
Possible	225	36.9%

Orchard Oriole

Icterus spurius

The spirited song of the Orchard Oriole is a widespread and surprisingly common summer sound across most of Kentucky. Mengel (1965) regarded the species as a common summer resident statewide except in the heavily forested portions of eastern Kentucky, where it was rather local and somewhat less numerous.

The Orchard Oriole is a bird of semi-open to open habitats with scattered trees. Although successional areas seem to be favored, large trees and mature forest edge are also used. The species is most frequent in rural areas, where it inhabits farmsteads, pastures, fencerows, roadway and utility corridors, and abandoned fields that are being invaded by young trees. A variety of other habitats are also used, including reclaimed surface mines, the margins of reservoirs, and suburban parks and yards. Although this oriole is primarily a bird of artificially created habitats today, it is occasionally found in natural openings, such as the riparian zones of rivers and streams and the shrubby borders of wetlands.

It is likely that Orchard Orioles were more locally distributed and less common in Kentucky before altered habitats became available. Audubon (1861) indicated that the species occurred in the valley of the Ohio River in the early 1800s, and considerable numbers may have occurred in naturally open situations across the state. Mature forest habitats, however, excluded the species from most regions, and clearing and fragmentation of forests have resulted in the creation of an abundance of suitable open and semi-open habitats. Consequently, the species probably occurs more widely across the state today than at any time.

Orchard Orioles winter primarily in the tropics (AOU 1983), and the first males typically return during the middle of April. Aggressive territorial behavior commences soon thereafter, and nest building is probably initiated by early May. According to Mengel (1965), clutches are completed from mid-May to the end of June, with a peak in clutch completion in late May. The species is apparently single-brooded. Family groups are conspicuous during June and early July, but these orioles become uncommon by early August, and they typically disappear by the end of the month. The average size of five clutches or broods reported by Mengel (1965) was 3.3 (range of 2–4).

Forest Cover

Value	% of Blocks	Avg Abund
All	72.8	2.2
1	77.2	2.1
2	89.5	2.3
3	80.4	2.3
4	53.2	2.0
5	28.8	1.4

Orchard Orioles generally avoid mature forest during the nesting season, although woodland edge is sometimes used. Nest trees are more typically situated alone or are scattered about yards and fields, or along fencerows. Many deciduous tree species are used, and a marked preference is not evident. Fine grass or other material is intricately woven together to form a neat, basket-shaped nest, which is suspended between twigs near the tips of branches so that it is concealed among the outer crown of leaves. The average height of 11 nests reported by Mengel (1965) was 29 feet (range of 10–60 feet), although more recently, nests have been reported only 6 feet (Croft and Stamm 1967) and 7 feet (KBBA data) above the ground.

Ron Austing

The atlas survey yielded records of Orchard Orioles in nearly 73% of priority blocks statewide, and 20 incidental observations were reported. Occurrence was greater than 75% across all of central and western Kentucky, but less than 50% in the Cumberland Plateau and the Cumberland Mountains. Average abundance was highest in the Mississippi Alluvial Plain and the East Gulf Coastal Plain and decreased eastward. Occurrence and average abundance were highest in open areas with at least some scattered forest, and lowest in predominantly forested areas.

About 21% of priority block records were for confirmed breeding. Although a number of active nests were located, most of the 113 confirmed records were based on the observation of recently fledged young. Adult orioles were also observed feeding young cowbirds on at least two occasions. Other confirmed records involved the observation of adults carrying food for young and nest building.

Orchard Orioles are reported in small to moderate numbers on most Kentucky BBS routes. The average number of individuals per BBS route for the periods 1966–91 and 1982–91 was 6.55 and 7.45, respectively. Trend analysis of these data reveals a nonsignificant increase of 1.1% per year for the period 1966–91, but a significant (p<.01) decrease of 3.6% per year for the period 1982–91.

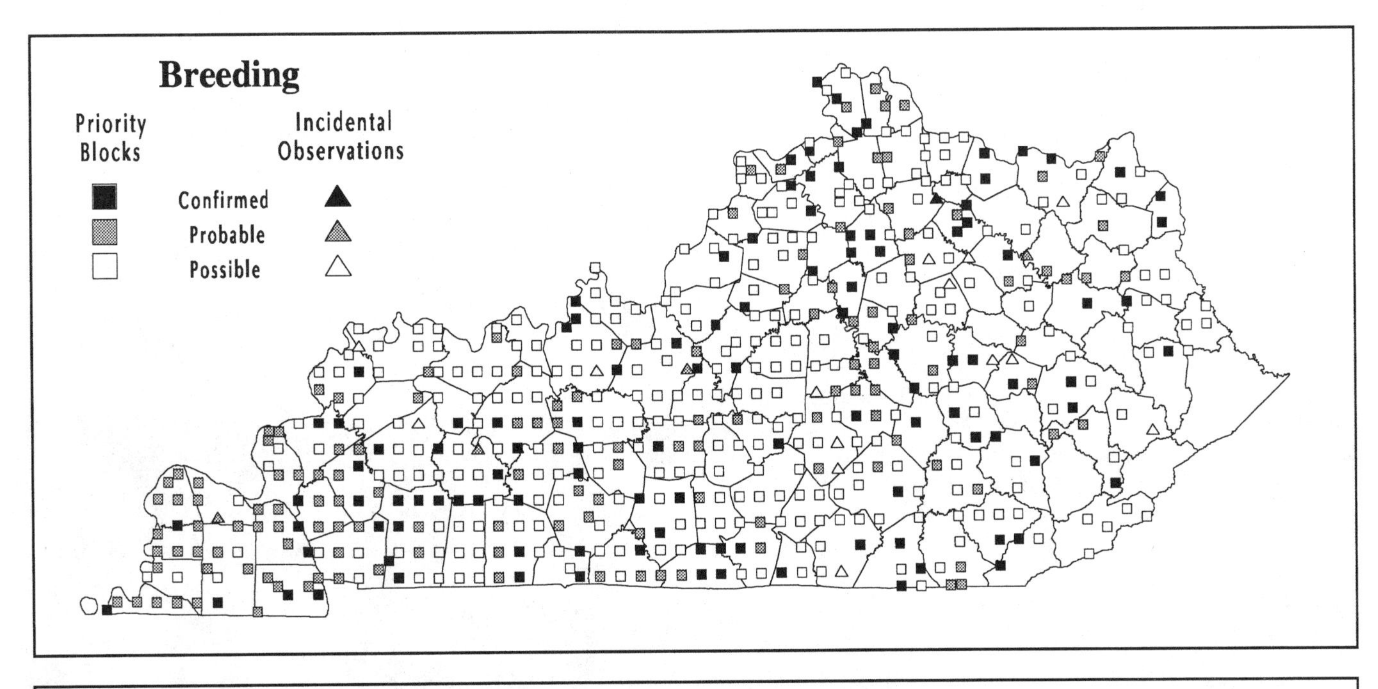

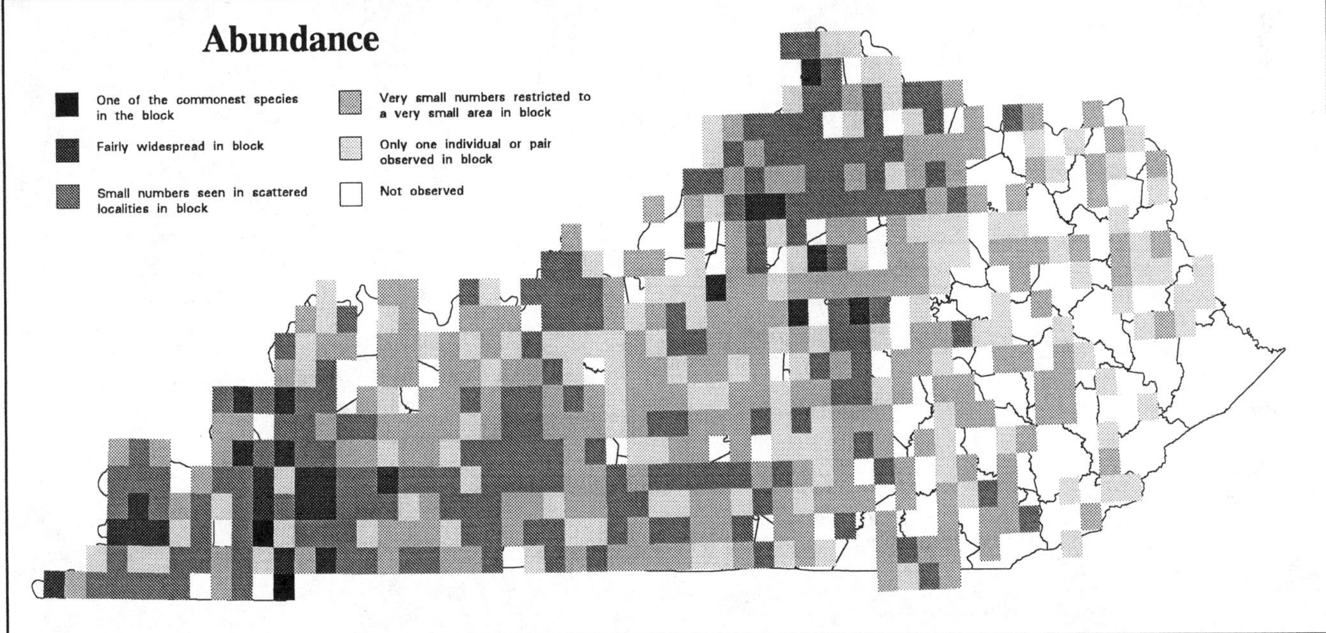

Analysis of Block Data by Physiographic Province Section

Physiographic Province Section	Total Blocks Surveyed	Blocks with Data	Avg Abund	% with Data	Section's % for State
Mississippi Alluvial Plain	14	12	2.8	85.7	2.3
East Gulf Coastal Plain	36	31	2.8	86.1	5.9
Highland Rim	139	128	2.2	92.1	24.2
Shawnee Hills	142	122	2.3	85.9	23.1
Blue Grass	204	156	2.2	76.5	29.5
Cumberland Plateau	173	74	1.7	42.8	14.0
Cumberland Mountains	19	6	1.3	31.6	1.1

Summary of Breeding Status

Number of Blocks in Which Species Was Recorded		
Total	**529**	**72.8%**
Confirmed	113	21.4%
Probable	139	26.3%
Possible	277	52.4%

Northern Oriole

Icterus galbula

Although the Northern Oriole's nesting range is typically considered to include all of Kentucky, the species occurs very locally throughout a large part of the state during the breeding season. Well distributed throughout the Blue Grass and portions of western Kentucky, the species is absent from most of the eastern and southern parts of the state. Mengel (1965) regarded this oriole as occurring statewide, but somewhat locally, and rare to uncommon in abundance.

The Northern Oriole is a bird of semi-open to open habitats with at least a scattering of large trees. The species typically avoids closed-canopy forest and is most conspicuous in rural areas where breeding pairs inhabit farmsteads, pastures, woodland borders, and stream corridors. These orioles also occur in other altered situations, including city parks, suburban yards, and the margins of reservoirs. Although the species is primarily a bird of artificial habitats today, substantial numbers occur locally along naturally open riparian corridors of larger rivers.

Prior to the 19th century this oriole likely was distributed very locally in Kentucky due to the extent of forestation. Audubon (1861) recorded the species along the Ohio River in the early 1800s, and substantial numbers may have occurred in open woodlands and along riparian corridors, especially in central and western Kentucky. After settlement the Northern Oriole likely became more widespread and common as forested areas were cleared and fragmented. Evidence summarized by Mengel (1965) suggests that the species was more numerous in the general region in the 1800s than today. In Kentucky, declines in abundance were noted during the first half of the 20th century in Calloway, Hopkins, and Warren counties (Mengel 1965). Reasons for these decreases are unknown.

Northern Orioles winter primarily in the tropics (AOU 1983), and they usually return to Kentucky during the last two weeks of April. Upon arrival, males begin noisily staking out their territories, and nest building has been noted before the beginning of May (Mengel 1965). Most clutches appear to be completed during the middle of May, and young are most conspicuous in nests, due to their calling, during the first half of June. After the young fledge, family groups forage in the vicinity for a few weeks, but most birds begin departing by early August, and the species is rare by mid-September. Kentucky data on clutch size are nearly absent, but Stamm (1969) included a report of a clutch of four eggs observed in Hopkins County in 1968. Harrison (1975) gives rangewide clutch size as typically 4, often 5, rarely 6.

Forest Cover

Value	% of Blocks	Avg Abund
All	35.5	1.7
1	48.5	1.8
2	51.1	1.7
3	42.2	2.2
4	9.7	1.5
5	—	—

Northern Orioles usually nest in large deciduous trees. They do not nest in forest interior, instead favoring trees that stand alone or are clustered loosely with others in open areas, in fencerows, or along woodland borders or streams. A variety of trees are used, but nests have been found most frequently in elms, sycamores, and willows (Mengel 1965). Nests are typically situated far out near the tips of higher branches. Fine grass and sometimes string or fishing line are intricately woven into a deep basket that is suspended, saclike, from two or more branches. The nest is lined with finer materials, including hair, fine grass, and cottony plant material (Harrison 1975). According to Mengel (1965), most nests seem to be built 25–40 feet above the ground.

Alvin E. Staffan

The atlas survey yielded records of Northern Orioles in nearly 36% of priority blocks, and 22 incidental observations were reported. Occurrence was greater than 50% only in the Mississippi Alluvial Plain, where the species occurs along riparian corridors in considerable numbers, and in the Blue Grass, where open farmland provides optimal habitat. A few Northern Orioles are probably present in the Cumberland Mountains, but none were recorded there during the atlas fieldwork. Average abundance was relatively low in all parts of the state. Occurrence was highest in areas with at least a good mix of open areas and forest, and it decreased substantially as the percentage of forest cover increased.

Less than 19% of priority block records were for confirmed breeding. A number of active nests were located, but the contents of none were determined. Other confirmed records were based on the observation of recently fledged young, adults carrying food, nest building, and used nests.

Northern Orioles are typically reported in small numbers on most Kentucky BBS routes. The average number of individuals per BBS route for the periods 1966–91 and 1982–91 was 1.23 and 2.40, respectively. Trend analysis of these data reveals significant increases of 3.7% per year for the period 1966–91 ($p<.01$) and 3.8% per year for the period 1982–91 ($p<0.1$). Reasons for these increases are not known, but they correspond with an increase that has occurred across much of the species's range (Robbins et al. 1986).

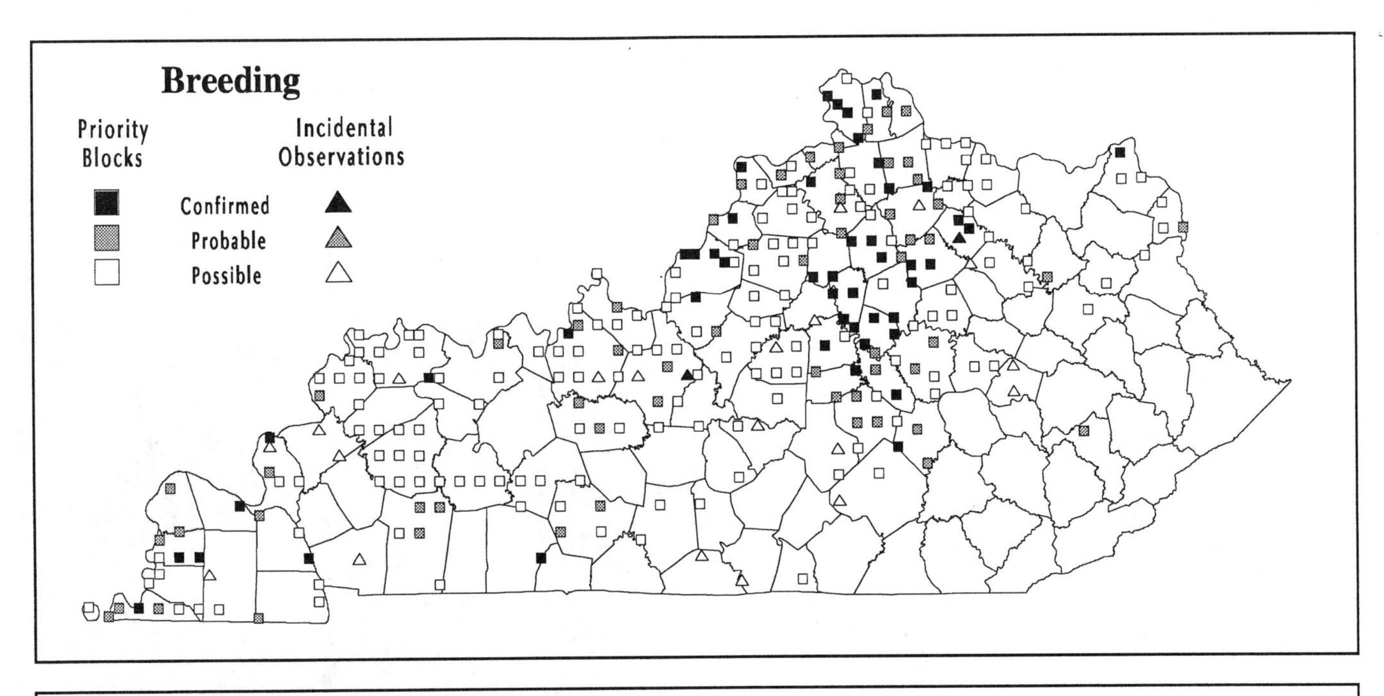

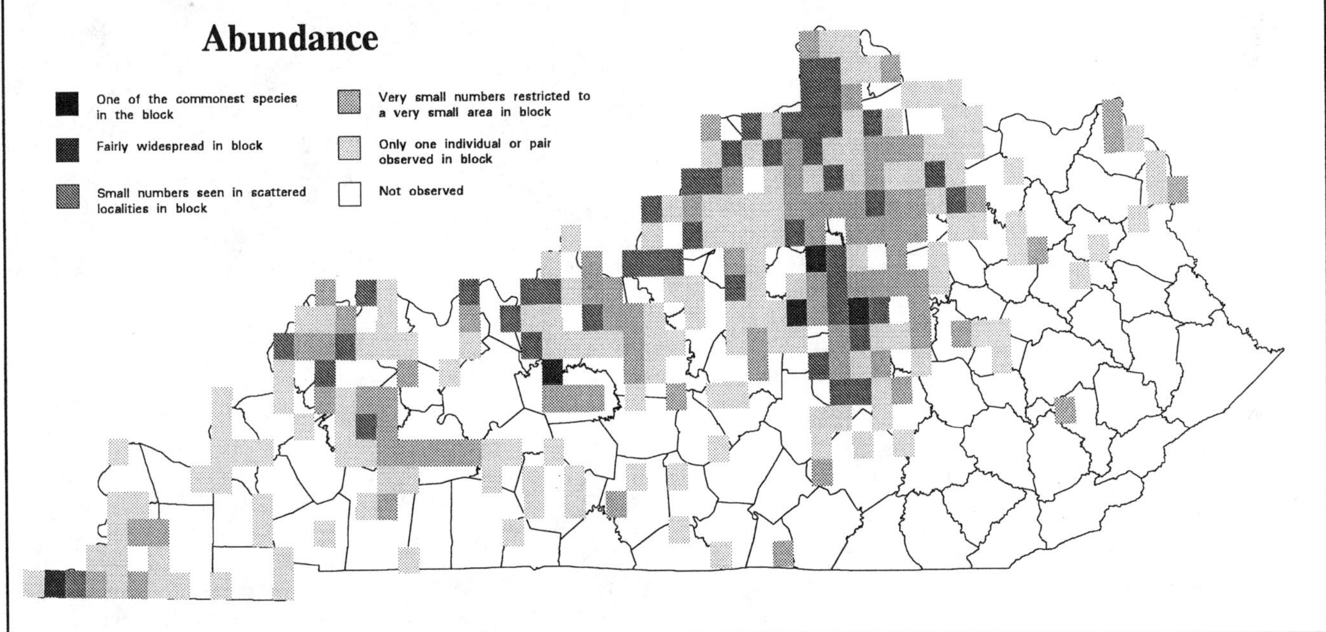

Analysis of Block Data by Physiographic Province Section

Physiographic Province Section	Total Blocks Surveyed	Blocks with Data	Avg Abund	% with Data	Section's % for State
Mississippi Alluvial Plain	14	12	1.5	85.7	4.6
East Gulf Coastal Plain	36	11	1.3	30.6	4.3
Highland Rim	139	30	1.5	21.6	11.6
Shawnee Hills	142	62	1.7	43.7	24.0
Blue Grass	204	133	1.9	65.2	51.6
Cumberland Plateau	173	10	1.3	5.8	3.9
Cumberland Mountains	19	-	-	-	-

Summary of Breeding Status

Number of Blocks in Which Species Was Recorded		
Total	**258**	**35.5%**
Confirmed	48	18.6%
Probable	59	22.9%
Possible	151	58.5%

Northern Oriole

House Finch

Carpodacus mexicanus

The House Finch is a recent addition to the breeding avifauna of Kentucky. Although the species is native to western North America, it was absent from the eastern United States until 1942, when a number of birds were released in the New York City area (Robbins et al. 1986). Soon thereafter the species became established, and it expanded rapidly to the south and west. During the late 1970s the first birds appeared in eastern Kentucky (Householder 1978; Elmore 1980), and by 1981 the first nesting had been documented (Barron 1981). House Finches are now found locally throughout the state, most frequently in association with cities and towns.

The House Finch is a bird of settled areas, whether they are urban parks, suburban yards, or rural farmland. The species is now frequent in larger cities and towns throughout Kentucky, but it remains somewhat less numerous and widespread in the western portion of the state. Away from settlement this finch is infrequently encountered, and it is typically absent in areas of extensive forest.

When the House Finch first appeared in Kentucky, it occurred primarily as a winter resident, presumably as a result of northern birds migrating south in the fall. Now that the species has become well established, many or most nesting birds appear to be resident, although some seasonal variation in abundance is still apparent. Males initiate courtship behavior during March, and spirited song seems incessant by early April. No intensive nesting study has been undertaken in Kentucky, but it appears that early clutches are often completed by mid-April. Later nests are at least occasionally reported into late June, and some young may not fledge until late July (Jackson and Jackson 1986). Late nestings suggest that at least some pairs raise two broods, although this is presently undocumented. Published Kentucky data on clutch size are limited to four nests: one with four eggs and three with five eggs (Barron 1981; Stickley 1983; Jackson and Jackson 1986; Stamm and Monroe 1991c). Harrison (1975) gives rangewide clutch size as 2–6, commonly 4 or 5.

House Finches are quite adaptable in choosing a nest site. Many nests are placed in evergreens, but the species also commonly nests in hanging flower baskets and on ledges under the eaves of homes and other buildings. These finches also nest occasionally in natural and artificial cavities, including old woodpecker holes, nest boxes, and nooks and crannies in buildings.

Forest Cover

Value	% of Blocks	Avg Abund
All	23.4	1.7
1	40.6	1.8
2	30.0	1.8
3	22.2	1.7
4	11.0	1.5
5	7.7	1.3

The nest is a shallow, somewhat loosely woven cup of dead grass, weed stems, and fine twigs and rootlets. It is lined with fine grass and other soft material (Jackson and Jackson 1986; J. Elmore, pers. comm.). In urban situations, a large amount of artificial material may be incorporated into the nest. In cavities, similar material is used, and a shallow cup is formed for the eggs. The heights of several nests reported in the literature and observed during atlas fieldwork have ranged from about 5 to 15 feet.

Philippe Roca

When the atlas project began, House Finches were just beginning to reach the western part of Kentucky, but by the time the fieldwork was completed, the species was being reported statewide. House Finches were recorded in more than 23% of priority blocks statewide, and 36 incidental observations were reported. Occurrence was generally related to the degree of settlement, being highest in the Blue Grass, where forested areas are limited and the land is heavily settled, and lowest in the Cumberland Plateau, where the percentage of forest cover is high and population density is low. As expected, occurrence and average abundance were highest in extensively cleared areas and decreased substantially as the amount of forest increased.

Despite the House Finch's conspicuousness in suburban and urban areas, less than 17% of priority block records were for confirmed breeding. Although several active nests were located, most of the 28 confirmed records were based on the observation of recently fledged young. Other confirmed records involved the observation of adults carrying food.

The House Finch is infrequently reported on most Kentucky BBS routes. Nonetheless, its statewide increase has been well documented by BBS surveys. The average number of individuals reported per BBS route for the periods 1966–91 and 1982–91 was 0.11 and 0.64, respectively. Trend analysis of these data reveals significant ($p<.05$) increases of 2.7% per year for the period 1966–91 and 17.4% per year for the period 1982–91.

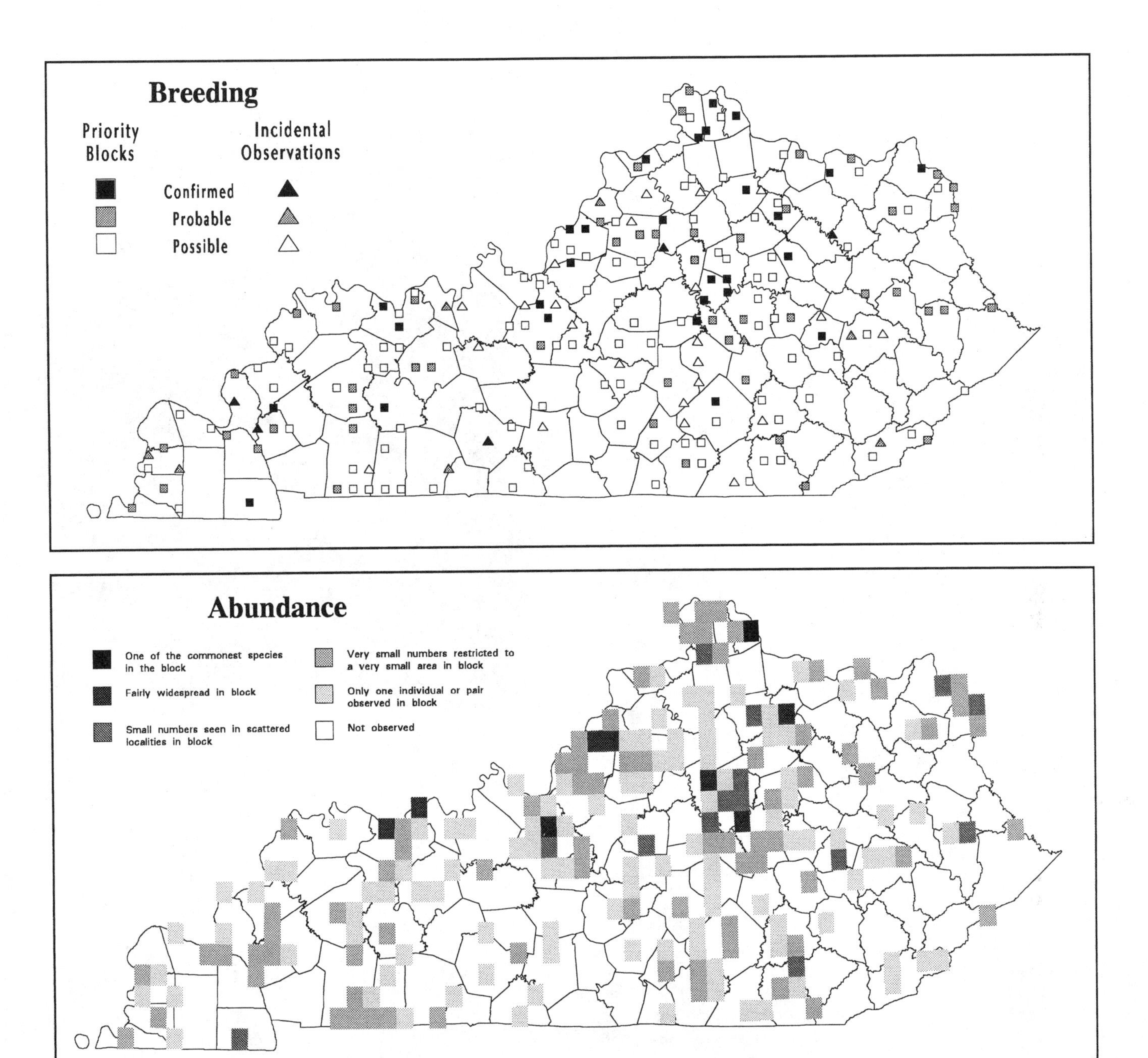

Analysis of Block Data by Physiographic Province Section

Physiographic Province Section	Total Blocks Surveyed	Blocks with Data	Avg Abund	% with Data	Section's % for State
Mississippi Alluvial Plain	14	4	1.5	28.6	2.4
East Gulf Coastal Plain	36	6	1.8	16.7	3.5
Highland Rim	139	34	1.7	24.5	20.0
Shawnee Hills	142	29	1.5	20.4	17.1
Blue Grass	204	67	1.9	32.8	39.4
Cumberland Plateau	173	26	1.7	15.0	15.3
Cumberland Mountains	19	4	1.3	21.1	2.4

Summary of Breeding Status

Number of Blocks in Which Species Was Recorded		
Total	**170**	**23.4%**
Confirmed	28	16.5%
Probable	50	29.4%
Possible	92	54.1%

House Finch

Pine Siskin

Carduelis pinus

The Pine Siskin normally breeds well to the north and west of Kentucky, primarily in the coniferous forests of Canada and the mountainous regions of the western United States (AOU 1983). Like some other members of the finch family, however, siskins occasionally nest far south of their normal range, especially after "invasion" years. Such nesting has been documented in Kentucky on at least one occasion (Palmer-Ball 1980).

Ron Austing

Pine Siskins are known primarily as winter residents in Kentucky, occurring from early October to early May. They fluctuate widely in abundance from year to year, occasionally being quite abundant and sometimes nearly absent. Although this small finch has always been rather irregular in occurrence, the species appears to have increased in recent years. Mengel (1965) noted that before about 1940 it was seldom recorded, but since that time siskins have become more regular in the state.

Before the atlas survey, the Pine Siskin had been documented as nesting only once, following the invasion winter of 1977–78 (Stamm 1978b; Palmer-Ball 1980). Nest construction and the incubation of a single egg were noted in late March 1978 in Jefferson County. Although this nest was abandoned, siskins remained in the area into early June, and there were a few reports of birds that resembled fledglings and possible immature birds being fed by adults (Stamm 1978b; K. Clay, pers. comm.; author's notes).

The 1978 nest was situated about 45 feet up in an introduced white pine of about 60 feet total height. The nest was a neatly woven cup of dead grass, lined with animal hair and placed at the base of a cluster of small branches near the end of a larger branch (Palmer-Ball 1980).

During the atlas survey, evidence of nesting was observed at the same Jefferson County location where nesting was attempted in 1978. On May 29–30, 1990, a juvenile siskin was observed begging and being fed by an adult bird on several occasions (Stamm and Monroe 1990b; author's notes). While the observation of fledglings being fed by adult birds may indicate nesting nearby, family groups composed of adults and partially dependent young may travel for many miles after fledging. For this reason, reports of nesting based on the observation of fledglings should be made with caution. Although the young bird observed in 1990 may have been raised some distance away, it was thought that the observation merited inclusion in the atlas results, because of the 1978 nesting attempt at the same locality.

These observations are similar to recent reports from Ohio (Peterjohn and Rice 1991), where small numbers of siskins have nested more regularly in recent years. As in Kentucky, most evidence of nesting in Ohio has come from suburban areas, where introduced species of conifers have been used for nesting and bird feeders have provided a good food supply. These reports indicate that Pine Siskins should be expected to nest occasionally in Kentucky, especially after invasion years.

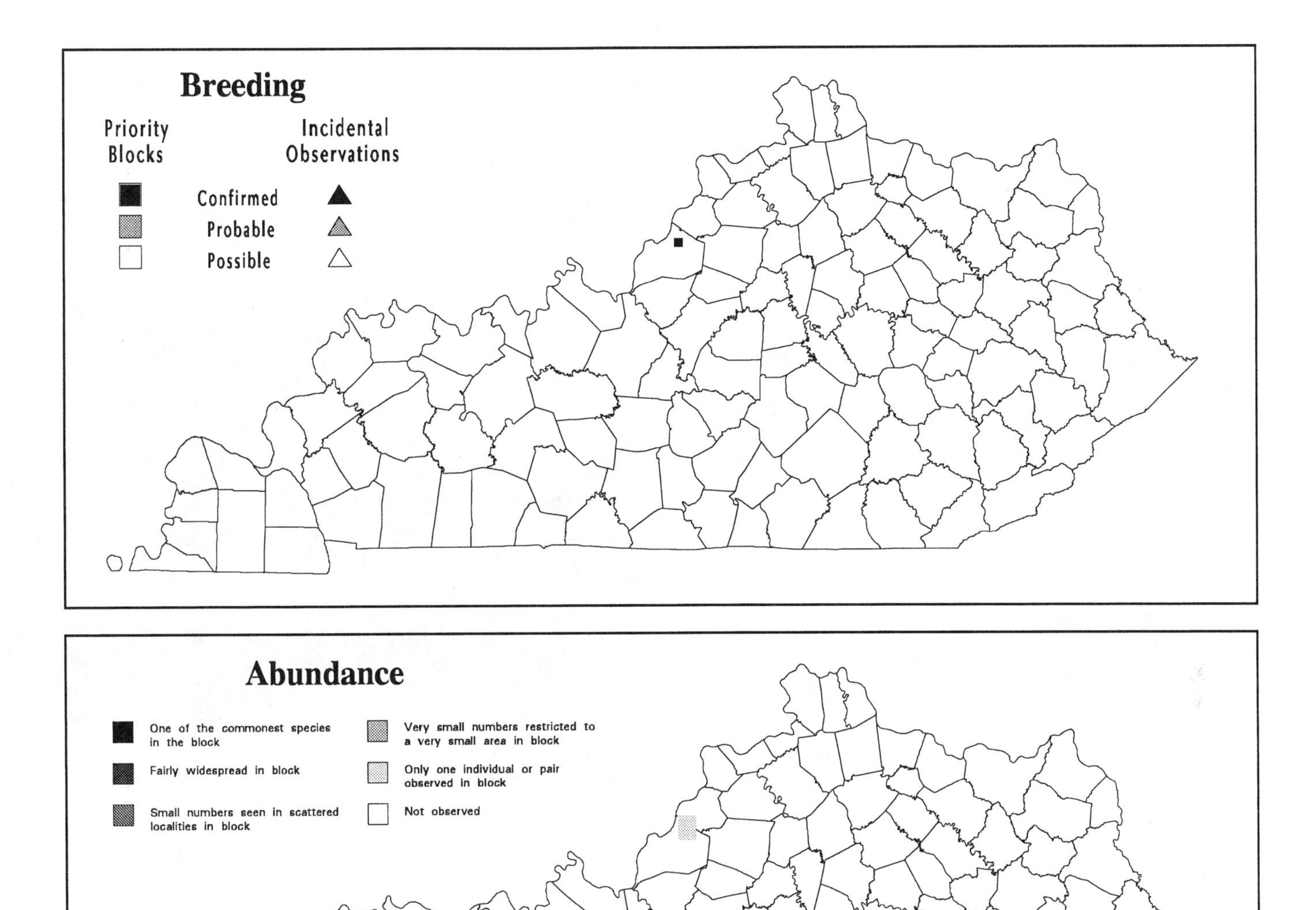

Analysis of Block Data by Physiographic Province Section

Physiographic Province Section	Total Blocks Surveyed	Blocks with Data	% with Data	Section's % for State
Mississippi Alluvial Plain	14	-	-	-
East Gulf Coastal Plain	36	-	-	-
Highland Rim	139	-	-	-
Shawnee Hills	142	-	-	-
Blue Grass	204	1	0.5	100.0
Cumberland Plateau	173	-	-	-
Cumberland Mountains	19	-	-	-

Summary of Breeding Status

Number of Blocks in Which Species Was Recorded		
Total	**1**	**0.1%**
Confirmed	1	100.0%
Probable	-	-
Possible	-	-

Pine Siskin

American Goldfinch

Carduelis tristis

The musical song of the American Goldfinch is a widespread and relatively common summer sound throughout much of Kentucky. The species is unique among the state's breeding birds in that its nesting season is so late, often not beginning until late July. Mengel (1965) regarded the species as a fairly common to common summer resident statewide.

The American Goldfinch occurs in a great variety of semi-open and open habitats. It is most frequent in areas dominated by early successional vegetation, whether the result of human activity or natural disturbance, and it typically avoids mature forest during the nesting season. The species primarily uses artificial situations, including abandoned fields, reverting clear-cuts, fencerows, reclaimed surface mines, rural yards, and woodland borders. Naturally occurring habitats are also used, including the margins of rivers, ponds, marshes, glades, and prairie remnants, as well as woodland openings created by fire or windstorms.

Although natural disturbance must have maintained an abundance of suitable early successional nesting habitat prior to settlement, goldfinches are certainly more widespread and numerous in Kentucky today. While some habitat has been lost as a result of conversion of natural openings to agricultural use and settlement, the clearing and fragmentation of once widespread forests have resulted in the creation of an abundance of suitable habitat. As a result, these colorful finches now occur in many areas where they were formerly excluded by the presence of mature forest.

Numbers of goldfinches vary somewhat throughout the year. In general, birds from farther north supplement the Kentucky population in winter, and many do not depart until early May. A few locally nesting goldfinches begin courtship activities in May, but it may be a month or more before nesting activity is under way in earnest. Although Mengel (1965) notes that early clutches are completed as early as mid-July, more recent reports describe earlier nestings. The spread of musk thistle, an early maturing exotic weed that is a source of food and nesting material for the bird, may have resulted in a shift in the timing of nesting activity by at least some birds where this plant is abundant. In general, a peak in clutch completion occurs in early August, and late clutches are completed by the first ten days of September (Mengel 1965). Kentucky data on clutch size are scarce, but the average size of four clutches reported by Mengel (1965) was 5.2 (range of 5–6). Harrison (1975) gives rangewide clutch size as 4–6, commonly 5.

Forest Cover

Value	% of Blocks	Avg Abund
All	96.6	2.8
1	90.0	2.7
2	96.8	2.8
3	97.8	2.8
4	100.0	2.9
5	92.3	2.6

The nest is usually placed in a small tree or shrub, often one standing in the open or along a fencerow. The nest is a compact cup, constructed of silky plant material and lined with soft plant down from thistles or cattails (Harrison 1975). It is typically placed in a sturdy, upright fork, although occasionally within the fork of a horizontal branch. The average height of six nests given by Mengel (1965) was 9.5 feet (range of 3–25 feet), although there is a more recent record of a nest 35 feet above the ground (Stamm 1978a).

Ron Austing

The atlas survey yielded records of American Goldfinches in nearly 97% of priority blocks statewide, and three incidental observations were reported. The species ranked 9th according to the number of priority block records, 16th by total abundance, and 24th by average abundance. Occurrence and average abundance were relatively uniform across the state. Occurrence was slightly lower only in highly cleared and predominantly forested areas.

Despite the goldfinch's overall abundance, only about 6% of priority block records were for confirmed breeding. Since most nesting of the species occurred after the majority of each season's fieldwork was done, it is not surprising that the confirmation rate was so low. Although several active nests were located, most of the 43 confirmed records were based on the observation of nest building and family groups including recently fledged young. The few remaining confirmed records involved the observation of adults carrying food and used nests.

American Goldfinches are recorded in small to moderate numbers on most Kentucky BBS routes. The average number of individuals per BBS route for the periods 1966–91 and 1982–91 was 8.67 and 9.79, respectively. According to these data, the species ranked 25th in abundance on BBS routes during the period 1982–91. Trend analysis of these data shows a slight, nonsignificant increase of 0.7% per year for the period 1966–91, but a significant ($p<.01$) increase of 5.3% per year for the period 1982–91. Reasons for the short-term increase are unknown.

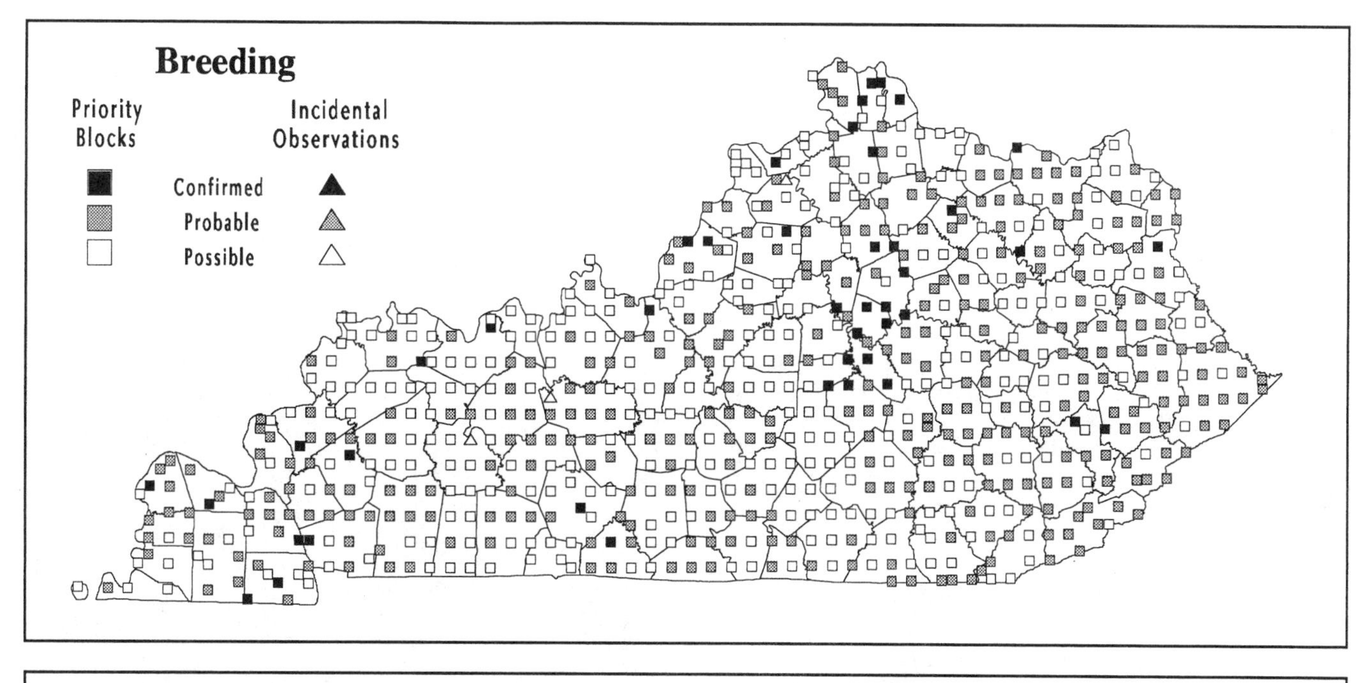

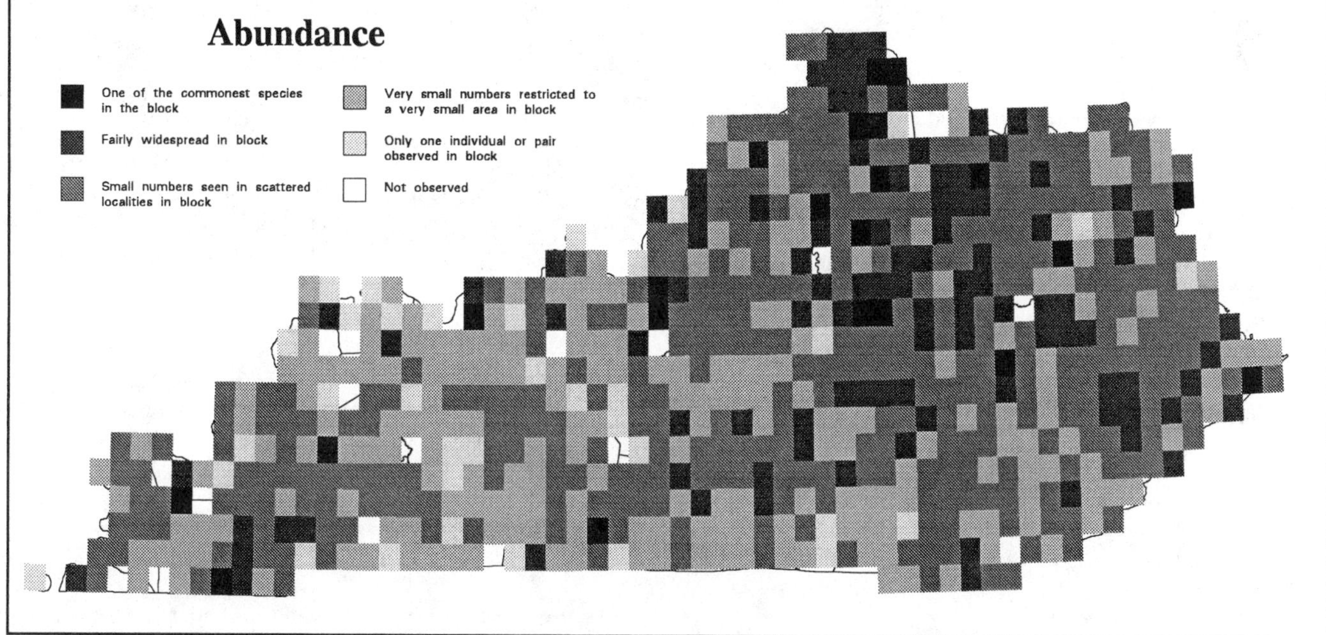

Analysis of Block Data by Physiographic Province Section

Physiographic Province Section	Total Blocks Surveyed	Blocks with Data	Avg Abund	% with Data	Section's % for State
Mississippi Alluvial Plain	14	12	2.7	85.7	1.7
East Gulf Coastal Plain	36	33	2.8	91.7	4.7
Highland Rim	139	136	2.6	97.8	19.4
Shawnee Hills	142	132	2.3	93.0	18.8
Blue Grass	204	200	3.2	98.0	28.5
Cumberland Plateau	173	172	2.8	99.4	24.5
Cumberland Mountains	19	17	2.9	89.5	2.4

Summary of Breeding Status

Number of Blocks in Which Species Was Recorded		
Total	**702**	**96.6%**
Confirmed	43	6.1%
Probable	347	49.4%
Possible	312	44.4%

American Goldfinch

House Sparrow

Passer domesticus

The House Sparrow is undoubtedly one of the most widely recognized of Kentucky's breeding birds. Fearless and common about most human dwellings, these sparrows and their monotonous chirping are familiar to most humans. Mengel (1965) regarded the species as a common to abundant permanent resident statewide in appropriate habitat, although less common in unsettled portions of eastern Kentucky.

Native to the Old World, House Sparrows were introduced at a number of locations in North America in the mid-1800s (AOU 1983). The species was released at Louisville sometime between 1865 and 1870, and it rapidly spread throughout the settled portions of the state, becoming common by about 1890 (Mengel 1965). By 1900 it was well established across Kentucky.

The House Sparrow is one of the most conspicuous birds of urban areas, and large numbers inhabit suburban parks and yards as well as rural farmland and settlement. Although the species is encountered primarily in and near settlement, it also occurs occasionally in other artificial situations, including roadway and utility corridors, where the birds use poles, signs, bridges, and vertical embankments for nesting.

House Sparrows are permanent residents in Kentucky. Males initiate courtship behavior in early spring, and nest building may be under way by mid-March. According to Mengel (1965), clutches are completed from late March to late July, with at least one peak of clutch completion in late April. The species is double-brooded, and late nests suggest that at least a few pairs may raise three broods. By mid-June large flocks of young birds have gathered to forage in sizable flocks, especially in farmland. Kentucky data on clutch size are scarce, but the average size of six clutches given by Mengel (1965) was 4.2 (range of 2–5). Harrison (1975) gives rangewide clutch size as 3–7, commonly 5.

Forest Cover

Value	% of Blocks	Avg Abund
All	89.5	3.2
1	99.0	4.0
2	99.5	3.7
3	96.5	3.1
4	77.3	2.4
5	42.3	1.8

House Sparrows usually nest in cavities, but in the absence of cavities they will build bulky, spherical nests among the branches of trees. These sparrows nest most often in nooks and crannies of buildings and barns, but they also readily accept nest boxes, and they will use old Cliff Swallow nests when these are present. The birds typically fill the entire cavity with dead grass, weed stalks, and pieces of trash and line it with finer grass and many feathers. Spherical nests are built of similar material but are constructed so that the entrance opens to one side. Nest sites may occur at just about any height, but most reported nests have been observed from 5 to 25 feet above the ground (Croft and Stamm 1967; Stamm and Croft 1968).

The atlas survey yielded records of House Sparrows in nearly 90% of priority blocks statewide, and 10 incidental observations were reported. The species ranked 19th according to the number of records in priority blocks, 10th by total abundance, and 8th by average abundance. Occurrence was lower than 85% only in the Cumberland Plateau and the Cumberland Mountains, where settlement is not as widespread. Average abundance varied similarly. These trends were closely related to the extent of settlement, which generally is inversely related to the percentage of forest cover.

Maslowski Wildlife Productions

Nearly 43% of priority block records were for confirmed breeding. By nesting in close proximity to humans, House Sparrows are relatively easy to confirm, and atlasers located a number of active nests. Family groups of House Sparrows that included recently fledged young were also conspicuous as they foraged around homes and barns. Other confirmed records were based on the observation of adults carrying food for young, nest building, and used nests.

House Sparrows are reported in substantial numbers on most Kentucky BBS routes. The average number of individuals recorded per BBS route for the periods 1966–91 and 1982–91 was 65.31 and 66.31, respectively. According to these data, the species ranked 4th in abundance on BBS routes during the period 1982–91. Trend analysis of these data reveals a nonsignificant decrease of 1.3% per year for the period 1966–91, but a significant ($p<.01$) decrease of 5.8% per year for the period 1982–91. The reasons for these declines are unclear, but they may include severe winter weather, changing farming practices (Robbins et al. 1986), and competition with the House Finch.

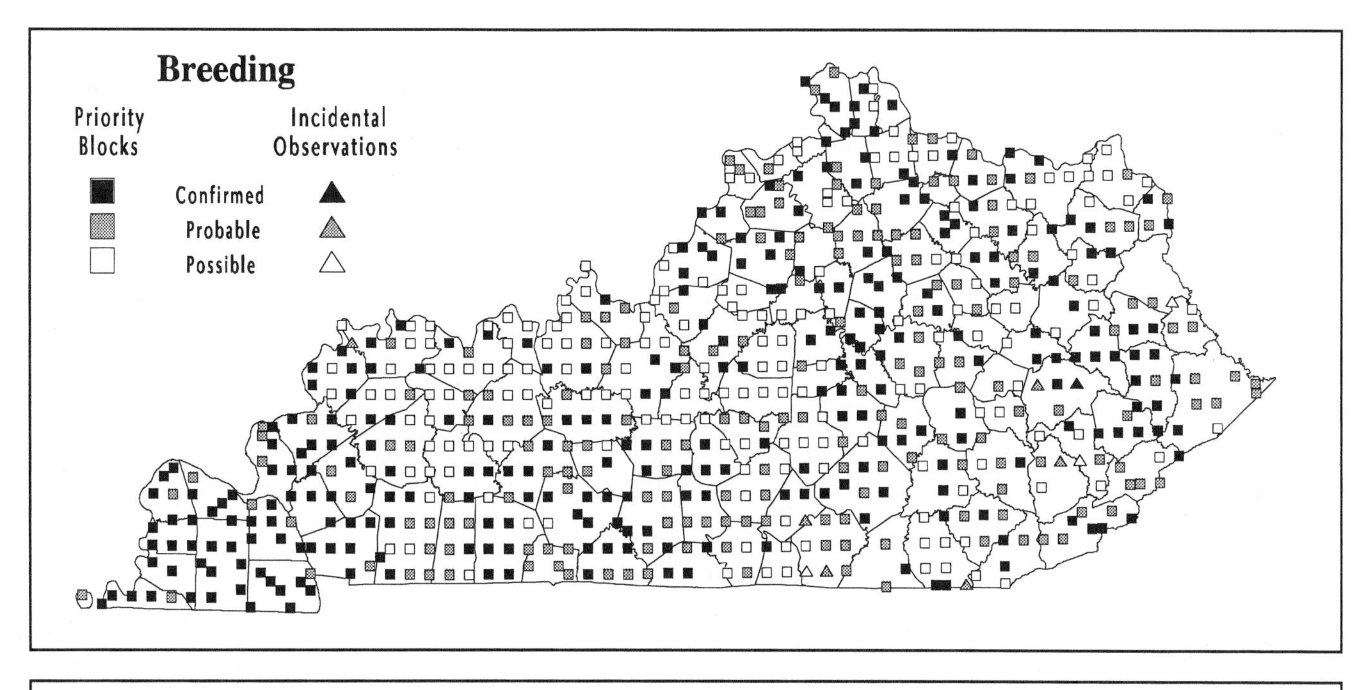

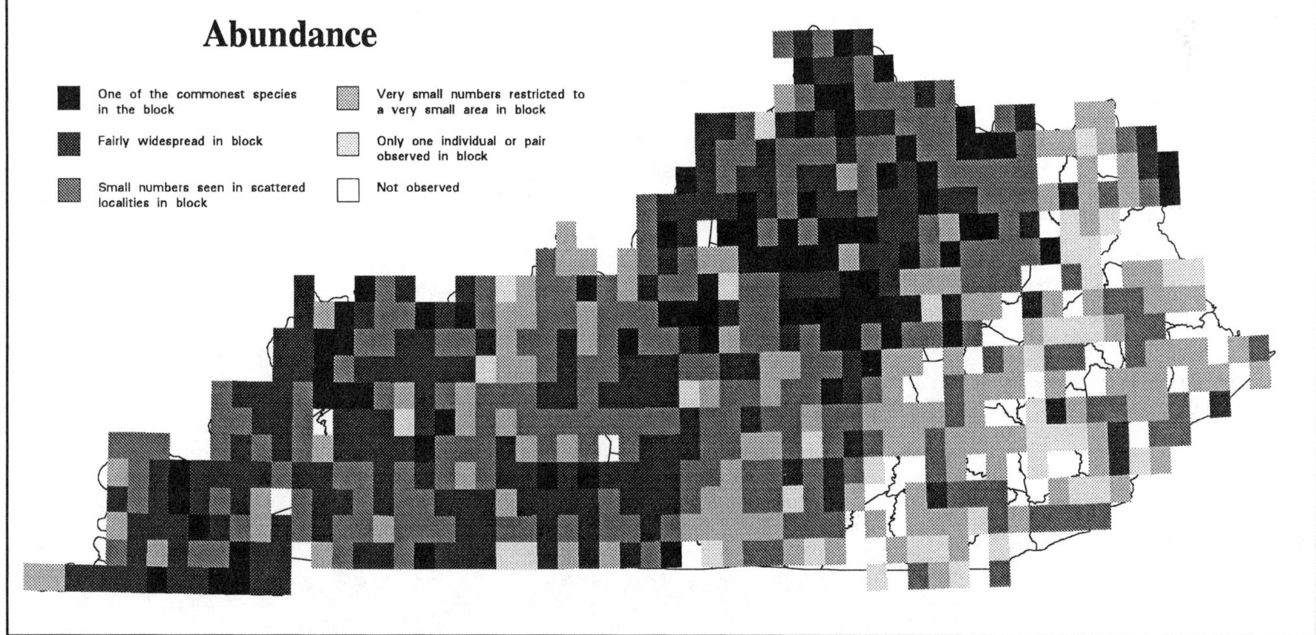

Analysis of Block Data by Physiographic Province Section

Physiographic Province Section	Total Blocks Surveyed	Blocks with Data	Avg Abund	% with Data	Section's % for State
Mississippi Alluvial Plain	14	12	3.1	85.7	1.8
East Gulf Coastal Plain	36	36	3.8	100.0	5.5
Highland Rim	139	135	3.2	97.1	20.7
Shawnee Hills	142	138	3.4	97.2	21.2
Blue Grass	204	201	3.6	98.5	30.9
Cumberland Plateau	173	117	2.3	67.6	18.0
Cumberland Mountains	19	12	2.2	63.2	1.8

Summary of Breeding Status

Number of Blocks in Which Species Was Recorded		
Total	**651**	**89.5%**
Confirmed	277	42.5%
Probable	196	30.1%
Possible	178	27.3%

House Sparrow

NONBREEDING SPECIES OBSERVED

Double-crested Cormorant

Phalacrocorax auritus

The Double-crested Cormorant was not documented to have nested in Kentucky before the 20th century, but as of the early 1950s Mengel (1965) knew of colonies at three localities. In 1936 a few pairs were observed nesting with Great Blue Herons and Great Egrets at each of two heronries along the lower Ohio River, one in western Henderson County and the other near Axe Lake, in Ballard County (Bacon and Monroe 1937). Subsequent information on the Axe Lake colony was not published, but on June 5, 1951, "several nests with eggs or young" were observed at the Henderson County heronry (Lovell 1951). The third colony was located near Bondurant, in Fulton County, and was referred to locally as Little Cranetown or Kentucky Cranetown (Mengel 1965). It was situated in swamp forest along the edge of a cypress slough on a northern arm of Reelfoot Lake, in what is now the Reelfoot NWR. In 1949 approximately 50 pairs of cormorants were nesting there with Anhingas, Great Egrets, and Great Blue Herons (Mengel 1965).

Unfortunately, cormorants have not been reported nesting in Kentucky since these reports. The effects of the accumulation of harmful pesticide residues, most notably from DDT, likely resulted in the species's disappearance as a breeding bird sometime before the early 1960s. Wiley (1964) reported that the Henderson County heronry was abandoned as of the summer of 1962, and Croft and Rowe (1966) alluded to abandonment of the Fulton County heronry as of June 1965.

During the 1960s and early 1970s cormorants were scarcely reported, with a few published accounts of transients representing the entire literature record (e.g., Stamm et al. 1967). The banning of DDT and other chlorinated hydrocarbon pesticides in the early 1970s, however, may have triggered a range-wide resurgence of cormorants. During the mid- to late 1970s nesting colonies on the Great Lakes began expanding dramatically (Ludwig 1985), and cormorants became much more common and widespread as transients. During the late 1980s summering individuals began to appear on the Ohio River in the Louisville area and on Kentucky Lake (Stamm 1987b, 1989c). Summering birds now occur regularly. Although these observations seem to represent nonbreeding individuals, the species may be on the verge of returning to formerly used sites along the Mississippi and lower Ohio Rivers.

Specific data on nesting in Kentucky are scarce, but observations of small young in nests in Fulton County on May 20, 1949, young in nests in Henderson County on June 16, 1936, and large young in nests in Henderson County on July 7, 1940, suggest that clutches may be completed as early as the latter part of April. Kentucky data on clutch size are lacking, but Ganier (1933) reported that clutches of three to five eggs were observed in nests in a colony elsewhere on Reelfoot Lake in Tennessee on April 26–May 29, 1919.

Although Double-crested Cormorants sometimes nest on the ground, all nests reported in Kentucky have been built in large trees. Furthermore, the species has been found nesting only in association with other waterbirds, typically Great Blue Herons. The nest usually is a bulky structure composed of sticks and debris and lined with finer material (Harrison 1975).

American Bittern

Botaurus lentiginosus

Very little is known about the breeding status of the American Bittern in Kentucky. The only definite nesting records were reported in Hopkins County from 1914 to 1943 (Bacon 1933; Mengel 1965). During that period immature birds too young to fly were noted at four separate locations: Loch Mary at Earlington, the Pond River bottoms, Atkinson Junction Lake, and Spring Lake at Madisonville. Most were observed in cattails and buttonbush along the margins of lakes.

Other than Bacon's observations, little if any evidence of nesting by this unusual wading bird has been published. American Bitterns have been reported from a few additional localities in summer, including the counties of Crittenden (Semple 1944), Fayette (Funkhouser 1925), Fulton (Pindar 1925), Henderson (Cooke 1913), and Warren (Wilson 1951). Although the species has been present during the breeding season, observers have not obtained evidence of nesting.

Throughout most of the American Bittern's breeding range, it is typically found in emergent wetland vegetation of marshes and the marshy borders of lakes and ponds. In addition, the species is sometimes found in weedy or grassy fields and meadows during migration, and such habitats are occasionally used for nesting in other parts of its range (Harrison 1975).

Although this bittern is an unusual find at any season, it is most often encountered during migration. The species typically winters in coastal marshes of the southeastern United States and southward (AOU 1983). A few birds are usually observed in Kentucky during early to mid-April, but they disappear by mid-May. Nesting data are so scarce that it is impossible to determine the limits of the normal breeding season, but the dates of observation of immature birds in Hopkins County ranged from June 11 to July 5 (Mengel 1965). Based on an incubation period of 24–28 days (Harrison 1975) and a period of at least three weeks from hatching to fledging, clutches were likely completed sometime during the middle of May.

During the atlas fieldwork American Bitterns were reported in suitable nesting habitat at four locations, all in the western third of the state: at a cattail marsh and pond on a reclaimed surface mine near Cool Springs in southern Ohio County, where one or two birds were seen on several occasions in April 1991 and May 1989 (Stamm 1989b; Stamm and Monroe 1991c); on extensive cattail and burreed marshes on the Sauerheber Unit of Sloughs WMA in western Henderson County, where single birds were flushed on several occasions in April 1986, 1988, 1989, and 1990 (Stamm 1986b, 1988c, 1989b; Stamm and Monroe 1990b); at a cattail-dominated marsh near Ferguson Spring in the Crooked Creek drainage of Land Between the Lakes, in Trigg County, where a bird was heard calling on April 9, 1989 (Stamm 1989b), and where three birds had been reported on April 22–23, 1982 (Stamm 1982c); and a marshy field on the Reelfoot NWR, in Fulton County, where two birds were flushed on April 20, 1986 (Stamm 1986b).

All these observations were made during the normal migratory period, and it is possible that some or all of the birds observed were simply spring migrants. These four locations are worthy of note, though, because of the suitability of habitat in combination with multiple records of sightings or calling.

While the American Bittern may have been overlooked in the past, it is likely that the species was never abundant as a nesting bird. These bitterns occur in greatest numbers where extensive marshes of emergent aquatic vegetation, especially cattails, are present. Even though suitable habitat was present in the state before settlement, such situations were not widely distributed. For this reason, it is likely that bitterns were never as common in Kentucky as in the Great Lakes and northern Great Plains, where suitable habitat was formerly abundant. It is possible that small numbers may have occurred in the state in canebrakes and other thick, grassy vegetation in upland areas, but documentation of this is lacking.

During the past several decades the American Bittern has declined substantially across much of its nesting range in the northern United States and southern Canada (R. Peterson 1980; Mumford and Keller 1984; Bohlen 1989; Peterjohn and Rice 1991). This decline can be attributed largely to the destruction of wetlands, but human alteration of some areas has actually resulted in the creation of suitable nesting habitat, especially where bottomland forest has been cleared and subsequently abandoned. Moreover, the return of the beaver has contributed immensely to the creation of suitable nesting habitat. These factors may assist the American Bittern in becoming reestablished as a nesting bird in the future.

Peregrine Falcon

Falco peregrinus

The historical breeding status of the Peregrine Falcon in Kentucky is obscure. Regional accounts from adjacent states suggest that the species nested sporadically throughout the Midwest (see Mengel 1965). Mengel (1965) considered it to have been a rare summer resident at one time, being distributed sporadically across at least the eastern third of the state. He based this belief on references to summer sightings at scattered localities across the state during the first half of the 20th century.

The most credible report of breeding was described to Mengel (1965) for a site in a cliffline along the Rockcastle River in Laurel County as of the late 1930s. Confirmed evidence of nesting also has been obtained from just across the Virginia line in the Cumberland Mountains (Mengel 1965) and at two nearby locations in Tennessee (Spofford 1947a, 1947b).

Although the species may have been distributed sporadically across Kentucky, sites most likely inhabited included clifflines in the Cumberland Mountains and in the Cliff Section of the Cumberland Plateau, bluffs along the Kentucky and Ohio Rivers, and hollow trees in the cypress swamps of far western Kentucky. It is also possible that Peregrines inhabited buildings in the larger cities (see Croft 1965; Mengel 1965).

The introduction of chlorinated hydrocarbon pesticides like DDT in the early 1960s resulted in a significant decline in numbers of some waterbirds and raptors through the early 1970s, and it is likely that any Peregrines nesting as of the late 1950s soon disappeared. A bird observed in a heavily forested area in Whitley County in mid-July 1965 was one of very few reported between about 1960 and the mid-1970s (Croft, Rowe, and Wiley 1965).

Since the mid-1980s interest in reestablishing nesting Peregrine Falcons into regions where they nested before the DDT era has resulted in the initiation of a major effort to raise young birds and release them at suitable sites throughout the Midwest and Appalachian regions. Recovery of natural populations from the effects of DDT, along with the success of these reintroduction efforts, have resulted in an increase in the number of observations of Peregrine Falcons in recent years.

During the atlas survey there were two reports of Peregrine Falcons during the breeding season. On June 17, 1989, an adult Peregrine Falcon was observed flying along and above White Rocks, a massive cliffline on the south face of Cumberland Mountain in Cumberland Gap National Historical Park, in Harlan County. Suitable nest sites on the cliffs there lie in Virginia, but the bird was observed over Kentucky. This is probably the site where Mengel (1965) noted a family group of three young and two adults on June 3, 1952.

The other report originated in central Kentucky, where at least one bird was observed in downtown Louisville during the summers of 1990 and 1991. Two birds were seen on a few occasions, and they even appeared to be engaged in courtship behavior. At least one of these birds is banded and likely originated from a reintroduction program in the northern United States or southern Canada. Observations suggest that the pair may be breeding in the immediate vicinity, but further evidence of nesting has not yet been obtained. It is expected that if efforts to reestablish the species continue to be successful, at least a few birds will begin to nest in Kentucky.

American Coot

Fulica americana

Although the American Coot is an abundant nesting bird throughout a large part of the United States and southern Canada, it has been reported nesting only a few times in Kentucky. In fact, only at the transient lakes in southern Warren County has attempted breeding been reported on more than one occasion.

Since the early 1900s coots have been observed summering or attempting to nest at both Chaney and McElroy Lakes on several occasions (Mengel 1965). Despite the regularity of the species's occurrence on the lakes when suitable habitat is present, however, it appears that successful nesting has never been documented. Fluctuating water levels have apparently always caused nestings to fail; sudden rises result in inundation of nests, and sudden drops lead to predation or desertion (Wilson 1940). Coots seem to have nested successfully only in Fayette County (Stamm 1969), where the species appears to summer on occasion at Lexington's water supply reservoirs (M. Flynn, pers. comm.). Wilson (1942) reported breeding from Crittenden and Union Counties, although no specific details have been provided. Coots have also been reported sporadically in summer from scattered locations throughout Kentucky; virtually all these reports, however, have involved single individuals, and most of the birds have been regarded as injured or nonbreeding.

During the breeding season, American Coots typically inhabit marshes and the marshy borders of lakes and ponds. Although the species is occasionally encountered in natural situations in Kentucky, most summer records come from reservoirs. It is unclear whether coots nested in the state before the 20th century, but a few birds probably summered on floodplain wetlands along the larger rivers. The type of vegetation that occurred naturally at transient lakes in Warren County is not known, but it may have been mature forest, unsuitable for nesting coots.

The limited information on nesting in Kentucky comes almost entirely from the transient lakes in Warren County. There, Wilson (1929, 1935) observed nests containing eggs on June 22, 1927, and deserted nests (two containing six eggs each) on June 19, 1935. The report of five young with an adult in Fayette County on April 28, 1968, has a remarkably early date (Stamm 1969), and most nesting must typically occur later in the season.

Two incidental records of American Coots were reported during the atlas survey. At the transient lakes in Warren County about six coots remained throughout the summer of 1989, but suitable nesting habitat was limited and the birds probably did not attempt to breed (Palmer-Ball and Boggs 1991). The only other report involved the observation of two coots that summered on a small cattail pond in southern Ohio County. Although the habitat was suitable at this site, further evidence of nesting was not observed.

Blue-winged × Golden-winged Warbler hybrids

Vermivora pinus × *chrysoptera*

Although hybridization between the Blue-winged and Golden-winged Warblers has traditionally occurred well to the north of Kentucky, recent observations suggest that the zone of hybrid occurrence may be shifting southward. In Ohio, for example, hybrid individuals have become more frequent in the southern half of the state in the last two decades (Peterjohn and Rice 1991).

In Kentucky, hybrids between the two species have been observed primarily during migration, with reports of Brewster's Warbler (*Vermivora "leuchobronchialis"*), the dominant form, outnumbering those of the recessive Lawrence's Warbler (*Vermivora "lawrencei"*) on the order of five to one. Most observations have been made from late April to early May, but there are a few records for mid- to late May and from late August to late September.

Before the mid-1980s these hybrid forms had been reported in summer on only one occasion. Mengel (1965) collected a male Brewster's Warbler about 12 miles southwest of London, in Laurel County, on June 15, 1952. Although neither a mate nor a nest could be located, this individual had enlarged testes and was heard singing for three days in a row, suggesting that it was breeding.

During the atlas survey hybrids of the two species were observed on three occasions, all on the Cumberland Plateau. On July 15, 1988, a male Brewster's Warbler was heard singing a song typical of a Golden-winged along Mill Creek near Mazie, in northern Lawrence County. This bird was in an area where Blue-winged Warblers were present, and nesting habitat was suitable. On July 9, 1991, a female or immature Lawrence's

Warbler was observed along the Grapevine Creek embayment of Fishtrap Lake, in Pike County. Although suitable habitat was present in the general area, forest surrounding the point of observation did not provide optimal habitat, and the bird was foraging with other small passerines. On July 31, 1991, a female or immature Brewster's Warbler was observed about two miles northwest of Hightop, in rural Laurel County. This individual was foraging high in trees with a mixed flock of small passerines and was not considered to be of local origin. Interestingly, the bird was observed in the general area where Mengel collected the singing male in 1952.

Although it is possible that all three reports could have pertained to locally nesting birds, none of the records were included in the atlas data set. All three birds were seen in July, when postbreeding Blue-winged Warblers begin turning up in nonbreeding areas across Kentucky. Furthermore, the Laurel and Pike County records involved individuals that were part of foraging groups of mixed passerines, which would not indicate territorial birds. The record most indicative of nesting would be the one of the Lawrence County male, but in the absence of further evidence of nesting, it was omitted as well.

While the atlas fieldwork did not yield substantial evidence of the interbreeding of Blue-winged and Golden-winged Warblers, it is possible that hybridization is occurring at a low level. Such interaction is most likely on the Cumberland Plateau, where all of the atlas observations of hybrid birds were obtained. The scarcity of Blue-winged Warblers in the Cumberland Mountains, where Golden-wingeds occur most frequently, minimizes the possibility of hybridization there. Likewise, the absence of Golden-winged Warblers in central and western Kentucky minimizes the possibility of hybridization west of the Cumberland Plateau.

EXTINCT AND EXTIRPATED BREEDING SPECIES

Anhinga

Anhinga anhinga

Anhingas were reported breeding in Kentucky at one locality in Fulton County until the early 1950s (Mengel 1965). Up to about 50 pairs nested at a mixed waterbird colony on the northern end of Reelfoot Lake, known as Little Cranetown or Kentucky Cranetown. No record of breeding has been reported since the DDT era, although nesting birds reappeared in western Tennessee in the late 1970s (Robinson 1990).

Northern Pintail

Anas acuta

The Northern Pintail was reported nesting on a large farm pond in northern Shelby County, in 1973 (Robinson 1974). This occurrence is considered accidental.

Lesser Scaup

Aythya affinis

A nest of this species was supposedly found in a small marsh near Carrollton, in Carroll County, in June 1950 (Webster 1951; Mengel 1965). This occurrence is considered accidental, and the report likely pertained to an injured bird.

American Swallow-tailed Kite

Elanoides forficatus

The Swallow-tailed Kite was once known as a regular summer resident along the floodplains of the larger rivers in central and western Kentucky (Audubon 1861). In the early 1800s Audubon found it nesting near the Falls of the Ohio, near Louisville, in Jefferson County. The species declined significantly during the late 1800s (Mengel 1965) and has not been reported in Kentucky in more than a century.

Greater Prairie-Chicken

Tympanuchus cupido

The Greater Prairie-Chicken was resident throughout the native prairies of southern and western Kentucky at the time of settlement (Audubon 1861; Mengel 1965). It was reported breeding at Bardstown and likely nested throughout the prairies (Mengel 1965). The species disappeared from Kentucky during the first half of the nineteenth century.

Virginia Rail

Rallus limicola

Audubon reported a Virginia Rail nest in Henderson County in the early 1800s (Mengel 1965), but more recent evidence of this species nesting in the state is lacking. Reference to nesting in Hopkins County in 1932 was not documented (Mengel 1965).

Upland Sandpiper

Bartramia longicauda

The Upland Sandpiper has been reported nesting in Kentucky only once. On June 4, 1950, a nest containing four eggs was found on the grounds of the Greater Cincinnati Airport, in Boone County (Kemsies et al. 1950). Otherwise, the species has been reported on only a few occasions during the breeding season, and it is likely that if breeding occurs at all, it is sporadic.

Black Tern

Chlidonias niger

The Black Tern was reported nesting near Louisville in the early 1800s, when Audubon (1861) observed up to 70 nests along the margins of a small pond (Mengel 1965). Wilson (1929) believed he found an egg of this species at McElroy Lake, in Warren County, in June 1927, and birds summered there in 1935 (Wilson 1935). Nesting, however, was not confirmed in either year. Suitable breeding habitat has been destroyed in Kentucky, and nesting is no longer likely.

Passenger Pigeon

Ectopistes migratorius

The Passenger Pigeon was widely known in Kentucky during the early 1800s. Large numbers passed through the state during migration, and breeding colonies were described from several localities (Audubon 1861; Mengel 1965). The species became very rare by about 1890, and it is now extinct (Mengel 1965).

Carolina Parakeet

Conuropsis carolinensis

The Carolina Parakeet was once a permanent resident in central and western Kentucky, being distributed mainly along the floodplains of the larger rivers (Audubon 1861). It probably disappeared from most of the state by the 1860s, and the species is now extinct (Mengel 1965). Although specific records are absent, evidence suggests that the species nested in the state.

Ivory-billed Woodpecker

Campephilus principalis

The Ivory-billed Woodpecker was reported by early ornithologists in the extensive floodplain forests along the Mississippi and lower Ohio Rivers during the nineteenth century (Audubon 1861; Mengel 1965). Within this limited range, it was probably resident throughout the year. The species was last seen in Kentucky in the late 1800s, and it is now considered to be nearing extinction (AOU 1983).

Bachman's Warbler

Vermivora bachmanii

The Bachman's Warbler was known to be a casual summer resident in Kentucky as of the early 1900s. Embody (1907) collected several birds and observed an active nest in swampy forest along Wolf Lick in northern Logan County in mid-May 1906. Otherwise, evidence of the species's occurrence is sketchy, with only a few unconfirmed reports (Mengel 1965). The Bachman's Warbler has not been reported for many years, and it is now considered to be possibly extinct (AOU 1983).

APPENDICES

APPENDIX A. KENTUCKY'S PHYSIOGRAPHIC PROVINCE SECTIONS

Kentucky lies within parts of three physiographic provinces: the Appalachian Plateaus, the Interior Low Plateau, and the Coastal Plain. These regions traditionally are split into the following subsections: the Appalachian Plateaus into the Cumberland Plateau and the Cumberland Mountains; the Interior Low Plateau into the Blue Grass, the Highland Rim, and the Shawnee Hills; and the Coastal Plain into the East Gulf Coastal Plain and the Mississippi Alluvial Plain (Fenneman 1938; see Figure 6). A brief description of the physiographic province sections of Kentucky follows. For more detailed physiographic information about these regions and specific material on the impacts on avifauna, see Mengel's *Birds of Kentucky* (1965). A series of photographs following this section (pp. 340-347) depict some of the characteristic features of the state's physiography and vegetation.

Mississippi Alluvial Plain

The Mississippi Alluvial Plain encompasses a relatively small portion of the westernmost part of Kentucky, but it extends southward through the Mississippi River floodplain to the Gulf Coast. In Kentucky the region is bounded on the west by the Mississippi River and on the north by the lower Ohio River. It includes the floodplains of these two large rivers, as well as those of the lower portions of their tributaries, primarily tributaries of the Mississippi, including Mayfield Creek, Obion Creek, and Bayou du Chien. Also included are islands within the Mississippi River and other lands deemed to be within Kentucky but that lie west of the main river channel. The Mississippi Alluvial Plain and the East Gulf Coastal Plain together constitute the Coastal Plain Physiographic Province in Kentucky, which is widely referred to as the Jackson Purchase. Mengel (1965) occasionally refers to the section as the Mississippi Lowlands. In this book the section is similar to that in Mengel's definition, but it has been expanded to include the floodplains of the lower portions of tributaries to the Mississippi River, as well as the floodplain of the lower Ohio River and the lower portion of the Clarks River. The area comprises only about 1% of Kentucky's land mass.

Presettlement vegetation was primarily bottomland forest and swamp, interspersed with floodplain sloughs. Along the larger rivers, broad riparian zones were maintained, which included strips of early successional vegetation along the shores and island margins. Almost no open habitats were not normally covered with water, but expansive sandbars were exposed along the river channels during periods of low flow.

Today, much of the Mississippi Alluvial Plain has been converted to row-crop agricultural use. Wheat, soybeans, and corn are the dominant crops. Settlement is primarily rural, and use of ground for pasture and hay is limited. Although some wetlands remain intact, many have been drained and cleared. Natural vegetation has been reduced to riparian strips and occasional blocks of fairly mature forest. Since the early 1900s beavers (*Castor canadensis*) have returned to the region in great abundance, resulting in an increase in the amount of shallow water wetlands in recent decades (Barbour and Davis 1974). Many of these new wetlands have flooded formerly used farmland, resulting in an increase in early successional swamp forest and marshes dominated by emergent herbaceous plants.

East Gulf Coastal Plain

The East Gulf Coastal Plain includes all of Kentucky west of the Tennessee River (Kentucky Lake) that is not part of the Mississippi Alluvial Plain. The region extends southward to the Gulf Coast, including much of western Tennessee, eastern Mississippi, eastern Louisiana, the southern half of Alabama, and the panhandle of Florida. The East Gulf Coastal Plain and the Mississippi Alluvial Plain together constitute the Coastal Plain Physiographic Province in Kentucky, which is widely referred to as the Jackson Purchase. The area comprises approximately 5% of Kentucky's land mass. The East Gulf Coastal Plain grades widely into the Mississippi Alluvial Plain, and the boundaries used in this book have been drawn arbitrarily on the

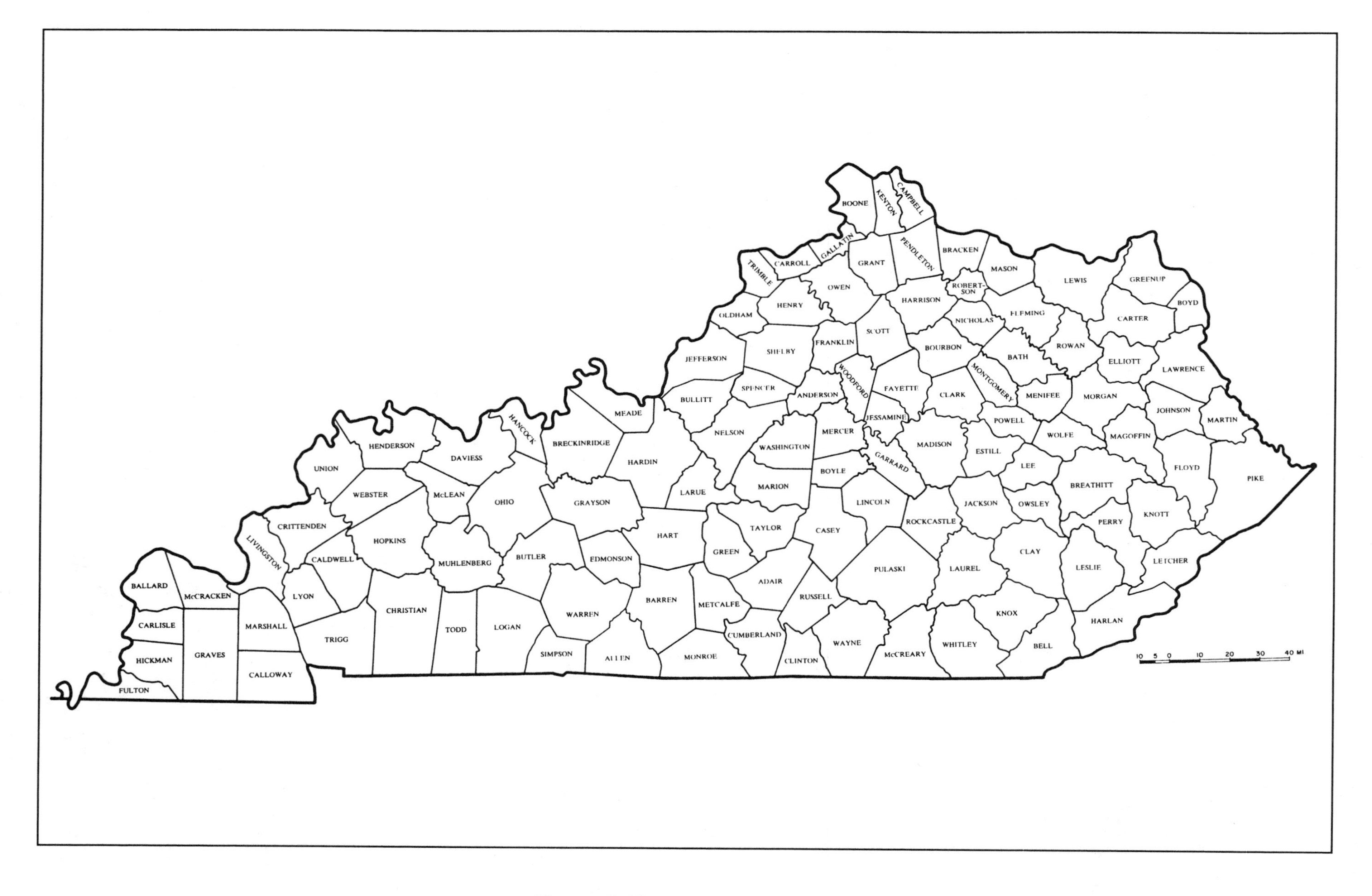

Figure 5. Kentucky's Counties

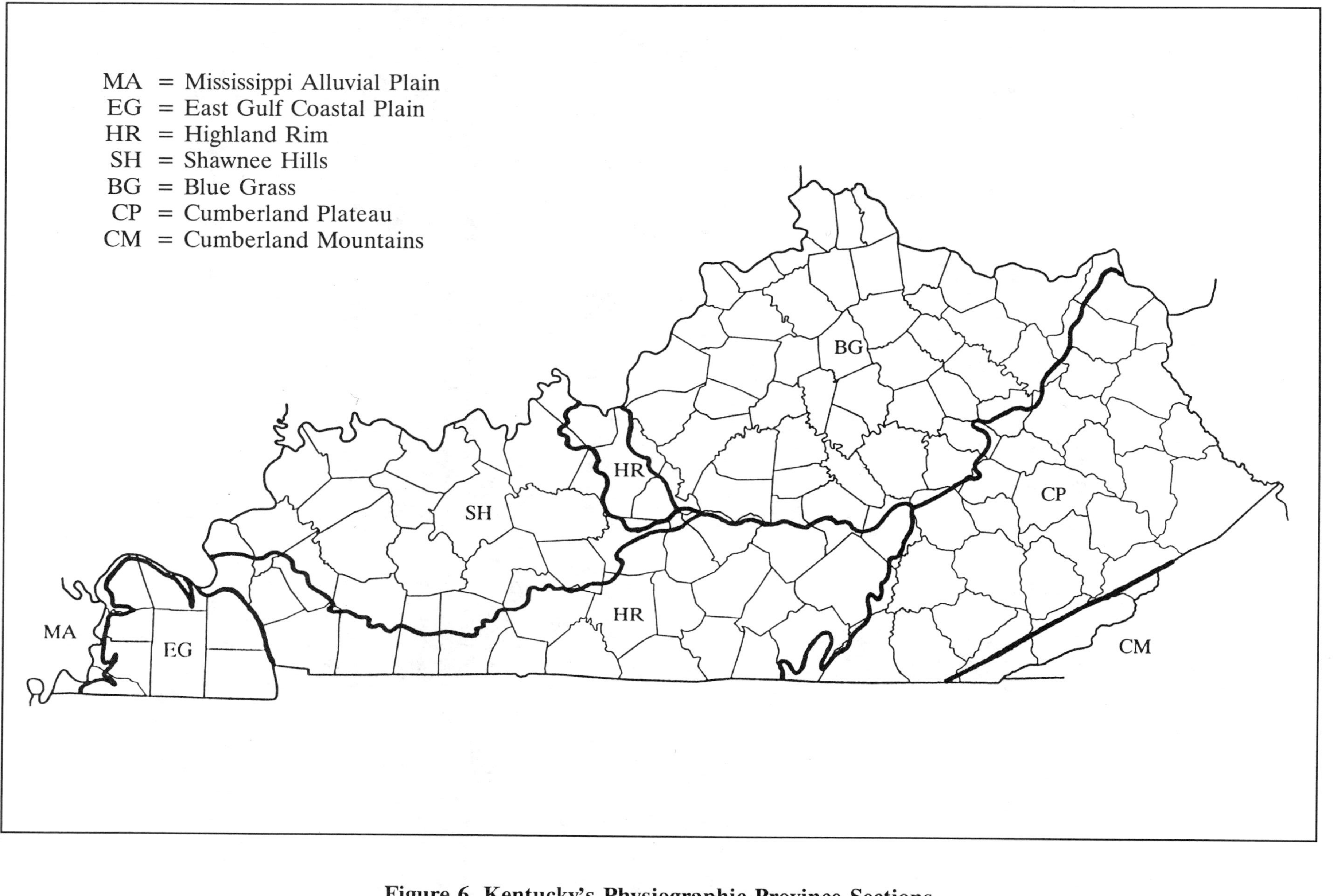

Figure 6. Kentucky's Physiographic Province Sections

tributaries of the larger rivers. In general, land lying above the floodplains of the major rivers and the lower portions of their tributaries, as well as all land (including floodplains) more than ten miles above stream mouths, are included in this section.

Before settlement, much of the East Gulf Coastal Plain was a mosaic of prairie and savannalike woodlands. Expansive areas of native grasses graded in and out of oak-hickory forest, and marshes may have been extensive along some stream floodplains. Mature closed-canopy forest was likely abundant only in the hills adjacent to the Tennessee River, where relief was greater, and in floodplains and swamps that occurred along streams. Closed-canopy oak flatwoods were also distributed locally on moist upland soils.

Today the area has been converted largely to agricultural use and settlement. Row crops of wheat, soybeans, and corn are grown widely, and much land is devoted to use as hayfields and pastures. Settlement is largely rural, but several cities are present. Forest cover is very sparse, the largest tracts lying on slopes along the major streams and among the hilly land near the Tennessee River. Timber companies have converted large areas of oak-hickory forest to monocultures of loblolly pine, and these pines occur in varying ages. Early successional habitat is scattered and not especially abundant.

Highland Rim

The Highland Rim is widely known as the Pennyroyal or Pennyrile. Less commonly it is called the Mississippian Plateau. The section comprises approximately 20% of Kentucky's land mass. The region extends from Kentucky southward across central Tennessee and into northern Alabama. In Kentucky it is bounded on the west by the Tennessee River (Kentucky Lake) and the Ohio River, on the northwest by the Shawnee Hills, on the northeast by the Blue Grass, and on the east by the Cumberland Plateau. The boundary with the Shawnee Hills is the Dripping Springs Escarpment, which is relatively well defined. A small subsection extends northward to the Ohio River, separating the Shawnee Hills from the western Blue Grass. This section includes the city of Elizabethtown. To the northeast, the division between the Highland Rim and the southern Blue Grass is a more or less well-defined escarpment known as Muldraugh's Hill, which dips sharply off the plateau and into the Blue Grass. The southern end of the west-facing Pottsville Escarpment serves as the section's eastern boundary with the Cumberland Plateau.

Before settlement, flatter portions of the Highland Rim were dominated by prairies and open-canopy forest known as barrens. This vegetation predominated across most of the western half of the section, including the northern Elizabethtown subsection. The region's relatively flat terrain and karstic geology enhanced the effects of fire, which was mainly responsible for maintaining these habitats. Forested areas in uplands were dominated by oak-hickory associations, and woodlands also were developed on some flatter lands, especially along streams. To the northeast and east, greater relief and less well-developed karst resulted in a higher degree of forest cover. Here, mesophytic forest predominated on lower slopes and higher terraces above floodplains of rivers and streams, where bottomland forest was well developed. It is not known to what extent the activities of Native Americans affected the landscape, especially in maintaining prairie grasslands with fire, but their impact may have been great.

Today the landscape of the Highland Rim has been highly altered. Native prairies and barrens have been nearly eliminated. Bottomland forest has been cleared, and most of the few wetlands have been drained and cleared. Forest cover remains relatively prominent only in areas of greater relief, primarily in the section's eastern transition to the Cumberland Plateau. Agricultural use of the land for row crops, hayfields, and pasture is considerable. Settlement is mostly rural, although several cities and many towns are situated within the region. The damming of the region's major rivers has resulted in the creation of several large reservoirs.

Shawnee Hills

The Shawnee Hills section is widely known as the Western Coal Field; Mengel (1965) frequently calls it the Western Highlands. The region comprises approximately 20% of Kentucky's land mass. It is bounded on the north by the Ohio River and is otherwise surrounded by the Highland Rim. The section extends northward into southern Illinois and Indiana. The boundary with the Highland Rim has been obscured at the eastern end by erosional patterns, but it is well defined to the south and west by the Dripping Springs Escarpment. Another feature, the Pottsville Escarpment, runs concentrically within the Dripping Spring Escarpment, creating greater relief and a higher degree of forest cover. The lowlands to the north, especially along the Ohio and lower Green rivers, are flatter.

Before settlement, most of the Shawnee Hills was dominated by forest. Upland areas and moderately dry slopes were covered with oak-hickory and other subxeric forest associations, some mixed with Virginia pine. Stream ravines and slopes graded to mesophytic forest, and in some areas mesic conditions fostered the dominance of forest similar to that found in ravines of eastern Kentucky, with species such as eastern hemlock, white pine, and umbrella magnolia. River floodplains were dominated by bottomland forest, and especially along the Ohio and lower Green rivers, many floodplain sloughs and extensive swamp forests were present. Open habitats were restricted to clifftops, the margins of floodplain sloughs, and drier slopes and flat areas in uplands, where fire may have maintained scattered prairie patches.

Today, much of the Shawnee Hills has been converted to agricultural use and settlement. Although much farmland supports row crops such as corn and soybeans, hayfields and pastures also occupy a significant amount of land. Settlement is primarily rural, although several cities are present, the largest occurring along the Ohio River. Surface mining has altered much of this section significantly. As a result of mining activities, expansive areas of upland forest and farmland have been replaced by grassland and secondary forest of varying ages. Most wetlands, too, have been cleared, drained, or otherwise degraded because of human activity. Forests are highly fragmented throughout most of the Shawnee Hills, although some large, contiguous tracts are present, especially along the Pottsville Escarpment and southward. Loblolly pines have been planted widely for pulpwood production in the western part of the region, and a few species of pines are common on reclaimed mines in the southern portion. Early successional habitats are more abundant here than in other parts of western Kentucky, in large part because of the greater relief and the amount of land left idle following mining activity.

Blue Grass

The Blue Grass is the largest physiographic province section in Kentucky, comprising nearly 30% of the state's land mass. It is bounded on the north by the Ohio River, on the west and south by the Highland Rim, and on the east by the Cumberland Plateau. Muldraugh's Hill serves as the boundary with the Highland Rim, while the Pottsville Escarpment defines the eastern boundary with the Cumberland Plateau. The section extends northward into southern Indiana and Ohio. The Kentucky River, deeply entrenched in a steep-sided gorge, bisects the Blue Grass. The section is commonly divided further into the Inner Blue Grass, the Outer Blue Grass, and the Knobs subsections.

The vegetation of the Inner Blue Grass before settlement was unique in Kentucky. Open-canopy forest with a herbaceous understory dominated the region, but unlike the open prairies and barrens of the western part of the state, these savannalike habitats apparently were maintained by a combination of fire, occasional drought, and great herds of bison (*Bos bison*) that roamed seasonally through the region. Thick stands of cane were scattered about, and closed-canopy forests may have been well developed only on slopes and floodplains along the major streams. Greater relief in the more rolling Outer Blue Grass and the well-defined Knobs subsections resulted in greater forest cover. Uplands were dominated by unbroken forests of oak-hickory associations, and some slopes supported rich, mesophytic forests. Small openings known as cedar glades were present on some drier slopes, and in some areas these prairielike habitats likely provided additional habitat for edge and grassland birds. Bottomland forests were well developed along the major rivers and streams, although wetlands were likely limited in extent.

Today, most of the Blue Grass has been highly altered by human activity. Although use of the land for row-crop agriculture is common, most land has been converted to hayfields and pastures. This is especially true of the Inner Blue Grass, where well-manicured horse farms are the predominant feature. Settlement is mostly rural, although the state's two largest cities, Louisville and Lexington, are located in the section, along with many smaller cities and towns. Native forests largely have been cleared, and most remnants are highly fragmented, although a few strips remain intact along the major rivers and within the less-developed portions of the Knobs, especially along the eastern transition to the Cumberland Plateau.

Cumberland Plateau

The Cumberland Plateau is sometimes called the Eastern Coal Field. It is Kentucky's second largest physiographic province section, comprising nearly 25% of the state's land mass. Its western boundary is defined by the Pottsville Escarpment, which descends toward the Blue Grass and the Highland Rim. To the north and east, the region extends into Ohio and West Virginia. To the southeast, the north slope of Pine Mountain provides a well-defined limit. Changes in elevation in this area are not significant, and most of the region lies below 2,000 feet. Some researchers have referred to the northern half of the section as the Allegheny Plateau, although this distinction is so obscure that it has not been used in this book. Of greater utility is recognition of the western Cliff Section (Braun 1950), which runs in a narrow band along the western boundary from Wolfe and Powell counties, southward through McCreary and Whitley counties. In this region, the topography is especially rugged, with extensive systems of clifflines and well-developed ravine forests.

Before settlement the Cumberland Plateau was almost completely forested. Open habitats apparently were restricted to riparian zones along the larger rivers, clifflines, and natural openings created by fire and severe winds. In addition, a few natural openings likely were maintained in areas of thin, rocky soil, especially on upland slopes. In some areas, it is believed that an open-canopy, mixed pine-hardwood community was maintained by fire (Braun 1950; Campbell et al. 1991). Forest vegetation varied from mixed pine-hardwood and subxeric hardwood associations in upland areas and drier slopes to rich mesophytic forest of lower slopes and ravines. Mesophytic associations, including magnolias and eastern hemlock, were common, especially in steep-sided ravines. Before the mid-1900s American chestnut was a common tree in the forests of eastern Kentucky, at least occasionally comprising up to 50% of the canopy (Braun 1950). Native pines were found in a variety of situations. Virginia, pitch, and shortleaf pines occurred in xeric to subxeric conditions, typically in association with hardwoods. In the Cliff Section, white pine occurred in a few mesic ravines. Mature bottomland forest dominated river floodplains, and early successional habitat was limited to the margins of natural openings.

Today the Cumberland Plateau remains primarily forested, although human impacts have been profound. Most of the floodplains have been cleared and settled or otherwise developed for human use. Surface mining for coal has resulted in the clearing of vast areas of forest in upland areas and on slopes. Most of the forest remaining has been logged, resulting in a significant shift in forest structure. Virgin forests often maintained a relatively open aspect in midstory levels as well as gaps in the canopy. Many second-growth forests now are composed of a closed canopy and a thick midstory, making them less suitable for some forest species such as Eastern Wood-Pewee and Great Crested Flycatcher. Whereas early successional and open habitats formerly occurred sporadically in this region, they are now relatively abundant. Recently reclaimed surface mines are dominated by herbaceous growth, and many neglected areas and logged sites have an abundance of early successional vegetation. Agricultural activity generally is restricted to the floodplains of the larger rivers and streams and is absent in many areas. Although some row-crop agriculture is practiced, most farmland is used for pasture, hay, and small patches of tobacco. Settlement is largely rural, occupying flatter lands along the rivers and streams, but several cities are located in the region.

Cumberland Mountains

The Cumberland Mountains physiographic province section lies along the state's southeastern border with Virginia. The section comprises only about 2% of Kentucky's land mass, but it includes some of the most interesting topography in the state. It is essentially a narrow sliver of mountain ranges, defined to the northwest by the fairly continuous ridge of Pine Mountain. The region extends south and east into adjacent portions of Tennessee, Virginia, and West Virginia. In Kentucky, the region includes three main mountain systems: Pine Mountain, including its north slope; Cumberland Mountain; and the Black Mountain range, including Black, Little Black, and Log Mountains. Elevation on Pine Mountain ranges up to about 3,250 feet, but Cumberland Mountain reaches just over 3,500 feet, and the summit of Black Mountain attains Kentucky's highest elevation, at 4,150 feet.

Before settlement the Cumberland Mountains were primarily covered in mature forest. Open habitat was restricted to the riparian corridors of the Cumberland River and its major tributaries, clifflines and rock outcrops, streamhead bogs, and natural openings created by fire and windstorms. It appears that relatively open pine savannas occurred in some xeric habitats, especially on the southeast-facing slope of Pine Mountain. Other mixed pine-hardwood associations were found on subxeric to xeric ridges and slopes. On lower slopes and ravines, mesophytic forest predominated, with a good mixture of eastern hemlock and northern hardwoods in more protected areas. American chestnut was once very common throughout the region. Floodplains of the rivers and streams were covered by mature bottomland forest.

Today much of the Cumberland Mountains section remains forested, but human manipulation of the landscape has greatly altered portions of the region. Flat land in the river and stream floodplains largely has been cleared for settlement. Surface mining has been extensive in parts of the Black Mountain range, creating open and early successional habitat. Logging of remaining forest has altered the forest structure, making it similar to that in the Cumberland Plateau. Settlement is largely restricted to rural strips along watercourses and the mouths of ravines, but two major cities are present. Likewise, agriculture is limited to the flatter land along rivers and streams, and it consists primarily of scattered patches of row crops, hay, and pastureland. A northern hardwood forest type unique to Kentucky occupies the highest slopes of Black Mountain. This forest is dominated by sugar maple, Ohio buckeye, yellow birch, and black cherry and occurs in areas that have not been disturbed by logging, fire, or mining.

Kentucky State Nature Preserves Commission

A water tupelo/bald cypress swamp at Axe Lake, in Ballard County. This wetland system is the state's largest remaining permanently inundated swamp. The area harbors Kentucky's only sizable nesting population of Brown Creepers, as well as an extensive heronry of Great Blue Herons and Great Egrets.

Brainard Palmer-Ball Jr.

A view of the Mississippi River from Chalk Bluff, in Hickman County. Natural edges along the banks and island shorelines provide an abundance of nesting habitat for a variety of species that use early successional vegetation. Such habitats were scarce in the region before most of the bottomland forests and swamps were cleared and drained.

Brainard Palmer-Ball Jr.

A bald cypress slough in Henderson County. Although greatly decreased by human alteration, similar wetlands are scattered across the western one-third of the state. Most occur along the major river floodplains from the Mississippi River northeastward along the Ohio and Green rivers to Daviess and Muhlenberg counties, respectively. Such wetlands are used by a variety of waterbirds, cavity nesters, and woodland and edge species.

Brainard Palmer-Ball Jr.

A sandbar on the Mississippi River in Carlisle County. The numbers and extent of such sandbars have been greatly diminished by navigation structures, especially on the Ohio River. These open habitats support the state's limited nesting population of Least Terns.

Marc Evans

A prairie remnant in Logan County. Such grassy openings are remnants of large expanses of native grassland that formerly occurred over much of the Highland Rim and East Gulf Coastal Plain. Virtually all such habitats have been converted to agricultural use.

Landon McKinney

Intensive agricultural use of land occurs extensively in the western two-thirds of Kentucky. Forested lands are often limited to isolated tracts and riparian corridors, as is shown here along the Green River in McLean County.

Brainard Palmer-Ball Jr.

The Knobs section of the Blue Grass is characterized by series of abrupt hills that rise out of otherwise relatively flat terrain dominated by agricultural use. These hills support deciduous forests with scattered patches of Virginia pine, and they provide the greatest amount of habitat for woodland species found in the Blue Grass.

Brainard Palmer-Ball Jr.

Exposed fossil beds at the Falls of the Ohio, in Jefferson County. This seasonally dry, rocky habitat with interspersed patches of sand, shrubs, and native grasses has provided the only traditional nesting habitat in the state for Spotted Sandpipers.

Tom Bloom

A horse farm in the Inner Blue Grass region. Similar habitat dominates much of the central portion of the Blue Grass. These intensively manicured farms provide little habitat for woodland and edge species, but the open habitats harbor populations of some grassland nesting species, such as Savannah Sparrows and Bobolinks.

Kentucky State Nature Preserves Commission

A Kentucky River bluff at Raven Run, in Fayette County. The Kentucky River has cut a deeply entrenched gorge through the heart of the Blue Grass, providing otherwise rare nesting habitat for woodland species. Protected sites along clifflines are commonly used by vultures and raptors for nesting.

Marc Evans

A hillside glade in Lewis County. Similar naturally open habitats are scattered across the state. Such areas formerly provided some of the only early successional habitat for species such as Prairie Warblers. Most have now been converted to pastures and homesites or have become overgrown with red cedars.

Brainard Palmer-Ball Jr.

Surface mining for coal has altered extensive areas in the Cumberland Mountains, Cumberland Plateau, and Shawnee Hills sections of the state. Within the first few years following reclamation, these grassy areas provide nesting habitat for species that use open ground, including Grasshopper Sparrows, Northern Harriers, and Short-eared Owls.

Deborah White

Big South Fork of Cumberland River, in McCreary County. Large river corridors in eastern Kentucky provide open habitats in a region otherwise naturally dominated by a variety of deciduous and mixed forest types.

Julian Campbell

Dumpling Rocks, in Pulaski County. Massive sandstone outcroppings add greatly to the diversity of nesting habitats available to birds in parts of eastern Kentucky, especially the Cliff Section of the Cumberland Plateau.

Marc Evans

Extensive second growth forests dominate much of the Cumberland Mountains of southeastern Kentucky. Large rock outcrops and cliffs scattered along the ridges provide nesting habitat for Common Ravens, while mesic ravines are inhabited by many woodland species, including Swainson's Warblers.

Marc Evans

Black Mountain, in Harlan County. Rich deciduous forest covers most of Kentucky's highest mountain, which reaches an elevation of approximately 4,150 feet. The northern hardwoods forest found at higher elevations provides habitat for several species seldom or never found nesting elsewhere in the state, including Veery, Blackburnian and Canada warblers, and Dark-eyed Junco.

APPENDIX B. COMMON AND SCIENTIFIC NAMES OF PLANTS NOTED IN THE TEXT

Common Name	Scientific Name
alder	*Alnus* spp.
ash	*Fraxinus* spp.
bald cypress	*Taxodium distichum*
basswood	*Tilia* spp.
(American) beech	*Fagus grandifolia*
black cherry	*Prunus serotina*
black locust	*Robinia pseudoacacia*
blackberry	*Rubus* sp.
blueberry	*Vaccinium* sp.
boxelder	*Acer negundo*
(Ohio) buckeye	*Aesculus glabra*
bulrush	*Scirpus* spp.
burreed	*Sparganium* spp.
buttonbush	*Cephalanthus occidentalis*
(giant) cane	*Arundinaria gigantea*
cattail	*Typha* spp.
(American) chestnut	*Castanea dentata*
cottonwood	*Populus deltoides*
crabapple	*Malus* sp.
(American) elm	*Ulmus americana*
fescue	*Festuca* spp.
greenbriar vine	*Smilax* sp.
hackberry	*Celtis occidentalis*
(eastern) hemlock	*Tsuga canadensis*
hickory	*Carya* spp.
(wild) hydrangea	*Hydrangea arborescens*
indigo bush	*Amorpha fruticosa*
loblolly pine	*Pinus taeda*
magnolias	*Magnolia* spp.
maple	*Acer* spp.
mountain laurel	*Kalmia latifolia*
mulberry	*Morus* spp.
musk thistle	*Carduus nutans*
oak	*Quercus* spp.
orchard grass	*Dactylis glomerata*
palmetto	*Serenoa repens*
pitch pine	*Pinus rigida*
red cedar	*Juniperus virginiana*
red maple	*Acer rubrum*
(great) rhododendron	*Rhododenron maximum*
sedge	*Carex* spp.
shortleaf pine	*Pinus echinata*
Spanish moss	*Tillandsia usneoides*
spruce	*Picea* spp.
sugar maple	*Acer saccharum*
sumac	*Rhus* spp.
(American) sycamore	*Platanus occidentalis*
thistle	*Cirsium* spp.
timothy grass	*Phleum pratense*
tulip poplar	*Liriodendron tulipfera*
umbrella magnolia	*Magnolia tripetala*
Virginia pine	*Pinus virginiana*
(black) walnut	*Juglans nigra*
water tupelo	*Nyssa aquatica*
wheat	*Triticum aestivum*
white pine	*Pinus strobus*
willow	*Salix* spp.
witch hazel	*Hamamelis virginiana*
yellow birch	*Betula lutea*

APPENDIX C. SAMPLE KBBA FIELD CARD

SPECIES	OB	PO	PR	CO	Ab

BLOCK COVERAGE

Trip No.	Date	Hours	# New Species
1.	3-10-85	1	3
2.	3-24-85	1	8
3.	4-14-85	2	19
4.	4-17-85	2	11
5.	4-27-85	4	27
6.	5-5-85	2	3
7.	5-11-85	3	4
8.	5-18-85	1.5	1
9.	5-25-85	2	3
10.	6-9-85	4.5	2
11.	6-27-85	2	1
12.	6-28-85	1	0
13.	6-30-85	4	1
14.	7-4-85	.5	0
15.	7-6-85	1	0
16.	8-4-85	2.5	0
17.	8-10-85	3	0
18.	8-18-85	1	0
19.			
20.			
TOTALS: HOURS 38		SPECIES 83	

FURTHER INSTRUCTIONS

Abundance codes to be used in the "Ab" column are: '1' only one pair observed, '2' very small numbers restricted to very small area, '3' small numbers seen in scattered localities, '4' fairly widespread in block, '5' one of the commonest nesting species in the block.

NOTE: '*' denotes species with uncertain breeding status that require completed 'Notable Species Location Form'.

If a species is encountered that does not appear on this list, immediately report possible nesting to your REGIONAL COORDINATOR.

NOTES

FIELD CARD

KENTUCKY

BREEDING BIRD ATLAS

Quadrangle

Quad name RISING SUN Region # 5

BLOCK (circle one) YEAR

nw ~~ne~~ cw ce sw se
A (B) C D E F 1985

Observer's Name: Lee McNeely

Address:

Phone:

BREEDING CRITERIA & CODES

OBserved
0 Species observed but not believed to be breeding in block

POssible
X Species heard or seen in breeding habitat

PRobable
A Agitated behavior/anxiety calls
P Pair seen
T Bird holding territory
C Courtship or copulation
N Visiting probable nest site
B Nest building by wrens or woodpeckers

COnfirmed
DD Distraction display
NB Nest building
UN Used nest
FL Recently fledged young
FS Parent with fecal sac
FY Parent with food for young
ON Adult leaving/entering nest site
NE Nest with eggs
NY Nest with young

SPECIES	OB	PO	PR	CO	Ab
Grebe, Pied-billed * .					
Cormorant, Double-cr *					
Bittern, American * .					
Least *					
Heron, Great Blue * .					
Egret, Great *					
Heron, Little Blue * .					
Egret, Cattle * . . .					
Heron, Green-backed .		X			1
Black-cr. Night * .					
Yellow-cr. Night *					
Goose, Canada					
Duck, Wood					
Mallard		X			1
Teal, Blue-winged * .					
Merganser, Hooded * .					
Vulture, Black					
Turkey		X			3
Osprey *					
Kite, Mississippi * .					
Harrier, Northern * .					
Hawk, Sharp-shinned *					
Cooper's *	0				1
Red-shouldered . . .					
Broad-winged		X			3
Red-tailed			P		3
Kestrel, American . .			P		1
Grouse, Ruffed					
Turkey, Wild					
Bobwhite, Northern . .				FL	3
Rail, King *					
Moorhen, Common * . .					
Coot, American * . . .					
Killdeer		X			3
Sandpiper, Spotted * .					
Upland *					
Woodcock, American . .			T		2
Tern, Least *					
Black *					

SPECIES	OB	PO	PR	CO	Ab
Dove, Rock			N		3
Mourning				NY	5
Cuckoo, Black-billed *		X			1
Yellow-billed . . .			T		3
Barn-Owl, Common . . .					
Screech-Owl, Eastern .		X			1
Owl, Great Horned . .			T		1
Barred		X			1
Nighthawk, Common . .	0				1
Chuck-will's-widow . .					
Whip-poor-will		X			2
Swift, Chimney		X			3
Hummingbird, Ruby-thr.		X			3
Kingfisher, Belted . .		X			3
Woodpecker, Red-hd. .					
Red-bellied			N		4
Downy		X			4
Hairy		X			3
Red-cockaded * . . .					
Flicker, Northern . .			T		4
Woodpecker, Pileated .		X			3
Wood-Pewee, Eastern .			T		4
Flycatcher, Acadian .				UN	3
Willow					
Least *					
Phoebe, Eastern . . .				NY	3
Flycatcher, Grt. Cr .			T		3
Kingbird, Eastern . .				FL	4
Lark, Horned					
Martin, Purple				ON	3
Swallow, Tree					
No. Rough-winged . .		X			3
Bank *					
Cliff *					
Barn				NB	4
Jay, Blue		X			4
Crow, American		X			4
Fish *					
Raven, Common * . . .					

SPECIES	OB	PO	PR	CO	Ab
Chickadee, Carolina .		X			4
Titmouse, Tufted . .		X			4
Nuthatch, White-br. .		X			3
Wren, Carolina . . .			T		3
Bewick's *					
House			T		4
Sedge *					
Gnatcatcher, Bl.-gr .				ON	4
Bluebird, Eastern . .			P		3
Veery *					
Thrush, Wood			T		3
Robin, American . . .				FL	5
Catbird, Gray				NB	4
Mockingbird, No. . .				NY	3
Thrasher, Brown . . .				NY	3
Waxwing, Cedar . . .		X			1
Shrike, Loggerhead .					
Starling, European .				FL	5
Vireo, White-eyed . .		X			3
Bell's *					
Solitary *					
Yellow-throated . .			T		3
Warbling			T		3
Red-eyed			T		4
Warbler, Blue-winged			T		3
Golden-winged * . .					
Parula, Northern . .					
Warbler, Yellow . . .			T		4
Chestnut-sided * .					
Black-thr. Blue * .					
Black-thr. Green					
Blackburnian * . .					
Yellow-throated . .			T		3
Pine					
Prairie					
Cerulean			T		3
Black-and-white . .					
Redstart, American .					
Warbler, Prothonotary					

SPECIES	OB	PO	PR	CO	Ab
Warbler, Worm-eating .			T		2
Swainson's					
Ovenbird			T		2
Waterthrush, Louisiana				NY	3
Warbler, Kentucky . .				NY	3
Mourning *					
Yellowthroat, Common .				FY	5
Warbler, Hooded . . .			T		2
Canada					
Chat, Yellow-breasted			T		3
Tanager, Summer . . .			T		3
Scarlet			T		3
Cardinal, Northern . .				NY	4
Grosbeak, Rose-br. * .					
Blue					
Bunting, Indigo . . .			T		5
Dickcissel					
Towhee, Rufous-sided .			T		4
Sparrow, Bachman's * .					
Chipping				FY	4
Field				NE	4
Vesper *					
Lark *					
Savannah *					
Grasshopper					
Henslow's *			T		1
Song			T		4
Bobolink					
Blackbird, Red-winged				NY	5
Meadowlark, Eastern .			T		4
Grackle, Common . . .				ON	5
Cowbird, Brown-headed				NE	4
Oriole, Orchard . . .			T		4
Northern			T		3
Finch, House					
Goldfinch, American .			T		4
Sparrow, House			P		5

LITERATURE CITED

Able, Kenneth P. 1967. Some recent observations from western Kentucky. Kentucky Warbler 43:27-34.

Alerich, Carol L. 1990. Forest statistics for Kentucky—1975 and 1988. United States Department of Agriculture, Northeastern Forest Experiment Station Resource Bulletin NE-117, Radnor, Pennsylvania.

Allaire, Pierre N. 1976. Nesting adaptations of bluebirds on surface mined lands. Kentucky Warbler 52:70-72.

————. 1979. The avifauna of reclaimed surface mined lands: Its composition and role in land use planning. Ph.D. diss., University of Louisville, Louisville, Kentucky.

————. 1981. Summer observations of birds on reclaimed surface coal mines in Breathitt, Bell, Pike, and Harlan Counties. Kentucky Warbler 57:51-54.

Allaire, Pierre N., William C. McComb, Wayne H. Davis, and Robert Brown. 1982. Short-eared Owls use reclaimed surface mine. Kentucky Warbler 58:58-59.

Allen, Jerry W. 1971. A Wood Duck survey at Land Between the Lakes. Kentucky Warbler 47:26-27.

————. 1972. Comments on a two-year Wood Duck survey. Kentucky Warbler 48:3-6.

Alsop, Fred J., III. 1971. An annotated list of birds observed in Hancock County, Kentucky. Kentucky Warbler 47:59-70.

Altsheler, Kay. 1962. Acadian Flycatcher nest used for two broods in one season. Kentucky Warbler 38:45-49.

American Ornithologists' Union. 1973. Thirty-second supplement to the AOU "Check-list of North American birds." Auk 90:411-19.

————. 1983. Check-list of North American birds. 6th ed. The Allen Press, Lawrence, Kansas.

————. 1985. Thirty-fifth supplement to the American Ornithologists' Union "Check-list of North American birds." Auk 102:680-86.

————. 1987. Thirty-sixth supplement to the American Ornithologists' Union "Check-list of North American birds." Auk 104:591-96.

————. 1989. Thirty-seventh supplement to the American Ornithologists' Union "Check-list of North American birds." Auk 106:532-38.

————. 1991. Thirty-eighth supplement to the American Ornithologists' Union "Check-list of North American birds." Auk 108:750-54.

————. 1993. Thirty-ninth supplement to the American Ornithologists' Union "Check-list of North American birds." Auk 110:675-82.

Anderson, D.W., J.J. Hickey, R.W. Risebrough, D.F. Hughes, and R.E. Christenson. 1969. Significance of chlorinated hydrocarbon residues to breeding pelicans and cormorants. Canadian Field Naturalist 83:91-112.

Anderson, William L., and Ronald E. Duzan. 1978. DDE residues and eggshell thinning in Loggerhead Shrikes. Wilson Bulletin 90:215-20.

Andrle, Robert F., and Janet R. Carroll, eds. 1988. The atlas of breeding birds in New York State. Cornell University Press, Ithaca, New York.

Audubon, John J. 1861. The birds of America, from drawings made in the United States and their territories. 7 vols. Roe Lockwood and Son, New York, New York.

Bacon, Brasher C. 1933. Water and wading birds of Hopkins County (in summary). Kentucky Warbler 9:13-15.

Bacon, Brasher C., and Burt L. Monroe. 1937. Birds of Kentucky. Kentucky Warbler 13:11-13.

Baker, Gerald F. 1943. Notes on the Wild Turkey in western Kentucky. Kentucky Warbler 19:25-27.

Barbour, Roger W. 1941. A preliminary list of the summer birds of the summit of Big Black Mountain. Kentucky Warbler 17:46-47.

————. 1951. The summer birds of Rowan and adjacent counties in eastern Kentucky. Kentucky Warbler 27:31-39.

Barbour, Roger W., and Wayne H. Davis. 1974. Mammals of Kentucky. University Press of Kentucky, Lexington, Kentucky.

Barron, Alan. 1981. First House Finch nest in Kentucky. Kentucky Warbler 57:64.

Beckham, Charles W. 1885. List of the birds of Nelson County, Kentucky. Kentucky Geological Survey. John D. Woods, Frankfort, Kentucky.

Bierly, Michael L. 1973. A flock of Mississippi Kites in western Kentucky. Kentucky Warbler 49:72.

Binnewies, Fred W. 1943. Ferry boat attracts Prothonotary Warblers. Kentucky Warbler 19:53-54.

Bohlen, H. David. 1978. An annotated check-list of the birds of Illinois. Illinois State Museum, Springfield, Illinois.

————. 1989. The birds of Illinois. Indiana University Press, Bloomington, Indiana.

Bowne, Ann H. 1972. Yellow Warbler feeds young cowbird. Kentucky Warbler 48:52.

Braun, E. Lucy. 1950. Deciduous forests of eastern North America. Blakiston Co., Philadephia, Pennsylvania.

Brauning, Daniel W., ed. 1992. Atlas of breeding birds in Pennsylvania. University of Pittsburgh Press, Pittsburgh, Pennsylvania.

Breiding, George H. 1947. A list of birds from Big Black Mountain. Kentucky Warbler 23:37-40.

Brittingham, Margaret C., and Stanley A. Temple. 1983. Have cowbirds caused forest songbirds to decline? Bioscience 33:31-35.

Buckelew, A.R., Jr., and G.A. Hall. 1994. The West Virginia breeding bird atlas. University of Pittsburgh Press, Pittsburgh, Pennsylvania.

Butler, Amos W. 1897. The birds of Indiana. Indiana Department of Geology and Natural Resources, 22d annual report. Pp. 515-1197.

Campbell, Julian N.N., David D. Taylor, Max E. Medley, and Allen C. Risk. 1991. Floristic and historical indications of fire-maintained, grassy pine-oak "barrens" before settlement in southeastern Kentucky. In Fire and the environment: Ecological and cultural perspectives, proceedings of an international symposium, ed. S.C. Nodvin and T.A. Waldrop, pp. 359-75. Southeastern Forest Experiment Station, Asheville, North Carolina.

Chapman, Frank M. 1968. The warblers of North America. Dover Publications, Inc., New York, New York.

Clark, Edith. 1962. A sight record of the Swainson's Warbler in eastern Kentucky. Kentucky Warbler 38:67-68.

Claus, Debra A., Wayne H. Davis, and William C. McComb. 1988. Bird use of eastern Kentucky surface mines. Kentucky Warbler 64:39-43.

Clay, Kathryn, and Herbert Clay. 1990. First record of nesting Northern Harriers in Hart County, Kentucky. Kentucky Warbler 66:99.

Coffey, Ben B., Jr. 1942. Fish Crow at Memphis. Migrant 13:42.

————. 1959. The season—Memphis. Migrant 30:36.

Colvin, B.A. 1985. Common Barn-Owl population decline in Ohio and the relationship to agricultural trends. Journal of Field Ornithology 56:224-35.

Cooke, A.S. 1973. Shell thinning in avian eggs by environmental pollutants. Environmental Pollution 4:85-152.

Cooke, Wells W. 1911. The migration of sparrows. Ninth paper. Bird-Lore 13:83-88.

————. 1913. Distribution and migration of North American herons and their allies. United States Department of Agriculture, Biological Survey Bureau Bulletin no. 45, pp. 1-70, May 24.

Cooper, Virginia. 1952. A five weeks' study of a hummingbird's nest. Kentucky Warbler 28:13-14.

Coskren, Dennis. 1979. Newly found localities for Swainson's Warbler in eastern Kentucky. Kentucky Warbler 55:65-66.

Croft, Joseph E. 1961. Traill's Flycatcher in Kentucky. Kentucky Warbler 37:63-70.

————. 1962. Nesting records of Traill's Flycatcher. Kentucky Warbler 38:59-61.

————. 1964. Traill's Flycatcher, 1963. Kentucky Warbler 40:27-28.

————. 1965. An urbanized Peregrine Falcon. Kentucky Warbler 41:15-16.

————. 1967. Summer records of the Blue Grosbeak. Kentucky Warbler 43:67-68.

————. 1969. Notes from the southeastern mountains. Kentucky Warbler 45:67-81.

————. 1970. Ravens in eastern Kentucky. Kentucky Warbler 46:21-22.

————. 1971. Notes from Cumberland Mountain. Kentucky Warbler 47:23-25.

————. 1972. Notes on some northern Kentucky field birds. Kentucky Warbler 48:39-42.

Croft, Joseph E., and Austin R. Lawrence. 1970. Birds of Fort Knox. Kentucky Warbler 46:59-69.

Croft, Joseph E., and William Rowe. 1966. Notes from Fulton County. Kentucky Warbler 42:23-26.

Croft, Joseph E., William Rowe, and R. Haven Wiley Jr. 1965. Notes from the Cumberland National Forest. Kentucky Warbler 41:3-4.

Croft, Joseph E., and Anne L. Stamm. 1964. Notes on the Chuck-will's-widow. Kentucky Warbler 40:31-32.

————. 1967. Kentucky nesting records, 1966. Kentucky Warbler 43:43-51.

Crosby, Gilbert T. 1972. Spread of the Cattle Egret in the Western Hemisphere. Bird Banding 43:205-12.

Davis, Wayne H., and Charles K. Smith. 1978. Birds of the higher mountains. Kentucky Warbler 54:73.

Davis, Wayne H., Charles K. Smith, Jarvis E. Hudson, and Greg Shields. 1980. Summer birds of Cumberland Gap National Historical Park. Kentucky Warbler 56:43-55.

DeLime, John. 1949. An Osprey's and a Bald Eagle's nest at Kentucky Lake. Kentucky Warbler 25:55.

Despard, Thomas L. 1954. Young of Spotted Sandpiper in Franklin County. Kentucky Warbler 30:62-63.

Dodge, Victor K. 1940. Swallows in the Bluegrass. Kentucky Warbler 16:22-23.

————. 1951. First nesting record of the Starling in Kentucky. Kentucky Warbler 27:41.

Dodson, Ronald G. 1976. Great Blue Heron rookery discovered. Kentucky Warbler 52:81.

————. 1977. Report of a Great Blue Heron rookery in Henderson County. Kentucky Warbler 53:40.

Dolbeer, Richard A. 1982. Migration patterns for age and sex classes of blackbirds and starlings. Journal of Field Ornithology 53:28-46.

Dubke, Kenneth H. 1966a. Blue Grosbeak nesting in Taylor County. Kentucky Warbler 42:55.

————. 1966b. Traill's Flycatcher in the Hodgenville area. Kentucky Warbler 42:60.

Durell, James, and David Yancy. 1990. Kentucky Bald Eagle count for 1990. Kentucky Warbler 66:63-64.

Eaden, Brenda, and Tony Eaden. 1988. Short-eared Owls (*Asio flammeus*). Kentucky Warbler 64:50.

Easterla, David A. 1965. Range extension of the Fish Crow in Missouri. Wilson Bulletin 77:297-98.

Ehrlich, Paul R., David S. Dobkin, and Darryl Wheye. 1988. The birder's handbook: A field guide to the natural history of North American birds. Simon and Schuster, Inc., New York, New York.

Elmore, Jackie B., Sr. 1980. House Finches in Pulaski County. Kentucky Warbler 56:21-22.

Embody, George C. 1907. Bachman's Warbler breeding in Logan County, Kentucky. Auk 24:41-42.

Evans, Sherri A. 1981. Ecology and behavior of the Mississippi Kite (*Ictinia mississippiensis*) in southern Illinois. M.A. thesis, Southern Illinois University, Carbondale, Illinois.

Fenneman, Nevin M. 1938. Physiography of the eastern United States. McGraw-Hill, New York, New York.

Fitzhugh, Henry, Jr. 1959. A yellow-crowned night heronry in Louisville. Kentucky Warbler 35:59-65.

———. 1961. A further note on Yellow-crowned Night Herons. Kentucky Warbler 37:20.

Ford, Robert P. 1987. Summary of recent Brown Creeper observations in west Tennessee. Migrant 58:50-51.

Fowler, Dale K., John R. MacGregor, Sherri A. Evans, and Lauren E. Schaaf. 1985. The Common Raven returns to Kentucky. American Birds 39:852-53.

Fuller, Tom. 1951. Nesting of the Wood Pewee. Kentucky Warbler 27:62-63.

Funkhouser, William D. 1925. Wild life in Kentucky. Kentucky Geological Survey, Frankfort, Kentucky.

Ganier, Albert F. 1933. A distributional list of the birds of Tennessee. Tennessee Avifauna no. 1. Tennessee Ornithological Society, Nashville, Tennessee.

———. 1935. Some King Rail nests in Kentucky. Kentucky Warbler 11:5.

———. 1951. The breeding herons of Tennessee. Migrant 22:1-8.

Gersbacher, Eva O. 1939. The heronries at Reelfoot Lake. Journal of the Tennessee Academy of Science 14:162-80.

Getz, L.L., J.E. Hofmann, B.J. Klatt, L. Verner, F.R. Cole, and R.L. Lindroth. 1987. Fourteen years of population fluctuations of *Microtus ochrogaster* and *Microtus pennsylvanicus* in east central Illinois. Canadian Journal of Zoology 65:1317-25.

Gill, F.B. 1980. Historical aspects of hybridization between Blue-winged and Golden-winged Warblers. Auk 97:1-18.

Graber, R.R., J.W. Graber, and E.L. Kirk. 1973. Illinois birds: Laniidae. Illinois Natural History Survey Biological Notes no. 83.

Gray, Willard. 1968a. Another nest of the Blue Grosbeak. Kentucky Warbler 44:51.

———. 1968b. Forty-fifth annual fall meeting. Kentucky Warbler 44:61-63.

Greene, William, Jr. 1978. Some observations from northeastern Kentucky. Kentucky Warbler 54:14.

Guthrie, Charles S. 1961. Barn Swallows use nest six consecutive years. Kentucky Warbler 37:80.

Hall, George A. 1976. The spring migration—April 1–May 31, 1976: Appalachian Region. American Birds 37:841-44.

———. 1983. West Virginia birds: Distribution and ecology. Special Publication no. 7. Carnegie Museum of Natural History, Pittsburgh, Pennsylvania.

Halverson, Emily. 1955. Notes on the nesting of the Yellow-crowned Night Heron near Caperton Swamp. Kentucky Warbler 31:64-66.

Hancock, James W. 1947. The Golden-winged Warbler in Hopkins County in summer. Kentucky Warbler 23:4.

———. 1951. Some nesting records for Hopkins County. Kentucky Warbler 27:9.

———. 1954. The breeding birds of Hopkins County. Kentucky Warbler 30:3-5, 19-25, 41-47.

———. 1966. A Pine Warbler nest at Pennyrile. Kentucky Warbler 42:57-58.

———. 1973. Nesting of the Song Sparrow in western Hopkins County. Kentucky Warbler 49:45.

Hardy, Frederick C. 1950. Ruffed Grouse nesting sites on the Laurel District Cumberland National Forest. Kentucky Warbler 26:2-4.

Hardy, John W. 1957. The Least Tern in the Mississippi Valley. Publications of the Museum of Michigan State University 1:1-60.

Harm, Ray. 1973. Notes from Bell County. Kentucky Warbler 49:16-17.

Harrison, Hal H. 1975. A field guide to birds' nests in the United States east of the Mississippi River. Houghton-Mifflin, Co., Boston, Massachusetts.

Hays, Rodney M. 1957. Cooperative breeding bird list for Kentucky, 1956. Kentucky Warbler 33:3-7.

Heilbrun, Lois H., and CBC regional editors. 1983. The eighty-third Audubon Christmas bird count—Breaks Interstate Park, Va.-Ky. American Birds 37:821.

Hendricks, William D., Landon E. McKinney, Brainard Palmer-Ball Jr., and Marc Evans. 1991. Biological inventory of the Jackson Purchase region of Kentucky. Kentucky State Nature Preserves Commission, Frankfort, Kentucky.

Hibbard, Claude W. 1935. Notes from Mammoth Cave National Park (proposed), Kentucky. Auk 52:465-66.

Hibbs, G.D. 1927. Notes from Cox's Creek, Nelson County. Kentucky Warbler 3:5-6.

Hodges, Carolyn Gay. 1992. Great Blue Herons nest in Pulaski County. Kentucky Warbler 68:51.

Hoover, Jeffrey, and Margaret Brittingham. 1993. Regional variation in cowbird parasitism of Wood Thrushes. Wilson Bulletin 105:228-38.

Householder, Jane. 1978. House Finches in Madison County. Kentucky Warbler 54:35-36.

Howell, Arthur H. 1910. Notes on the summer birds of Kentucky and Tennessee. Auk 27:295-304.

Hudson, Jarvis. 1971. Some notes on the birds of Lilley's Woods. Kentucky Warbler 47:27-28.

Jackson, Jerome A., Ray Weeks, and Patricia Shindala. 1976. The present status and future of Red-cockaded Woodpeckers in Kentucky. Kentucky Warbler 52:75-80.

Jackson, Violet, and Wilbur Jackson. 1986. House Finch builds nest in a hanging fern plant. Kentucky Warbler 62:20.

Johnson, George. 1980. Cooper's Hawk nest in Barren County. Kentucky Warbler 56:21.

Jones, Howard P. 1966. Nest of a Parula Warbler near Frankfort. Kentucky Warbler 42:56-57.

———. 1969. The Common Grackle: A nesting study. Kentucky Warbler 45:3-8.

———. 1975. Red-headed Woodpeckers may be increasing in Franklin County. Kentucky Warbler 51:82-83.

Kain, Teta, ed. 1987. Virginia's birdlife: An annotated checklist. 2d ed. Virginia Avifauna no. 3. Virginia Society of Ornithology, Lynchburg, Virginia.

Keller, Charles E., Shirley A. Keller, and Timothy C. Keller. 1986. Indiana birds and their haunts. 2d ed. Indiana University Press, Bloomington, Kentucky.

Kemsies, Emerson, William H. Mers, and Worth Randle. 1950. First Kentucky breeding record for Upland Plover. Kentucky Warbler 26:49.

Kemsies, Emerson, and Worth Randle. 1953. Birds of southwestern Ohio. Edward Brothers, Ann Arbor, Michigan.

Kentucky State Nature Preserves Commission. 1992. Endangered, threatened, and special concern plants and animals of Kentucky. Frankfort, Kentucky.

Kleber, John E., ed. 1992. The Kentucky encyclopedia. University Press of Kentucky, Lexington, Kentucky.

Kleen, Vernon M. 1973. The nesting season—June 1–July 31, 1973: Middlewestern prairie region. American Birds 27:874-78.

———. 1975. The nesting season—June 1–July 31, 1975: Middlewestern prairie region. American Birds 29:978-82.

Kleen, Vernon M., and Lee Bush. 1973. The fall migration, August 16–November 30, 1972: Middlewestern prairie region. American Birds 27:66-70.

Lancaster, L.Y., and Gordon Wilson. 1964. Blue Grosbeak breeding at Bowling Green. Kentucky Warbler 40:54-55.

Langdon, Frank W. 1879. A revised list of Cincinnati birds. Journal of the Cincinnati Society of Natural History 1:167-93.

Larson, Edwin R. 1970. A nesting study of the Common Nighthawk. Kentucky Warbler 46:3-6.

———. 1973. Killdeer nesting on roof tops. Kentucky Warbler 49:44-45.

———. 1979. A follow-up of nesting Red-cockaded Woodpeckers in Laurel County. Kentucky Warbler 55:67.

Lovell, Harvey B. 1942. The nesting of the starling in Kentucky. Kentucky Warbler 18:29-34.

———. 1948. The Pine Warbler in Kentucky. Kentucky Warbler 24:33-39.

———. 1950a. Breeding birds of Big Black Mountain. Kentucky Warbler 26:57-66.

————. 1950b. Some breeding records from eastern Kentucky. Auk 67:106-8.

————. 1951. Breeding bird list for Kentucky, 1951. Kentucky Warbler 27:58-62.

————. 1952. Birds in a Kentucky swamp. Kentucky Warbler 28:53-54.

Lowe, Richard L. 1980. Bald eagle restoration efforts at LBL. Kentucky Warbler 56:71.

Ludwig, J.P. 1985. Decline, resurgence, and population dynamics of Michigan and Great Lakes Double-crested Cormorants. Jack-Pine Warbler 62:91-102.

Martin, William H. 1989. The role and history of fire in the Daniel Boone National Forest. Report prepared for the Daniel Boone National Forest by the Division of Natural Areas, Eastern Kentucky University, Richmond, Kentucky.

Mason, Wayne M. 1977. Cliff Swallows nesting on Barren River Reservoir. Kentucky Warbler 53:40-41.

————. 1978. Blue-winged Warbler nest in Barren County. Kentucky Warbler 54:53.

Mason, Wayne M., and Blaine R. Ferrell. 1983. Cedar Waxwings nesting in Warren County. Kentucky Warbler 59:59.

Mayfield, Al H. 1956. Nest of the Yellow-crowned Night Heron in the Lexington area. Kentucky Warbler 32:62.

McNeely, Lee K. 1987. A Broad-winged Hawk's nest in northern Kentucky. Kentucky Warbler 63:67.

————. 1988. Observations of nesting Common Barn-Owls in Boone County. Kentucky Warbler 64:23-24.

McNeely, Lee K., Joe and Kathy Caminiti, and Tommy Stephens. 1991. Summering Rose-breasted Grosbeaks in Boone County. Kentucky Warbler 67:86.

Mengel, Robert M. 1965. The birds of Kentucky. American Ornithologists' Union Monograph no. 3. The Allen Press, Lawrence, Kansas.

Monroe, Burt L. 1935. Birds of a Jefferson County marsh. Kentucky Warbler 11:20-22.

————. 1946. Nest of Cedar Waxwing at Louisville. Kentucky Warbler 22:45-46.

————. 1947. Hooded Merganser at Louisville in summer. Kentucky Warbler 23:57-60.

Monroe, Burt L., and Burt L. Monroe Jr. 1961. Birds of the Louisville region. Kentucky Warbler 37:23-42.

Monroe, Burt L., Jr. 1976. Birds of the Louisville region. Kentucky Warbler 52:39-64.

————. 1978. Analysis of Kentucky's breeding birds: Declining species. Kentucky Warbler 54:19-26.

————. 1979. Analysis of Kentucky's breeding birds: Increasing species. Kentucky Warbler 55:23-28.

Monroe, Burt L., Jr., and Kenneth P. Able. 1968. Recent additions to the avifauna of Kentucky. Kentucky Warbler 44:55-57.

Monroe, Burt L., Jr., and Lois S. Cronholm. 1977. Blackbird study. Final report submitted to the Kentucky Environmental Quality Commission. Frankfort, Kentucky.

Moore, Suzanne C. 1969. Vesper Sparrow nesting in Franklin County. Kentucky Warbler 45:56-57.

Morse, John S. 1948. Nest of Woodcock in Marshall County. Kentucky Warbler 24:41.

Mumford, Russell E., and Charles E. Keller. 1984. The birds of Indiana. Indiana University Press, Bloomington, Indiana.

Murphy, Ginger A. 1982. 1981 status of the Red-cockaded Woodpecker on the Daniel Boone National Forest. Kentucky Warbler 58:43-47.

Nelson, Edward W. 1877. Notes upon birds observed in southern Illinois, between July 17 and September 4, 1875. Bulletin of Essex Institute 9:32-65.

Nelson, Lee K. 1973. Woodcock nesting records. Kentucky Warbler 49:43-44.

———. 1981. Mourning Dove nesting study. Kentucky Warbler 57:31-34.

Nelson, Ray. 1959. Wild Turkeys at Mammoth Cave National Park. Kentucky Warbler 34:13-14.

Nice, Margaret M. 1964. Studies in the life history of the Song Sparrow. 2 vols. Dover Publications, Inc., New York, New York.

Nicholson, Charles P. 1981. Nesting of the Bell's Vireo in Kentucky. Kentucky Warbler 57:77-79.

Palmer-Ball, Brainard, Jr. 1980. Some bird observations in the Louisville area. Kentucky Warbler 56:37-39.

———. 1991a. Current status of colonial nesting waterbirds in Kentucky. Journal of the Tennessee Academy of Science 66:211-14.

———. 1991b. Nesting of Pied-billed Grebe at Chaney Lake. Kentucky Warbler 67:86-87.

———. 1992. Late nesting of House Wrens in Franklin County. Kentucky Warbler 68:63-64.

———. 1993. A natural nest site of the Bewick's Wren. Kentucky Warbler 69:64-65.

———. 1995. 1994 survey of Least Tern nesting colonies in Kentucky. Kentucky Warbler 71:5-8.

Palmer-Ball, Brainard, Jr., and Alan Barron. 1982. Western Kentucky observations. Kentucky Warbler 58:75-81.

———. 1990. Notes on breeding birds of the reclaimed surface mines in western Kentucky. Kentucky Warbler 66:73-80.

Palmer-Ball, Brainard, Jr., and Gary Boggs. 1991. Return of the Woodburn lakes, 1989: Parts I and II. Kentucky Warbler 67:33-45 (pt. I), 60-66 (pt. II).

Palmer-Ball, Brainard, Jr., and Sherri A. Evans. 1986. Nesting of herons on Shippingport Island. Kentucky Warbler 62:75-76.

Palmer-Ball, Brainard, Jr., and Wendell Haag. 1989. First reported nesting of Brown Creeper in Kentucky. Kentucky Warbler 65:77-78.

Palmer-Ball, Brainard, Jr., and Richard R. Hannan. 1986. Resident and migrant bird study, McAlpine Locks and Dam vicinity. Report submitted to the United States Army Corps of Engineers, Louisville District, by the Kentucky State Nature Preserves Commission, Frankfort, Kentucky.

Pasikowski, James. 1972. An August nest of the Wood Thrush. Kentucky Warbler 48:33.

Pearson, William, and Juanelle Pearson. 1985. Black Vultures nesting in an abandoned building. Kentucky Warbler 61:49-51.

Perring, F.H., and S.M. Walters, eds. 1962. Atlas of the British flora. T. Nelson, London, England.

Peterjohn, Bruce G. 1989. The birds of Ohio. Indiana University Press, Bloomington, Indiana.

Peterjohn, Bruce G., and Dan L. Rice. 1991. The Ohio breeding bird atlas. Ohio Department of Natural Resources, Division of Natural Areas and Preserves, Columbus, Ohio.

Peterson, Clell T. 1970. Cliff Swallows breeding in Land Between the Lakes area. Kentucky Warbler 46:7-9.

————. 1973. Cliff Swallows in western Kentucky. Kentucky Warbler 49:63-65.

————. 1980. Cedar Waxwings nest in western Kentucky. Kentucky Warbler 56:65.

Peterson, Roger Tory. 1969. Population trends of Ospreys in the northeastern United States. In Peregrine Falcon populations, their biology and decline, ed. Joseph J. Hickey, pp. 333-337. University of Wisconsin Press, Madison, Wisconsin.

————. 1980. A field guide to the birds east of the Mississippi River. Houghton-Mifflin Co., Boston, Massachusetts.

Phillips, John H. 1973. Screech Owl utilization of Wood Duck boxes in Madison County, Kentucky. Kentucky Warbler 49:51-53.

Pindar, Leon O. 1925. Birds of Fulton County, Kentucky. Wilson Bulletin 37:77-88, 163-69.

Powell, A.L. 1961. Some birds of the Owensboro lakes—III. Kentucky Warbler 37:4-7.

Ridgway, Robert. 1873. The prairie birds of southern Illinois. American Naturalist 7:197-203.

Rippy, Peggy T. 1974. Late nesting of Eastern Bluebird. Kentucky Warbler 50:18.

Ritchison, Gary. 1984. The Blue Grosbeak in Kentucky. Kentucky Warbler 60:29-31.

Robbins, Chandler S., Danny Bystrak, and Paul H. Geissler. 1986. The breeding bird survey: Its first fifteen years, 1965-1979. United States Department of the Interior, Fish and Wildlife Service. Resource Publication 157. Washington, D.C.

Robbins, Chandler S., John W. Fitzpatrick, and Paul B. Hamel. 1992. A warbler in trouble: *Dendroica cerulea.* In Ecology and conservation of neotropical migrant landbirds, ed. John M. Hagan III and David W. Johnston, pp. 549-562. Smithsonian Institution Press. Washington, D.C.

Robbins, Chandler S., John R. Sauer, Russell S. Greenberg, and Sam Droege. 1989. Population declines in North American birds that migrate to the neotropics. Proceedings of the National Academy of Science 86:7658-62.

Robbins, Mark B., and David A. Easterla. 1992. Birds of Missouri, their distribution and abundance. University of Missouri Press, Columbia, Missouri.

Robinson, John C. 1989. A concentration of Bewick's Wrens in Stewart County, Tennessee. Migrant 60:1-3.

————. 1990. An annotated checklist of the birds of Tennessee. University of Tennessee Press, Knoxville, Tennessee.

Robinson, Thane S. 1965. A preliminary list of the breeding birds of the lower Cumberland River Valley. Kentucky Warbler 41:59-62.

————. 1974. Pintail Duck nesting in Kentucky. Kentucky Warbler 50:18.

Rowe, William. 1964. Short-billed Marsh Wrens in Meade County. Kentucky Warbler 40:29-31.

————. 1967. Parula Warblers in Meade County. Kentucky Warbler 43:63-64.

Russell, Dan M. 1954. Woodcock studies. Kentucky Warbler 30:58-61.

Russell, Marvin, Ernest Beal, Herbert Shadowen, and Gordon Wilson. 1967. Blue Grosbeak and Cattle Egret. Kentucky Warbler 43:67.

Schmaltz, Jeffrey E. 1981. Past and present status of the Red-cockaded Woodpecker on the Daniel Boone National Forest, Kentucky. Kentucky Warbler 57:3-7.

Schneider, Evelyn. 1944. The summer range of the Chuck-will's-widow in Kentucky. Kentucky Warbler 20:13-19.

Semple, Sue W. 1944. Bittern meets poet. Kentucky Warbler 20:53-54.

Sharrock, J.T.R. 1976. The atlas of breeding birds in Britain and Ireland. T. and A.D. Poyser, Staffordshire, England.

Smith, Charles K., and Wayne H. Davis. 1979. Raven and Osprey in southeastern Kentucky. Kentucky Warbler 55:19-20.

Smith, Jerry R. 1950. A heron rookery at the Falls of the Ohio. Kentucky Warbler 26:6-8.

Smith, John W. 1986. Interim report. 1986 survey of the interior Least Tern on the Mississippi River (Cape Girardeau, Missouri, to Island No. 20, Tennessee). Sept. 10, 1986. Missouri Department of Conservation, Columbia, Missouri.

Smith, John W., and Rochelle B. Renken. 1990. Final report: Surveys and investigations projects. Endangered Species Project no. SE-01-19. April 16, 1990. Missouri Department of Conservation, Columbia, Missouri.

Smith, Lawrence D. 1972. Yellow Warbler parasitized by cowbird. Kentucky Warbler 48:33.

Spofford, Walter R. 1947a. Another tree-nesting Peregrine Falcon record for Tennessee. Migrant 18:60.

————. 1947b. A successful nesting of Peregrine Falcon with three adults present. Migrant 18:49-51.

Stamm, Anne L. 1951a. The breeding of the House Wren in Kentucky. Kentucky Warbler 27:47-56.

————. 1951b. Breeding status of the Cedar Waxwing in Kentucky. Kentucky Warbler 27:7-8.

————. 1961a. An August nest of the Eastern Meadowlark. Kentucky Warbler 37:60.

————. 1961b. Some summer notes from the Kentucky mountains. Kentucky Warbler 37:70-74.

————. 1962. Notes on the Dickcissel in the Louisville area. Kentucky Warbler 38:62-65.

————. 1963a. High nest of the Mourning Dove. Kentucky Warbler 39:12.

————. 1963b. Some June observations at Mammoth Cave. Kentucky Warbler 39:46-48.

————. 1964. High nest of the robin. Kentucky Warbler 40:56.

————. 1965. Notes on the incubation and nestling period of the mockingbird. Kentucky Warbler 41:48.

————. 1966. A nest of a Spotted Sandpiper at the Falls of the Ohio. Kentucky Warbler 42:3-4.

————. 1969. K.O.S. nest-card report, 1968. Kentucky Warbler 45:30-31.

————. 1970. Comments on the 1969 Kentucky nest-card program. Kentucky Warbler 46:46-47.

————. 1975. High nest of the Blue Jay. Kentucky Warbler 51:63.

————. 1976. Tree Swallows nesting in Jefferson County. Kentucky Warbler 52:13-14.

————. 1977. K.O.S. nest-record report, 1976. Kentucky Warbler 53:23-25.

————. 1978a. The nesting season, summer 1978. Kentucky Warbler 54:60-64.

————. 1978b. The spring season of 1978. Kentucky Warbler 54:42-47.

————. 1979a. The nesting season, summer 1979. Kentucky Warbler 55:55-58.

————. 1979b. The spring season of 1979. Kentucky Warbler 55:46-50.

————. 1980. The nesting season, summer 1980. Kentucky Warbler 56:78-82.

————. 1981a. The fall migration season, 1980. Kentucky Warbler 57:18-24.

————. 1981b. The nesting season, summer 1981. Kentucky Warbler 57:71-75.

————. 1981c. The spring migration of 1981. Kentucky Warbler 57:54-60.

————. 1982a. The fall migration season, 1981. Kentucky Warbler 58:13-19.

————. 1982b. The nesting season, summer 1982. Kentucky Warbler 58:81-86.

————. 1982c. The spring migration season, 1982. Kentucky Warbler 58:48-55.

————. 1983. The nesting season, summer 1983. Kentucky Warbler 59:51-54.

————. 1984a. The fall migration season, 1983. Kentucky Warbler 60:14-20.

————. 1984b. The nesting season, summer 1984. Kentucky Warbler 60:51-56.

————. 1984c. The spring season, 1984. Kentucky Warbler 60:40-46.

————. 1985. The nesting season, summer 1985. Kentucky Warbler 61:55-60.

————. 1986a. The nesting season, summer 1986. Kentucky Warbler 62:65-70.

————. 1986b. The spring migration of 1986. Kentucky Warbler 62:41-46.

————. 1987a. The fall migration season, 1986. Kentucky Warbler 63:5-12.

————. 1987b. The nesting season, summer 1987. Kentucky Warbler 63:58-62.

————. 1987c. The spring migration of 1987. Kentucky Warbler 63:47-53.

————. 1988a. The fall migration season, 1987. Kentucky Warbler 64:15-20.

————. 1988b. The nesting season, summer 1988. Kentucky Warbler 64:61-67.

————. 1988c. The spring season of 1988. Kentucky Warbler 64:43-50.

————. 1989a. The nesting season, summer 1989. Kentucky Warbler 65:83-92.

————. 1989b. The spring season of 1989. Kentucky Warbler 65:61-72.

————. 1990. The fall migration season, 1989. Kentucky Warbler 66:3-14.

Stamm, Anne L., Leonard C. Brecher, and Joseph E. Croft. 1967. Recent studies at the Falls of the Ohio. Kentucky Warbler 43:3-12.

Stamm, Anne L., Leonard C. Brecher, and Harvey B. Lovell. 1960. The 1959 autumn season at the Falls of the Ohio. Kentucky Warbler 36:3-8.

Stamm, Anne L., and Kathryn W. Clay. 1989. First breeding record of the Short-eared Owl in Kentucky. Kentucky Warbler 65:75-76.

Stamm, Anne L., and Joseph E. Croft. 1968. Kentucky nesting records, 1967. Kentucky Warbler 44:23-30.

Stamm, Anne L., and Howard P. Jones. 1966. Kentucky nesting records, 1965. Kentucky Warbler 42:39-43.

Stamm, Anne L., and Dorothea McConnell. 1971. Savannah Sparrows breeding in Oldham County. Kentucky Warbler 47:45.

Stamm, Anne L., and Burt L. Monroe Jr. 1990a. The nesting season, summer 1990. Kentucky Warbler 66:83-93.

———. 1990b. The spring season, 1990. Kentucky Warbler 66:52-62.

———. 1991a. The fall migration season, 1990. Kentucky Warbler 67:3-10.

———. 1991b. The nesting season, 1991. Kentucky Warbler 67:75-81.

———. 1991c. The spring season, 1991. Kentucky Warbler 67:51-60.

———. 1992. The nesting season, 1992. Kentucky Warbler 68:55-58.

Stamm, Anne L., and Mabel Slack. 1955. Notes on the nesting of the Grasshopper Sparrow at Louisville. Kentucky Warbler 31:52-54.

———. 1957. Nest of the Black and White Warbler in Bernheim Forest. Kentucky Warbler 33:70-71.

Stamm, Frederick W. 1960. September nesting record of the Black-crowned Night-Heron. Kentucky Warbler 36:33.

———. 1968. A nest of the Least Tern. Kentucky Warbler 44:49-51.

Steilberg, Robert H. 1949. Yellow-crowned Night Heron at Louisville. Kentucky Warbler 25:16.

Stevenson, Thomas E. 1985. Observations of nesting Barn Owls (*Tyto alba*). Kentucky Warbler 61:51-52.

Stickley, Allen R., Jr. 1983. House Finch nests in Warren County. Kentucky Warbler 59:47.

Terres, John. 1980. The Audubon Society encyclopedia of North American Birds. Alfred A. Knopf, Inc., New York, New York.

Thomas, E.S. 1932. Chuck-will's-widow, a new bird for Ohio. Auk 49:479.

Thomas, Julia. 1982. Colony observations on Lake Barkley. Kentucky Warbler 58:35.

Triquet, Alexis M., William C. McComb, and Jeffrey D. Sole. 1988. Ruffed Grouse drumming sites in eastern Kentucky. Kentucky Warbler 64:54-60.

Twedt, Daniel J., and Robert S. Oddo. 1984. Breeding biology of the starling in Kentucky. Kentucky Warbler 60:35-40.

United States Congress. 1983. The Endangered Species Act as amended by Public Law 97-304 (The Endangered Species Act Amendments of 1982). U.S. Government Printing Office, Washington, D.C.

United States Department of Agriculture. 1978. The forest resources of Kentucky. United States Department of Agriculture, Northeastern Forest Experiment Station Resource Bulletin NE-54. Broomall, Pennsylvania.

United States Fish and Wildlife Service. 1982. Endangered and threatened wildlife and plants: Review of vertebrate wildlife for listing as endangered or threatened species. Federal Register 47:58454-60.

———. 1990. Recovery plan for the interior population of the Least Tern (*Sterna antillarum*). United States Fish and Wildlife Service, Twin Cities, Minnesota.

———. 1991. Endangered and threatened wildlife and plants: Animal candidate review for listing as endangered or threatened species, proposed rule. Federal Register 56:58804-36.

———. 1993. Endangered and threatened wildlife and plants. 50 CFR 17.11 and 17.12. United States Department of the Interior, Washington, D.C.

Van Arsdall, C. Alex. 1949. A list of the breeding birds of Mercer County, Kentucky. Kentucky Warbler 25:21-29.

Warren, M.L., Jr., W.H. Davis, R.R. Hannan, M. Evans, D.L. Batch, B.D. Anderson, B. Palmer-Ball Jr., J.R. MacGregor, R.R. Cicerello, R. Athey, B.A. Branson, G.J. Fallo, B.M. Burr, M.E. Medley, and J.M. Baskin. 1986. Endangered, threatened, and rare plants and animals of Kentucky. Transactions of the Kentucky Academy of Science 47:83-98.

Webster, Conley. 1960. Nests of the Yellow-crowned Night Heron near Lexington. Kentucky Warbler 36:30.

Webster, J. Dan. 1951. An interesting swamp in Carroll County. Kentucky Warbler 27:21-22.

Wetmore, Alexander. 1940. Notes on the birds of Kentucky. Proceedings of the U.S. National Museum 88:529-74.

Whitt, A.L. 1969. Distribution of the Dickcissel in Madison County. Kentucky Warbler 45:29-30.

———. 1974. Current status of the Red-cockaded Woodpecker. Kentucky Warbler 50:51-53.

———. 1977. Notes on the status of the Carolina Wren in six eastern Kentucky counties. Kentucky Warbler 53:42.

Widmann, Otto. 1907. A preliminary catalog of the birds of Missouri. Transactions of the Academy of Science of St. Louis 17:1-288.

Wiley, R. Haven, Jr. 1964. Report on some Kentucky heronries. Kentucky Warbler 40:3-5.

———. 1970. Audubon's Kentucky birds, including the Ivory-billed Woodpecker. Kentucky Warbler 46:27-36.

Wiley, R. Haven, Jr., and Joseph E. Croft. 1964. Ecological notes on Henslow's Sparrows near Louisville. Kentucky Warbler 40:39-41.

Wilson, Gordon. 1922. A real Sycamore Warbler. Wilson Bulletin 34:119.

———. 1923. Birds of Calloway County, Kentucky. Wilson Bulletin 35:129-36.

———. 1929. Bird life of a transient lake in Kentucky. Wilson Bulletin 41:177-85.

———. 1935. The McElroy Farm—season of 1935. Kentucky Warbler 11:22-23.

———. 1940. The McElroy Farm—a study of a transient lake. Kentucky Warbler 16:13-21.

———. 1942. Breeding birds of Kentucky—a composite list. Kentucky Warbler 18:17-25.

———. 1947. Some restricted habitats in southern Kentucky. Kentucky Warbler 23:61-63.

———. 1951. The Woodburn lakes since 1939. Kentucky Warbler 27:1-6.

———. 1960. Sight record of Cattle Egret at Bowling Green. Kentucky Warbler 36:72.

———. 1961. Additions to "Birds of the Mammoth Cave National Park." Kentucky Warbler 37:7-9.

———. 1962a. Birds of south-central Kentucky. Kentucky Warbler 38:3-24.

———. 1962b. The tragedy of the bluebirds. Kentucky Warbler 38:49-50.

———. 1967. The Blue Grosbeak in south-central Kentucky. Kentucky Warbler 43:59-60.

———. 1969. Additions to "Birds of south-central Kentucky." Kentucky Warbler 45:32-39.

Wilson, Gordon, and Anne L. Stamm. 1960. The 1959-60 winter season and its effect on Kentucky birdlife. Kentucky Warbler 36:39-43.

Wright, Albert H. 1915. Early records of the Wild Turkey: Part IV. Auk 32:207-24.

Young, James B. 1973. A Mourning Dove fledgling in early March. Kentucky Warbler 49:46.

INDEX TO BIRD NAMES